Jeep Grand Cherokee Automotive Repair Manual

by Larry Warren
and John H Haynes
Member of the Guild of Motoring Writers

Models covered:

All Jeep Grand Cherokee models
1993 through 2000

(9E12 - 50025)

Haynes Publishing Group
Sparkford Nr Yeovil
Somerset BA22 7JJ England

Haynes North America, Inc
861 Lawrence Drive
Newbury Park
California 91320 USA

Acknowledgements

We are grateful to the DaimlerChrysler Corporation for assistance with technical information and certain illustrations. Technical writers who contributed to this project include Eric Godfrey, Bob Henderson, Rob Maddox, Mike Stubblefield, Jay Storer and Jeff Kibler.

A book in the Haynes Automotive Repair Manual Series

Printed in the U.S.A.

ISBN 1 56392 394 7

Library of Congress Control Number 00-104991

00-384

Contents

Haynes mechanic, author and photographer with 1993 Jeep Grand Cherokee

About this manual

Its purpose

The purpose of this manual is to help you get the best value from your vehicle. It can do so in several ways. It can help you decide what work must be done, even if you choose to have it done by a dealer service department or a repair shop; it provides information and procedures for routine maintenance and servicing; and it offers diagnostic and repair procedures to follow when trouble occurs.

We hope you use the manual to tackle the work yourself. For many simpler jobs, doing it yourself may be quicker than arranging an appointment to get the vehicle into a shop and making the trips to leave it and pick it up. More importantly, a lot of money can be saved by avoiding the expense the shop must pass on to you to cover its labor and overhead costs. An added benefit is the sense of satisfaction and accomplishment that you feel after doing the job yourself.

Using the manual

The manual is divided into Chapters. Each Chapter is divided into numbered Sections, which are headed in bold type between horizontal lines. Each Section consists of consecutively numbered paragraphs.

At the beginning of each numbered Section you will be referred to any illustrations which apply to the procedures in that Section. The reference numbers used in illustration captions pinpoint the pertinent Section and the Step within that Section. That is, illustration 3.2 means the illustration refers to Section 3 and Step (or paragraph) 2 within that Section.

Procedures, once described in the text, are not normally repeated. When it's necessary to refer to another Chapter, the reference will be given as Chapter and Section number. Cross references given without use of the word "Chapter" apply to Sections and/or paragraphs in the same Chapter. For example, "see Section 8" means in the same Chapter.

References to the left or right side of the vehicle assume you are sitting in the driver's seat, facing forward.

Even though we have prepared this manual with extreme care, neither the publisher nor the author can accept responsibility for any errors in, or omissions from, the information given.

NOTE

A **Note** provides information necessary to properly complete a procedure or information which will make the procedure easier to understand.

CAUTION

A **Caution** provides a special procedure or special steps which must be taken while completing the procedure where the Caution is found. Not heeding a Caution can result in damage to the assembly being worked on.

WARNING

A **Warning** provides a special procedure or special steps which must be taken while completing the procedure where the Warning is found. Not heeding a Warning can result in personal injury.

Introduction to the Jeep Grand Cherokee

Chassis layout is conventional, with the engine mounted at the front and the power being transmitted through either a manual or automatic transmission to a driveshaft and solid rear axle on 2WD models. On 4WD models a transfer case transmits power to the front axle by way of a driveshaft. Transmissions used are a five-speed overdrive manual and four-speed overdrive automatic.

The suspension on these models features solid axles with coil springs and control arms front and rear. 2WD vehicles use a beam-type front axle while 4WD vehicles use a live-axle, similar in layout to the rear axle assembly.

All models are equipped with power assisted front disc and either drum or disc rear brakes. An Anti-lock Braking System (ABS) is used on all models.

Vehicle identification numbers

Vehicle identification numbers

Modifications are a continuing and unpublicized process in vehicle manufacturing. Since spare parts manuals and lists are compiled on a numerical basis, the individual vehicle numbers are necessary to correctly identify the component required.

Vehicle Identification Number (VIN)

This very important identification number is located on a plate attached to the left side dashboard just inside the windshield **(see illustration)**. The VIN also appears on the Vehicle Certificate of Title and Registration. It contains information such as where and when the vehicle was manufactured, engine type, the model year and the body style.

VIN engine and model year codes

Two particularly important pieces of information found in the VIN are the engine code and the model year code. Counting from the left, the engine code letter designation is the 8th character and the model year code letter designation is the 10th character **(see illustration)**.

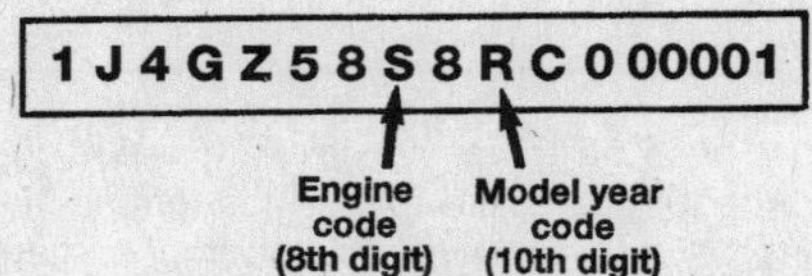

The engine code is the 8th character of the VIN number; the model code is the 10th character

On the models covered by this manual the engine codes are:

S ..... 4.0L inline six-cylinder engine
Y ..... 5.2L V8 engine
Z ..... 5.9L V8 engine
N ..... 4.7L V8 engine

On models covered by this manual the model year codes are:

P 1993
R 1994
S 1995
T 1996
V 1997
W 1998
X 1999
Y 2000

Vehicle Certification Label

The Vehicle Certification label (VC label is attached to the rear of the left (driver's side)

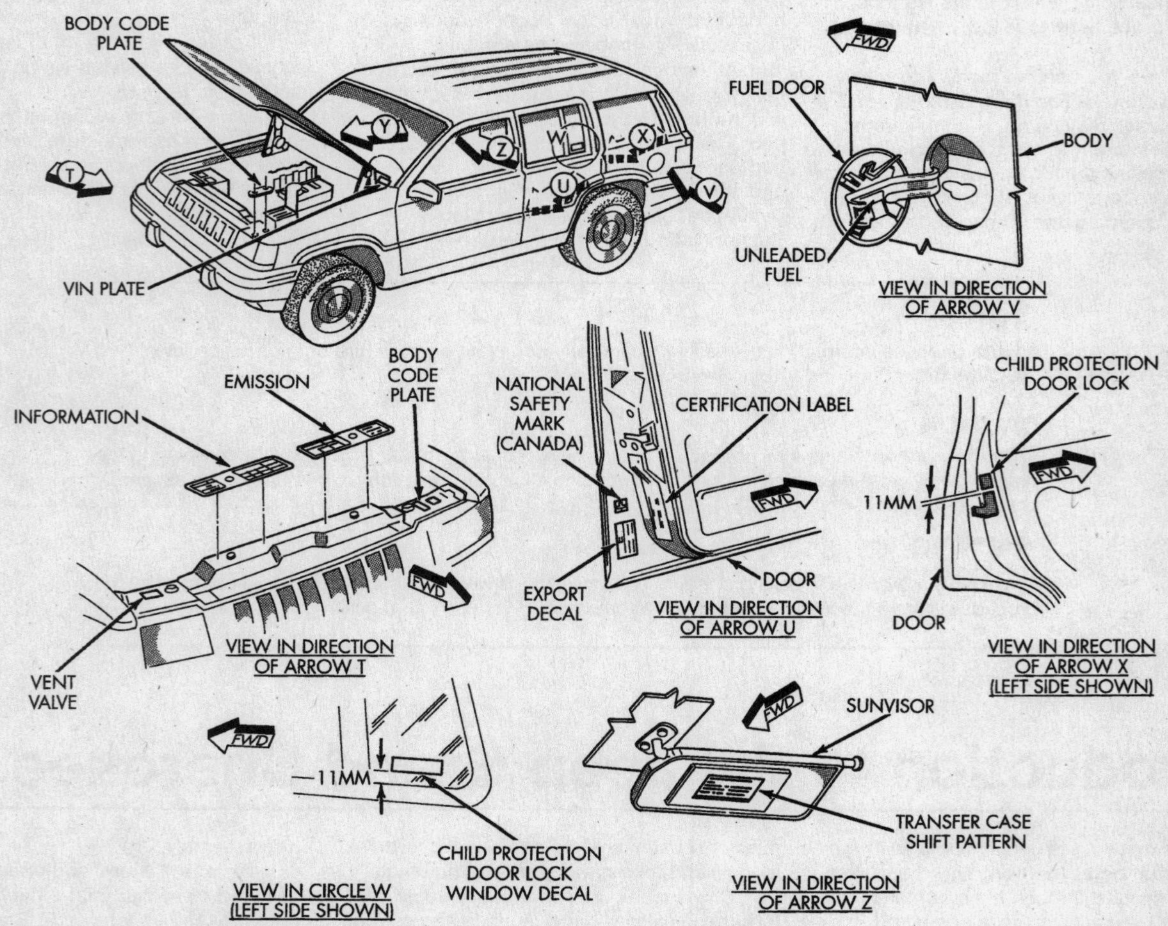

The Vehicle Identification Number (VIN), Vehicle Certification and other label locations

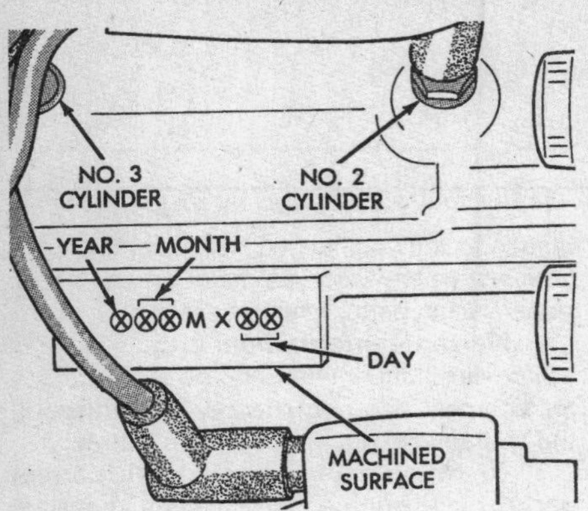

Six-cylinder engine ID number location

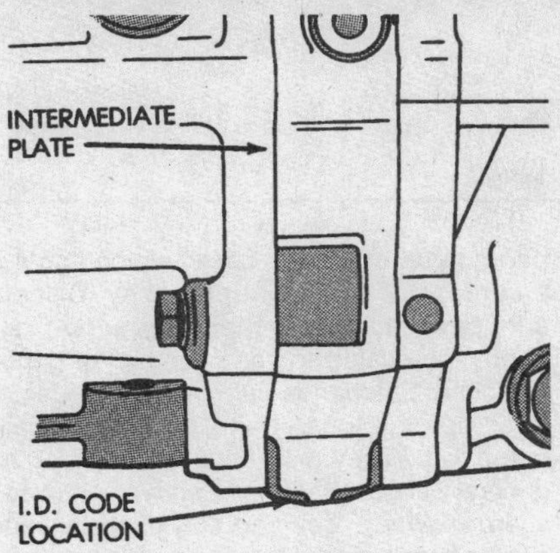

Manual transmission identification number location

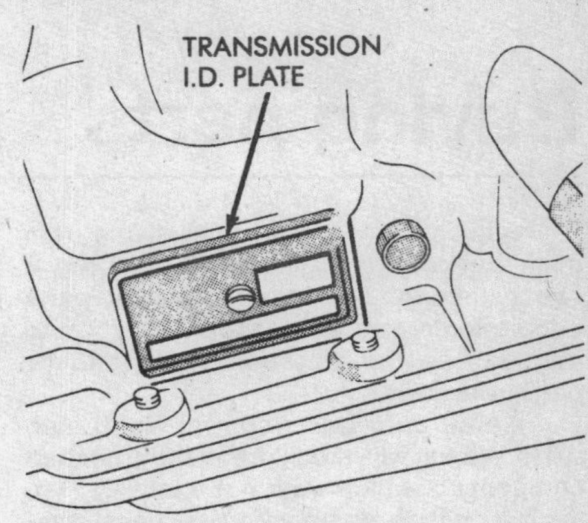

1993 model automatic transmission identification plate location

door. The upper half of the label contains the name of the manufacturer, the month and year of production, the Gross Vehicle Weight Rating (GVWR), the Gross Axle Weight Rating (GAWR) and the certification statement.

The VC label also contains the VIN number, which is used for warranty identification of the vehicle, and provides such information as manufacturer, type of restraint system, body type, engine, transmission, model year and vehicle serial number.

Engine identification numbers

On six-cylinder engines, build date information can be found stamped on a machined pad located on the right side of the cylinder block between the number 2 and 3 cylinders **(see illustration)**.

On 5.2L/5.9L V8 engines, the serial number is stamped on a machined pad on the left front corner of the cylinder block.

On 4.7L V8 engines, the serial number is located on the right front side of the engine block.

Manual transmission identification number

The manual transmission identification number and serial numbers can be found on a tag on the side of the transmission case near the drain plug **(see illustration)**.

Automatic transmission identification numbers

On 1993 models, a plate with the automatic transmission serial number, build date and other information is attached to the right rear side of the case **(see illustration)**. On 1994 and later models, the information is stamped on the left side of the case, just above the fluid pan gasket **(see illustration)**.

Transfer case identification number

The transfer case identification plate is attached to the left rear side of the case **(see illustration)**.

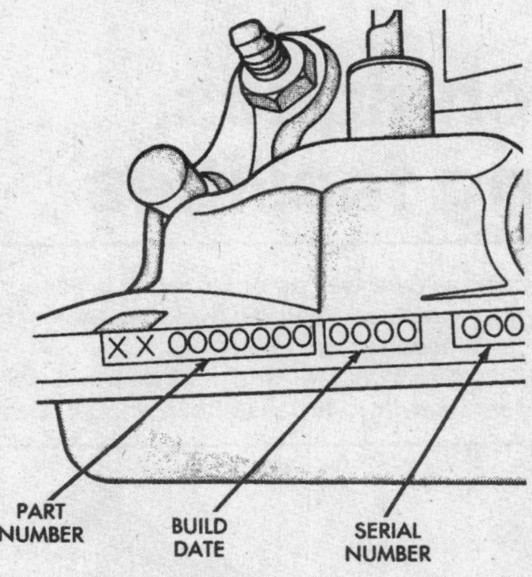

1994 and later model automatic transmission identification number location

Transfer case identification plate

Axle identification numbers

On both front and rear axles the identification number is located on a tag attached to the differential cover **(see illustration)**.

Vehicle Emissions Control Information (VECI) label

This label is found under the hood, usually on the radiator support (see Chapter 6).

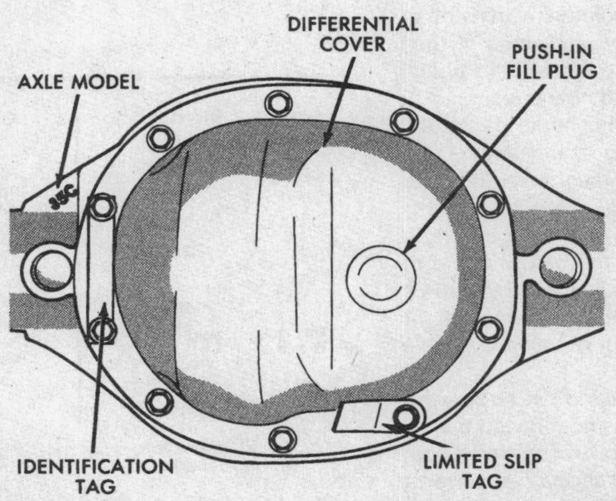

Axle identification tag location

Buying parts

Replacement parts are available from many sources, which generally fall into one of two categories - authorized dealer parts departments and independent retail auto parts stores. Our advice concerning these parts is as follows:

Retail auto parts stores: Good auto parts stores will stock frequently needed components which wear out relatively fast, such as clutch components, exhaust systems, brake parts, tune-up parts, etc. These stores often supply new or reconditioned parts on an exchange basis, which can save a considerable amount of money. Discount auto parts stores are often very good places to buy materials and parts needed for general vehicle maintenance such as oil, grease, filters, spark plugs, belts, touch-up paint, bulbs, etc. They also usually sell tools and general accessories, have convenient hours, charge lower prices and can often be found not far from home.

Authorized dealer parts department: This is the best source for parts which are unique to the vehicle and not generally available elsewhere (such as major engine parts, transmission parts, trim pieces, etc.).

Warranty information: If the vehicle is still covered under warranty, be sure that any replacement parts purchased - regardless of the source - do not invalidate the warranty!

To be sure of obtaining the correct parts, have engine and chassis numbers available and, if possible, take the old parts along for positive identification.

Maintenance techniques, tools and working facilities

Maintenance techniques

There are a number of techniques involved in maintenance and repair that will be referred to throughout this manual. Application of these techniques will enable the home mechanic to be more efficient, better organized and capable of performing the various tasks properly, which will ensure that the repair job is thorough and complete.

Fasteners

Fasteners are nuts, bolts, studs and screws used to hold two or more parts together. There are a few things to keep in mind when working with fasteners. Almost all of them use a locking device of some type, either a lockwasher, locknut, locking tab or thread adhesive. All threaded fasteners should be clean and straight, with undamaged threads and undamaged corners on the hex head where the wrench fits. Develop the habit of replacing all damaged nuts and bolts with new ones. Special locknuts with nylon or fiber inserts can only be used once. If they are removed, they lose their locking ability and must be replaced with new ones.

Rusted nuts and bolts should be treated with a penetrating fluid to ease removal and prevent breakage. Some mechanics use turpentine in a spout-type oil can, which works quite well. After applying the rust penetrant, let it work for a few minutes before trying to loosen the nut or bolt. Badly rusted fasteners may have to be chiseled or sawed off or removed with a special nut breaker, available at tool stores.

If a bolt or stud breaks off in an assembly, it can be drilled and removed with a special tool commonly available for this purpose. Most automotive machine shops can perform this task, as well as other repair procedures, such as the repair of threaded holes that have been stripped out.

Flat washers and lockwashers, when removed from an assembly, should always be replaced exactly as removed. Replace any damaged washers with new ones. Never use a lockwasher on any soft metal surface (such as aluminum), thin sheet metal or plastic.

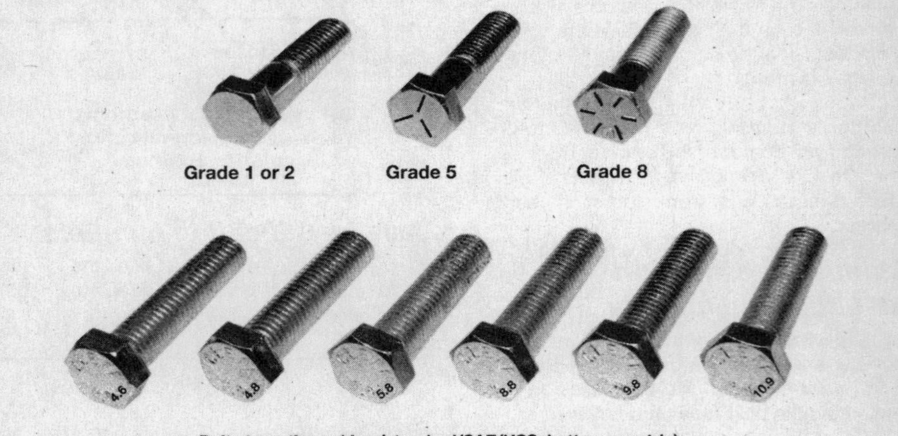

Bolt strength marking (standard/SAE/USS; bottom - metric)

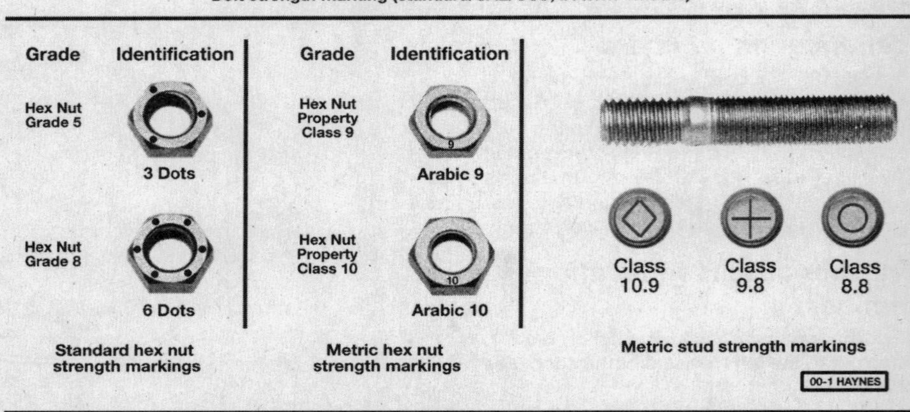

Fastener sizes

For a number of reasons, automobile manufacturers are making wider and wider use of metric fasteners. Therefore, it is important to be able to tell the difference between standard (sometimes called U.S. or SAE) and metric hardware, since they cannot be interchanged.

All bolts, whether standard or metric, are sized according to diameter, thread pitch and length. For example, a standard 1/2 - 13 x 1 bolt is 1/2 inch in diameter, has 13 threads per inch and is 1 inch long. An M12 - 1.75 x 25 metric bolt is 12 mm in diameter, has a thread pitch of 1.75 mm (the distance between threads) and is 25 mm long. The two bolts are nearly identical, and easily confused, but they are not interchangeable.

In addition to the differences in diameter, thread pitch and length, metric and standard bolts can also be distinguished by examining the bolt heads. To begin with, the distance across the flats on a standard bolt head is measured in inches, while the same dimension on a metric bolt is sized in millimeters (the same is true for nuts). As a result, a standard wrench should not be used on a metric bolt and a metric wrench should not be used on a standard bolt. Also, most standard bolts have slashes radiating out from the center of the head to denote the grade or strength of the bolt, which is an indication of the amount of torque that can be applied to it. The greater the number of slashes, the greater the strength of the bolt. Grades 0 through 5 are commonly used on automobiles. Metric bolts have a property class (grade) number, rather than a slash, molded into their heads to indicate bolt strength. In this case, the higher the number, the stronger the bolt. Property class numbers 8.8, 9.8 and 10.9 are commonly used on automobiles.

Strength markings can also be used to distinguish standard hex nuts from metric hex nuts. Many standard nuts have dots stamped into one side, while metric nuts are marked with a number. The greater the number of dots, or the higher the number, the greater the strength of the nut.

Metric studs are also marked on their ends according to property class (grade). Larger studs are numbered (the same as metric bolts), while smaller studs carry a geometric code to denote grade.

It should be noted that many fasteners, especially Grades 0 through 2, have no distinguishing marks on them. When such is the case, the only way to determine whether it is standard or metric is to measure the thread pitch or compare it to a known fastener of the same size.

Standard fasteners are often referred to as SAE, as opposed to metric. However, it should be noted that SAE technically refers to a non-metric fine thread fastener only. Coarse thread non-metric fasteners are referred to as USS sizes.

Since fasteners of the same size (both standard and metric) may have different

Metric thread sizes	Ft-lbs	Nm
M-6	6 to 9	9 to 12
M-8	14 to 21	19 to 28
M-10	28 to 40	38 to 54
M-12	50 to 71	68 to 96
M-14	80 to 140	109 to 154
Pipe thread sizes		
1/8	5 to 8	7 to 10
1/4	12 to 18	17 to 24
3/8	22 to 33	30 to 44
1/2	25 to 35	34 to 47
U.S. thread sizes		
1/4 - 20	6 to 9	9 to 12
5/16 - 18	12 to 18	17 to 24
5/16 - 24	14 to 20	19 to 27
3/8 - 16	22 to 32	30 to 43
3/8 - 24	27 to 38	37 to 51
7/16 - 14	40 to 55	55 to 74
7/16 - 20	40 to 60	55 to 81
1/2 - 13	55 to 80	75 to 108

Standard (SAE and USS) bolt dimensions/grade marks

- *G* *Grade marks (bolt strength)*
- *L* *Length (in inches)*
- *T* *Thread pitch (number of threads per inch)*
- *D* *Nominal diameter (in inches)*

Metric bolt dimensions/grade marks

- *P* *Property class (bolt strength)*
- *L* *Length (in millimeters)*
- *T* *Thread pitch (distance between threads in millimeters)*
- *D* *Diameter*

strength ratings, be sure to reinstall any bolts, studs or nuts removed from your vehicle in their original locations. Also, when replacing a fastener with a new one, make sure that the new one has a strength rating equal to or greater than the original.

Tightening sequences and procedures

Most threaded fasteners should be tightened to a specific torque value (torque is the twisting force applied to a threaded component such as a nut or bolt). Overtightening the fastener can weaken it and cause it to break, while undertightening can cause it to eventually come loose. Bolts, screws and studs, depending on the material they are made of and their thread diameters, have specific torque values, many of which are noted in the Specifications at the beginning of each Chapter. Be sure to follow the torque recommendations closely. For fasteners not assigned a specific torque, a general torque value chart is presented here as a guide. These torque values are for dry (unlubricated) fasteners threaded into steel or cast iron (not aluminum). As was previously mentioned, the size and grade of a fastener determine the amount of torque that can safely be applied to it. The figures listed here are approximate for Grade 2 and Grade 3 fasteners. Higher grades can tolerate higher torque values.

Fasteners laid out in a pattern, such as cylinder head bolts, oil pan bolts, differential cover bolts, etc., must be loosened or tightened in sequence to avoid warping the com-

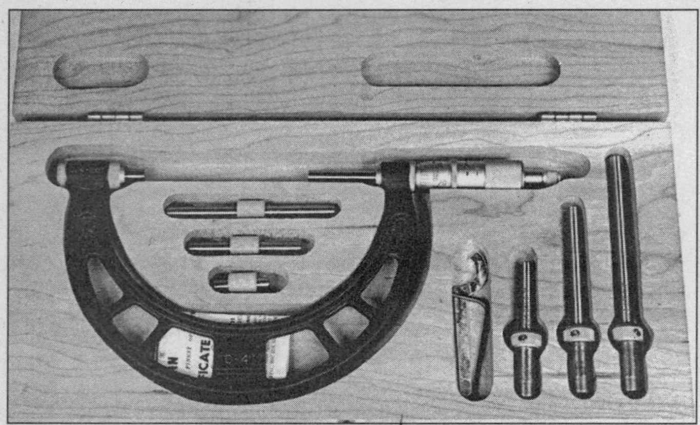

Micrometer set

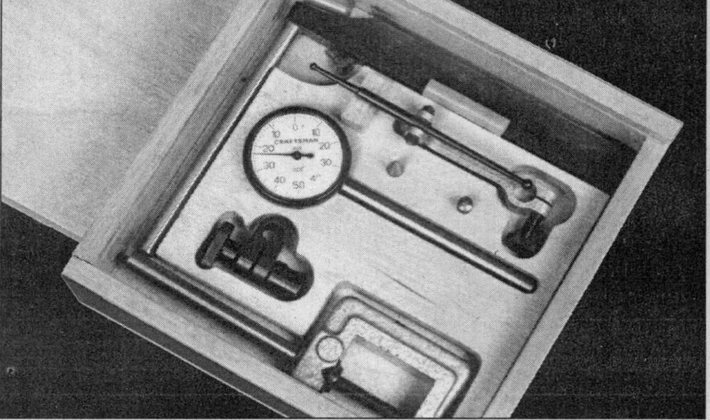

Dial indicator set

ponent. This sequence will normally be shown in the appropriate Chapter. If a specific pattern is not given, the following procedures can be used to prevent warping.

Initially, the bolts or nuts should be assembled finger-tight only. Next, they should be tightened one full turn each, in a criss-cross or diagonal pattern. After each one has been tightened one full turn, return to the first one and tighten them all one-half turn, following the same pattern. Finally, tighten each of them one-quarter turn at a time until each fastener has been tightened to the proper torque. To loosen and remove the fasteners, the procedure would be reversed.

Component disassembly

Component disassembly should be done with care and purpose to help ensure that the parts go back together properly. Always keep track of the sequence in which parts are removed. Make note of special characteristics or marks on parts that can be installed more than one way, such as a grooved thrust washer on a shaft. It is a good idea to lay the disassembled parts out on a clean surface in the order that they were removed. It may also be helpful to make sketches or take instant photos of components before removal.

When removing fasteners from a component, keep track of their locations. Sometimes threading a bolt back in a part, or putting the washers and nut back on a stud, can prevent mix-ups later. If nuts and bolts cannot be returned to their original locations, they should be kept in a compartmented box or a series of small boxes. A cupcake or muffin tin is ideal for this purpose, since each cavity can hold the bolts and nuts from a particular area (i.e. oil pan bolts, valve cover bolts, engine mount bolts, etc.). A pan of this type is especially helpful when working on assemblies with very small parts, such as the carburetor, alternator, valve train or interior dash and trim pieces. The cavities can be marked with paint or tape to identify the contents.

Whenever wiring looms, harnesses or connectors are separated, it is a good idea to identify the two halves with numbered pieces of masking tape so they can be easily reconnected.

Gasket sealing surfaces

Throughout any vehicle, gaskets are used to seal the mating surfaces between two parts and keep lubricants, fluids, vacuum or pressure contained in an assembly.

Many times these gaskets are coated with a liquid or paste-type gasket sealing compound before assembly. Age, heat and pressure can sometimes cause the two parts to stick together so tightly that they are very difficult to separate. Often, the assembly can be loosened by striking it with a soft-face hammer near the mating surfaces. A regular hammer can be used if a block of wood is placed between the hammer and the part. Do not hammer on cast parts or parts that could be easily damaged. With any particularly stubborn part, always recheck to make sure that every fastener has been removed.

Avoid using a screwdriver or bar to pry apart an assembly, as they can easily mar the gasket sealing surfaces of the parts, which must remain smooth. If prying is absolutely necessary, use an old broom handle, but keep in mind that extra clean up will be necessary if the wood splinters.

After the parts are separated, the old gasket must be carefully scraped off and the gasket surfaces cleaned. Stubborn gasket material can be soaked with rust penetrant or treated with a special chemical to soften it so it can be easily scraped off. A scraper can be fashioned from a piece of copper tubing by flattening and sharpening one end. Copper is recommended because it is usually softer than the surfaces to be scraped, which reduces the chance of gouging the part. Some gaskets can be removed with a wire brush, but regardless of the method used, the mating surfaces must be left clean and smooth. If for some reason the gasket surface is gouged, then a gasket sealer thick enough to fill scratches will have to be used during reassembly of the components. For most applications, a non-drying (or semi-drying) gasket sealer should be used.

Hose removal tips

Warning: *If the vehicle is equipped with air conditioning, do not disconnect any of the A/C hoses without first having the system depressurized by a dealer service department or a service station.*

Hose removal precautions closely parallel gasket removal precautions. Avoid scratching or gouging the surface that the hose mates against or the connection may leak. This is especially true for radiator hoses. Because of various chemical reactions, the rubber in hoses can bond itself to the metal spigot that the hose fits over. To remove a hose, first loosen the hose clamps that secure it to the spigot. Then, with slip-joint pliers, grab the hose at the clamp and rotate it around the spigot. Work it back and forth until it is completely free, then pull it off. Silicone or other lubricants will ease removal if they can be applied between the hose and the outside of the spigot. Apply the same lubricant to the inside of the hose and the outside of the spigot to simplify installation.

As a last resort (and if the hose is to be replaced with a new one anyway), the rubber can be slit with a knife and the hose peeled from the spigot. If this must be done, be careful that the metal connection is not damaged.

If a hose clamp is broken or damaged, do not reuse it. Wire-type clamps usually weaken with age, so it is a good idea to replace them with screw-type clamps whenever a hose is removed.

Tools

A selection of good tools is a basic requirement for anyone who plans to maintain and repair his or her own vehicle. For the owner who has few tools, the initial investment might seem high, but when compared to the spiraling costs of professional auto maintenance and repair, it is a wise one.

To help the owner decide which tools are needed to perform the tasks detailed in this manual, the following tool lists are offered: *Maintenance and minor repair, Repair/overhaul* and *Special.*

The newcomer to practical mechanics

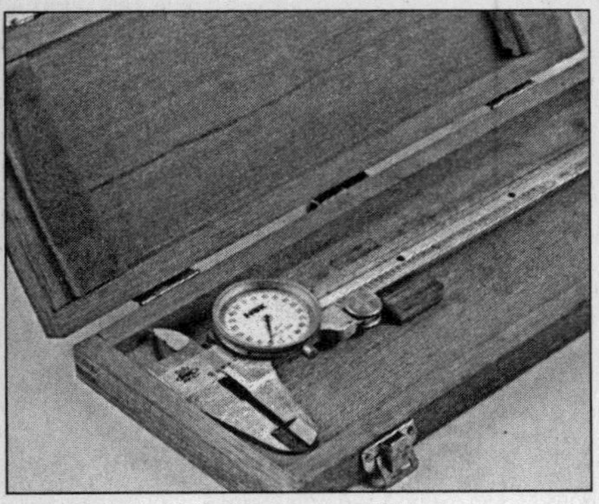

Dial caliper

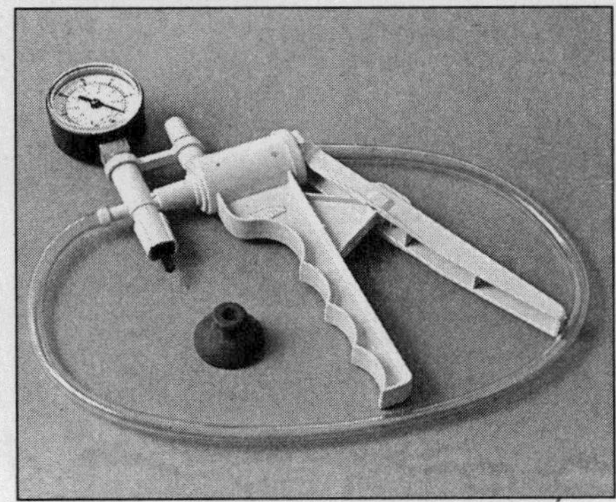

Hand-operated vacuum pump

Timing light

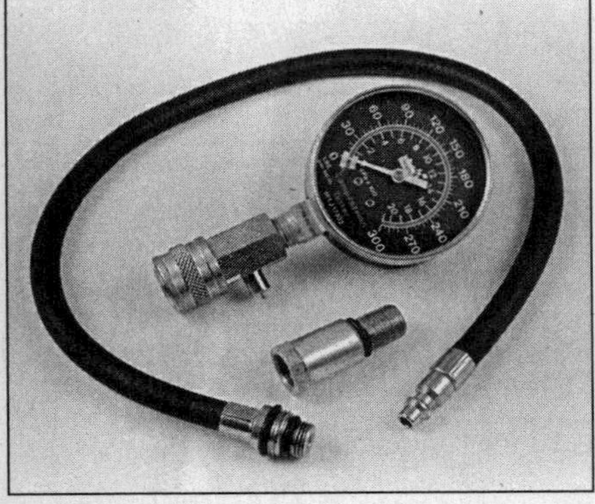

Compression gauge with spark plug hole adapter

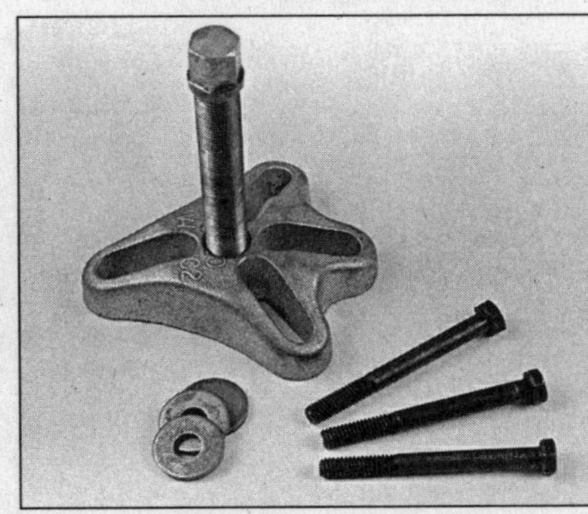

Damper/steering wheel puller

General purpose puller

Hydraulic lifter removal tool

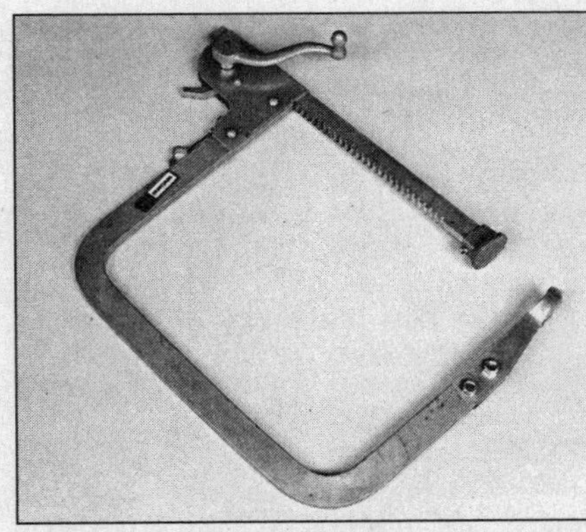

Valve spring compressor

Valve spring compressor

Ridge reamer

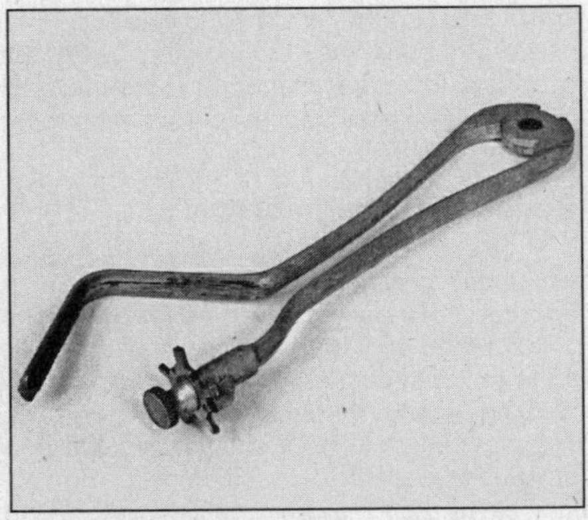

Piston ring groove cleaning tool

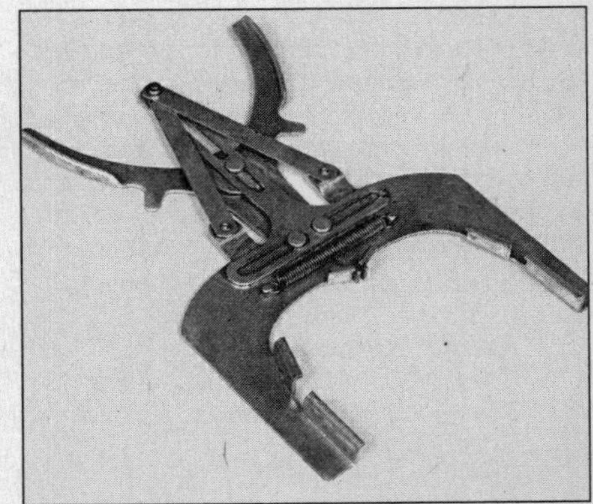

Ring removal/installation tool

Ring compressor

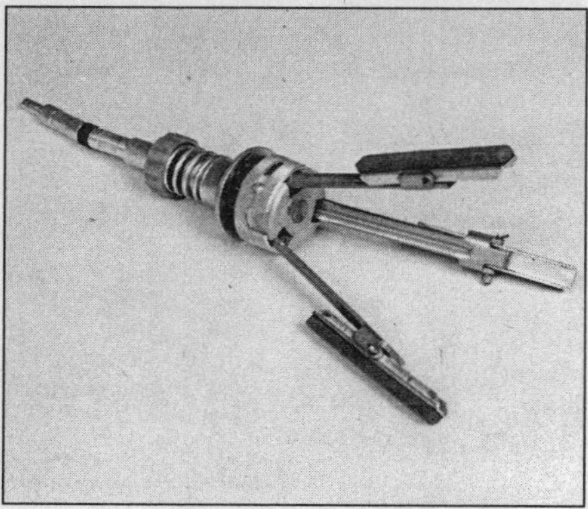
Cylinder hone

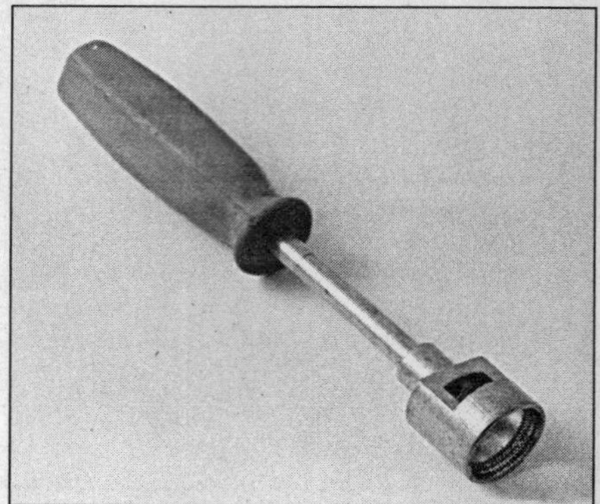
Brake hold-down spring tool

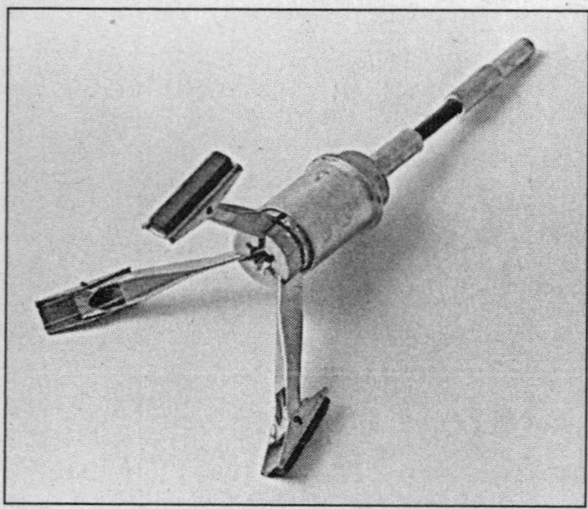
Brake cylinder hone

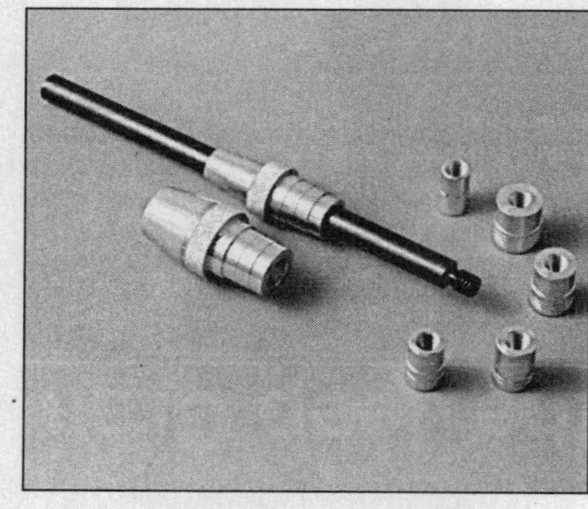
Clutch plate alignment tool

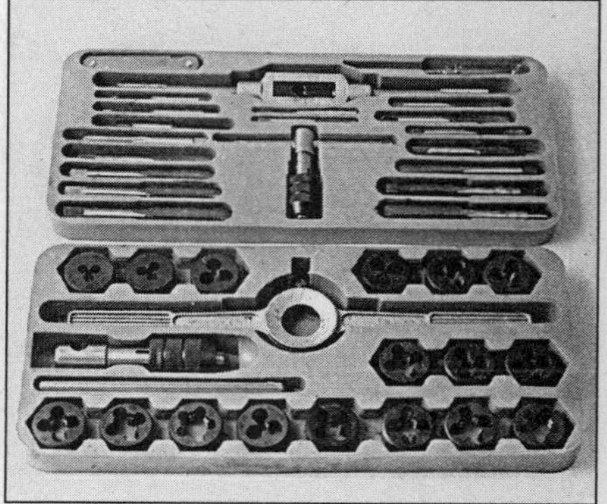
Tap and die set

should start off with the *maintenance and minor repair* tool kit, which is adequate for the simpler jobs performed on a vehicle. Then, as confidence and experience grow, the owner can tackle more difficult tasks, buying additional tools as they are needed. Eventually the basic kit will be expanded into the *repair and overhaul* tool set. Over a period of time, the experienced do-it-yourselfer will assemble a tool set complete enough for most repair and overhaul procedures and will add tools from the special category when it is felt that the expense is justified by the frequency of use.

Maintenance and minor repair tool kit

The tools in this list should be considered the minimum required for performance of routine maintenance, servicing and minor repair work. We recommend the purchase of combination wrenches (box-end and open-end combined in one wrench). While more expensive than open end wrenches, they offer the advantages of both types of wrench.

Combination wrench set (1/4-inch to 1 inch or 6 mm to 19 mm)
Adjustable wrench, 8 inch
Spark plug wrench with rubber insert
Spark plug gap adjusting tool
Feeler gauge set
Brake bleeder wrench
Standard screwdriver (5/16-inch x 6 inch)
Phillips screwdriver (No. 2 x 6 inch)
Combination pliers - 6 inch
Hacksaw and assortment of blades
Tire pressure gauge
Grease gun
Oil can
Fine emery cloth
Wire brush
Battery post and cable cleaning tool
Oil filter wrench
Funnel (medium size)
Safety goggles
Jackstands (2)
Drain pan

Note: *If basic tune-ups are going to be part of routine maintenance, it will be necessary to purchase a good quality stroboscopic timing light and combination tachometer/dwell meter. Although they are included in the list of special tools, it is mentioned here because they are absolutely necessary for tuning most vehicles properly.*

Repair and overhaul tool set

These tools are essential for anyone who plans to perform major repairs and are in addition to those in the maintenance and minor repair tool kit. Included is a comprehensive set of sockets which, though expensive, are invaluable because of their versatility, especially when various extensions and drives are available. We recommend the 1/2-inch drive over the 3/8-inch drive. Although the larger drive is bulky and more expensive, it has the capacity of accepting a very wide range of large sockets. Ideally, however, the mechanic should have a 3/8-inch drive set and a 1/2-inch drive set.

Socket set(s)
Reversible ratchet
Extension - 10 inch
Universal joint
Torque wrench (same size drive as sockets)
Ball peen hammer - 8 ounce
Soft-face hammer (plastic/rubber)
Standard screwdriver (1/4-inch x 6 inch)
Standard screwdriver (stubby - 5/16-inch)
Phillips screwdriver (No. 3 x 8 inch)
Phillips screwdriver (stubby - No. 2)
Pliers - vise grip
Pliers - lineman's
Pliers - needle nose
Pliers - snap-ring (internal and external)
Cold chisel - 1/2-inch
Scribe
Scraper (made from flattened copper tubing)
Centerpunch
Pin punches (1/16, 1/8, 3/16-inch)
Steel rule/straightedge - 12 inch
Allen wrench set (1/8 to 3/8-inch or 4 mm to 10 mm)
A selection of files
Wire brush (large)
Jackstands (second set)
Jack (scissor or hydraulic type)

Note: *Another tool which is often useful is an electric drill with a chuck capacity of 3/8-inch and a set of good quality drill bits.*

Special tools

The tools in this list include those which are not used regularly, are expensive to buy, or which need to be used in accordance with their manufacturer's instructions. Unless these tools will be used frequently, it is not very economical to purchase many of them. A consideration would be to split the cost and use between yourself and a friend or friends. In addition, most of these tools can be obtained from a tool rental shop on a temporary basis.

This list primarily contains only those tools and instruments widely available to the public, and not those special tools produced by the vehicle manufacturer for distribution to dealer service departments. Occasionally, references to the manufacturer's special tools are included in the text of this manual. Generally, an alternative method of doing the job without the special tool is offered. However, sometimes there is no alternative to their use. Where this is the case, and the tool cannot be purchased or borrowed, the work should be turned over to the dealer service department or an automotive repair shop.

Valve spring compressor
Piston ring groove cleaning tool
Piston ring compressor
Piston ring installation tool
Cylinder compression gauge
Cylinder ridge reamer
Cylinder surfacing hone
Cylinder bore gauge
Micrometers and/or dial calipers
Hydraulic lifter removal tool
Balljoint separator
Universal-type puller
Impact screwdriver
Dial indicator set
Stroboscopic timing light (inductive pick-up)
Hand operated vacuum/pressure pump
Tachometer/dwell meter
Universal electrical multimeter
Cable hoist
Brake spring removal and installation tools
Floor jack

Buying tools

For the do-it-yourselfer who is just starting to get involved in vehicle maintenance and repair, there are a number of options available when purchasing tools. If maintenance and minor repair is the extent of the work to be done, the purchase of individual tools is satisfactory. If, on the other hand, extensive work is planned, it would be a good idea to purchase a modest tool set from one of the large retail chain stores. A set can usually be bought at a substantial savings over the individual tool prices, and they often come with a tool box. As additional tools are needed, add-on sets, individual tools and a larger tool box can be purchased to expand the tool selection. Building a tool set gradually allows the cost of the tools to be spread over a longer period of time and gives the mechanic the freedom to choose only those tools that will actually be used.

Tool stores will often be the only source of some of the special tools that are needed, but regardless of where tools are bought, try to avoid cheap ones, especially when buying screwdrivers and sockets, because they won't last very long. The expense involved in replacing cheap tools will eventually be greater than the initial cost of quality tools.

Care and maintenance of tools

Good tools are expensive, so it makes sense to treat them with respect. Keep them clean and in usable condition and store them properly when not in use. Always wipe off any dirt, grease or metal chips before putting them away. Never leave tools lying around in the work area. Upon completion of a job, always check closely under the hood for tools that may have been left there so they won't get lost during a test drive.

Some tools, such as screwdrivers, pliers, wrenches and sockets, can be hung on a panel mounted on the garage or workshop wall, while others should be kept in a tool box or tray. Measuring instruments, gauges, meters, etc. must be carefully stored where they cannot be damaged by weather or impact from other tools.

When tools are used with care and stored properly, they will last a very long time. Even with the best of care, though, tools will wear out if used frequently. When a tool is damaged or worn out, replace it. Subsequent jobs will be safer and more enjoyable if you do.

How to repair damaged threads

Sometimes, the internal threads of a nut or bolt hole can become stripped, usually from overtightening. Stripping threads is an all-too-common occurrence, especially when working with aluminum parts, because aluminum is so soft that it easily strips out.

Usually, external or internal threads are only partially stripped. After they've been cleaned up with a tap or die, they'll still work. Sometimes, however, threads are badly damaged. When this happens, you've got three choices:

1) *Drill and tap the hole to the next suitable oversize and install a larger diameter bolt, screw or stud.*
2) *Drill and tap the hole to accept a threaded plug, then drill and tap the plug to the original screw size. You can also buy a plug already threaded to the original size. Then you simply drill a hole to the specified size, then run the threaded plug into the hole with a bolt and jam nut. Once the plug is fully seated, remove the jam nut and bolt.*
3) *The third method uses a patented thread repair kit like Heli-Coil or Slimsert. These easy-to-use kits are designed to repair damaged threads in straight-through holes and blind holes. Both are available as kits which can handle a variety of sizes and thread patterns. Drill the hole, then tap it with the special included tap. Install the Heli-Coil and the hole is back to its original diameter and thread pitch.*

Regardless of which method you use, be sure to proceed calmly and carefully. A little impatience or carelessness during one of these relatively simple procedures can ruin your whole day's work and cost you a bundle if you wreck an expensive part.

Working facilities

Not to be overlooked when discussing tools is the workshop. If anything more than routine maintenance is to be carried out, some sort of suitable work area is essential.

It is understood, and appreciated, that many home mechanics do not have a good workshop or garage available, and end up removing an engine or doing major repairs outside. It is recommended, however, that the overhaul or repair be completed under the cover of a roof.

A clean, flat workbench or table of comfortable working height is an absolute necessity. The workbench should be equipped with a vise that has a jaw opening of at least four inches.

As mentioned previously, some clean, dry storage space is also required for tools, as well as the lubricants, fluids, cleaning solvents, etc. which soon become necessary.

Sometimes waste oil and fluids, drained from the engine or cooling system during normal maintenance or repairs, present a disposal problem. To avoid pouring them on the ground or into a sewage system, pour the used fluids into large containers, seal them with caps and take them to an authorized disposal site or recycling center. Plastic jugs, such as old antifreeze containers, are ideal for this purpose.

Always keep a supply of old newspapers and clean rags available. Old towels are excellent for mopping up spills. Many mechanics use rolls of paper towels for most work because they are readily available and disposable. To help keep the area under the vehicle clean, a large cardboard box can be cut open and flattened to protect the garage or shop floor.

Whenever working over a painted surface, such as when leaning over a fender to service something under the hood, always cover it with an old blanket or bedspread to protect the finish. Vinyl covered pads, made especially for this purpose, are available at auto parts stores.

Booster battery (jump) starting

Observe the following precautions when using a booster battery to start a vehicle:

a) Before connecting the booster battery, make sure the ignition switch is in the Off position.
b) Turn off the lights, heater and other electrical loads.
c) Your eyes should be shielded. Safety goggles are a good idea.
d) Make sure the booster battery is the same voltage as the dead one in the vehicle.
e) The two vehicles MUST NOT TOUCH each other.
f) Make sure the transmission is in Neutral (manual transaxle) or Park (automatic transaxle).
g) If the booster battery is not a maintenance-free type, remove the vent caps and lay a cloth over the vent holes.

Connect the red jumper cable to the positive (+) terminals of each battery.

Connect one end of the black cable to the negative (-) terminal of the booster battery. The other end of this cable should be connected to a good ground on the engine block **(see illustration)**. Make sure the cable will not come into contact with the fan, drivebelts or other moving parts of the engine.

Start the engine using the booster battery, then, with the engine running at idle speed, disconnect the jumper cables in the reverse order of connection

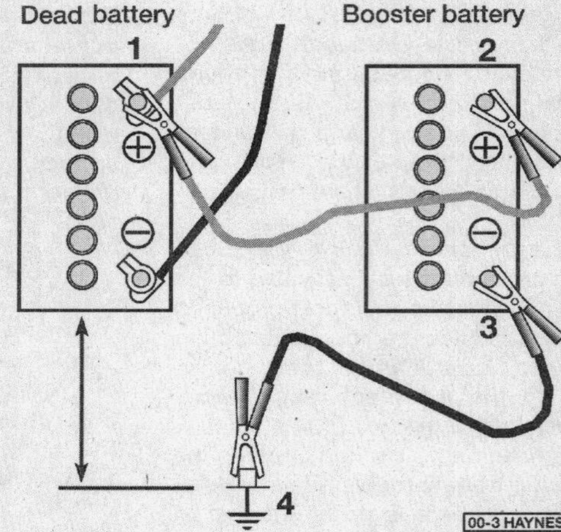

Make the booster battery cable connections in the numerical order shown (note that the negative cable of the booster battery is NOT attached to the negative terminal of the dead battery)

Jacking and towing

Jacking

The jack supplied with the vehicle should only be used for raising the vehicle when changing a tire or placing jackstands under the frame. NEVER work under the vehicle or start the engine when the vehicle supported only by a jack.

The vehicle should be parked on level ground with the wheels blocked, the parking brake applied and the transmission in Park (automatic) or Reverse (manual). If the vehicle is parked alongside the roadway, or in any other hazardous situation, turn on the emergency hazard flashers. If a tire is to be changed, loosen the lug nuts one-half turn before raising off the ground.

Place the jack under the vehicle in the indicated position **(see illustration)**. Operate the jack with a slow, smooth motion until the wheel is raised off the ground. Remove the lug nuts, pull off the wheel, install the spare and thread the lug nuts back on with the beveled side facing in. Tighten the lug nuts snugly, lower the vehicle until some weight is on the wheel, tighten them completely in a criss-cross pattern and remove the jack. Note that some spare tires are designed for temporary use only - don't exceed the recommended speed, mileage or other restriction instructions accompanying the spare.

Towing

Equipment specifically designed for towing should be used and attached to the main structural members of the vehicle. Optional tow hooks may be attached to the frame at both ends of the vehicle; they are intended for emergency use only, for rescuing a stranded vehicle. Do not use the tow hooks for highway towing. Stand clear when using tow straps or chains, they may break causing serious injury.

Safety is a major consideration when towing and all applicable state and local laws must be obeyed. In addition to a tow bar, a safety chain must be used for all towing.

Four-wheel drive vehicles may be towed with all four wheels on the ground with no speed or mileage restrictions. Turn the ignition key to the OFF position to unlock the steering column, shift the transmission in gear (manual) or Park (automatic) and shift the transfer case to Neutral.

Two-wheel drive vehicles may be towed with four wheels on the ground for a distance of 15 miles or less, as long as the speed doesn't exceed 30 mph. If the vehicle has to be towed more than 15 miles, place the rear wheels on a towing dolly.

If any vehicle is to be towed with the front wheels on the ground and the rear wheels raised, the ignition key must be turned to the OFF position to unlock the steering column and a steering wheel clamping device designed for towing must be used or damage to the steering column lock may occur.

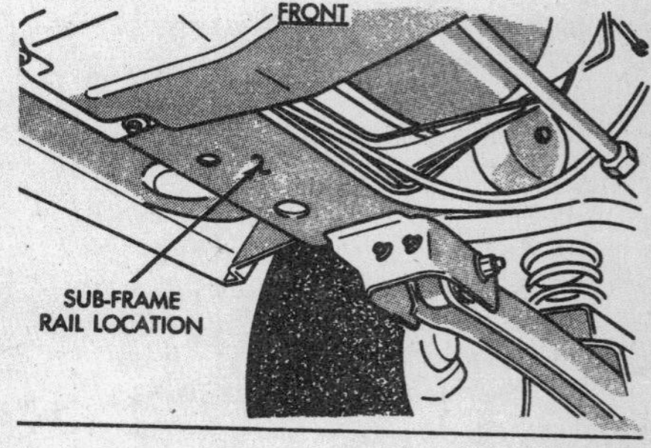

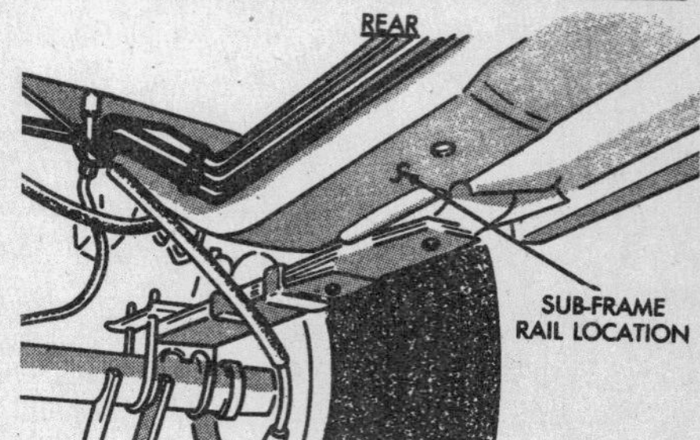

Front and rear jacking points

Automotive chemicals and lubricants

A number of automotive chemicals and lubricants are available for use during vehicle maintenance and repair. They include a wide variety of products ranging from cleaning solvents and degreasers to lubricants and protective sprays for rubber, plastic and vinyl.

Cleaners

Carburetor cleaner and choke cleaner is a strong solvent for gum, varnish and carbon. Most carburetor cleaners leave a dry-type lubricant film which will not harden or gum up. Because of this film it is not recommended for use on electrical components.

Brake system cleaner is used to remove grease and brake fluid from the brake system, where clean surfaces are absolutely necessary. It leaves no residue and often eliminates brake squeal caused by contaminants.

Electrical cleaner removes oxidation, corrosion and carbon deposits from electrical contacts, restoring full current flow. It can also be used to clean spark plugs, carburetor jets, voltage regulators and other parts where an oil-free surface is desired.

Demoisturants remove water and moisture from electrical components such as alternators, voltage regulators, electrical connectors and fuse blocks. They are non-conductive, non-corrosive and non-flammable.

Degreasers are heavy-duty solvents used to remove grease from the outside of the engine and from chassis components. They can be sprayed or brushed on and, depending on the type, are rinsed off either with water or solvent.

Lubricants

Motor oil is the lubricant formulated for use in engines. It normally contains a wide variety of additives to prevent corrosion and reduce foaming and wear. Motor oil comes in various weights (viscosity ratings) from 0 to 50. The recommended weight of the oil depends on the season, temperature and the demands on the engine. Light oil is used in cold climates and under light load conditions. Heavy oil is used in hot climates and where high loads are encountered. Multi-viscosity oils are designed to have characteristics of both light and heavy oils and are available in a number of weights from 5W-20 to 20W-50.

Gear oil is designed to be used in differentials, manual transmissions and other areas where high-temperature lubrication is required.

Chassis and wheel bearing grease is a heavy grease used where increased loads and friction are encountered, such as for wheel bearings, balljoints, tie-rod ends and universal joints.

High-temperature wheel bearing grease is designed to withstand the extreme temperatures encountered by wheel bearings in disc brake equipped vehicles. It usually contains molybdenum disulfide (moly), which is a dry-type lubricant.

White grease is a heavy grease for metal-to-metal applications where water is a problem. White grease stays soft under both low and high temperatures (usually from -100 to +190-degrees F), and will not wash off or dilute in the presence of water.

Assembly lube is a special extreme pressure lubricant, usually containing moly, used to lubricate high-load parts (such as main and rod bearings and cam lobes) for initial start-up of a new engine. The assembly lube lubricates the parts without being squeezed out or washed away until the engine oiling system begins to function.

Silicone lubricants are used to protect rubber, plastic, vinyl and nylon parts.

Graphite lubricants are used where oils cannot be used due to contamination problems, such as in locks. The dry graphite will lubricate metal parts while remaining uncontaminated by dirt, water, oil or acids. It is electrically conductive and will not foul electrical contacts in locks such as the ignition switch.

Moly penetrants loosen and lubricate frozen, rusted and corroded fasteners and prevent future rusting or freezing.

Heat-sink grease is a special electrically non-conductive grease that is used for mounting electronic ignition modules where it is essential that heat is transferred away from the module.

Sealants

RTV sealant is one of the most widely used gasket compounds. Made from silicone, RTV is air curing, it seals, bonds, waterproofs, fills surface irregularities, remains flexible, doesn't shrink, is relatively easy to remove, and is used as a supplementary sealer with almost all low and medium temperature gaskets.

Anaerobic sealant is much like RTV in that it can be used either to seal gaskets or to form gaskets by itself. It remains flexible, is solvent resistant and fills surface imperfections. The difference between an anaerobic sealant and an RTV-type sealant is in the curing. RTV cures when exposed to air, while an anaerobic sealant cures only in the absence of air. This means that an anaerobic sealant cures only after the assembly of parts, sealing them together.

Thread and pipe sealant is used for sealing hydraulic and pneumatic fittings and vacuum lines. It is usually made from a Teflon compound, and comes in a spray, a paint-on liquid and as a wrap-around tape.

Chemicals

Anti-seize compound prevents seizing, galling, cold welding, rust and corrosion in fasteners. High-temperature anti-seize, usually made with copper and graphite lubricants, is used for exhaust system and exhaust manifold bolts.

Anaerobic locking compounds are used to keep fasteners from vibrating or working loose and cure only after installation, in the absence of air. Medium strength locking compound is used for small nuts, bolts and screws that may be removed later. High-strength locking compound is for large nuts, bolts and studs which aren't removed on a regular basis.

Oil additives range from viscosity index improvers to chemical treatments that claim to reduce internal engine friction. It should be noted that most oil manufacturers caution against using additives with their oils.

Gas additives perform several functions, depending on their chemical makeup. They usually contain solvents that help dissolve gum and varnish that build up on carburetor, fuel injection and intake parts. They also serve to break down carbon deposits that form on the inside surfaces of the combustion chambers. Some additives contain upper cylinder lubricants for valves and piston rings, and others contain chemicals to remove condensation from the gas tank.

Miscellaneous

Brake fluid is specially formulated hydraulic fluid that can withstand the heat and pressure encountered in brake systems. Care must be taken so this fluid does not come in contact with painted surfaces or plastics. An opened container should always be resealed to prevent contamination by water or dirt.

Weatherstrip adhesive is used to bond weatherstripping around doors, windows and trunk lids. It is sometimes used to attach trim pieces.

Undercoating is a petroleum-based, tar-like substance that is designed to protect metal surfaces on the underside of the vehicle from corrosion. It also acts as a sound-deadening agent by insulating the bottom of the vehicle.

Waxes and polishes are used to help protect painted and plated surfaces from the weather. Different types of paint may require the use of different types of wax and polish. Some polishes utilize a chemical or abrasive cleaner to help remove the top layer of oxidized (dull) paint on older vehicles. In recent years many non-wax polishes that contain a wide variety of chemicals such as polymers and silicones have been introduced. These non-wax polishes are usually easier to apply and last longer than conventional waxes and polishes.

Conversion factors

Length (distance)					
Inches (in)	X 25.4	= Millimetres (mm)	X 0.0394	= Inches (in)
Feet (ft)	X 0.305	= Metres (m)	X 3.281	= Feet (ft)
Miles	X 1.609	= Kilometres (km)	X 0.621	= Miles
Volume (capacity)				
Cubic inches (cu in; in³)	X 16.387	= Cubic centimetres (cc; cm³)	X 0.061	= Cubic inches (cu in; in³)
Imperial pints (Imp pt)	X 0.568	= Litres (l)	X 1.76	= Imperial pints (Imp pt)
Imperial quarts (Imp qt)	X 1.137	= Litres (l)	X 0.88	= Imperial quarts (Imp qt)
Imperial quarts (Imp qt)	X 1.201	= US quarts (US qt)	X 0.833	= Imperial quarts (Imp qt)
US quarts (US qt)	X 0.946	= Litres (l)	X 1.057	= US quarts (US qt)
Imperial gallons (Imp gal)	X 4.546	= Litres (l)	X 0.22	= Imperial gallons (Imp gal)
Imperial gallons (Imp gal)	X 1.201	= US gallons (US gal)	X 0.833	= Imperial gallons (Imp gal)
US gallons (US gal)	X 3.785	= Litres (l)	X 0.264	= US gallons (US gal)
Mass (weight)				
Ounces (oz)	X 28.35	= Grams (g)	X 0.035	= Ounces (oz)
Pounds (lb)	X 0.454	= Kilograms (kg)	X 2.205	= Pounds (lb)
Force				
Ounces-force (ozf; oz)	X 0.278	= Newtons (N)	X 3.6	= Ounces-force (ozf; oz)
Pounds-force (lbf; lb)	X 4.448	= Newtons (N)	X 0.225	= Pounds-force (lbf; lb)
Newtons (N)	X 0.1	= Kilograms-force (kgf; kg)	X 9.81	= Newtons (N)
Pressure				
Pounds-force per square inch (psi; lbf/in²; lb/in²)	X 0.070	= Kilograms-force per square centimetre (kgf/cm²; kg/cm²)	X 14.223	= Pounds-force per square inch (psi; lbf/in²; lb/in²)
Pounds-force per square inch (psi; lbf/in²; lb/in²)	X 0.068	= Atmospheres (atm)	X 14.696	= Pounds-force per square inch (psi; lbf/in²; lb/in²)
Pounds-force per square inch (psi; lbf/in²; lb/in²)	X 0.069	= Bars	X 14.5	= Pounds-force per square inch (psi; lbf/in²; lb/in²)
Pounds-force per square inch (psi; lbf/in²; lb/in²)	X 6.895	= Kilopascals (kPa)	X 0.145	= Pounds-force per square inch (psi; lbf/in²; lb/in²)
Kilopascals (kPa)	X 0.01	= Kilograms-force per square centimetre (kgf/cm²; kg/cm²)	X 98.1	= Kilopascals (kPa)
Torque (moment of force)				
Pounds-force inches (lbf in; lb in)	X 1.152	= Kilograms-force centimetre (kgf cm; kg cm)	X 0.868	= Pounds-force inches (lbf in; lb in)
Pounds-force inches (lbf in; lb in)	X 0.113	= Newton metres (Nm)	X 8.85	= Pounds-force inches (lbf in; lb in)
Pounds-force inches (lbf in; lb in)	X 0.083	= Pounds-force feet (lbf ft; lb ft)	X 12	= Pounds-force inches (lbf in; lb in)
Pounds-force feet (lbf ft; lb ft)	X 0.138	= Kilograms-force metres (kgf m; kg m)	X 7.233	= Pounds-force feet (lbf ft; lb ft)
Pounds-force feet (lbf ft; lb ft)	X 1.356	= Newton metres (Nm)	X 0.738	= Pounds-force feet (lbf ft; lb ft)
Newton metres (Nm)	X 0.102	= Kilograms-force metres (kgf m; kg m)	X 9.804	= Newton metres (Nm)
Vacuum				
Inches mercury (in. Hg)	X 3.377	= Kilopascals (kPa)	X 0.2961	= Inches mercury
Inches mercury (in. Hg)	X 25.4	= Millimeters mercury (mm Hg)	X 0.0394	= Inches mercury
Power				
Horsepower (hp)	X 745.7	= Watts (W)	X 0.0013	= Horsepower (hp)
Velocity (speed)				
Miles per hour (miles/hr; mph)	X 1.609	= Kilometres per hour (km/hr; kph)	X 0.621	= Miles per hour (miles/hr; mph)
Fuel consumption*				
Miles per gallon, Imperial (mpg)	X 0.354	= Kilometres per litre (km/l)	X 2.825	= Miles per gallon, Imperial (mpg)
Miles per gallon, US (mpg)	X 0.425	= Kilometres per litre (km/l)	X 2.352	= Miles per gallon, US (mpg)

Temperature

Degrees Fahrenheit = (°C x 1.8) + 32

Degrees Celsius (Degrees Centigrade; °C) = (°F - 32) x 0.56

It is common practice to convert from miles per gallon (mpg) to litres/100 kilometres (l/100km), where mpg (Imperial) x l/100 km = 282 and mpg (US) x l/100 km = 235

Safety first!

Regardless of how enthusiastic you may be about getting on with the job at hand, take the time to ensure that your safety is not jeopardized. A moment's lack of attention can result in an accident, as can failure to observe certain simple safety precautions. The possibility of an accident will always exist, and the following points should not be considered a comprehensive list of all dangers. Rather, they are intended to make you aware of the risks and to encourage a safety conscious approach to all work you carry out on your vehicle.

Essential DOs and DON'Ts

DON'T rely on a jack when working under the vehicle. Always use approved jackstands to support the weight of the vehicle and place them under the recommended lift or support points.

DON'T attempt to loosen extremely tight fasteners (i.e. wheel lug nuts) while the vehicle is on a jack - it may fall.

DON'T start the engine without first making sure that the transmission is in Neutral (or Park where applicable) and the parking brake is set.

DON'T remove the radiator cap from a hot cooling system - let it cool or cover it with a cloth and release the pressure gradually.

DON'T attempt to drain the engine oil until you are sure it has cooled to the point that it will not burn you.

DON'T touch any part of the engine or exhaust system until it has cooled sufficiently to avoid burns.

DON'T siphon toxic liquids such as gasoline, antifreeze and brake fluid by mouth, or allow them to remain on your skin.

DON'T inhale brake lining dust - it is potentially hazardous (see *Asbestos* below).

DON'T allow spilled oil or grease to remain on the floor - wipe it up before someone slips on it.

DON'T use loose fitting wrenches or other tools which may slip and cause injury.

DON'T push on wrenches when loosening or tightening nuts or bolts. Always try to pull the wrench toward you. If the situation calls for pushing the wrench away, push with an open hand to avoid scraped knuckles if the wrench should slip.

DON'T attempt to lift a heavy component alone - get someone to help you.

DON'T rush or take unsafe shortcuts to finish a job.

DON'T allow children or animals in or around the vehicle while you are working on it.

DO wear eye protection when using power tools such as a drill, sander, bench grinder, etc. and when working under a vehicle.

DO keep loose clothing and long hair well out of the way of moving parts.

DO make sure that any hoist used has a safe working load rating adequate for the job.

DO get someone to check on you periodically when working alone on a vehicle.

DO carry out work in a logical sequence and make sure that everything is correctly assembled and tightened.

DO keep chemicals and fluids tightly capped and out of the reach of children and pets.

DO remember that your vehicle's safety affects that of yourself and others. If in doubt on any point, get professional advice.

Asbestos

Certain friction, insulating, sealing, and other products - such as brake linings, brake bands, clutch linings, torque converters, gaskets, etc. - may contain asbestos. Extreme care must be taken to avoid inhalation of dust from such products, since it is hazardous to health. If in doubt, assume that they do contain asbestos.

Fire

Remember at all times that gasoline is highly flammable. Never smoke or have any kind of open flame around when working on a vehicle. But the risk does not end there. A spark caused by an electrical short circuit, by two metal surfaces contacting each other, or even by static electricity built up in your body under certain conditions, can ignite gasoline vapors, which in a confined space are highly explosive. Do not, under any circumstances, use gasoline for cleaning parts. Use an approved safety solvent.

Always disconnect the battery ground (-) cable at the battery before working on any part of the fuel system or electrical system. Never risk spilling fuel on a hot engine or exhaust component. It is strongly recommended that a fire extinguisher suitable for use on fuel and electrical fires be kept handy in the garage or workshop at all times. Never try to extinguish a fuel or electrical fire with water.

Fumes

Certain fumes are highly toxic and can quickly cause unconsciousness and even death if inhaled to any extent. Gasoline vapor falls into this category, as do the vapors from some cleaning solvents. Any draining or pouring of such volatile fluids should be done in a well ventilated area.

When using cleaning fluids and solvents, read the instructions on the container carefully. Never use materials from unmarked containers.

Never run the engine in an enclosed space, such as a garage. Exhaust fumes contain carbon monoxide, which is extremely poisonous. If you need to run the engine, always do so in the open air, or at least have the rear of the vehicle outside the work area.

If you are fortunate enough to have the use of an inspection pit, never drain or pour gasoline and never run the engine while the vehicle is over the pit. The fumes, being heavier than air, will concentrate in the pit with possibly lethal results.

The battery

Never create a spark or allow a bare light bulb near a battery. They normally give off a certain amount of hydrogen gas, which is highly explosive.

Always disconnect the battery ground (-) cable at the battery before working on the fuel or electrical systems.

If possible, loosen the filler caps or cover when charging the battery from an external source (this does not apply to sealed or maintenance-free batteries). Do not charge at an excessive rate or the battery may burst.

Take care when adding water to a non maintenance-free battery and when carrying a battery. The electrolyte, even when diluted, is very corrosive and should not be allowed to contact clothing or skin.

Always wear eye protection when cleaning the battery to prevent the caustic deposits from entering your eyes.

Household current

When using an electric power tool, inspection light, etc., which operates on household current, always make sure that the tool is correctly connected to its plug and that, where necessary, it is properly grounded. Do not use such items in damp conditions and, again, do not create a spark or apply excessive heat in the vicinity of fuel or fuel vapor.

Secondary ignition system voltage

A severe electric shock can result from touching certain parts of the ignition system (such as the spark plug wires) when the engine is running or being cranked, particularly if components are damp or the insulation is defective. In the case of an electronic ignition system, the secondary system voltage is much higher and could prove fatal.

Troubleshooting

Contents

This section provides an easy reference guide to the more common problems which may occur during the operation of your vehicle. These problems and possible causes are grouped under various components or systems; i.e. Engine, Cooling System, etc., and also refer to the Chapter and/or Section which deals with the problem.

Remember that successful troubleshooting is not a mysterious black art practiced only by professional mechanics. It's simply the result of a bit of knowledge combined with an intelligent, systematic approach to the problem. Always work by a process of elimination, starting with the simplest solution and working through to the most complex - and never overlook the obvious. Anyone can forget to fill the gas tank or leave the lights on overnight, so don't assume that you are above such oversights.

Finally, always get clear in your mind why a problem has occurred and take steps to ensure that it doesn't happen again. If the electrical system fails because of a poor connection, check all other connections in the system to make sure that they don't fail as well. If a particular fuse continues to blow, find out why - don't just go on replacing fuses. Remember, failure of a small component can often be indicative of potential failure or incorrect functioning of a more important component or system.

Engine

1 Engine will not rotate when attempting to start

1 Battery terminal connections loose or corroded. Check the cable terminals at the battery. Tighten the cable or remove corrosion as necessary.
2 Battery discharged or faulty. If the cable connections are clean and tight on the battery posts, turn the key to the On position and switch on the headlights and/or windshield wipers. If they fail to function, the battery is discharged.
3 Automatic transmission not completely engaged in Park or Neutral or clutch pedal not completely depressed.
4 Broken, loose or disconnected wiring in the starting circuit. Inspect all wiring and connectors at the battery, starter solenoid and ignition switch.
5 Starter motor pinion jammed in flywheel ring gear. If manual transmission, place transmission in gear and rock the vehicle to manually turn the engine. Remove starter and inspect pinion and flywheel at earliest convenience (Chapter 5).
6 Starter solenoid faulty (Chapter 5).
7 Starter motor faulty (Chapter 5).
8 Ignition switch faulty (Chapter 12).

2 Engine rotates but will not start

1 Fuel tank empty.
2 Fault in the fuel injection system (Chapter 4).
3 Battery discharged (engine rotates slowly). Check the operation of electrical components as described in the previous Section.
4 Battery terminal connections loose or corroded (see previous Section).
5 Fuel pump faulty (Chapter 4).
6 Excessive moisture on, or damage to, ignition components (see Chapter 5).
7 Worn, faulty or incorrectly gapped spark plugs (Chapter 1).
8 Broken, loose or disconnected wiring in the starting circuit (see previous Section).
9 Broken, loose or disconnected wires at the ignition coil (Chapter 5).

3 Starter motor operates without rotating engine

1 Starter pinion sticking. Remove the starter (Chapter 5) and inspect.
2 Starter pinion or flywheel teeth worn or broken. Remove the flywheel/driveplate access cover and inspect.

4 Engine hard to start when cold

1 Battery discharged or low. Check as described in Section 1.
2 Fault in the fuel or electrical systems (Chapters 4 and 5).

5 Engine hard to start when hot

1 Air filter clogged (Chapter 1).
2 Fault in the fuel or electrical systems (Chapters 4 and 5).
3 Fuel not reaching the injectors (see Chapter 4).

6 Starter motor noisy or excessively rough in engagement

1 Pinion or flywheel gear teeth worn or broken. Remove the cover at the rear of the engine (if equipped) and inspect.
2 Starter motor mounting bolts loose or missing.

7 Engine starts but stops immediately

1 Loose or faulty electrical connections at distributor, coil or alternator.
2 Fault in the fuel or electrical systems (Chapters 4 and 5).
3 Vacuum leak at the gasket surfaces of the intake manifold or throttle body. Make sure all mounting bolts/nuts are tightened securely and all vacuum hoses connected to the manifold are positioned properly and in good condition.

8 Engine lopes while idling or idles erratically

1 Vacuum leakage. Check the mounting bolts/nuts at the throttle body and intake manifold for tightness. Make sure all vacuum hoses are connected and in good condition. Use a stethoscope or a length of fuel hose held against your ear to listen for vacuum leaks while the engine is running. A hissing sound will be heard. A soapy water solution will also detect leaks.
2 Fault in the fuel or electrical systems (Chapters 4 and 5).
3 Plugged PCV valve or CCV hose (see Chapters 1 and 6).
4 Air filter clogged (Chapter 1).
5 Fuel pump not delivering sufficient fuel to the fuel injectors (see Chapter 4).
6 Leaking head gasket. Perform a compression check (Chapter 2).
7 Camshaft lobes worn (Chapter 2).

9 Engine misses at idle speed

1 Spark plugs worn, fouled or not gapped properly (Chapter 1).
2 Fault in the fuel or electrical systems (Chapters 4 and 5).
3 Faulty spark plug wires (Chapter 1).
4 Vacuum leaks at intake or hose connections. Check as described in Section 8.
5 Uneven or low cylinder compression. Check compression as described in Chapter 1.

10 Engine misses throughout driving speed range

1 Fuel filter clogged and/or impurities in the fuel system (Chapter 1).
2 Faulty or incorrectly gapped spark plugs (Chapter 1).
3 Fault in the fuel or electrical systems (Chapters 4 and 5).
4 Defective spark plug wires (Chapter 1).
5 Faulty emissions system components (Chapter 6).
6 Low or uneven cylinder compression pressures. Remove the spark plugs and test the compression with a gauge (Chapter 2).
7 Weak or faulty ignition system (Chapter 5).
8 Vacuum leaks at the throttle body, intake manifold or vacuum hoses (see Section 8).

11 Engine stalls

1 Idle speed incorrect. Refer to the VECI label.
2 Fuel filter clogged and/or water and impurities in the fuel system (Chapter 1).
3 Fault in the fuel system or sensors (Chapters 4 and 6).
4 Faulty emissions system components (Chapter 6).
5 Faulty or incorrectly gapped spark plugs (Chapter 1). Also check the spark plug wires (Chapter 1).
6 Vacuum leak at the throttle body, intake manifold or vacuum hoses. Check as described in Section 8.

12 Engine lacks power

1 Fault in the fuel or electrical systems (Chapters 4 and 5).
2 Faulty or incorrectly gapped spark plugs (Chapter 1).
3 Faulty coil (Chapter 5).
4 Brakes binding (Chapter 1).
5 Automatic transmission fluid level incorrect (Chapter 1).
6 Clutch slipping (Chapter 8).
7 Fuel filter clogged and/or impurities in the fuel system (Chapter 1).
8 Emissions control system not functioning properly (Chapter 6).
9 Use of substandard fuel. Fill the tank with the proper octane fuel.
10 Low or uneven cylinder compression pressures. Test with a compression tester, which will detect leaking valves and/or a blown head gasket (Chapter 2).

13 Engine backfires

1 Emissions system not functioning properly (Chapter 6).
2 Fault in the fuel or electrical systems (Chapters 4 and 5).
3 Faulty secondary ignition system (cracked spark plug insulator or faulty plug wires) (Chapters 1 and 5).
4 Vacuum leak at the throttle body, intake manifold or vacuum hoses. Check as described in Section 8.
5 Valves sticking (Chapter 2).
6 Crossed plug wires (Chapter 1).
7 Valve timing incorrect (Chapter 2).

14 Pinging or knocking engine sounds during acceleration or uphill

1 Incorrect grade of fuel. Fill the tank with fuel of the proper octane rating.
2 Fault in the fuel or electrical systems (Chapters 4 and 5).
3 Improper spark plugs. Check the plug type against the VECI label located in the engine compartment. Also check the plugs and wires for damage (Chapter 1).
4 Faulty emissions system (Chapter 6).
5 Vacuum leak. Check as described in Section 9.

15 Engine diesels (continues to run) after switching off

1 Fault in the fuel, electrical or emissions/engine control systems (Chapters 4, 5 and 6).
2 Excessive engine operating temperature. Probable causes of this are a low coolant level (see Chapter 1), malfunctioning thermostat, clogged radiator or faulty water pump (see Chapter 3).

Engine electrical system

16 Battery will not hold a charge

1 Alternator drivebelt defective or not adjusted properly (Chapter 1).
2 Electrolyte level low or battery discharged (Chapter 1).
3 Battery terminals loose or corroded (Chapter 1).
4 Alternator not charging properly (Chapter 5).
5 Loose, broken or faulty wiring in the charging circuit (Chapter 5).
6 Short in the vehicle wiring causing a continuous drain on the battery (refer to Chapter 12 and the Wiring Diagrams).
7 Battery defective internally.

17 Ignition light fails to go out

1 Fault in the alternator or charging circuit (Chapter 5).
2 Alternator drivebelt defective or not properly adjusted (Chapter 1).

18 Ignition light fails to come on when key is turned on

1 Instrument cluster warning light bulb defective (Chapter 12).
2 Alternator faulty (Chapter 5).
3 Fault in the instrument cluster printed circuit, dashboard wiring or bulb holder (Chapter 12).

Fuel system

19 Excessive fuel consumption

1 Dirty or clogged air filter element (Chapter 1).
2 Emissions system not functioning properly (Chapter 6).
3 Fault in the fuel or electrical systems (Chapters 4 and 5).
4 Low tire pressure or incorrect tire size (Chapter 1).

20 Fuel leakage and/or fuel odor

1 Leak in a fuel feed or vent line (Chapter 4).
2 Tank overfilled. Fill only to automatic shut-off.
3 Evaporative emissions system canister clogged (Chapter 6).
4 Vapor leaks from system lines (Chapter 4).

Cooling system

21 Overheating

1 Insufficient coolant in the system (Chapter 1).
2 Water pump drivebelt defective or not adjusted properly (Chapter 1).
3 Radiator core blocked or radiator grille dirty and restricted (see Chapter 3).
4 Thermostat faulty (Chapter 3).
5 Fan blades broken or cracked (Chapter 3).
6 Radiator cap not maintaining proper pressure. Have the cap pressure tested by a gas station or repair shop.

22 Overcooling

1 Thermostat faulty (Chapter 3).
2 Inaccurate temperature gauge (Chapter 12).

23 External coolant leakage

1 Deteriorated or damaged hoses or loose clamps. Replace hoses and/or tighten the clamps at the hose connections (Chapter 1).
2 Water pump seals defective. If this is the case, water will drip from the weep hole in the water pump body (Chapter 3).
3 Leakage from the radiator core or side tank(s). This will require the radiator to be professionally repaired (see Chapter 3 for removal procedures).
4 Engine drain plug leaking (Chapter 1) or water jacket core plugs leaking (see Chapter 2).

24 Internal coolant leakage

Note: *Internal coolant leaks can usually be*

detected by examining the oil. Check the dipstick and inside of the valve cover for water deposits and an oil consistency like that of a milkshake.

1 Leaking cylinder head gasket. Have the cooling system pressure tested.
2 Cracked cylinder bore or cylinder head. Dismantle the engine and inspect (Chapter 2).

25 Coolant loss

1 Too much coolant in the system (Chapter 1).
2 Coolant boiling away due to overheating (see Section 15).
3 External or internal leakage (see Sections 23 and 24).
4 Faulty radiator cap. Have the cap pressure tested.

26 Poor coolant circulation

1 Worn water pump impeller vanes (see Chapter 3).
2 Restriction in the cooling system. Drain, flush and refill the system (Chapter 1). If necessary, remove the radiator (Chapter 3) and have it reverse flushed.
3 Water pump drivebelt defective or not adjusted properly (Chapter 1).
4 Thermostat sticking (Chapter 3).
5 Drivebelt incorrectly routed, causing the pump to turn backwards (Chapter 1).

Clutch

27 Fails to release (pedal pressed to the floor - shift lever does not move freely in and out of Reverse)

1 Leak in the clutch hydraulic system. Check the master cylinder, slave cylinder and lines (Chapter 8).
2 Clutch plate warped, damaged or stuck to transmission input shaft splines (Chapter 8).
3 Defective pressure plate (Chapter 8)

28 Clutch slips (engine speed increases with no increase in vehicle speed)

1 Clutch plate oil soaked or lining worn. Remove clutch (Chapter 8) and inspect.
2 Clutch plate not seated. It may take 30 or 40 normal starts for a new one to seat.
3 Pressure plate glazed or worn (Chapter 8).
4 Flywheel glazed (Chapter 2)

29 Grabbing (chattering) as clutch is engaged

1 Oil on clutch plate lining. Remove (Chapter 8) and inspect. Correct any leakage source.
2 Worn or loose engine or transmission mounts. These units move slightly when the clutch is released. Inspect the mounts and bolts (Chapter 2).
3 Worn splines on clutch plate hub. Remove the clutch components (Chapter 8) and inspect.
4 Warped pressure plate or flywheel. Remove the clutch components and inspect.

30 Squeal or rumble with clutch fully engaged (pedal released)

1 Release bearing binding on transmission bearing retainer. Remove clutch components (Chapter 8) and check bearing. Remove any burrs or nicks; clean and relubricate bearing retainer before installing.

31 Squeal or rumble with clutch fully disengaged (pedal depressed)

1 Worn, defective or broken release bearing (Chapter 8).
2 Worn or broken pressure plate springs (or diaphragm fingers) (Chapter 8).
3 Worn, defective or dry pilot bearing (Chapter 8).

32 Clutch pedal stays on floor when disengaged

1 Release bearing binding (Chapter 8).
2 Make sure proper pedal stop (bumper) is installed.
3 Defective pressure plate (Chapter 8).

Manual transmission

Note: *All the following references are in Chapter 7, unless noted.*

33 Noisy in Neutral with engine running

1 Input shaft bearing worn.
2 Damaged main drive gear bearing.
3 Worn countershaft bearings.
4 Worn or damaged countershaft endplay shims.

34 Noisy in all gears

1 Any of the above causes, and/or:
2 Insufficient lubricant (see the checking procedures in Chapter 1).

35 Noisy in one particular gear

1 Worn, damaged or chipped gear teeth for that particular gear.
2 Worn or damaged synchronizer for that particular gear.

36 Slips out of high gear

1 Transmission loose on clutch housing.
2 Dirt between the transmission case and engine or misalignment of the transmission (Chapter 7).

37 Difficulty in engaging gears

1 Clutch not releasing completely (see clutch adjustment in Chapter 1).
2 Loose, damaged or out-of-adjustment shift linkage. Make a thorough inspection, replacing parts as necessary (Chapter 7).

38 Oil leakage

1 Excessive amount of lubricant in the transmission (see Chapter 1 for correct checking procedures). Drain lubricant as required.
2 Transmission oil seal or speedometer oil seal in need of replacement (Chapter 7).

Automatic transmission

Note: *Due to the complexity of the automatic transmission, it's difficult for the home mechanic to properly diagnose and service this component. For problems other than the following, the vehicle should be taken to a dealer service department or a transmission shop.*

39 General shift mechanism problems

1 Chapter 7B deals with checking and adjusting the shift cable on automatic transmissions. Common problems which may be attributed to poorly adjusted cable are:

a) *Engine starting in gears other than Park or Neutral.*
b) *Indicator on shifter pointing to a gear other than the one actually being selected.*
c) *Vehicle moves when in Park.*

2 Refer to Chapter 7B to adjust the cable.

40 Transmission will not downshift with accelerator pedal pressed to the floor

Throttle valve (TV) cable misadjusted (Chapter 7B).

41 Transmission slips, shifts rough, is noisy or has no drive in forward or reverse gears

1 There are many probable causes for the above problems, but the home mechanic should be concerned with only one possibility - fluid level.
2 Before taking the vehicle to a repair shop, check the level and condition of the fluid as described in Chapter 1. Correct fluid level as necessary or change the fluid and filter if needed. If the problem persists, have a professional diagnose the probable cause.

42 Fluid leakage

1 Automatic transmission fluid is a deep red color. Fluid leaks should not be confused with engine oil, which can easily be blown by air flow to the transmission.
2 To pinpoint a leak, first remove all built-up dirt and grime from around the transmission. Degreasing agents and/or steam cleaning will achieve this. With the underside clean, drive the vehicle at low speeds so air flow will not blow the leak far from its source. Raise the vehicle and determine where the leak is coming from. Common areas of leakage are:

a) ***Pan:*** *Tighten the mounting bolts and/or replace the pan gasket as necessary (see Chapter 1).*
b) ***Filler pipe:*** *Replace the rubber seal where the pipe enters the transmission case.*
c) ***Transmission oil lines:*** *Tighten the connectors where the lines enter the transmission case and/or replace the lines.*
d) ***Vent pipe:*** *Transmission overfilled and/or water in fluid (see checking procedures, Chapter 1).*
e) ***Speedometer connector:*** *Replace the O-ring where the speedometer sensor enters the transmission case (Chapter 7B).*

Transfer case

43 Transfer case is difficult to shift into the desired range

1 Speed may be too great to permit engagement. Stop the vehicle and shift into the desired range.
2 Shift linkage loose, bent or binding (early models). Check the linkage for damage or wear and replace or lubricate as necessary (Chapter 7C).
3 Shift cable misadjusted or binding (later models) (Chapter 7C).
4 If the vehicle has been driven on a paved surface for some time, the driveline torque can make shifting difficult. Stop and shift into two-wheel drive on paved or hard surfaces.
5 Insufficient or incorrect grade of lubricant. Drain and refill the transfer case with the specified lubricant. (Chapter 1).
6 Worn or damaged internal components. Disassembly and overhaul of the transfer case may be necessary.

44 Transfer case noisy in all gears

Insufficient or incorrect grade of lubricant. Drain and refill (Chapter 1).

45 Noisy or jumps out of four-wheel drive Low range

1 Transfer case not fully engaged. Stop the vehicle, shift into Neutral and then engage 4L.
2 Shift linkage loose, worn or binding (early models). Tighten, repair or lubricate linkage as necessary (Chapter 7C).
3 Shift fork cracked, inserts worn or fork binding on the rail.
4 Shift cable misadjusted (later models) (Chapter 7C).

46 Lubricant leaks from the vent or output shaft seals

1 Transfer case is overfilled. Drain to the proper level (Chapter 1).
2 Vent is clogged or jammed closed. Clear or replace the vent.
3 Output shaft seal incorrectly installed or damaged. Replace the seal and check contact surfaces for nicks and scoring.

Driveshaft

47 Oil leak at seal end of driveshaft

Defective transmission or transfer case oil seal. See Chapter 7B for replacement procedures. While this is done, check the splined yoke for burrs or a rough condition which may be damaging the seal. Burrs can be removed with crocus cloth or a fine whetstone.

48 Knock or clunk when the transmission is under initial load (just after transmission is put into gear)

1 Loose or disconnected rear suspension components. Check all mounting bolts, nuts and bushings (see Chapter 10).
2 Loose driveshaft bolts. Inspect all bolts and nuts and tighten them to the specified torque.
3 Worn or damaged universal joint bearings. Check for wear (see Chapter 8).

49 Metallic grinding sound consistent with vehicle speed.

Pronounced wear in the universal joint bearings. Check as described in Chapter 8.

50 Vibration

Note: *Before assuming that the driveshaft is at fault, make sure the tires are perfectly balanced and perform the following test.*
1 Install a tachometer inside the vehicle to monitor engine speed as the vehicle is driven. Drive the vehicle and note the engine speed at which the vibration (roughness) is most pronounced. Now shift the transmission to a different gear and bring the engine speed to the same point.
2 If the vibration occurs at the same engine speed (rpm) regardless of which gear the transmission is in, the driveshaft is NOT at fault since the driveshaft speed varies.
3 If the vibration decreases or is eliminated when the transmission is in a different gear at the same engine speed, refer to the following probable causes.
4 Bent or dented driveshaft. Inspect and replace as necessary (see Chapter 8).
5 Undercoating or built-up dirt, etc. on the driveshaft. Clean the shaft thoroughly and recheck.
6 Worn universal joint bearings. Remove and inspect (see Chapter 8).
7 Driveshaft and/or companion flange out of balance. Check for missing weights on the shaft. Remove the driveshaft (see Chapter 8) and reinstall 180-degrees from original position, then retest. Have the driveshaft professionally balanced if the problem persists.

Axles

51 Noise

1 Road noise. No corrective procedures available.
2 Tire noise. Inspect tires and check tire pressures (Chapter 1).
3 Rear wheel bearings loose, worn or damaged (Chapter 8).

52 Vibration

See probable causes under *Driveshaft*. Proceed under the guidelines listed for the driveshaft. If the problem persists, check the rear wheel bearings by raising the rear of the vehicle and spinning the rear wheels by hand. Listen for evidence of rough (noisy) bearings. Remove and inspect (see Chapter 8).

53 Oil leakage

1 Pinion seal damaged (see Chapter 8).

2 Axleshaft oil seals damaged (see Chapter 8).
3 Differential inspection cover leaking. Tighten the bolts or replace the gasket as required (see Chapters 1 and 8).

Brakes

Note: *Before assuming that a brake problem exists, make sure that the tires are in good condition and inflated properly (see Chapter 1), that the front end alignment is correct and that the vehicle is not loaded with weight in an unequal manner.*

54 Vehicle pulls to one side during braking

1 Defective, damaged or oil contaminated disc brake pads or shoes on one side. Inspect as described in Chapter 9.
2 Excessive wear of brake shoe or pad material or drum/disc on one side. Inspect and correct as necessary.
3 Defective drum brake or caliper assembly. Remove the drum or caliper and inspect for a stuck piston or other damage (Chapter 9).
4 Inadequate lubrication of front brake caliper slide pins or bushings. Remove caliper and lubricate the pins or bushings (Chapter 9).

55 Noise (high-pitched squeal with the brakes applied)

1 Disc brake pads worn out. The noise comes from the wear sensor rubbing against the disc (does not apply to all vehicles) or the actual pad backing plate itself if the material is completely worn away. Replace the pads with new ones immediately (Chapter 9). If the pad material has worn completely away, the brake discs should be inspected for damage as described in Chapter 9.
2 Linings contaminated with dirt or grease. Replace pads or shoes.
3 Incorrect linings. Replace with correct linings.

56 Excessive brake pedal travel

1 Partial brake system failure. Inspect the entire system (Chapter 9) and correct as required.
2 Insufficient fluid in the master cylinder. Check (Chapter 1), add fluid and bleed the system if necessary (Chapter 9).
3 Rear brakes not adjusting properly. Make a series of starts and stops while the vehicle is in Reverse. If this does not correct the situation, remove the drums and inspect the self-adjusters (Chapter 9).

57 Brake pedal feels spongy when depressed

1 Air in the hydraulic lines. Bleed the brake system (Chapter 9).
2 Faulty flexible hoses. Inspect all system hoses and lines. Replace parts as necessary.
3 Master cylinder mounting bolts/nuts loose.
4 Master cylinder defective (Chapter 9).

58 Excessive effort required to stop vehicle

1 Power brake booster not operating properly (Chapter 9).
2 Excessively worn linings or pads. Inspect and replace if necessary (Chapter 9).
3 One or more caliper pistons or wheel cylinders seized or sticking. Inspect and rebuild as required (Chapter 9).
4 Brake linings or pads contaminated with oil or grease. Inspect and replace as required (Chapter 9).
5 New pads or shoes installed and not yet seated. It will take a while for the new material to seat against the drum (or disc).

59 Pedal travels to the floor with little resistance

1 Little or no fluid in the master cylinder reservoir caused by leaking wheel cylinder(s), leaking caliper piston(s), loose, damaged or disconnected brake lines. Inspect the entire system and correct as necessary.
2 Worn master cylinder seals (Chapter 9).

60 Brake pedal pulsates during brake application

1 Caliper improperly installed. Remove and inspect (Chapter 9).
2 Disc or drum defective. Check for excessive lateral runout and parallelism (discs) or excessive out-of-roundness (drums). Have the disc or drum resurfaced or replace it with a new one (Chapter 9).

Suspension and steering systems

61 Vehicle pulls to one side

1 Tire pressures uneven (Chapter 1).
2 Defective tire (Chapter 1).
3 Excessive wear in suspension or steering components (Chapter 10).
4 Front end in need of alignment.
5 Front brakes dragging. Inspect the brakes as described in Chapter 9.

62 Shimmy, shake or vibration

1 Tire or wheel out-of-balance or out-of-round. Have professionally balanced.
2 Loose, worn or out-of-adjustment rear wheel bearings (Chapter 1).
3 Shock absorbers and/or suspension components worn or damaged (Chapter 10).

63 Excessive pitching and/or rolling around corners or during braking

1 Defective shock absorbers. Replace as a set (Chapter 10).
2 Weak springs and/or suspension components. Inspect as described in Chapter 10.
3 Broken stabilizer bar link (Chapter 10).

64 Excessively stiff steering

1 Lack of fluid in power steering fluid reservoir (Chapter 1).
2 Incorrect tire pressures (Chapter 1).
3 Lack of lubrication at steering joints (see Chapter 1).
4 Front end out of alignment.
5 Lack of power assistance (see Section 66).

65 Excessive play in steering

1 Loose front wheel bearings (Chapter 1 and 10).
2 Excessive wear in suspension or steering components (Chapter 10).
3 Steering gearbox damaged or out of adjustment (Chapter 10).

66 Lack of power assistance

1 Steering pump drivebelt faulty or not adjusted properly (Chapter 1).
2 Fluid level low (Chapter 1).
3 Hoses or lines restricted. Inspect and replace parts as necessary.
4 Air in power steering system. Bleed the system (Chapter 10).

67 Excessive tire wear (not specific to one area)

1 Incorrect tire pressures (Chapter 1).
2 Tires out-of-balance. Have professionally balanced.
3 Wheels damaged. Inspect and replace as necessary.
4 Suspension or steering components excessively worn (Chapter 10).

68 Excessive tire wear on outside edge

1 Inflation pressures incorrect (Chapter 1).
2 Excessive speed in turns.
3 Front end alignment incorrect (excessive toe-in). Have professionally aligned.

69 Excessive tire wear on inside edge

1 Inflation pressures incorrect (Chapter 1).
2 Front end alignment incorrect (toe-out). Have professionally aligned.
3 Loose or damaged steering components (Chapter 10).

70 Tire tread worn in one place

1 Tires out-of-balance.
2 Damaged or buckled wheel. Inspect and replace if necessary.
3 Defective tire (Chapter 1).

Chapter 1
Tune-up and routine maintenance

Contents

Specifications

Recommended lubricants and fluids

Note: *Listed here are manufacturer recommendations at the time this manual was written. Manufacturers occasionally upgrade their fluid and lubricant specifications, so check with your local auto parts store for current recommendations.*

Engine oil type	API grade “certified for gasoline engines”
Engine oil viscosity	See accompanying chart

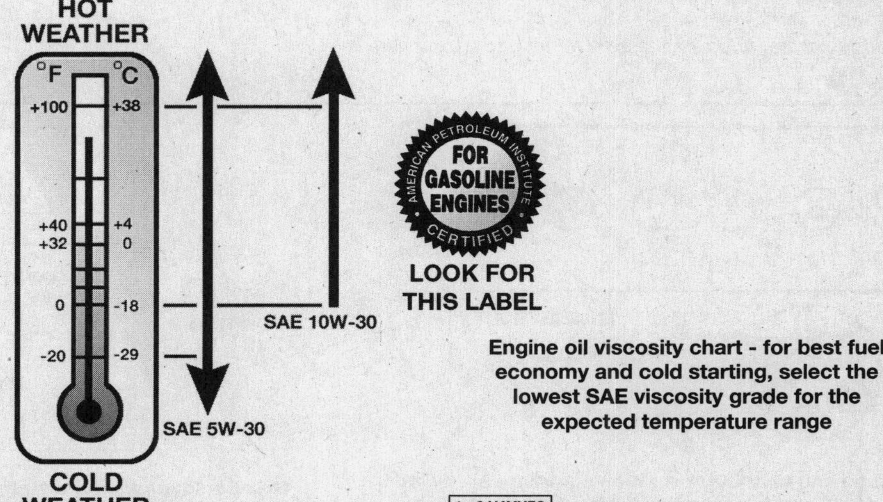

Engine oil viscosity chart - for best fuel economy and cold starting, select the lowest SAE viscosity grade for the expected temperature range

Recommended lubricants and fluids (continued)

Automatic transmission fluid type	
1993	Mercon/Dexron III automatic transmission fluid or equivalent
1994 and later	Mopar Plus 3 Type 7176 ATF or equivalent
Manual transmission lubricant type	SAE 75W-90 API GL-5 gear lubricant
Transfer case lubricant type	Dexron III/Mercon automatic transmission fluid
Differential lubricant type	
Normal operation	
Front axle	SAE 75W-90 API GL-5 gear lubricant
Rear axle	SAE 90W API GL-5 gear lubricant
Trailer towing package rear axle	SAE 75W-140 synthetic gear lubricant
Limited slip differential	Add MOPAR Friction Modifier
Brake fluid type	DOT 3 brake fluid
Power steering fluid	Mopar power steering fluid or equivalent
Manual steering gear lubricant type	SAE 90 API GL-5 hypoid gear lubricant
Chassis grease type	NLGI no. 2 EP chassis grease

Capacities*

Engine oil (with filter change)	
Six-cylinder engine	6 qts
V8 engines	
5.2L/5.9L	5 qts
4.7L	6 qts
Cooling system	
Six-cylinder engine	
1993 through 1999	12.7 qts
2000	11.7 qts
V8 engine	
1993 through 1998	12.6 qts
1999	12.2 qts
2000	11.7 qts
Automatic transmission (drain and refill)	Capacities vary - see Section 26 for procedure
Manual transmission	3.2 qts
Transfer case	
NP 231 (Command-Trac)	2.5 pts
NP 242/NV 242 (Selec-Trac)	2.9 pts
NP 249/NV 249 (Quadra-Trac)	2.5 pts
NV 247	2.5 pts
Differential	
Front	
Model 30	3.1 pts
Model 181	2.5 pts
Model 186	2.5 pts
Rear	
Model 35	3.1 pts
Model 194	3.5 pts
Model 216	4.7 pts
Model 198	3.5 pts
Model 226	4.7 pts

** All capacities approximate. Add as necessary to bring to appropriate level.*

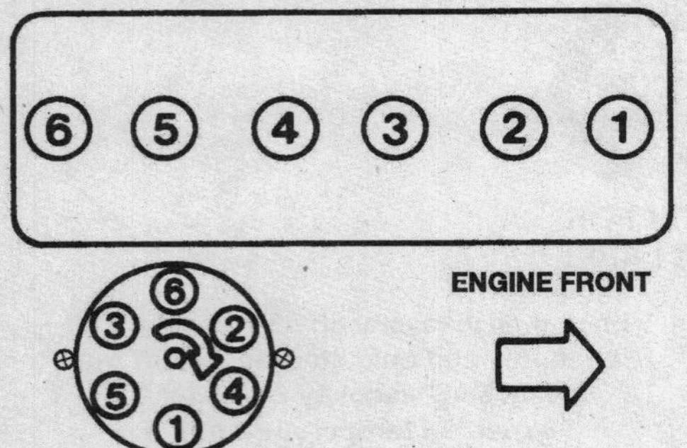

Cylinder location and distributor rotation diagram - six cylinder engine (1999 and later models have no distributor)

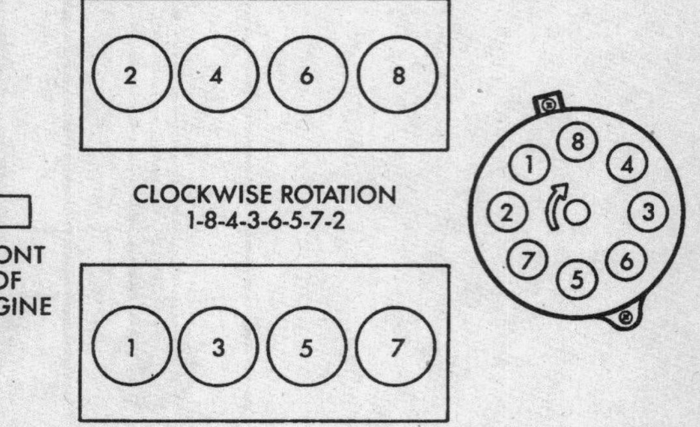

Cylinder location and distributor rotation diagram - V8 engines (1999 and later models have no distributor)

Ignition system

Spark plug type and gap	
Six-cylinder engine	
1993 through 1998	Champion RC12LYC or equivalent @ 0.035 inch
1999 and later	Champion RC12ECC or equivalent @ 0.035 inch
4.7L V8 engine (1999 and later)	Champion RC12MCC4 or equivalent @ 0.040 inch
5.2L V8 engine	
1993 through 1996	Champion RC12YC or equivalent @ 0.035 inch
1997	Champion RC12LC4 or equivalent @ 0.035 inch
1998	Champion RC12LC4 or equivalent @ 0.040 inch
5.9L V8 engine	
1997	Champion RC12YC or equivalent @ 0.035 inch
1998	Champion RC12LC4 or equivalent @ 0.040 inch
Firing order	
Six-cylinder engine	1-5-3-6-2-4
V8 engines	1-8-4-3-6-5-7-2

General

Disc brake pad lining thickness (minimum)	1/8-inch
Drum brake shoe lining thickness (minimum)	
Bonded shoes	1/16-inch
Riveted shoes (thickness above rivets)	1/32-inch
Drivebelt tension	
Six-cylinder engine	
1993 through 1998 (measured with special tool)	
New belt	180 to 200 lbs
Used belt	140 to 160 lbs
1999 and later	Self-adjusting
V8 engines	Self-adjusting

Automatic transmission band adjustment

1994 and later*	
Front band	
42RE	Tighten to 72 in-lbs, back-off 3-5/8 turns
42RH	Tighten to 72 in-lbs, back-off 2-1/2 turns
44RE	Tighten to 72 in-lbs, back-off 2-1/4 turns
46RH	Tighten to 72 in-lbs, back-off 2-1/2 turns
46RE	Tighten to 72 in-lbs, back off 2-7/8 turns
Rear band	
42RE and 44RE	Tighten to 72 in-lbs, back-off 4 turns
42RH	Tighten to 72 in-lbs, back-off 4 turns
44RE	Tighten to 72 in-lbs, back-off 4 turns
46RH and 46RE	Tighten to 72 in-lbs, back-off 2 turns

**Except 1999 and later 45RFE, which does not have band adjustments*

Torque specifications

	Ft-lbs (unless otherwise indicated)
Front band adjusting screw locknut	30
Rear band adjusting screw locknut	25
Automatic transmission pan bolts	144 in-lbs
Differential cover bolts	35
Transfer case fill/drain plug	30 to 40
Spark plugs	25 to 30
Engine oil drain plug	20
Oxygen sensor	20
Wheel lug nuts	80 to 110

Engine compartment component locations - 5.2L/5.9L V8 engines

1. *PCV valve*
2. *Automatic transmission fluid dipstick location (not visible)*
3. *Brake fluid reservoir*
4. *Windshield washer fluid reservoir*
5. *Engine oil filler cap*
6. *Air filter housing*
7. *Radiator hose*
8. *Power steering fluid reservoir*
9. *Serpentine drivebelt routing decal*
10. *Drivebelt*
11. *Engine oil dipstick*
12. *Radiator cap*
13. *Battery*
14. *Relay box*
15. *Coolant reservoir*

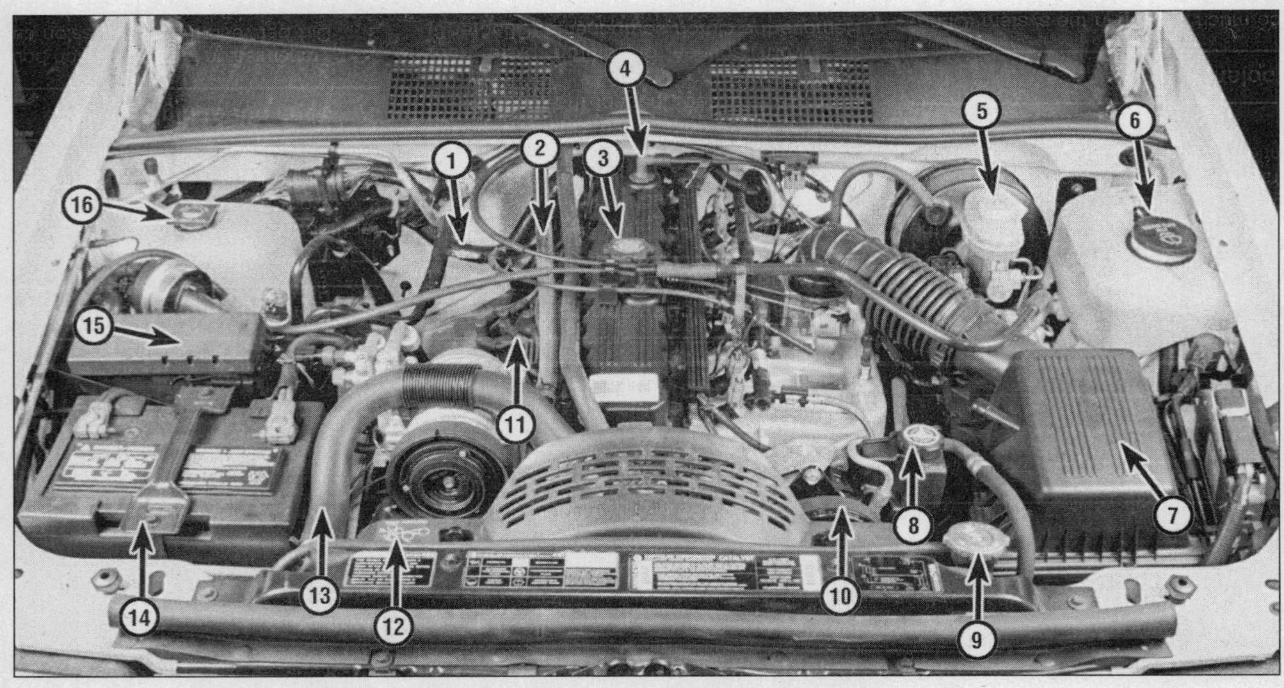

Engine compartment component locations - six cylinder engine

1. *Engine oil dipstick*
2. *Automatic transmission fluid dipstick location (not visible)*
3. *Engine oil filler cap*
4. *CCV system hose*
5. *Brake fluid reservoir*
6. *Windshield washer fluid reservoir*
7. *Air filter housing*
8. *Power steering fluid reservoir*
9. *Radiator cap*
10. *Drivebelt*
11. *Distributor cap and wires*
12. *Serpentine drivebelt routing decal*
13. *Radiator hose*
14. *Battery*
15. *Relay box*
16. *Coolant reservoir*

Engine compartment component locations - 4.7L V8 engine

1 *Engine oil dipstick*
2 *Automatic transmission fluid dipstick*
3 *Brake fluid reservoir*
4 *Power steering fluid reservoir*
5 *Windshield washer fluid reservoir*
6 *Air filter housing*
7 *Upper radiator hose*
8 *Ignition coil(s)/spark plugs (left bank)*
9 *Alternator*
10 *Drivebelt*
11 *Engine oil filler cap*
12 *Radiator cap*
13 *Battery*
14 *Power distribution center*
15 *Coolant reservoir*

Typical engine compartment underside component locations

1 *Steering damper (4WD models)*
2 *Steering grease fitting*
3 *Front differential check/fill plug*
4 *Driveaxle boot (4WD models)*
5 *Driveshaft grease fitting (4WD models)*
6 *Automatic transmission fluid pan*
7 *Engine oil pan*
8 *Exhaust system*
9 *Front disc brake*

Typical rear underside component locations

1 *Rear disc brake*
2 *Fuel filter*
3 *Differential check/fill plug*
4 *Rear shock absorber*
5 *Exhaust system*
6 *Driveshaft rear universal joint*

1 Jeep Cherokee Maintenance schedule

Every 250 miles or weekly, whichever comes first

Check the engine oil level (Section 4)
Check the engine coolant level (Section 4)
Check the windshield washer fluid level (Section 4)
Check the brake and clutch fluid levels (Section 4)
Check the tires and tire pressures (Section 5)
Check the automatic transmission fluid level (Section 6)

Every 3000 miles or 3 months, whichever comes first

All items listed above, plus . . .
Check the power steering fluid level (Section 7)
Change the engine oil and filter (Section 8)
Check and service the battery (Section 9)
Check the cooling system (Section 10)
Inspect and replace, if necessary, all underhood hoses (Section 11)
Inspect and replace, if necessary, the windshield wiper blades (Section 12)

Every 7500 miles or 6 months, whichever comes first

Inspect the suspension and steering components (Section 13)
Inspect the exhaust system (Section 14)
Check the manual transmission lubricant (Section 15)
Check the transfer case lubricant level (4WD models) (Section 16)
Check the differential lubricant level (Section 17)
Rotate the tires (Section 18)
Check the brakes (Section 19) *
Inspect the fuel system (Section 20)
Check the engine drivebelt (Section 21)
Check the seat belts (Section 22)
Check the Park/Neutral Position (PNP) switch (Section 23)

Every 15,000 miles or 12 months, whichever comes first

Lubricate the chassis components (Section 24) *

Every 30,000 miles or 24 months, whichever comes first

All items listed above, plus . . .
Replace the air filter (Section 25)
Change the automatic transmission fluid and filter (Section 26)**
Adjust the automatic transmission bands, if necessary (except 1993 models) (see note in Section 27)**
Change the manual transmission lubricant (Section 28)
Change the transfer case lubricant (4WD models) (Section 29)
Change the differential lubricant (Section 30)
Service the cooling system (drain, flush and refill) (Section 31)
Check the Positive Crankcase Ventilation (PCV) system (V8 engines) (Section 32)
Check the Crankcase Ventilation (CCV) system (6-cylinder engines) (Section 33)
Check the evaporative emissions control (Section 34)
Clean the crankcase inlet filter (V8 models) (Section 35)
Replace the fuel filter (Section 36)
Replace the spark plugs (Section 37)
Inspect the spark plug wires, distributor cap and rotor (Section 38)

Every 60,000 miles or 24 months, whichever comes first

Replace the drivebelt (Section 21)

* *This item is affected by "severe" operating conditions, as described below. If the vehicle is operated under severe conditions, perform all maintenance indicated with an asterisk (*) at 3000 mile/three-month intervals. Severe conditions exist if you mainly operate the vehicle . . .*
in dusty areas
towing a trailer
idling for extended periods and/or driving at low speeds when outside temperatures remain below freezing and most trips are less than four miles long

** *If operated under one or more of the following conditions, change the automatic transmission fluid and adjust the bands every 15,000 miles:*
in heavy city traffic where the outside temperature regularly reaches 90-degrees F or higher
in hilly or mountainous terrain
frequent trailer pulling

2 Introduction

This Chapter is designed to help the home mechanic maintain Jeep Grand Cherokee models with the goals of maximum performance, economy, safety and reliability in mind.

Included is a master maintenance schedule, followed by procedures dealing specifically with each item on the schedule. Visual checks, adjustments, component replacement and other helpful items are included. Refer to the accompanying illustrations of the engine compartment and the underside of the vehicle for the locations of various components.

Servicing your vehicle in accordance with the mileage/time maintenance schedule and the step-by-step procedures will result in a planned maintenance program that should produce a long and reliable service life. Keep in mind that it's a comprehensive plan, so maintaining some items but not others at the specified intervals will not produce the same results.

As you service your vehicle, you will discover that many of the procedures can - and should - be grouped together because of the nature of the particular procedure you're performing or because of the close proximity of two otherwise unrelated components to one another.

For example, if the vehicle is raised for chassis lubrication, you should inspect the exhaust, suspension, steering and fuel systems while you're under the vehicle. When you're rotating the tires, it makes good sense to check the brakes since the wheels are already removed. Finally, let's suppose you have to borrow or rent a torque wrench. Even if you only need it to tighten the spark plugs, you might as well check the torque of as many critical fasteners as time allows.

The first step in this maintenance program is to prepare yourself before the actual work begins. Read through all the procedures you're planning to do, then gather up all the parts and tools needed. If it looks like you might run into problems during a particular job, seek advice from a mechanic or an experienced do-it-yourselfer.

The following maintenance intervals are based on the assumption the vehicle owner will be doing the maintenance or service work, as opposed to having a dealer service department do the work. Although the time/mileage intervals are loosely based on factory recommendations, most have been shortened to ensure, for example, that such items as lubricants and fluids are checked/changed at intervals that promote maximum engine/driveline service life. Also, subject to the preference of the individual owner interested in keeping the vehicle in peak condition at all times, and with the vehicle's ultimate resale in mind, many of the maintenance procedures may be performed more often than recommended in the following schedule. We encourage such owner initiative.

When the vehicle is new it should be serviced initially by a factory authorized dealer service department to protect the factory warranty. In many cases the initial maintenance check is done at no cost to the owner (check with your dealer service department for additional information).

3 Tune-up general information

The term tune-up is used in this manual to represent a combination of individual operations rather than one specific procedure.

If, from the time the vehicle is new, the routine maintenance schedule is followed closely and frequent checks are made of fluid levels and high wear items, as suggested throughout this manual, the engine will be kept in relatively good running condition and the need for additional work will be minimized.

More likely than not, however, there will be times when the engine is running poorly due to lack of regular maintenance. This is even more likely if a used vehicle, which has not received regular and frequent maintenance checks, is purchased. In such cases, an engine tune-up will be needed outside of the regular routine maintenance intervals.

The first step in any tune-up or diagnostic procedure to help correct a poor running engine is a cylinder compression check. A compression check (see Chapter 2, Part D) will help determine the condition of internal engine components and should be used as a guide for tune-up and repair procedures. If, for instance, the compression check indicates serious internal engine wear, a conventional tune-up won't improve the performance of the engine and would be a waste of time and money. Because of its importance, the compression check should be done by someone with the right equipment and the knowledge to use it properly.

The following procedures are those most often needed to bring a generally poor running engine back into a proper state of tune.

Minor tune-up

- Check all engine related fluids (Section 4)
- Clean, inspect and test the battery (Section 9)
- Check the cooling system (Section 10)
- Check all underhood hoses (Section 11)
- Check the drivebelt (Section 21)
- Check the air filter (Section 25)
- Check the PCV valve (V8 engine) (Section 32)
- Check the CCV system (6-cylinder engine) (Section 33)
- Replace the spark plugs (Section 37)
- Inspect the spark plug and coil wires (Section 38)
- Inspect the distributor cap and rotor (Section 38)

4.2 The engine oil dipstick is located at the front of the engine on 5.2L/5.9L V8 engines and at the rear of the engine on 4.7L V8 engines

Major tune-up

All items listed under Minor tune-up, plus . . .

- Check the fuel system (Section 20)
- Replace the air filter (Section 25)
- Replace the spark plug wires (Section 38)
- Replace the distributor cap and rotor (Section 38)
- Check the ignition system (Chapter 5)
- Check the charging system (Chapter 5)

4 Fluid level checks (every 250 miles or weekly)

Note: *The following are fluid level checks to be done on a 250 mile or weekly basis. Additional fluid level checks can be found in specific maintenance procedures which follow. Regardless of intervals, be alert to fluid leaks under the vehicle which would indicate a fault to be corrected immediately.*

1 Fluids are an essential part of the lubrication, cooling, brake, clutch and windshield washer systems. Because the fluids gradually become depleted and/or contaminated during normal operation of the vehicle, they must be periodically replenished. See *Recommended lubricants and fluids* at the beginning of this Chapter before adding fluid to any of the following components. **Note:** *The vehicle must be on level ground when fluid levels are checked.*

Engine oil

Refer to illustrations 4.2, 4.4 and 4.6

2 The engine oil level is checked with a dipstick that extends through a tube and into the oil pan at the bottom of the engine **(see illustration)**.

3 The oil level should be checked before the vehicle has been driven, or about 15 minutes after the engine has been shut off. If the oil is checked immediately after driving the vehicle, some of the oil will remain in the upper engine components, resulting in an inaccurate reading on the dipstick.

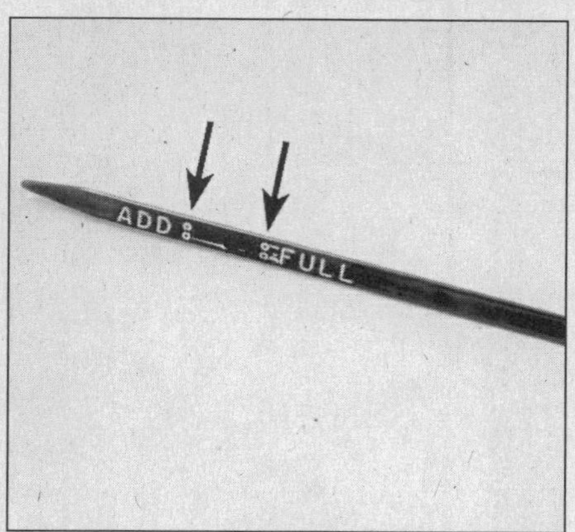

4.4 The oil level must be maintained between the marks at all times - it takes one quart of oil to raise the level from the ADD to the FULL mark

4 Pull the dipstick out of the tube and wipe all the oil from the end with a clean rag or paper towel. Insert the clean dipstick all the way back into the tube, then pull it out again. Note the oil at the end of the dipstick. Add oil as necessary to keep the level between the ADD and FULL marks on the dipstick **(see illustration)**.

5 Do not overfill the engine by adding too much oil since this may result in oil fouled spark plugs, oil leaks or oil seal failures.

6 Oil is added to the engine after removing a cap from the valve cover **(see illustration)**. A funnel may help to reduce spills.

7 Checking the oil level is an important preventive maintenance step. A consistently low oil level indicates oil leakage through damaged seals, defective gaskets or past worn rings or valve guides. If the oil looks milky or has water droplets in it, the cylinder head gasket(s) may be blown or the head(s) or block may be cracked. The engine should be checked immediately. The condition of the oil should also be checked. Whenever you check the oil level, slide your thumb and index finger up the dipstick before wiping off the oil. If you see small dirt or metal particles clinging to the dipstick, the oil should be changed (see Section 8).

Engine coolant

Refer to illustration 4.8

Warning: *Do not allow antifreeze to come in contact with your skin or painted surfaces of the vehicle. Flush contaminated areas immediately with plenty of water. Don't store new coolant or leave old coolant lying around where it's accessible to children or pets - they're attracted by its sweet smell. Ingestion of even a small amount of coolant can be fatal! Wipe up garage floor and drip pan coolant spills immediately. Keep antifreeze containers covered and repair leaks in the cooling system as soon as they are noted.*

8 All vehicles covered by this manual are equipped with a pressurized coolant recovery system. A white plastic coolant reservoir

4.6 Oil is added to the engine after removing the oil filler cap from the valve cover (or the oil filler tube on 4.7L engines) - always make sure the area around the opening is clean before removing the cap to prevent dirt from contaminating the engine (5.2L/5.9L engine shown)

located in the engine compartment is connected by a hose to the radiator filler neck **(see illustration)**. If the engine overheats, coolant escapes through a valve in the radiator cap and travels through the hose into the reservoir. As the engine cools, the coolant is automatically drawn back into the cooling system to maintain the correct level. **Warning:** *Do not remove the radiator cap to check the coolant level when the engine is warm.*

9 The coolant level in the reservoir should be checked regularly. The level in the reservoir varies with the temperature of the engine. When the engine is cold, the coolant level should be at or slightly above the ADD mark on the reservoir. Once the engine has warmed up, the level should be at or near the FULL mark. If it isn't, allow the engine to cool, then remove the cap from the reservoir and add a 50/50 mixture of ethylene glycol-based antifreeze and water.

10 Drive the vehicle and recheck the coolant level. If only a small amount of coolant is required to bring the system up to the proper level, water can be used. However, repeated additions of water will dilute the antifreeze and water solution. In order to maintain the proper ratio of antifreeze and water, always top up the coolant level with the correct mixture. An empty plastic milk jug or bleach bottle makes an excellent container for mixing coolant. Do not use rust inhibitors or additives.

11 If the coolant level drops consistently, there may be a leak in the system. Inspect the radiator, hoses, filler cap, drain plugs and water pump (see Section 10). If no leaks are noted, have the radiator cap pressure tested by a service station.

12 If you have to remove the radiator cap, wait until the engine has cooled, then wrap a thick cloth around the cap and turn it to the first stop. If coolant or steam escapes, let the engine cool down longer, then remove the cap.

4.8 The coolant reservoir is located in the right rear corner of the engine compartment - keep the level near the Full mark on the reservoir

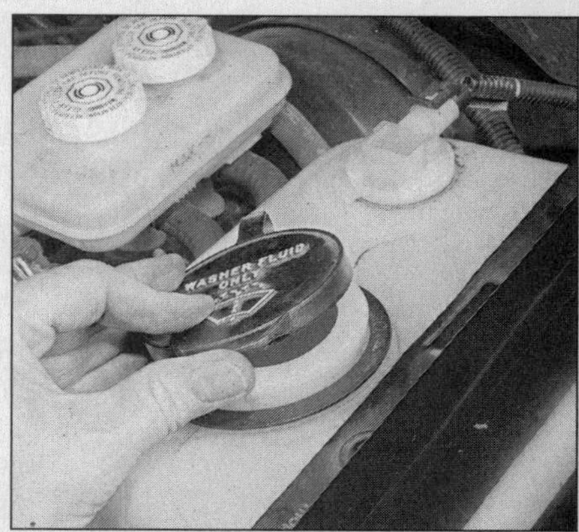

4.14 The windshield washer fluid reservoir is located on the left side of the engine compartment

13 Check the condition of the coolant as well. It should be relatively clear. If it's brown or rust colored, the system should be drained, flushed and refilled. Even if the coolant appears to be normal, the corrosion inhibitors wear out, so it must be replaced at the specified intervals.

Windshield washer fluid

Refer to illustration 4.14

14 Fluid for the windshield washer system is located in a plastic reservoir in the engine compartment **(see illustration)**.

15 In milder climates, plain water can be used in the reservoir, but it should be kept no more than 2/3 full to allow for expansion if the water freezes. In colder climates, use windshield washer system antifreeze, available at any auto parts store, to lower the freezing point of the fluid. Mix the antifreeze with water in accordance with the manufacturer's directions on the container. **Caution:** *Don't use cooling system antifreeze - it will damage the vehicle's paint.*

16 To help prevent icing in cold weather, warm the windshield with the defroster before using the washer.

4.17 Remove the cell caps to check the electrolyte level in the battery - if the level is low, add distilled water only

4.19 The brake fluid level is easily checked visually - remove the reservoir cap(s) to add fluid

Battery electrolyte

Refer to illustration 4.17

17 Most vehicles with which this manual is concerned are equipped with a battery which is permanently sealed (except for vent holes) and has no filler caps. Water doesn't have to be added to these batteries at any time. If a maintenance-type battery is installed, the caps on the top of the battery should be removed periodically to check for a low electrolyte level **(see illustration)**. This check is most critical during the warm summer months.

Brake and clutch fluid

Refer to illustration 4.19

18 The brake and (if equipped) clutch master cylinder is mounted on the upper left of the engine compartment firewall.

19 The fluid inside can be checked looking at the translucent plastic reservoir **(see illustration)**. Be sure to wipe the top of the reservoir cover with a clean rag to prevent contamination of the brake and/or clutch system before removing the cover.

20 When adding fluid, pour it carefully into the reservoir to avoid spilling it on surrounding painted surfaces. Be sure the specified fluid is used, since mixing different types of brake fluid can cause damage to the system. See *Recommended lubricants and fluids* at the front of this Chapter or your owner's manual. **Warning:** *Brake fluid can harm your eyes and damage painted surfaces, so use extreme caution when handling or pouring it. Do not use brake fluid that has been standing open or is more than one year old. Brake fluid absorbs moisture from the air. Excess moisture can cause a dangerous loss of brake performance.*

21 At this time, the fluid and master cylinder can be inspected for contamination. The system should be drained and refilled if deposits, dirt particles or water droplets are seen in the fluid.

22 After filling the reservoir to the proper level, make sure the cover or cap is on tight to prevent fluid leakage.

23 The brake fluid level in the master cylinder will drop slightly as the pads at the front wheels wear down during normal operation. If the master cylinder requires repeated additions to keep it at the proper level, it's an indication of leakage in the brake system, which should be corrected immediately. Check all brake lines and connections (see Section 19 for more information).

24 If, upon checking the master cylinder fluid level, you discover one or both reservoirs empty or nearly empty, the brake system should be bled (Chapter 9).

5 Tire and tire pressure checks (every 250 miles or weekly)

Refer to illustrations 5.2, 5.3, 5.4a, 5.4b and 5.8

1 Periodic inspection of the tires may spare you the inconvenience of being stranded with a flat tire. It can also provide you with vital information regarding possible problems in the steering and suspension systems before major damage occurs.

2 The original tires on this vehicle are equipped with 1/2-inch wide wear bands that will appear when tread depth reaches 1/16-inch, at which point the tires can be considered worn out. Tread wear can be monitored with a simple, inexpensive device known as a tread depth indicator **(see illustration)**.

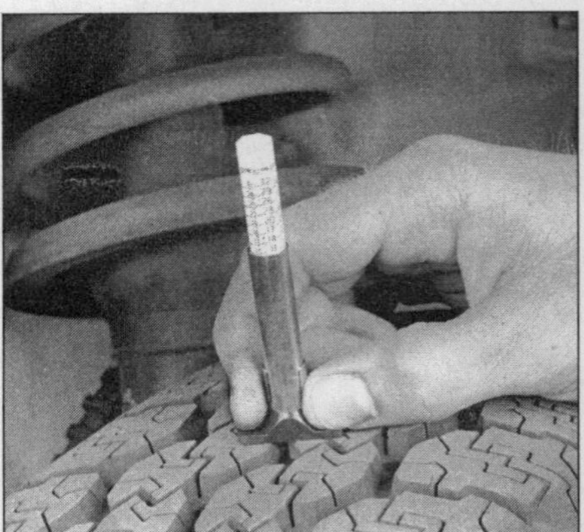

5.2 Use a tire tread depth indicator to monitor tire wear - they are available at auto parts stores and service stations and cost very little

3 Note any abnormal tread wear **(see illustration)**. Tread pattern irregularities such as cupping, flat spots and more wear on one side than the other are indications of front end alignment and/or balance problems. If any of these conditions are noted, take the vehicle to a tire shop or service station to correct the problem.

4 Look closely for cuts, punctures and embedded nails or tacks. Sometimes a tire will hold air pressure for a short time or leak down very slowly after a nail has embedded itself in the tread. If a slow leak persists, check the valve stem core to make sure it's tight **(see illustration)**. Examine the tread for an object that may have embedded itself in the tire or for a "plug" that may have begun to leak (radial tire punctures are repaired with a plug that's installed in a puncture). If a puncture is suspected, it can be easily verified by spraying a solution of soapy water onto the puncture area **(see illustration)**. The soapy solution will bubble if there's a leak. Unless the puncture is unusually large, a tire shop or service station can usually repair the tire.

5 Carefully inspect the inner sidewall of each tire for evidence of brake fluid leakage. If you see any, inspect the brakes immediately.

6 Correct air pressure adds miles to the lifespan of the tires, improves mileage and enhances overall ride quality. Tire pressure cannot be accurately estimated by looking at a tire, especially if it's a radial. A tire pressure gauge is essential. Keep an accurate gauge in the vehicle. The pressure gauges attached to the nozzles of air hoses at gas stations are often inaccurate.

UNDERINFLATION

CUPPING

Cupping may be caused by:

- Underinflation and/or mechanical irregularities such as out-of-balance condition of wheel and/or tire, and bent or damaged wheel.
- Loose or worn steering tie-rod or steering idler arm.
- Loose, damaged or worn front suspension parts.

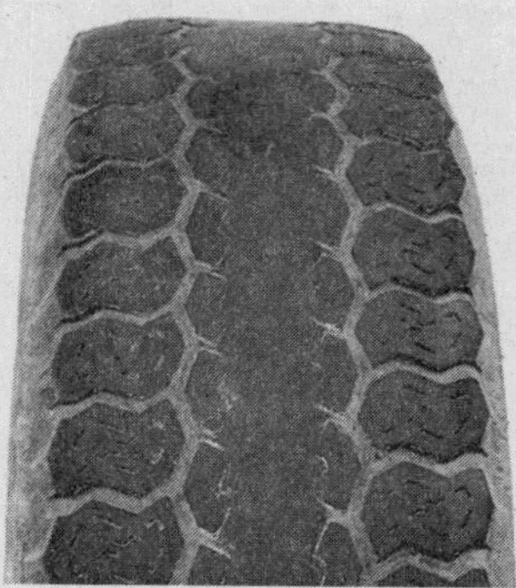

OVERINFLATION

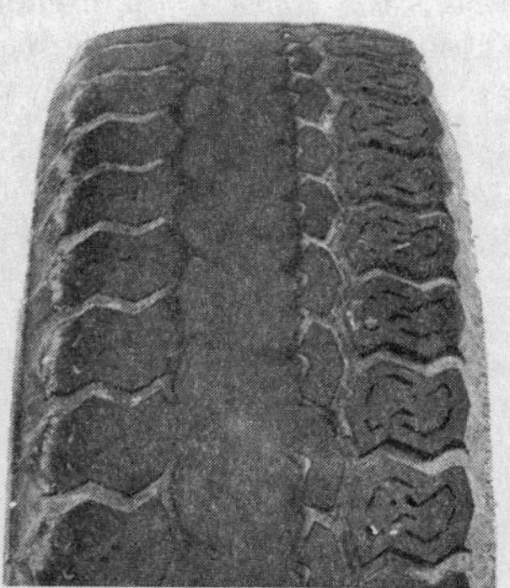

INCORRECT TOE-IN OR EXTREME CAMBER

FEATHERING DUE TO MISALIGNMENT

5.3 This chart will help you determine the condition of the tires, the probable cause(s) of abnormal wear and the corrective action necessary

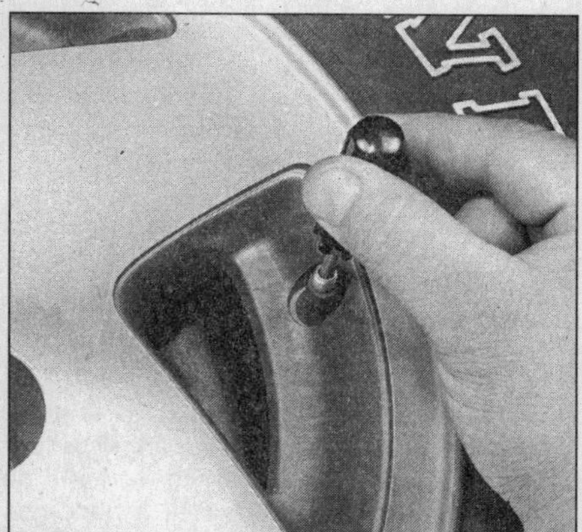

5.4a If a tire loses air on a steady basis, check the valve stem core first to make sure it's snug (special inexpensive wrenches are commonly available at auto parts stores)

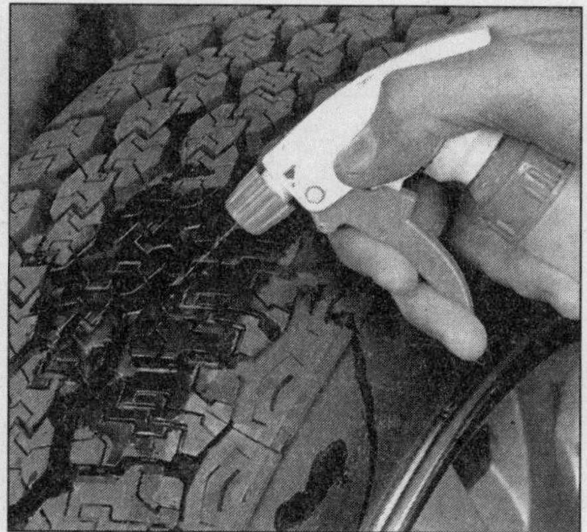

5.4b If the valve stem core is tight, raise the corner of the vehicle with the low tire and spray a soapy water solution onto the tread as the tire is turned slowly - leaks will cause small bubbles to appear

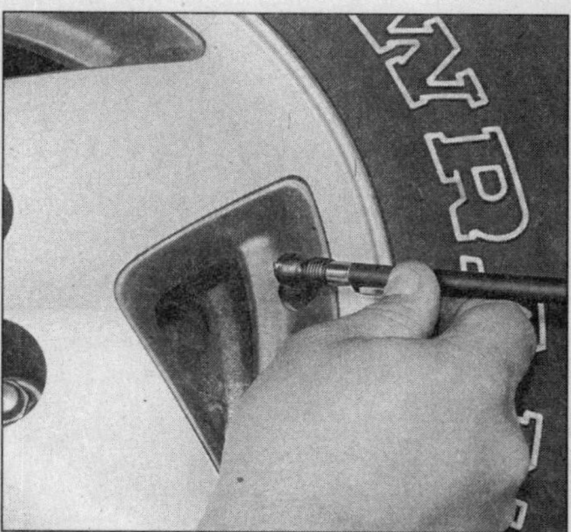

5.8 To extend the life of the tires, check the air pressure at least once a week with an accurate gauge (don't forget the spare!)

7 Always check tire pressure when the tires are cold. Cold, in this case, means the vehicle has not been driven over a mile in the three hours preceding a tire pressure check. A pressure rise of four to eight pounds is not uncommon once the tires are warm.

8 Unscrew the valve cap protruding from the wheel or hubcap and push the gauge firmly onto the valve stem **(see illustration)**. Note the reading on the gauge and compare the figure to the recommended tire pressure shown on the placard on the driver's side door pillar. Be sure to reinstall the valve cap to keep dirt and moisture out of the valve stem mechanism. Check all four tires and, if necessary, add enough air to bring them up to the recommended pressure.

9 Don't forget to keep the spare tire inflated to the specified pressure (refer to your owner's manual or the tire sidewall).

6 Automatic transmission fluid level check (every 250 miles or weekly)

Refer to illustrations 6.3, 6.6 and 6.7

1 The automatic transmission fluid level should be carefully maintained. Low fluid level can lead to slipping or loss of drive, while overfilling can cause foaming and loss of fluid.

6.3 The automatic transmission dipstick is located at the rear of the engine compartment (5.2L/5.9L engine shown; on six-cylinder and 4.7L V8 engines it's on the passenger's side of the vehicle)

6.6 Use a funnel like this one (available at auto parts stores) inserted into the transmission filler tube to add fluid

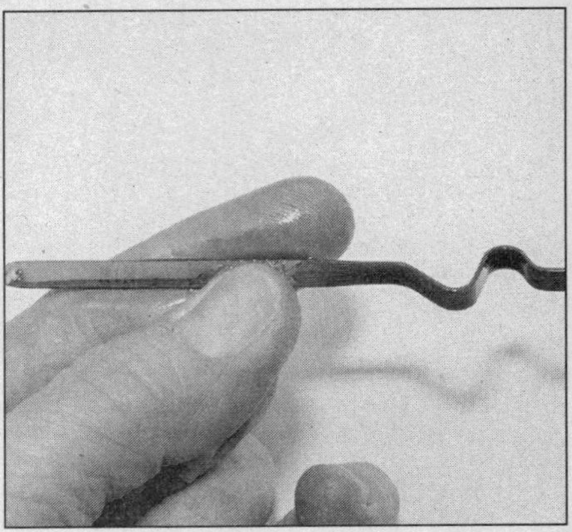

6.7 Check the condition of the fluid - if it isn't red in color it should be changed

2 With the parking brake set, start the engine, then move the shift lever through all the gear ranges, ending in Neutral. The fluid level must be checked with the vehicle level and the engine running at idle. **Note:** *Incorrect fluid level readings will result if the vehicle has just been driven at high speeds for an extended period, in hot weather in city traffic, or if it has been pulling a trailer. If any of these conditions apply, wait until the fluid has cooled (about 30 minutes).*

3 With the transmission at normal operating temperature, remove the dipstick from the filler tube. The dipstick is located at the rear of the engine compartment **(see illustration)**.

4 Determine if the fluid is warm or hot. Wipe the fluid from the dipstick with a clean rag and push it back into the filler tube until the cap seats.

5 Pull the dipstick out again and note the fluid level.

6 If the fluid is warm, the level should be between the two dimples. If it's hot, the level should be in the crosshatched area, near the MAX line. If additional fluid is required, add it directly into the tube using a funnel **(see illustration)**. It takes about one pint to raise the level from the bottom of the crosshatched area to the MAX line with a hot transmission, so add the fluid a little at a time and keep checking the level until it's correct.

7 The condition of the fluid should also be checked along with the level. If the fluid at the end of the dipstick is a dark reddish-brown color, or if it smells burned, it should be changed **(see illustration)**. If you are in doubt about the condition of the fluid, purchase some new fluid and compare the two for color and smell.

7 Power steering fluid level check (every 3000 miles or 3 months)

Refer to illustrations 7.5 and 7.6

1 Unlike manual steering, the power steering system relies on fluid which may, over a period of time, require replenishing.

2 The fluid reservoir for the power steering pump is located on the pump body at the front of the engine.

3 For the check, the front wheels should be pointed straight ahead and the engine should be off.

4 Use a clean rag to wipe off the reservoir cap and the area around the cap. This will help prevent any foreign matter from entering the reservoir during the check.

5 Twist off the cap and determine the temperature of the fluid **(see illustration)**.

6 Wipe off the fluid with a clean rag, reinsert the dipstick, then withdraw it and read the fluid level. The fluid should be at the proper level, depending on whether it was checked hot or cold **(see illustration)**. Never allow the fluid level to drop below the lower mark on the dipstick.

7 If additional fluid is required, pour the specified type directly into the reservoir, using a funnel to prevent spills.

8 If the reservoir requires frequent fluid additions, all power steering hoses, hose connections and the power steering pump should be carefully checked for leaks.

7.5 The power steering fluid dipstick is located in the power steering pump reservoir - turn the cap counterclockwise to remove it

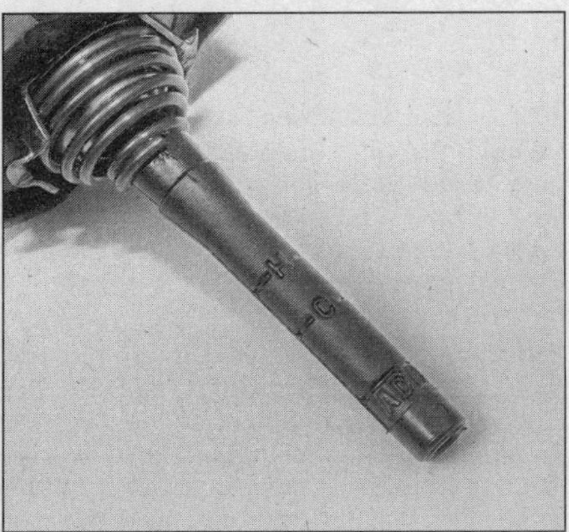

7.6 The power steering fluid dipstick has marks on it so the fluid can be checked hot or cold

8 Engine oil and filter change (every 3000 miles or 3 months)

Refer to illustrations 8.3, 8.9, 8.14 and 8.18

1 Frequent oil changes are the most important preventive maintenance procedures that can be done by the home mechanic. As engine oil ages, it becomes diluted and contaminated, which leads to premature engine wear.

2 Although some sources recommend oil filter changes every other oil change, we feel that the minimal cost of an oil filter and the relative ease with which it is installed dictate that a new filter be installed every time the oil is changed.

3 Gather together all necessary tools and materials before beginning this procedure **(see illustration)**.

4 You should have plenty of clean rags and newspapers handy to mop up any spills. Access to the under side of the vehicle may be improved if the vehicle can be lifted on a hoist, driven onto ramps or supported by jackstands. **Warning:** *Do not work under a*

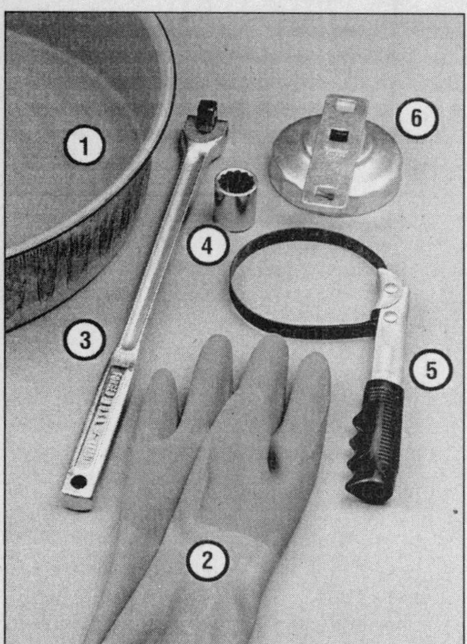

8.3 These tools are required when changing the engine oil and filter

1. ***Drain pan*** *- It should be fairly shallow in depth, but wide to prevent spills*
2. ***Rubber gloves*** *- When removing the drain plug and filter, you will get oil on your hands (the gloves will prevent burns)*
3. ***Breaker bar*** *- Sometimes the oil drain plug is tight, and a long breaker bar is needed to loosen it*
4. ***Socket*** *- To be used with the breaker bar or a ratchet (must be the correct size to fit the drain plug - six-point preferred)*
5. ***Filter wrench*** *- This is a metal band-type wrench, which requires clearance around the filter to be effective*
6. ***Filter wrench*** *- This type fits on the bottom of the filter and can be turned with a ratchet or breaker bar (different-size wrenches are available for different types of filters)*

vehicle which is supported only by a bumper, hydraulic or scissors-type jack.

5 If this is your first oil change, get under the vehicle and familiarize yourself with the locations of the oil drain plug and the oil filter. The engine and exhaust components will be warm during the actual work, so note how they are situated to avoid touching them when working under the vehicle.

6 Warm the engine to normal operating temperature. If the new oil or any tools are needed, use this warm-up time to gather everything necessary for the job. The correct type of oil for your application can be found in *Recommended lubricants and fluids* at the beginning of this Chapter.

7 With the engine oil warm (warm engine oil will drain better and more built-up sludge will be removed with it), raise and support the vehicle. Make sure it's safely supported!

8 Move all necessary tools, rags and newspapers under the vehicle. Set the drain pan under the drain plug. Keep in mind that the oil will initially flow from the pan with some force; position the pan accordingly.

9 Being careful not to touch any of the hot exhaust components, use a wrench to remove the drain plug near the bottom of the oil pan **(see illustration)**. Depending on how hot the oil is, you may want to wear gloves while removing the plug the final few turns.

10 Allow the old oil to drain into the pan. It may be necessary to move the pan as the oil flow slows to a trickle.

11 After all the oil has drained, wipe off the drain plug with a clean rag. Small metal particles may cling to the plug and would immediately contaminate the new oil.

12 Clean the area around the drain plug opening and reinstall the plug. Tighten the plug securely with the wrench. If a torque wrench is available, use it to tighten the plug to the torque listed in this Chapter's Specifications.

13 Move the drain pan into position under the oil filter.

14 Use the filter wrench to loosen the oil filter **(see illustration)**.

15 Completely unscrew the old filter. Be careful; it's full of oil. Empty the oil inside the filter into the drain pan, then lower the filter (V8 engines) or lift the filter out from above (six-cylinder engines).

16 Compare the old filter with the new one to make sure they're the same type.

17 Use a clean rag to remove all oil, dirt and sludge from the area where the oil filter mounts to the engine. Check the old filter to make sure the rubber gasket isn't stuck to the engine. If the gasket is stuck to the engine, remove it.

18 Apply a light coat of clean oil to the rubber gasket on the new oil filter **(see illustration)**.

19 Attach the new filter to the engine, following the tightening directions printed on the filter canister or packing box. Most filter manufacturers recommend against using a filter wrench due to the possibility of overtightening and damage to the seal.

20 Remove all tools, rags, etc. from under the vehicle, being careful not to spill the oil in the drain pan, then lower the vehicle.

21 Move to the engine compartment and locate the oil filler cap.

22 Pour the fresh oil through the filler opening. A funnel may be helpful.

23 Pour four quarts of fresh oil into the engine. Wait a few minutes to allow the oil to drain into the pan, then check the level on the oil dipstick (see Section 4 if necessary). If the oil level is above the ADD mark, start the engine and allow the new oil to circulate.

24 Run the engine for only about a minute and then shut it off. Immediately look under the vehicle and check for leaks at the oil pan drain plug and around the oil filter. If either is leaking, tighten with a bit more force.

25 With the new oil circulated and the filter now completely full, recheck the level on the dipstick and add more oil as necessary.

26 During the first few trips after an oil change, make it a point to check frequently for leaks and proper oil level.

27 The old oil drained from the engine cannot be reused in its present state and should be disposed of. Check with your local auto parts store, disposal facility or environmental

1

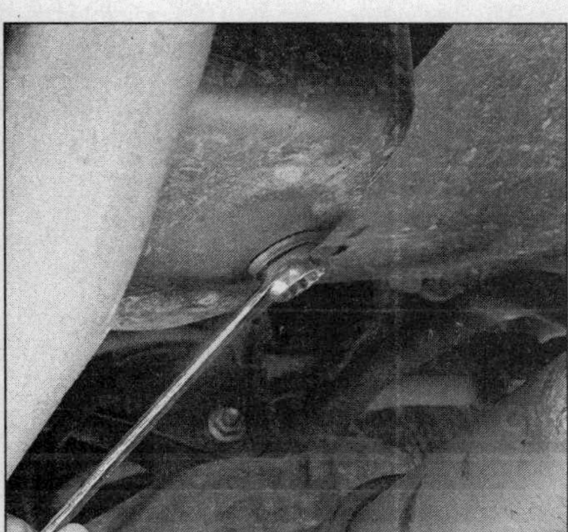

8.9 Use a proper size box-end wrench or socket to remove the oil drain plug and avoid rounding it off

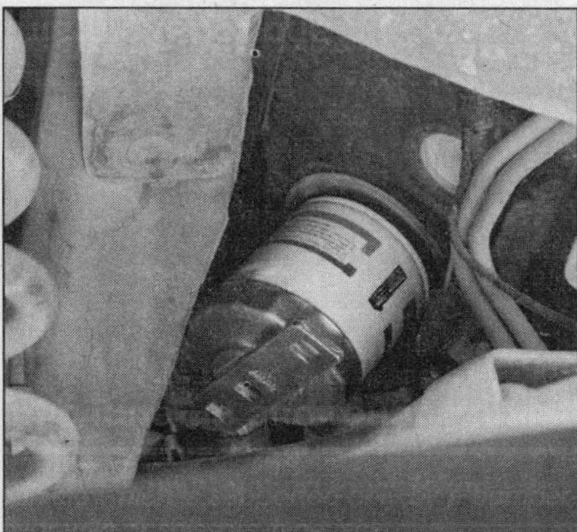

8.14 Since the oil filter is on very tight, you'll need a special wrench for removal - DO NOT use the wrench to tighten the new filter

8.18 Lubricate the oil filter gasket with clean engine oil before installing the filter on the engine

agency to see if they will accept the oil for recycling. After the oil has cooled it can be drained into a container (capped plastic jugs, topped bottles, milk cartons, etc.) for transport to one of these disposal sites. Don't dispose of the oil by pouring it on the ground or down a drain!

Service reminder - resetting

28 Some models are equipped with a service reminder feature in the Vehicle Information Center (VIC). This feature displays a reminder when an oil change service is due to be performed. After servicing is complete, reset the service reminder using the following procedure:

- *a) Turn the ignition switch ON.*
- *b) Momentarily press and release the SELECT button on the VIC.*
- *c) Hold down the SET button for at least two seconds.*
- *d) You will hear a "beep" and the number of miles to the next service will be indicated.*

9 Battery check, maintenance and charging (every 3000 miles or 3 months)

Refer to illustrations 9.1, 9.6a, 9.6b, 9.7a and 9.7b

Warning: *Certain precautions must be followed when checking and servicing the battery. Hydrogen gas, which is highly flammable, is always present in the battery cells, so keep lighted tobacco and all other open flames and sparks away from the battery. The electrolyte inside the battery is actually dilute sulfuric acid, which will cause injury if splashed on your skin or in your eyes. It will also ruin clothes and painted surfaces. When removing the battery cables, always detach the negative cable first and hook it up last!*

Maintenance

1 A routine preventive maintenance program for the battery in your vehicle is the only way to ensure quick and reliable starts. But before performing any battery maintenance, make sure that you have the proper equipment necessary to work safely around the battery **(see illustration)**.

2 There are also several precautions that should be taken whenever battery maintenance is performed. Before servicing the battery, always turn the engine and all accessories off and disconnect the cable from the negative terminal of the battery.

3 The battery produces hydrogen gas, which is both flammable and explosive. Never create a spark, smoke or light a match around the battery. Always charge the battery in a ventilated area.

4 Electrolyte contains poisonous and corrosive sulfuric acid. Do not allow it to get in your eyes, on your skin or your clothes. Never ingest it. Wear protective safety glasses when working near the battery. Keep children away from the battery.

9.1 Tools and materials required for battery maintenance

1 ***Face shield/safety goggles*** *- When removing corrosion with a brush, the acidic particles can easily fly up into your eyes*

2 ***Baking soda*** *- A solution of baking soda and water can be used to neutralize corrosion*

3 ***Petroleum jelly*** *- A layer of this on the battery posts will help prevent corrosion*

4 ***Battery post/cable cleaner*** *- This wire brush cleaning tool will remove all traces of corrosion from the battery posts and cable clamps*

5 ***Treated felt washers*** *- Placing one of these on each post, directly under the cable clamps, will help prevent corrosion*

6 ***Puller*** *- Sometimes the cable clamps are very difficult to pull off the posts, even after the nut/bolt has been completely loosened. This tool pulls the clamp straight up and off the post without damage*

7 ***Battery post/cable cleaner*** *- Here is another cleaning tool which is a slightly different version of Number 4 above, but it does the same thing*

8 ***Rubber gloves*** *- Another safety item to consider when servicing the battery; remember that's acid inside the battery!*

5 Note the external condition of the battery. If the positive terminal and cable clamp on your vehicle's battery is equipped with a rubber protector, make sure that it's not torn or damaged. It should completely cover the terminal. Look for any corroded or loose connections, cracks in the case or cover or loose hold-down clamps. Also check the entire length of each cable for cracks and frayed conductors.

9.6a Battery terminal corrosion usually appears as light, fluffy powder

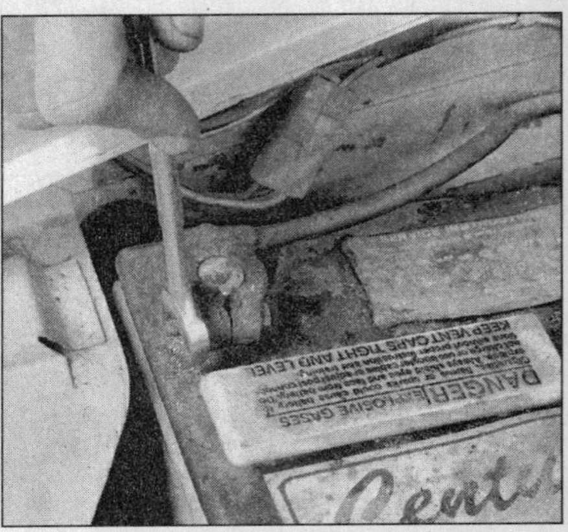

9.6b Removing the cable from a battery post with a wrench - sometimes special battery pliers are required for this procedure if corrosion has caused deterioration of the nut hex (always remove the ground cable first and hook it up last!)

6 If corrosion, which looks like white, fluffy deposits **(see illustration)** is evident, particularly around the terminals, the battery should be removed for cleaning. Loosen the cable clamp bolts with a wrench, being careful to remove the ground cable first, and slide them off the terminals **(see illustration)**. Then disconnect the hold-down clamp bolt and nut, remove the clamp and lift the battery from the engine compartment.

7 Clean the cable clamps thoroughly with a battery brush or a terminal cleaner and a solution of warm water and baking soda **(see illustration)**. Wash the terminals and the top of the battery case with the same solution but make sure that the solution doesn't get into the battery. When cleaning the cables, terminals and battery top, wear safety goggles and rubber gloves to prevent any solution from coming in contact with your eyes or hands. Wear old clothes too - even diluted, sulfuric acid splashed onto clothes will burn holes in them. If the terminals have been extensively corroded, clean them up with a terminal cleaner **(see illustration)**. Thoroughly wash all cleaned areas with plain water.

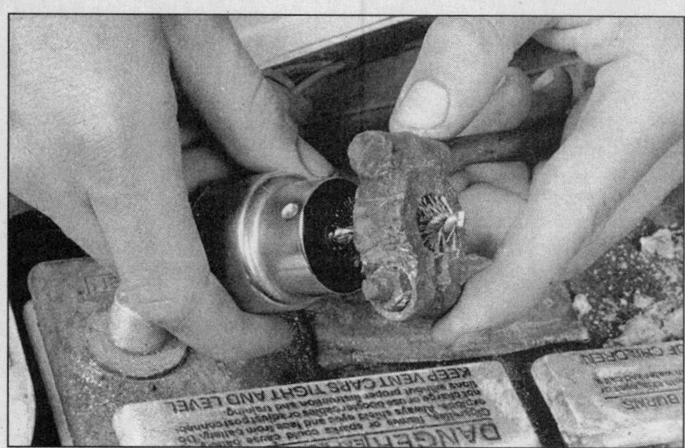

9.7a When cleaning the cable clamps, all corrosion must be removed (the inside of the clamp is tapered to match the taper on the post, so don't remove too much material)

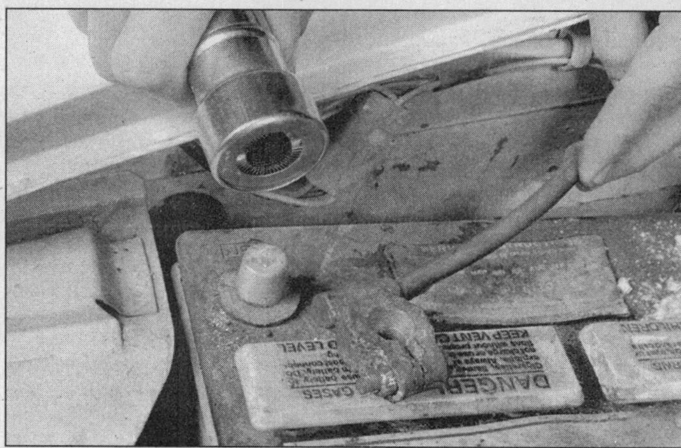

9.7b Regardless of the type of tool used on the battery posts, a clean, shiny surface should be the result

8 Make sure that the battery tray is in good condition and the hold-down clamp bolts are tight. If the battery is removed from the tray, make sure no parts remain in the bottom of the tray when the battery is reinstalled. When reinstalling the hold-down clamp bolts, do not overtighten them.

9 Any metal parts of the vehicle damaged by corrosion should be covered with a zinc-based primer, then painted.

10 Information on removing and installing the battery can be found in Chapter 5. Information on jump starting can be found at the front of this manual. For more detailed battery checking procedures, refer to the *Haynes Automotive Electrical Manual*.

Charging

Warning: *When batteries are being charged, hydrogen gas, which is very explosive and flammable, is produced. Do not smoke or allow open flames near a charging or a recently charged battery. Wear eye protection when near the battery during charging. Also, make sure the charger is unplugged before connecting or disconnecting the battery from the charger.*

Note: *The manufacturer recommends the battery be removed from the vehicle for charging because the gas that escapes during this procedure can damage the paint. Fast charging with the battery cables connected can result in damage to the electrical system.*

11 Slow-rate charging is the best way to restore a battery that's discharged to the point where it will not start the engine. It's also a good way to maintain the battery charge in a vehicle that's only driven a few miles between starts. Maintaining the battery charge is particularly important in the winter when the battery must work harder to start the engine and electrical accessories that drain the battery are in greater use.

12 It's best to use a one or two-amp battery charger (sometimes called a "trickle" charger). They are the safest and put the least strain on the battery. They are also the least expensive. For a faster charge, you can use a higher amperage charger, but don't use one rated more than 1/10th the amp/hour rating of the battery. Rapid boost charges that claim to restore the power of the battery in one to two hours are hardest on the battery and can damage batteries not in good condition. This type of charging should only be used in emergency situations.

13 The average time necessary to charge a battery should be listed in the instructions that come with the charger. As a general rule, a trickle charger will charge a battery in 12 to 16 hours.

14 Remove all the cell caps (if equipped) and cover the holes with a clean cloth to prevent spattering electrolyte. Disconnect the negative battery cable and hook the battery charger cable clamps up to the battery posts (positive to positive, negative to negative), then plug in the charger. Make sure it is set at 12-volts if it has a selector switch.

15 If you're using a charger with a rate higher than two amps, check the battery regularly during charging to make sure it doesn't overheat. If you're using a trickle charger, you can safely let the battery charge overnight after you've checked it regularly for the first couple of hours.

16 If the battery has removable cell caps, measure the specific gravity with a hydrometer every hour during the last few hours of the charging cycle. Hydrometers are available inexpensively from auto parts stores - follow the instructions that come with the hydrometer. Consider the battery charged when there's no change in the specific gravity reading for two hours and the electrolyte in the cells is gassing (bubbling) freely. The specific gravity reading from each cell should be very close to the others. If not, the battery probably has a bad cell(s).

17 Some batteries with sealed tops have built-in hydrometers on the top that indicate the state of charge by the color displayed in the hydrometer window. Normally, a bright-colored hydrometer indicates a full charge and a dark hydrometer indicates the battery still needs charging.

18 If the battery has a sealed top and no built-in hydrometer, you can hook up a digital voltmeter across the battery terminals to check the charge. A fully charged battery should read 12.6 volts or higher after the surface charge has been removed.

19 Further information on the battery and jump starting can be found in Chapter 5 and at the front of this manual.

10 Cooling system check (every 3000 miles or 3 months)

Refer to illustration 10.4

1 Many major engine failures can be attributed to a faulty cooling system. If the vehicle is equipped with an automatic transmission, the cooling system also cools the transmission fluid and thus plays an important role in prolonging transmission life.

2 The cooling system should be checked with the engine cold. Do this before the vehicle is driven for the day or after it has been shut off for at least three hours.

3 Remove the radiator cap by turning it to the left until it reaches a stop. If you hear a hissing sound (indicating there is still pressure in the system), wait until this stops. Now press down on the cap with the palm of your hand and continue turning to the left until the cap can be removed. Thoroughly clean the cap, inside and out, with clean water. Also clean the filler neck on the radiator. All traces of corrosion should be removed. The coolant inside the radiator should be relatively transparent. If it is rust colored, the system should be drained and refilled (see Section 31). If the coolant level is not up to the top, add additional antifreeze/coolant mixture (see Section 4).

4 Carefully check the large upper and lower radiator hoses along with the smaller diameter heater hoses which run from the engine to the firewall. Inspect each hose along its entire length, replacing any hose which is cracked, swollen or shows signs of

deterioration. Cracks may become more apparent if the hose is squeezed **(see illustration)**. Regardless of condition, it's a good idea to replace hoses with new ones every two years.

5 Make sure all hose connections are tight. A leak in the cooling system will usually show up as white or rust colored deposits on the areas adjoining the leak. If wire-type clamps are used at the ends of the hoses, it may be a good idea to replace them with more secure screw-type clamps.

6 Use compressed air or a soft brush to remove bugs, leaves, etc. from the front of the radiator or air conditioning condenser. Be careful not to damage the delicate cooling fins or cut yourself on them.

7 Every other inspection, or at the first indication of cooling system problems, have the cap and system pressure tested. If you don't have a pressure tester, most gas stations and repair shops will do this for a minimal charge.

Check for a chafed area that could fail prematurely.

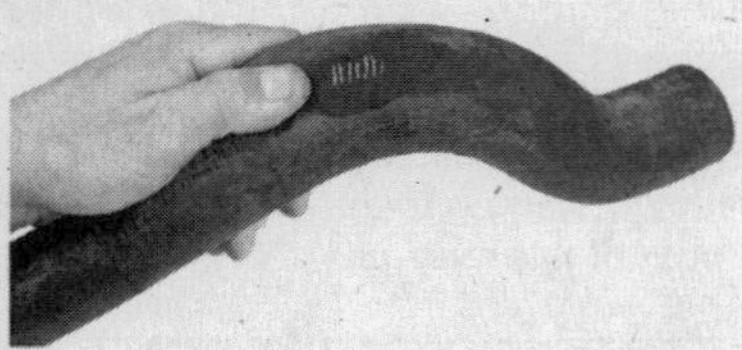

Check for a soft area indicating the hose has deteriorated inside.

Overtightening the clamp on a hardened hose will damage the hose and cause a leak.

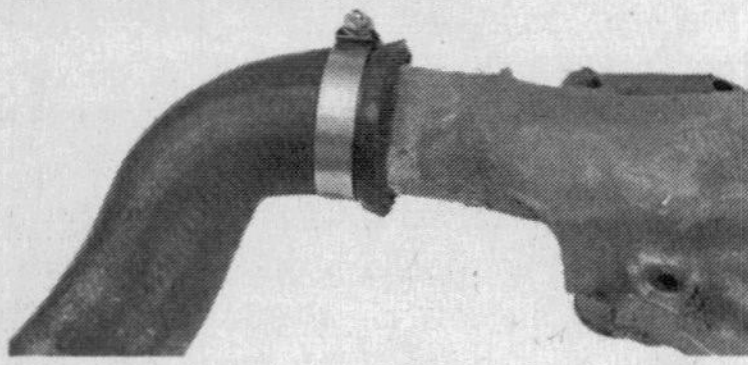

Check each hose for swelling and oil-soaked ends. Cracks and breaks can be located by squeezing the hose.

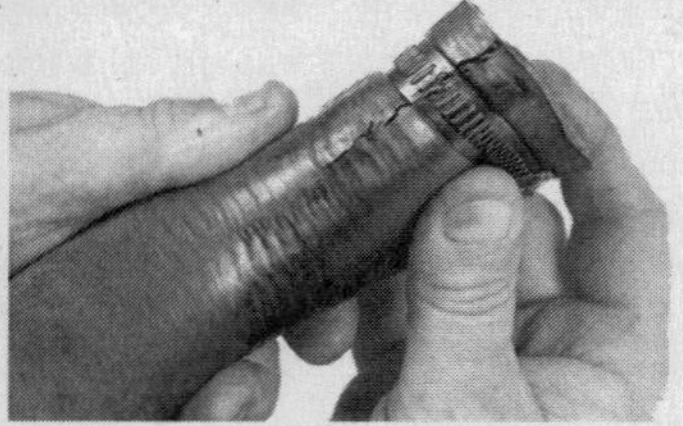

10.4 Hoses, like drivebelts, have a habit of failing at the worst possible time - to prevent the inconvenience of a blown radiator or heater hose, inspect them carefully as shown here

11 Underhood hose check and replacement (every 3000 miles or 3 months)

Caution: *Replacement of air conditioning hoses must be left to a dealer service department or air conditioning shop that has the equipment to depressurize the system safely. Never remove air conditioning components or hoses until the system has been depressurized.*

General

1 High temperatures in the engine compartment can cause the deterioration of the rubber and plastic hoses used for engine, accessory and emission systems operation. Periodic inspection should be made for cracks, loose clamps, material hardening and leaks. Information specific to the cooling system hoses can be found in Section 10.

2 Some, but not all, hoses are secured to the fittings with clamps. Where clamps are used, check to be sure they haven't lost their tension, allowing the hose to leak. If clamps aren't used, make sure the hose has not expanded and/or hardened where it slips over the fitting, allowing it to leak.

Vacuum hoses

3 It's quite common for vacuum hoses, especially those in the emissions system, to be color coded or identified by colored stripes molded into them. Various systems require hoses with different wall thickness, collapse resistance and temperature resistance. When replacing hoses, be sure the new ones are made of the same material.

4 Often the only effective way to check a hose is to remove it completely from the vehicle. If more than one hose is removed, be sure to label the hoses and fittings to ensure correct installation.

5 When checking vacuum hoses, be sure to include any plastic T-fittings in the check. Inspect the fittings for cracks and the hose where it fits over the fitting for distortion, which could cause leakage.

6 A small piece of vacuum hose (1/4-inch inside diameter) can be used as a stethoscope to detect vacuum leaks. Hold one end of the hose to your ear and probe around vacuum hoses and fittings, listening for the "hissing" sound characteristic of a vacuum leak. **Warning:** *When probing with the vacuum hose stethoscope, be very careful not to come into contact with moving engine components such as the drivebelt, cooling fan, etc.*

Fuel hose

Warning: *There are certain precautions which must be taken when inspecting or servicing fuel system components. Work in a well ventilated area and do not allow open flames (cigarettes, appliance pilot lights, etc.) or bare light bulbs near the work area. Mop up any spills immediately and do not store fuel soaked rags where they could ignite. On fuel-injected models, the fuel system is under high pressure, so if any fuel lines are to be disconnected, the pressure in the system must be relieved first* (see Chapter 4 for more information).

7 Check all rubber fuel lines for deterioration and chafing. Check especially for cracks in areas where the hose bends and just before fittings, such as where a hose attaches to the fuel filter.

8 High quality fuel line, specifically designed for high-pressure fuel injection applications, should be used for fuel line replacement. Never, under any circumstances, use unreinforced vacuum line, clear plastic tubing or water hose for fuel lines.

9 Spring-type clamps are commonly used on fuel lines. These clamps often lose their tension over a period of time, and can be "sprung" during removal. Replace all spring-type clamps with screw clamps whenever a hose is replaced.

Metal lines

10 Sections of metal line are often used in the fuel system. Check carefully to be sure the line has not been bent or crimped and that cracks have not started in the line.

11 If a section of metal fuel line must be replaced, only seamless steel tubing should be used, since copper and aluminum tubing don't have the strength necessary to withstand normal engine vibration.

12 Check the metal brake lines where they enter the master cylinder and brake proportioning unit (if used) for cracks in the lines or loose fittings. Any sign of brake fluid leakage calls for an immediate thorough inspection of the brake system.

12 Wiper blade inspection and replacement (every 3000 miles or 3 months)

Refer to illustrations 12.6, 12.7 and 12.8

1 The wiper and blade assembly should be inspected periodically for damage, loose components and cracked or worn blade elements.

2 Road film can build up on the wiper blades and affect their efficiency, so they should be washed regularly with a mild detergent solution.

3 The action of the wiping mechanism can loosen the bolts, nuts and fasteners, so they should be checked and tightened, as necessary, at the same time the wiper blades are checked.

4 If the wiper blade elements (sometimes called inserts) are cracked, worn or warped, they should be replaced with new ones.

5 Pull the wiper blade/arm assembly away from the glass.

6 On early model rear wipers, lift the release lever and slide the blade assembly off the pin on the wiper arm and replace the assembly as a unit **(see illustration)**.

7 On windshield wipers and later rear window wipers, depress the retaining tab and slide the blade assembly down and out of the hook in the wiper end **(see illustration)**.

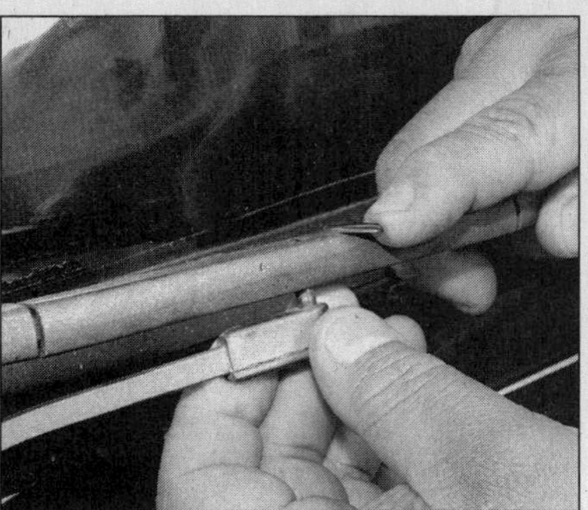

12.6 On earlier model rear wipers, lift the release lever up and slide the blade off the pin

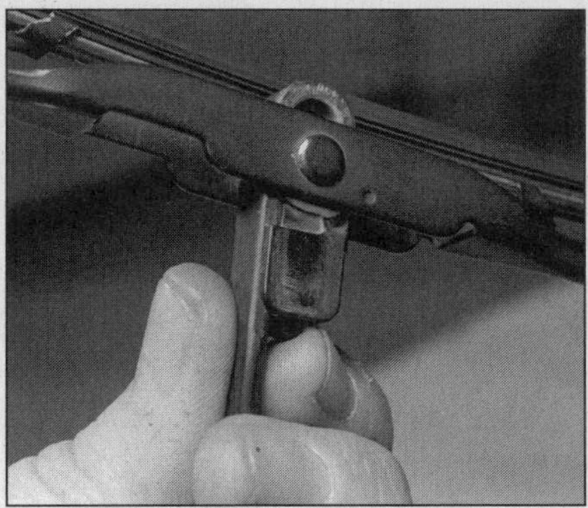

12.7 On later models press the retaining tab in and slide the wiper assembly down and out of the hook in the end of the wiper arm

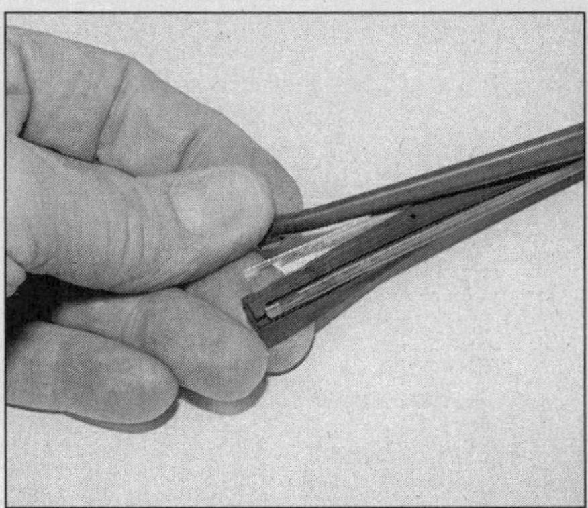

12.8 Detach the end of the element and use needle-nose pliers to pull out the support rods

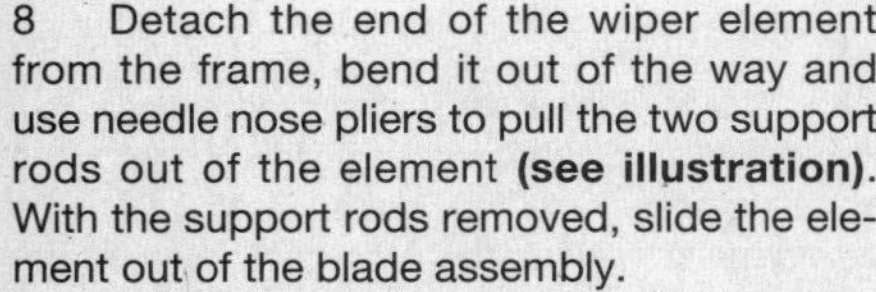

8 Detach the end of the wiper element from the frame, bend it out of the way and use needle nose pliers to pull the two support rods out of the element **(see illustration)**. With the support rods removed, slide the element out of the blade assembly.

9 Compare the new element with the old for length, design, etc.

10 Slide the new element into place and insert the support rods.

11 Reinstall the blade assembly on the arm, wet the windshield and check for proper operation.

13 Suspension and steering check (every 7500 miles or 6 months)

Refer to illustrations 13.4 and 13.6

1 Indications of a fault in these systems are excessive play in the steering wheel before the front wheels react, excessive sway around corners, body movement over rough roads or binding at some point as the steering wheel is turned.

2 Raise the front of the vehicle periodically, support it securely on jackstands and visually check the suspension and steering components for wear.

3 Check the wheel bearings. Do this by spinning the front wheels. Listen for any abnormal noises and watch to make sure the wheel spins true (doesn't wobble). Grab the top and bottom of the tire and pull in-and-out on it. Notice any movement which would indicate a loose wheel bearing assembly. If the bearings are suspect, refer to Chapter 10 for more information.

4 From under the vehicle, check for loose bolts, broken or disconnected parts and deteriorated rubber bushings on all suspension and steering components **(see illustration)**. Look for fluid leaking from the steering gear assembly. Check the power steering hoses, belts and connections for leaks. Inspect the shock absorbers for fluid leaks, indicating the need for replacement.

5 Have an assistant turn the steering wheel from side-to-side and check the steering components for free movement, chafing and binding. If the steering doesn't react with the movement of the steering wheel, try to determine where the slack is located.

6 The front axle Constant Velocity (CV) joints on 4WD models are protected by rubber boots. Check the boots for cuts, wear and signs of leaking grease **(see illustration)**. The boots must be replaced if they are damaged, otherwise the CV joint will be contaminated and eventually fail (see Chapter 8).

14 Exhaust system check (every 7500 miles or 6 months)

Refer to illustration 14.4

1 With the engine cold (at least three hours after the vehicle has been driven), check the complete exhaust system from the manifold to the end of the tailpipe. Be careful around the catalytic converter, which may be hot even after three hours. The inspection should be done with the vehicle on a hoist to permit unrestricted access. If a hoist isn't available, raise the vehicle and support it securely on jackstands.

2 Check the exhaust pipes and connections for signs of leakage and/or corrosion indicating a potential failure. Make sure that all brackets and hangers are in good condition and tight.

3 Inspect the underside of the body for holes, corrosion, open seams, etc. which may allow exhaust gasses to enter the passenger compartment. Seal all body openings with silicone sealant or body putty.

4 Rattles and other noises can often be traced to the exhaust system, especially the hangers, mounts and heat shields **(see illustration)**. Try to move the pipes, mufflers and catalytic converter. If the components can come in contact with the body or suspension parts, secure the exhaust system with new brackets and hangers.

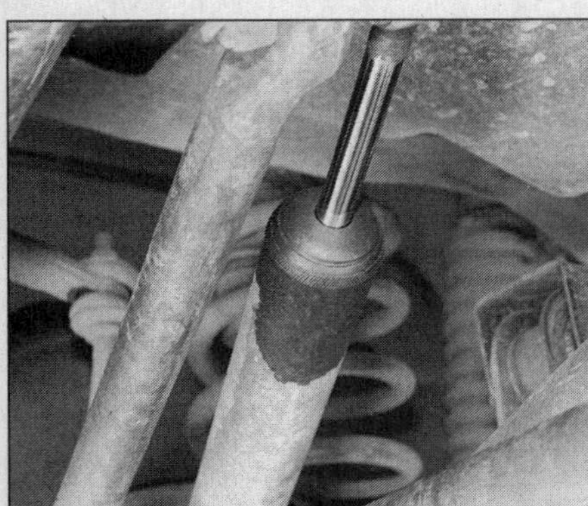

13.4 Inspect the suspension for damage - this steering damper is leaking fluid

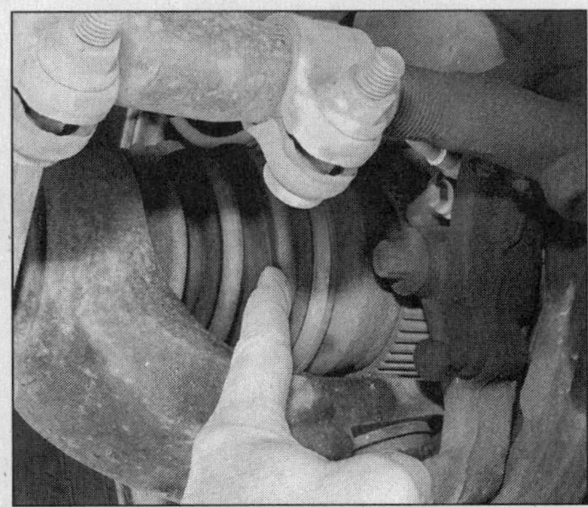

13.6 Push in on the CV joint boots and check for damage and leaking grease

14.4 Inspect the rubber exhaust system mounts for damage

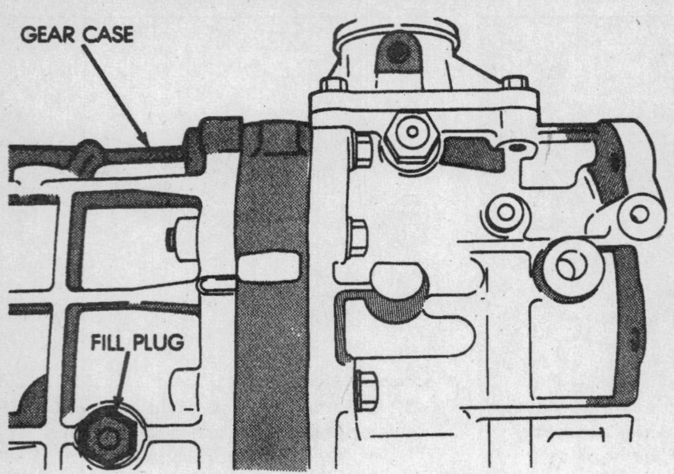

15.2 The manual transmission fill plug is located on the side of the case

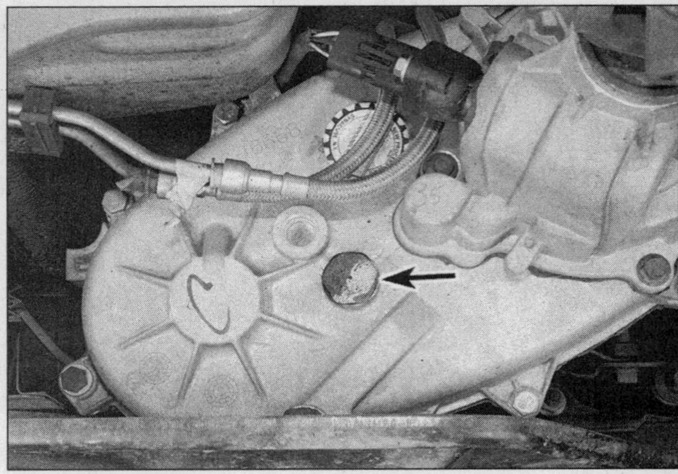

16.1 The 4WD transfer case fill plug (arrow) is located on the rear side of the housing

15 Manual transmission lubricant level check (every 7500 miles or 6 months)

Refer to illustration 15.2

1 The manual transmission has a filler plug which must be removed to check the lubricant level. If the vehicle is raised to gain access to the plug, be sure to support it safely on jackstands - DO NOT crawl under a vehicle which is supported only by a jack! Be sure the vehicle is level or the check may be inaccurate.

2 Unscrew the plug from the transmission **(see illustration)** and use a finger to reach inside the housing to determine the lubricant level. The level should be at or near the bottom of the plug hole.

3 If it isn't, add the recommended lubricant through the plug hole with a pump or squeeze bottle.

4 Install and tighten the plug and check for leaks after the first few miles of driving.

16 Transfer case lubricant level check (4WD models) (every 7500 miles or 6 months)

Refer to illustration 16.1

1 The transfer case lubricant level is checked by removing the upper plug located in the back of the case **(see illustration)**.

2 Use a finger to reach inside the housing to determine the lubricant level. The lubricant level should be just at the bottom of the hole. If not, add the appropriate lubricant through the opening.

17 Differential lubricant level check (every 7500 miles or 6 months)

Refer to illustrations 17.2a, 17.2b and 17.3

Note: *4WD vehicles have two differentials - be sure to check the lubricant level in both differentials.*

1 The filler plug on the front differential is a threaded plug, while the rear differential has a rubber press-in filler plug. Both must be removed to check the lubricant level. If the vehicle is raised to gain access to the plug, be sure to support it safely on jackstands - DO NOT crawl under the vehicle when it's supported only by the jack. Be sure the vehicle is level or the check may not be accurate.

2 Remove the plug from the filler hole in the differential housing cover **(see illustrations)**.

3 The lubricant level should be at the bottom of the filler hole **(see illustration)**. If not, use a pump or squeeze bottle to add the recommended lubricant until it just starts to run out of the opening. On some models a tag is located in the area of the plug which gives information regarding lubricant type.

4 Install the plug securely into the filler hole.

17.2a On a front differential, remove the inspection plug with a 3/8-drive ratchet

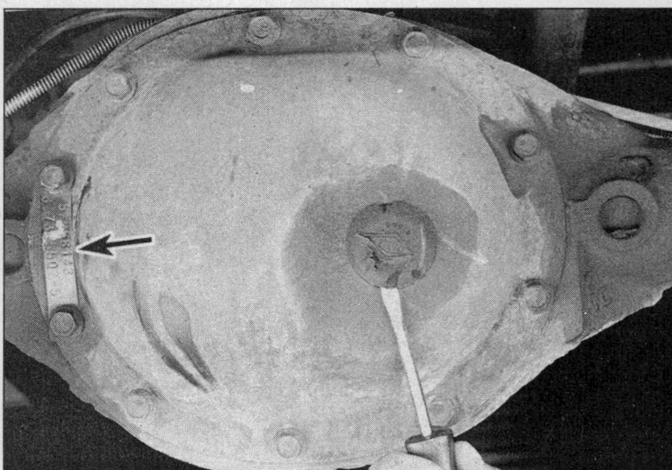

17.2b On rear differentials, use a screwdriver to pry out the rubber plug - note the tag (arrow) which contains important information on the differential

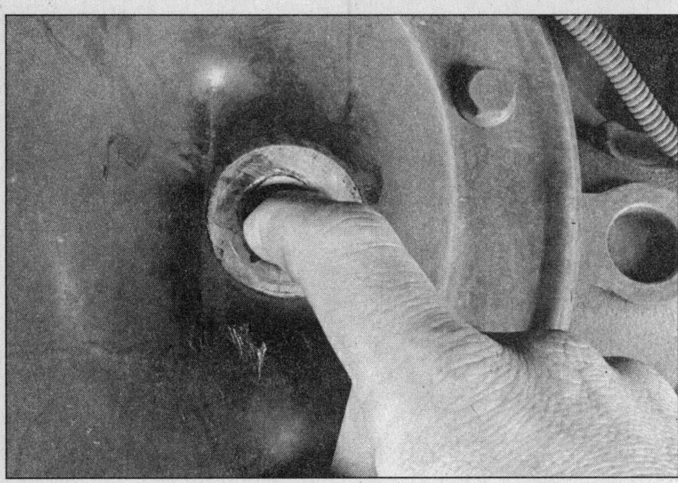

17.3 Use your finger as a dipstick to check the lubricant level

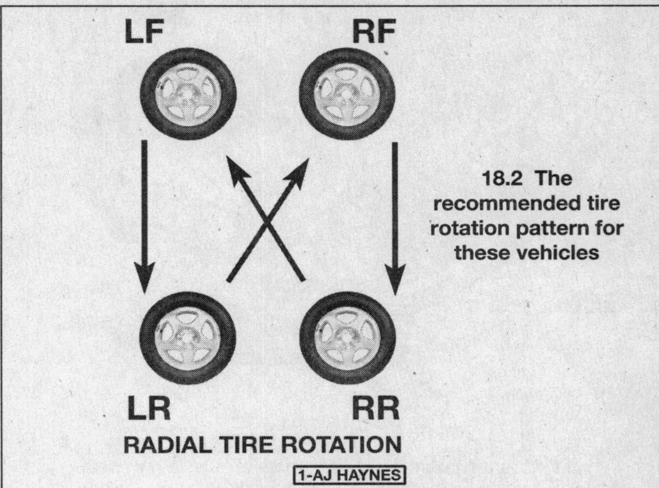

18.2 The recommended tire rotation pattern for these vehicles

18 Tire rotation (every 7500 miles or 6 months)

Refer to illustration 18.2

1 The tires should be rotated at the specified intervals and whenever uneven wear is noticed.

2 Refer to the accompanying illustration for the preferred tire rotation pattern **(see illustration)**.

3 Refer to the information in *Jacking and towing* at the front of this manual for the proper procedures to follow when raising the vehicle and changing a tire. If the brakes are to be checked, don't apply the parking brake as stated. Make sure the tires are blocked to prevent the vehicle from rolling as it's raised.

4 Preferably, the entire vehicle should be raised at the same time. This can be done on a hoist or by jacking up each corner and then lowering the vehicle onto jackstands placed under the frame rails. Always use four jackstands and make sure the vehicle is safely supported.

5 After rotation, check and adjust the tire pressures as necessary and be sure to properly tighten the lug nuts.

19 Brake check (every 7500 miles or 6 months)

Note: *For detailed information of the brake system, refer to Chapter 9.*

Warning: *Brake system dust is hazardous to your health. Wear an approved filtering mask whenever working on the brakes. DO NOT blow it out with compressed air, inhale it or use gasoline or solvents to remove it. Use brake system cleaner only.*

1 In addition to the specified intervals, the brakes should be inspected every time the wheels are removed or whenever a defect is suspected.

2 To check the brakes, raise the vehicle and place it securely on jackstands. Remove the wheels (see *Jacking and towing* at the front of the manual, if necessary).

Disc brakes

Refer to illustration 19.5

3 Disc brakes are used on the front wheels on all models as well as the rear wheels on some models. Extensive damage to the discs can occur if the pads are not replaced when needed.

19.5 You will find an inspection hole like this in most calipers - placing a ruler across the hole should enable you to determine the thickness of remaining pad material

4 The disc brake calipers, which contain the pads, are visible with the wheels removed. There is an outer pad and an inner pad in each caliper. All pads should be inspected.

5 Each caliper has a "window" or opening to inspect the pads. Check the thickness of the pad lining by looking into the caliper at each end and down through the inspection window at the top of the housing **(see illustration)**. If the lining material is less than the thickness listed in this Chapter's Specifications, the pads should be replaced. **Note:** *The metal backing plate is not included in this measurement.*

6 If you're unsure about the exact thickness of the remaining lining material, remove the pads for further inspection or replacement (refer to Chapter 9).

7 Before installing the wheels, check for leakage and/or damage (cracks, splitting, etc.) around the brake hose connections. Replace the hose or fittings as necessary, referring to Chapter 9.

8 Check the condition of the disc. Look for score marks, deep scratches and burned spots. If these conditions exist, the hub/disc assembly should be removed for servicing (see Chapter 9).

Drum brakes

Refer to illustrations 19.11, 19.13 and 19.15

9 Remove the drum by pulling it off the axle and brake assembly. If this proves difficult, make sure the parking brake is released, then squirt penetrating oil around the center hub areas. Allow the oil to soak in and try to pull the drum off again.

10 If the drum still cannot be pulled off, the self-adjusting mechanism may have to be retracted slightly. This is done by first removing the small plug from the backing plate.

11 With the plug removed, insert a thin screwdriver and lift the adjusting lever off the star wheel, then use an adjusting tool or screwdriver to back off the star wheel several turns **(see illustration)**. This will move the brake shoes away from the drum. If the drum

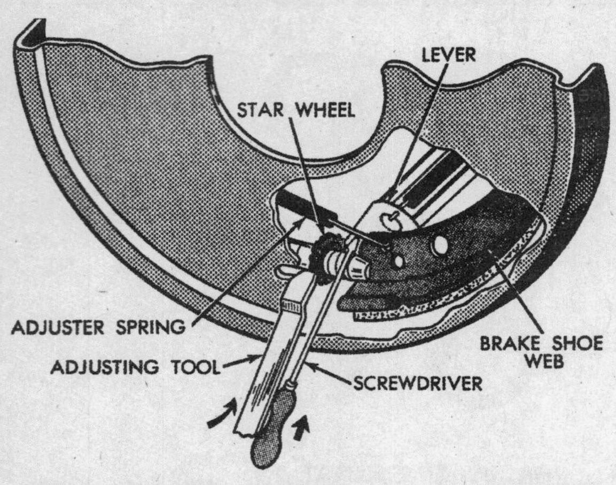

19.11 Use a thin screwdriver to push the lever away, then use an adjusting tool or another screwdriver to back off the star wheel

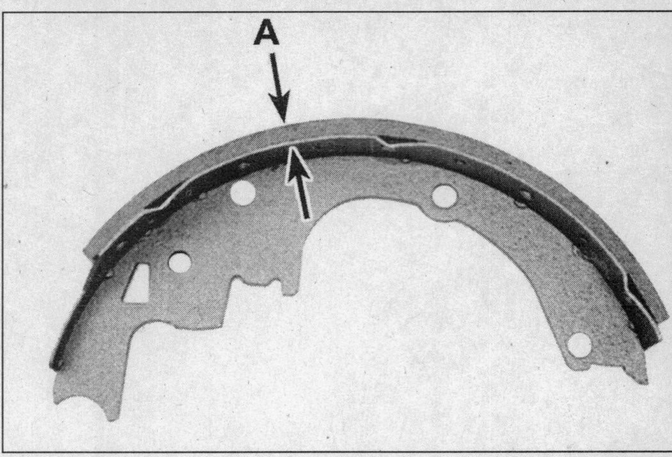

19.13 If the lining is bonded to the brake shoe, measure the lining thickness from the outer surface to the metal shoe, as shown here; if the lining is riveted to the shoe, measure from the lining outer surface to the rivet head

still won't pull off, tap around its inner circumference with a soft-face hammer.

12 With the drum removed, do not touch any brake dust (see the **Warning** at the beginning of this Section).

13 Note the thickness of the lining material on both the front and rear brake shoes. If the material has worn away to within 1/32-inch of the recessed rivets or 1/16-inch of the metal backing, the shoes should be replaced **(see illustration)**. The shoes should also be replaced if they're cracked, glazed (shiny surface) or contaminated with brake fluid.

14 Make sure that all the brake assembly springs are connected and in good condition.

15 Check the brake components for any signs of fluid leakage. Carefully pry back the rubber cups on the wheel cylinders located at the top of the brake shoes **(see illustration)**. Any leakage is an indication that the wheel cylinders should be overhauled immediately (see Chapter 9). Also check brake hoses and connections for signs of leakage.

16 Wipe the inside of the drum with a clean rag and brake system cleaner. Again, be careful not to breathe the dangerous asbestos dust.

19.15 To check for wheel cylinder leakage, use a small screwdriver to pry the boot away from the cylinder

17 Check the inside of the drum for cracks, score marks, deep scratches and hard spots, which will appear as small discolorations. If these imperfections cannot be removed with fine emery cloth, the drum must be taken to a machine shop equipped to resurface the drums.

18 If after the inspection process all parts are in good working condition, reinstall the brake drum.

19 Install the wheels and lower the vehicle.

Parking brake

20 The parking brake is operated by a hand lever and locks the rear brake system. The easiest, and perhaps most obvious method of periodically checking the operation of the parking brake assembly is to stop the vehicle on a steep hill with the parking brake set and the transmission in Neutral. If the parking brake cannot prevent the vehicle from rolling, adjust it (see Chapter 9).

20 Fuel system check (every 7500 miles or 6 months)

Warning: *Gasoline is extremely flammable, so take extra precautions when working on any part of the fuel system. Don't smoke or allow open flames or bare light bulbs in or near the work area, and don't work in a garage where a gas-type appliance (such as a water heater or clothes dryer) is present. Since gasoline is carcinogenic, wear latex gloves when there's a possibility of being exposed to fuel, and, if you spill fuel on your skin, rinse it off immediately with soap and water. Have a Class B fire extinguisher on hand.*

1 The fuel tank is located under the rear of the vehicle.

2 The fuel system is most easily checked with the vehicle raised on a hoist so the components underneath the vehicle are readily visible and accessible.

3 If the smell of gasoline is noticed while driving or after the vehicle has been in the sun, the system should be thoroughly inspected immediately.

4 Remove the gas tank cap and check for damage, corrosion and an unbroken sealing imprint on the gasket. Replace the cap with a new one if necessary.

5 With the vehicle raised, check the fuel tank and filler neck for punctures, cracks and other damage. The connection between the filler neck and the tank is especially critical. Sometimes a rubber filler neck will leak due to loose clamps or deteriorated rubber; problems a home mechanic can usually rectify. **Warning:** *Do not, under any circumstances, try to repair a fuel tank yourself (except rubber components). A welding torch or any open flame can easily cause the fuel vapors to explode if the proper precautions are not taken!*

6 Carefully check all rubber hoses and metal lines leading away from the fuel tank. Look for loose connections, deteriorated hoses, crimped lines and other damage. Follow the lines to the front of the vehicle, carefully inspecting them all the way. Repair or replace damaged sections as necessary.

7 If a fuel odor is still evident after the inspection, refer to Section 34.

21 Drivebelt check, adjustment and replacement (every 7500 miles or 6 months)

Refer to illustrations 21.3, 21.4, 21.5, 21.6, 21.7a, 21.7b, 21.8a, 21.8b and 21.8c

Check

1 The drivebelt, also referred to as a V-ribbed belt or simply "fan" belt, is located at the front of the engine. The condition and proper adjustment of the belt is critical to the operation of the engine. Because of the composition and the high stresses to which it is subjected, the drivebelt stretches and deteri-

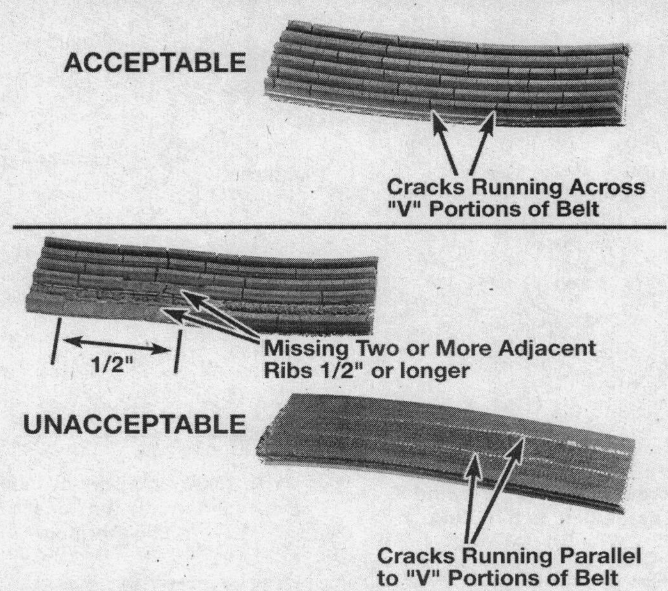

21.3 Small cracks on the underside of V-ribbed belts are acceptable - lengthwise cracks, or missing pieces that cause the belt to make noise are cause for replacement

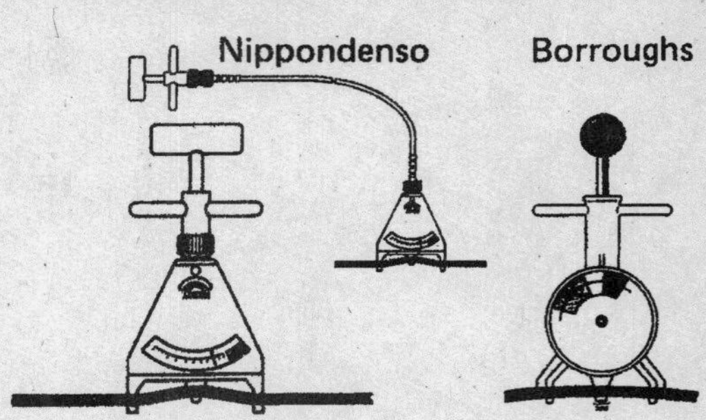

21.4 If you are able to borrow or rent either a Nippondenso or Burroughs belt tension gauge, this is how it is installed on the belt - compare the reading on the scale with the specified drivebelt tension (six-cylinder models without an automatic adjuster only)

orates as it gets older. It must therefore be periodically inspected.

2 On these models a single serpentine drivebelt driving the alternator, power steering pump and air conditioning compressor, if used.

3 With the engine off, open the hood and locate the drivebelt at the front of the engine. With a flashlight, check the belt for separation of the ribs from the adhesive rubber, cracking or separation of the ribs, and torn or worn ribs or cracks in the inner ridges of the ribs **(see illustration)**. Both sides of the belt should be inspected, which means you will have to twist the belt to check the underside. Use your fingers to feel the belt where you can't see it. If any of the above conditions are evident, replace the belt (go to Step 7 or 8).

Adjustment

4 The tension of the drivebelt on six-cylinder engines without an automatic adjuster is checked by using a belt tension gauge at a distance halfway between the idler and power steering pump pulley **(see illustration)**. Compare the tension gauge readings with those in the Specifications Section at the beginning of this Chapter.

5 Adjust the tension by loosening the power steering pump pivot, locknut and lockbolts and turn the adjusting bolt counterclockwise to loosen the belt and clockwise to tighten it **(see illustration)**. Repeat the procedure until the drivebelt tension is correct and tighten the locknut and bolts.

6 On some six-cylinder engines and all V8 models, the belt tension is controlled by a spring-loaded tensioner and the belt does not require periodic adjustment. If the indexing arrow on the back of the tensioner is more than 1/8-th of an inch of the indexing mark on the tensioner housing, the belt must be replaced **(see illustration)**.

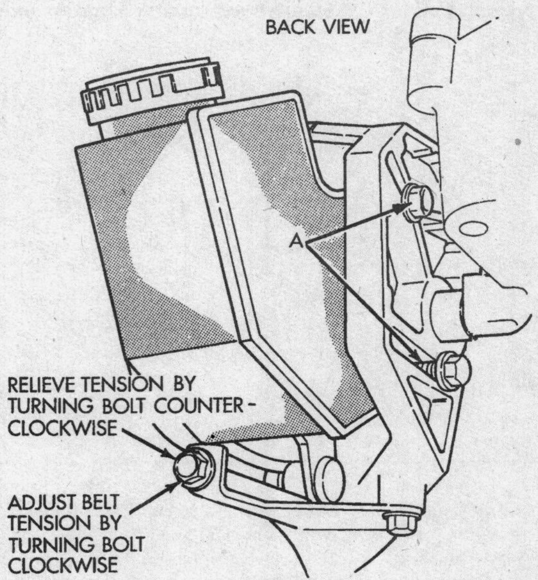

21.5 On six-cylinder engines without an automatic adjuster, loosen the power steering pump bolts and turn the tensioner adjusting bolt to tighten or loosen the drivebelt

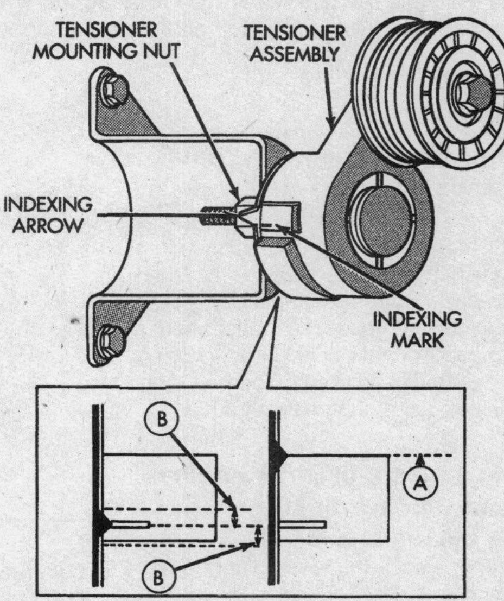

21.6 The serpentine drivebelt tensioner used on later six-cylinder engines and all V8 models automatically keeps the proper tension on the drivebelt, but it does have limits - the indexing arrow on the tensioner housing (B) must not move beyond point A on the tensioner housing

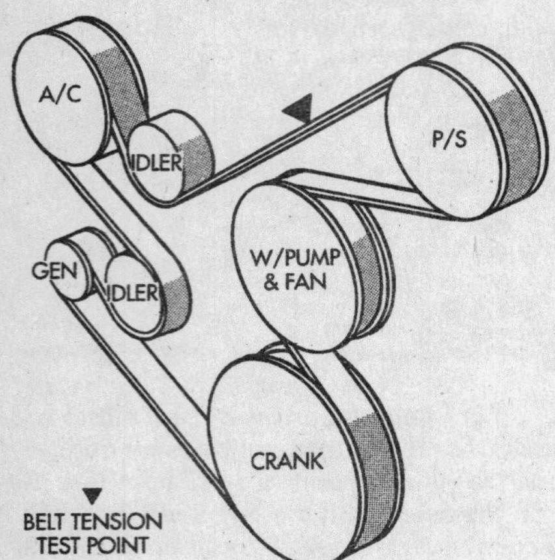

21.7a 1995 and earlier model six-cylinder engine serpentine belt routing layout

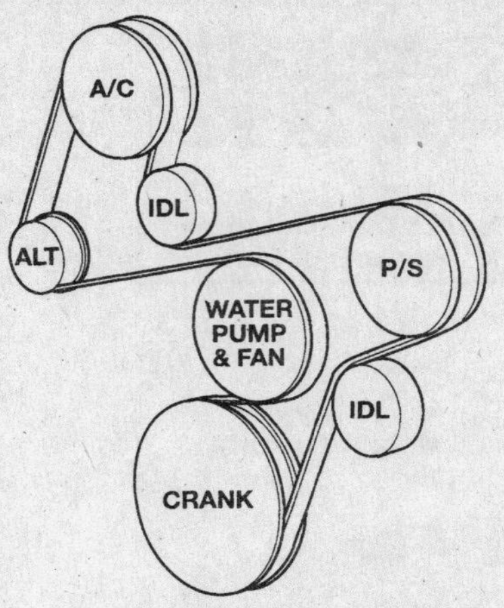

21.7b 1996 through 1998 six-cylinder engine serpentine belt routing

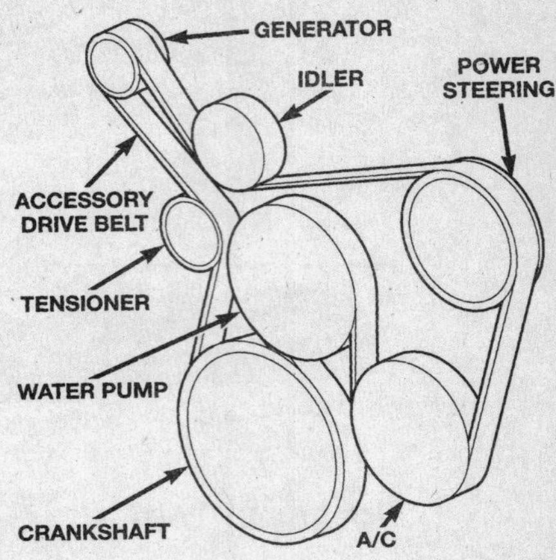

21.7c 2000 model six-cylinder engines have an automatic tensioner and different belt routing

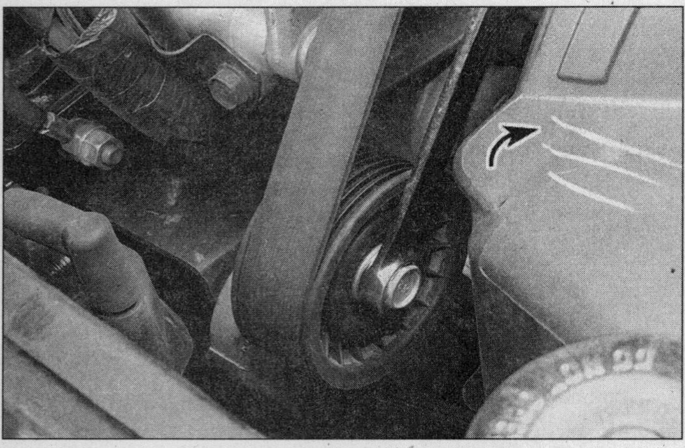

21.8a Release the belt tension by rotating the tensioner clockwise using a wrench on the tensioner bolt

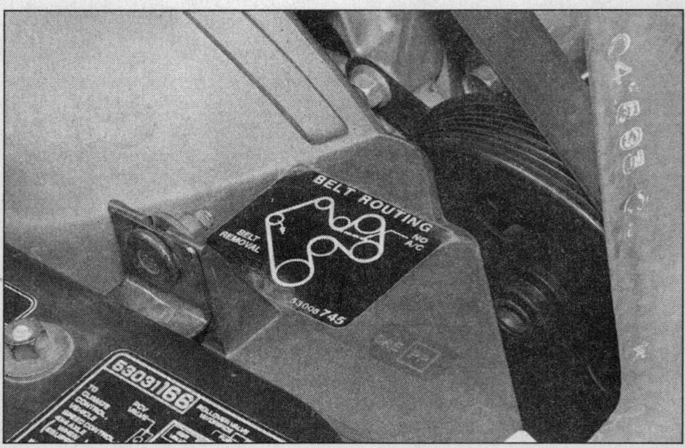

21.8b A drivebelt routing diagram is found on the radiator brace or shroud (5.2L/5.9L drivebelt routing diagram shown)

Replacement

Inline six-cylinder engine (without an automatic adjuster)

7 To replace a belt, follow the above procedures for drivebelt adjustment but slip the belt off the crankshaft pulley and remove it. When installing the new belt, make sure it is routed properly because improper routing can cause the water pump to rotate backwards, causing overheating **(see illustrations)**. The belt routing diagram is located on the fan shroud.

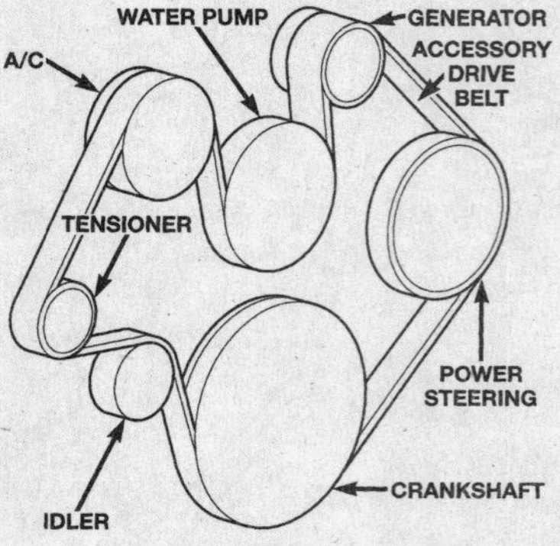

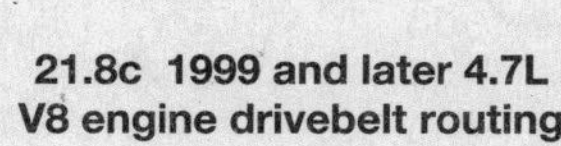

21.8c 1999 and later 4.7L V8 engine drivebelt routing

V8 engine (and six-cylinder engines with an automatic adjuster)

8 Use a wrench to rotate the tensioner clockwise and release the tension on the belt **(see illustration)**. The tensioner will swing down out of the way once the tension of the belt is released. Remove the belt and carefully release the tensioner. Route the new belt over the pulleys, again rotating the tensioner to allow the belt to be installed, then release the belt tensioner. When installing the new belt, make sure it is routed properly because improper routing can cause the water pump to rotate backwards, causing overheating. The belt routing diagram is located on the fan shroud **(see illustrations)**.

All engines

9 Take the old belt to the parts store in order to make a direct comparison for length, width and design.

10 When replacing the drivebelt, make sure it is routed correctly on the pulley or ribbed grooves in the pulleys. It is essential that the belt be properly centered.

11 Adjust the belt(s) in accordance with the procedure outlined above.

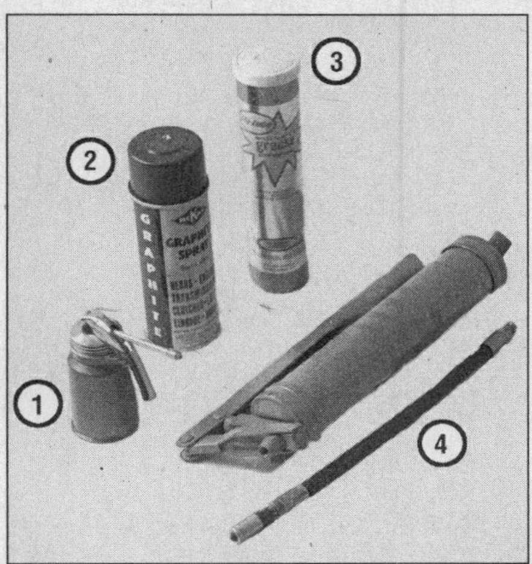

24.1 Materials required for chassis and body lubrication

1 ***Engine oil*** *- Light engine oil in a can like this can be used for door and hood hinges*
2 ***Graphite spray*** *- Used to lubricate lock cylinders*
3 ***Grease*** *- Grease, in a variety of types and weights, is available for use in a grease gun. Check the Specification for your requirements*
4 ***Grease gun*** *- A common grease gun, shown here with a detachable hose and nozzle, is needed for chassis lubrication. After use, clean it thoroughly!*

22 Seat belt check (every 7500 miles or 6 months)

1 Check the seat belts, buckles, latch plates and guide loops for any obvious damage or signs of wear.
2 On later models, make sure the seat belt reminder light comes on when the key is turned on.
3 The seat belts are designed to lock up during a sudden stop or impact, yet allow free movement during normal driving. The retractors should hold the belt against your chest while driving and rewind the belt when the buckle is unlatched.
4 If any of the above checks reveal problems with the seat belt system, replace parts as necessary.

23 Park/Neutral Position (PNP) switch check (models with automatic transmission) (every 7500 miles or 6 months)

Warning: *During the following checks there is a chance the vehicle could lunge forward, possibly causing damage or injuries. Allow plenty of room around the vehicle, apply the parking brake firmly and press down on the brake pedal during the checks.*
1 The Park/Neutral Position (PNP) switch prevents the engine from being cranked unless the gear selector is in Park or Neutral.
2 Try to start the vehicle in each gear. The engine should crank only in Park or Neutral. If it cranks in any other gear, the PNP switch is faulty or in need of adjustment (see Chapter 7 Part B).
3 Make sure the steering column lock allows the key to go into the Lock position only when the shift lever is in Park.
4 The ignition key should come out only in the Lock position.

24 Chassis lubrication (every 15,000 miles or 12 months)

Refer to illustrations 24.1, 24.6a, 24.6b and 24.8
1 Refer to *Recommended lubricants and fluids* at the front of this Chapter to obtain the necessary grease, etc. You'll also need a grease gun **(see illustration)**. Occasionally plugs will be installed rather than grease fittings. If so, grease fittings will have to be purchased and installed. Some models may have sealed tie-rod ends and balljoints with no plugs or fittings for lubrication.
2 Look under the vehicle and locate the grease fittings.
3 For easier access under the vehicle, raise it with a jack and place jackstands under the frame. Make sure it's safely supported by the stands. If the wheels are to be removed at this interval for tire rotation or brake inspection, loosen the lug nuts slightly while the vehicle is still on the ground.
4 Before beginning, force a little grease out of the nozzle to remove any dirt from the end of the gun. Wipe the nozzle clean with a rag.
5 With the grease gun and plenty of clean rags, crawl under the vehicle and begin lubricating the components.
6 Wipe one of the grease fitting nipples clean and push the nozzle firmly over it. Pump the gun until the component is completely lubricated. On balljoints and steering connections, stop pumping when the rubber seal is firm to the touch **(see illustrations)**. Do not pump too much grease into the fitting as it could rupture the seal. For all other suspension and steering components, continue pumping grease into the fitting until it oozes out of the joint between the two components. If it escapes around the grease gun nozzle, the nipple is clogged or the nozzle is not completely seated on the fitting. Resecure the gun nozzle to the fitting and try again. If necessary, replace the fitting with a new one.
7 Wipe the excess grease from the components and the grease fitting. Repeat the procedure for the remaining fittings.
8 Lubricate the driveshaft slip joint by pumping grease into the fitting until it can be seen coming out of the slip joint **(see illustration)**.
9 Lubricate the transfer case shift mechanism contact surfaces with clean engine oil.
10 Also clean and lubricate the parking brake cable, along with the cable guides and levers. This can be done by smearing some of the chassis grease onto the cable and its related parts with your fingers.
11 On 1999 and later models, the brake caliper slide pins should be cleaned and lubricated regularly, especially when the vehicle is used off-road or in dusty conditions. See Chapter 9 for the procedure.

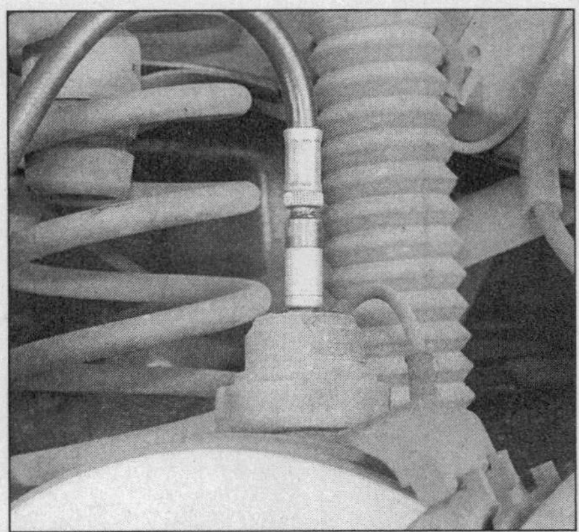

24.6a The upper balljoint grease fitting is accessible after removing the front wheel

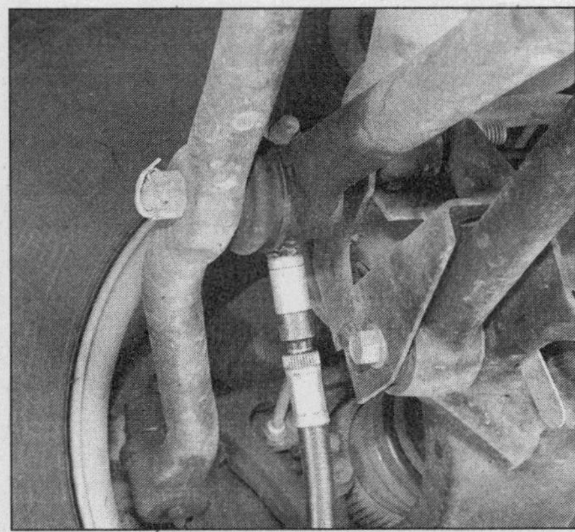

24.6b Pump the grease into the fitting until the rubber seal is firm to the touch

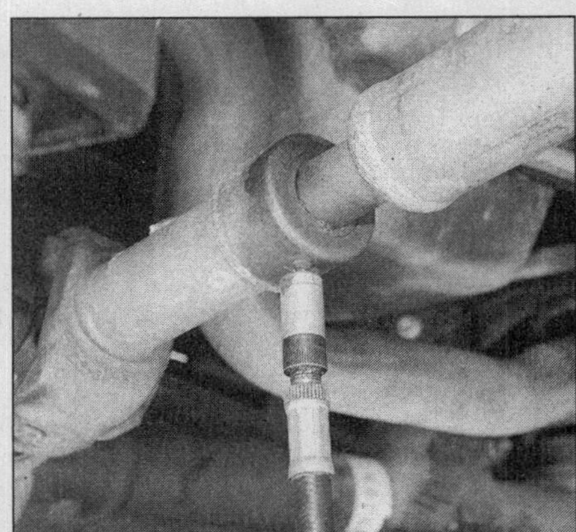

24.8 The front driveshaft slip joint grease fitting is located on the collar - pump grease into it until it comes out the collar

25.3 Release the four air cleaner housing retaining clips

25.4 Lift the cover and slide the element out of the air cleaner housing

25 Air filter replacement (every 30,000 miles or 24 months)

Refer to illustrations 25.3 and 25.4

1 At the specified intervals, the air filter element should be replaced with a new one.

2 The filter is located in the left front corner of the engine compartment.

3 Detach the four spring clips **(see illustration)**.

4 Lift the air filter housing cover up and slide the element out of the housing **(see illustration)**.

5 Wipe out the inside of the air cleaner housing with a clean rag.

6 Place the new filter element in the air cleaner housing. Make sure it seats properly in the bottom of the housing.

7 Install the housing cover, making sure it is seated, then secure it with the clips.

26 Automatic transmission fluid and filter change (every 30,000 miles or 24 months)

Refer to illustrations 26.6, 26.9, 26.10a, 26.10b and 26.12

1 At the specified intervals, the transmission fluid should be drained and replaced. Since the fluid will remain hot long after driving, perform this procedure only after the engine has cooled down completely. On 1994 and later models, the manufacturer also recommends adjusting the transmission bands at this time, since this procedure requires removing the fluid pan (see Section 27).

2 Before beginning work, purchase the specified transmission fluid (see *Recommended lubricants and fluids* at the front of this Chapter) and a new filter.

3 Other tools necessary for this job include a floor jack, jackstands to support the vehicle in a raised position, a drain pan capable of holding at least eight pints, newspapers and clean rags.

4 Raise the vehicle and support it securely on jackstands.

5 Place the drain pan underneath the transmission pan. Remove the rear and side pan mounting bolts, but only loosen the front pan bolts approximately four turns.

6 Carefully pry the transmission pan loose with a screwdriver, allowing the fluid to drain **(see illustration)**.

7 Remove the remaining bolts, pan and gasket. Carefully clean the gasket surface of the transmission to remove all traces of the old gasket and sealant.

8 Drain the fluid from the transmission pan, clean the pan with solvent and dry it with compressed air, if available.

9 Remove the filter retaining screws from the valve body and remove the filter **(see illustration)**.

10 Once the pan and filter are off, you can get a good idea of the condition of your transmission by thoroughly inspecting the bottom of the pan, the filter and the fluid. Although normally bright red, transmission fluid may turn dark red or brown during normal use. If you find the fluid very dark col-

26.6 With the front bolts in place but loose, pull the rear of the pan down to drain the fluid

26.9 Remove the filter screws - a special Torx-head wrench is necessary

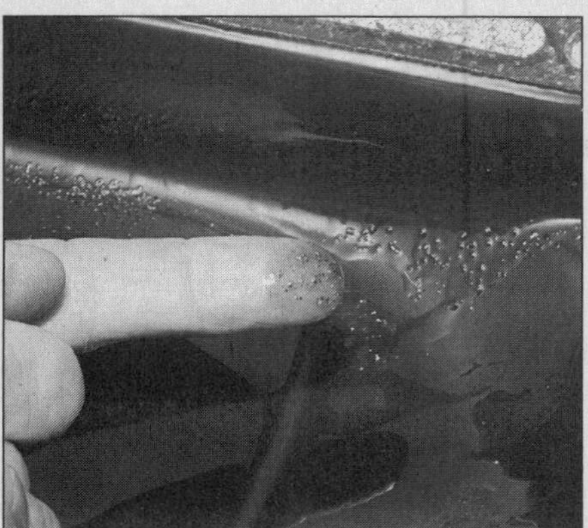
26.10a Metal flakes found at the bottom of the pan could be a sign of internal transmission damage

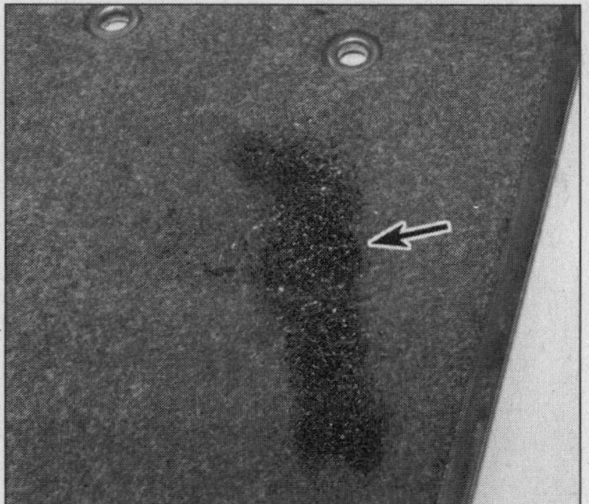
26.10b Inspect the filter for pieces of metal or clutch material (arrow)

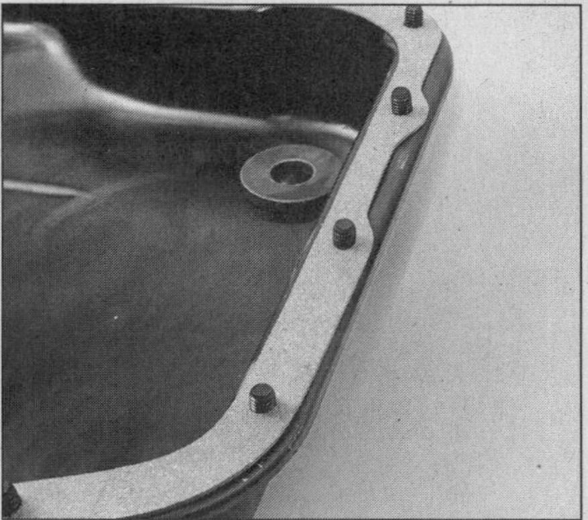
26.12 Place a new gasket in position on the pan and install the bolts to hold it in place

ored, or if it smells burned, it usually indicates the transmission has been overheated. If you find small pieces of metal or clutch material in the pan or filter, it indicates wear or damage have occurred to the internal parts or clutches **(see illustrations)**. If you have any concerns about the condition of your transmission based on what you find in the fluid, pan and filter, it's a good idea to take your vehicle to your dealer or a transmission shop for further evaluation.

11 Use a gasket scraper to remove any traces of old gasket material remaining on the valve body. **Note:** *Be very careful not to gouge the delicate aluminum gasket surface on the valve body.* Install a new filter and gasket.

12 Make sure the gasket surface on the transmission pan is clean, then install the magnet (if equipped) and a new gasket on the pan **(see illustration)**. Put the pan in place against the transmission and, working around the pan, tighten each bolt a little at a time to the torque listed in this Chapter's Specifications.

13 Lower the vehicle and add approximately two quarts of the specified type of automatic transmission fluid through the filler tube (see Section 6).

14 With the transmission in Park and the parking brake set, run the engine at a fast idle, but don't race it.

15 Move the gear selector through each range and back to Park. Check the fluid level and add fluid, if necessary, until the level is within the correct range on the dipstick.

16 Check under the vehicle for leaks during the first few trips. Check the fluid level again when the transmission is hot (see Section 6).

27 Automatic transmission band adjustment (1994 and later models)

Refer to illustrations 27.2 and 27.7

Note: *The band adjustment is not a regularly scheduled maintenance item. In the event of rough shifting, follow the procedure for band adjustment along with the fluid change.*

Front band (kickdown)

1 The transmission bands should be adjusted as necessary when the transmission fluid and filter are being replaced (see Section 26). **Note:** *The 45RFE transmission in 1999 and later models does not require band adjustments.*

2 The front band adjusting screw is located on the left side of the transmission **(see illustration)**.

3 Raise the front of the vehicle and support it securely on jackstands.

4 Loosen the adjusting screw locknut approximately five turns, then loosen the adjusting screw a few turns. Make sure the adjusting screw turns freely, with no binding; lubricate it with penetrating oil if necessary.

5 Tighten the adjusting screw to the torque listed in the Specifications Section at the beginning of this Chapter, then back it off the number of turns listed in the Specifications. Hold the screw from turning, then tighten the locknut to the torque listed in this Chapter's Specifications.

Rear band (low-reverse)

6 To gain access to the rear band, the fluid pan must be removed (see Section 26).

7 Loosen the adjusting screw locknut and back it off four turns **(see illustration)**. Make sure the screw turns freely in the lever.

8 Tighten the adjusting screw to the torque listed in the Specifications section at the beginning of this Chapter, then back it off the number of turns listed in the Specifications. Hold the screw from turning, then tighten the locknut to the torque listed in this Chapter's Specifications.

9 Install the transmission fluid pan (see Section 26).

27.2 Location of the front band adjusting screw (arrow)

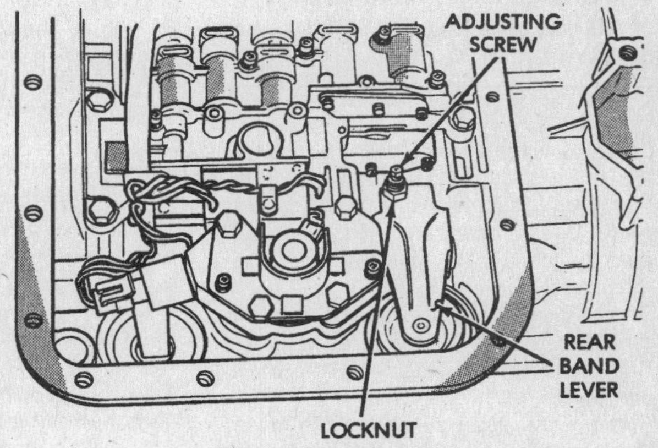

27.7 Rear band adjusting screw details

29.4 Use a wrench to remove the transfer case drain plug

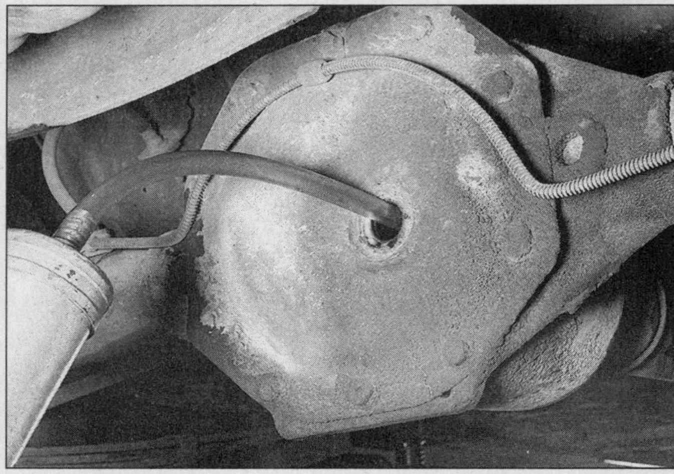

30.6 This is the easiest way to remove the lubricant: work the end of the hose to the bottom of the differential housing and draw out the old lubricant with a hand pump

28 Manual transmission lubricant change (every 30,000 miles or 24 months)

1 Raise the vehicle and support it securely on jackstands.

2 Move a drain pan, rags, newspapers and wrenches under the transmission.

3 Remove the transmission drain plug at the bottom of the case and allow the lubricant to drain into the pan.

4 After the lubricant has drained completely, reinstall the plug and tighten it securely.

5 Remove the fill plug from the side of the transmission case. Using a hand pump, syringe or funnel, fill the transmission with the specified lubricant until it begins to leak out through the hole. Reinstall the fill plug and tighten it securely.

6 Lower the vehicle.

7 Drive the vehicle for a short distance, then check the drain and fill plugs for leakage.

29 Transfer case lubricant change (4WD models) (every 30,000 miles or 24 months)

Refer to illustration 29.4

1 Drive the vehicle for at least 15 minutes to warm the lubricant in the case. Perform this warm-up procedure with 4WD engaged, if possible. Use all gears, including Reverse, to ensure the lubricant is sufficiently warm to drain completely.

2 Raise the vehicle and support it securely on jackstands.

3 Remove the filler plug from the case **(see illustration 16.1)**.

4 Remove the drain plug from the lower part of the case and allow the old lubricant to drain completely **(see illustration)**.

30.8a Remove the bolts from the lower edge of the cover . . .

5 Carefully clean and install the drain plug after the case is completely drained. Tighten the plug to the torque listed in this Chapter's Specifications.

6 Fill the case with the specified lubricant until it is level with the lower edge of the filler hole.

7 Install the filler plug and tighten it securely.

8 Drive the vehicle for a short distance and recheck the lubricant level. In some instances a small amount of additional lubricant will have to be added.

30 Differential lubricant change (every 30,000 miles or 24 months)

Drain

Refer to illustration 30.6, 30.8a, 30.8b, 30.8c and 30.10

1 This procedure should be performed after the vehicle has been driven so the lubricant will be warm and therefore flow out of the differential more easily.

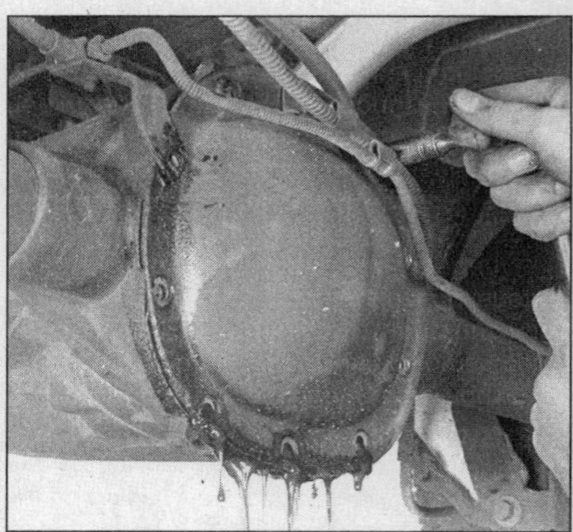

30.8b . . . then loosen the top bolts and let the lubricant drain out

2 Raise the vehicle and support it securely on jackstands.

3 The easiest way to drain the differential is to remove the lubricant through the filler plug hole with a suction pump. If the differential cover gasket is leaking, it will be necessary to remove the cover to drain the lubricant (which will also allow you to inspect the differential).

Changing the lubricant with a suction pump

4 Remove the filler plug from the differential (see Section 17).

5 Insert the flexible hose.

6 Work the hose down to the bottom of the differential housing and pump the lubricant out **(see illustration)**.

Changing lubricant by removing the cover

7 Move a drain pan, rags, newspapers and wrenches under the vehicle.

8 Remove the bolts on the lower half of the cover. Loosen the bolts on the upper half and use them to loosely retain the cover.

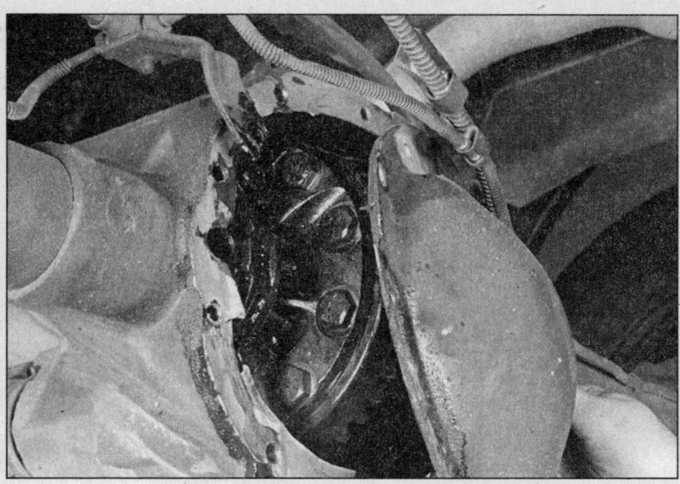

30.8c After the lubricant has drained, remove the bolts and the cover

30.10 Carefully scrape the old gasket material off to ensure a leak-free seal

Allow the oil to drain into the pan, then completely remove the cover **(see illustrations)**.

9 Using a lint-free rag, clean the inside of the cover and the accessible areas of the differential housing. As this is done, check for chipped gears and metal particles in the lubricant, indicating that the differential should be more thoroughly inspected and/or repaired.

10 Thoroughly clean the gasket mating surfaces of the differential housing and the cover plate. Use a gasket scraper or putty knife to remove all traces of the old gasket **(see illustration)**.

11 Apply a thin layer of RTV sealant to the cover flange, then press a new gasket into position on the cover. Make sure the bolt holes align properly.

Refill

12 Use a hand pump, syringe or funnel to fill the differential housing with the specified lubricant until it's level with the bottom of the plug hole.

13 Install the fill plug and tighten it securely.

31 Cooling system servicing (draining, flushing and refilling) (every 30,000 miles or 24 months)

Warning: *Do not allow antifreeze to come in contact with your skin or painted surfaces of the vehicle. Rinse off spills immediately with plenty of water. Antifreeze is highly toxic if ingested. Never leave antifreeze lying around in an open container or in puddles on the floor; children and pets are attracted by it's sweet smell and may drink it. Check with local authorities about disposing of used antifreeze. Many communities have collection centers which will see that antifreeze is disposed of safely.*

Draining

Refer to illustrations 31.4a and 31.4b

1 Periodically, the cooling system should be drained, flushed and refilled to replenish the antifreeze mixture and prevent formation of rust and corrosion, which can impair the performance of the cooling system and cause engine damage. When the cooling system is serviced, all hoses and the radiator cap should be checked and replaced if necessary.

2 Apply the parking brake and block the wheels. **Warning:** *If the vehicle has just been driven, wait several hours to allow the engine to cool down before beginning this procedure.*

3 Move a large container under the radiator drain to catch the coolant. The radiator drain plug on six-cylinder models is located on the lower right side of the radiator while on V8 models it is on the left. Attach a 3/8-inch diameter hose to the drain fitting (if possible) to direct the coolant into the container, then open the drain fitting (a pair of pliers may be required to turn it). Remove the radiator cap.

4 Allow the radiator to drain, then close the drain fitting and move the container under the engine block drain plugs - there's one on each side of the block on V8 engines and one on the right side only on six-cylinder engines **(see illustrations)**. Remove the plugs and

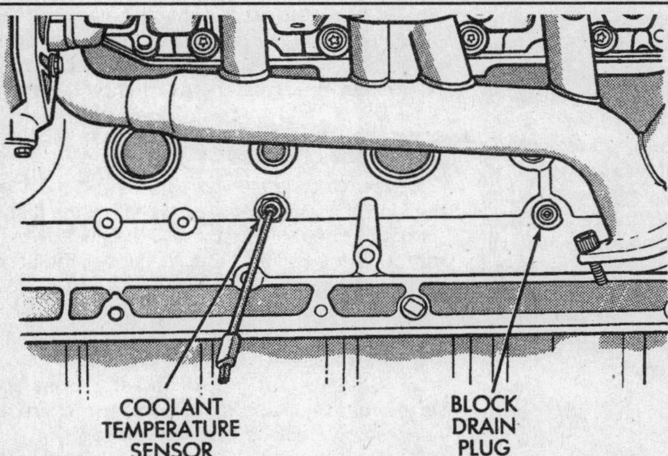

31.4a On six-cylinder engines, the block drain plug is located on the left side, below the exhaust manifold

31.4b V8 engine cylinder block drain location - there is one on each side of the block

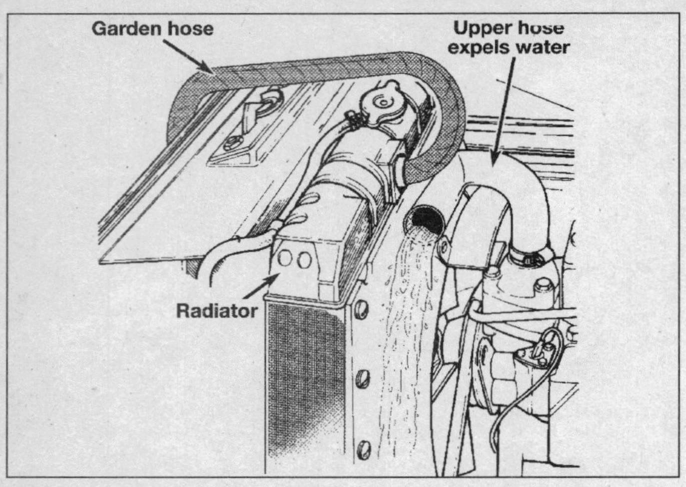

31.9 With the thermostat removed, disconnect the upper radiator hose and flush the radiator and engine block with a garden hose

31.14 Location of the cooling system bleed plug (1) - 4.7L V8 engines

32.1 The PCV valve (arrow) on 5.2L/5.9L engines is pressed into the right side valve cover

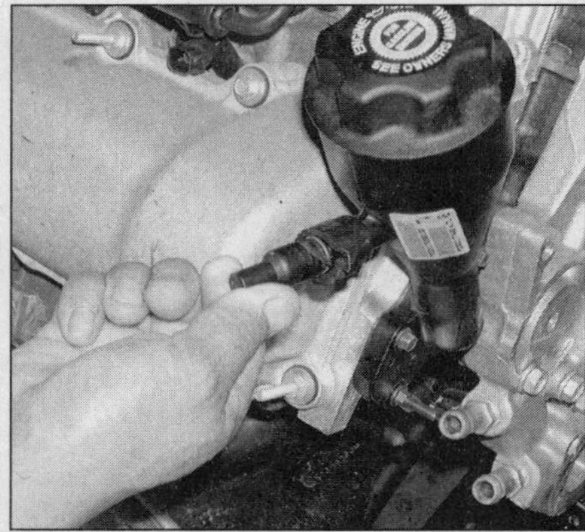

32.2 On 4.7L engines, rotate the PCV valve 90 degrees counterclockwise and pull straight out to remove it

allow the coolant in the block to drain. **Note:** *Frequently, the coolant will not drain from the block after the plug is removed. This is due to a rust layer that has built up behind the plug. Insert a Phillips screwdriver into the hole to break the rust barrier.*

5 While the coolant is draining, check the condition of the radiator hoses, heater hoses and clamps (refer to Section 11 if necessary). Replace any damaged clamps or hoses.

6 Remove the coolant reservoir (see Chapter 3) and pour the coolant into the container.

Flushing

Refer to illustration 31.9

7 Once the system is completely drained, remove the thermostat from the engine (see Chapter 3). Then reinstall the thermostat housing without the thermostat. This will allow the system to be thoroughly flushed.

8 Tighten the radiator drain plug, if not already done. Turn your heating system controls to Hot, so that the heater core will be flushed at the same time as the rest of the cooling system.

9 Disconnect the upper radiator hose, then place a garden hose in the upper radiator inlet and flush the system until the water runs clear at the upper radiator hose **(see illustration)**.

10 In severe cases of contamination or clogging of the radiator, remove the radiator (see Chapter 3) and have a radiator repair facility clean and repair it if necessary.

11 Many deposits can be removed by the chemical action of a cleaner available at auto parts stores. Follow the procedure outlined in the manufacturer's instructions. **Note:** *When the coolant is regularly drained and the system refilled with the correct antifreeze/water mixture, there should be no need to use chemical cleaners or descalers.*

Refilling

Refer to illustration 31.14

12 To refill the system, install the thermostat and reconnect any radiator hoses.

13 Place the heater temperature control in the maximum heat position.

14 If you're working on a 4.7L V8 engine, remove the air bleed plug from the casting where the upper radiator hose connects to the engine **(see illustration)**.

15 Make sure to use the proper coolant listed in this Chapter's Specifications. Slowly fill the radiator with the recommended mixture of antifreeze and water to the base of the filler neck or, on 4.7L V8 engines, until coolant begins to flow from the bleed hole (then install the plug). Then add coolant to the reservoir until it reaches the FULL mark. Wait five minutes and recheck the coolant level in the radiator, adding if necessary.

4.7L V8 engine

16 Install the radiator cap and the reservoir cap. Start the engine, rev it up to 3000 rpm for ten seconds, then turn it off.

17 Remove the radiator cap and unscrew the bleed plug. Add coolant to the radiator again until coolant flows from the bleed hole.

18 Apply a film of RTV sealant to the threads of the bleed plug, then install the plug and tighten it securely.

19 Check the level of the coolant in the radiator, adding as necessary. Install the radiator cap. Start the engine, allow it to reach normal operating temperature and check for leaks.

All other engines

20 Leave the reservoir cap off and run the engine in a well-ventilated area until the thermostat opens (coolant will begin flowing through the radiator and the upper radiator hose will become hot).

21 Turn the engine off and let it cool. Add more coolant mixture to bring the level back up to the FULL level on the reservoir.

22 Squeeze the upper radiator hose to expel air, then add more coolant mixture if necessary. Replace the reservoir cap.

23 Start the engine, allow it to reach normal operating temperature and check for leaks.

32 Positive Crankcase Ventilation (PCV) valve check and replacement (V8 engines) (every 30,000 miles or 24 months)

Refer to illustrations 32.1 and 32.2

1 The PCV valve on 5.2L and 5.9L engines is located in the right valve cover **(see illustration)**. The PCV valve on the 4.7L engine is located in the neck of the oil filler tube which is mounted to the front of the right cylinder head.

2 With the engine idling at normal operating temperature, pull the valve (with hose attached) from the rubber grommet. On 4.7L engines, it will be necessary to remove the vacuum hose, rotate the valve 90 degrees counterclockwise, then pull straight up to remove it from the filler neck and reconnect the vacuum hose **(see illustration)**.

3 Place your finger over the valve opening. If there's no vacuum at the valve, check for a plugged hose, manifold port, or the valve itself. Replace any plugged or deteriorated hoses.

4 Turn off the engine and shake the PCV valve, listening for a rattle. If the valve doesn't rattle, replace it with a new one.

5 To replace the valve, pull it from the end of the hose, noting its installed position.

6 When purchasing a replacement PCV valve, make sure it's for your particular vehicle and engine size. Compare the old valve with the new one to make sure they're the same.

7 On 5.2L and 5.9L engines, inspect the rubber grommet on the valve cover for damage and hardening. Replace it with a new one if necessary. On 4.7L engines, check to make sure the new PCV valve is equipped with an O-ring. If not, remove the O-ring from the old valve and install it on the new one. Be sure the O-ring is in good condition before doing so or vacuum leaks may occur.

8 Install the PCV valve securely into position on the valve cover (5.2L/5.9L engines) or the oil filler tube (4.7L engine). On 4.7L engines it will be necessary to rotate the valve 90 degrees clockwise to lock it into place on the oil filler tube.

9 Push the vacuum hose over the end of the valve until it's seated.

33 Crankcase Ventilation (CCV) hose check, cleaning and replacement (six-cylinder engines) (every 30,000 miles or 24 months)

Refer to illustration 33.4

1 Six-cylinder models are equipped with a CCV system which performs the same function as the PCV system but uses a rubber fitting with a molded-in orifice that is pressed into a hole in the valve cover. The fitting is connected to the intake manifold by a plastic hose.

2 With the engine idling at normal operating temperature, pull the CCV hose from the rubber grommet.

3 Place your finger over the hose opening. If there's no vacuum, check for a plugged hose, manifold port, or valve. Replace any plugged or deteriorated hoses.

4 If there is no vacuum at the end of the hose, turn off the engine, remove the fitting and clean the hose with solvent. Clean the fitting orifice if it's plugged **(see illustration)**. If the fitting or hose are cracked or deteriorated, replace them with new ones.

5 Install the fitting and hose securely in the valve cover.

34 Evaporative emissions control system check (every 30,000 miles or 24 months)

Refer to illustration 34.2

1 The function of the evaporative emissions control system is to draw fuel vapors from the gas tank and fuel system, store them in a charcoal canister and route them to the intake manifold during normal engine operation.

2 The most common symptom of a fault in the evaporative emissions system is a strong fuel odor in the engine compartment. If a fuel odor is detected, inspect the charcoal canister, located in the left front corner of the engine compartment **(see illustration)**. Check the canister and all hoses for damage and deterioration.

3 The evaporative emissions control system is explained in more detail in Chapter 6.

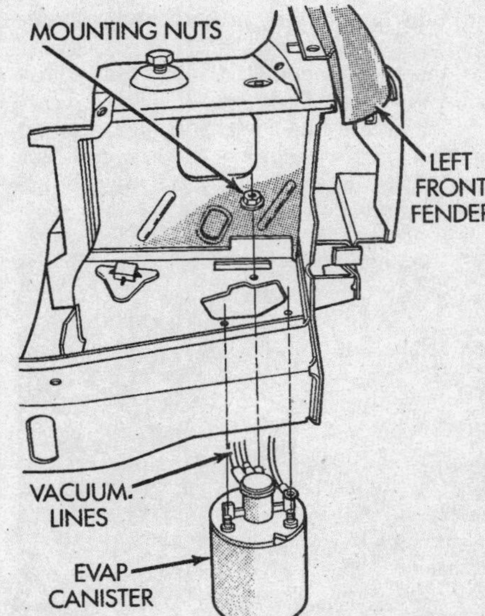

34.2 The evaporative emissions canister is located in the left front corner of the engine compartment - check the hoses and connections for damage

33.4 A paper clip can be used to clean the CCV system orifice

1

35 Crankcase inlet filter cleaning (5.2L/5.9L engines) (every 30,000 miles or 24 months)

Refer to illustration 35.3

1 Disconnect the hose and remove the crankcase inlet filter from the valve cover.

2 Wash the inside of the filter with solvent.

3 Lubricate the filter by pouring clean engine oil into the large opening and allowing it to drain out through the smaller (inlet) opening **(see illustration)**.

4 Reinstall the filter in the valve cover and connect the hose.

36 Fuel filter replacement (1993 through 1996 models) (every 30,000 miles or 24 months)

Refer to illustration 36.4

Warning: *Gasoline is extremely flammable, so take extra precautions when working on any part of the fuel system. Don't smoke or allow open flames or bare light bulbs in or near the work area, and don't work in a garage where a gas-type appliance (such as a water heater or*

35.3 Lubricate the filter by pouring clean engine oil into the large opening until it drains out the smaller opening

36.4 Squeeze the white plastic tabs together to detach the fuel lines from the filter - be careful because some residual fuel could leak out

clothes dryer) is present. Since gasoline is carcinogenic, wear latex gloves when there's a possibility of being exposed to fuel, and, if you spill fuel on your skin, rinse it off immediately with soap and water. Have a Class B fire extinguisher on hand.

Note: *Only 1993 through 1996 models use a replaceable inline fuel filter. The manufacturer does not suggest periodic fuel filter replacement on 1997 and later models. Since the fuel filter is part of the fuel pump assembly inside the fuel tank on 1997 and later models, refer to Chapter 4 if fuel contamination makes it necessary to replace the fuel filter.*

1 Relieve the fuel system pressure (see Chapter 4).

2 The fuel filter is mounted under the left side of the vehicle, near the fuel tank.

3 With the exhaust system cold, place a container or rags under the filter.

4 Squeeze the white plastic clips on the fuel line connectors and detach the fuel lines from the filter **(see illustration)**.

5 Remove the retaining bolt, lower the clamp retaining strap and separate the filter from the bracket.

6 Transfer the plastic retainers for the quick-connect fittings to the new filter. To remove them, gently pry them off the filter using two small screwdrivers. Replace them with new ones if they are cracked or damaged in any way.

7 Install the new filter, making sure it is facing the proper direction.

8 Install the fuel lines, pressing them on the filter until an audible click is heard. Pull back on the fuel line to verify that the connection is secure.

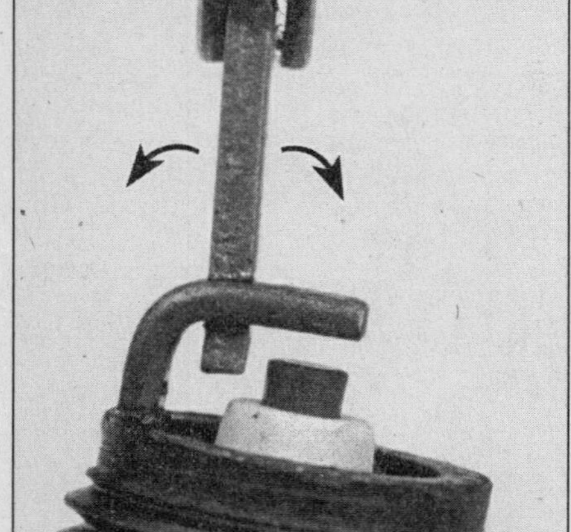

37.5b To change the gap, bend the side electrode only, as indicated by the arrows, and be very careful not to crack or chip the porcelain insulator surrounding the center electrode

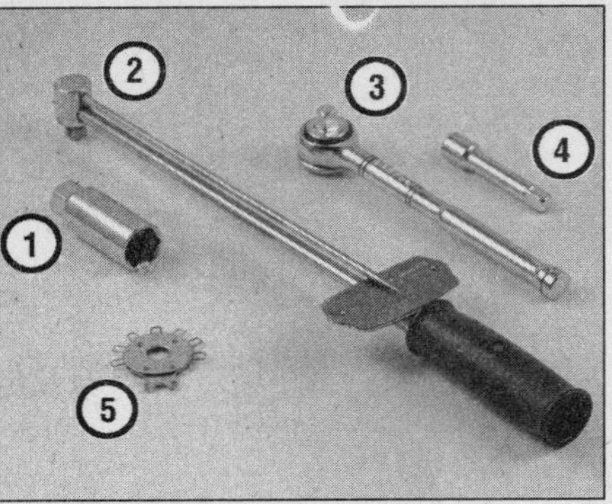

37.2 Tools required for spark plugs replacement

1 ***Spark plug socket*** *- This will have special padding inside to protect the spark plug's porcelain insulator*

2 ***Torque wrench*** *- Although not mandatory, using this tool is the best way to ensure the plugs are tightened properly*

3 ***Ratchet*** *- Standard hand tool to fit the spark plug socket*

4 ***Extension*** *- Depending on model and accessories, you may need special extensions and universal joints to reach one or more of the plugs*

5 ***Spark plug gap gauge*** *- This gauge for checking the gap comes in a variety of styles. Make sure the gap for your engine is included*

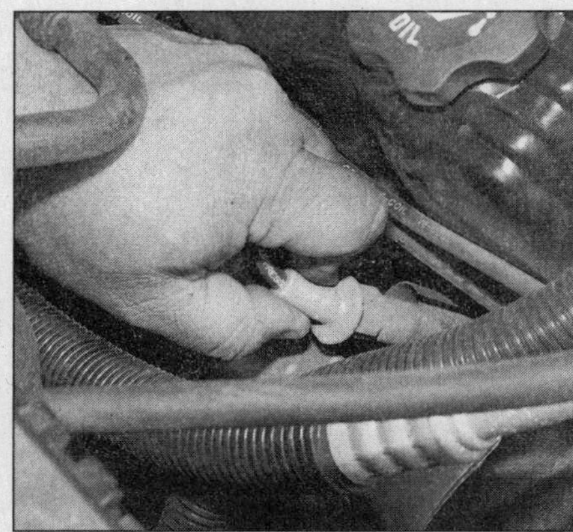

37.6a When removing the spark plug wires, pull only on the boot and twist it back-and-forth

37.5a Spark plug manufacturers recommend using a wire-type gauge when checking the gap - if the wire does not slide between the electrodes with a slight drag, adjustment is required

37 Spark plug replacement (every 30,000 miles or 24 months)

Refer to illustrations 37.2, 37.5a, 37.5b, 37.6a, 37.6b, 37.6c, 37.8 and 37.10

1 Open the hood and label each spark plug wire to ensure proper installation (this Step does not apply to 1999 and later models).

2 In most cases, the tools necessary for spark plug replacement include a spark plug socket which fits onto a ratchet (spark plug sockets are padded inside to prevent damage to the porcelain insulators on the new plugs), various extensions and a gap gauge to check and adjust the gaps on the new plugs **(see illustration)**. A special plug wire removal tool is available for separating the wire boots from the spark plugs, but it isn't absolutely necessary. A torque wrench should be used to tighten the new plugs.

3 The best approach when replacing the spark plugs is to purchase the new ones in advance, adjust them to the proper gap and replace them one at a time. When buying the new spark plugs, be sure to obtain the correct plug type for your particular engine. This information can be found on the *Emission Control Information* label located under the hood, in the factory owner's manual and the Specifications at the front of this Chapter. If differences exist between the plug specified on the emissions label and in the owner's manual, assume that the emissions label is correct.

4 Allow the engine to cool completely before attempting to remove any of the plugs. While you're waiting for the engine to cool, check the new plugs for defects and adjust the gaps.

5 The gap is checked by inserting the proper-thickness gauge between the electrodes at the tip of the plug **(see illustration)**. The gap between the electrodes should be the same as the one specified on the *Emissions Control Information* label or in this

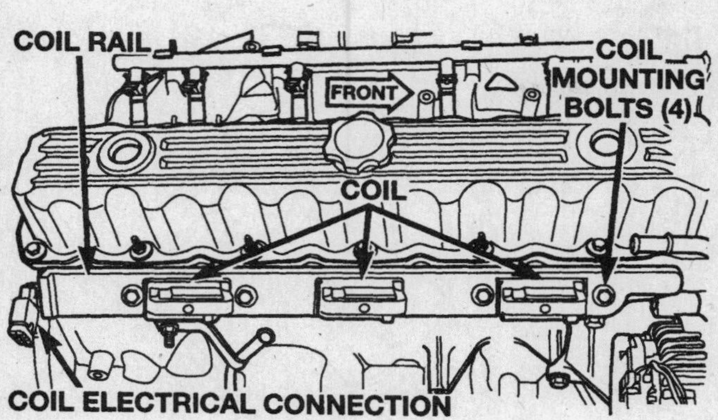

37.6b On 1999 and later six-cylinder engines, remove the coil mounting bolts and the coil-pack, which mounts directly over the spark plugs

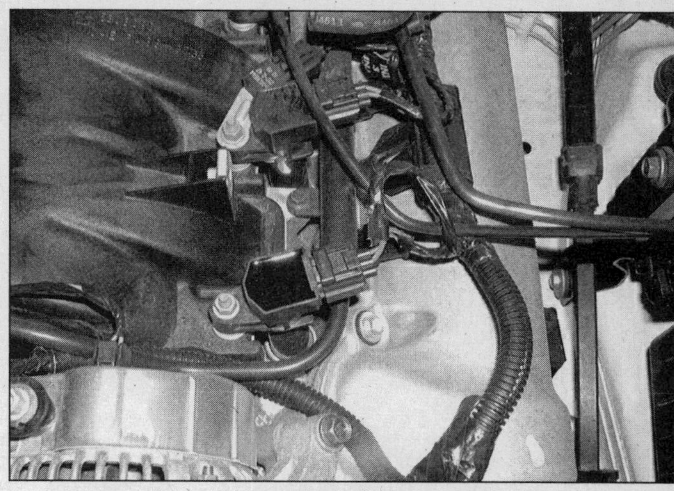

37.6c On 4.7L V8 engines, remove the mounting nut(s) (arrow) and the individual coil(s) over each spark plug

1

Chapter's Specifications. The wire should just slide between the electrodes with a slight amount of drag. If the gap is incorrect, use the adjuster on the gauge body to bend the curved side electrode slightly until the proper gap is obtained **(see illustration)**. If the side electrode is not exactly over the center electrode, bend it with the adjuster until it is. Check for cracks in the porcelain insulator (if any are found, the plug should not be used).

6 With the engine cool, remove the spark plug wire from one spark plug (1998 and earlier models only). Pull only on the boot at the end of the wire - do not pull on the wire **(see illustration)**. A plug wire removal tool should be used if available. On 1999 and later models, remove the coil pack (six-cylinder engines) or the individual coils (4.7L V8 engine) **(see illustrations)**.

7 If compressed air is available, use it to blow any dirt or foreign material away from the spark plug hole. The idea here is to eliminate the possibility of debris falling into the cylinder as the spark plug is removed.

8 Place the spark plug socket over the plug and remove it from the engine by turning it in a counterclockwise direction **(see illustration)**.

9 Compare the spark plug with the chart on the inside back cover of this manual to get an indication of the general running condition of the engine.

10 Thread one of the new plugs into the hole until you can no longer turn it with your fingers, then tighten it with a torque wrench (if available) or the ratchet. It's a good idea to slip a short length of rubber hose over the end of the plug to use as a tool to thread it into place **(see illustration)**. The hose will grip the plug well enough to turn it, but will start to slip if the plug begins to cross-thread in the hole - this will prevent damaged threads and the accompanying repair costs.

11 Before pushing the spark plug wire onto the end of the plug, inspect it following the procedures outlined in Section 38.

12 Attach the plug wire to the new spark plug, again using a twisting motion on the boot until it's seated on the spark plug.

13 Repeat the procedure for the remaining spark plugs, replacing them one at a time to prevent mixing up the spark plug wires.

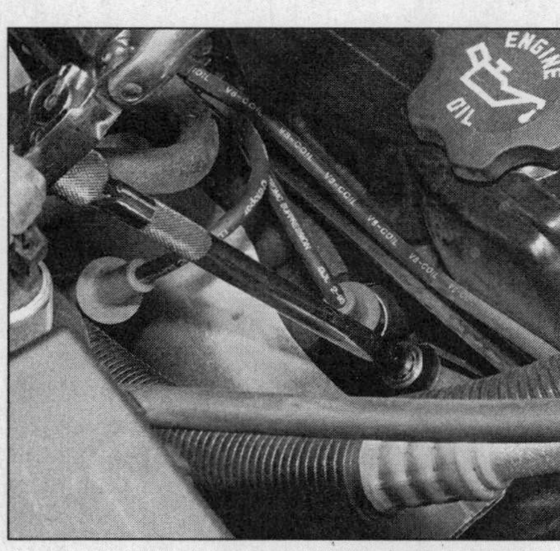

37.8 Use a socket and extension to unscrew the spark plugs

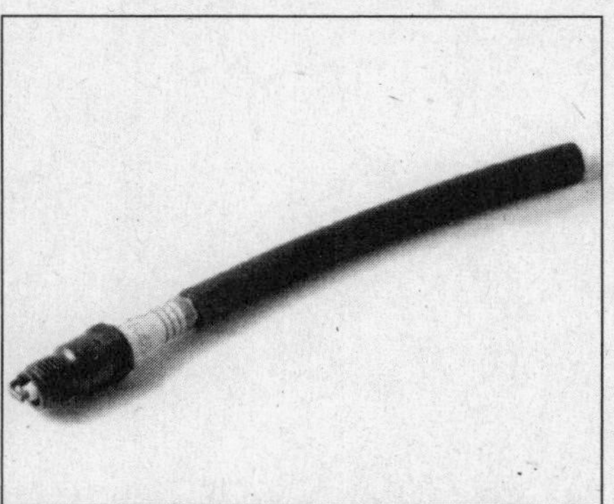

37.10 A length of snug-fitting 3/8-inch ID rubber hose will save time and prevent damaged threads when installing the spark plugs

38 Spark plug wires, distributor cap and rotor check and replacement (1993 through 1998 models) (every 30,000 miles or 24 months)

Refer to illustrations 38.10 and 38.13

1 The spark plug wires should be checked at the recommended intervals and whenever new spark plugs are installed in the engine.

2 The wires should be inspected one at a time to prevent mixing up the order, which is essential for proper engine operation.

3 Disconnect the plug wire from one spark plug. To do this, grab the rubber boot, twist slightly and pull the wire free. Do not pull on the wire itself, only on the rubber boot.

4 Check inside the boot for corrosion, which will look like a white crusty powder. Push the wire and boot back onto the end of the spark plug. It should be a tight fit on the plug. If it isn't, remove the wire and use a pair of pliers to carefully crimp the metal connector inside the boot until it fits securely on the end of the spark plug.

5 Using a clean rag, wipe the entire length of the wire to remove any built-up dirt and grease. Once the wire is clean, check for holes, burned areas, cracks and other damage. Don't bend the wire excessively or the conductor inside might break.

6 Disconnect the wire from the distributor cap. Pull the wire straight out of the cap. Pull only on the rubber boot during removal. Check for corrosion and a tight fit in the same manner as the spark plug end. Reattach the wire to the distributor cap.

7 Check the remaining spark plug wires one at a time, making sure they are securely fastened at the distributor and the spark plug when the check is complete.

8 If new spark plug wires are required, purchase a new set for your specific engine model. Wire sets are available pre-cut, with the rubber boots already installed. Remove

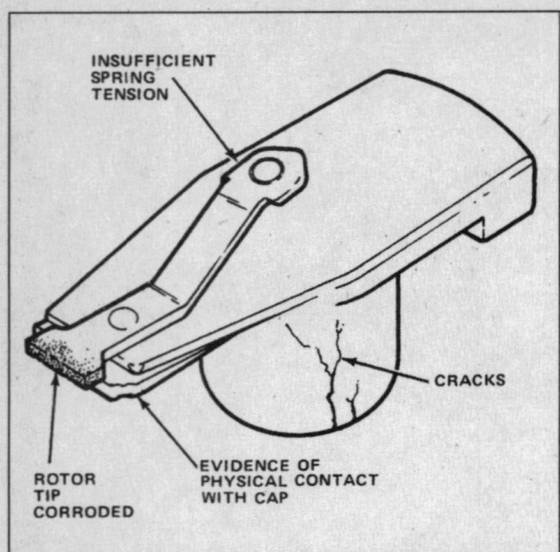

38.10 The ignition rotor should be checked for wear and corrosion as indicated here (if in doubt about its condition, buy a new one)

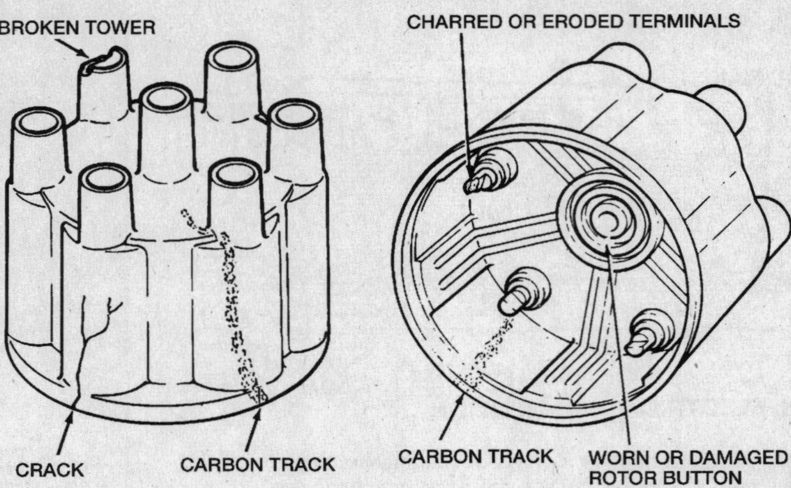

38.13 Shown here are some of the common defects to look for when inspecting the distributor cap (if in doubt about its condition, install a new one)

and replace the wires one at a time to avoid mix-ups in the firing order. The wire routing is extremely important, so be sure to note exactly how each wire is situated before removing it.

9 Remove the distributor screws. On V8 engines, it will be necessary to disconnect the EGR valve transducer vacuum lines and electrical connectors for access. Pull up on the cap, with the wires attached, to separate it from the distributor, then position it to one side.

10 The rotor is now visible on the end of the distributor shaft. Check it carefully for cracks and carbon tracks. Make sure the center terminal spring tension is adequate and look for corrosion and wear on the rotor tip **(see illustration)**. If in doubt about its condition, replace it with a new one.

11 If replacement is required, detach the rotor from the shaft and install a new one. The rotor is a press fit on the shaft and can be pried or pulled off.

12 The rotor is indexed to the shaft so it can only be installed one way. It has an internal key that must line up with a slot in the end of the shaft (or vice versa).

13 Check the distributor cap for carbon tracks, cracks and other damage. Closely examine the terminals on the inside of the cap for excessive corrosion and damage **(see illustration)**. Slight deposits are normal. Again, if in doubt about the condition of the cap, replace it with a new one. Be sure to apply a small dab of silicone dielectric grease to each terminal before installing the cap. Also, make sure the carbon brush (center terminal) is correctly installed in the cap - a wide gap between the brush and rotor will result in rotor burn-through and/or damage to the distributor cap.

14 To replace the cap, simply separate it from the distributor and transfer the spark plug wires, one at a time, to the new cap. Be very careful not to mix up the wires!

15 Reattach the cap to the distributor, then install the screws to hold it in place.

Chapter 2 Part A
Inline six-cylinder engine

Contents

Specifications

General

Displacement	242 cu. in (4.0 liters)
Cylinder numbers (front-to-rear)	1-2-3-4-5-6
Firing order	1-5-3-6-2-4

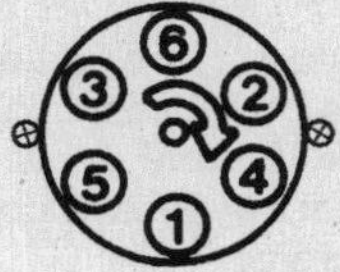

Cylinder location and distributor rotation - distributor used on 1998 and earlier models only

Camshaft

Lobe lift	
1999 and earlier	
Intake and exhaust	0.2530 inch
2000	
Intake	0.2545 inch
Exhaust	0.2590 inch
Endplay	N/A
Journal diameter	
No. 1	2.029 to 2.030 inches
No. 2	2.019 to 2.020 inches
No. 3	2.009 to 2.010 inches
No. 4	1.999 to 2.000 inches
Journal-to-bearing (oil) clearance	0.001 to 0.003 inch

Oil pump

Gear-to cover clearance	0.002 to 0.006 inch
Gear-to-body clearance	0.002 to 0.004 inch

Torque specifications

	Ft-lbs (unless otherwise indicated)
Camshaft sprocket bolt	
1998 and earlier	80
1999 and later	50
Crankshaft pulley-to-vibration damper bolts	20
Cylinder head bolts	
Step A	22
Step B	45
Step C	
Bolt no. 11	100
All other bolts	110
Flywheel/driveplate bolts	105
Intake and exhaust manifold retaining bolts and nuts	
No. 6 and 7 exhaust manifold bolts	17
All other bolts/nuts	24
Oil pan mounting bolts	
1998 and earlier	
1/4-20	120 in-lbs
5/16-18	156 in-lbs
1999 and later	
1/4-20	84 in-lbs
5/16-18	132 in-lbs
Oil pump mounting bolt	
Short	120 in-lbs
Long	204 in-lbs
Rear main bearing cap bolts	80
Rocker arm bolts	21
Valve cover mounting bolts	
1993	75 in-lbs
1994 and later	85 in-lbs
Tensioner bracket-to-block bolts	14
Timing chain cover-to-block	
1/4-20	60 in-lbs
5/16-18	192 in-lbs
Vibration damper center bolt (lubricated)	80

1 General information

This Part of Chapter 2 is devoted to in-vehicle repair procedures for the 4.0L inline six-cylinder engine. All information concerning engine removal and installation and engine block and cylinder head overhaul can be found in Part D of this Chapter.

The following repair procedures are based on the assumption that the engine is installed in the vehicle. If the engine has been removed from the vehicle and mounted on a stand, many of the steps outlined in this Part of Chapter 2 will not apply.

The Specifications included in this Part of Chapter 2 apply only to the procedures contained in this Part. Part D of Chapter 2 contains the Specifications necessary for cylinder head and engine block rebuilding.

2 Repair operations possible with the engine in the vehicle

Many major repair operations can be accomplished without removing the engine from the vehicle.

Clean the engine compartment and the exterior of the engine with some type of pressure washer before any work is done. It will make the job easier and help keep dirt out of the internal areas of the engine.

Remove the hood, if necessary, to improve access to the engine as repairs are performed (refer to Chapter 11 if necessary).

If vacuum, exhaust, oil or coolant leaks develop, indicating a need for gasket or seal replacement, the repairs can generally be made with the engine in the vehicle. The intake and exhaust manifold gaskets, timing cover gasket, oil pan gasket, crankshaft oil seals and cylinder head gaskets are all accessible with the engine in place.

Exterior engine components, such as the intake and exhaust manifolds, the oil pan (and the oil pump), the water pump, the starter motor, the alternator, the distributor and the fuel system components can be removed for repair with the engine in place.

Since the cylinder heads can be removed without pulling the engine, valve component servicing can also be accomplished with the engine in the vehicle. Replacement of the camshaft and timing chain and sprockets is also possible with the engine in the vehicle.

In extreme cases caused by a lack of necessary equipment, repair or replacement of piston rings, pistons, connecting rods and rod bearings is possible with the engine in the vehicle. However, this practice is not recommended because of the cleaning and preparation work that must be done to the components involved.

3 Top Dead Center (TDC) for number one piston - locating

Refer to illustration 3.6

1 Top Dead Center (TDC) is the highest point in the cylinder that each piston reaches as it travels up-and-down when the crankshaft turns. Each piston reaches TDC on the compression stroke and again on the exhaust stroke, but TDC generally refers to piston position on the compression stroke. The timing marks on the vibration damper/crankshaft pulley installed on the front of the crankshaft are referenced to the number one piston at TDC on the compression stroke.

2 Positioning the piston(s) at TDC is an essential part of procedures such as timing belt and sprocket replacement.

3 In order to bring any piston to TDC, the crankshaft must be turned using one of the methods outlined below. When looking at the front end of the engine, normal crankshaft

3.6 Align the notch (arrow) on the vibration damper with the "0" on the timing chain cover

4.1 Location of the crankcase ventilation hoses (arrows)

rotation is clockwise. **Warning:** *Before beginning this procedure, be sure to place the transmission in Neutral, set the parking brake and remove the ignition key.*

a) *The preferred method is to turn the crankshaft with a large socket and breaker bar attached to the large bolt threaded into the center of the crankshaft pulley.*
b) *A remote starter switch, which may save some time, can also be used. Attach the switch leads to the S (switch) and B (battery) terminals on the starter motor. Once the piston is close to TDC, use a socket and breaker bar as described in the previous paragraph.*
c) *If an assistant is available to turn the ignition switch to the Start position in short bursts, you can get the piston close to TDC without a remote starter switch. Use a socket and breaker bar as described in Paragraph a) to complete the procedure.*

4 Disable the ignition system by disconnecting the primary electrical connectors at the ignition coil (see Chapter 5).

5 Remove the spark plugs (see Chapter 1) and install a compression gauge in the number one cylinder. Turn the crankshaft clockwise with a socket and breaker bar as described above.

6 When the piston approaches TDC, compression will be noted on the compression gauge. Continue turning the crankshaft until the notch in the crankshaft damper is aligned with the O or TDC mark on the front cover **(see illustration)**. At this point number one cylinder is at TDC on the compression stroke. If the marks are aligned but there was no compression, the piston was on the exhaust stroke. Continue rotating the crankshaft 360-degrees (1-turn). **Note:** *If a compression gauge is not available, you can simply place your finger over the spark plug hole and feel for compression as the engine is rotated. Once compression at the No.1 spark plug hole is noted the remainder of the Step is the same.*

9 After the number one piston has been positioned at TDC on the compression stroke, TDC for any of the remaining cylinders can be located by turning the crankshaft 120 degrees and following the firing order (refer to the Specifications). Rotating the engine 120 degrees past TDC #1 will position the engine at TDC compression for cylinder #5.

4 Valve cover - removal and installation

Refer to illustrations 4.1, 4.3 and 4.4

1 Remove the crankcase ventilation tube and hose from the valve cover **(see illustration)**.

2 Remove the wire loom clips, noting the locations of their studs for reinstallation.

3 Remove the accelerator cable, kickdown cable and cruise-control cable from the snap-in clips on top of the valve cover **(see illustration)**, disconnect them at the throttle body and lay them aside.

4 Remove the valve cover retaining bolts and lift the cover off. If the cover is stuck, tap on it gently with a soft-face mallet **(see illustration)**. Do not pry on the gasket flange.

4.3 Disconnect the accelerator, kickdown and cruise cables from the throttle body, remove them from the valve cover clips (arrow) and lay them aside

4.4 Use a soft-face mallet to break the cover loose - DO NOT pry between the cover and head

5.2 Alternating between the intake and exhaust rocker arms, loosening each bolt 1/4-turn at a time

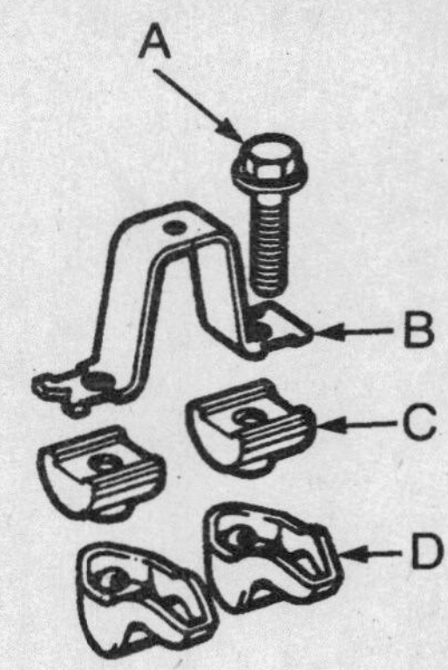

5.3 Rocker arm components

A Bolt
B Bridge
C Fulcrum
D Rocker arm

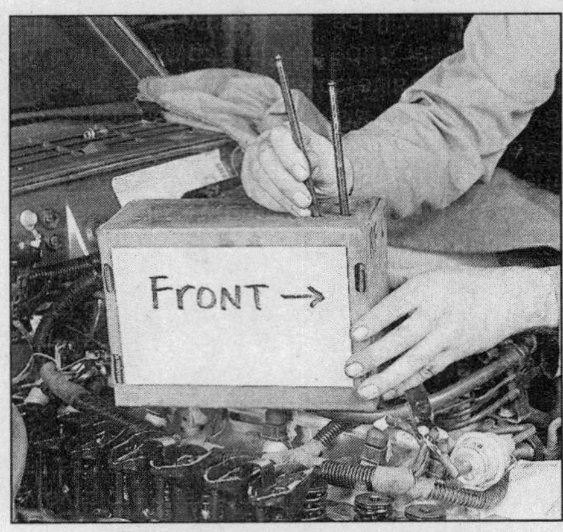

5.4 Store the pushrods in a box like this to ensure reinstallation in the same location

5 Clean the sealing surfaces, removing any traces of oil with lacquer thinner or acetone and a clean rag. **Note:** *These covers have a reusable, moulded rubber gasket that is attached to the cover.*

6 Small cracks in the pre-cured gasket are allowable and can be repaired by applying RTV sealer to the cracked area before the cover is installed.

7 Install the cover and bolts. Tighten the bolts to the torque listed in this Chapter's Specifications.

8 Reinstall the crankcase breather hoses and cruise control servo, if equipped.

9 Run the engine and check for oil leaks.

5 Rocker arms and pushrods - removal, inspection and installation

Refer to illustrations 5.2, 5.3 and 5.4

Removal

1 Remove the valve cover (see Section 4).

2 Beginning at the front of the cylinder head, loosen and remove the rocker arm mounting bolts in pairs **(see illustration)**.

3 Remove the rocker arms, bridges and fulcrums **(see illustration)** and store them with their respective mounting bolts. Store each set of rocker arm components separately in a marked plastic bag to ensure they are reinstalled in their original locations. The bridges may be reinstalled in any location.

4 Remove the pushrods and store them separately to make sure they don't get mixed up during installation **(see illustration)**.

Inspection

5 Inspect each rocker arm for wear, cracks and other damage, especially where the pushrods and valve stems contact the rocker arm faces. Inspect the fulcrum seat in each rocker arm and the fulcrum faces. Look for galling, stress cracks and unusual wear patterns. If the rocker arms are worn or damaged, replace them with new ones and install new fulcrums as well.

6 Make sure the oil hole at the pushrod end of each rocker arm is open.

7 Inspect the pushrods for cracks and excessive wear at the ends. Roll each pushrod across a piece of plate (flat) glass to see if it's bent (if it wobbles, it's bent).

Installation

8 Lubricate the lower end of each pushrod with clean engine oil or engine assembly lube and install it in its original location. Make sure each pushrod seats completely in the lifter socket.

9 Bring the number one piston to top dead center on the compression stroke (Section 3).

10 Apply clean engine oil or engine assembly lube to the ends of the valve stems and the upper ends of the pushrods before placing the rocker arms in position.

11 Apply clean engine oil or engine assembly lube to the fulcrums to prevent damage to the mating surfaces before engine oil pressure builds up. Install the rocker arms, fulcrums, bridges and bolts in their original locations. Tighten the bolts to the torque listed in this Chapter's Specifications.

12 Install the valve cover (see Section 4).

13 Start the engine, listen for unusual valve train noises and check for oil leaks at the valve cover joint.

6 Valve springs, retainers and seals - replacement

Refer to illustrations 6.4, 6.8, 6.10 and 6.17

Note: *Broken valve springs and defective valve stem seals can be replaced without removing the cylinder heads. Two special tools and a compressed air source are normally required to perform this operation, so read through this Section carefully and rent or buy the tools before beginning the job. If compressed air isn't available, a length of nylon rope can be used to keep the valves from falling into the cylinder during this procedure.*

1 Remove the valve cover from the cylinder head (see Section 4).

2 Remove the spark plug from the cylinder which has the defective component. If all of the valve stem seals are being replaced, all of the spark plugs should be removed.

3 Turn the crankshaft until the piston in the affected cylinder is at Top Dead Center on the compression stroke (see Section 3). If you're replacing all of the valve stem seals, begin with cylinder number one and work on the valves for one cylinder at a time. Move from cylinder-to-cylinder following the firing order sequence (see this Chapter's Specifications).

4 Thread an adapter into the spark plug hole **(see illustration)** and connect an air hose from a compressed air source to it. Most auto parts stores can supply the air hose adapter. **Note:** *Many cylinder compression gauges utilize a screw-in fitting that may work with your air hose quick-disconnect fitting.*

5 Remove the rocker arms, bridges and fulcrums for the cylinder with the defective parts and pull out the pushrods. If all of the valve stem seals are being replaced, all of the rocker arms, bridges, fulcrums and pushrods should be removed (see Section 5).

6 Apply compressed air to the cylinder. **Warning:** *The piston may be forced down by compressed air, causing the crankshaft to turn suddenly. If the wrench used when positioning the number one piston at TDC is still attached to the bolt in the crankshaft nose, it could cause damage or injury when the crankshaft moves.*

7 If you don't have access to compressed air, an alternative method can be used. Position the piston at a point approximately 45-degrees before TDC on the compression stroke, then feed a long piece of nylon rope through the spark plug hole until it fills the combustion chamber. Be sure to leave the end of the rope hanging out of the engine so it can be removed easily. Use a large ratchet and socket to rotate the crankshaft in the normal direction of rotation until slight resistance is felt.

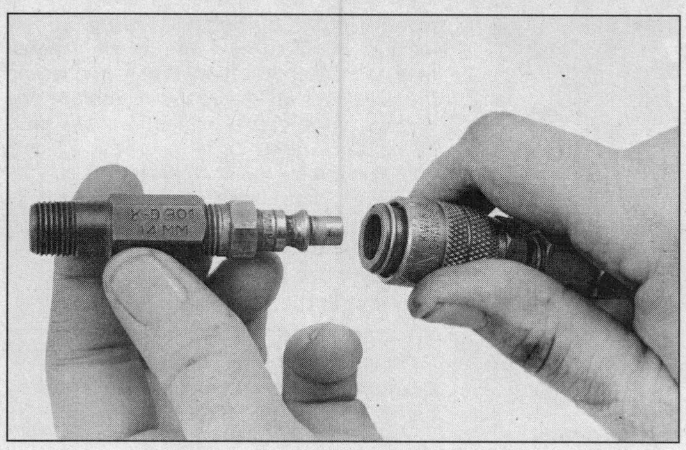

6.4 This is what the air-hose adapter that threads into the spark plug hole looks like - they're commonly available in auto parts stores

6.8 Once the spring is compressed, the keepers can be removed with a small magnet or needle-nose pliers (a magnet is preferred to prevent dropping the keepers)

8 Stuff shop rags into the cylinder-head holes near the valves to prevent parts and tools from falling into the engine, then use a valve spring compressor to compress the spring. Remove the keepers with small needle-nose pliers or a magnet **(see illustration)**. **Note:** *A couple of different types of tools are available for compressing the valve springs with the head in place. One type, shown here, grips the lower spring coils and presses on the retainer as the knob is turned, while the other type utilizes the rocker arm bolt for leverage. Both types work very well, although the lever type is usually less expensive.*

9 The valves should be held in place by the air pressure. If the valve faces or seats are in poor condition, leaks may prevent air pressure from retaining the valves - if that's the case, the head will have to be removed for repair.

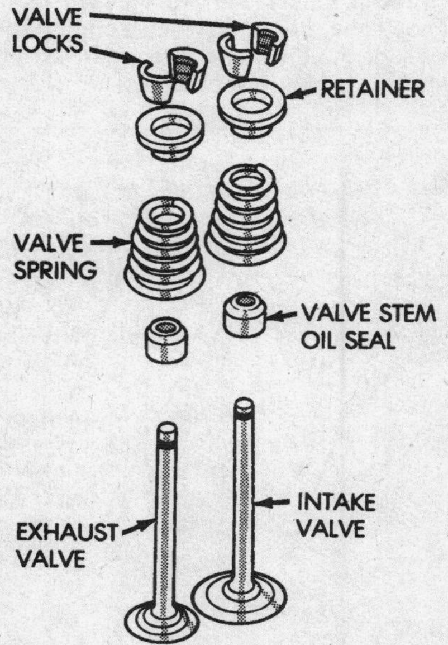

6.10 Valves, springs, seals and related components

10 Remove the spring retainer or rotator, sleeve (used on some intake valves) and valve spring assembly, then remove the valve guide seal **(see illustration)**.

11 Wrap a rubber band or tape around the top of the valve stem so the valve won't fall into the combustion chamber, then release the air pressure. **Note:** *If a rope was used instead of air pressure, turn the crankshaft slightly in the direction opposite normal rotation.*

12 Inspect the valve stem for damage. Rotate the valve in the guide and check the end for eccentric movement, which would indicate that the valve is bent.

13 Move the valve up-and-down in the guide and make sure it doesn't bind. If the valve stem binds, either the valve is bent or the guide is damaged. In either case, the head will have to be removed for repair.

14 Reapply air pressure to the cylinder to retain the valve in the closed position, then remove the tape or rubber band from the valve stem. If a rope was used instead of air pressure, rotate the crankshaft in the normal direction of rotation until slight resistance is felt.

15 Lubricate the valve stem with engine oil and the valve stem tip with engine assembly lube, then install a new guide seal, lightly pushing the seal over the valve stem until it seats on the valve guide boss. **Caution:** *Install seals carefully to prevent damage from the sharp edges of the valve keeper grooves.*

6.17 Apply a small dab of grease to the keepers before installation - it will hold them in place on the valve stem as the spring is released

16 Install the spring in position over the valve.

17 Install the valve spring retainer or rotator. Compress the valve spring and carefully position the keepers in the groove. Apply a small dab of grease to the inside of each keeper to hold it in place **(see illustration)**.

18 Remove the pressure from the spring tool and make sure the keepers are seated.

19 Disconnect the air hose and remove the adapter from the spark plug hole. If a rope was used in place of air pressure, pull it out of the cylinder.

20 Install the pushrods, rocker arms, pivots and bridges (see Section 5).

21 Install the spark plugs and connect the spark plug wires (see Chapter 1).

22 Install the valve cover (see Section 4).

23 Start and run the engine, then check for oil leaks and unusual sounds coming from the valve cover area.

7 Intake/exhaust manifolds - removal and installation

Refer to illustrations 7.4, 7.5, 7.10 and 7.18

Warning: *Allow the engine to cool completely before following this procedure.*

Note: *Since the intake and exhaust manifolds share a common gasket, they must be removed and replaced at the same time. All studs and bolts should be soaked with penetrating oil before beginning manifold removal.*

Removal

1 Disconnect the cable from the negative battery terminal.

2 Apply penetrating oil to the threads of the exhaust manifold attaching studs and the exhaust pipe-to-manifold attaching bolts.

3 Remove the air cleaner assembly (see

7.4 Disconnect the accelerator, transmission and cruise control cables (arrows) from the throttle body

7.5 Label the connections before detaching them

Chapter 4).

4 Detach the accelerator cable, transmission control cable (automatic transmission-equipped models), and cruise-control cables (if equipped) from the throttle body **(see illustration)**.

5 Label and then disconnect all vacuum and electrical connectors on the intake manifold **(see illustration)**.

6 Relieve the fuel pressure and then disconnect the fuel supply and return lines from the fuel rail assembly (see Chapter 4). Cap the open ends.

7 Loosen the serpentine drivebelt (see Chapter 1).

8 Remove the power steering pump and bracket from the intake manifold and set it aside without disconnecting the hoses. Be sure to leave the pump in an upright position so fluid won't spill.

9 Remove the fuel rail and injectors (see Chapter 4).

10 Remove the intake manifold bolts/nuts **(see illustration)**.

11 Pull the manifold away from the engine slightly to disengage it from the locating dowels in the cylinder head, then lift the manifold out of the engine compartment.

12 Remove the bolts and nuts that secure the exhaust manifold to the exhaust pipe and detach the pipe from the manifold. Discard the exhaust manifold seal ring.

13 Remove the exhaust manifold.

Installation

14 Thoroughly clean the gasket mating surfaces, removing all traces of old gasket material.

15 If the manifold is being replaced, make sure all the fittings, etc. are transferred to the replacement manifold.

16 Position a new gasket on the cylinder head, using the locating dowels to hold it in place. Install the exhaust manifold and hand tighten the nuts.

17 Position the intake manifold loosely on the cylinder head.

18 Install the manifold retaining bolts and tighten all fasteners in sequence **(see illustration)** to the torque listed in this Chapter's Specifications. Note that the end bolts require a different torque.

19 Reinstall the remaining parts in the reverse order of removal, using a new exhaust manifold seal ring at the exhaust pipe connection. **Caution:** *Before connecting the fuel lines to the fuel rail, replace the O-rings in the quick-connect fuel line couplings* (see Chapter 4).

20 Run the engine and check for fuel, vacuum and exhaust leaks.

8 Cylinder head - removal and installation

Refer to illustrations 8.11, 8.19 and 8.22

Caution: *Allow the engine to cool completely before following this procedure.*

Removal

1 Disconnect the cable from the negative battery terminal.

2 Drain the coolant from the radiator and the engine block (see Chapter 1).

3 Remove the air cleaner assembly (see Chapter 4).

4 Detach the fuel pipe and vacuum-advance hose.

5 Remove the valve cover (see Section 4).

6 Remove the rocker arms and pushrods (see Section 5).

7 Unbolt the power-steering pump bracket (if equipped) and set the pump aside without disconnecting the hoses. Leave the pump upright so fluid doesn't spill.

8 Remove the intake and exhaust manifolds (see Sections 7 and 8).

9 Remove the bracket on the cylinder head that supports the idler pulley for the air-conditioning compressor drivebelt (if equipped).

10 Loosen the alternator drivebelt and remove the alternator bracket/cylinder head mounting bolt.

11 Disconnect the wiring harness and unbolt the air-conditioning compressor (if equipped) without disconnecting the refrigerant hoses. Set the compressor aside and

7.10 Four of the intake manifold mounting fasteners (arrows) can be accessed from the top - you must reach below the manifold to remove the remaining four

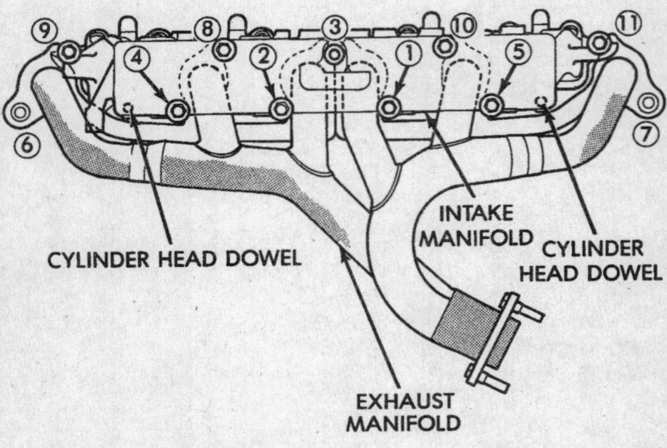

7.18 Intake/exhaust manifold bolt/nut tightening sequence

8.11 **Set the compressor aside with the refrigerant lines still attached - then remove the upper bracket bolts (arrows)**

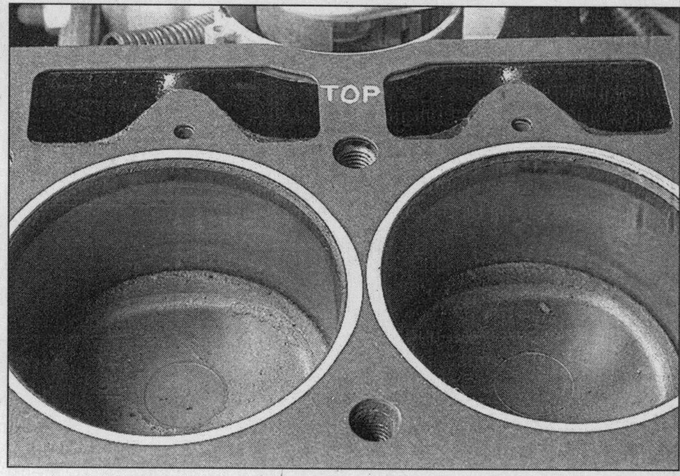

8.19 **Install the head gasket with the TOP mark facing up**

remove the upper two bolts from the bracket **(see illustration)**.

12 On 1998 and earlier models, label the spark plug wires and remove the distributor cap with the wires attached, then remove the ignition coil and bracket assembly. On 1999 and later models, disconnect the electrical connector from the coil rail and remove the coil rail from the engine (see Chapter 5).

13 Remove the spark plugs as described in Chapter 1.

14 Disconnect the wire from the coolant temperature sensor, which is at the front of the engine in the thermostat housing (see Chapter 3). Also disconnect the battery ground cable from the right side of the engine compartment.

15 Unclip the heater hoses along the passenger side of the valve cover, undo the hose clamps at the front and lay the hoses out of the way.

16 Remove the cylinder head bolts and lift the head off the engine. If the head sticks to the engine, insert a prybar into an exhaust port and pry gently to break the seal. **Note:** *Bolt number 14* **(see illustration 8.22)** *cannot be removed completely until the head is moved forward. Pull the bolt up as far as you can and hold it up by applying masking tape around it, then pull the head forward and remove the bolt.*

Installation

17 Thoroughly clean the gasket mating surfaces, removing all traces of old gasket material. Stuff shop towels into each cylinder so material doesn't fall in.

18 Inspect the head for cracks and warpage. See Chapter 2, Part D, for cylinder head servicing information. Use a straightedge against the head and the block to check for warpage. If a feeler gauge of greater than .008-in. fits under the straightedge (when measuring the whole length of head or block) the head or block must be resurfaced. If you are replacing the cylinder head, be sure to transfer all fittings, etc. to the new head.

19 Position the new head gasket on the engine block with the word TOP facing up **(see illustration)**. Do not use any sealing compound on the gasket.

20 Apply masking tape to bolt number 14 as during removal, and install it in the head. Install the cylinder head on the engine block, and remove the tape from the bolt.

21 Coat the threads of bolt number 11 **(see illustration 8.22)** with Loctite 592 sealant (or equivalent) and install the head bolts hand tight. **Note:** *The head bolts can be reused only once (original use plus one reuse). Clean each bolt head and mark it with a dab of paint. If the bolts in your engine are already marked with paint, replace them with new ones.*

22 Tighten the cylinder head bolts in sequence **(see illustration)**, according to the following procedure:

Step A - tighten all bolts (1 through 14) in sequence to the specified torque for Step A.

Step B - tighten all bolts in sequence to the specified torque for Step B.

Step C - tighten all bolts to the specified torque for Step C. **Caution:** *In Step C, bolt no. 11 is tightened to a lower torque than the rest of the bolts. Do not overtighten it. Tighten bolt no. 11 to the specified torque for Step C.*

23 Reinstall the remaining components in the reverse order of removal.

24 Add coolant and run the engine, checking for proper operation and coolant and oil leaks.

9 Crankshaft front oil seal - replacement

Refer to illustrations 9.6, 9.7, 9.9 and 9.10

1 Disconnect the cable from the negative battery terminal.

2 Raise the front of the vehicle and support it securely on jackstands.

3 Remove the shield mounted below the front of the engine.

4 Remove the drivebelts (see Chapter 1).

5 Remove the radiator and cooling fan(s) as described in Chapter 3.

6 Remove the vibration damper retaining bolt and washer. **Note:** *To prevent the crankshaft from rotating, place two 5/16 x 1-1/2 inch long bolts into the damper holes and hold a prybar between them* **(see illustra-**

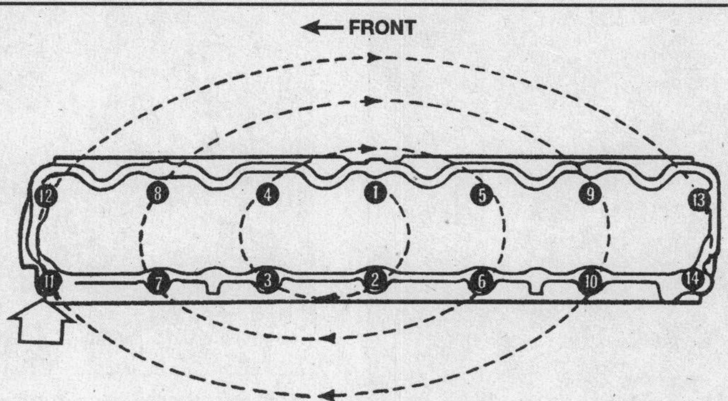

8.22 **Cylinder head bolt tightening sequence - be sure to coat the threads of bolt no. 11 (arrow) with Loctite 592 sealant (or equivalent)**

9.6 Install two bolts in the damper and use them to keep the crankshaft from rotating

9.7 Use a vibration damper removal tool such as this one - do not use a gear puller with jaws; it will damage the damper

9.9 Pry the old seal out with a seal removal tool (shown here) or a screwdriver

tion). *Rotate the crankshaft until the bar contacts the frame.*

7 Using a vibration damper removal tool **(see illustration)**, pull the damper off the crankshaft.

8 Clean and inspect the area on the center hub of the damper where the front oil seal contacts it. Minor imperfections can be cleaned up with emery cloth. If there is a groove worn in the hub, replace the vibration damper or have a machine shop install a special sleeve on the hub to restore the contact surface.

9 Carefully pry the oil seal out of the timing chain cover with a seal removal tool or screwdriver **(see illustration)**. Don't scratch the cover bore or damage the crankshaft in the process (if the crankshaft is damaged the new seal will end up leaking).

10 Clean the bore in the cover and coat the outer edge of the new seal with engine oil or multi-purpose grease. Using a socket with an outside diameter slightly smaller than the outside diameter of the seal, carefully drive the new seal into place with a hammer **(see illustration)**. If a socket isn't available, a short section of large-diameter pipe will work. Check the seal after installation to be sure that the spring didn't pop out of place.

11 Apply clean engine oil to the seal contact surface of the damper hub.

9.10 Gently drive the new seal into place with a hammer and large socket

12 Align the key slot of the vibration damper hub with the crankshaft key and tap the damper onto the crankshaft with a soft-face mallet.

13 Install the vibration damper bolt and tighten it to the torque listed in this Chapter's Specifications.

14 Reinstall the remaining parts in the reverse order of removal.

15 Run the engine and check for leaks.

10 Timing cover, chain and sprockets - inspection, removal and installation

Cover removal

1 Disconnect the cable from the negative battery terminal.

2 Remove the fan, fan shroud and water pump pulley (see Chapter 3).

3 Remove the crankshaft vibration damper (see Section 9).

4 Remove the air conditioning compressor and alternator and set them aside without disconnecting the wiring or hoses. Remove any brackets connected to the timing chain cover.

5 Remove the oil pan-to-timing chain cover bolts and timing chain cover-to-engine block bolts.

6 Separate the timing chain cover from the engine. If necessary, tap on it gently with a soft-face mallet to break the seal. Temporarily stuff a rag into the oil pan opening to prevent entry of debris.

7 Clean the mating surfaces of the timing chain cover and engine block, removing all traces of oil and old gasket material.

Timing chain inspection

Refer to illustration 10.8

8 Slip the oil slinger off the crankshaft **(see illustration)** and reinstall the vibration damper bolt. Using this bolt, rotate the crankshaft clockwise just enough to take up the slack on one side of the chain.

9 Establish a reference point on the block. Move the slack side of the chain from side-to-side with your fingers and measure the movement. The difference between the two measurements is the deflection. If the deflection exceeds 1/2-inch, replace the timing chain and sprockets.

10.8 Slip the oil slinger off the crankshaft, noting that the cupped side faces away from the engine, and reinstall the vibration damper bolt

10.10 With the number one piston at Top Dead Center, the timing chain sprocket index dots (arrows) are directly opposite each other

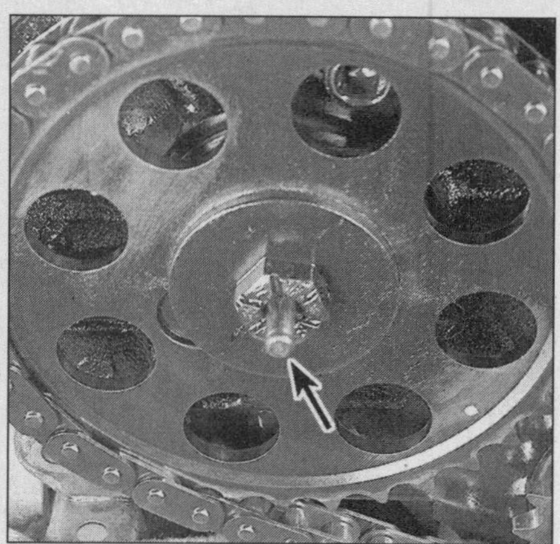
10.11 Remove the thrust pin and spring (arrow), then remove the bolt in the center of the camshaft sprocket - put a large screwdriver through one of the holes in the sprocket to keep it from turning while loosening the bolt

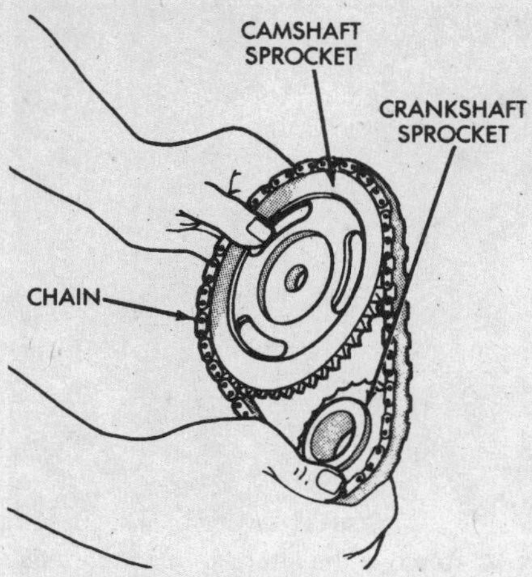

10.12 Remove and install the chain and sprockets as an assembly

10.13a Note that the engine side of the camshaft sprocket has a hole (arrow) . . .

Timing chain removal

Refer to illustrations 10.10, 10.11 and 10.12

10 Align the sprocket timing marks **(see illustration)**.

11 Remove the camshaft thrust pin and spring and the sprocket-retaining bolt and washer **(see illustration)**.

12 Pull the crankshaft sprocket, camshaft sprocket and timing chain off as an assembly **(see illustration)**. **Caution:** *Do not turn the crankshaft or camshaft while the timing chain is removed.*

Installation

Refer to illustrations 10.13a, 10.13b, 10.16 and 10.22

13 Be sure the crankshaft key is still pointing up. Note the locations of the locating dowel on the camshaft and the corresponding hole in the cam sprocket **(see illustrations)**.

14 Pre-assemble the timing chain, crankshaft sprocket and camshaft sprocket with the timing marks aligned and facing out. Slip the assembly onto the engine in such a way that a line drawn through the timing marks will also pass through the centers of the sprockets.

15 Install the camshaft sprocket bolt and tighten it to the torque listed in this Chapter's Specifications. Reinstall the thrust pin and spring.

16 To verify the correct installation of the timing chain, turn the crankshaft clockwise until the camshaft sprocket timing mark is at the one o'clock position. This positions the crankshaft sprocket timing mark where the adjacent tooth meshes with the chain at the three o'clock position. There must be 15 chain pins between the sprocket timing marks **(see illustration)**.

17 Install the crankshaft oil slinger on the crankshaft with the cupped side facing out.

18 Check for tears and deformation of the one-piece oil pan gasket.

19 If the oil pan gasket is damaged, it must be replaced before installing the timing chain cover. It is suggested that the oil pan gasket always be replaced to ensure a leak free seal at the timing cover to oil pan joint (see Section 12).

20 Apply a thin layer of RTV sealant to both sides of the new cover gasket and the corners of the pan and block, then position the new cover gasket on the engine. The sealant will hold it in place. **Note:** *If the oil pan has been removed for replacement of the pan gasket, install the front cover before the oil pan and it's gaskets are installed.*

21 Position the timing chain cover on the engine block.

22 Use the vibration damper to center the timing chain cover **(see illustration)**. Leave the old oil seal in place, as it may be damaged during installation.

23 Install the timing chain cover-to-block and oil pan-to-cover bolts and tighten them to the torque listed in this Chapter's Specifications.

24 Replace the crankshaft front oil seal (see Section 9).

25 With the key inserted in the crankshaft, install the vibration damper as described in Section 9.

26 Reinstall the remaining components in the reverse order of removal.

27 Run the engine and check for oil leaks.

10.13b . . . for the camshaft locating dowel (arrow) - be sure they are aligned properly during installation

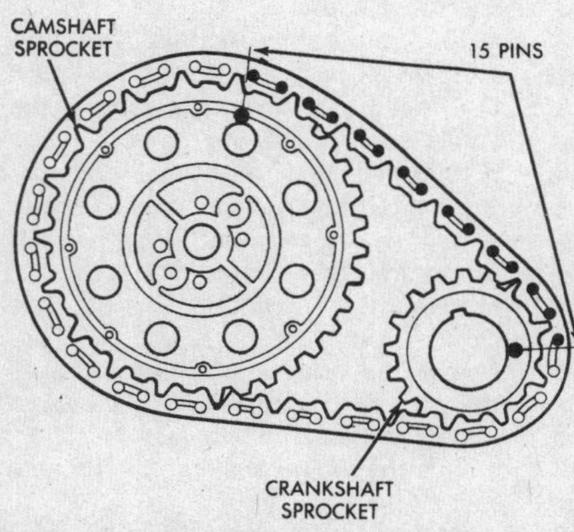

10.16 To verify correct installation, turn the crankshaft clockwise until the timing marks are positioned as shown and count the pins between the marks

10.22 Use the vibration damper to center the timing chain cover during installation

11.3 When checking the camshaft lobe lift, the dial indicator plunger must be positioned directly above the pushrod

11.16 Remove the lifters with a magnetic pick-up tool (shown here) or a special lifter removal tool

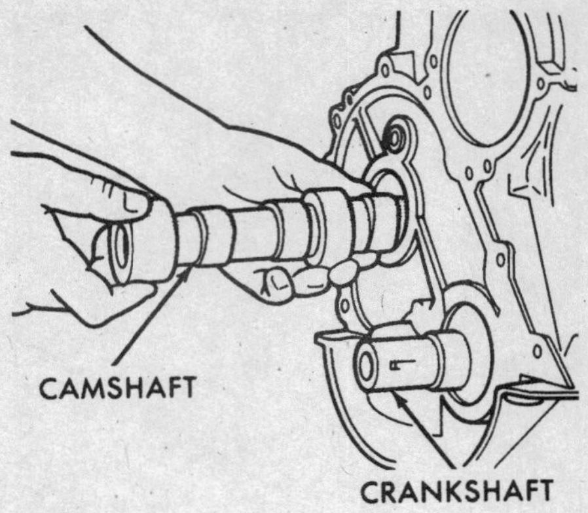

11.18 Support the camshaft near the block as you slowly withdraw it - it helps to use a long bolt in the sprocket-bolt hole as a handle

11 Camshaft, bearings and lifters - removal, inspection and installation

Camshaft lobe lift check

Refer to illustration 11.3

1 In order to determine the extent of cam lobe wear, the lobe lift should be checked prior to camshaft removal. Remove the valve cover (see Section 4).

2 Position the number one piston at TDC on the compression stroke (see Section 3). Remove the rocker arms (see Section 5).

3 Beginning with the number one cylinder, mount a dial-indicator on the engine and position the plunger in-line with the pushrod and resting on the end of the pushrod **(see illustration)**.

4 Zero the dial indicator, then very slowly turn the camshaft in the normal direction of rotation until the indicator needle stops and begins to move in the opposite direction. The point at which it stops indicates maximum cam lobe lift.

5 Record this figure for future reference, then reposition the piston at TDC on the compression stroke.

6 Move the dial indicator to the next number one cylinder pushrod and repeat the check. Be sure to record the results.

7 Repeat the check for the remaining camshaft lobes. Since each piston must be at TDC on the compression stroke for this procedure, work from cylinder-to-cylinder following the firing order sequence.

8 After the check is complete, compare the results to this Chapter's Specifications. If camshaft lobe lift is less than specified, cam lobe wear has occurred and a new camshaft should be installed.

Camshaft removal

Refer to illustrations 11.16 and 11.18

Warning 1: *These models are equipped with airbags. The airbag is armed and can deploy (inflate) anytime the battery is connected. To prevent accidental deployment (and possible injury), disconnect the negative battery cable whenever working near airbag components, including crash sensors that may be located near the grille/front bumper area. After the battery is disconnected, wait at least 2 minutes before beginning work (the system has a back-up capacitor that must fully discharge). For more information see Chapter 12.*

Warning 2: *The air conditioning system is under high pressure. Do not loosen any hose fittings or remove any components until after the system has been discharged. Air conditioning refrigerant should be properly discharged into an EPA-approved recovery/recycling unit at a dealer service department or an automotive air conditioning repair facility. Always wear eye protection when disconnecting air conditioning system fittings.*

9 If equipped with air conditioning, have the air conditioning system discharged at a dealer service department or automotive air conditioning repair facility. Disconnect the cable from the negative battery terminal.

10 Drain the cooling system, remove the radiator, fan and fan shroud (see Chapter 3).

11 Remove the grille (see Chapter 11).

12 On air-conditioned models, remove the condenser (see Chapter 3).

13 On 1998 and earlier models, label and remove the spark plug wires from the spark plugs. On 1999 and later models, remove the ignition coils (see Chapter 5).

14 On 1998 and earlier models, remove the distributor (see Chapter 5). On 1999 and later models, remove the camshaft position sensor/oil pump drive (see chapter 6).

15 Remove the cylinder head (see Section 8).

16 Remove the valve lifters **(see illustration)** and store them separately so they can be reinstalled in the same bores.

17 Remove the timing chain and sprockets (see Section 10).

18 Carefully pull the camshaft out. Temporarily install a long bolt, if necessary, to use as a handle. Support the cam so the lobes don't nick or gouge the bearings as the cam is withdrawn **(see illustration)**.

Inspection

Refer to illustration 11.20

19 After the camshaft has been removed from the engine, cleaned with solvent and dried, inspect the bearing journals for uneven wear, pitting and evidence of seizure. If the journals are damaged, the bearing inserts in the block are probably damaged also. Both the camshaft and bearings will have to be replaced.

20 If the bearing journals are in good condition, measure them with a micrometer and record the measurements **(see illustration)**. Measure each journal at several locations around its circumference. If you get different measurements at different locations, the journal is out of round.

21 Check the inside diameter of each camshaft bearing with a telescoping gauge and measure the gauge with a micrometer. Subtract each cam journal diameter from the corresponding camshaft bearing inside diam-

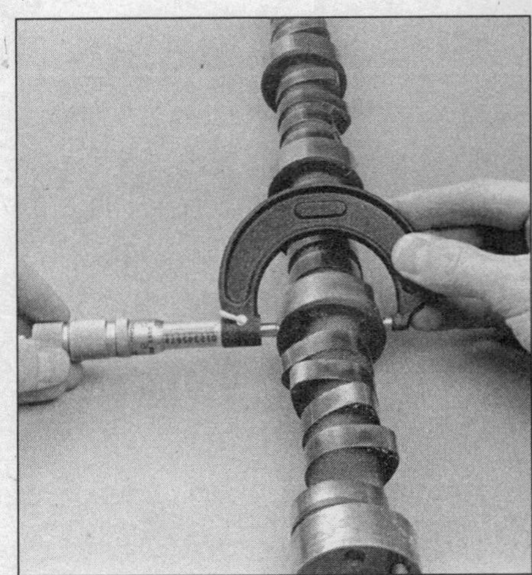

11.20 Use a micrometer to accurately measure the lobes and bearing journals of the camshaft

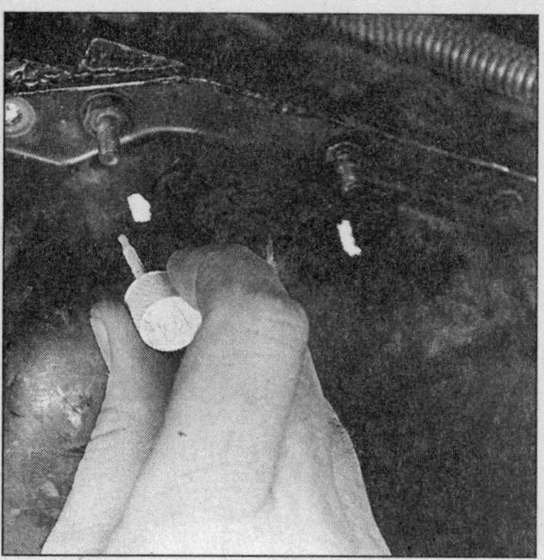

12.11 Mark the locations of the studs with paint to ensure proper reassembly

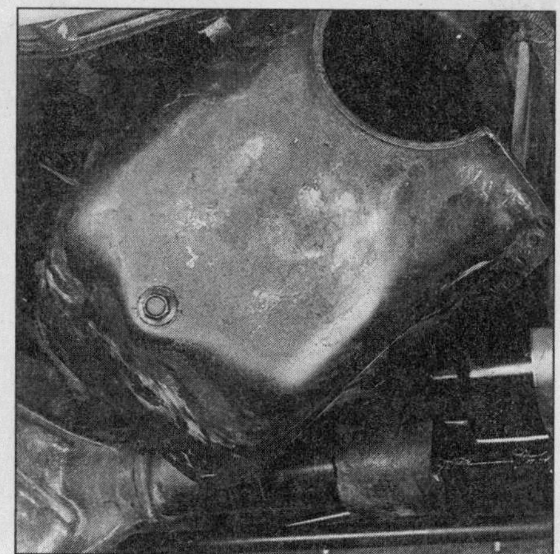

12.13 Remove the oil pan from the rear by sliding it outbetween the axle and bellhousing

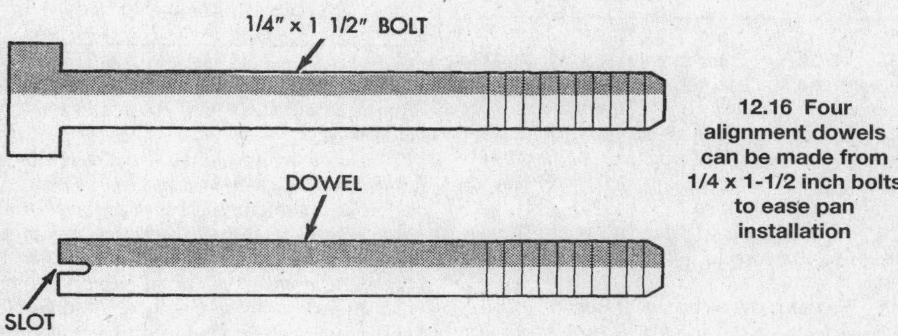

12.16 Four alignment dowels can be made from 1/4 x 1-1/2 inch bolts to ease pan installation

eter to obtain the bearing oil clearance.

22 Compare the clearance for each bearing to the specifications. If it is excessive, for any of the bearings, have new bearings installed by an automotive machine shop.

23 Inspect the distributor drive gear for wear. Replace the camshaft if the gear is worn.

24 Inspect the camshaft lobes for heat discoloration, score marks, chipped areas, pitting and uneven wear. If the lobes are in good condition and if the lobe lift measurements are as specified, the camshaft can be reused.

25 Camshaft bearing replacement requires special tools and expertise that place it outside the scope of the home mechanic. Remove the engine and take the block to an automotive machine shop to ensure the job is done correctly. **Note:** *If the camshaft appears to have been rubbing hard against the timing chain cover, first check the camshaft thrust pin and spring and then examine the oil pressure relief holes in the rear cam journal to make sure they are open.*

Installation

26 Lubricate the camshaft journals and lobes with engine assembly lube.

27 Slide the camshaft into the engine. Support the cam near the block and be careful not to scrape or nick the bearings.

28 The rest of the installation procedure is the reverse of removal. **Note:** *Camshafts and lifters develop a wear pattern that makes them compatible. If a new camshaft has been installed, new lifters must be installed as well.*

29 Before starting and running the engine, change the oil and filter (see Chapter 1).

30 If a new camshaft and lifters have been installed, the engine should be brought to operating temperature and run at a fast idle for 15 to 20 minutes to "break in" the components. Then change the oil and filter once more.

12 Oil pan - removal and installation

Refer to illustrations 12.11, 12.13 and 12.16

Removal

1 Disconnect the cable from the negative battery terminal.

2 Raise the vehicle and support it securely on jackstands.

3 Remove the shield from under the front of the engine.

4 Drain the oil and replace the oil filter (see Chapter 1).

5 Disconnect the exhaust pipe at the manifold and the exhaust hanger near the catalytic converter, allow the pipe to hang down.

6 Remove the starter motor (see Chapter 5).

7 Remove the flywheel inspection cover from the front of the transmission.

8 Disconnect the oil level sensor, if equipped.

9 Position a jack under the vibration damper and raise it just enough to take tension off the engine mounts. Place a wood block between the jack head and the engine to protect the damper.

10 Remove the engine mount through-bolts and raise the engine to obtain enough clearance to remove the oil pan.

11 Mark the locations of the oil pan mounting studs **(see illustration)**.

12 Remove the oil pan mounting bolts/studs and carefully separate the pan from the engine block. If the pan sticks to the block, tap the side of the pan gently with a soft-face mallet.

13 Remove the oil pan by sliding it out to the rear **(see illustration)**.

2A

Installation

14 Thoroughly clean the mating surfaces, removing all traces of oil and old gasket material.

15 Check the oil pan flange for distortion and warpage. Straighten the flange by placing the distorted area on a wood block and pounding it flat with a hammer.

16 Prepare four pan alignment dowels from 1/4-inch bolts 1-1/2-inches long. Cut off the bolt heads and slot the ends with a hacksaw **(see illustration)**.

17 Install the dowels in the two pan-bolt holes in the front cover, and the rearmost pair of 1/4-inch pan-bolt holes in the block.

18 Install the one-piece oil pan gasket to the block and front cover, over the four alignment dowels.

19 With several 1/4-inch pan bolts and a socket close at hand, position the oil pan over the alignment dowels and against the block, then insert and hand-tighten several bolts.

20 Install all of the 1/4-inch bolts and remove the dowels, replacing them with bolts.

21 Install the 5/16-inch bolts, and tighten all bolts to the torque listed in this Chapter's Specifications, working from the center out in several steps.

22 Reinstall the remaining parts in the reverse order of removal.

23 Check that the drain plug is tight and then add the amount of oil specified in Chapter 1.

24 Run the engine and check for oil leaks.

13 Oil pump - removal, inspection and installation

Refer to illustrations 13.2, 13.5 and 13.7

Removal

1 Remove the oil pan (see Section 12).

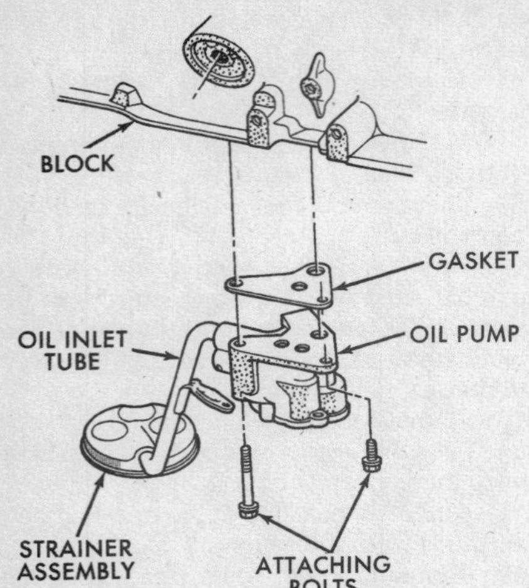

13.2 Oil pump installation details

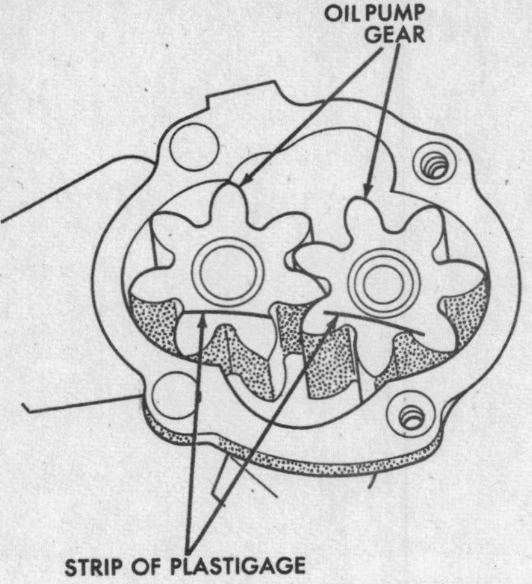

13.5 Place two strips of Plastigage a shown and reinstall the cover - remove the cover and compare the width of the crushed Plastigage against the scale

13.7 Using a feeler gauge, check the clearance between the oil pump gears and the pump body

2 Remove the two oil pump attaching bolts from the engine block **(see illustration)**.

3 Detach the oil pump and strainer assembly from the block.

Inspection

4 Remove the oil pump cover from the body.

5 Position two strips of Plastigage on the oil pump gears **(see illustration),** reinstall the cover and tighten it to 70 in-lbs.

6 Remove the cover and compare the width of the Plastigage with the scale on the Plastigage envelope. The clearance should be no greater than that listed in this Chapter's Specifications.

7 Using feeler gauges, measure the gear-to-body clearance **(see illustration)**. The clearance should be no greater than that listed in this Chapter's Specifications.

Installation

8 If the pump is defective, replace it with a new one. If there is any doubt about its condition, install a new oil pump - don't reuse the original or attempt to rebuild it.

9 Install the pump with a new gasket, turning the shaft so the gear tang mates with the slot on the lower end of the distributor or oil pump drive. The oil pump should slide easily into place. If it doesn't, pull it off and turn the tang until it's aligned with the distributor or oil pump drive.

10 Install the pump attaching bolts. Tighten them to the torque listed in this Chapter's Specifications.

11 Reinstall the oil pan (see Section 12).

12 Add oil, run the engine and check for leaks.

14 Flywheel/driveplate - removal and installation

Refer to illustrations 14.3 and 14.4

Removal

1 Raise the vehicle and support it securely on jackstands, then refer to Chapter 7A or 7B and remove the transmission. If it's leaking, now would be a very good time to replace the front pump seal/O-ring (automatic transmission only).

2 Remove the pressure plate and clutch disc (see Chapter 8) (manual transmission equipped vehicles). Now is a good time to check/replace the clutch components and pilot bearing.

3 Use paint or a center punch to make alignment marks on the flywheel/driveplate and crankshaft to ensure correct alignment during reinstallation **(see illustration)**.

4 Remove the bolts that secure the flywheel/driveplate to the crankshaft **(see illustration)**. If the crankshaft turns, hold the flywheel with a prybar or wedge a screwdriver into the ring gear teeth.

5 Remove the flywheel/driveplate from the crankshaft. Since the flywheel is fairly heavy, be sure to support it while removing the last bolt. **Warning:** *The edges of the ring gear can be very sharp - wear gloves or rags to protect your hands.*

14.3 Before removing the flywheel, index it to the crankshaft with a paint mark

14.4 Use a prybar to hold the flywheel while loosening the flywheel bolts

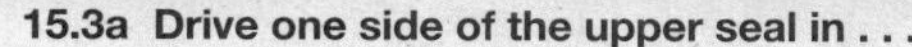

15.3a Drive one side of the upper seal in . . .

15.3b . . . until the other side protrudes far enough to be grasped with needle-nose pliers and pulled out

Installation

6 Clean the flywheel to remove grease and oil. Inspect the surface for cracks, rivet grooves, burned areas and score marks. Light scoring can be removed with emery cloth. Check for cracked and broken ring gear teeth or a loose ring gear. Lay the flywheel on a flat surface and use a straightedge to check for warpage. If any defects are evident, have the flywheel resurfaced at a machine shop.

7 Clean and inspect the mating surfaces of the flywheel/driveplate and the crankshaft. If the crankshaft rear seal is leaking, replace it before reinstalling the flywheel/driveplate (see Section 15).

8 Position the flywheel/driveplate against the crankshaft. Be sure to align the marks made during removal. Note that some engines have an alignment dowel or staggered bolt holes to ensure correct installation. Before installing the bolts, apply thread locking compound to the threads.

9 Wedge a screwdriver into the ring gear teeth to keep the flywheel/driveplate from turning **(see illustration 14.4)** as you tighten the bolts to the torque listed in this Chapter's Specifications, in several steps and in a criss-cross pattern. **Note:** *It helps to mark each bolt head with a chalk mark as it is tightened. When all bolts have three marks (using three steps), the torque sequence is complete.*

10 The remainder of installation is the reverse of the removal procedure.

15 Rear main oil seal - replacement

Refer to illustrations 15.3a, 15.3b, 15.5 and 15.8

1 Remove the oil pan (see Section 12). On 1999 and later models, remove the main bearing cap brace.

2 Remove the rear main bearing cap and pry the old seal half out of the bearing cap with a small screwdriver.

3 Carefully drive one side of the old upper main seal in with a small brass punch and a hammer until it protrudes sufficiently from the other side of the engine block to be gripped with needle-nose pliers and removed **(see illustrations)**. Use great care to avoid damaging the crankshaft.

4 Thoroughly clean the main bearing cap and the rear of the block/crankshaft, removing all traces of oil and old sealer.

5 Coat the lip of the new upper seal with engine oil; coat the outside surface with liquid soap **(see illustration)**.

6 Insert the upper seal into the groove in the engine block with the lip facing forward.

7 Apply a 0.090 inch bead of RTV sealant to the oil seal extension grooves on each side of the lower bearing cap. Then coat the outside surface of the new lower seal with liquid soap and press the seal into place in the cap with the lip facing forward. Do not apply RTV sealant or soap to the seal lip.

8 Apply a 0.125 bead of RTV sealant to the chamfered edges of the rear main bearing cap and to the top of the oil seal extensions after the seal has been installed in the rear cap **(see illustration)**.

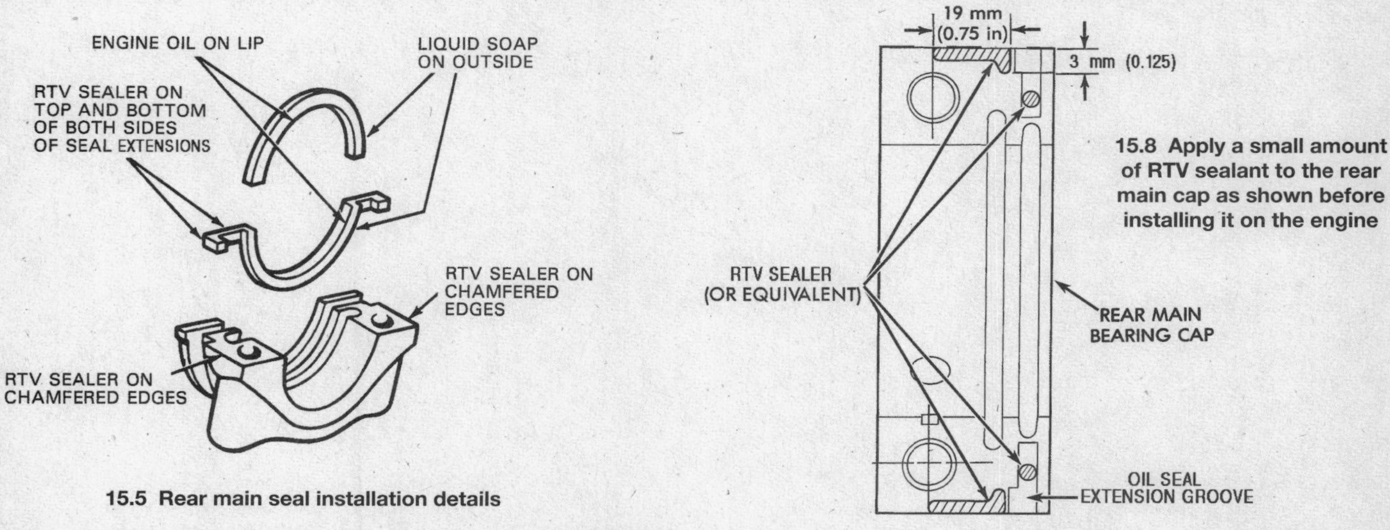

15.5 Rear main seal installation details

15.8 Apply a small amount of RTV sealant to the rear main cap as shown before installing it on the engine

9 Apply a film of engine oil to the seal lip and install the cap. Tighten the bolts to the torque listed in this Chapter's Specifications. **Caution:** *Do not apply an excessive amount of sealant to the cylinder block mating surfaces of the rear main bearing cap. Doing this will alter the bearing-to-journal clearance.*
10 Reinstall the remaining components in the reverse order of removal.
11 Add the amount of oil specified in Chapter 1, run the engine and check for oil leaks.

16 Engine mounts - check and replacement

Refer to illustration 16.9

1 Engine mounts seldom require attention, but broken or deteriorated mounts should be replaced immediately or the added strain placed on the driveline components may cause damage or wear.

Check

2 During the check, the engine must be raised slightly to remove the weight from the mounts.
3 Raise the vehicle and support it securely on jackstands, then position a jack under the engine oil pan. Place a large wood block between the jack head and the oil pan, then carefully raise the engine just enough to take the weight off the mounts. **Warning:** *DO NOT place any part of your body under the engine when it's supported only by a jack!*
4 Check the mounts to see if the rubber is cracked, hardened or separated from the metal plates. Sometimes the rubber will split right down the center.
5 Check for relative movement between the mount plates and the engine or frame (use a large screwdriver or prybar to attempt to move the mounts). If movement is noted, lower the engine and tighten the mount fasteners.
6 Rubber preservative should be applied to the mounts to slow deterioration.

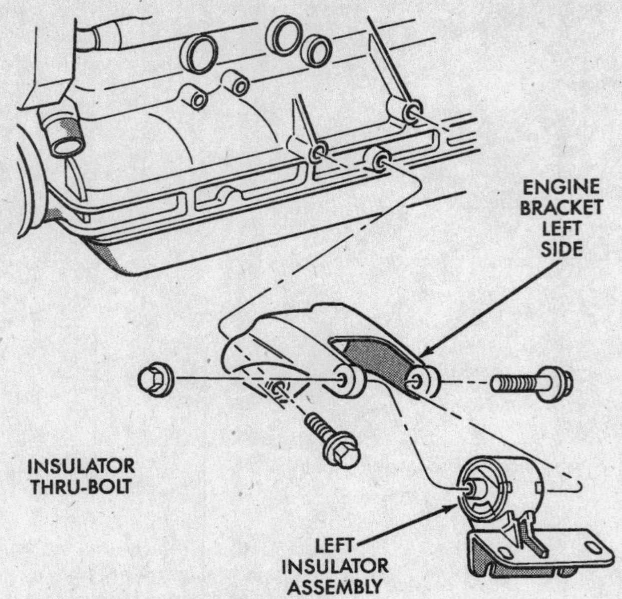

16.9 Typical six-cylinder engine mount details (driver's side shown)

Replacement

7 Disconnect the cable from the negative battery terminal. Raise the vehicle and support it securely on jackstands.
8 Raise the engine slightly with a jack or hoist (make sure the fan doesn't hit the radiator or shroud).
9 Loosen the nut on the through-bolt and remove the bolts that secure the mount to the engine or the frame **(see illustration)**. **Note:** *On 1999 and later models, the engine mounts are bolted to the engine, not the frame as on previous models.*
10 Remove the through-bolt and nut and detach the mount.
11 Installation is the reverse of removal. Use thread locking compound on the mount bolts and be sure to tighten them securely.

Chapter 2 Part B
5.2L and 5.9L V8 engines

Contents

Specifications

General

Displacement	
5.2 liter	318 cubic inches
5.9 liter	360 cubic inches
Cylinder numbers (front-to-rear)	
Left (driver's) side	1-3-5-7
Right (passenger's) side	2-4-6-8
Firing order	1-8-4-3-6-5-7-2

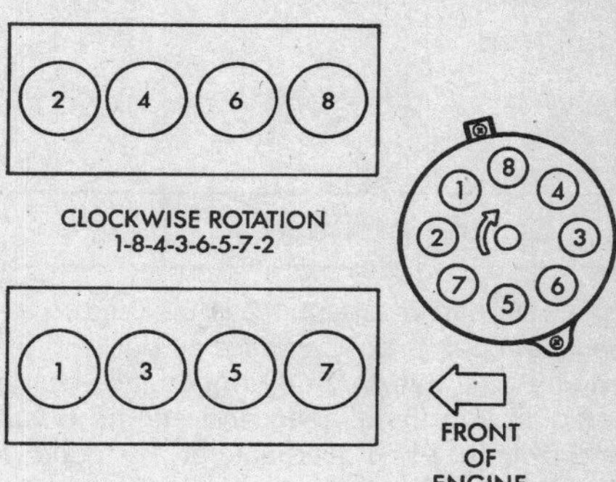

Cylinder location and distributor rotation diagram

Camshaft

Journal diameters	
No. 1 (front of engine)	1.998 to 1.999 inches
No. 2	1.982 to 1.983 inches
No. 3	1.967 to 1.968 inches
No. 4	1.951 to 1.952 inches
No. 5 (rear of engine)	1.560 to 1.561 inches
Bearing inside diameter	
No. 1	2.000 to 2.001 inches
No. 2	1.984 to 1.985 inches
No. 3	1.969 to 1.970 inches
No. 4	1.953 to 1.954 inches
No. 5	1.562 to 1.563 inches
Bearing oil clearance	0.001 to 0.003 inch
End play	
Standard	0.002 to 0.010 inch
Service limit	0.010 inch
Lobe lift	
5.2L (intake and exhaust)	0.2880 inch
5.9L	
1997 (intake and exhaust)	0.2734 inch
1998	
Intake	0.2734 inch
Exhaust	0.2780 inch

Oil pump

Oil pump cover surface variation	0.0015 inch
Outer rotor thickness	0.825 inch
Inner rotor thickness	0.825 inch
Rotor-to-cover clearance	0.004 inch
Outer rotor-to-body clearance	0.014 inch
Rotor tip clearance	0.008 inch

Torque specifications*

Torque specifications*	Ft-lbs (unless otherwise indicated)
Camshaft sprocket bolt	50
Camshaft thrust plate bolts	18
Cylinder head bolts	
Step 1	50
Step 2	105
Exhaust manifold bolts	20
Exhaust manifold nuts	15
Flywheel/driveplate bolts	105
Plenum pan bolts	
Step 1	24 in-lbs
Step 2	48 in-lbs
Step 3	84 in-lbs
Intake manifold bolts	
Step 1	72 in-lbs
Step 2	144 in-lbs
Oil pan bolts	18
Oil pump mounting bolts	30
Rocker arm bolts	21
Timing chain cover bolts	30
Main bearing cap bolts	85
Pulley-to-damper bolts	200 in-lbs
Valve cover	
Bolts	95 in-lbs
Studs	115 in-lbs
Vibration damper-to-crankshaft bolt	135

***Note:** *Refer to Part D for additional specifications*

1 General information

This part of Chapter 2 is devoted to in-vehicle repair procedures for all pushrod V8 engines. All information concerning engine removal and installation and engine block and cylinder head overhaul can be found in Part D of this Chapter.

Since the repair procedures included in this Part are based on the assumption that the engine is still installed in the vehicle, if they are being used during a complete engine overhaul (with the engine already out of the vehicle and on a stand) many of the steps included here will not apply.

The specifications included in this Part of Chapter 2 apply only to the procedures found here. The specifications necessary for rebuilding the block and cylinder heads are included in Part D.

The engines covered by this chapter include the 5.2L (318 cubic inches) and the 5.9L (360 cubic inches) overhead valve pushrod V8 with hydraulic roller lifters and electronic fuel injection. The 5.9L engine is basically the same as the 5.2L engine with a larger bore and longer stroke. Service procedures, and specifications, for both engines are identical unless otherwise noted.

3.1 Location of timing marks for TDC

2 Repair operations possible with the engine in the vehicle

Many major repair operations can be accomplished without removing the engine from the vehicle.

Clean the engine compartment and the exterior of the engine with some type of pressure washer before any work is done. A clean engine will make the job easier and will help keep dirt out of the internal areas of the engine.

Depending on the components involved, it may be a good idea to remove the hood to improve access to the engine as repairs are performed (refer to Chapter 11 if necessary).

If oil or coolant leaks develop, indicating a need for gasket or seal replacement, the repairs can generally be made with the engine in the vehicle. The oil pan gasket, the cylinder head gaskets, intake and exhaust manifold gaskets, timing cover gaskets and the crankshaft oil seals are all accessible with the engine in place.

Exterior engine components, such as the water pump, the starter motor, the alternator, the distributor and the carburetor, as well as the intake and exhaust manifolds, can be removed for repair with the engine in place.

Since the cylinder heads can be removed without pulling the engine, valve component servicing can also be accomplished with the engine in the vehicle.

Replacement of, repairs to, or inspection of the timing chain and sprockets and the oil pump are all possible with the engine in place.

In extreme cases caused by a lack of necessary equipment, repair or replacement of piston rings, pistons, connecting rods and rod bearings is possible with the engine in the vehicle. However, this practice is not recommended because of the cleaning and preparation work that must be done to the components involved.

3 Top Dead Center (TDC) for number one piston - locating

Refer to illustration 3.1

See Chapter 2, Part A for this procedure, but **refer to the illustrations** included in this Section.

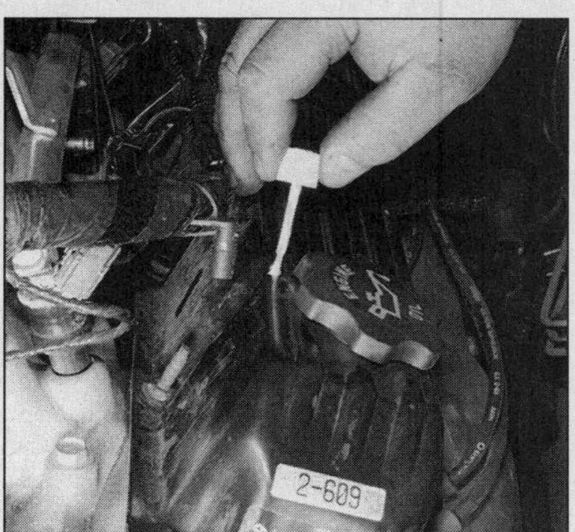

4.7 Mark the valve covers to indicate the location of studs

4.8 After striking the valve cover with a rubber mallet to break it loose, pull it straight up

4 Valve covers - removal and installation

Removal

Refer to illustrations 4.7, and 4.8

1 Disconnect the cable from the negative battery terminal.

2 Remove the air filter assembly (see Chapter 4).

3 Remove the crankcase ventilation and evaporative control hoses.

4 Remove the coolant tube bracket from the left valve cover.

5 Pulling on the boots only, remove the spark plug wires from the plugs. It may be helpful to number the wires with a felt pen before disconnecting them.

6 Remove the plastic spark plug wire holders from the valve cover studs and position the holders/wires out of the way. **Caution:** *Pull straight up on the plastic holders - they can easily break if pulled at an angle.*

7 Remove the valve cover mounting bolts. **Note:** *Many of the fasteners are studs, not bolts - mark the valve covers at each stud so they can be installed in the proper location* **(see illustration).**

8 Remove the valve cover **(see illustration). Note:** *If the cover is stuck to the head, bump the cover with a block of wood and a hammer to release it. If it still will not come loose, try to slip a flexible putty knife between the head and cover to break the seal. Don't pry at the cover-to-head joint, as damage to the sealing surface and cover flange will result and oil leaks will develop.*

Installation

9 The mating surfaces of each cylinder head and valve cover must be perfectly clean when the covers are installed. Use a gasket scraper to remove all traces of sealant or old gasket, then wipe the mating surfaces with a cloth saturated with lacquer thinner or acetone. If there is sealant or oil on the mating surfaces when the cover is installed, oil leaks may develop. **Note:** *The steel-backed silicone gaskets can be reused if they haven't been damaged.*

10 Make sure any threaded holes are clean. Run a tap into them to remove corrosion and restore damaged threads.

11 Mate the gaskets to the heads before installing the covers.

12 Carefully position the cover on the head and install the bolts/studs.

13 Tighten the bolts/studs in three steps to the torque listed in this Chapter's Specifications. **Caution:** *DON'T over-tighten the valve cover bolts.*

14 The remaining installation steps are the reverse of removal.

15 Start the engine and check carefully for oil leaks as the engine warms up.

5 Rocker arms and pushrods - removal, inspection and installation

Removal

Refer to illustrations 5.2 and 5.3

1 Remove the valve covers from the cylinder heads (see Section 4).

2 Remove the rocker arm bolts and pivots **(see illustration)**, placing them on the bench or in a container that will keep them in the proper order.

3 Remove the pushrods and store them separately to make sure they don't get mixed up during installation **(see illustration). Caution:** *All valve train components must go back in their original position. Organize and store the parts in a way that they won't get mixed up.*

Inspection

4 Check each rocker arm for wear, cracks and other damage, especially where the pushrods and valve stems contact the rocker arm faces. **Note:** *Keep in mind that there is no valve adjustment on these engines, so excessive wear or damage in the valve train can easily result in excessive valve clearance, which in turn will cause valve "tapping" or "clattering" noises when the engine is running.*

5.2 Remove the rocker-arm bolts, pivots and rocker arms

5.3 A perforated piece of cardboard can be used to store the pushrods to ensure that they're reinstalled in their original locations - note the arrow indicating the front of the engine

5.7 Before reinstalling pushrods and rocker arms, rotate the engine until the "V8" mark lines up with the TDC mark on the cover

5 Make sure the oil-feed holes in each pushrod are not plugged so that each rocker arm gets proper lubrication.

6 Inspect the pushrods for cracks and excessive wear at the ends. Roll each pushrod across a piece of plate glass to see if it is bent (if it wobbles, it is bent). Replace the pushrods if any of these conditions are present.

Installation

Refer to illustration 5.7

7 Rotate the engine until the V8 mark on the damper aligns with the TDC mark on the timing cover **(see illustration)**, which is 147 degrees after TDC for the number one piston. This represents a neutral point in the engine's directional rotation and the point at which the rocker arms can be tightened.

8 Lubricate the lower end of each pushrod with clean engine oil or moly-base grease and install them in their original locations. Make sure each pushrod seats completely in the lifter socket.

9 Apply moly-base grease to the ends of the valve stems and the upper ends of the pushrods before positioning the rocker arms onto the studs on the engine.

10 Install the pivot assemblies and rocker-arm bolts. Tighten the bolts to the torque listed in this Chapter's Specifications. **Caution:** *Do not rotate the engine for at least five minutes after installing the pushrods and rocker arms - the hydraulic roller tappets must "bleed down."*

11 Refer to Section 4 and install the valve covers. Start the engine, listen for unusual valve train noises and check for oil leaks at the valve cover joints. **Caution:** *Do not run the engine above a fast idle until all the hydraulic lifters have filled with oil and become quiet again.*

6 Valve springs, retainers and seals - replacement

Refer to illustrations 6.1a and 6.1b

This procedure is essentially the same as described in Chapter 2 Part A. Use the procedures for valve cover and rocker arm removal from this Chapter. Intake and exhaust seals may be different, so be sure to install the correct seal on the valve stem **(see illustrations)**.

7 Intake manifold - removal and installation

Removal

Refer to illustrations 7.5, 7.7, 7.8, 7.9, 7.12 and 7.18

1 Disconnect the cable from the negative battery terminal.

2 Drain the cooling system (see Chapter 1).

3 Remove the air filter assembly.

6.1a Be sure to install the seals on the correct valve stems

1 *Exhaust seal (shield)*
2 *Intake seal*

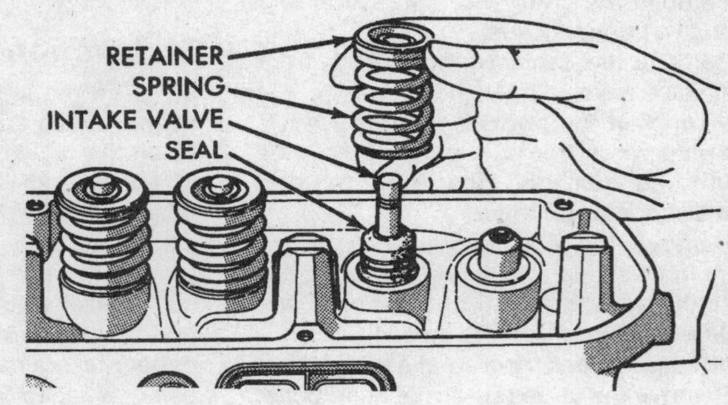

6.1b Valve spring, retainer and seal installation details

7.5 With a wrench on the tensioner bolt, rotate the tensioner clockwise until the belt can be slipped off the idler pulley

7.7 Access the power steering pump mounting bolts through the holes in the pulley

7.8 One bracket bolt is hidden under the idler pulley - remove the pulley, the bolts and the bracket

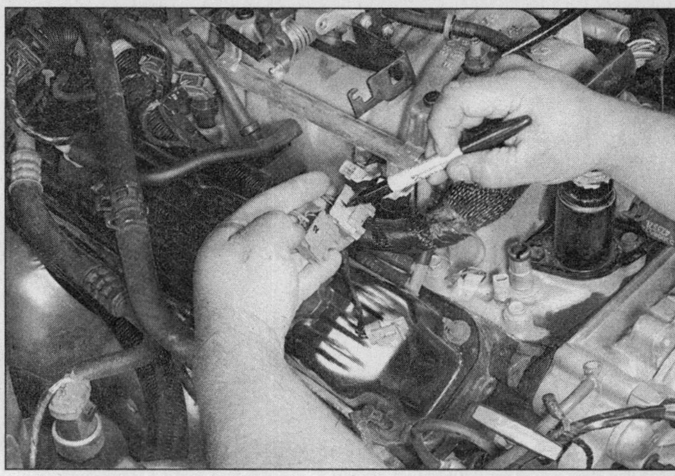

7.9 Disconnect each fuel injector electrical connector and label it clearly so it can be installed on the proper injector during reassembly

4 Relieve the fuel system pressure (see Chapter 4).

5 Rotate the accessory belt tensioner over enough to slip the serpentine belt off the idler pulley **(see illustration)** and remove the belt.

6 Remove the alternator (see Chapter 5). On air-conditioned models, remove the air-conditioning compressor and set it aside. **Caution:** *Do not disconnect the refrigerant lines.*

7 Remove the three power-steering-pump mounting bolts **(see illustration)**. Lay the pump aside, in an upright position, to avoid spilling fluid.

8 Remove the idler pulley (it covers access to one of the bracket bolts) **(see illustration)**. Remove the bolts securing the alternator/compressor bracket to the front of the engine and remove the bracket.

9 Label and then disconnect the electrical connector to each fuel injector **(see illustration)**. Disconnect any vacuum hoses attached to the intake manifold or throttle body.

10 Disconnect any remaining electrical connectors connected to the intake manifold or throttle body and pull the engine wiring harness up and over the intake manifold to the rear of the engine.

11 Remove the fuel rails and injectors as an assembly (see Chapter 4). The two fuel rails can be pulled straight up with the injectors still attached, but it will take some force to dislodge the injectors from the intake manifold.

12 Remove the upper radiator hose from the engine, then disconnect the heater hose and water pump bypass hose from the intake manifold **(see illustration).**

13 Disconnect the accelerator linkage (see Chapter 4) and, if equipped, the cruise control linkage. On models equipped with an automatic transmission, disconnect the kickdown cable at the throttle body.

14 Remove the two distributor-cap-retaining screws (use a 1/4-inch-drive socket and proper-length extension). Disconnect the spark plug wires from the spark plugs (see Chapter 1). Feed the distributor cap and spark plug wires from behind the intake manifold and out of the engine compartment.

15 Remove the vacuum-distribution block from the driver's side of the manifold and move it to the opposite side of the engine.

16 Remove the two bolts securing the EGR tube to the passenger-side exhaust manifold (see Section 8).

17 Remove the EGR valve from the back of the intake manifold, this will allow access to the EGR tube fitting threaded into the rear of the intake manifold, just below the EGR valve. Remove the EGR tube.

18 Remove the intake manifold with the throttle body attached **(see illustration)**. As the manifold is lifted from the engine, be sure to check for and disconnect anything still attached to the manifold. **Caution:** *Do not pry between the block and manifold or the heads and manifold or damage to the gasket-sealing surfaces may result and vacuum leaks could develop.*

7.12 Disconnect the heater hose from the manifold, and the water-pump bypass hose

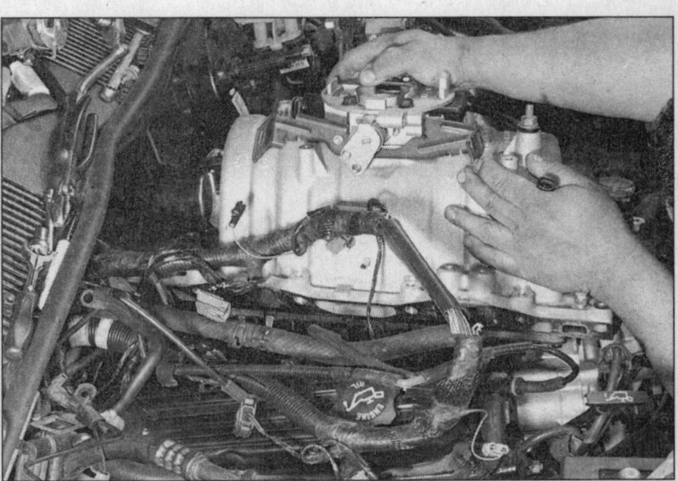

7.18 The intake manifold is bulky and heavy - watch for wires or hoses hanging up as you remove the manifold

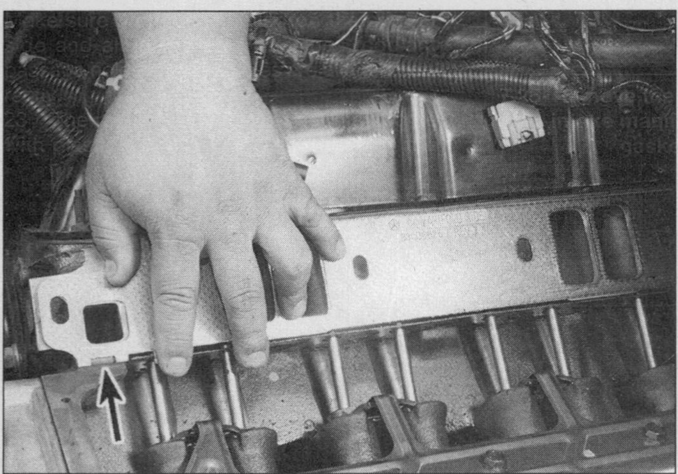

7.22 The gaskets must be installed on the proper side; some may be marked left or right or "manifold side," but align the gasket with the ports and make sure the side gasket's cutouts clear the tabs on the head gasket (arrow) front and rear

7.23 Apply RTV sealant into the "dowel" holes (arrows) for the end gaskets

Installation

Refer to illustrations 7.22, 7.23, 7.25, 7.26, 7.27 and 7.29

Note: *The mating surfaces of the cylinder heads, block and manifold must be perfectly clean when the manifold is installed. Gasket removal solvents in aerosol cans are available at most auto parts stores and may be helpful when removing old gasket material that is stuck to the heads and manifold. Be sure to follow the directions printed on the container, and follow cautions on ventilation and handling.*

19 Scrape the carbon deposits from the EGR passage. Use a gasket scraper to remove all traces of sealant and old gasket material, then wipe the mating surfaces with a cloth saturated with lacquer thinner or acetone. If there is old sealant or oil on the mating surfaces when the manifold is installed, oil or vacuum leaks may develop. Cover the lifter valley with shop rags to keep debris out of the engine. Use a vacuum cleaner to remove any gasket material that falls into the intake ports in the heads.

20 Use a tap of the correct size to chase the threads in the bolt holes, then use compressed air (if available) to remove the debris from the holes. **Warning:** *Wear safety glasses or a face shield to protect your eyes when using compressed air.*

21 Apply a thin coat of RTV sealant to the cylinder-head side of the new intake manifold gaskets.

22 Position the side gaskets on the cylinder heads. Look to see if the gaskets are marked LT for left or RT for right **(see illustration)**, usually one side is marked "manifold" side. If they are, this will ensure proper installation. Make sure they are installed on the correct side and all intake port openings, coolant passage holes and bolt holes are aligned correctly.

23 The replacement gasket set may come with plastic dowels that fit into holes in the end gaskets and the front and rear of the block. If dowels are included, install them. If there are no dowels, squirt some RTV sealant into the holes in the block **(see illustration)**. Apply a thin, uniform coating of quick-dry cement to the intake manifold end seals and the cylinder block contact surfaces and press the gaskets down onto the block, allowing the RTV to squish up through the holes in the gaskets. Refer to the instructions with the gasket set for further information.

24 If the engine has more than 50,000 miles on it and you have a complete intake gasket set, now is a good time to replace the gasket between the manifold and the throttle body.

25 Bolted to the bottom of the manifold is the plenum pan. If necessary, remove the pan **(see illustration)** and clean the gasket-sealing surfaces thoroughly.

26 Reassemble the pan to the manifold, using a new gasket, and hand-start all the bolts. Tighten the bolts to the torque listed in this Chapter's Specifications in three steps, and in the sequence shown **(see illustration).**

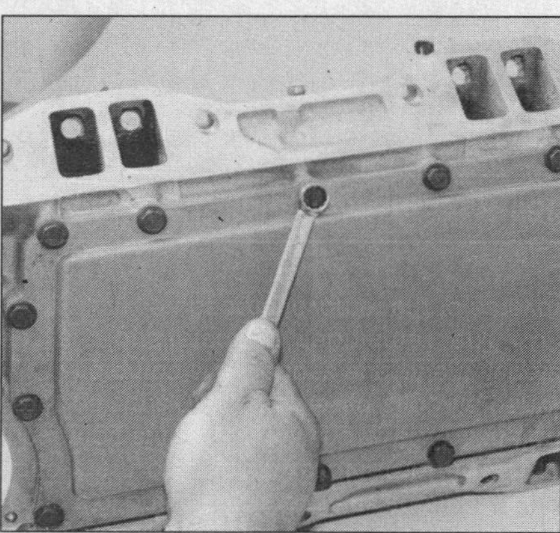

7.25 Remove the plenum pan under the intake manifold and clean the gasket surfaces

27 Apply a dab of RTV sealant to the four corners of the intake gaskets on the engine **(see illustration)** but do not overdo it. The RTV should be only slightly higher than the front and rear gaskets.

28 Carefully set the manifold in place, **Caution:** *Do not disturb the gaskets and DO NOT move the manifold fore-and-aft after it contacts the front and rear seals or the gaskets will be pushed out of place and you may not notice the problem until you see an oil leak later.*

29 Install all the intake manifold bolts and hand-tighten them. Then tighten the bolts to the torque listed in this Chapter's Specifications in two steps, in the sequence shown **(see illustration)**. **Note:** *The two front and two rearmost bolts are nearest the water pas-*

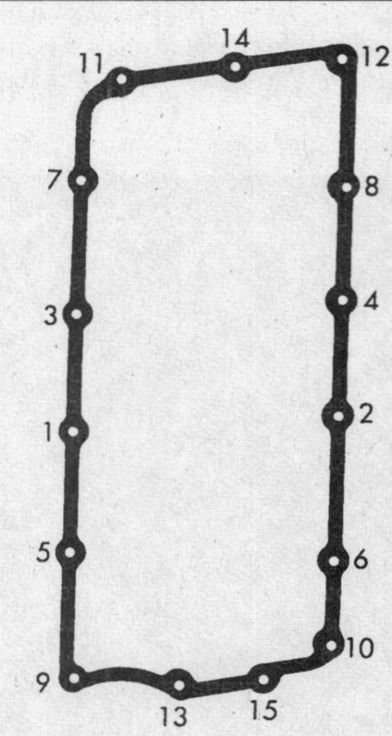

7.26 Tightening sequence for the plenum pan

7.27 Apply a small bead of RTV sealant to the four corners, where the end gaskets meet the side gaskets

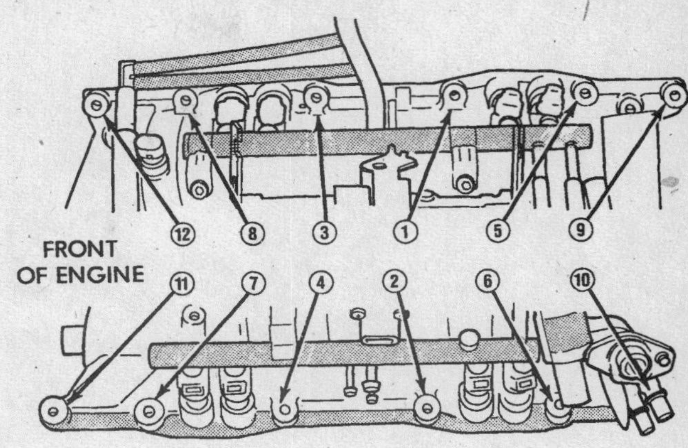

7.29 Intake manifold bolt tightening sequence

sages and should have sealant applied to the threads before installation.

30 The remaining installation steps are the reverse of removal. Start the engine and check carefully for oil, vacuum or coolant leaks at the intake manifold joints.

8 Exhaust manifolds - removal and installation

Removal

Refer to illustrations 8.3, 8.6, 8.7 and 8.8

Warning: *Allow the engine to cool completely before beginning this procedure.*

1 Disconnect the cable from the negative battery terminal.

2 Disconnect the spark plug wires and remove the spark plugs (see Chapter 1).

3 Remove the nuts securing the exhaust heat shields, and remove the shields **(see illustration)**.

4 Disconnect the EGR tube from the intake manifold.

5 Firmly apply the parking brake and block the rear wheels. Raise the front of the vehicle and support it securely on jackstands.

6 Disconnect the exhaust pipe from the manifold outlet **(see illustration)**. **Note:** *Soak the bolts with penetrating oil, as necessary, to remove frozen exhaust-attaching bolts.*

7 On the passenger-side manifold, remove the two bolts retaining the EGR tube and remove the tube **(see illustration)**.

8 Loosen the outer fasteners first, then the center ones and separate the manifold from the head **(see illustration)**. **Note:** *Many of the exhaust manifold fasteners are studs, so mark their location before removal to ensure they're installed in the correct location. Also, the studs that retain the heat shields use large spacer washers (between the studs and the shields), save these washers and mark their location.*

Installation

9 Installation is the reverse of the removal procedure. Clean the manifold and head gasket surfaces. Scrape any remains of old gasket and corrosion from the EGR tube flange and its mating surface on the passenger-side manifold, and install a new EGR tube gasket.

10 Install the fasteners. Tighten them to the torque listed in this Chapter's Specifications. Work from the center to the ends and approach the final torque in three steps.

11 Apply anti-seize compound to the exhaust manifold-to-exhaust pipe bolts and tighten them securely.

8.3 Remove the exhaust manifold heat shield

8.6 Remove the exhaust pipe-to-manifold bolts

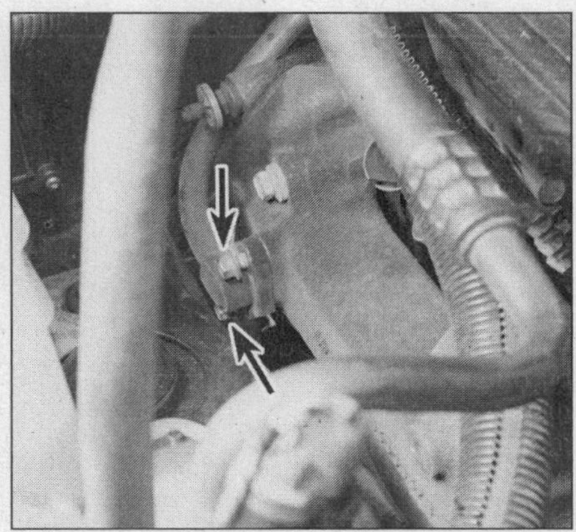
8.7 Remove the two bolts (arrows) retaining the EGR tube to the passenger-side exhaust manifold

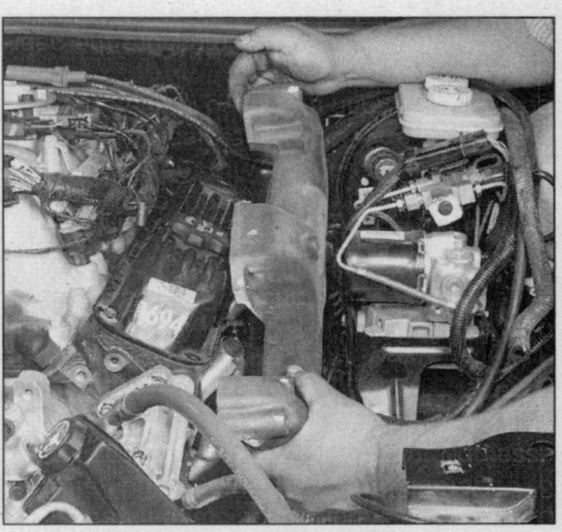
8.8 Mark the location of studs and spacers, then remove the manifold

9.6 Following the reverse order of the tightening sequence, loosen all the head bolts

9.7 After all the bolts have been removed get a good hold on the cylinder head and remove it from the engine block

9 Cylinder heads - removal and installation

Removal

Refer to illustrations 9.6 and 9.7

1 Disconnect the cable from the negative battery terminal.

2 Drain the cooling system (see Chapter 1).

3 Remove the valve covers (see Section 4). Remove the intake manifold (see Section 7).

4 Detach both exhaust manifolds from the cylinder heads (see Section 8).

5 Remove the rocker arm assemblies and pushrods (see Section 5). **Caution:** *keep all the parts in order so they are reinstalled in their original locations*.

6 Following the reverse order of the tightening sequence **(see illustration 9.16),** loosen the head bolts in 1/4-turn increments using a breaker-bar **(see illustration)** until they can be removed by hand. There are two lengths of headbolts, the shorter ones are the outside bolts.

7 Lift the heads off the engine **(see illustration)**. If resistance is felt, do not pry between the head and block as damage to the mating surfaces will result. To dislodge the head, place a block of wood against the end of it and strike the wood block with a hammer, or lift on a casting protrusion Store the heads on blocks of wood to prevent damage to the gasket sealing surfaces.

8 Cylinder head disassembly and inspection procedures are covered in detail in Chapter 2, Part D.

Installation

Refer to illustrations 9.10, 9.12, 9.13 and 9.16

9 The mating surfaces of the cylinder heads and block must be perfectly clean when the heads are installed. Gasket removal solvents are available at auto parts stores and may prove helpful.

10 Use a gasket scraper to remove all traces of carbon and old gasket material, then wipe the mating surfaces with a cloth saturated with lacquer thinner or acetone. If there is oil on the mating surfaces when the heads are installed, the gaskets may not seal correctly and leaks may develop. When working on the block, cover the lifter valley with shop rags to keep debris out of the engine **(see illustration)**. Use a vacuum cleaner to remove any debris that falls into the cylinders.

11 Check the block and head mating surfaces for nicks, deep scratches and other damage. If damage is slight, it can be removed with emery cloth. If it is excessive, machining may be the only alternative.

12 Use a tap of the correct size to chase the threads in the head bolt holes in the block. Mount each bolt in a vise and run a die down the threads to remove corrosion and restore the threads **(see illustration)**. Dirt, corrosion, sealant and damaged threads will affect torque readings.

13 Position the new gaskets over the dowels in the block **(see illustration)**.

14 Carefully position the heads on the block without disturbing the gaskets.

15 Before installing the head bolts, apply a light coat of clean engine oil to the threads.

16 Install the bolts in their original locations and tighten them finger tight. Following the recommended sequence **(see illustration)**,

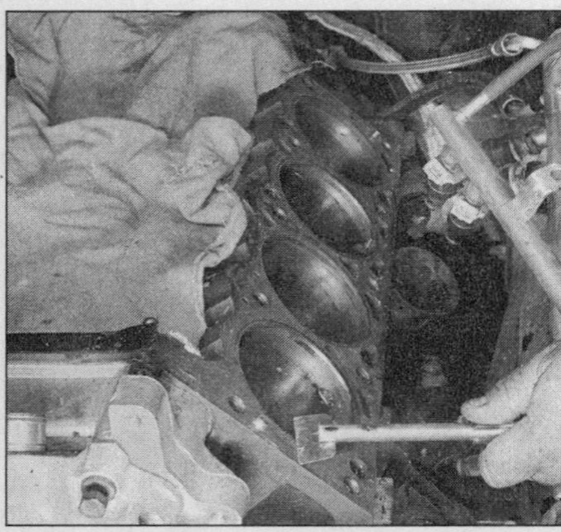

9.10 Remove all traces of old gasket material - keep the intake valley covered with hop rags

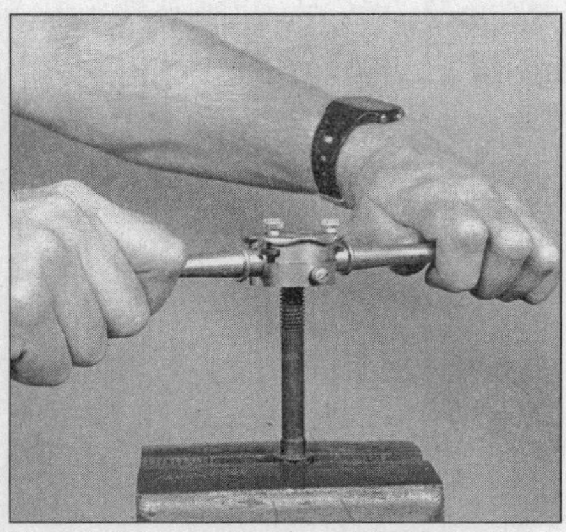

9.12 A die should be used to remove sealant and corrosion from the bolt threads prior to installation

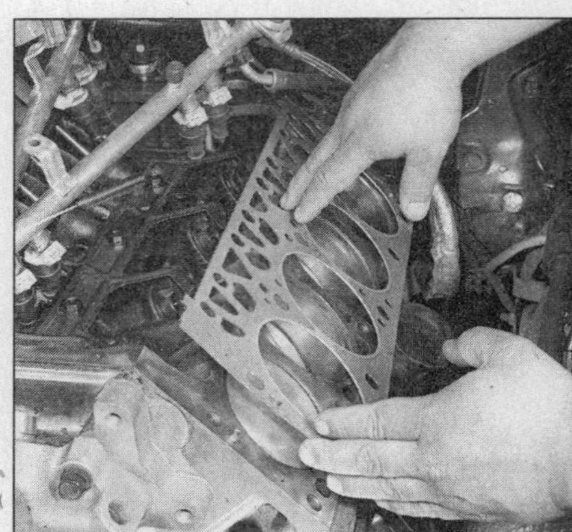

9.13 Install the new head gasket over the dowels at each end cylinder bank

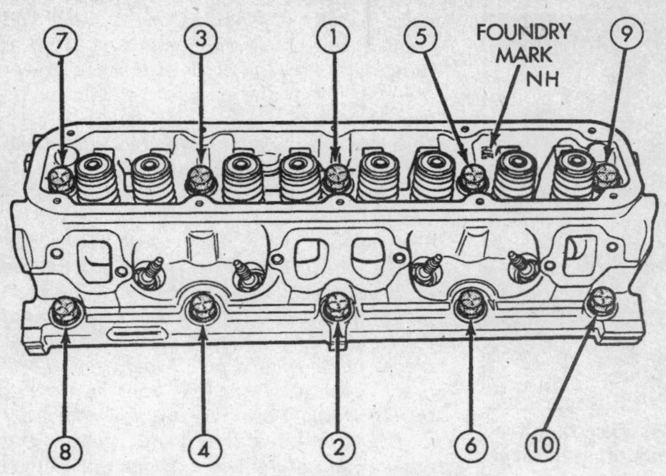

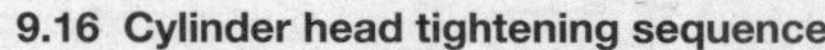

9.16 Cylinder head tightening sequence

10.4 Remove the crankshaft pulley bolts (arrows) and the large vibration damper bolt in the center

tighten the bolts in several steps to the torque listed in this Chapter's Specifications.

17 The remaining installation steps are the reverse of removal. As mentioned in Section 5, after tightening the rocker arms, allow the lifters five minutes to "bleed down" before starting the engine.

18 Add coolant and change the oil and filter (see Chapter 1). Start the engine and check for proper operation and for coolant or oil leaks. **Caution:** *Do not run the engine above a fast idle until all the hydraulic lifters have filled with oil and become quiet again.*

10 Crankshaft front oil seal - replacement

Removal

Refer to illustrations 10.4, 10.6 and 10.7

1 Disconnect the cable from the negative battery terminal.

2 Unbolt the fan shroud and move it back over the cooling fan.

3 Pry the accessory drivebelt tensioner clockwise until the serpentine belt can be removed from the idler pulley.

4 Remove the bolts **(see illustration)** and separate the crankshaft pulley from the vibration damper.

5 Remove the large vibration damper bolt. To keep the crankshaft from turning, remove the starter (see Chapter 5) and have an assistant wedge a large screwdriver against the ring gear teeth.

6 Use a vibration damper puller (commonly available from auto parts stores) to detach the vibration damper **(see illustration)**. **Caution:** *Do not use a puller with jaws that grip the outer edge of the damper. The puller must be the type that utilizes bolts to apply force to the damper hub only, and make sure that any adapter on the front of the tool is larger than the hole in the front of the crankshaft.*

7 Carefully pry the seal out of the cover with a seal puller or a large screwdriver **(see illustration)**. **Caution:** *Be careful not to scratch, gouge or distort the area that the seal fits into or a leak will develop.*

Installation

Refer to illustrations 10.9 and 10.10

8 Clean the bore to remove any old seal material and corrosion. Position the new seal in the bore with the seal lip (usually the side with the spring) facing IN (toward the inside of the engine). A small amount of oil applied to the outer edge of the new seal will make installation easier - don't overdo it!

9 Drive the seal into the bore with a large socket and hammer until it's completely seated **(see illustration)**. Select a socket that's the same outside diameter as the seal and make sure the new seal is pressed into place until it bottoms against the cover flange.

10 Check the surface of the damper that the oil seal rides on. If the surface has been grooved from prolonged contact with the seal, a press-on sleeve may be available renewing the damper sealing surface **(see illustration)**. These sleeves are commonly available from auto parts stores.

11 Lubricate the seal lips with engine oil and reinstall the vibration damper.

10.6 Use a bolt-on-type puller to remove the vibration damper

10.7 Use a seal puller, being careful not to nick the crankshaft surface

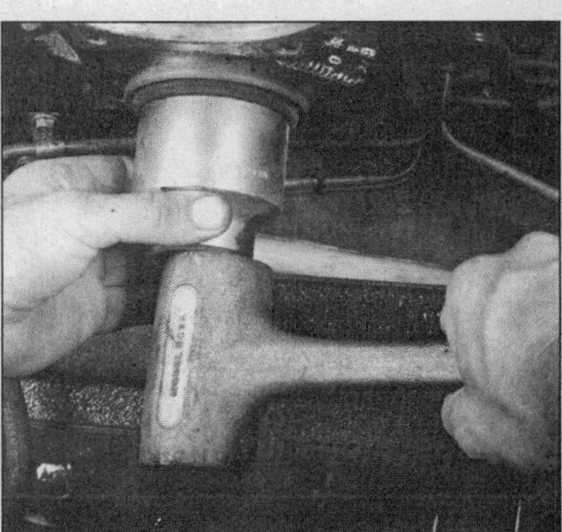

10.9 Drive the new front seal into place with a large socket and hammer

10.10 If the sealing surface of the damper has a wear groove from contact with the seal, repair sleeves, like the one shown here, are available at auto parts stores - when pressed in place, these sleeves provide a fresh new surface for the seal and can prevent oil leaks

11 Timing cover, chain and sprockets - removal, inspection and installation

Timing chain slack check (cover on engine)

Note: *This procedure will allow you to check the amount of slack in the timing chain without removing the timing chain cover from the engine.*

1 Disconnect the cable from the negative battery terminal.

2 Place the engine at TDC (Top Dead Center) (see Section 3) for number one cylinder.

3 Keeping it on the compression stroke, place number one at about 30-degrees before TDC (BTDC).

4 Remove the distributor cap (see Chapter 1).

11.11 Remove the accessories, then the alternator/compressor mounting bracket

5 You must rotate the crankshaft very slowly by hand for accuracy on this check, so use a 1/2-inch-drive breaker bar, extension and correct-size socket on the large bolt in the center of the crankshaft pulley.

6 Turn the crankshaft slowly clockwise until the number one piston is at TDC. This will take up the slack on the left side of the timing chain. Do not go past the timing mark and reverse direction.

7 Mark the position of the distributor rotor on the distributor housing.

8 Slowly turn the crankshaft counterclockwise until the slightest movement is seen at the distributor rotor. Stop and note how far the crankshaft has moved away from the TDC mark by looking at the ignition timing marks.

9 If the mark has moved more than 10 degrees, the timing chain is probably worn excessively. Remove the timing chain cover for a more accurate check.

Cover removal

Refer to illustrations 11.11 and 11.14

10 Detach accessories such as the power steering pump, alternator and air conditioning compressor that block access to the timing chain cover. Leave the hoses connected to the air conditioning compressor and power steering pump and secure the units aside (see Section 7).

11 Remove the accessory bracket from the front of the engine **(see illustration)**.

12 Remove the water pump (see Chapter 3).

13 Remove the front two oil pan bolts, that thread into the timing chain cover - they're most easily accessed from below. **Note:** *Even though this procedure can be done without the removal of the oil pan, it is likely the one-piece, metal-reinforced oil pan gasket will break when removing the front cover. If it doesn't break, it could be coated with sealant in the front-cover area and reused, but replacement of the entire oil pan gasket (*see Section 13*) is the only guarantee of a leak-free installation.*

14 Remove the remaining mounting bolts and separate the timing chain cover from the block and oil pan **(see illustration)**. The cover may be stuck. If so, use a putty knife to break the gasket seal. The cover is easily damaged, so DO NOT attempt to pry it off.

15 Remove the oil slinger from the end of the crankshaft, if equipped. Note how it's installed so you can reinstall it the same way on reassembly. It's a good idea to replace the crankshaft front seal at this time (see Section 10).

Timing chain and gear inspection (cover removed from engine)

Refer to illustration 11.16

16 Attach a socket and torque wrench to the camshaft sprocket bolt **(see illustration)** and apply force in the normal direction of crankshaft rotation (30 ft-lbs if the cylinder head is still in position complete with rocker arms, or 15 ft-lbs if the cylinder head has been removed). Don't allow the crankshaft to rotate. Using a ruler, note the position of a

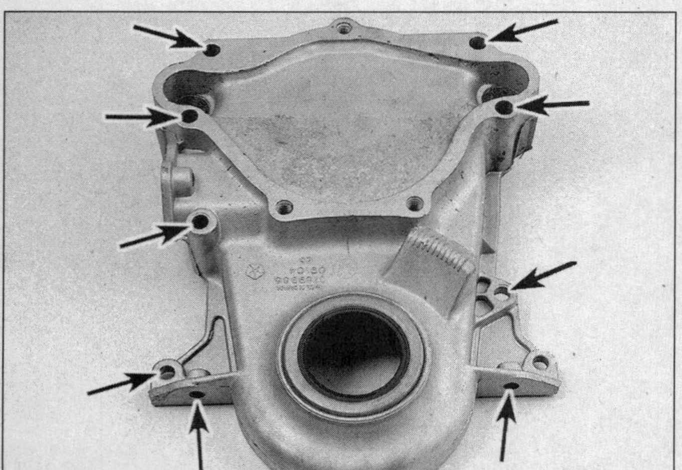

11.14 Timing chain cover bolt hole locations

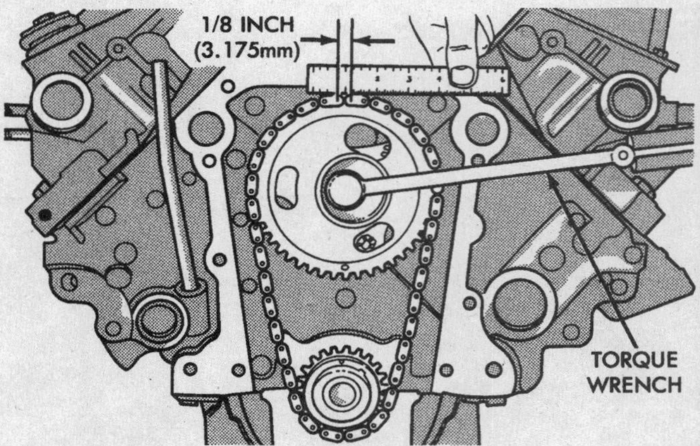

11.16 Position a ruler over the chain to measure slack while applying pressure with a torque wrench to the camshaft sprocket bolt

11.18 Align the timing marks on the crankshaft and camshaft sprockets

11.19a The camshaft and crankshaft sprockets can be removed with a two or three-jaw puller. . .

11.19b . . . or two screwdrivers

chain linkpin on the ruler and reverse the motion of the torque wrench. Check the movement against the ruler and observe the total amount of chain movement. If it exceeds 1/8-inch, a new timing chain will be required. **Note:** *Whenever a new timing chain is required, the entire set (chain, camshaft sprocket and crankshaft sprocket) must be replaced as an assembly.*

17 Inspect the camshaft and crankshaft sprockets for damage or wear.

Chain removal

Refer to illustrations 11.18, 11.19a and 11.19b

18 Be sure the timing marks are aligned **(see illustration)** and remove the bolt from the camshaft sprocket.

19 The sprockets on the camshaft and crankshaft can be removed with a two or three-jaw puller or by using two screwdrivers **(see illustrations)**, but be careful not to damage the threads in the end of the crankshaft.

Installation

Refer to illustrations 11.22a and 11.22b

20 Use a gasket scraper to remove all traces of old gasket material and sealant from the cover and engine block. Stuff a shop rag into the opening at the front of the oil pan to keep debris out of the engine. Remove the rubber semi-circular seal from the front of the oil pan. Wipe the cover and block sealing surfaces with a cloth saturated with lacquer thinner or acetone.

21 If you're installing a new crankshaft sprocket, be sure to align the keyway in the crankshaft sprocket with the Woodruff key in the end of the crankshaft. **Note:** *Timing chains must be replaced as a set with the camshaft and crankshaft sprockets. Never put a new chain on old sprockets.* Align the sprocket with the Woodruff key and press the sprocket onto the crankshaft with the vibration damper bolt, a large socket and some washers or tap it gently into place until it is completely seated. **Caution:** *If resistance is encountered, do not hammer the sprocket onto the crankshaft. It may eventually move onto the shaft, but it may be cracked in the process and fail later, causing extensive engine damage.*

22 Loop the new chain over the camshaft sprocket, then position the sprocket with the timing mark at the bottom **(see illustration)**. Mesh the chain with the crankshaft sprocket and position the camshaft sprocket on the camshaft. If necessary, turn the camshaft so the key fits into the sprocket keyway with the timing mark in the 6 o'clock position **(see illustration)**. When the chain is installed, the timing marks MUST align as shown **(see illustration 11.18)**. **Note:** *The number six piston is at TDC on the compression stroke when the timing marks are aligned.*

23 Apply a thread-locking compound to the camshaft sprocket bolt threads, and tighten the bolt to the torque listed in this Chapter's Specifications.

24 Lubricate the chain with clean engine oil.

25 If the crankshaft front oil seal has been leaking, refer to Section 10 and install a new one. If the seal is being replaced with the timing chain cover removed, support the cover on two blocks of wood and drive the seal out from the back with a hammer and punch. **Caution:** *Be careful not to scratch, gouge or distort the area that the seal fits into or a leak will develop.* Drive a new one in with a hammer and a socket the size of the seal.

26 Check for tears or deformation of the one-piece, metal-reinforced oil pan gasket before installing the timing cover. If the gas-

2B

11.22a Slip the chain and camshaft sprocket in place over the crankshaft sprocket with the timing mark (arrow) at the bottom

11.22b The keyway must align with the key (arrows) with the timing mark at 6 o'clock

12.3 When checking the camshaft lobe lift, the dial indicator plunger must be positioned directly above and in-line with the pushrod

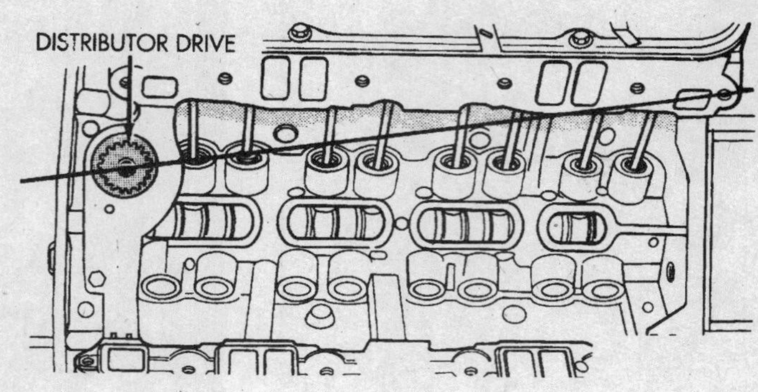

12.12 Pull up on the distributor drive gear to remove it - upon installation be sure the slot is aligned as shown with the No. 1 piston at TDC

ket is damaged, it must be replaced (see Section 13) before reinstalling the timing cover.

27 Apply a thin layer of RTV sealant to both sides of the new cover gasket and the corners of the pan and block, then position the new cover gasket on the engine. The sealant will hold it in place. **Note:** *If the oil pan has been removed for replacement of the pan gasket, install the front cover before the oil pan and its gaskets are installed.*

28 Install the timing chain cover on the block. Install all front cover bolts, except the ones that also secure the water pump, tightening them finger-tight.

29 Install the water pump with a new gasket and tighten the timing chain cover bolts to the torque listed in this Chapter's Specifications.

30 Install the oil pan bolts, bringing the oil pan up against the timing chain cover.

31 Lubricate the oil seal contact surface of the vibration damper hub with clean engine oil, then install the damper on the end of the crankshaft. The keyway in the damper must be aligned with the Woodruff key in the crankshaft nose. If the damper cannot be seated by hand, slip the large washer over the bolt, install the bolt and tighten it to pull the damper into place. Tighten the bolt to the torque listed in this Chapter's Specifications.

32 The remaining installation steps are the reverse of removal.

33 Add coolant and check the oil level. Run the engine and check for oil and coolant leaks.

12 Camshaft and lifters - removal, inspection and installation

Warning 1: *These models are equipped with airbags. The airbag is armed and can deploy (inflate) anytime the battery is connected. To prevent accidental deployment (and possible injury), disconnect the negative battery cable whenever working near airbag components or crash sensors which may be located in the grille area. After the battery is disconnected, wait at least 2 minutes before beginning work (the system has a back-up capacitor that must fully discharge). For more information see Chapter 12.*

Warning 2: *The air conditioning system is under high pressure. Do not loosen any hose fittings or remove any components until after the system has been discharged. Air conditioning refrigerant should be properly discharged into an EPA-approved recovery/recycling unit at a dealer service department or an automotive air conditioning repair facility. Always wear eye protection when disconnecting air conditioning system fittings.*

Note: *Camshaft lobe lift and endplay should be checked before removal of the camshaft* (see Steps 1 through 10*).*

Camshaft lobe lift check

Refer to illustration 12.3

1 In order to determine the extent of cam lobe wear, the lobe lift should be checked prior to camshaft removal.

2 Remove the valve covers (see Section 4), the spark plugs (see Chapter 1) and the rocker arms (see Section 5).

3 Beginning with the number one cylinder, mount a dial indicator on the engine and position the plunger against the top surface of the push rod. Position the number one cylinder at TDC on the compression stroke (see Section 3). The plunger should be directly above and in line with the pushrod **(see illustration)**.

4 Zero the dial indicator, then very slowly turn the crankshaft in the normal direction of rotation (clockwise) until the indicator needle stops and begins to move in the opposite direction. The point at which it stops indicates maximum cam lobe lift.

5 Record this figure for future reference, then reposition the piston at TDC on the compression stroke again.

6 Move the dial indicator to the other number one cylinder pushrod and repeat the check. Be sure to record the results.

7 Repeat the check for the remaining camshaft lobes. Since each piston must be at TDC on the compression stroke for this procedure, work from cylinder-to-cylinder following the firing order sequence.

8 After the check is complete, compare the lobe lift results to the specifications in this Chapter. If the lobe lift is 0.003 inch less than specified, cam lobe wear has occurred and a new camshaft should be installed.

Camshaft endplay

9 Check the camshaft end play by placing a dial indicator with the stem in line with the camshaft and position the plunger against a flat surface on the upper timing chain gear. Push the camshaft all the way to the rear and zero the dial indicator. Next, pry the camshaft to the front as far as possible and check the reading on the dial indicator. The distance it moves is the endplay. If it's greater than the Specifications listed in Chapter, check the camshaft thrust plate for wear. If the thrust plate is worn it must be replaced.

Removal

Refer to illustrations 12.12, 12.13, 12.14, 12.16, 12.17 and 12.18

Note: *Have the air-conditioning system discharged by a dealer service department or automotive air conditioning repair facility before beginning the camshaft removal procedure.*

10 Refer to the appropriate Sections and remove the intake manifold, valve covers, rocker arms, pushrods and the timing chain.

11 Remove the radiator and air conditioning condenser (see Chapter 3).

12 Remove the distributor (see Chapter 5) and the distributor drive gear **(see illustration)**.

13 Before removing the lifters and guides note the marks showing which side of the lifter faces the lifter "valley." If they aren't marked, apply some paint dabs before removing the guides **(see illustration)**. **Caution:** *The lifters must be installed the same way to aim the oil feed holes properly.*

12.13 Note the paint marks (arrows) that indicate which side of the roller lifters face the valley - apply marks if none are visible

12.14 Remove the three bolts (arrows) and lift out the lifter guide retainer

14 Remove the lifter guide retainer **(see illustration)**.

15 There are several ways to extract the lifters from the bores. A special tool designed to grip and remove lifters is manufactured by many tool companies and is widely available, but it may not be required in every case. On newer engines without a lot of varnish buildup, the lifters can often be removed with a small magnet or even with your fingers. A machinist's scribe with a bent end can be used to pull the lifters out by positioning the point under the retainer ring inside the top of each lifter. **Caution:** *Do not use pliers to remove the lifters unless you intend to replace them with new ones. The pliers will damage the precision machined and hardened lifters, rendering them useless.*

12.16 Remove the lifter and the lifter guide (arrow) which keeps the roller lifters from rotating in their bores

16 Remove the lifter guides and lifters **(see illustration)**. Store the lifters in a clearly labeled box to ensure that they are reinstalled in their original locations.

17 Unbolt and remove the camshaft thrust plate and oil tab **(see illustration)**. Note how the oil tab is installed so you can return it to its original location on reassembly.

18 Thread a long bolt into the camshaft sprocket bolt hole to use as a handle when removing the camshaft from the block **(see illustration)**.

19 Carefully pull the camshaft out. Support the cam near the block so the lobes do not nick or gouge the bearings as it is withdrawn. Inspect the camshaft, bearings and lifters as described below.

Inspection

Refer to illustrations 12.21 and 12.24

20 After the camshaft has been removed from the engine, cleaned with solvent and dried, inspect the bearing journals for uneven wear, pitting and evidence of seizure. If the journals are damaged, the bearing inserts in the block are probably damaged as well. Both the camshaft and bearings will have to be replaced. **Note:** *Camshaft bearing replacement requires special tools and expertise that place it beyond the scope of this manual. If the bearings are bad, that the engine should be removed and the block taken to an automotive machine shop to ensure that the job is done correctly.*

21 Measure the bearing journals with a micrometer to determine if they are excessively worn or out-of-round **(see illustration)**.

12.17 Remove the camshaft thrust plate and oil tab

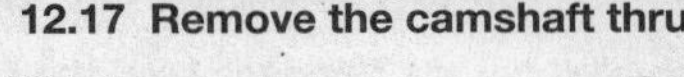
1 *Oil tab bolt*
2 *Thrust plate bolt*

12.18 Thread a long bolt into the camshaft sprocket bolt hole to use as a handle - as the camshaft is being removed, support it near the block so the lobes do not nick or gouge the bearings

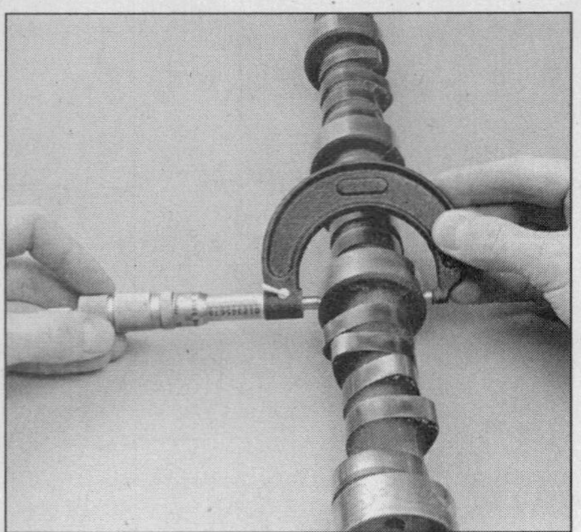

12.21 Check the diameter of each camshaft bearing journal to pinpoint excessive wear and out-of-round conditions

12.24 The roller on the roller lifters must turn freely - check for wear and excessive play as well

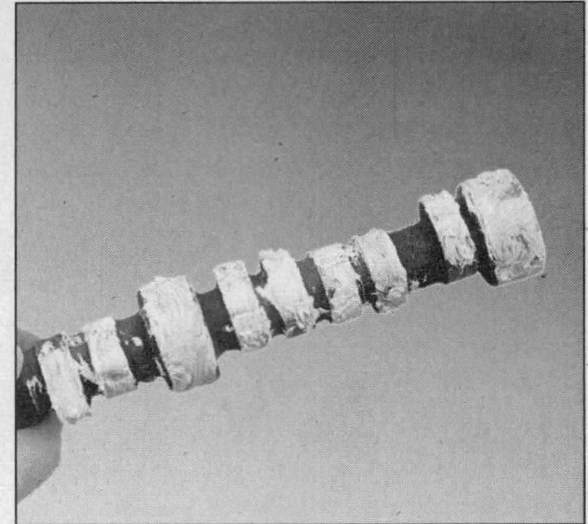

12.25 Apply engine assembly lube to the cam lobes and bearing journals before installing the camshaft

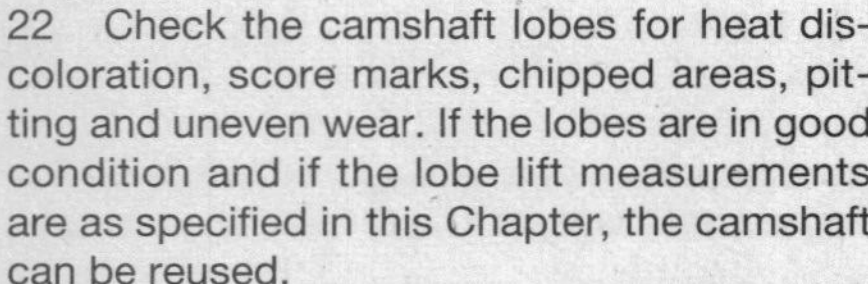

22 Check the camshaft lobes for heat discoloration, score marks, chipped areas, pitting and uneven wear. If the lobes are in good condition and if the lobe lift measurements are as specified in this Chapter, the camshaft can be reused.

23 With conventional, flat-tappet lifters, the camshaft must be replaced whenever lifters are replaced, and vice-versa, due to the wear pattern. However, used roller lifters in good shape can be transferred to a new cam if necessary, but should be installed in the same bores as they were originally.

24 After cleaning thoroughly, check the rollers carefully for wear and damage and make sure they turn freely without excessive play **(see illustration)**.

Installation

Refer to illustration 12.25

25 Lubricate the camshaft bearing journals and cam lobes with camshaft installation lube **(see illustration)**.

26 Slide the camshaft slowly and gently into the engine. Support the cam near the block and be careful not to scrape or nick the bearings. Only install the camshaft far enough to allow the installation of the camshaft thrust plate. Pushing it in too far could dislodge the camshaft plug at the rear of the engine, causing an oil leak.

27 Install with the camshaft thrust plate as it was originally **(see illustration 12.17)**.

28 Align the timing marks on the crankshaft and camshaft sprockets, install the timing chain and sprockets (see Section 11), then recheck the camshaft endplay as described in Step 9.

29 With the timing chain marks aligned, rotate the engine 360 degrees to bring the No.1 piston to TDC on the compression stroke. Install the distributor drive gear as **shown in illustration 12.12. Note:** *Removing the spark plugs will allow the engine to rotate easier if they're not already removed.*

30 Lubricate the lifters with clean engine oil and install them in the block. If the original lifters are being reinstalled, be sure to return them to their original locations, and with the paint marks facing the valley and the oil-feed holes on the side of the lifter body facing UP, away from the crankshaft. Install the lifter guides and the lifter guide retainer. **Note:** *The lifter guides must be installed with their arrows pointing toward the camshaft.*

31 Refer to the appropriate Sections and install the timing chain cover, vibration damper, pushrods and rocker arms.

32 The remaining installation steps are the reverse of removal.

33 Change the oil and install a new oil filter (see Chapter 1).

34 Start the engine, check for oil pressure and leaks. **Caution:** *Do not run the engine above a fast idle until all the hydraulic lifters have filled with oil and become quiet again.*

35 If a new camshaft and lifters have been installed, the engine should be brought to operating temperature and run at a fast idle for 15 to 20 minutes to "break in" the new components. Change the oil and filter again after 500 miles of operation.

13 Oil pan - removal and installation

Removal

Refer to illustrations 13.2, 13.3, 13.8, 13.9 and 13.11

1 Disconnect the cable from the negative battery terminal. Raise the vehicle and support the chassis securely on jackstands (see Chapter 1). Place the jackstands on the frame so the front axle hangs down to the limit of its travel.

13.2 Remove the front skid-plate - use a long extension through these holes to reach the rear bolts

13.3 The rubber splash-shield is easily removed by popping out the plastic retainers (arrow) with a screwdriver

13.8 Support the transmission or transfer case with a jack, unbolt the crossmember (arrow) and allow the exhaust system to drop down

13.9 Disconnect the oil-level sensor (A) and the nut retaining the transmission cooling lines (B)

2 Remove the bolts retaining the front skid-plate in place. The rear bolts are reached with a long extension through access holes **(see illustration)**.
3 Use a blunt screwdriver, pop the plastic retainers from the flexible splash-shield **(see illustration)**.
4 Drain the engine oil and remove the oil filter (see Chapter 1).
5 Remove the starter motor and lay it aside (see Chapter 5).
6 Disconnect the oxygen sensor from the exhaust pipe, remove the bolts/nuts and separate the exhaust pipe from both exhaust manifolds (see Section 8).
7 Remove the oil dipstick.
8 Place a jack under the transmission or transfer case to take the weight off the transmission crossmember. Unbolt the crossmember from the frame and from the transmission mount **(see illustration)**. This allows the complete exhaust system to drop down.
9 Disconnect the oil-level sensor and the bracket retaining the transmission cooling lines to the pan rail **(see illustration)**.
10 Remove all the oil pan bolts, then lower the pan from the engine. Strike the pan with a rubber mallet to break the gasket seal. **Caution:** *Before using force on the oil pan, be sure all the bolts have been removed*.
11 Twist the pan sideways and down to remove it to the rear **(see illustration)**.

Installation

Refer to illustration 13.16

12 Thoroughly clean the mounting surfaces of the oil pan and engine block of old gasket material and sealer.
13 If the oil pan is distorted at the bolt-hole areas, straighten the flange by supporting it from below on a 1x4 wood block and tapping the bolt holes with the rounded end of a ball-peen hammer. Wipe the gasket surfaces clean with a rag soaked in lacquer thinner or acetone.
14 Prepare four pan alignment dowels from 5/16-inch bolts 1-1/2-inches long. Cut off the bolt heads and slot the ends with a hacksaw.
15 Install the four alignment dowels into the frontmost and rearmost pairs of pan bolt holes in the block. Apply RTV sealant on both sides where the pan meets the rear main cap, and up front at the junction of the front cover and block.
16 Lift the pan into position with its gasket in place, slipping it over the alignment dowels and being careful not to disturb the gasket. Install several bolts finger tight **(see illustration)**.
17 Check that the gasket isn't sticking out anywhere around the block's perimeter. When all the bolts are in place, replace the alignment dowels with four pan bolts.
18 Starting at the center and alternating from side-to-side towards the ends, tighten the bolts to the torque listed in this Chapter's Specifications.
19 The remainder of the installation procedure is the reverse of removal.
20 Add the proper type and quantity of oil (see Chapter 1) and a new oil filter. Start the engine and check for leaks before placing the vehicle back in service.

13.11 Take your time and don't force anything - the oil pan can be removed from under the vehicle

13.16 Place the pan and gasket into position over the alignment studs (arrow) and install several pan bolts finger-tight

14.2 Remove the bolts (arrows) and lower the oil pump

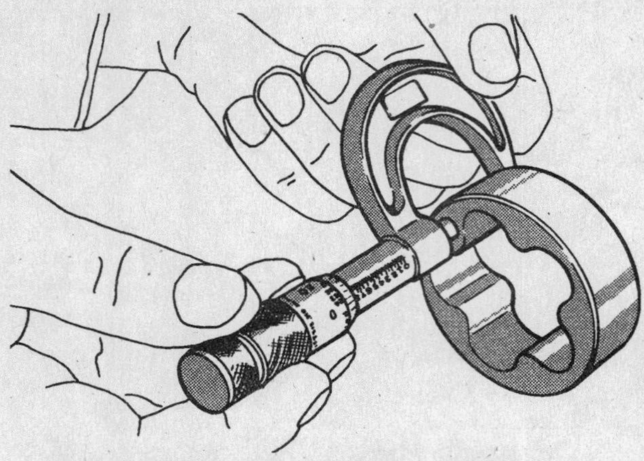
14.6 Measure the outer oil-pump rotor thickness and compare it to Specifications

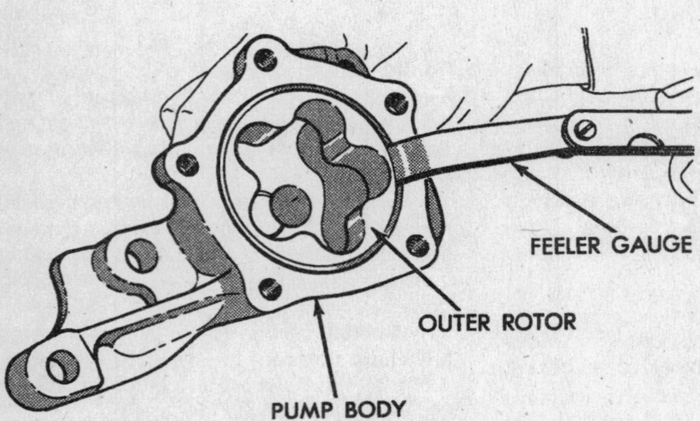

14.7a Measure the clearance between the outer rotor and the pump body

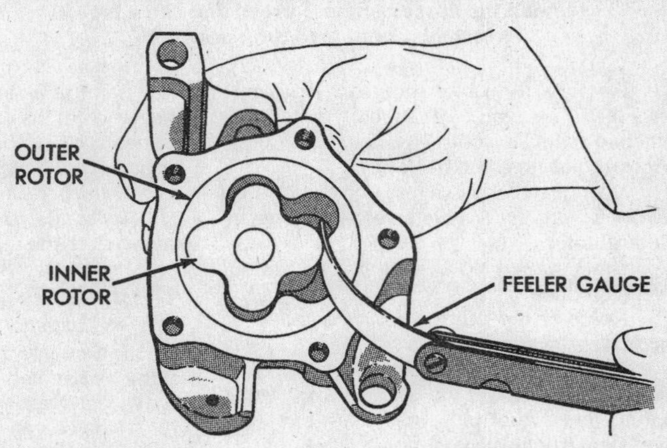

14.7b Measure the clearance between the rotor tips

14 Oil pump - removal, inspection and installation

Refer to illustration 14.2

Removal

1 Unbolt and lower the oil pan (see Section 13).

2 While supporting the oil pump, remove the pump-to-rear main bearing cap bolts **(see illustration)**.

3 Lower the pump and pickup screen assembly from the vehicle and clean the gasket surfaces.

Inspection

Refer to illustrations 14.6, 14.7a, 14.7b and 14.7c

4 Remove the oil pump cover.

5 Place a straightedge across the inner surface of the oil pump cover and try to insert a 0.0015-inch feeler gauge between the straightedge and pump cover. If the gauge fits, the oil pump assembly should be replaced.

6 Remove the outer rotor **(see illustration)** and measure its thickness with a micrometer. If it is less than the rotor thickness listed in this Chapter's Specifications, the pump assembly should be replaced. The inner rotor should meet this same minimum thickness.

7 With feeler gauges, measure the clearance between the outer rotor and the body, between the inner and outer rotor tips, and the clearance between the rotors and the cover **(see illustrations).** Compare these measurements to this Chapter's Specifications. If any clearance is greater than specified, replace the pump assembly.

Installation

Refer to illustration 14.10

8 Thread the oil pickup tube and screen into the oil pump and tighten it so that the pick-up screen is parallel with the bottom of the oil pan. **Caution:** *Be absolutely certain that the pickup screen is properly tightened so that no air can be sucked into the oiling system at this connection.*

9 Prime the pump by pouring clean engine oil into the pickup tube.

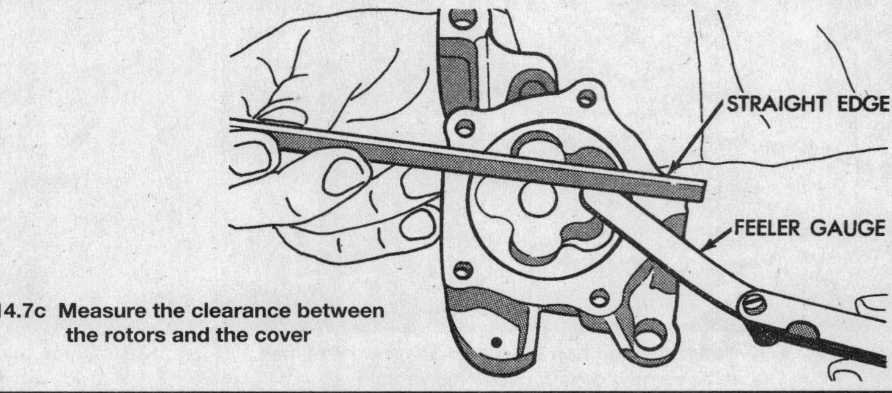

14.7c Measure the clearance between the rotors and the cover

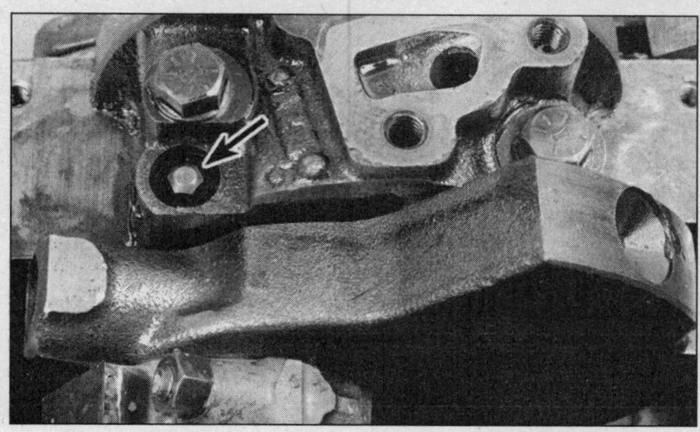
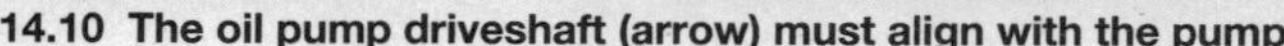

14.10 The oil pump driveshaft (arrow) must align with the pump

16.3 Remove the two bolts (arrows) and the rear main cap

10 Position the pump on the engine with a new gasket and make sure the oil pump driveshaft is aligned with the oil pump **(see illustration)**.
11 Install the mounting bolts and tighten them to the torque listed in this Chapter's Specifications.
12 Install the oil pan and add oil.
13 Run the engine and check for oil pressure and leaks.

15 Flywheel/driveplate - removal and installation

Refer to Chapter 2, Part A for this procedure, but be sure to use the torque value listed in this Chapter's Specifications.

16 Rear main oil seal - replacement

Refer to illustrations 16.3 and 16.10
1 Remove the oil pan (see Section 13).
2 Remove the oil pump (see Section 14).
3 Remove the bolts and detach the rear main bearing cap from the engine **(see illustration)**.
4 Remove the lower half of the oil seal from the bearing cap or retainer.
5 Loosen the main cap bolts on the number 2 and number 3 main caps.
6 Use a small brass punch to carefully press on one end of the upper half of the existing seal until the other side protrudes, then pull the seal out from the other side with needle-nose pliers. Clean the bearing cap and engine block surfaces carefully to degrease them and remove any sealant.
7 Lightly oil the lips of the new (neoprene rubber) crankshaft seals. **Caution:** *Always wipe the crankshaft surface clean, then oil it lightly before installing a new seal.*
8 Rotate a new seal half into cylinder block with the seal lip facing toward the front of the engine (white or yellow paint mark facing the rear of the engine). **Caution:** *Hold your thumb firmly against the outside diameter of the seal as you're rotating it into place. This will prevent the seal outside diameter from being shaved from contact with the sharp edge of the engine block. If the seal gets shaved, you may wind up with an oil leak.*
9 Place the other seal half in the bearing cap with the paint mark toward the rear of the engine.
10 Apply a 0.20-inch drop of Loctite 515, or equivalent sealant on both sides of the rear main cap as indicated **(see illustration)**.
11 Install the rear main bearing cap with cleaned and oiled bolts and tighten them to the torque listed in this Chapter's Specifications. Tighten the remainder of the main bearing caps to the torque listed in this Chapter's Specifications. Apply RTV sealant at the junction of the cap and block **(see illustration 16.10)**.
12 Install the oil pump and oil pan.
13 The remainder of installation is the reverse of removal. Fill the pan with oil, run the engine and check for leaks.

17 Engine mounts - check and replacement

Refer to illustrations 17.1

Refer to Chapter 2, Part A for this procedure, but use the **accompanying illustration** for V8 engine mount locations and descriptions.

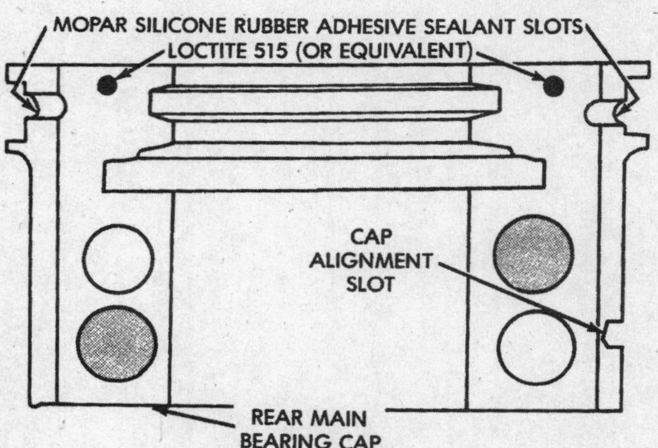

16.10 Apply Loctite 515, or equivalent, to either side of the cap as shown, and RTV sealant at the block/cap junction after the cap is installed

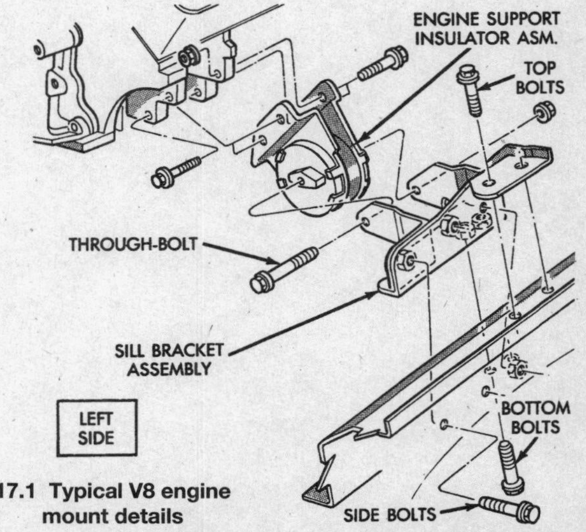

17.1 Typical V8 engine mount details

Notes

Chapter 2 Part C
4.7L V8 engine

Contents

Specifications

General

Displacement	287 cubic inches
Bore and stroke	3.66 x 3.40 inches
Cylinder numbers (front to rear)	
Left bank	1-3-5-7
Right bank	2-4-6-8
Firing order	1-8-4-3-6-5-7-2

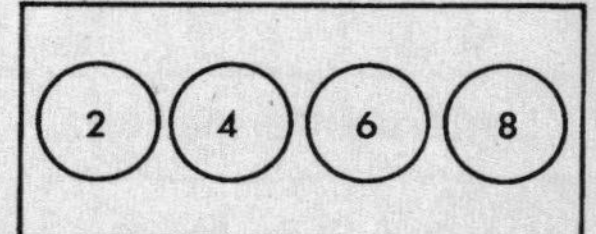

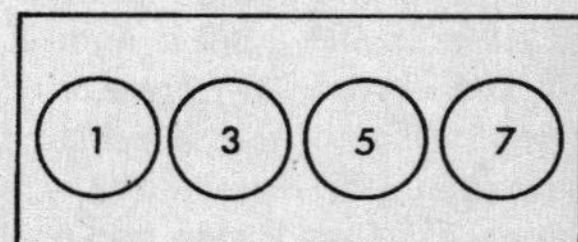

Cylinder identification diagram

Camshaft

Endplay	0.003 to 0.0079 inch
Valve lift	
Intake	0.443 inch
Exhaust	0.4292 inch
Camshaft bearing oil clearance	
Standard	0.001 to 0.0026 inch
Service limit	0.0026 inch
Camshaft journal diameter	1.0227 to 1.0235 inch
Camshaft bore diameter	1.0245 to 1.0252 inch

Timing chain

Idler gear endplay	0.004 to 0.010 inch

Oil pump

Cover warpage limit	0.001 inch
Inner and outer rotor thickness	0.4756 inch
Outer rotor diameter (minimum)	3.3843 inches
Outer rotor-to-housing clearance (maximum)	0.0186 inch
Inner rotor-to-outer rotor lobe clearance	0.006 inch
Oil pump housing-to-rotor side clearance	0.0014 to 0.0038 inch

Torque specifications

	Ft-lbs (unless otherwise indicated)
Camshaft sprocket bolts (non-oiled)	90
Camshaft bearing cap bolts	100 in-lbs
Crankshaft pulley/vibration damper bolt	130
Cylinder head bolts	
Step one	
11 mm bolts (1 through 10)	15
Step two	
11 mm bolts (1 through 10)	35
Step three	
8 mm bolts (11 through 14)	18
Step four	
11 mm bolts (1 through 10)	Turn an additional 90 degrees
Step five	
8 mm bolts (11 through 14)	22
Driveplate bolts	45
Engine mount bracket-to-block bolts	45
Engine mount through bolt/nut	45
Exhaust manifold bolts	18
Exhaust manifold heat shield nuts	
Step 1	72 in-lbs
Step 2	Loosen an additional 45 degrees
Intake manifold bolts	105 in-lbs
Oil pan bolts	132 in-lbs
Oil pan drain plug	25
Oil pick-up tube mounting bolt/nut	21
Oil pump mounting bolts	21
Oil pump cover screws	105 in-lbs
Timing chain cover bolts	40
Timing chain guide bolts	21
Timing chain guide access plugs	15
Timing chain idler sprocket bolt	25
Timing chain tensioner arm pivot bolt	150 in-lbs
Timing chain tensioner (secondary)	21
Timing chain tensioner (primary)	21
Transmission-to-oil pan support brace	40
Valve cover bolts	105 in-lbs
Water outlet housing	105 in-lbs

1 General information

This Part of Chapter 2 is devoted to in-vehicle repair procedures for the 4.7L V8 engine. These engines utilize a cast iron engine block with eight cylinders arranged in a "V" shape at a 90-degree angle between the two banks. The overhead camshaft aluminum cylinder heads are equipped with replaceable valve guides and seats. Stamped steel rocker arms with an integral roller bearing actuate the valves.

All information concerning engine removal and installation and engine block and cylinder head overhaul can be found in Part D of this Chapter.

The following repair procedures are based on the assumption that the engine is installed in the vehicle. If the engine has been removed from the vehicle and mounted on a stand, many of the steps outlined in this Part of Chapter 2 will not apply.

The Specifications included in this Part of Chapter 2 apply only to the procedures contained in this Part. Part D of Chapter 2 contains the Specifications necessary for cylinder head and engine block rebuilding.

2 Repair operations possible with the engine in the vehicle

Many major repair operations can be accomplished without removing the engine from the vehicle.

Clean the engine compartment and the exterior of the engine with some type of degreaser before any work is done. It will make the job easier and help keep dirt out of the internal areas of the engine.

Depending on the components involved,

3.5 A compression gauge can be used in the number one plug hole to assist in finding TDC

3.8 Align the groove in the damper with the TDC mark on the timing chain cover

it may be helpful to remove the hood to improve access to the engine as repairs are performed (refer to Chapter 11, if necessary). Cover the fenders to prevent damage to the paint. Special pads are available, but an old bedspread or blanket will also work.

If vacuum, exhaust, oil or coolant leaks develop, indicating a need for gasket or seal replacement, the repairs can generally be made with the engine in the vehicle. The intake and exhaust manifold gaskets, oil pan gasket, crankshaft oil seals and cylinder head gaskets are all accessible with the engine in place.

Exterior engine components, such as the intake and exhaust manifolds, the oil pan, the oil pump, the water pump (see Chapter 3), the starter motor, the alternator and the fuel system components (see Chapter 4) can be removed for repair with the engine in place.

Since the cylinder heads can be removed without pulling the engine, valve component servicing can also be accomplished with the engine in the vehicle. Replacement of the camshafts, timing chains and sprockets are also possible with the engine in the vehicle.

In extreme cases caused by a lack of necessary equipment, repair or replacement of piston rings, pistons, connecting rods and rod bearings is possible with the engine in the vehicle. However, this practice is not recommended because of the cleaning and preparation work that must be done to the components involved.

3 Top Dead Center (TDC) for number one piston - locating

Refer to illustrations 3.5 and 3.8

1 Top Dead Center (TDC) is the highest point in the cylinder that each piston reaches as it travels up the cylinder bore. Each piston reaches TDC on the compression stroke and again on the exhaust stroke, but TDC generally refers to piston position on the compression stroke.

2 Positioning the piston(s) at TDC is an essential part of many procedures such as valve timing, camshaft and timing chain/sprocket removal.

3 Before beginning this procedure, be sure to place the transmission in Neutral and apply the parking brake or block the rear wheels. Also, disable the ignition system by disconnecting the primary electrical connectors at the ignition coil packs and remove the spark plugs (see Chapter 1).

4 In order to bring any piston to TDC, the crankshaft must be turned using one of the methods outlined below. When looking at the front of the engine, normal crankshaft rotation is clockwise.

a) The preferred method is to turn the crankshaft with a socket and ratchet attached to the bolt threaded into the front of the crankshaft. Apply pressure on the bolt in a clockwise direction only. Never turn the bolt counterclockwise.

b) A remote starter switch, which may save some time, can also be used. Follow the instructions included with the switch. Once the piston is close to TDC, use a socket and ratchet as described in the previous paragraph.

c) If an assistant is available to turn the ignition switch to the Start position in short bursts, you can get the piston close to TDC without a remote starter switch. Make sure your assistant is out of the vehicle, away from the ignition switch, then use a socket and ratchet as described in Paragraph (a) to complete the procedure.

5 Install a compression pressure gauge in the number one spark plug hole (refer to Chapter 2D). It should be a gauge with a screw-in fitting and a hose at least six inches long **(see illustration)**.

6 Rotate the crankshaft using one of the methods described above while observing for pressure on the compression gauge. The moment the gauge shows pressure indicates that the number one cylinder has begun the compression stroke.

7 Once the compression stroke has begun, TDC for the compression stroke is reached by bringing the piston to the top of the cylinder.

8 Continue turning the crankshaft until the notch in the crankshaft damper is aligned with the TDC mark on the timing chain cover **(see illustration)**. At this point, the number one cylinder is at TDC on the compression stroke. If the marks are aligned but there was no compression, the piston was on the exhaust stroke. Continue rotating the crankshaft 360-degrees (1-turn). **Note:** *If a compression gauge is not available, you can simply place a blunt object over the spark plug hole and listen for compression as the engine is rotated. Once compression at the No.1 spark plug hole is noted the remainder of the Step is the same.*

9 After the number one piston has been positioned at TDC on the compression stroke, TDC for any of the remaining cylinders can be located by turning the crankshaft 90 degrees and following the firing order (refer to the Specifications). Rotating the engine 90 degrees past TDC #1 will put the engine at TDC compression for cylinder #8.

4 Valve cover - removal and installation

Refer to illustrations 4.4, 4.5, 4.9, 4.10, 4.11 and 4.12

Removal

1 Disconnect the cable from the negative terminal of the battery.

2 Remove the air intake duct and the throttle body resonator (see Chapter 4).

3 Detach the electrical connector(s) from the ignition coils and the fuel injectors on the side from which the valve cover is to be

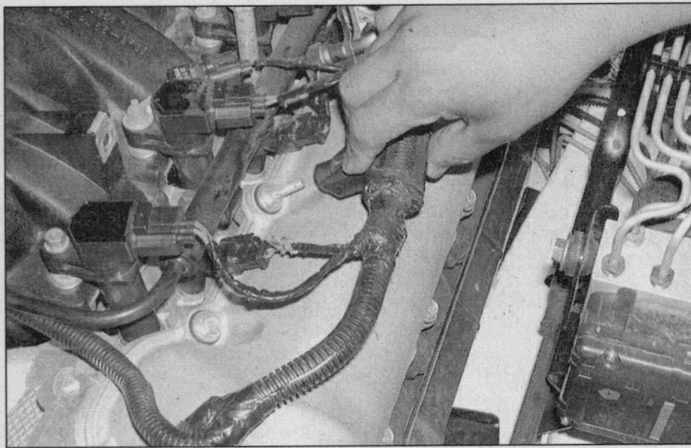

4.4 Detach the wiring harness from the valve cover studs and position it aside

4.5 Left valve cover mounting details - arrows indicate stud locations

4.9 When working on the right valve cover, it will be necessary to position aside the accumulator (A), the heater hoses and bracket (B) and the fuel injection wiring harness (C)

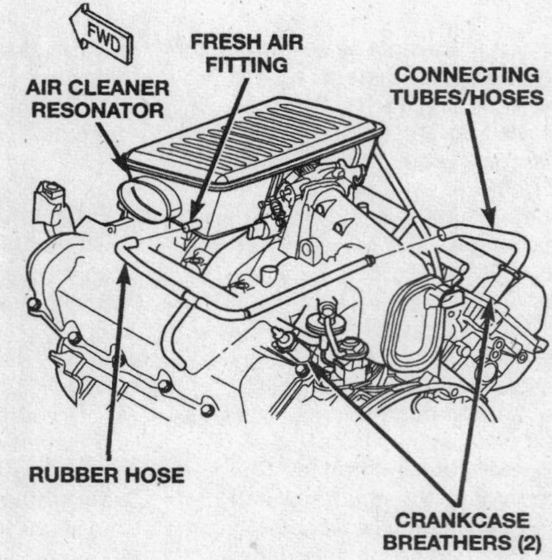

4.10 Location of the crankcase breather valves

removed. If both valve covers are to be removed, disconnect all of the connectors from the ignition coils and the fuel injectors.

Left valve cover

4 Unclip the fuel injector wiring harness from the studs on the valve cover and position it aside **(see illustration)**.

5 Remove the valve cover studs and bolts **(see illustration)**. Make a note of the stud locations before removal to ensure correct positioning during installation.

6 Detach the valve cover. **Note:** *If the cover sticks to the cylinder head, use a block of wood and a hammer to dislodge it. If the cover still won't come loose, pry on it carefully, but don't distort the sealing flange.*

Right valve cover

7 Detach the power distribution center from the battery tray and position it aside. Make sure to disconnect the cable from the positive terminal of the battery before doing so, then remove the battery tray from the engine compartment (see Chapter 5).

8 Partially drain the cooling system until the level is below heater hoses (see Chapter 1 if necessary), then remove the heater hoses from the engine and the retaining clips and position them aside.

9 Remove the engine drivebelt (see Chapter 1), then unbolt the air conditioning compressor from the engine and position it aside without disconnecting the refrigerant lines. Also loosen the air conditioning accumulator bracket **(see illustration)**.

10 Detach the breather hose and filter assembly from the rear of the right cylinder head **(see illustration)**.

11 Remove the PCV hose from the oil fill tube, then remove the oil fill tube with the PCV valve attached from the cylinder head **(see illustration)**.

12 Remove the valve cover studs/nuts and bolts **(see illustration)**. Make a note of the stud locations before removal to ensure correct positioning during installation.

13 Detach the valve cover. **Note:** *If the cover sticks to the cylinder head, use a block of wood and a hammer to dislodge it. If the cover still won't come loose, pry on it carefully, but don't distort the sealing flange.*

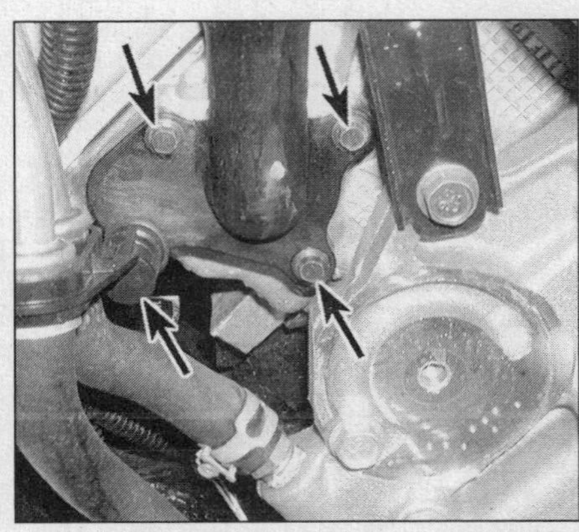

4.11 Oil filler tube mounting bolts/nut (arrows) - detach the heater hose bracket from the stud to access the nut

4.12 Right valve cover mounting details - arrows indicate stud locations

5.5 Using a special type valve spring compressor, depress the valve spring just enough to remove the rocker arm

Installation

14 The mating surfaces of each cylinder head and valve cover must be perfectly clean when the covers are installed. Use a gasket scraper to remove all traces of sealant or old gasket material, then clean the mating surfaces with lacquer thinner or acetone (if there's sealant or oil on the mating surfaces when the cover is installed, oil leaks may develop). Be extra careful not to nick or gouge the mating surfaces with the scraper. **Caution:** *Do not use harsh cleaners when cleaning the valve covers or damage to the covers may occur.*

15 Clean the mounting bolt or stud threads with a wire brush if necessary to remove any corrosion and restore damaged threads. Use a tap to clean the threaded holes in the heads.

16 Place the valve cover and new gasket in position, then install the stud/nuts and bolts in the correct location from which they where removed. Tighten the bolts in several steps to the torque listed in this Chapter's Specifications.

17 Complete the installation by reversing the removal procedure. Start the engine and check carefully for oil leaks.

5 Rocker arms and hydraulic lash adjusters - removal, inspection and installation

Refer to illustrations 5.5, 5.7 and 5.8

Note 1: *A special valve spring compressor available from most aftermarket specialty tool manufacturers will be required for this procedure. The only other alternative to accomplishing this task without the use of this special tool is to remove the timing chains and the camshafts, which requires major disassembly of the engine and surrounding components.*

Note 2: *This engine is a non-freewheeling engine and the pistons must be down in the cylinder bore before the valve and spring assembly can be compressed to allow rocker arm removal.*

1 Before beginning this procedure, be sure to place the transmission in Park and apply the parking brake or block the rear wheels. Also, disable the ignition system by disconnecting the primary electrical connectors at the ignition coils and remove the spark plugs (see Chapter 1).

2 Remove the valve cover(s) (see Section 4).

3 Before the rocker arms and lash adjusters are removed, arrange to label and store them, so they can be kept separate and reinstalled on the same valve they were removed from.

4 Rotate the engine with a socket and ratchet in a clockwise direction only by the crankshaft pulley/vibration damper bolt until the piston(s) are positioned correctly to remove the rocker arms from the corresponding cylinders as follows:

a) *Remove the rocker arms from cylinders* **No. 2 and 8** *with the No.1 piston at TDC on the compression stroke* (see Section 3).
b) *Remove the rocker arms from cylinders* **No. 3 and 5** *with the No. 1 piston at TDC on the exhaust stroke. Rotate the crankshaft exactly 360 degrees (1-turn) from TDC on the compression stroke to bring the No.1 piston to TDC on the exhaust stroke. This is also verified by observing that the "V8" marks on the camshaft sprocket are pointing straight upward.*
c) *Remove the rocker arms from cylinders* **No. 4 and 6** *with the No. 3 piston at TDC on the compression stroke.*
d) *Remove the rocker arms from cylinders* **No. 1 and 7** *with the No. 2 piston at TDC on the compression stroke.*

5 Hook the valve spring compressor around the base of the camshaft. Depress the valve spring just enough to release tension on the rocker arm to be removed. Once tension on the rocker arm is relieved, the rocker arm can be removed by simply pulling it out **(see illustration)**.

6 If you're removing or replacing only a few of the rockers arms or lash adjusters, locate the cylinder number of the rocker arm or lash adjuster you wish to remove in Step 4, then rotate the crankshaft to the corresponding position. Remember to keep the rocker arm and lash adjuster for each valve together so they can be reinstalled in the same locations. Refer to Section 3 as necessary to help position the designated cylinder at TDC.

5.7 Pull the lash adjuster up and out of its bore to remove it from the cylinder head

7 Once the rocker arms are removed, the lash adjusters can be pulled out of the cylinder head and stored with the corresponding rocker arm **(see illustration)**.

8 Inspect each rocker arm for wear,

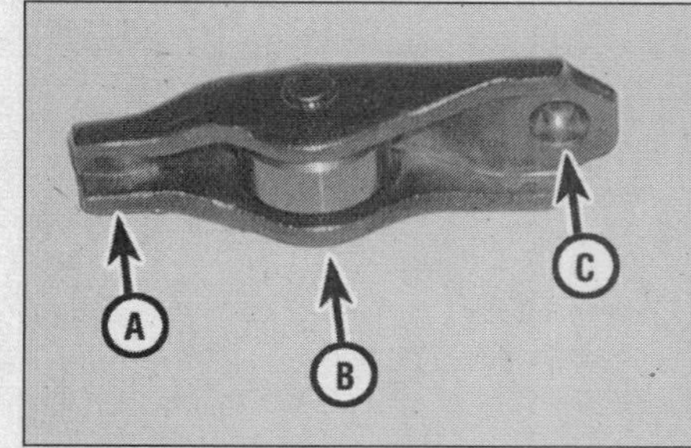

5.8 Inspect the rocker arms at the following locations

A Valve stem seat
B Roller
C Lash adjuster pocket

cracks and other damage. Make sure the rollers turn freely and show no signs of wear, also check the pivot area for wear, cracks and galling **(see illustration)**.

9 Inspect the lash adjuster contact surfaces wear or damage. Make sure the lash adjusters move up and down freely in their bores on the cylinder head without excessive side to side play.

10 Installation is the reverse of removal with the following exceptions: Always install the lash adjuster first and make sure they're at least partially full of oil before installation. This is indicated by little or no lash adjuster plunger travel.

6 Valve springs, retainers and seals - replacement

Refer to illustrations 6.5, 6.7a, 6.7b, 6.8, 6.13 and 6.15

Note: *Broken valve springs and defective valve stem seals can be replaced without removing the cylinder heads. Two special tools and a compressed air source are normally required to perform this operation, so read through this Section carefully. The special-type of valve spring compressor required for rocker arm and valve spring removal may not be available at all tool rental yards, so check on the availability before beginning the job.*

1 Remove the valve cover(s) (see Section 4).

2 Refer to Section 5 and remove the rocker arms from the affected cylinder head.

3 Remove the spark plug (see Chapter 1) from the cylinder that has the defective component. If all of the valve stem seals are being replaced, all of the spark plugs and rockers arms should be removed.

4 Turn the crankshaft until the piston in the affected cylinder is at Top Dead Center on the compression stroke (refer to Section 3). If you're replacing all of the valve stem seals, begin with cylinder number one and work on the valves for one cylinder at a time. Move from cylinder-to-cylinder following the firing order sequence (see this Chapter's Specifications).

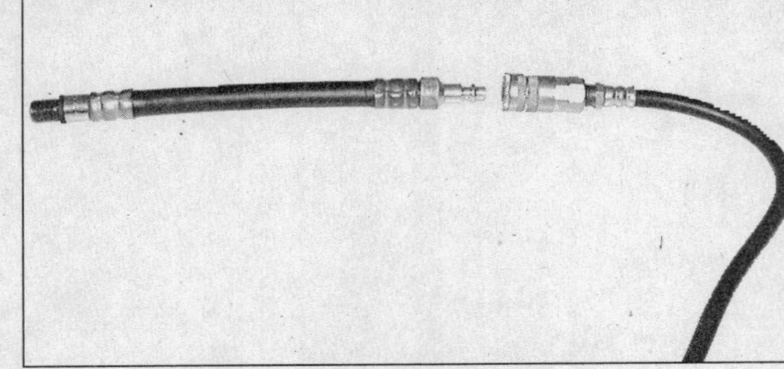

6.5 You'll need an air hose adapter this long to reach down into the spark plug tubes - they're commonly available from auto parts stores

5 Thread a long adapter into the spark plug hole and connect an air hose from a compressed air source to it **(see illustration)**. Most auto parts stores can supply the air hose adapter. **Note:** *Because of the length of the spark plug wells, it will be necessary to use a long spark plug adapter with a length of hose attached (as used on many cylinder compression gauges) utilizing a quick-disconnect fitting to hook to your air source.*

6 Apply 90 to 100 psi of compressed air to the cylinder. **Warning:** *The piston may be forced down by the compressed air, causing the crankshaft to turn suddenly. If the wrench used when positioning the number one piston at TDC is still attached to the bolt in the crankshaft nose, it could cause damage or injury when the crankshaft moves.*

7 Stuff shop rags into the cylinder head holes around the valves to prevent parts and tools from falling into the engine, then use a valve spring compressor to compress the spring. Remove the valve stem locks with small needle-nose pliers or a magnet **(see illustrations)**. **Note:** *The valves should be held in place by the air pressure. If the valve faces or seats are in poor condition, leaks may prevent air pressure from retaining the valves. If the valves cannot hold air, the cylinder head should be removed and taken to a machine shop for a valve job.*

8 Remove the spring retainer and valve spring, then remove the valve stem seal **(see illustration)**.

9 Wrap a rubber band or tape around the top of the valve stem so the valve won't fall into the combustion chamber, then release the air pressure.

10 Inspect the valve stem for damage. Rotate the valve in the guide and check the end for eccentric movement, which would indicate that the valve is bent.

11 Move the valve up-and-down in the guide and make sure it doesn't bind. If the valve stem binds, either the valve is bent or the guide is damaged. In either case, the head will have to be removed for repair.

12 Reapply air pressure to the cylinder to retain the valve in the closed position, then remove the tape or rubber band from the valve stem.

13 Lubricate the valve stems with engine oil and install the valve spring seat/valve seal assembly over the top of the valves stems. Using the stem of the valves as a guide, slide the seals down to the top of each valve guide. Using a hammer and a deep socket or seal installation tool, gently tap each seal into place until it's completely seated on the guide **(see illustration)**. Don't twist or cock the seals during installation or they won't seal properly on the valve stems. Make sure the garter spring (if equipped) is still in place around the top of the seal.

14 Install the spring and retainer in position over the valve with the colored mark (if

6.7a Compress the valve spring enough to release the valve stem locks . . .

6.7b . . . and lift them out with a magnet or needle-nose pliers

6.8 A pair of pliers will be required to remove the valve stem seal from the valve guide

6.13 Using a deep socket and hammer, gently tap the new seals onto the valve guide only until seated

equipped) on the spring facing UP. Compress the valve spring assembly only enough to install the keepers in the valve stem.
15 Position the keepers in the valve stem groove. Apply a small dab of grease to the inside of each keeper to hold it in place if necessary **(see illustration)**. Remove the pressure from the spring tool and make sure the keepers are seated.
16 Disconnect the air hose and remove the adapter from the spark plug hole.
17 Repeat the above procedure on the remaining cylinders, following the firing order sequence (see the Specifications). Bring each piston to top dead center on the compression stroke before applying air pressure.
18 Refer to Section 5 and install the rocker arms.
19 Refer to Section 4 and install the valve covers.
20 Install the spark plug(s) and the ignition coils referring to the appropriate sections as necessary.
21 Start and run the engine, then check for oil leaks and unusual sounds coming from the valve cover area.

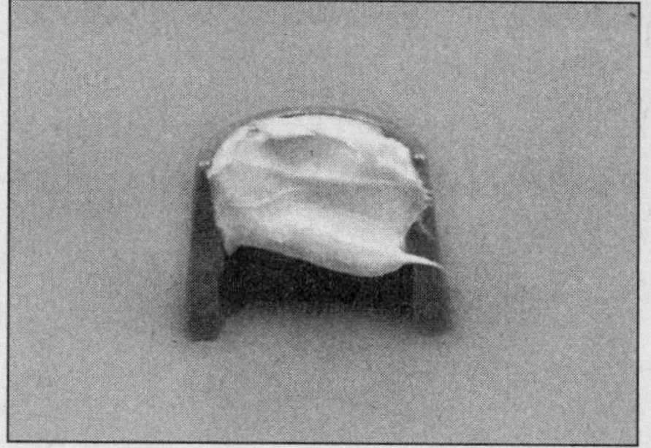

6.15 Apply a small dab of grease to each valve stem lock as shown here before installation - it will hold them in place on the valve stem as the spring is released

7 Timing chain and sprockets - removal, inspection and installation

Note 1: *Special tools are necessary to complete this procedure. Read through the entire procedure and obtain the special tools before beginning work.*
Note 2: *The 4.7L engine utilizes three timing chains to produce proper valve timing. The primary timing chain runs around the crankshaft sprocket and the idler gear sprocket. This chain synchronizes the crankshaft and pistons with the idler gear, while two secondary timing chains run around the rear of the idler gear and up to the camshaft sprockets to synchronize the valve timing with the crankshaft. Refer to Step 19 for in-vehicle timing chain inspection procedures and to Chapter 2D for idler gear endplay check prior to timing chain removal.*

Removal

Refer to illustrations 7.7, 7.11, 7.13, 7.14, 7.15, 7.16a, 7.16b, 7.17 and 7.18
1 Disconnect the cable from the negative terminal of the battery.
2 Drain the cooling system (see Chapter 1).

7.7 Drivebelt tensioner retaining bolt (arrow)

3 Refer to Chapter 3 and remove the cooling fan, the accessory drivebelt and the fan shroud from the engine compartment.
4 Detach the heater hoses and the lower radiator hose from the timing chain cover and position them aside.
5 Unbolt the power steering pump and set it aside without disconnecting the fluid lines (see Chapter 10).
6 Remove the alternator (see Chapter 5). Also unbolt the air conditioning compressor (if equipped) and position it aside without disconnecting the refrigerant lines.
7 Remove the accessory drivebelt tensioner from the timing chain cover **(see illustration)**.
8 Remove the valve covers (see Section 4) and the spark plugs (see Chapter 1).
9 Remove all of the rocker arms following the procedure outlined in Section 5. **Note:** *This step is not absolutely necessary, but it will help make alignment of the camshaft sprockets easier upon installation and also eliminate any possibility of the pistons contacting the valves during this procedure, since the 4.7L V8 is an interference engine.*
10 Remove the camshaft position sensor from the right cylinder head (see Chapter 6).
11 Position the number one piston at TDC on the exhaust stroke (see Section 3). Visu-

2C

7.11 When the engine is at TDC on the exhaust stroke, the crankshaft pulley/vibration damper is aligned with the mark on the timing chain cover and the "V8" marks on the camshaft sprockets should be pointing to the 12 o'clock position

7.13 Timing chain cover retaining bolts

7.14 Locking the primary timing chain tensioner in the retracted position

7.15 Secondary timing chain tensioner mounting bolts (arrows)

7.16a Removing the left camshaft sprocket bolt while holding the sprocket with a pin spanner wrench - note the chain guide access plug

7.16b The right camshaft sprocket is identified by the camshaft position sensor ring (arrow) which is fastened to the rear of the sprocket - be extremely careful not to damage or place a magnetic object of any kind near the camshaft position sensor ring or a no start condition may occur after installation

ally confirm the engine is at TDC on the exhaust stroke, by verifying that the timing mark on the crankshaft pulley/vibration damper is aligned with the mark on the timing chain cover and the "V8" marks on the camshaft sprockets are pointing straight up in the 12 o'clock position **(see illustration 3.8 and the accompanying illustration)**.

12 Remove the crankshaft pulley (see Section 12).

13 Remove the timing chain cover and the water pump as an assembly **(see illustration)**. Note that various types and sizes of bolts are used. They must be reinstalled in their original locations. Mark each bolt or make a sketch to help remember where they go.

14 Cover the oil pan opening with shop rags to prevent any components from falling into the engine. Collapse the primary timing chain tensioner with a pair of locking pliers and install a locking pin into the holes in the tensioner body to keep it in the retracted position **(see illustration)**.

15 Remove the secondary timing chain tensioners **(see illustration)**.

16 Remove the camshaft sprocket retaining bolts **(see illustrations)**. Pull the camshaft sprockets off the camshaft hubs one at a time. Lower the sprocket(s) into the cylinder head opening until the chain can be displaced from around the sprocket, then remove the camshaft sprockets from the engine and let the secondary chains fall down between the timing chain guides. **Warning 1:** *If the rocker arms* **were not** *removed as suggested in Step 9, it will be necessary to hold the camshafts from rotating with a set of locking pliers while the sprocket is being removed. Work on one camshaft and sprocket at a time starting with the left sprocket and proceeding to the right sprocket. After the sprocket is removed from the camshaft(s), let the camshaft slowly rotate to its neutral position. This is typically 15 degrees clockwise on the left camshaft sprocket and 45 degrees counterclockwise on the right camshaft. Pressure from the valve springs will make the camshafts rotate as the sprockets are removed. Sudden movement of the camshafts may allow the valves to strike the pistons.* **Warning 2:** *Never install the locking pliers on a camshaft lobe as damage the camshaft will occur. When using locking pliers, always rotate the camshaft with locking pliers by the base shaft.* **Warning 3:** *Do not rotate the crankshaft or camshafts separately after the secondary timing chains are loosened or removed with the rocker arms installed in the engine as piston or valve damage may occur. The only exception to this rule is when the camshafts must be rotated slightly, to realign the camshaft sprockets with the camshafts during installation.* **Caution:** *The right camshaft sprocket is identified by the camshaft position sensor ring which is fas-*

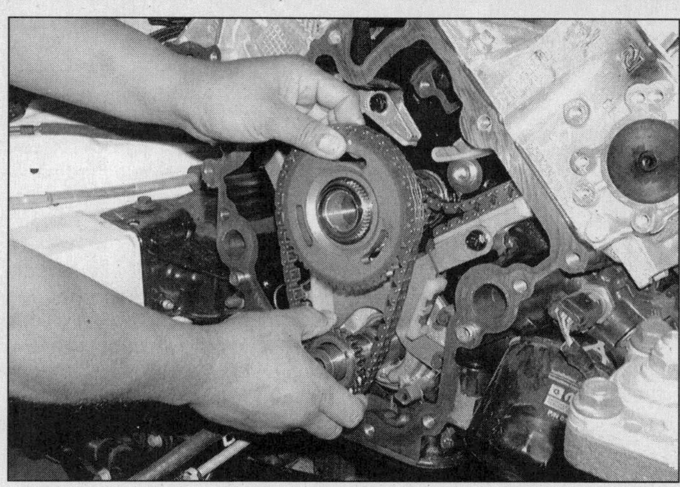

7.17 Remove the primary timing chain and the secondary chains as an assembly from the engine

7.18 Timing chain guide and tensioner arm pivot bolts (arrows)

tened to the rear of the sprocket. Be extremely careful not to damage or place a magnetic object of any kind near the camshaft position sensor ring or a no start condition may occur after installation.

17 Remove the idler sprocket bolt, then detach the idler sprocket, the crankshaft sprocket, the primary timing chain and the secondary timing chains as an assembly **(see illustration)**.

18 Remove the cylinder head access plugs **(see illustrations 7.16a and 7.16b)**. Also remove the oil fill tube (if not already removed) from the front of the right cylinder head. Detach the timing chain guides and tensioner arms **(see illustration)**.

Inspection

Refer to illustrations 7.19a and 7.19b

19 Inspect the camshaft and crankshaft sprockets for wear on the teeth and keyways. Inspect the chains for cracks or excessive wear of the rollers. Inspect the facing of the chain guides and tensioner arm for excessive wear. If any of the components show signs of excessive wear or the chain guides are grooved in excess of 0.039 inch deep they must be replaced. If any of the timing chain guides are excessively grooved or melted, the tensioner lube jet may be clogged. Be sure to remove the jet and blow compressed air through it to remove any debris or foreign material **(see illustration)**. Also inspect the idler sprocket bushing and spline joint for wear (see Chapter 2D). **Note:** *Secondary timing chain stretch can be checked by removing the timing chain cover and the water pump assembly, then rotate the engine clockwise until the pistons in the secondary tensioners reach their maximum travel or extension. Using a machinist's ruler or a dial caliper, measure the piston protrusion or extension from the stepped ledge on the piston to the tensioner housing on each tensioner* **(see illustration)**. *If the maximum extension of either tensioner piston exceeds 0.590 inch, the secondary timing chains are worn beyond their limits and should be replaced.*

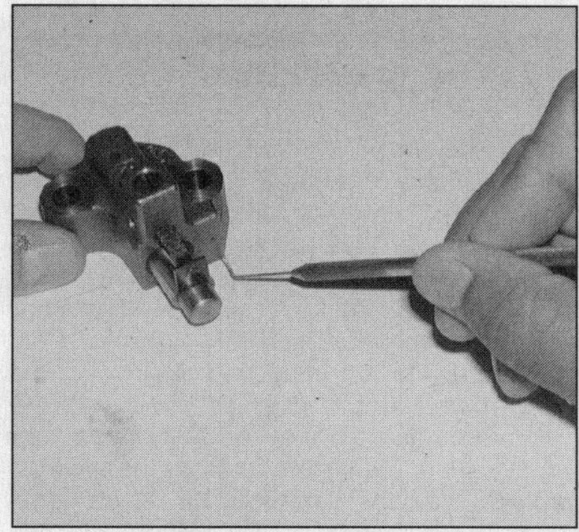

7.19a If excessive wear on the chain guides is evident, check the oil jet on the side of each secondary tensioner for a clogging

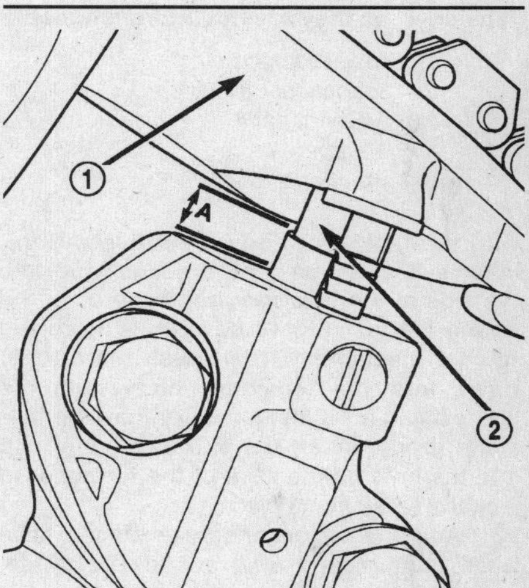

7.19b Measure the piston protrusion of the secondary tensioners to check the stretch of the secondary timing chains

1 *Secondary tensioner arm*
2 *Secondary tensioner piston*

Installation

Refer to illustrations 7.21, 7.22, 7.25, 7.26, 7.28, 7.33, 7.41a and 7.41b

20 If removed, install the timing chain guides and tensioner pivot arms back onto the engine and tighten the bolts to the torque listed in this Chapter's Specifications. Apply several drops of medium strength thread-locking compound to the tensioner pivot arm bolts before installing them. Note that the silver bolts retain the guides to the cylinder head and the black colored bolts retain the guides to the engine block.

21 If the primary timing chain tensioner was removed or replaced install it back onto the engine in the locked position and tighten the bolts to the torque listed in this Chapter's Specifications **(see illustration)**.

22 Working on one secondary timing chain tensioner at a time, compress the tensioner piston in a vise until the stepped edge is flush

7.21 Primary timing chain tensioner/oil pump mounting bolts

2C

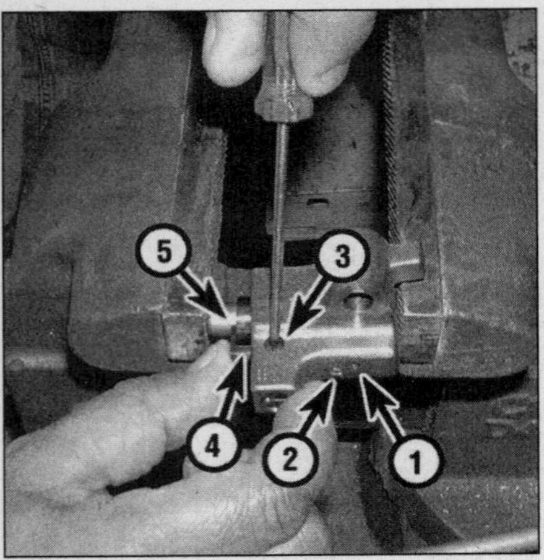

7.22 Locking the secondary tensioner(s) in the retracted position - note the "identification" mark on the side of the tensioner, as they are not interchangeable

1 *Insert locking pin*
2 *Identification mark*
3 *Ratchet pawl*
4 *Ratchet*
5 *Tensioner piston*

with the tensioner body **(see illustration)**. Insert a small scribe or other suitable tool into the side of the tensioner body and push the spring loaded ratchet pawl away from the ratchet mechanism, then push the ratchet down into the tensioner body until it's approximately 0.080 inch away from the tensioner body. Insert the end of a paper clip into the hole on the front of the tensioner to lock the tensioner in place.

23 After the two secondary tensioners have been compressed and locked into place, install them on the engine and tighten the bolts to the torque listed in this Chapter's Specifications. Make sure the tensioner with the "R" mark is installed on the right (passenger side) secondary chain and the tensioner with "L" mark is installed on the left (driver's side) secondary chain. The secondary chain tensioners **cannot** be switched with one another. Also make sure the plate behind the left secondary chain tensioner is installed correctly.

24 If you purchased a new timing chain verify that you have the correct timing chain for your vehicle by counting the number of links the chain has and comparing the new chain with the old chain. Also compare the position of the colored links in the new chain with the position of the colored links in the old chain.

25 Note that the idler sprocket has three drive gears incorporated into it. The front or forward facing gear (the largest of the three) is for the primary timing chain, the second or middle gear is for the left secondary chain and the third or rear gear is for the right secondary sprocket **(see illustration)**. The next 5 Steps will involve assembling the timing chains onto the idler sprocket on a workbench.

26 Place the idler sprocket on a workbench

7.25 Side view of the idler gear

1 *Idler sprocket assembly*
2 *Left camshaft (secondary) chain teeth*
3 *Right camshaft (secondary) chain teeth*
4 *Primary timing chain teeth*

with the mark on the front in the 12 o'clock position. Loop the right camshaft chain over the rear gear (farthest away from the primary chain gear) on the idler sprocket and position it so that the two plated links on the chain are visible through the lower (4 o'clock) window in the idler sprocket **(see illustration)**.

27 Loop the left camshaft chain over the front of the idler sprocket and position it over the middle gear so that the two plated links on the chain are visible through the lower (8 o'clock) window in the idler sprocket. **Note:** *After the left chain is in position, the two plated links on the right chain will no longer be visible through the 4 o'clock window in the idler sprocket, so be sure that the right camshaft chain is installed correctly before installing the left camshaft chain.*

28 After the secondary (camshaft) chains have been installed properly on the idler sprocket, install the special secondary chain holding tool onto the idler sprocket **(see illustration)**. This tool serves as a third hand, to secure the camshaft chains to the idler sprocket during the installation of the idler sprocket onto the engine.

29 Install the primary timing chain onto the primary chain gear of the idler sprocket and align the double plated links with the mark on the front of the sprocket. The mark on the idler sprocket should still be in the 12 o'clock position.

30 Insert the teeth of the crankshaft sprocket into the primary timing chain with the mark on the crankshaft sprocket pointing down in the 6 o'clock position and aligned with the single plated link on the chain.

31 Lubricate the idler shaft and bushing with clean engine oil.

32 Install the idler sprocket, the crankshaft sprocket and the timing chains with the chain holding tool as an assembly onto the engine. Slide the crankshaft sprocket over the keyway on the crankshaft and position the idler sprocket partially over the idler shaft (just enough to hold the primary chain in place). Then feed the secondary chains up through the chain guides and the cylinder head. **Note:** *It may be easier to bend a hook in the end of a coat hanger to help pull the secondary chains up through the timing chain guides and the cylinder head opening.*

33 Loop the secondary chains over the camshaft hubs and secure them with rubber bands to remove the slack from the chains, then push the idler gear, primary timing chain and the crankshaft sprocket assembly back on to the engine until they're fully seated

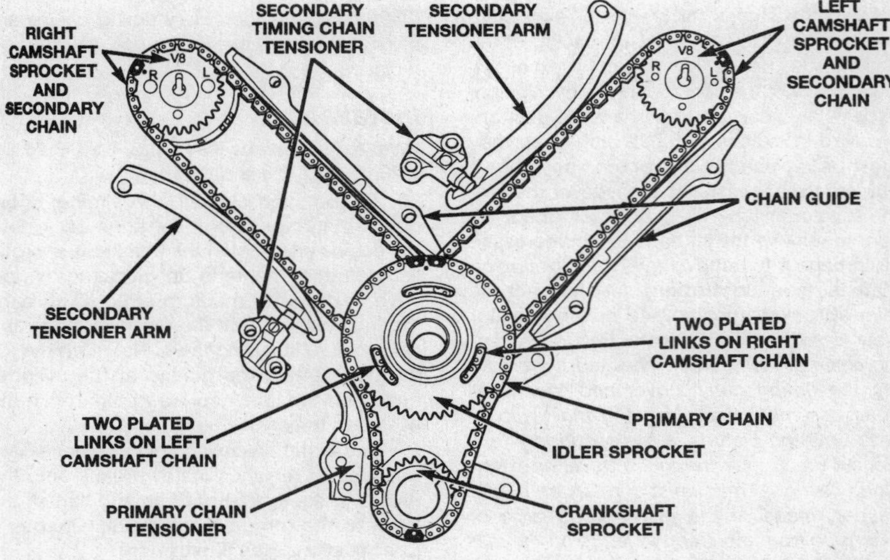

7.26 Timing chain installation details and related components

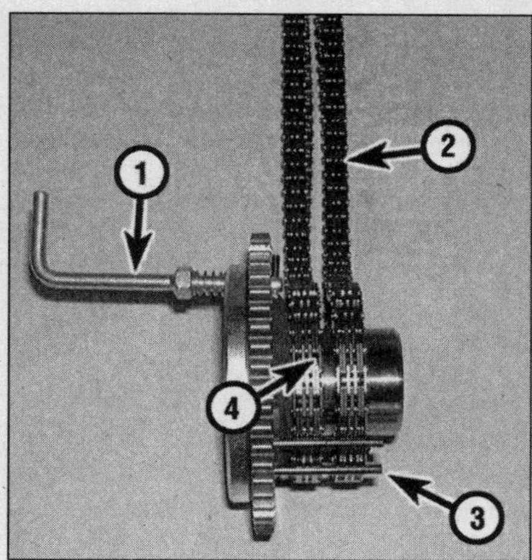

7.28 Install the secondary chain holding tool onto the idler sprocket with the plated links on the right camshaft chain in the 4 o'clock position and the left camshaft chain in the 8 o'clock position

1 *Secondary chain holding tool*
2 *Right camshaft timing chain*
3 *Secondary chain holding tool retaining pins*
4 *Left camshaft timing chain*

against the block **(see illustration).**

34 Align the "L" mark on left camshaft sprocket with the plated link on the left camshaft chain and position the camshaft sprocket over the camshaft hub (**see illustration 7.26**). The camshaft may have to be rotated slightly to align the dowel pin on the camshaft with the slot on the sprocket.

35 Align the "R" mark on right camshaft sprocket with the plated link on the right camshaft chain and position the camshaft sprocket over the camshaft hub. The camshaft may have to be rotated slightly to align the dowel pin on the camshaft with the slot on the sprocket. **Note:** *If the rocker arms* ***were not*** *removed as suggested in Step 9, it will be necessary to rotate and hold the camshafts with a set of locking pliers while the sprocket is being installed. This is typically 15 degrees counterclockwise on the left camshaft sprocket and 45 degrees clockwise on the right camshaft (the exact opposite of removal). Work on one camshaft and sprocket at a time starting with the left sprocket and proceeding to the right sprocket and never rotate the camshaft by a camshaft lobe or damage to the camshaft will occur.*

7.41b Tightening the right camshaft sprocket bolt

7.33 Partially install the idler sprocket, the crankshaft sprocket, the timing chains and the chain holding tool as an assembly onto the engine with the marks aligned as shown - feed the secondary chains up through the timing chain guides and loop them over the camshaft hubs, then push the primary chain assembly the rest of the way on the engine until it's seated against the block

36 Thoroughly clean the camshaft and idler sprocket bolts. Make sure all oil is removed from the bolt threads before installation, as over-tightening of bolts may occur if oil is not removed, then lubricate the bolt washers with small amount of clean engine oil making sure not to get oil on the threads.

37 Install the camshaft sprocket bolts finger tight. Remove the secondary timing chain holding tool from the idler sprocket, then install the idler sprocket retaining bolt and tighten it to the torque listed in this Chapter's Specifications.

38 Verify that all the plated timing chain links are aligned as shown in **illustration 7.26.**

39 Remove the locking pins from the primary timing chain tensioner and the secondary timing chain tensioners. **Caution:** *Do not manually extend the tensioners by hand; doing so will only over extend the tensioners and lead to premature timing chain wear.*

40 Rotate the engine two complete revolutions and reverify the position of the timing marks again. The idler sprocket mark should be located in the 12 o'clock position and the crankshaft sprocket mark should be located in the 6 o'clock position with the "V8" marks on the camshaft sprockets located in the 12 o'clock position.

41 Using a spanner wrench to hold the sprockets from turning, tighten the camshaft sprocket bolts to the torque listed in this Chapter's Specifications **(see illustrations).**

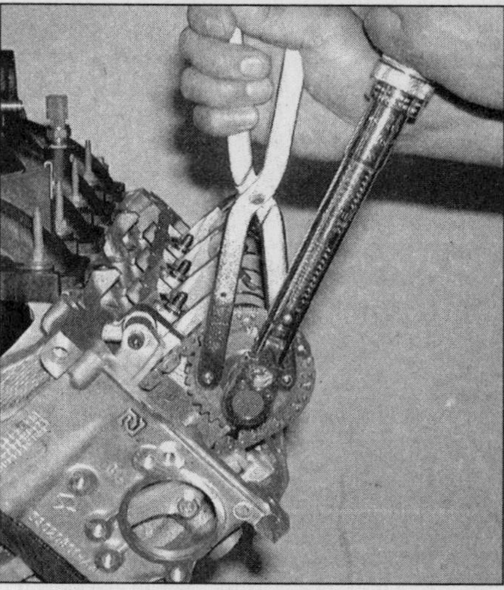

7.41a Tightening the left camshaft sprocket bolt

42 Remove all traces of old sealant or gasket material from the timing chain cover and the engine block.

43 Place the timing cover and gasket in position on the engine and install the bolts in their original locations and tighten the bolts to the torque listed in this Chapter's Specifications.

44 The remainder of the installation is the reverse of removal. Be sure to use pipe sealant on the cylinder head plugs to prevent oil leaks.

2C

8 Camshafts - removal, inspection and installation

Note 1: *Special tools are necessary to complete this procedure. Read through the entire procedure and obtain the special tools before beginning work.*

Note 2: *The camshafts should always be thoroughly inspected before installation and camshaft endplay should always be checked prior to camshaft removal.*

Removal

Refer to illustrations 8.7a, 8.7b and 8.9

1 Disconnect the cable from the negative terminal of the battery.

2 Remove the valve covers (see Section 4).

3 Remove the rocker arms (see Section 5).

4 Rotate the engine with a socket and ratchet (in a clockwise direction only) by the crankshaft pulley/vibration damper bolt until the "V8" marks on the camshaft sprockets are located in the 12 o'clock position **(see illustration 7.11).**

5 Using a permanent marker, apply alignment marks to the secondary timing chain links on either side of the "V8" marks on both camshaft sprockets to help aid the installation process (4 marks total).

6 Using a spanner wrench to hold the

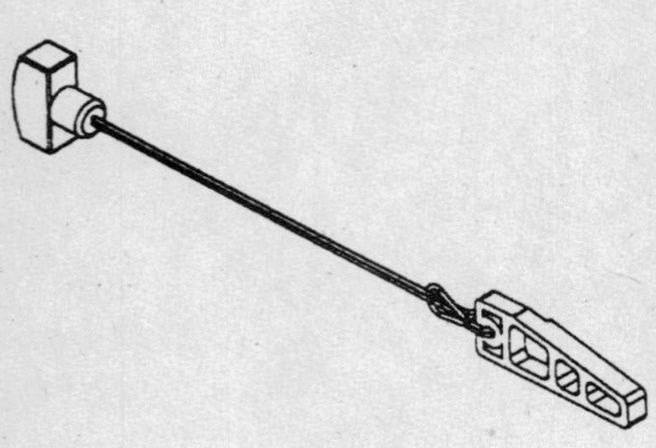

8.7a Timing chain tensioner wedge

camshaft sprockets from turning, loosen the camshaft sprocket bolts several turns, then retighten the bolts by hand until they're snug up against the sprocket. If the camshaft sprockets have rotated during the bolt loosening process, rotate the engine clockwise until the "V8" marks on the cam sprockets are realigned in the 12 o'clock position.

7 Install a timing chain tensioner wedge through the opening in the top of the cylinder head and force the wedge down between the narrowest section of the secondary chain **(see illustrations)**. If both camshafts are to removed, two timing chain wedges will be necessary (one for the left camshaft chain and one for the right camshaft chain). The wedge is used to secure the chain and the secondary tensioner in place while the camshaft is removed. **Caution 1:** *Failure to use a timing chain wedge will allow the secondary tensioner to over-extend and require removal of the timing chain cover to reset the tensioners.* **Caution 2:** *Never force the wedge past the narrowest section of the secondary timing chain as damage to the tensioner will occur.* If a timing chain wedge is not available, they may be fabricated using a block of wood that is 3/8 to 1/2-inch thick and a piece

8.7b The wedge is pushed down between the chain strands to secure the secondary chain and the tensioner in place while the camshaft is removed - this wedge is fabricated from a block of wood and a piece of wire

of wire to pull the wedge out of the cylinder head after installation.

8 Remove the camshaft sprocket retaining bolt(s) and detach the camshaft sprocket(s) from the camshaft hub(s). Disengage the camshaft chain(s) from the sprocket(s) and remove the camshaft sprocket(s) from the engine.

9 Verify the markings on the camshaft bearing caps. The caps should be marked from 1 to 5 with arrow marks on the caps indicating the front of the engine **(see illustration)**. If both camshafts are being removed, use a permanent marker to mark each bearing cap on the right cylinder head with an "R" and each bearing cap on the left cylinder head with an "L" to indicate from which cylinder head they came from. Loosen the camshaft bearing caps in two or three steps, in the reverse order of the tightening sequence **(see illustration 8.19)**. **Caution:** *Keep the caps in order. They must go back in the same location they were removed from.*

8.9 Verify that the camshaft bearing caps are marked to ensure correct reinstallation - do not mix-up the caps from the left cylinder head with the caps from the right cylinder head

10 Detach the bearing caps, then remove the camshaft(s) from the cylinder head. Mark the camshaft(s) "Left" or "Right" to indicate which cylinder head it came from.

11 Inspect the camshafts as described in Steps 12 through 16. Also inspect the camshaft secondary sprockets for wear on the teeth. Inspect the chains for cracks or excessive wear of the rollers. If any of the components show signs of excessive wear they must be replaced.

Inspection

Refer to illustrations 8.12, 8.13 and 8.16

12 Visually check the camshaft bearing surfaces for pitting, score marks, galling and abnormal wear. If the bearing surfaces are damaged, the cylinder head may have to be replaced **(see illustration)**.

13 Measure the outside diameter of each camshaft bearing journal and record your measurements **(see illustration)**. Compare them to the journal outside diameter specified in this Chapter, then measure the inside diameter of each corresponding camshaft bearing and record the measurements. Subtract each cam journal outside diameter from its respective cam bearing bore inside diameter to determine the oil clearance for each bearing. Compare the results to the specified journal-to-bearing clearance. If any of the measurements fall outside the standard specified wear limits in this Chapter, either the camshaft or the cylinder head, or both, must be replaced.

14 Check camshaft runout by placing the camshaft back into the cylinder head and set up a dial indicator on the center journal. Zero the dial indicator. Turn the camshaft slowly and note the dial indicator readings. Runout should not exceed .0010 inch. If the measured runout exceeds the specified runout, replace the camshaft.

15 Check the camshaft endplay by placing a dial indicator with the stem in line with the camshaft and touching the snout. Push the camshaft all the way to the rear and zero the

8.12 Inspect the cam bearing surfaces in each cylinder head for pits, score marks and abnormal wear - if wear or damage is noted, the cylinder head must be replaced

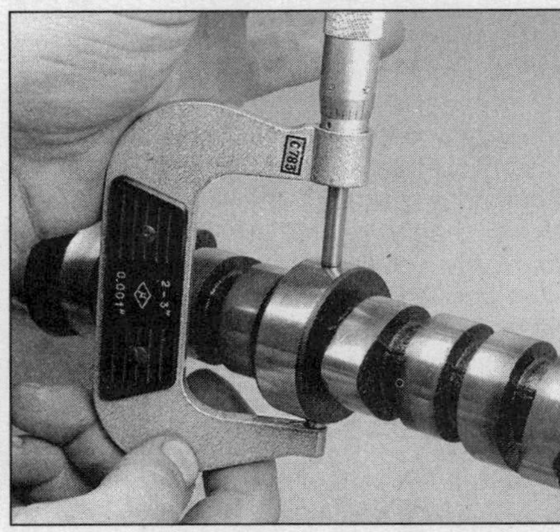

8.13 Measure the outside diameter of each camshaft journal and the inside diameter of each bearing to determine the oil clearance measurement

8.16 Measuring cam lobe height with a micrometer, make sure you move the micrometer to get the highest reading (top of cam lobe)

8.19 Camshaft bearing cap TIGHTENING sequence

dial indicator. Next, pry the camshaft to the front as far as possible and check the reading on the dial indicator. The distance it moves is the endplay. If it's greater than the Specifications listed in this Chapter, check the bearing caps for wear. If the bearing caps are worn, the cylinder head must be replaced.

16 Compare the camshaft lobe height by measuring each lobe with a micrometer **(see illustration)**. Measure of each of the intake lobes and write the measurements and relative positions down on a piece of paper. Then measure of each of the exhaust lobes and record the measurements and relative positions also. This will let you compare all of the intake lobes to one another and all of the exhaust lobes to one another. If the difference between the lobes exceeds 0.005 inch the camshaft should be replaced. Do not compare intake lobe heights to exhaust lobe heights, as lobe lift may be different. Only compare intake lobes to intake lobes and exhaust lobes to exhaust lobes for this comparison.

Installation

Refer to illustration 8.19

17 Apply moly-based engine assembly lubricant to the camshaft lobes and journals and install the camshaft(s) into the cylinder head with the dowel pins in the 12 o'clock position. If the old camshafts are being used, make sure they're installed in the exact location from which they came.

18 Install the bearing caps and bolts and tighten them hand tight.

19 Tighten the bearing cap bolts in several steps, to the torque listed in this Chapter's Specifications, using the proper tightening sequence **(see illustration)**.

20 Engage the camshaft sprocket teeth with the camshaft drive chain links so that the "V8" mark on the sprocket(s) is between the two marks made in Step 5 during removal, then position the sprocket over the dowel on the camshaft hub. At this point the chain marks and the "V8" marks on the sprocket should be pointing up in the 12 o'clock position.

21 Thoroughly clean the camshaft sprocket bolts. Make sure all oil is removed from the bolt threads before installation, as over-tightening of bolts may occur if oil is not removed, then lubricate the bolt washers with small amount of clean engine oil making sure not to get oil on the threads.

22 Install the camshaft sprocket bolts and tighten them to the torque listed in this Chapter's Specifications **(see illustration 7.41a and 7.41b)**.

23 Remove the timing chain wedge(s).

24 Install the rocker arms (see Section 5).

25 The remainder of installation is the reverse of removal.

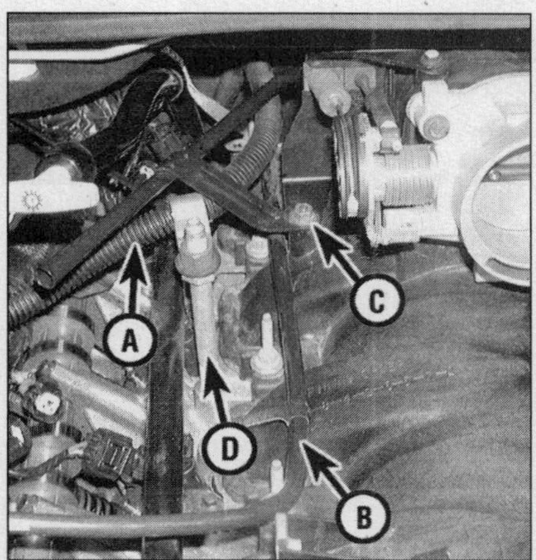

9.4 Label and disconnect the wiring harnesses (A), the vacuum hoses (B), the oil dipstick tube (C) and the hood-to-cowl seal (D)

9 Intake manifold - removal and installation

Warning: *The engine must be completely cool before beginning this procedure.*

Removal

Refer to illustration 9.4

1 Relieve the fuel pressure (see Chapter 4).

2 Disconnect the cable from the negative terminal of the battery.

3 Refer to Chapter 4 and remove the air intake duct from the throttle body.

4 Label and disconnect the vacuum hoses, the electrical connectors and the ground straps attached to the intake manifold and the throttle body **(see illustration)**. Also remove the engine oil dipstick, the hood-to-cowl seal and the right engine lifting stud. When removing the oil dipstick, detach the nut from the stud on the intake manifold, then follow the dipstick tube to the rear of the engine block and remove the bolt securing the tube at the back of the block. Pull the dipstick tube up and out of the engine block to remove it.

5 Detach the throttle cable and the cruise control cable from the throttle body and the throttle cable bracket (see Chapter 4).

6 Remove the accessory drivebelt (see Chapter 1).

7 Remove the alternator and the ignition coils (see Chapter 5).

8 Unbolt the air conditioning compressor and position it aside without disconnecting the refrigerant lines (see Chapter 3).

9 Remove the fuel rails and injectors, then remove the throttle body (see Chapter 4).

10 Refer to Chapter 1 and partially drain the cooling system, then refer to Chapter 6 and remove the coolant temperature sensor. This step is necessary to allow clearance for intake manifold as it is removed.

11 Loosen the manifold mounting bolts/nuts in 1/4-turn increments until they can be removed by hand in the reverse order of the tightening sequence **(see illustration 9.17)**.

12 The manifold will probably be stuck to the cylinder heads and force may be required to break the gasket seal. **Caution:** *Don't pry between the manifold and the heads or damage to the gasket sealing surfaces may occur, leading to vacuum leaks.*

13 Label and detach any remaining hoses which would interfere with the removal of the intake manifold.

14 Lift the manifold up level with the vehicle and remove it from the engine. Interference with the cowl will be apparent, but with careful maneuvering the manifold can be removed.

Installation

Refer to illustration 9.17

15 Clean and inspect the intake manifold to cylinder head sealing surfaces. Inspect the

9.17 Intake manifold TIGHTENING sequence

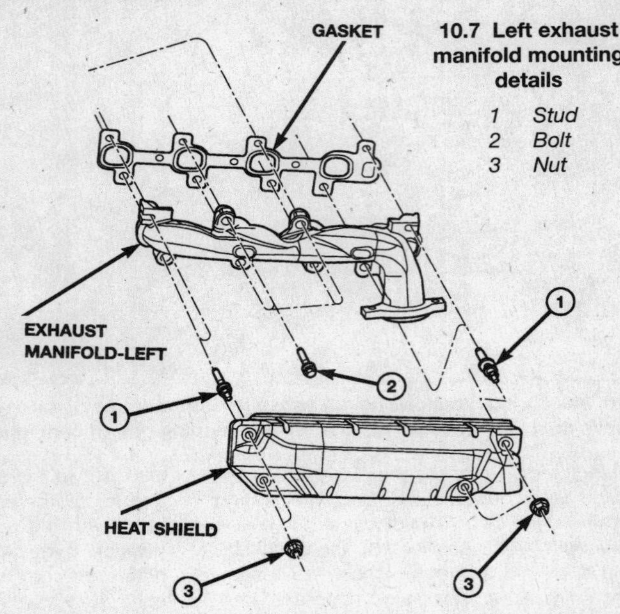

10.7 Left exhaust manifold mounting details

1 *Stud*
2 *Bolt*
3 *Nut*

gaskets on the manifold for tears or cracks replacing them if necessary. The gaskets can be reused if not damaged.

16 Position the manifold on the engine making sure the gaskets and manifold are aligned correctly over the cylinder heads, then install the intake manifold bolts hand tight.

17 Following the recommended tightening sequence, tighten the bolts, in several steps, to the torque listed in this Chapter's Specifications **(see illustration)**.

18 The remainder of the installation is the reverse of the removal procedure. Fill the cooling system, run the engine and check for fuel, vacuum and coolant leaks.

10 Exhaust manifold - removal and installation

Warning: *The engine must be completely cool before beginning this procedure.*

Removal

Refer to illustrations 10.7 and 10.12

1 Disconnect the cable from the negative terminal of the battery.

2 Block the rear wheels, set the parking brake, raise the front of the vehicle and support it securely on jackstands. Working below the vehicle, disconnect the electrical connectors from the oxygen sensors. Unbolt the front Y-pipe from both manifolds, then detach the Y-pipe from the exhaust system (see Chapter 4). **Note:** *It will be necessary to detach the center exhaust system mount from the transmission and loosen the exhaust pipe clamp to accomplish this task.*

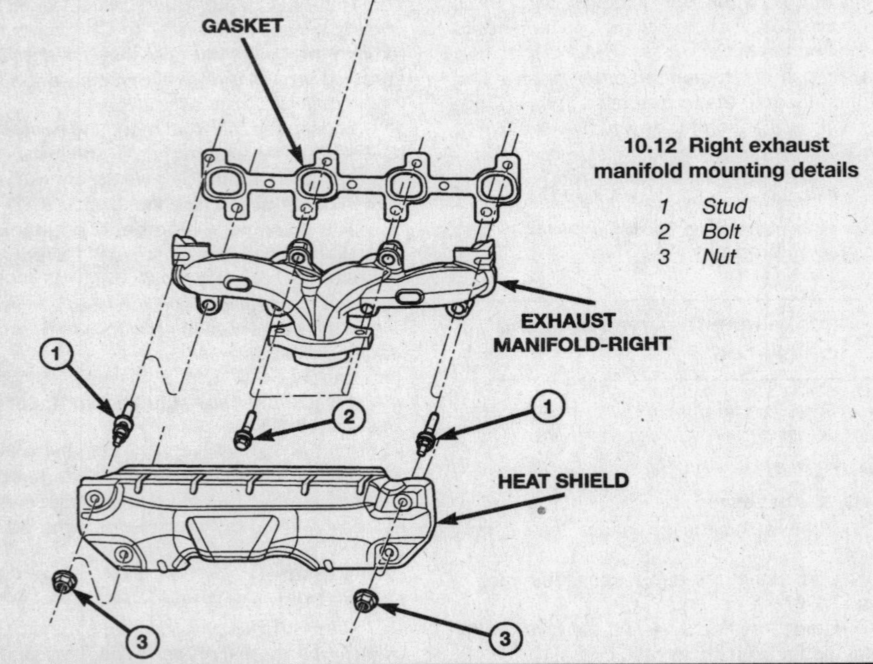

10.12 Right exhaust manifold mounting details

1 *Stud*
2 *Bolt*
3 *Nut*

Left manifold

3 Remove the mounting bolts at the rear of the heat shield.

4 Working in the engine compartment, remove the air intake duct and the throttle body resonator (see Chapter 4).

5 Remove the mounting bolts at the front of the heat shield and detach the heat shield from the manifold.

6 Remove the upper exhaust manifold bolts.

7 Working below the vehicle again, remove the lower manifold bolts and lower the manifold from the vehicle **(see illustration)**.

Right manifold

8 Remove the starter (see Chapter 5).

9 Working in the engine compartment, detach the power distribution center from the battery tray and position it aside. Make sure to disconnect the cable from the positive terminal of the battery before doing so, then remove the battery and the battery tray from the engine compartment (see Chapter 5).

10 Partially drain the cooling system until the level is below heater hoses (see Chapter 1 if necessary), then remove the heater hoses from the engine and the retaining clips and position them aside.

11 Remove the engine drivebelt (see Chapter 1), then unbolt the air conditioning compressor from the engine and position it aside without disconnecting the refrigerant lines. Also loosen the air conditioning accumulator

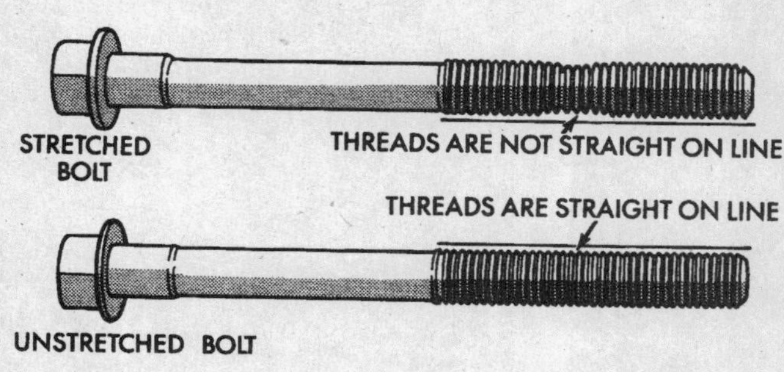

11.13 Place a metal ruler or straight edge against the cylinder head bolt threads to check if the bolts have stretched

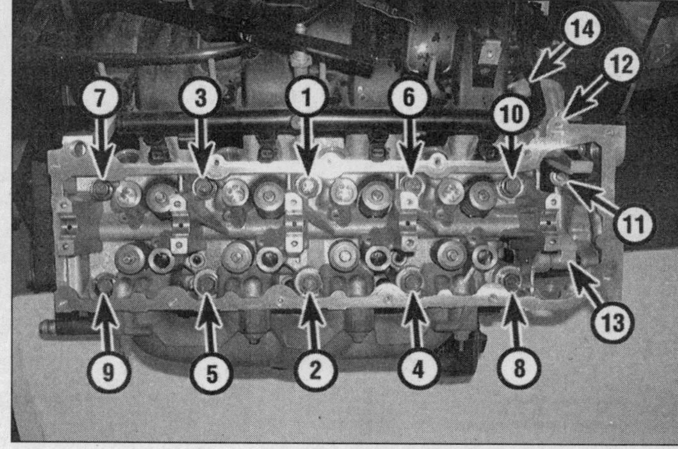

11.19a Cylinder head TIGHTENING sequence

bracket **(see illustration 4.9)**.

12 Remove the exhaust manifold heat shield mounting bolts **(see illustration)** and lift the heat shields from the engine compartment.

13 Remove the upper manifold mounting bolts.

14 Working below the vehicle, detach the lower manifold mounting bolts and lower manifold from the vehicle.

Installation

15 Clean the mating surfaces to remove all traces of old gasket material, then inspect the manifold for distortion and cracks. Warpage can be checked with a precision straightedge held against the mating flange. If a feeler gauge thicker than 0.030-inch can be inserted between the straightedge and flange surface, take the manifold to an automotive machine shop for resurfacing.

16 Place the exhaust manifold in position with a new gasket and install the mounting bolts finger tight.

17 Starting in the middle and working out toward the ends, tighten the mounting bolts in several increments, to the torque listed in this Chapter's Specifications.

18 Install the remaining components in the reverse order of removal. **Caution:** *Do not overtighten the exhaust manifold heat shields as distortion and cracks may occur.*

19 Start the engine and check for exhaust leaks between the manifold and cylinder head and between the manifold and exhaust pipe.

11 Cylinder head - removal and installation

Caution: *The engine must be completely cool before beginning this procedure.*

Note: *The following procedure describes how to remove the cylinder heads with the camshaft(s) and the exhaust manifold(s) still attached to the cylinder head.*

Removal

1 Refer to Section 7 and remove the timing chains, sprockets and the timing chain guides.

2 Remove the intake manifold (see Section 9).

3 Raise the front of the vehicle and support it securely on jackstands.

4 Working below the vehicle, disconnect the electrical connector(s) from the oxygen sensor(s). Unbolt the front Y-pipe(s) from the exhaust manifolds.

5 Label and remove any remaining items attached to the cylinder head, such as coolant fittings, ground straps, cables, hoses, wires or brackets.

6 Using a breaker bar and the appropriate sized socket, loosen the cylinder head bolts in 1/4-turn increments until they can be removed by hand. Loosen the bolts in the reverse order of the tightening sequence **(see illustration 11.19a)** to avoid warping or cracking the head.

7 Lift the cylinder head off the engine block with the camshaft in place and the exhaust manifold attached. If it's stuck, very carefully pry up at the front end of the cylinder head, beyond the gasket surface, at a casting protrusion.

8 Remove all external components from the head to allow for thorough cleaning and inspection. **Note:** *See Chapter 2, Part D, for cylinder head inspection and servicing procedures.*

Installation

Refer to illustrations 11.13, 11.19a and 11.19b

9 The mating surfaces of the cylinder head and block must be perfectly clean when the head is installed.

10 Use a gasket scraper to remove all traces of carbon and old gasket material from the cylinder head and engine block being careful not to gouge the aluminum, then clean the mating surfaces with lacquer thinner or acetone. If there's oil on the mating surfaces when the head is installed, the gasket may not seal correctly and leaks could develop. When working on the block, stuff the cylinders with clean shop rags to keep out debris. Use a vacuum cleaner to remove material that falls into the cylinders.

11 Check the block and head mating surfaces for nicks, deep scratches and other damage. If damage is slight, it can be removed with a file; if it's excessive, machining may be the only alternative.

12 Use a tap of the correct size to chase the threads in the head bolt holes, then clean the holes with compressed air - make sure that nothing remains in the holes. **Warning:** *Wear eye protection when using compressed air!*

13 Check each cylinder head bolt for stretching **(see illustration)**. If the diameter of the bolt threads has necked down anywhere in the threaded area the bolts have exceeded the maximum amount of stretch and will need to be replaced.

14 Check the cylinder head for warpage (see Chapter 2D). Check the head gasket, intake and exhaust manifold surfaces.

15 Install any components that were removed from the head such as the lash adjusters, the exhaust manifold and the camshaft back onto the cylinder head.

16 Position the new cylinder head gasket over the dowel pins on the block noting which direction on the gasket faces up.

17 Carefully set the head over the dowels on the block without disturbing the gasket.

18 Before installing the 11 mm head bolts (bolts 1 through 10), apply a small amount of clean engine oil to the threads and hardened washers (if equipped). The chamfered side of the washers must face the bolt heads. Before installing the 8 mm head bolts (bolts 11 through 14), apply a small amount of thread sealant to the bolt threads.

19 Install the bolts in their original locations and tighten them finger tight. Then tighten all the bolts in five steps, following the proper sequence to the torque and angle of rotation listed in this Chapter's Specifications **(see illustrations)**.

20 Install the timing chain guides, the timing chains and the timing chain sprockets as described in Section 7. The remaining instal-

11.19b Using an angle measurement gauge during the final stages of tightening

lation steps are the reverse of removal.

21 Refill the cooling system and check the engine oil, adding if necessary (see Chapter 1).

22 Start the engine and check for oil and coolant leaks.

12 Crankshaft pulley/vibration damper - removal and installation

Refer to illustrations 12.4, 12.5 and 12.6

1 Disconnect the cable from the negative terminal of the battery.

2 Refer to Chapter 3 and remove the cooling fans and shroud assembly.

3 Remove the drivebelts (see Chapter 1) and position the belt tensioner away from the crankshaft pulley.

4 Use a strap wrench around the crankshaft pulley to hold it while using a breaker bar and socket to remove the crankshaft pulley center bolt **(see illustration)**.

5 Pull the damper off the crankshaft with a

12.4 Use a strap wrench to hold the crankshaft pulley while removing the center bolt (a chain-type wrench may be used if you wrap a section of old drivebelt or a rag around the crankshaft pulley first)

puller **(see illustration)**. **Caution:** *The jaws of the puller must only contact the hub of the pulley - not the outer ring.* **Note:** *A long Allen-head bolt should be inserted into the crankshaft nose for the puller's tapered tip to push against to prevent damage to the crankshaft threads.*

6 Check the surface on the pulley hub that the oil seal rides on. If the surface has been grooved from long-time contact with the seal, a press-on sleeve may be available to renew the sealing surface **(see illustration)**. This sleeve is pressed into place with a hammer and a block of wood and is commonly available at auto parts stores for various applications.

7 Lubricate the pulley hub with clean engine oil. Align the slot in the pulley with the key on the crankshaft and push the crankshaft pulley on the crankshaft as far as it will go. Use a vibration damper installation tool to press the pulley the rest of the way onto the crankshaft.

12.5 The use of a three jaw puller will be necessary to remove the crankshaft pulley - always place the puller jaws around the pulley hub, not the outer ring

8 Install the crankshaft pulley retaining bolt and tighten it to the torque listed in this Chapter's Specifications.

9 The remainder of installation is the reverse of the removal.

13 Crankshaft front oil seal - replacement

Refer to illustrations 13.2 and 13.4

1 Remove the crankshaft pulley from the engine (see Section 12).

2 Carefully pry the seal out of the cover with a seal removal tool or a large screwdriver **(see illustration)**. **Caution:** *Be careful not to scratch, gouge or distort the area that the seal fits into or an oil leak will develop.*

3 Clean the bore to remove any old seal material and corrosion. Position the new seal in the bore with the seal lip (usually the side with the spring) facing IN (toward the engine). A small amount of oil applied to the outer edge of the new seal will make installation

12.6 If the sealing surface of the pulley hub has a wear groove from contact with the seal, repair sleeves are available at most auto parts stores

13.2 Pry the seal out very carefully with a seal removal tool or screwdriver, being careful not to nick or gouge the seal bore or the crankshaft

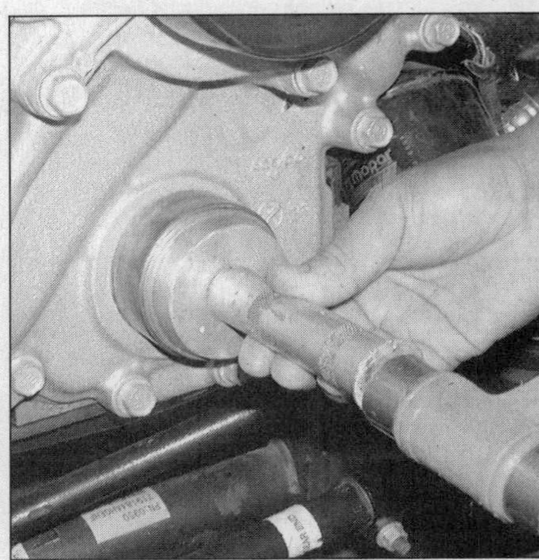

13.4 Use a seal driver or large-diameter socket to drive the new seal into the cover

14.5 Remove the exhaust Y-pipe (A) and the transmission-to-oil pan support brace (B), then position the transmission cooler lines (C) aside

14.7 If the oil pan is stuck to the gasket, gently tap on the side of the oil pan to break the gasket seal

easier - but don't overdo it!

4 Drive the seal into the bore with a seal driver or a large socket and hammer until it's completely seated **(see illustration)**. Select a socket that's the same outside diameter as the seal and make sure the new seal is pressed into place until it bottoms against the cover flange.

5 Lubricate the seal lips with engine oil and reinstall the crankshaft pulley.

6 The remainder of installation is the reverse of the removal. Run the engine and check for oil leaks.

14 Oil pan - removal and installation

Removal

Refer to illustrations 14.5 and 14.7

1 Disconnect the cable from the negative terminal of the battery.

2 Apply the parking brake and block the rear wheels. Raise the front of the vehicle and place it securely on jackstands. Drain the engine oil (see Chapter 1).

3 Disconnect the electrical connectors from the oxygen sensors. Unbolt the front Y-pipe from both manifolds, then detach the Y-pipe from the exhaust system (see Chapter 4). **Note:** *It will be necessary to detach the center exhaust system mount from the transmission and loosen the exhaust pipe clamp to accomplish this task.*

4 Remove the transmission oil cooler lines from the right side of the oil pan.

5 Remove the transmission-to-oil pan support brace at the rear of the pan **(see illustration)**.

6 Remove the starter (see Chapter 5). Also detach any clips securing the transmission oil cooler lines that would interfere with removal of the oil pan.

7 Remove the bolts and nuts, noting the stud locations, then carefully separate the oil pan from the block. Don't pry between the block and the pan or damage to the sealing surfaces and gasket could occur and oil leaks may develop. Instead, tap on the side of the oil pan with a rubber mallet if necessary to break the gasket seal **(see illustration)**.

8 Remove the oil pump pick-up tube **(see illustrations 15.3a and 15.3b)** and detach the oil pan gasket/windage tray from the engine block.

Installation

Refer to illustration 14.13

9 Clean the oil pan with solvent and remove any gasket material from the block and the pan mating surfaces. Clean the mating surfaces with lacquer thinner or acetone and make sure the bolt holes in the block are clear. Check the oil pan flange for distortion, particularly around the bolt holes.

10 Inspect the oil pan gasket for cuts and tears, replacing it if necessary. If the gasket is in good condition it can be reused. **Note:** *The oil pan gasket and the windage tray are a one piece design, therefore must be replaced together.*

11 Position the oil pan gasket/windage tray in place on the block and install the oil pump pick-up tube. (see Section 15). Always use a new O-ring on the pick-up tube and tighten the pick-up tube-to-oil pump bolt first, when installing it.

12 Place the oil pan in position on the block and install the nuts/bolts.

13 After the fasteners are installed, tighten all the bolts in several steps, following the proper sequence to the torque listed in this Chapter's Specifications **(see illustration)**.

14 Place the transmission-to-oil pan support brace in position and install the vertically mounted bolts. Torque the vertical mounted bolts to 10 in-lbs, then install the horizontally mounted bolts. Torque the horizontal mounted bolts to 40 ft-lbs, then retorque the vertical mounted bolts to 40 ft-lbs **(see illustration 14.5)**.

15 The remaining steps are the reverse of the removal procedure.

16 Refill the engine with oil (see Chapter 1), replace the filter, run it until normal operating temperature is reached and check for leaks.

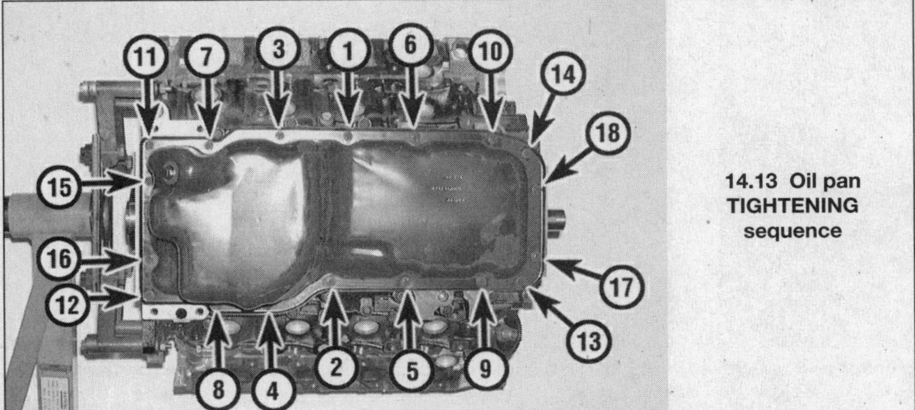

14.13 Oil pan TIGHTENING sequence

15 Oil pump - removal, inspection and installation

Removal

Refer to illustrations 15.3a, 15.3b and 15.4

1 Refer to Section 7 and remove the timing chains and sprockets.

2 Remove the oil pan (see Section 14).

2C

15.3a Remove the bolt (arrow) securing oil pump pick-up tube to the oil pump - when installing the pick up tube always tighten this bolt first

15.3b Remove the nuts (arrows) securing the oil pump pick-up to the main caps and remove the pick-up tube - be prepared for the oil pan gasket/windage tray to come off at the same time

15.4 Oil pump housing/primary timing chain tensioner retaining bolts (arrows)

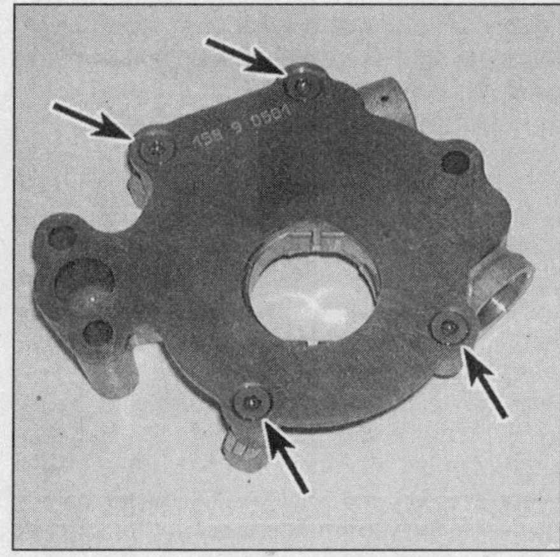

15.6 Remove the screws (arrows) and lift the cover off

3 Remove the oil pump pick-up tube **(see illustrations)**.

4 Remove the bolts and detach the primary timing chain tensioner the front of the oil pump**(see illustration)**.

5 Gently pry the oil pump housing outward enough to clear the flats on the crankshaft and remove it from the engine.

Inspection

Refer to illustrations 15.6, 15.8a 15.8b, 15.8c, 15.8d, 15.8e and 15.8f

6 Remove the screws holding the front cover on the oil pump housing **(see illustration)**.

7 Clean all components with solvent, then inspect them for wear and damage. **Caution:** *The oil pressure relief valve and spring are an integral part of the oil pump housing. Removal of the relief valve and spring from the oil pump housing will damage the oil pump and require replacement of the entire oil pump assembly.*

8 Check the clearance of the following oil pump components with a feeler gauge and a micrometer or dial caliper **(see illustrations)** and compare the measurements to the clearance listed in this Chapter's Specifications:

a) *Cover flatness*
b) *Outer rotor diameter and thickness*
c) *Inner rotor thickness*
d) *Outer rotor-to-body clearance*
e) *Inner rotor-to-outer rotor tip clearance*

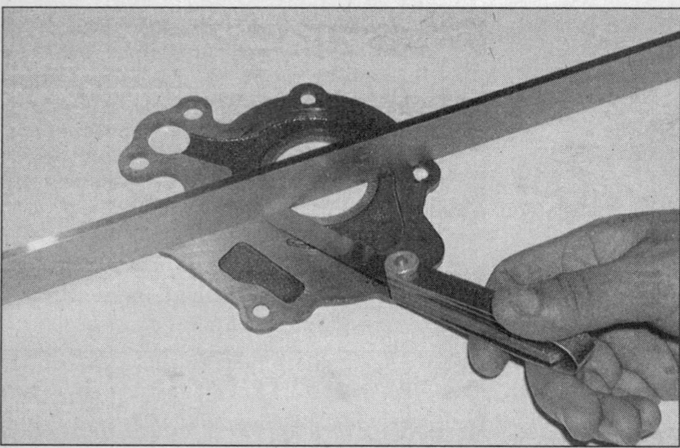

15.8a Place a straightedge across the oil pump cover and check it for warpage with a feeler gauge

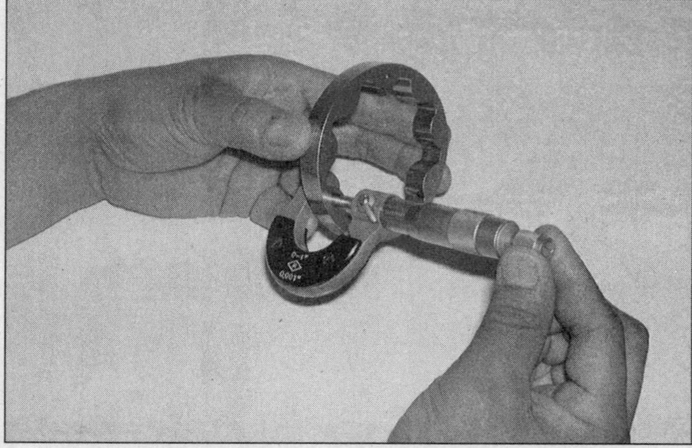

15.8b Use a micrometer or dial caliper to check the thickness and the diameter of the outer rotor

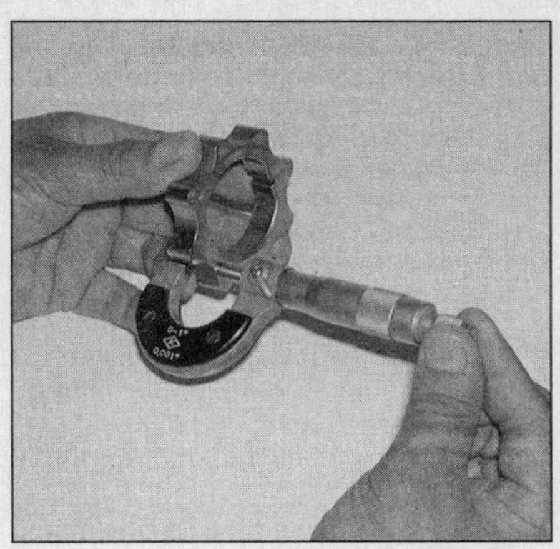
15.8c Use a micrometer or dial caliper to check the thickness of the inner rotor

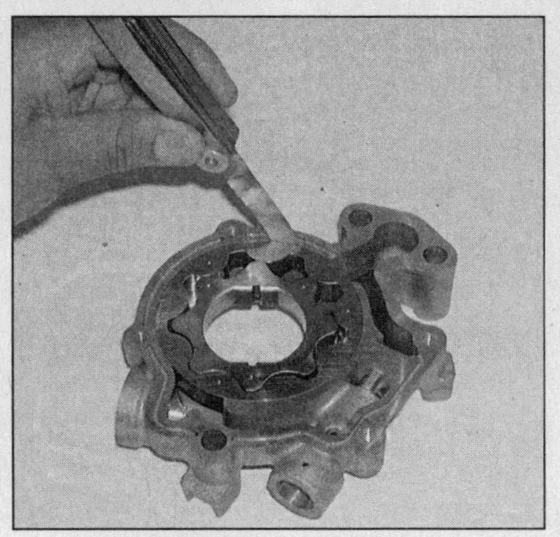
15.8d Check the outer rotor-to-housing clearance

15.8e Check the clearance between the tips of the inner and outer rotors

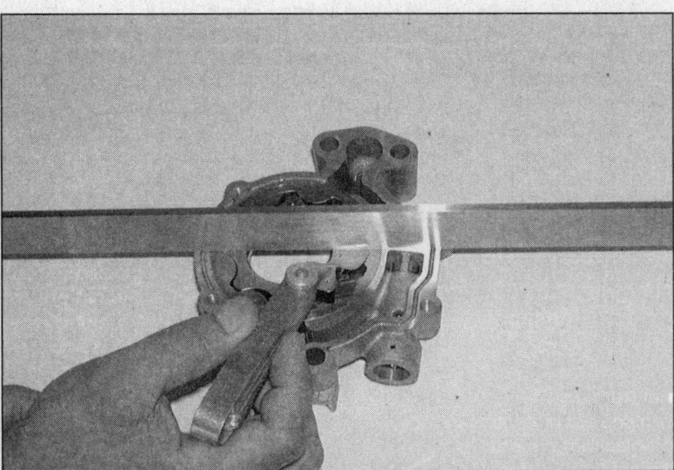
15.8f Using a straightedge and feeler gauge, check the side clearance between the surface of the oil pump and the inner and outer rotors

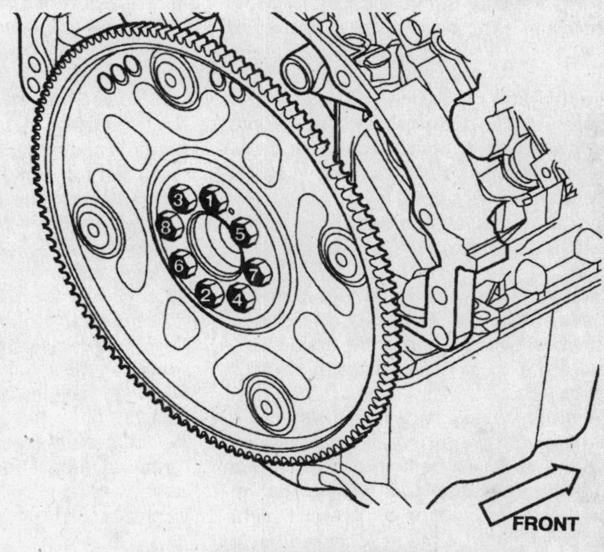

16.6 Driveplate TIGHTENING sequence

2C

f) Cover-to-inner rotor side clearance
g) Cover-to-outer rotor side clearance

If any clearance is excessive, replace the entire oil pump assembly.

9 **Note:** *Pack the pump with petroleum jelly to prime it.* Assemble the oil pump and tighten all fasteners to the torque listed in this Chapter's Specifications. Install the oil pressure regulator valve, spring and washer, then tighten the oil pressure regulator valve cap.

Installation

10 To install the pump, turn the flats in the rotor so they align with the flats on the crankshaft and push the oil pump back into position against the block.

11 Position the primary timing chain tensioner over the oil pump and install the pump-to-block bolts. Tighten the oil pump/primary timing chain tensioner bolts to the torque listed in this Chapter's Specifications.

12 The remainder of installation is the reverse of removal.

16 Driveplate - removal and installation

Refer to illustration 16.6

1 Raise the vehicle and support it securely on jackstands, then refer to Chapter 7B and remove the transmission. **Warning:** *The engine must be supported from above with an engine hoist or three-bar support fixture before working underneath the vehicle with the transmission removed.*

2 Now would be a good time to check and replace the transmission front pump seal.

3 Use paint or a center-punch to make alignment marks on the driveplate and crankshaft to ensure correct alignment during reinstallation.

4 Remove the bolts that secure the driveplate to the crankshaft. If the crankshaft turns, jam a large screwdriver or prybar through the driveplate to keep the crankshaft from turning, then remove the mounting bolts.

5 Pull straight back on the driveplate to detach it from the crankshaft.

6 Installation is the reverse of removal. Be sure to align the matching paint marks. Use thread locking compound on the bolt threads and tighten them in several steps, following the proper sequence to the torque listed in this Chapter's Specifications **(see illustration)**.

17 Rear main oil seal - replacement

Refer to illustrations 17.2 and 17.5

1 These models use a one-piece rear main seal that is sandwiched between the engine block and the lower main bearing cap assembly, or "bed plate" as it's often referred to. Replacing this seal requires removal of the transmission, torque converter and driveplate. Refer to Chapter 7 for the transmission removal procedures.

2 The seal can be removed by prying it out of the engine block with a screwdriver, being careful not to nick the crankshaft surface

17.2 Pry the seal out very carefully with a seal removal tool or screwdriver - if the crankshaft is damaged, the new seal will leak!

17.5 The rear oil seal can be pressed into place with a seal installation tool, a section of pipe or a blunt object shown here - in any case be sure the seal is installed squarely into the seal bore and flush with the rear of the engine block

(see illustration). Wrap the screwdriver tip with tape to avoid damage. Be sure to note how far it's recessed into the housing bore before removal so the new seal can be installed to the same depth.

3 Thoroughly clean the seal bore in the block with a shop towel. Remove all traces of oil and dirt.

4 Lubricate the outside diameter of the seal and the seal lip with clean engine oil and install the seal over the end of the crankshaft. Make sure the lip of the seal points toward the engine.

5 Preferably, a seal installation tool (available at most auto parts store) should be used to press the new seal back into place. If the proper seal installation tool is unavailable, use a large socket, section of pipe or a blunt tool and carefully drive the new seal squarely into the seal bore and flush with the rear of the engine block **(see illustration)**.

6 The remainder of installation is the reverse of the removal procedure.

18 Engine mounts - check and replacement

Refer to illustration 18.12

1 There are three powertrain mounts on the vehicles covered by this manual; left and right engine mounts attached to the engine block and to the frame and a rear mount attached to the transmission and the frame. The rear transmission mount is covered in Chapter 7B. Engine mounts seldom require attention, but broken or deteriorated mounts should be replaced immediately or the added strain placed on the driveline components may cause damage or wear.

Check

2 During the check, the engine must be raised slightly to remove the weight from the mounts.

3 Raise the vehicle and support it securely on jackstands, then position a jack under the engine oil pan. Place a large wood block between the jack head and the oil pan, then carefully raise the engine just enough to take the weight off the mounts. **Warning:** *DO NOT place any part of your body under the engine when it's supported only by a jack!*

4 Check for relative movement between the inner and outer portions of the mount (use a large screwdriver or prybar to attempt to move the mounts). If movement is noted, lower the engine and tighten the mount fasteners.

5 Check the mounts to see if the rubber is cracked, hardened or separated from the metal casing which would indicate a need for replacement.

6 Rubber preservative should be applied to the mounts to slow deterioration.

Replacement

7 Disconnect the cable from the negative terminal of the battery.

8 Remove the engine cooling fan and shroud (see Chapter 3).

9 Raise the front of the vehicle and support it securely on jackstands.

10 Support the engine with a lifting device from above. **Caution:** *Do not connect the lifting device to the intake manifold.* Raise the engine just enough to take the weight off the engine mounts. If you're removing the driver's side engine mount, removal of the oil filter will be necessary.

11 Working below the vehicle, remove the engine mount-to-frame support bracket through-bolt.

12 Remove the mount-to-engine block bolts, then raise the engine until the mounts can be maneuvered past the frame bracket and removed from the vehicle **(see illustration)**.

13 To remove the engine mount support brackets, simply remove the bolts from the frame and remove the brackets from the engine compartment.

14 After the engine mounts have been installed onto the engine, lower the engine while guiding the engine mount and through-bolt into the frame support bracket. Install the through-bolt nut and tighten it to the torque listed in this Chapter's Specifications.

15 The remainder of installation is the reverse of removal. Remove the engine hoist and the jackstands and lower the vehicle.

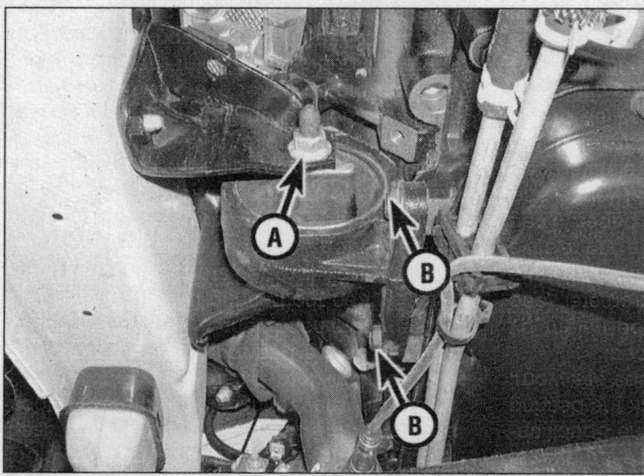

18.12 The engine mounts are secured by the through-bolt (A) and the mount-to-engine block bolts (B) - the two upper mount to block bolts are not visible in this photo

Chapter 2 Part D
General engine overhaul procedures

Contents

2D

Specifications

4.0L inline six-cylinder engine

General

Displacement	4.0 liters (258 cubic inches)
Cylinder compression pressure	120 to 150 psi
Maximum variation between cylinders	30 psi (from the highest reading)
Oil pressure	
At idle (600 rpm)	13 psi
Above 1600 rpm	37 to 75 psi

Engine block

Bore diameter	3.875 to 3.877 inches
Taper and out-of-round (maximum)	0.001 inch
Deck (head gasket surface) warpage limit	0.002 inch per 6 inches
Lifter bore diameter	0.9055 to 0.9065 inch

Cylinder head and valves

Head warpage limit	0.002 inch per 6 inches
Minimum valve margin	1/32 inch
Valve seat angle, intake and exhaust	44.5-degrees
Valve face angle, intake and exhaust	
1999 and earlier	45-degrees
2000	46.5-degrees
Valve stem diameter	0.0311 to 0.312 inch
Valve stem-to-guide clearance	0.001 to 0.003 inch
Valve spring free length	
1993 through 1995	1.967 inches
1996 and later	1.876 to 1.890 inches
Valve spring installed height	1.64 inches
Valve length	
Intake	4.822 to 4.837 inches
Exhaust	4.837 to 4.852 inches

4.0L inline six-cylinder engine (continued)

Crankshaft and connecting rods

Rod journal diameter	2.0934 to 2.0955 inches
Rod bearing oil clearance	
Desired	0.0015 to 0.002 inch
Allowable	0.001 to 0.003 inch
Connecting rod side clearance (endplay)	0.010 to 0.019 inch
Main bearing journal diameter	
No.1 thru 6	2.4996 to 2.5001 inches
No. 7	2.4980 to 2.4995 inches
Main bearing oil clearance	
Desired	0.002 inch
Allowable	0.001 to 0.0025 inch
Crankshaft endplay (at thrust bearing)	0.0015 to 0.0065 inch
Maximum taper and out-of-round (all journals)	0.0005 inch

Pistons and rings

Piston-to-bore clearance	
1998 and earlier	
Desired	0.0013 to 0.0015 inch
Allowable	0.0008 to 0.0015 inch
1999 and later	see Sections 16 and 18
Piston ring end gap	
Compression rings	
1993 through 1995	0.010 to 0.020 inch
1996 and later	
Top compression ring	0.0090 to 0.0240 inch
Second compression ring	0.0190 to 0.0380 inch
Oil control ring (steel rails)	
1993 through 1995	0.010 to 0.025 inch
1996 and later	0.010 to 0.060 inch
Piston ring side clearance	
Compression rings	
Desired	0.0017 inch
Allowable	0.0017 to 0.0032 inch
Oil control ring	
Desired	0.003 inch
Allowable	
1993 through 1995	0.001 to 0.0095 inch
1996 and later	0.0024 to 0.0083 inch

Torque specifications *

	Ft-lbs (unless otherwise indicated)
Main bearing cap bolts	80
Connecting rod cap nuts	33

* **Note:** *Refer to Part A for additional torque specifications.*

5.2L and 5.9L V8 engines

General

Displacement	
5.2L	5.2 liters (318 cubic inches)
5.9L	5.9 liters (360 cubic inches)
Cylinder compression pressure	120 to 150 psi
Maximum variation between cylinders	30 psi (from the highest reading)
Oil pressure	
At idle (600 rpm)	10 psi (approximately)
At 3000 rpm	30 to 80 psi

Engine block

Bore diameter	
5.2L	3.910 to 3.914 inches
5.9L	4.000 to 4.002 inches
Taper (maximum)	0.010 inch
Out-of-round (maximum)	0.005 inch
Deck (head gasket surface) warpage limit	0.002 inch per 6 inches
Lifter bore diameter	0.9051 to 0.9059 inch

Cylinder head and valves

Head warpage limit	0.002 inch per 6 inches
Minimum valve margin	3/64 inch
Valve seat angle, intake and exhaust	44.25 to 44.75-degrees
Valve face angle, intake and exhaust	43.25 to 43.75-degrees
Valve stem diameter	0.0311 to 0.312 inch
Valve stem-to-guide clearance	0.001 to 0.003 inch
Valve spring free length	1.967 inches
Valve spring installed height	1.64 inches
Valve length	
5.2L	
Intake	4.893 to 4.918 inches
Exhaust	4.907 to 4.932 inches
5.9L	
Intake	4.969 to 4.994 inches
Exhaust	4.978 to 5.012 inches

Crankshaft and connecting rods

Rod journal diameter	2.124 to 2.125 inch
Rod bearing oil clearance	
Desired	0.0005 to 0.0022 inch
Allowable	0.003 inch
Connecting rod side clearance (endplay)	0.006 to 0.014 inch
Main bearing journal diameter	2.4995 to 2.5005 inch
Main bearing oil clearance	
Number 1	0.005 to 0.0015 inch
Numbers 2, 3, 4, and 5	
Desired	0.005 to 0.0020 inch
Allowable	0.0025 inch
Crankshaft endplay	0.002 to 0.007 inch
Maximum taper and out-of-round (all journals)	0.001 inch

Pistons and rings

Piston-to-bore clearance	0.0005 to 0.0015 inch
Piston ring end gap	
5.2L	
Compression rings	0.010 to 0.020 inch
Oil control ring (steel rails)	0.010 to 0.050 inch
5.9L	
Compression rings	
Top ring	0.012 to 0.022 inch
Second ring	0.022 to 0.031 inch
Oil control ring (steel rails)	0.015 to 0.055 inch
Piston ring side clearance	
5.2L	
Compression rings	0.0015 to 0.0030 inch
Oil control ring (steel rails)	0.002 to 0.008 inch
5.9L	
Compression rings	0.0029 to 0.0038 inch
Oil control ring (steel rails)	0.0097 inch maximum

Torque specifications*

	Ft-lbs (unless otherwise indicated)
Main bearing cap bolts	85
Connecting rod cap nuts	45

* **Note:** *Refer to Part B for additional torque specifications.*

4.7L V8 engine

General

Displacement	4.7 liters (287 cubic inches)
Cylinder compression pressure	140 to 180 psi
Maximum variation between cylinders	30 psi (from the highest reading)
Oil pressure	
At curb idle	4 psi minimum
At 3000 rpm	25 to 80 psi

2D

4.7L V8 engine (continued)

Engine block

Bore diameter	3.6616 to 3.6622 inches
Taper (maximum)	0.002 inch
Out-of-round (maximum)	0.003 inch
Deck (head gasket surface) warpage limit	0.002 inch per 6 inches

Cylinder head and valves

Head warpage limit	0.002 inch per 6 inches
Minimum valve margin (typical)	1/32 inch
Valve seat angle, intake and exhaust	44.5-degrees
Valve face angle, intake and exhaust	45-degrees
Valve stem diameter	
Intake	0.2729 to 0.2739 inch
Exhaust	0.2717 to 0.2718 inch
Valve stem-to-guide clearance	
Intake	0.0008 to 0.0028 inch
Exhaust	0.0019 to 0.0039 inch
Valve spring free length	
1999	1.870 inches
2000	1.9134 inches
Valve spring installed height	
Intake	1.613 inches
Exhaust	1.606 inches
Valve length	
Intake	
1999	4.4539 to 4.4839 inches
2000	4.4666 to 4.4965 inches
Exhaust	4.5244 to 4.5543 inches

Crankshaft and connecting rods

Rod journal diameter	2.0076 to 2.0082 inches
Rod bearing oil clearance**	0.0004 to 0.0019 inch
Connecting rod side clearance (endplay)	0.004 to 0.0138 inch
Main bearing journal diameter	2.4996 to 2.5005 inches
Main bearing oil clearance**	0.0002 to 0.0013 inch
Crankshaft endplay (at thrust bearing)	0.0021 to 0.0112 inch
Taper (maximum)	0.0004 inch
Out-of-round (maximum)	0.0002 inch

Pistons and rings

Piston-to-bore clearance	See Sections 16 and 18
Piston ring end gap	
Top compression ring	0.015 to 0.025 inch
Second compression ring	0.015 to 0.025 inch
Oil control ring (steel rails)	
1999	0.010 to 0.050 inch
2000	0.010 to 0.030 inch
Piston ring side clearance	
Top compression ring	0.002 to 0.0037 inch
Second compression ring	0.0016 to 0.0031 inch
Oil control ring (steel rails)	0.0007 to 0.0091 inch

Torque specifications *

	Ft-lbs (unless otherwise indicated)
Bedplate bolts **(see illustration 22.24)**	
Step one	
Tighten bolts 1 through 12	40
Step two	
Tighten bolts 13 through 22	25 in-lbs
Step three	
Tighten bolts 13 through 22	Turn an additional 90 degrees
Step four	
Tighten bolts 23 through 28	20
Connecting rod cap bolts	
Step one	20
Step two	Turn an additional 90 degrees

* **Note:** *Refer to Part C for additional torque specifications.*

** **Note:** *These are the manufacturers specifications for new engines at the time this manual was written. Typically during a rebuild, a suitable tolerance for main and rod bearing oil clearances is 0.001 to 0.002 inch.*

2.4a Remove the oil pressure sending unit (arrow) . . .

2.4b . . . and connect a gauge to check oil pressure - inline six-cylinder engine shown; the sending unit on 5.2L/5.9L V8 engines is on the top of the block, behind the intake manifold

2.4c On 4.7L V8 engines, the oil pressure sending unit is located at the front of the block, just above the oil filter

1 General information

Included in this portion of Chapter 2 are the general overhaul procedures for the cylinder head(s) and internal engine components.

The information ranges from advice concerning preparation for an overhaul and the purchase of replacement parts to detailed, step-by-step procedures covering removal and installation of internal engine components and the inspection of parts.

The following Sections have been written based on the assumption that the engine has been removed from the vehicle. For information concerning in-vehicle engine repair, as well as removal and installation of the external components necessary for the overhaul, see Sections 5 and 8, and Chapters 2A, 2B and 2C.

The Specifications included in this Part are only those necessary for the inspection and overhaul procedures which follow. Refer to Parts A, B or C for additional Specifications.

2 Engine overhaul - general information

Refer to illustrations 2.4a, 2.4b and 2.4c

It's not always easy to determine when, or if, an engine should be completely overhauled, as a number of factors must be considered.

High mileage is not necessarily an indication that an overhaul is needed, while low mileage doesn't preclude the need for an overhaul. Frequency of servicing is probably the most important consideration. An engine that's had regular and frequent oil and filter changes, as well as other required maintenance, will most likely give many thousands of miles of reliable service. Conversely, a neglected engine may require an overhaul very early in its life.

Excessive oil consumption is an indication that piston rings, valve seals and/or valve guides are in need of attention. Make sure that oil leaks aren't responsible before deciding that the rings and/or guides are bad. Perform a cylinder compression check to determine the extent of the work required (see Section 3).

Check the oil pressure with a gauge installed in place of the oil pressure sending unit **(see illustrations)** and compare it to the Specifications. If it's extremely low, the bearings and/or oil pump are probably worn out.

Loss of power, rough running, knocking or metallic engine noises, excessive valve train noise and high fuel consumption rates may also point to the need for an overhaul, especially if they're all present at the same time. If a complete tune-up doesn't remedy the situation, major mechanical work is the only solution.

An engine overhaul involves restoring the internal parts to the specifications of a new engine. During an overhaul, the piston rings are replaced and the cylinder walls are reconditioned (rebored and/or honed). If a rebore is done by an automotive machine shop, new oversize pistons will also be installed. The main bearings, connecting rod bearings and camshaft bearings are generally replaced with new ones and, if necessary, the crankshaft may be reground to restore the journals. Generally, the valves are serviced as well, since they're usually in less-than-perfect condition at this point.

While the engine is being overhauled, other components, such as the distributor, starter and alternator, can be rebuilt as well. The end result should be a like new engine that will give many thousands of trouble free miles. **Note:** *Critical cooling system components such as the hoses, drivebelts, thermostat and water pump MUST be replaced with new parts when an engine is overhauled.* There are engine rebuilders who will not guarantee their work if the radiator hasn't been professionally cleaned at the time of rebuilt-engine installation. The radiator should be checked carefully to ensure that it isn't clogged or leaking (see Chapter 3). Also, we don't recommend overhauling the oil pump - always install a new one when an engine is rebuilt.

Before beginning the engine overhaul, read through the entire procedure to familiarize yourself with the scope and requirements of the job. Overhauling an engine isn't difficult if you have the right equipment and follow the instructions carefully, but it is time-consuming. Plan on the vehicle being tied up for a minimum of two weeks, especially if parts must be taken to an automotive machine shop for repair or reconditioning. Check on availability of parts and make sure that any necessary special tools and equipment are obtained in advance.

Most work can be done with typical hand tools, although a number of precision measuring tools are required for inspecting parts to determine if they must be replaced. Often an automotive machine shop will handle the inspection of parts and offer advice concerning reconditioning and replacement. **Note:** *Always wait until the engine has been completely disassembled and all components, especially the engine block, have been inspected before deciding what service and repair operations must be performed by an automotive machine shop.*

Since the block's condition will be the major factor to consider when determining whether to overhaul the original engine or buy a rebuilt one, never purchase parts or have machine work done on other components until the block has been thoroughly inspected. As a general rule, time is the primary cost of an overhaul, so it doesn't pay to install worn or substandard parts.

As a final note, to ensure maximum life and minimum trouble from a rebuilt engine, everything must be assembled with care in a spotlessly-clean environment.

3 Compression check

Refer to illustration 3.6

1 A compression check will tell you what mechanical condition the upper end (pistons, rings, valves, head gaskets) of your engine is in. Specifically, it can tell you if the compression is down due to leakage caused by worn piston rings, defective valves and seats or a blown head gasket. **Note:** *The engine must be at normal operating temperature and the battery must be fully charged for this check.*

2 Disable the ignition system by unplugging the primary (low voltage) electrical connector(s) from the ignition coil(s). The fuel pump circuit should also be disabled (see Chapter 4, Section 2).

3 Clean the area around the spark plugs before you remove them (compressed air should be used, if available). The idea is to prevent dirt from getting into the cylinders as the compression check is being done.

4 Remove all of the spark plugs from the engine (see Chapter 1).

5 Block the throttle wide open.

6 Install the compression gauge in the number one spark plug hole **(see illustration)**.

7 Crank the engine over at least seven compression strokes and watch the gauge. The compression should build up quickly in a healthy engine. Low compression on the first stroke, followed by gradually increasing pressure on successive strokes, indicates worn piston rings. A low compression reading on the first stroke, which doesn't build up during successive strokes, indicates leaking valves or a blown head gasket (a cracked head could also be the cause). Deposits on the undersides of the valve heads can also cause low compression. Record the highest gauge reading obtained.

8 Repeat the procedure for the remaining cylinders and compare the results to this Chapter's Specifications.

9 Add some engine oil (about three squirts from a plunger-type oil can) to each cylinder, through the spark plug hole, and repeat the test.

10 If the compression increases after the oil is added, the piston rings are definitely worn. If the compression doesn't increase significantly, the leakage is occurring at the valves or head gasket. Leakage past the valves may be caused by burned valve seats and/or faces or warped, cracked or bent valves.

11 If two adjacent cylinders have equally-low compression, there's a strong possibility that the head gasket between them is blown. The appearance of coolant in the combustion chambers or the crankcase would verify this condition.

12 If one cylinder is about 20 percent lower than the others, and the engine has a slightly rough idle, a worn exhaust lobe on the camshaft could be the cause.

13 If the compression is unusually high, the combustion chambers are probably coated with carbon deposits. If that's the case, the cylinder head(s) should be removed and decarbonized.

14 If compression is way down or varies greatly between cylinders, it would be a good idea to have a leak-down test performed by an automotive repair shop. This test will pinpoint exactly where the leakage is occurring and how severe it is.

3.6 A compression gauge with a threaded fitting for the spark plug hole is preferred over the type that requires hand pressure to maintain the seal

4.4 An inexpensive vacuum gauge can tell you a lot about an engine's condition

4 Vacuum gauge diagnostic checks

Refer to illustrations 4.4 and 4.6

A vacuum gauge provides valuable information about the condition of internal engine components. You can check for worn rings or cylinder walls, leaking head or intake manifold gaskets, restricted exhaust, stuck or burned valves, weak valve springs, improper ignition or valve timing and ignition problems.

Unfortunately, vacuum gauge readings are easy to misinterpret, so they should be used in conjunction with other tests to confirm the diagnosis.

Both the absolute readings and the rate of needle movement are important for accurate interpretation. Most gauges measure vacuum in inches of mercury (in-Hg). The following references to vacuum assume the diagnosis is being performed at sea level. As elevation increases (or atmospheric pressure decreases), the reading will decrease. For every 1,000 foot increase in elevation above approximately 2000 feet, the gauge readings will decrease about one inch of mercury.

Connect the vacuum gauge directly to intake manifold vacuum, not to ported (throttle body) vacuum **(see illustration)**. Be sure no hoses are left disconnected during the test or false readings will result.

Before you begin the test, allow the engine to warm up completely. Block the wheels and set the parking brake. With the transmission in Park, start the engine and allow it to run at normal idle speed. **Warning:** *Carefully inspect the fan blades for cracks or damage before starting the engine. Keep your hands and the vacuum gauge clear of the fan and do not stand in front of the vehicle or in line with the fan when the engine is running.*

Read the vacuum gauge; an average, healthy engine should normally produce about 17 to 22 inches of vacuum with a fairly steady needle **(see illustration)**. Refer to the following vacuum gauge readings and what they indicate about the engine's condition:

a) *A low steady reading usually indicates a leaking gasket between the intake manifold and throttle body, a leaky vacuum hose, late ignition timing or incorrect camshaft timing. Check ignition timing with a timing light and eliminate all other possible causes, utilizing the tests provided in this Chapter before you remove the timing chain cover to check the timing marks.*
b) *If the reading is three to eight inches below normal and it fluctuates at that low reading, suspect an intake manifold gasket leak at an intake port or a faulty fuel injector.*
c) *If the needle has regular drops of about two-to-four inches at a steady rate, the valves are probably leaking. Perform a compression check or leak-down test to confirm this.*
d) *An irregular drop or down-flick of the needle can be caused by a sticking valve or an ignition misfire. Perform a compression check or leak-down test and read the spark plugs.*
e) *A rapid vibration of about four inches Hg vibration at idle combined with exhaust smoke indicates worn valve guides. Perform a leak-down test to confirm this. If the rapid vibration occurs with an increase in engine speed, check for a leaking intake manifold gasket or head gasket, weak valve springs, burned valves or ignition misfire.*
f) *A slight fluctuation, say one inch up and down, may mean ignition problems.*

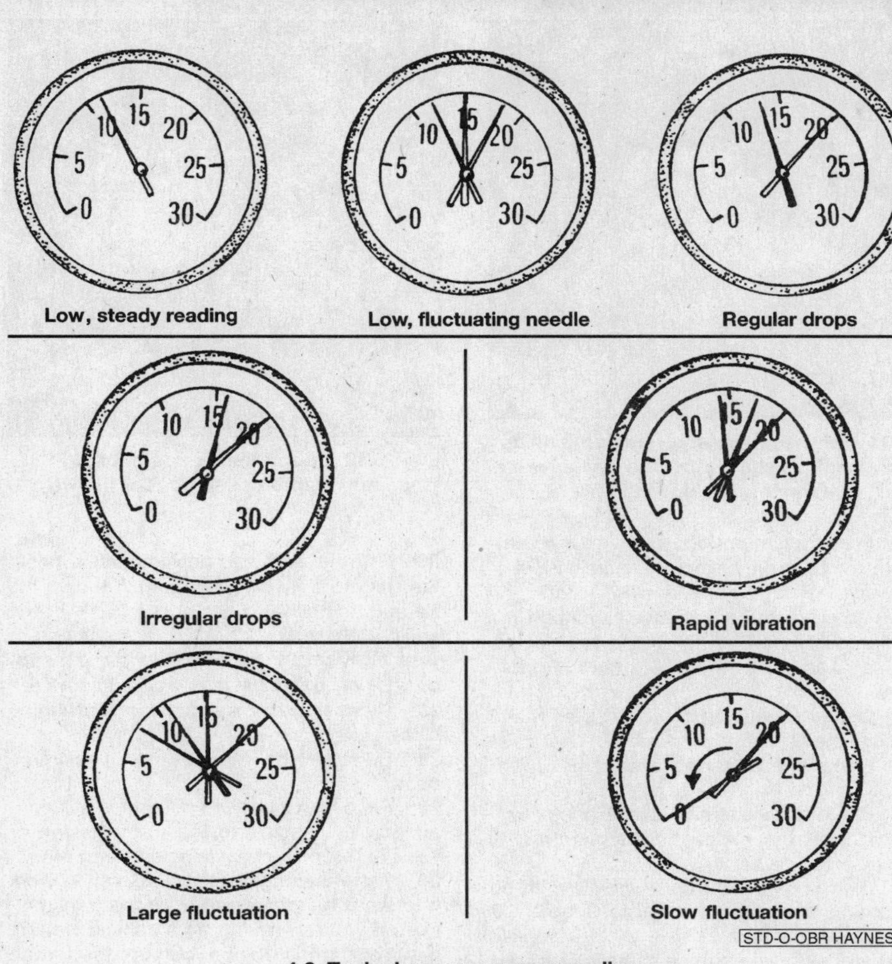

4.6 Typical vacuum gauge readings

Check all the usual tune-up items and, if necessary, run the engine on an ignition analyzer.

- g) *If there is a large fluctuation, perform a compression or leak-down test to look for a weak or dead cylinder or a blown head gasket.*
- h) *If the needle moves slowly through a wide range, check for a clogged PCV system, incorrect idle fuel mixture, carburetor/throttle body or intake manifold gasket leaks.*
- i) *Check for a slow return after revving the engine by quickly snapping the throttle open until the engine reaches about 2,500 rpm and let it shut. Normally the reading should drop to near zero, rise above normal idle reading (about 5 in.-Hg over) and then return to the previous idle reading. If the vacuum returns slowly and doesn't peak when the throttle is snapped shut, the rings may be worn. If there is a long delay, look for a restricted exhaust system (often the muffler or catalytic converter). An easy way to check this is to temporarily disconnect the exhaust ahead of the suspected part and re-test.*

5 Engine removal - methods and precautions

If you've decided that an engine must be removed for overhaul or major repair work, several preliminary steps should be taken.

Locating a suitable place to work is extremely important. Adequate work space, along with storage space for the vehicle, will be needed. If a shop or garage isn't available, at the very least a flat, level, clean work surface made of concrete or asphalt is required.

Cleaning the engine compartment and engine before beginning the removal procedure will help keep your tools and your hands clean.

An engine hoist or A-frame will also be necessary. Make sure the equipment is rated in excess of the combined weight of the engine and accessories. Safety is of primary importance, considering the potential hazards involved in lifting the engine out of the vehicle.

If the engine is being removed by a novice, a helper should be available. Advice and aid from someone more experienced would also be helpful. There are many instances when one person cannot simultaneously perform all of the operations required when lifting the engine out of the vehicle.

Plan the operation ahead of time. Arrange for or obtain all of the tools and equipment you'll need prior to beginning the job. Some of the equipment necessary to perform engine removal and installation safely and with relative ease are (in addition to an engine hoist) a heavy duty floor jack, complete sets of wrenches and sockets as described in the front of this manual, wooden blocks and plenty of rags and cleaning solvent for mopping up spilled oil, coolant and gasoline. If the hoist must be rented, make sure that you arrange for it in advance and perform all of the operations possible without it beforehand. This will save you money and time.

Plan for the vehicle to be out of use for quite a while. A machine shop will be required to perform some of the work which the do-it-yourselfer can't accomplish without special equipment. These shops often have a busy schedule, so it would be a good idea to consult them before removing the engine in order to accurately estimate the amount of time required to rebuild or repair components that may need work.

Always be extremely careful when removing and installing the engine. Serious injury can result from careless actions. Plan ahead, take your time and a job of this nature, although major, can be accomplished successfully.

6 Engine - removal and installation

Refer to illustrations 6.5, 6.11, 6.12, 6.20 and 6.24

Warning 1: *The air conditioning system is under high pressure! Have a dealer service department or service station discharge the system before disconnecting any system hoses or fittings.*

Warning 2: *Wait until the engine is completely cool before beginning this procedure.*

Removal

1 Refer to Chapter 4 and relieve the fuel system pressure, then disconnect the negative cable from the battery. **Warning:** *These models are equipped with airbags. The airbag is armed and can deploy (inflate) anytime the battery is connected. To prevent accidental deployment (and possible injury), disconnect the negative battery cable whenever working near airbag components, such as the crash sensors at the front of the vehicle. After the battery is disconnected, wait at least 2 minutes before beginning work (the system has a back-up capacitor that must fully discharge). For more information* see Chapter 12.

2 Cover the fenders and cowl and remove the hood (see Chapter 11). Special pads are available to protect the fenders, but an old bedspread or blanket will also work.

3 Remove the air cleaner assembly (see Chapter 4).

4 Drain the cooling system (see Chapter 1).

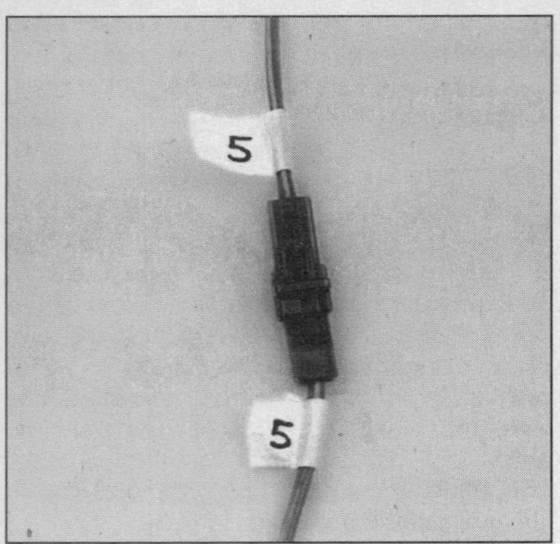

6.5 Label both ends of each wire before unplugging the connector

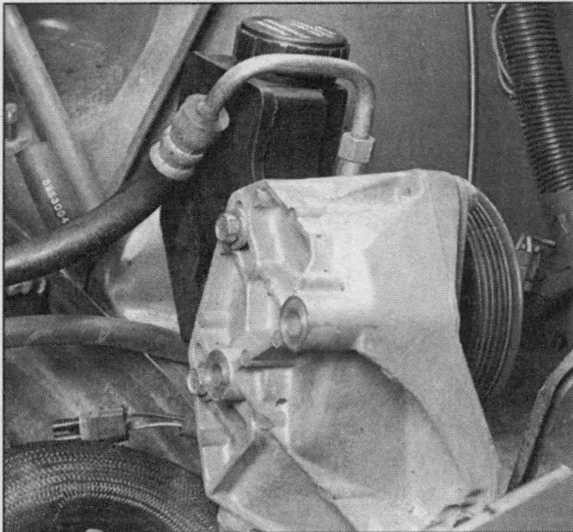

6.11 Set the power steering pump aside with the lines still connected - be sure it's upright so fluid won't spill

6.12 Unbolt the air conditioning compressor and set it out of the way

5 Label the vacuum lines, emissions system hoses, wiring connectors, ground strap and fuel lines, to ensure correct reinstallation **(see illustration)**, then detach them. If there's any possibility of confusion, make a sketch of the engine compartment and clearly label the lines, hoses and wires.

6 Label and detach all coolant hoses from the engine.

7 Remove the drivebelt(s) (see Chapter 1).

8 Remove the cooling fan, shroud and radiator (see Chapter 3).

9 **Warning:** *Gasoline is extremely flammable, so extra precautions must be taken when working on any part of the fuel system. DO NOT smoke or allow open flames or bare light bulbs near the vehicle. Also, don't work in a garage if a gas-type appliance is present.* Disconnect the fuel lines running from the engine to the chassis (see Chapter 4). Plug or cap all open fittings/lines.

10 Disconnect the throttle cable (and TV cable/cruise control cable, if equipped) from the engine (see Chapter 4).

11 Unbolt the power steering pump (see Chapter 10). Leave the lines/hoses attached **(see illustration)** and make sure the pump is kept in an upright position in the engine compartment (use wire or rope to restrain it out of the way).

12 On air-conditioned vehicles, unbolt the compressor (see Chapter 3) and set it aside. Do not disconnect the hoses **(see illustration)**.

13 Drain the engine oil (see Chapter 1) and remove the oil filter.

14 Remove the starter motor (see Chapter 5).

15 Remove the alternator (see Chapter 5).

16 Unbolt the exhaust system from the engine (see Chapter 4).

17 If you're working on a vehicle with an automatic transmission, refer to Chapter 7B and remove the torque converter-to-driveplate fasteners. Also unclip the automatic transmission fluid cooler lines from any studs on the engine oil pan, on models so equipped.

18 Support the transmission with a jack. Position a block of wood between the jack and transmission to prevent damage to the transmission. Special transmission jacks with safety chains are available - use one if possible.

19 Attach an engine sling or a length of chain to the lifting brackets on the engine.

20 Roll the hoist into position and connect the sling to it **(see illustration)**. Take up the slack in the sling or chain, but don't lift the engine. **Warning:** *DO NOT place any part of your body under the engine when it's supported only by a hoist or other lifting device.*

21 Remove the transmission-to-engine block bolts.

22 Remove the engine mount-to-frame bolts.

23 Recheck to be sure nothing is still connecting the engine to the transmission or vehicle. Disconnect anything still remaining.

24 Raise the engine slightly. Carefully work it forward to separate it from the transmission. If you're working on a vehicle with an automatic transmission, be sure the torque converter stays in the transmission (clamp a pair of vise-grips to the housing to keep the converter from sliding out). If you're working on a vehicle with a manual transmission, the input shaft must be completely disengaged from the clutch. Slowly raise the engine out of the engine compartment **(see illustration)**. Check carefully to make sure nothing is hanging up.

6.20 4.7L V8 engines are equipped with engine lifting studs (arrows) that protrude from the cylinder heads and the front of the water pump

6.24 Pull the engine forward as far as possible to clear the transmission, then lift the engine high enough to clear the body

25 Remove the flywheel/driveplate and mount the engine on an engine stand.

Installation

26 Check the engine and transmission mounts. If they're worn or damaged, replace them.
27 If you're working on a vehicle with a manual transmission, install the clutch and pressure plate (Chapter 7A). Now is a good time to install a new clutch.
28 Carefully lower the engine into the engine compartment - make sure the engine mounts line up.
29 If you're working on a vehicle with an automatic transmission, guide the torque converter into the crankshaft following the procedure outlined in Chapter 7B.
30 If you're working on a vehicle with a manual transmission, apply a dab of high-temperature grease to the input shaft and guide it into the crankshaft pilot bearing until the bellhousing is flush with the engine block.
31 Install the transmission-to-engine bolts and tighten them securely. **Caution:** *DO NOT use the bolts to force the transmission and engine together!*
32 Reinstall the remaining components in the reverse order of removal.
33 Add coolant, oil, power steering and transmission fluid as needed. If you're working on a vehicle with a 4.7L V8 engine, be sure to properly bleed the cooling system (see Chapter 1).
34 Run the engine and check for leaks and proper operation of all accessories, then install the hood and test drive the vehicle.
35 Have the air conditioning system recharged and leak tested.

7 Engine rebuilding alternatives

The do-it-yourselfer is faced with a number of options when performing an engine overhaul. The decision to replace the engine block, piston/connecting rod assemblies and crankshaft depends on a number of factors, with the number one consideration being the condition of the block. Other considerations are cost, access to machine shop facilities, parts availability, time required to complete the project and the extent of prior mechanical experience on the part of the do-it-yourselfer. **Note:** *The costs of alternatives described in this Section can vary, depending upon quality of parts, machine work required and the necessary tools and equipment to correctly do the work. Many automotive parts stores carry complete (long and short block) assemblies in addition to individual repair parts. Consult the local parts store on price and availability to make the final repair/replace decision.*

Some of the rebuilding alternatives include:

Individual parts - If the inspection procedures reveal that the engine block and most engine components are in reusable condition, purchasing individual parts may be the most economical alternative. The block, crankshaft and piston/connecting rod assemblies should all be inspected carefully. Even if the block shows little wear, the cylinder bores should be surface honed.

Short block - A short block consists of an engine block with a crankshaft, camshaft (cam-in-block engines) and piston/connecting rod assemblies already installed. All new bearings are incorporated and all clearances will be correct. The existing valve train components, cylinder head(s) and external parts can be bolted to the short block with little or no machine-shop work necessary.

Long block - A long block consists of a short block plus an oil pump, oil pan, cylinder head(s), valve cover(s) and valve train components, timing sprockets and chain or gears and timing cover. All components are installed with new bearings, seals and gaskets incorporated throughout. The installation of manifolds and external parts is all that's necessary.

Give careful thought to which alternative is best for you and discuss the situation with local automotive machine shops, auto parts dealers and experienced rebuilders before ordering or purchasing replacement parts.

8 Engine overhaul - disassembly sequence

1 It's much easier to disassemble and work on the engine if it's mounted on a portable engine stand. A stand can often be rented quite cheaply from an equipment rental yard. Before the engine is mounted on a stand, the flywheel/driveplate should be removed from the engine.
2 If a stand isn't available, it's possible to disassemble the engine with it blocked up on the floor. Be extra careful not to tip or drop the engine when working without a stand.
3 If you're going to obtain a rebuilt engine, all external components must come off first to be transferred to the replacement engine, just as they will if you're doing a complete engine overhaul yourself. These include:

Alternator and brackets
Emissions control components
Spark plug wires and spark plugs
Ignition coil(s) and distributor (if equipped)
Thermostat and housing cover
Water pump
Fuel injection components
Intake/exhaust manifold(s)
Oil filter
Engine mounts
Clutch and flywheel/driveplate

Note: *When removing the external components from the engine, pay close attention to details that may be helpful or important during installation. Note the installed position of gaskets, seals, spacers, pins, brackets, washers, bolts and other small items.*
4 If you're obtaining a short block, which consists of the engine block, crankshaft, pistons and connecting rods all assembled, then the cylinder head(s), oil pan and oil pump will have to be removed as well. See *Engine rebuilding alternatives* for additional information regarding the different possibilities to be considered.
5 If you're planning a complete overhaul, the engine must be disassembled and the internal components removed in the following order:

Six-cylinder engine and 5.2L/5.9L V8 engines

Valve cover(s)
Intake and exhaust manifolds
Rocker arms and pushrods
Cylinder head(s)
Valve lifters
Timing cover
Timing chain and sprockets
Camshaft
Oil pan
Oil pump
Piston/connecting rod assemblies
Crankshaft and main bearings

4.7L V8 engines

Valve cover(s)
Intake and exhaust manifolds
Rocker arms and hydraulic lash adjusters
Timing cover
Timing chains and sprockets
Camshaft(s)
Cylinder head(s)
Oil pan
Oil pick-up tube and windage tray
Oil pump
Piston/connecting rod assemblies
Bedplate
Crankshaft and main bearings

6 Before beginning the disassembly and overhaul procedures, make sure the following items are available. Also, refer to Section 21 for a list of tools and materials needed for engine reassembly.

Common hand tools
Small cardboard boxes or plastic bags for storing parts
Gasket scraper
Ridge reamer
Vibration damper puller
Micrometers
Telescoping gauges
Dial indicator set
Valve spring compressor
Cylinder surfacing hone
Piston ring groove-cleaning tool
Electric drill motor
Tap and die set
Wire brushes
Oil gallery brushes
Cleaning solvent

9 Cylinder head - disassembly

Refer to illustrations 9.2, 9.3 and 9.4

Note: *New and rebuilt cylinder heads are commonly available for most engines at deal-*

9.2 A small plastic bag, with an appropriate label, can be used to store the valve train components so they can be kept together and reinstalled in the correct guide

9.3 Use a valve spring compressor to compress the spring, then remove the keepers with needle-nose pliers or a small magnet

9.4 If the valve won't pull through the guide, deburr the edge of the stem and the area around the top of the keeper groove with a file

erships and auto parts stores. Due to the fact that some specialized tools are necessary for the disassembly and inspection procedures, and some parts may not be readily available, it may be more practical and economical for the home mechanic to purchase replacement head(s) rather than taking the time to disassemble, inspect and recondition the original(s).

1 Cylinder head disassembly involves removal of the intake and exhaust valves and related components. Make sure to remove all the valve actuating components before disassembling the cylinder head. Valve actuating components consist of: the rocker arms, rocker arm bolts and pivots on six-cylinder and 5.2L/5.9L V8 engines and the rocker arms, camshafts and hydraulic lash adjusters on 4.7L V8 engines. Label the parts or store them separately so they can be reinstalled in their original locations.

2 Before the valves are removed, arrange to label and store them, along with their related components, so they can be kept separate and reinstalled in the same valve guides they were removed from **(see illustration)**.

3 Compress the springs on the first valve with a spring compressor and remove the keepers **(see illustration)**. Carefully release the valve spring compressor and remove the retainer, the spring and the spring seat (if used).

4 Pull the valve out of the head, then remove the oil seal from the guide. If the valve binds in the guide (won't pull through), push it back into the head and deburr the area around the keeper groove with a fine file or whetstone **(see illustration)**.

5 Repeat the procedure for the remaining valves. Remember to keep all the parts for each valve together so they can be reinstalled in the same locations.

6 Once the valves and related components have been removed and stored in an organized manner, the head should be thoroughly cleaned and inspected. If a complete engine overhaul is being done, finish the engine disassembly procedures before beginning the cylinder head cleaning and inspection process.

10 Cylinder head - cleaning and inspection

1 Thorough cleaning of the cylinder heads and related valve train components, followed by a detailed inspection, will enable you to decide how much valve service work must be done during the engine overhaul. **Note:** *If the engine was severely overheated, the cylinder heads are probably warped (see Step 12).*

Cleaning

2 Scrape all traces of old gasket material and sealing compound off the cylinder head gasket, intake manifold and exhaust manifold sealing surfaces. Be very careful not to gouge the cylinder head. Special gasket removal solvents that soften gaskets and make removal much easier are available at auto parts stores.

3 Remove all built up scale from the coolant passages.

4 Run a stiff wire brush through the various holes to remove deposits that may have formed in them.

5 Run an appropriate-size tap into each of the threaded holes to remove corrosion and thread sealant that may be present. If compressed air is available, use it to clear the holes of debris produced by this operation. **Warning:** *Wear eye protection when using compressed air!*

6 Clean the combustion chambers with a brass wire brush and solvent if carbon has accumulated.

7 Clean the cylinder head with solvent and dry it thoroughly. Compressed air will speed the drying process and ensure that all holes and recessed areas are clean. **Note:** *Decarbonizing chemicals are available and may prove very useful when cleaning cylinder heads and valve train components. They are very caustic and should be used with caution. Be sure to follow the instructions on the container.*

8 Clean the rocker arms with solvent and dry them thoroughly (don't mix them up during the cleaning process). Compressed air will speed the drying process and can be used to clean out the oil passages.

9 Clean all the valve springs, spring seats, valve stem locks and retainers with solvent and dry them thoroughly. Do the components from one valve at a time to avoid mixing up the parts.

10 Scrape off any heavy deposits that may have formed on the valves, then use a motorized wire brush to remove deposits from the valve heads and stems. Again, make sure the valves don't get mixed up.

Inspection

Note: *Be sure to perform all of the following inspection procedures before concluding that machine shop work is required. Make a list of the items that need attention.*

Cylinder head

Refer to illustrations 10.12 and 10.14

11 Inspect the heads very carefully for cracks, evidence of coolant leakage and other damage. If cracks are found, check with an automotive machine shop concerning repair. If repair isn't possible, a new cylinder head should be obtained.

12 Using a straightedge and feeler gauge, check the cylinder head gasket mating surface for warpage **(see illustration)**. If the warpage exceeds the limit specified in this Chapter, it can be resurfaced at an automotive machine shop. **Note:** *The manufacturer recommends replacing the cylinder heads on 4.7L V8 engines if the warpage exceeds the specified amount. Since these engines are relatively new at the time this manual was written, be sure to consult your local machine shop for alternative solutions before replacing the cylinder heads.*

10.12 Check the cylinder head surface for warpage by trying to slip a feeler gauge under the straightedge - see the Specifications for the maximum warpage and use a feeler gauge of that size

13 Examine the valve seats in each of the combustion chambers. If they're pitted, cracked or burned, the cylinder head will require valve service that's beyond the scope of the home mechanic.

14 Check the valve stem-to-guide clearance with a small hole gauge and micrometer **(see illustration)**, then measure the valve stem diameter with a micrometer and subtract it from the valve guide inside diameter to obtain the stem-to-guide clearance. When using a small hole gauge or telescoping snap gauge, insert the gauge to the middle portion of the valve guide (where wear should be minimal) and tighten the gauge. Move the gauge up and down in the guide. If the guide isn't worn the clearance should be equal from top to bottom. Loose areas indicate that the guide is tapered. If the measurement exceeds the stem-to-guide clearance limit found in this Chapter's Specifications, the valve guides should be resized or replaced. After this is done, if there's still some doubt regarding the condition of the valve guides they should be checked by an automotive machine shop (the cost should be minimal).

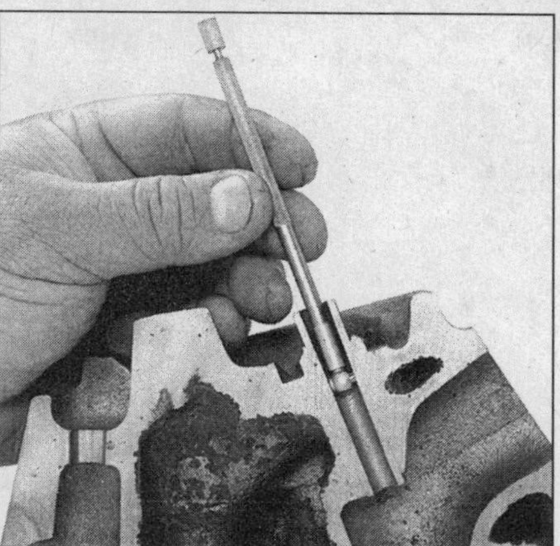

10.14 Use a small hole gage to determine the inside diameter of the valve guides (the gauge is then measured with a micrometer

Valves

Refer to illustrations 10.15 and 10.16

15 Carefully inspect each valve face for uneven wear, deformation, cracks, pits and burned areas **(see illustration)**. Check the valve stem for scuffing and galling and the neck for cracks. Rotate the valve and check for any obvious indication that it's bent. Look for pits and excessive wear on the end of the stem. The presence of any of these conditions indicates the need for valve service by an automotive machine shop.

16 Measure the margin width on each valve **(see illustration)**. Any valve with a margin narrower than specified in this Chapter will have to be replaced with a new one.

Valve components

Refer to illustrations 10.17 and 10.18

17 Check each valve spring for wear (on the ends) and pits. Measure the free length and compare it to the Specifications in this Chapter **(see illustration)**. Any springs that are shorter than specified have sagged and should not be reused. The tension of all springs should be checked with a special fixture before deciding that they're suitable for use in a rebuilt engine (take the springs to an automotive machine shop for this check).

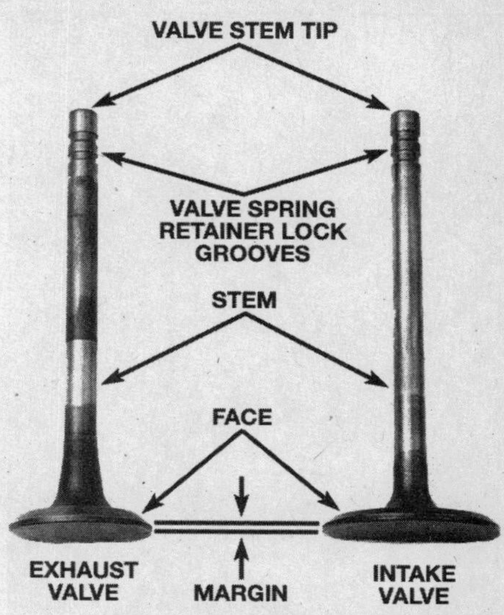

10.15 Check for valve wear at the points shown here

18 Stand each spring on a flat surface and check it for squareness **(see illustration)**. If any of the springs are distorted or sagged, replace all of them with new parts.

19 Check the spring retainers and valve stem keepers for obvious wear and cracks. Any questionable parts should be replaced with new ones, as extensive damage will occur if they fail during engine operation.

Camshaft, bearings, rocker arms and hydraulic lash adjusters

20 On six-cylinder and 5.2L/5.9L V8 engines, refer to Chapter 2A or 2B for the camshaft, camshaft bearing and the rocker arm inspection procedures. On 4.7L engines, refer to Chapter 2C for the camshaft, rocker arms and the hydraulic lash adjuster inspection procedures. Be sure to inspect the camshaft bearing journals in the cylinder head (4.7L only) before the cylinder head is sent to the machine shop to have the valves

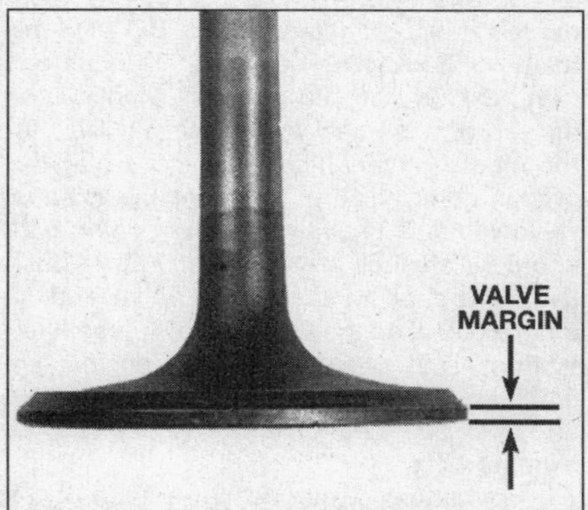

10.16 Valve margin width must be as specified (if no margin exists, the valve cannot be reused)

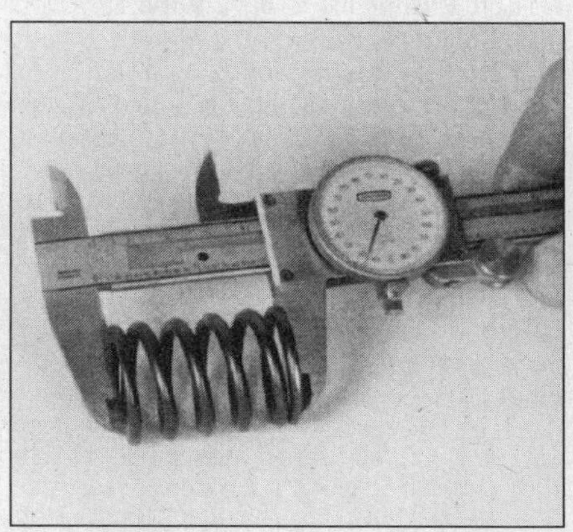

10.17 Measure the length of each valve spring with a dial or vernier caliper

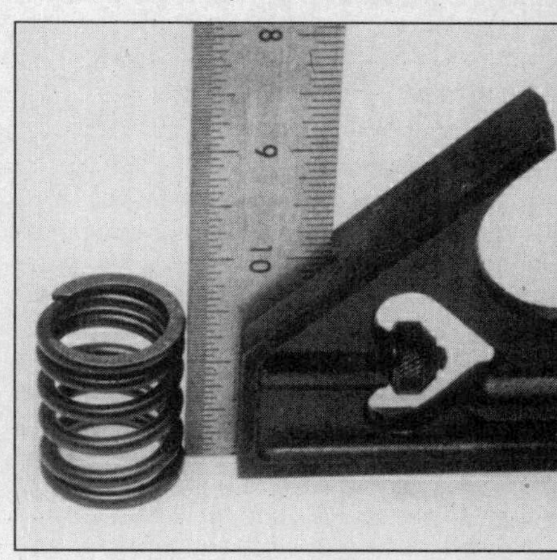

10.18 Check each spring for squareness

11.3a The valve seats are ground at the machine shop using a machine like this one - the precisely-shaped grinding stone is centered on a mandrel in the new valve guide, making the new seat concentric to the guide

serviced. If the journals are gouged or scored the cylinder head will have to be replaced regardless of the condition of the valves and related components.

21 Any damaged or excessively worn parts must be replaced with new ones.

22 If the inspection process indicates that the valve components are in generally poor condition and worn beyond the limits specified, which is usually the case in an engine that's being overhauled, reassemble the valves in the cylinder head and refer to Section 11 for valve servicing recommendations.

11 Valves - servicing

Refer to illustrations 11.3a and 11.3b

1 Because of the complex nature of the job and the special tools and equipment needed, servicing of the valves, the valve seats and the valve guides, commonly known as a valve job, should be done by a professional.

2 The home mechanic can remove and disassemble the heads, do the initial cleaning and inspection, then reassemble and deliver them to a dealer service department or an automotive machine shop for the actual service work. Doing the inspection will enable you to see what condition the cylinder head and valvetrain components are in and will

11.3b A machine like this mills a new, flat surface on heads that are warped

ensure that you know what work and new parts are required when dealing with an automotive machine shop.

3 The dealer service department, or automotive machine shop, will remove the valves and springs, recondition or replace the valves and valve seats, recondition or replace the valve guides, check and replace the valve springs, spring retainers and keepers (as necessary), replace the valve seals with new ones, reassemble the valve components and make sure the installed spring height is correct. The cylinder head gasket surface will also be resurfaced if it's warped **(see illustrations)**.

4 After the valve job has been performed by a professional, the cylinder head will be in like-new condition. When the cylinder head is returned, be sure to clean it again before installation on the engine to remove any metal particles and abrasive grit that may still be present from the valve service or cylinder head resurfacing operations. Use compressed air, if available, to blow out all the oil holes and passages.

12 Cylinder head - reassembly

Refer to illustrations 12.6, 12.9 and 12.10

1 Regardless of whether or not a head was sent to an automotive repair shop for valve servicing, make sure it's clean before beginning reassembly.

2 If a head was sent out for valve servicing, the valves and related components will already be in place. Begin the reassembly procedure with Step 8.

3 Beginning at one end of the head, lubricate and install the first valve. Apply moly-base grease or clean engine oil to the valve stem.

4 Repeat the procedure for the remaining valves. Be sure to return the components to their original locations - don't mix them up!

5 Valve stem seals are manufactured in several different varieties. The first type of seal is called a Positive type seal. This type of seal stays in its installed position while the

12.6 Valve seal installation details - 5.2L/5.9L engines

1 Exhaust valve seal (umbrella type)
2 Intake valve seal (positive type)

valve moves up and down so it can meter a precise amount of oil to the valve guide. Positive type seals use a small press fit between the seal and the valve guide to hold the seal in place. Always apply a small amount of oil to the top of the valve guide and valve stem when installing positive type seals, this will help ease the installation process and possible misalignment of the seal. The second type of seal is called an Umbrella seal. This type of seal is simply placed over the valve stem and pushed down to the top of the valve guide, when the valve opens for first time the seal is pushed up on the valve stem (the distance of the valve lift) where it stays for the remainder of its service life. Umbrella seals will then move up and down with the valve as it opens and closes, shrouding oil that is supplied to the rocker arms from going directly down into the valve guide. Most valve seal kits come equipped with a plastic installation tool which must be inserted over the end of valve stem during installation of the seal. Once the seal installation tool is inserted over the valve stem, the valve seal can easily slide past the valve keeper grooves preventing premature damage to the seal.

6 Intake and exhaust valves on most models require different seals - DO NOT mix them up! Six-cylinder engines are equipped with positive type seals and are identified on the top of each seal by the initials "INT" for the intake valve which is typically black or gray in color and by "EXH" for the exhaust valve which is typically brown in color. 5.2L and 5.9L engines are equipped with a positive type seal on the intake valve and an umbrella seal on the exhaust valve **(see illustration)**. The valve seals on 4.7L engines are common for the intake and the exhaust valve and can be used on either the intake or exhaust valve.

7 Install new valve seals on each of the intake and exhaust valves. Gently push each seal into place until it's completely seated on the guide. Don't twist or cock the seals during installation or they won't seal properly on

12.9 Apply a small dab of grease to each keeper as shown here before installation - it will hold them in place on the valve stem as the spring is released

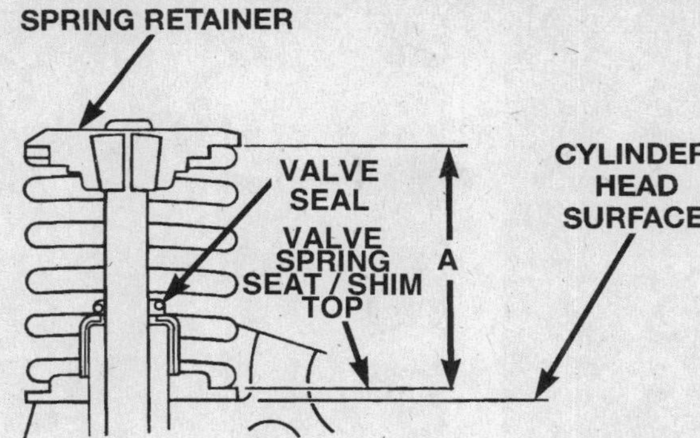

12.10 Be sure to check the valve spring installed height (the distance from the top of the seat/shims to the bottom of the retainer)

the valve stems. **Note:** *Some positive type seals may require the use a small mallet and a seal installer or deep socket to gently tap the seal onto the boss on the head.*

8 Place the valve springs and retainers in place on the head and apply pressure with the spring compressor.

9 Position the keepers in the upper groove, then slowly release the compressor and make sure the keepers seat properly. Apply a small dab of grease to each keeper to hold it in place if necessary **(see illustration)**.

10 Check the installed valve spring height with a ruler graduated in 1/64-inch increments or a dial caliper. If the head was sent out for service work, the installed height should be correct (but don't automatically assume that it is). The measurement is taken from the top of each spring seat or shim(s) to the bottom of the retainer **(see illustration)**. If the height is greater than specified, shims can be added under the springs to correct it. **Caution:** *Don't, under any circumstances, shim the springs to the point where the installed height is less than specified.*

11 Install the remaining cylinder head components in their original locations on the cylinder head. Make sure to apply moly-base grease to the rocker arm contact points before installation of the rocker arms. **Note:** *On 4.7L engines, it is best to install the rocker arms after the cylinder heads/camshafts and the timing chains have been installed on the engine, then follow the rocker arm installation procedure in Chapter 2C.*

13 Pistons/connecting rods - removal

Refer to illustrations 13.1, 13.3, 13.4, 13.6a, 13.6b and 13.9

Note: *Prior to removing the piston/connecting rod assemblies, remove the cylinder head(s), the oil pan and the oil pump by referring to the appropriate Sections in Chapter 2A, 2B or 2C.*

1 Use your fingernail to feel if a ridge has formed at the upper limit of ring travel (about 1/4-inch down from the top of each cylinder). If carbon deposits or cylinder wear have produced ridges, they must be completely removed with a special tool **(see illustration)**. Follow the manufacturer's instructions provided with the tool. Failure to remove the ridges before attempting to remove the piston/connecting rod assemblies may result in piston breakage.

13.1 A ridge reamer is required to remove the ridge from the top of each cylinder - do this before removing the pistons!

2 After the cylinder ridges have been removed, turn the engine upside-down so the crankshaft is facing up.

3 Before the connecting rods are removed, check the endplay with feeler gauges. Slide them between the first connecting rod and the crankshaft throw until the play is removed **(see illustration)**. The endplay is equal to the thickness of the feeler gauge(s). If the endplay exceeds the service limit, new connecting rods will be required. If new rods (or a new crankshaft) are installed, the endplay may fall under the specified minimum (if it does, the rods will have to be machined to restore it - consult an automotive machine shop for advice if necessary). Repeat the procedure for the remaining connecting rods.

4 Check the connecting rods and caps for identification marks. If they aren't plainly marked, use a small center punch to make the appropriate number of indentations on each rod and cap (1, 2, 3, etc., depending on the engine type and cylinder they're associated with). **Caution:** *DO NOT use a marking punch to mark the caps on 4.7L engines, as damage to the rod cap will occur. Use a marking pen only to identify the rod caps* **(see illustration)**.

5 Loosen each of the connecting rod cap nuts or bolts 1/2-turn at a time until they can

13.3 Check the connecting rod side clearance (endplay) with a feeler gauge as shown here

13.4 On 4.7L engines, use a marking pen only to identify each connecting rod - using any type of punch will damage the connecting rod cap

13.6a On six-cylinder and 5.2L/5.9L engines, slip a short section of rubber hose over the end of each rod bolt to prevent damage to the cylinder walls as the piston and connecting rod assemblies are removed and installed

13.6b On 4.7L V8 engines, it will be necessary to fabricate a connecting rod guide tool before removing and installing the piston and connecting rod assemblies

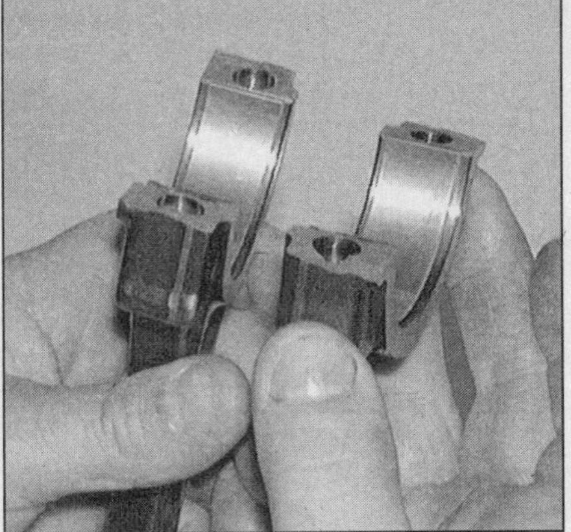

13.9 The rod cap parting line on 4.7L V8 engines is fractured at the factory (not machined) - therefore it is imperative that the rod caps be placed in the exact same position from which they were removed

be removed by hand. Remove the number one connecting rod cap and bearing insert. Don't drop the bearing insert out of the cap.

6 Slip a short length of plastic or rubber hose over each connecting rod cap bolt to protect the crankshaft journal and cylinder wall as the piston is removed **(see illustration)**. **Note:** *On 4.7L V8 engines the rod bolts are removed with the caps, so it may be helpful to make a set of connecting rod guides. Find several bolts that fit the rods at your local hardware store, or a couple of the old rod bolts, since new rod bolts must be used during installation anyway. Cut the heads of the bolts off with a hacksaw and slip a short length of hose over the end of each bolt to make a pair of connecting rod guides. Screw the guides into the connecting rods during removal and installation of the pistons/connecting rods from the engine block* **(see illustration)**.

7 Remove the bearing insert and push the connecting rod/piston assembly out through the top of the engine. Use a wooden hammer handle to push on the upper bearing surface in the connecting rod. If resistance is felt, double-check to make sure that all of the ridge was removed from the cylinder.

8 Repeat the procedure for the remaining cylinders.

9 After removal, reassemble the connecting rod caps and bearing inserts in their respective connecting rods and install the cap nuts finger tight **(see illustration)**. Leaving the old bearing inserts in place until reassembly will help prevent the connecting rod bearing surfaces from being accidentally nicked or gouged.

10 Don't separate the pistons from the connecting rods (see Section 18 for additional information).

14 Crankshaft - removal

Refer to illustrations 14.1, 14.3, 14.4 and 14.5

Note: *The crankshaft can be removed only after the engine has been removed from the vehicle. It's assumed that the flywheel or driveplate, vibration damper, timing chain(s), oil pan, oil pump and piston/connecting rod assemblies have already been removed.*

1 Before the crankshaft is removed, check the endplay. Mount a dial indicator with the stem in line with the crankshaft and touching one of the crank throws or the nose of the crank **(see illustration)**.

2 Push the crankshaft all the way to the rear and zero the dial indicator. Next, pry the crankshaft to the front as far as possible and check the reading on the dial indicator. The distance that it moves is the endplay. If it's greater than specified, check the crankshaft thrust surfaces for wear. If no wear is evident, new main bearings should correct the endplay.

3 If a dial indicator isn't available, feeler gauges can be used. Gently pry or push the crankshaft all the way to the front of the engine. Slip feeler gauges between the crankshaft and the front face of the thrust main bearing to determine the clearance **(see illustration)**.

4 Check the main bearing caps to see if they're marked to indicate their locations. They should be numbered consecutively from the front of the engine to the rear **(see illustration)**. If they aren't, mark them with

14.1 Checking the crankshaft endplay with a dial indicator

14.3 Checking crankshaft endplay with a feeler gauge

14.4 Main caps are generally numbered front to rear (arrow) - on the 5.2L/5.9L V8 the numbers are on the passenger side of the caps

14.5 On 4.7L V8 engines, pry the bedplate off the engine block at the casting protrusion only, then remove the bedplate to access the crankshaft - be careful not to drop the lower bearing inserts when removing the bedplate

15.1 The core plugs should be removed carefully by tapping in one side, then pulling out the opposite side with locking pliers

number-stamping dies or a center punch. Main bearing caps often have a cast-in arrow, which points to the front of the engine. If not, mark a paint arrow before removing the main caps. **Note:** *This step is not necessary on 4.7L V8 engines since the main bearing caps are an integral part of the bedplate which can only be installed one way.* In either case loosen the main bearing cap bolts 1/4-turn at a time in the reverse order of the tightening sequence, until they can be removed by hand (see Section 22). Note if any stud bolts are used and make sure they're returned to their original locations when the crankshaft is reinstalled.

5 Gently tap the caps with a soft-face hammer, then separate them from the engine block. If necessary, use the bolts as levers to remove the caps. Try not to drop the bearing inserts if they come out with the caps. **Note:** *On 4.7L V8 engines it will be necessary to pry the bedplate off the engine block* **(see illustration)**.

6 Carefully lift the crankshaft out of the engine. It may be a good idea to have an assistant available, since the crankshaft is quite heavy. With the bearing inserts in place in the engine block and main bearing caps, return the caps to their respective locations on the engine block and tighten the bolts finger tight.

15 Engine block - cleaning

Refer to illustrations 15.1, 15.5a, 15.5b, 15.9 and 15.11

1 Most engine core plugs are corroded firmly in place on an engine that has been in service for years. Use a small punch at the edge of the plug and carefully drive that edge in with a hammer **(see illustration)**. Drive one edge in only far enough for the opposite side to protrude far enough to grab it with locking pliers and pull out. **Caution:** *The core plugs (also known as freeze or soft plugs) may be difficult or impossible to retrieve if they're driven completely into the block coolant passages.*

2 Using a gasket scraper, remove all traces of gasket material from the engine block. Be very careful not to nick or gouge the gasket sealing surfaces.

3 Remove the main bearing caps and separate the bearing inserts from the caps and the engine block. Tag the bearings, indicating which cylinder they were removed from and whether they were in the cap or the block, then set them aside.

4 Remove all of the threaded oil gallery plugs from the block. The plugs are usually very tight - they may have to be drilled out and the holes retapped. Use new plugs when the engine is reassembled.

5 On 4.7L V8 engines, inspect the idler shaft at the front of the engine block for damage. Even if damage to the idler shaft is not evident, it a good idea to remove the idler shaft from the engine block before the block is hot tanked, since oil is fed from the main oil gallery through the idler shaft to oil the idler gear **(see illustrations)**. This will prevent any dirt and debris from lodging itself in the idler shaft oil holes during the hot tanking process. Be sure to clean the oil holes in the idler shaft before installing it back onto the engine block. Also note that the idler shaft oil holes can be installed in any position.

6 The engine should be taken to an automotive machine shop to be steam cleaned or hot tanked, although with patience it can be cleaned at home.

7 After the block is returned, clean all oil holes and oil galleries one more time. Brushes specifically designed for this purpose are available at most auto parts stores. Flush the passages with warm water until the water runs clear, dry the block thoroughly and wipe all machined surfaces with a light, rust-preventive oil. If you have access to compressed air, use it to speed the drying process and to blow out all the oil holes and

15.5a Tap the end of the idler shaft to accept a slide hammer tool - the factory has predrilled the end of the idler shaft to accept a 12mm x 1.75 tap

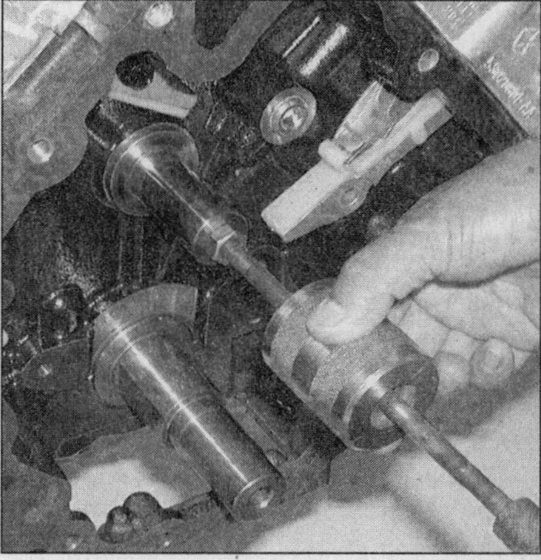

15.5b Use a slide hammer to pull the idler shaft from the block - when installing the idler shaft use the idler gear retaining bolt and washer to pull the shaft back into the block

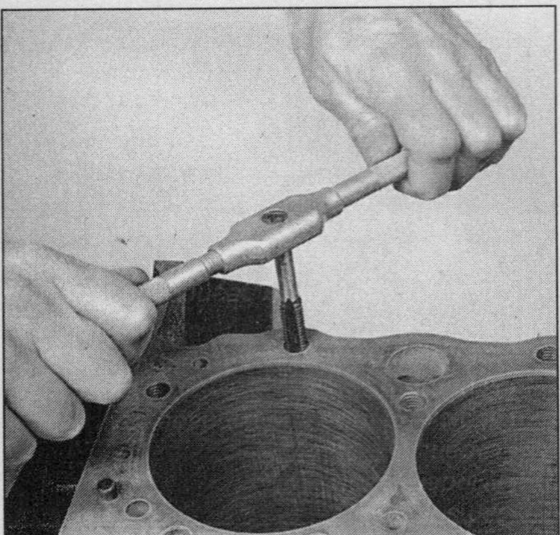
15.9 All bolt holes in the block, particularly the main bearing cap and head bolt holes, should be cleaned with the proper-sized tap

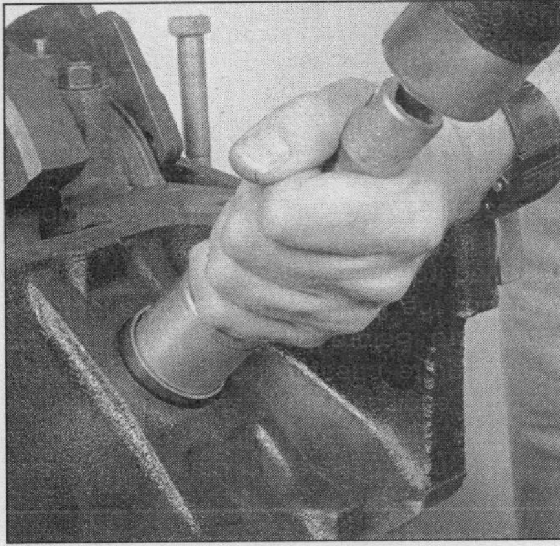
15.11 A large socket on an extension can be used to drive the new core plugs into the block - go slowly and stop when the plug is evenly flush with the block

galleries. **Warning:** *Wear eye protection when using compressed air!*

8 If the block isn't extremely dirty or sludged up, you can do an adequate cleaning job with hot soapy water and a stiff brush. Take plenty of time and do a thorough job. Regardless of the cleaning method used, be sure to clean all oil holes and galleries very thoroughly, dry the block completely and coat all machined surfaces with light oil.

9 The threaded holes in the block must be clean to ensure accurate torque readings during reassembly. Run the proper size tap into each of the holes to remove rust, corrosion, thread sealant or sludge and restore damaged threads **(see illustration)**. If possible, use compressed air to clear the holes of debris produced by this operation. **Warning:** *Wear eye protection when using compressed air!* Now is a good time to clean the threads on the head bolts and the main bearing cap bolts as well.

10 Reinstall the main bearing caps and tighten the bolts finger tight.

11 After coating the sealing surfaces of the new core plugs with core plug sealant, install them in the engine block **(see illustration)**. Make sure they're driven in straight and seated properly or leakage could result. Special tools are available for this purpose, but a large socket, with an outside diameter that will just slip into the core plug, a 1/2-inch drive extension and a hammer will work just as well.

12 Apply non-hardening sealant (such as Permatex no. 2 or Teflon pipe sealant) to the new oil gallery plugs and thread them into the holes in the block. Make sure they're tightened securely.

13 If the engine isn't going to be reassembled right away, cover it with a large plastic trash bag to keep it clean.

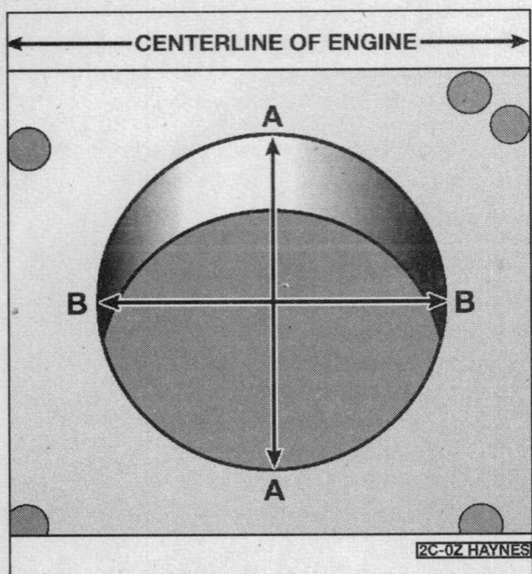

16.4a Measure the diameter of each cylinder at point A and at point B in three places, just under the wear ridge, at the center of the bore and at the bottom of the bore - out-of-round is the difference in diameter between point A and B at any place in the cylinder - taper is the difference in diameter between the top and the bottom of the cylinder

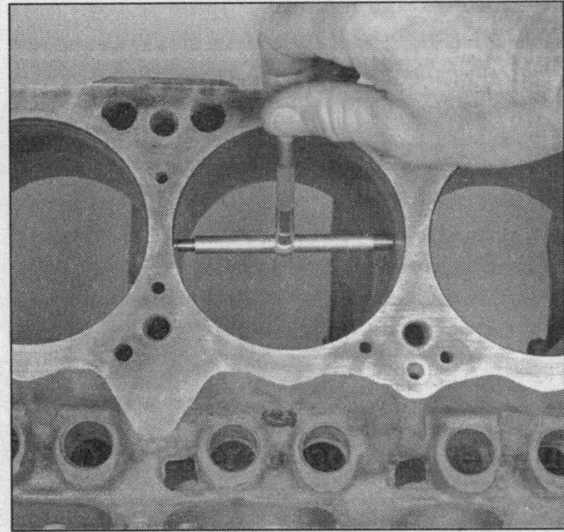
16.4b The ability to "feel" when the telescoping gauge is at the correct point will be developed over time, so work slowly and repeat the check until you're satisfied the bore measurement is accurate

16 Engine block - inspection

Refer to illustrations 16.4a, 16.4b, 16.4c, 16.13a, 16.13b, 16.13c and 16.13d

1 Before the block is inspected, it should be cleaned as described in Section 15.

2 Visually check the block for cracks, rust and corrosion. Look for stripped threads in the threaded holes. It's also a good idea to have the block checked for hidden cracks by an automotive machine shop that has the special equipment to do this type of work. If defects are found, have the block repaired, if possible, or replaced.

3 Check the cylinder bores for scuffing and scoring.

4 Measure the diameter of each cylinder at the top (just under the ridge area), center and bottom of the cylinder bore, parallel to the crankshaft axis **(see illustrations)**.

5 Next, measure each cylinder's diameter at the same three locations across the crankshaft axis. Compare the results to this Chapter's Specifications. **Caution:** *The cylinder bores on 1998 six-cylinder engines and all 1999 and later engines must be accurately measured with a dial bore gauge (that can measure 0.0001 inch) since these engines are equipped with moly-coated pistons. Piston-to-bore clearance is especially critical on engines with moly-coated pistons, so have an automotive machine shop perform this task if a dial bore gauge is not available.*

6 If the required precision measuring tools aren't available, the piston-to-cylinder clearances on 1998 and earlier engines can be obtained, though not quite as accurately, using feeler-gauge stock. Feeler gauge stock comes in 12-inch lengths and various thicknesses and is generally available at auto parts stores.

7 To check the clearance, select a feeler gauge and slip it into the cylinder along with the matching piston. The piston must be positioned exactly as it normally would be. The feeler gauge must be between the piston and cylinder on one of the thrust faces (90-degrees to the piston-pin bore).

8 The piston should slip through the cylinder (with the feeler gauge in place) with moderate pressure.

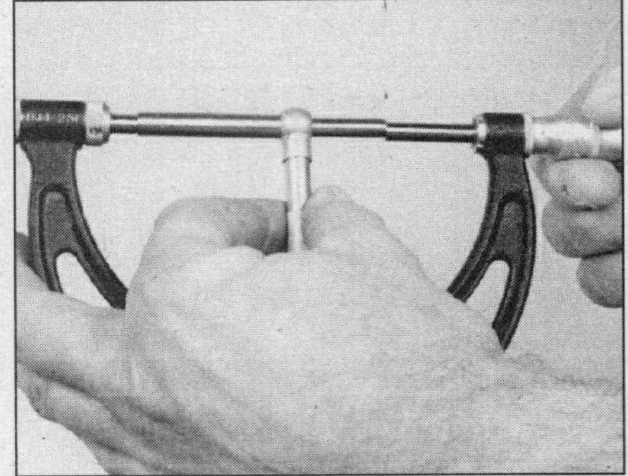
16.4c The gauge is then measured with a micrometer to determine the bore size

9 If it falls through or slides through easily, the clearance is excessive and a new piston will be required. If the piston binds at the lower end of the cylinder and is loose toward the top, the cylinder is tapered. If tight spots are encountered as the piston/feeler gauge is rotated in the cylinder, the cylinder is out-of-round.

10 Repeat the procedure for the remaining pistons and cylinders.

11 If the cylinder walls are badly scuffed or scored, or if they're out-of-round or tapered beyond the limits given in the specifications, have the engine block rebored and honed at an automotive machine shop. If a rebore is done, oversize pistons and rings will be required. **Note:** *The factory recommends replacing the engine block on 4.7L V8 engines if the cylinder bore is out of specifications, however check with your local machine shop for alternative possibilities.*

12 If the cylinders are in reasonably-good condition and not worn to the outside of the limits, and if the piston-to-cylinder clearances can be maintained properly, then they don't have to be rebored. Honing is all that's necessary (see Section 17).

13 Some engines are produced with oversize or undersize components, such as oversize cylinder bores, undersize main bearing

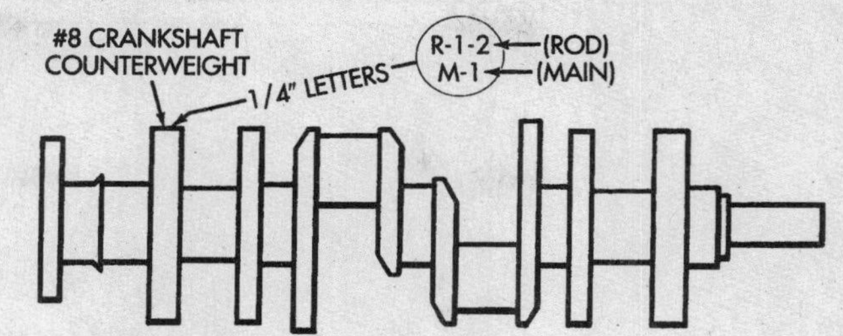

16.13a 5.2L/5.9L V8 crankshafts may have letters/numbers stamped into the number eight counterweight to indicate if it has undersize rod (R) or main (M) journals

CONDITION	IDENTIFICATION	LOCATION OF IDENTIFICATION
CRANKSHAFT JOURNALS (UNDERSIZE) 0.0254 mm (0.001 in.)	R or M M-2-3 etc. (indicating no. 2 and 3 main bearing journal) and/or R-1-4 etc. (indicating no. 1 and 4 connecting rod journal)	Milled flat on no. 8 crankshaft counterweight.
HYDRAULIC TAPPETS (OVERSIZE) 0.2032 mm (0.008 in.)	♦	Diamond-shaped stamp top pad – front of engine and flat ground on outside surface of each O/S tappet bore.
VALVE STEMS (OVERSIZE) 0.127 mm (0.005 in.)	X	Milled pad adjacent to two tapped holes (3/8 in.) on each end of cylinder head.

16.13b 5.2L/5.9L V8 engine oversize/undersize markings

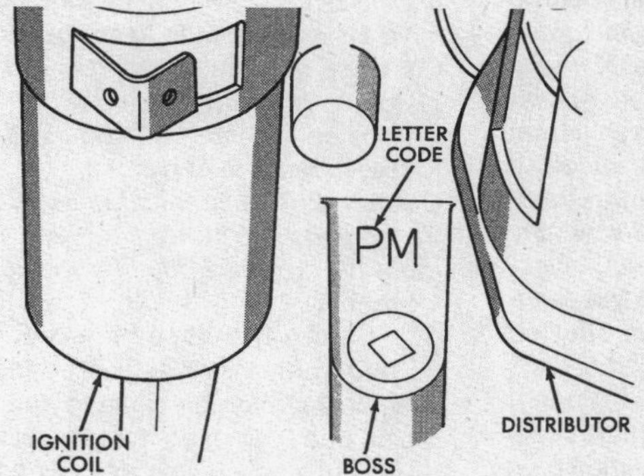

CODE	COMPONENT	UNDERSIZE
P	One or more connecting rod bearing journals	0.254 mm (0.010 in)
M	All crankshaft main bearing journals	0.254 mm (0.010 in)
PM	All crankshaft main bearing journals and one or more connecting rod journals	0.254 mm (0.010 in)
CODE	**COMPONENT**	**OVERSIZE**
B	All cylinder bores	0.254 mm (0.010 in)
C	All camshaft bearing bores	0.254 mm (0.010 in)

16.13c The oversize/undersize codes on six-cylinder engines are on a pad between the distributor and ignition coil

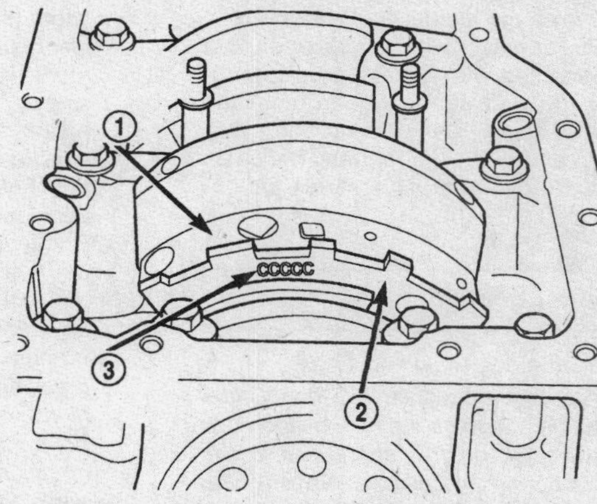

GRADE MARKING	SIZE mm (in.)	FOR USE WITH JOURNAL SIZE
A	0.008 mm U/S (0.0004 in.) U/S	63.488–63.496 mm (2.4996–2.4999 in.)
B	NOMINAL	63.496–63.504 mm (2.4999–2.5002 in.)
C	0.008 mm O/S (0.0004 in.) O/S	63.504–63.512 mm (2.5002–2.5005 in.)

16.13d Undersize and oversize markings on 4.7L V8 engines can be found on the crankshaft position sensor reluctor ring

1 *Rear crankshaft counterweight*
2 *Reluctor ring*
3 *Main bearing markings*

17.3a A "bottle brush" hone is the easiest type of hone to use

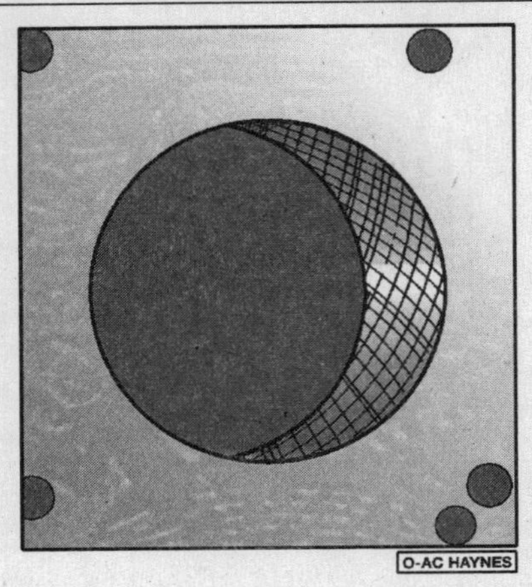

17.3b The cylinder hone should leave a smooth, crosshatch pattern with the lines intersecting at approximately a 60-degree angle

and connecting rod journals, or oversize camshaft bearing bores. These engines are identified by a letter or symbol code stamped on the engine block, crankshaft or lifter valley. On 5.2L/5.9L engines, undersize crankshaft journals are indicated by a letter "R" or "M" on the number 8 crank counterweight **(see illustrations)**. A designation of "M-2-3" would indicate .001-inch-undersize journals on mains 2 and 3, while an R prefix indicates which rod journals are undersize.

14 Earlier engines with oversize tappet bores are marked with a stamped diamond-shape on the pad at the top front of the engine, while an "X" stamped into the milled pad on the end of a cylinder head indicates that there are oversize valve stems (0.005-inch).

17 Cylinder honing

Refer to illustrations 17.3a and 17.3b

1 Prior to engine reassembly, the cylinder bores must be honed so the new piston rings will seat correctly and provide the best possible combustion-chamber seal. **Note:** *If you don't have the tools or don't want to tackle the honing operation, most automotive machine shops will do it for a reasonable fee.*

2 Before honing the cylinders, install the main bearing caps and tighten the bolts to the torque listed in this Chapter's Specifications.

3 Two types of cylinder hones are commonly available - the flex hone or "bottle brush" type and the more traditional surfacing hone with spring-loaded stones. Both will do the job, but for the less experienced mechanic the "bottle brush" hone will probably be easier to use. You'll also need some kerosene or honing oil, rags and an electric drill motor. Proceed as follows:

a) *Mount the hone in the drill motor, compress the stones (if applicable) and slip it into the first cylinder* **(see illustration)***. Be sure to wear safety goggles or a face shield!*

b) *Lubricate the cylinder with plenty of honing oil or kerosene, turn on the drill and move the hone up-and-down in the cylinder at a pace that will produce a fine crosshatch pattern on the cylinder walls. Ideally, the crosshatch lines should intersect at approximately a 60-degree angle* **(see illustration)***. Be sure to use plenty of lubricant and don't take off any more material than is absolutely necessary to produce the desired finish.* **Note:** *Piston ring manufacturers may specify a smaller crosshatch angle than the traditional 60-degrees - read and follow any instructions included with the new rings.*

c) *Don't withdraw the hone from the cylinder while it's running. Instead, shut off the drill and continue moving the hone up-and-down in the cylinder until it comes to a complete stop, then compress the stones and withdraw the hone. If you're using a "bottle brush" type hone, stop the drill motor, then turn the chuck in the normal direction of rotation while withdrawing the hone from the cylinder.*

d) *Wipe the oil out of the cylinder and repeat the procedure for the remaining cylinders.*

4 After the honing job is complete, chamfer the top edges of the cylinder bores with a small file so the rings won't catch when the pistons are installed. Be very careful not to nick the cylinder walls with the end of the file.

5 The entire engine block must be washed again very thoroughly with warm, soapy water to remove all traces of the abrasive grit produced during the honing operation. **Note:** *The bores can be considered clean when a lint-free white cloth - dampened with clean engine oil - used to wipe them out doesn't pick up any more honing residue, which will show up as gray areas on the cloth. Be sure to run a brush through all oil holes and galleries and flush them with running water.*

6 After rinsing, dry the block and apply a coat of light rust preventive oil to all machined surfaces. Wrap the block in a plastic trash bag to keep it clean and set it aside until reassembly.

18 Pistons/connecting rods - inspection

Refer to illustrations 18.4a, 18.4b, 18.8, 18.10 and 18.11

Caution 1: *The 1998 six-cylinder engines and all 1999 and later engines are equipped with pistons that are moly-coated after the final machining process. Measuring a coated piston will not provide an accurate measurement due to the thickness of the coating. To establish the piston-to-cylinder bore clearance in these engines it is mandatory to use a dial bore gauge capable of reading 0.0001 inch increments to accurately measure the cylinder bore. Moly-coated pistons should always be replaced in engines with high mileage, since the coating wears off in time and excessive piston-to-bore clearance will result. Once the cylinder bores are accurately measured, the measurements can be taken to a dealer service department or a local machine shop to select the proper size pistons for your engine block. The new pistons may seem to have a slight press fit in the bores, but this is OK since the moly coating is not part of the piston-to-bore fitting procedure. Tin-coated pistons should not be used as replacements for the moly-coated pistons.*

Caution 2: *The connecting rod bolts on 4.7L V8 engines are torque-to-yield design and must be replaced anytime the rod bolts are loosened.*

1 Before the inspection process can be carried out, the piston/connecting rod assemblies must be cleaned and the original piston rings removed from the pistons. **Note:** *Always use new piston rings when the engine is reassembled.*

2 Using a piston ring installation tool, carefully remove the rings from the pistons. Be careful not to nick or gouge the pistons in the process.

3 Scrape all traces of carbon from the top of the piston. A hand-held wire brush or a piece of fine emery cloth can be used once the majority of the deposits have been scraped away. Do not, under any circumstances, use a wire brush mounted in a drill motor to remove deposits from the pistons. The piston material is soft and may be eroded away by the wire brush.

4 Use a piston ring groove cleaning tool to remove carbon deposits from the ring grooves. If a tool isn't available, a piece broken off the old ring will do the job. Be very careful to remove only the carbon deposits - don't remove any metal and do not nick or scratch the sides of the ring grooves **(see illustrations)**.

5 Once the deposits have been removed, clean the piston/rod assemblies with solvent

18.4a The piston ring grooves can be cleaned with a special tool, as shown here . . .

18.4b . . . or a section of a broken ring

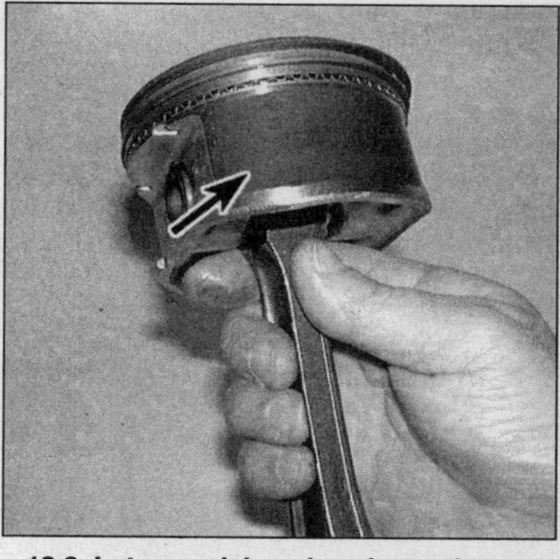

18.8 Later model engines have pistons with a moly coating (arrow) applied to the piston skirt - check the coating for excessive scoring and flaking - never use abrasive materials to clean the coating or damage to the piston will occur

and dry them with compressed air (if available). Make sure the oil return holes in the back sides of the ring grooves are clear.

6 If the pistons and cylinder walls aren't damaged or worn excessively, and if the engine block is not rebored, new pistons won't be necessary. Normal piston wear appears as even vertical wear on the piston thrust surfaces and slight looseness of the top ring in its groove. New piston rings, however, should always be used when an engine is rebuilt.

7 Carefully inspect each piston for cracks around the skirt, at the pin bosses and at the ring lands.

8 Look for scoring and scuffing on the thrust faces of the skirt, holes in the piston crown and burned areas at the edge of the crown. If the skirt is scored or scuffed, the engine may have been suffering from overheating and/or abnormal combustion, which caused excessively high operating temperatures. Such pistons should be replaced **(see illustration)**. The cooling and lubrication systems should be checked thoroughly. A hole in the piston crown is an indication that abnormal combustion (preignition) was occurring. Burned areas at the edge of the piston crown are usually evidence of spark knock (detonation). If any of the above problems exist, the causes must be corrected or the damage will occur again. The causes may include intake air leaks, incorrect fuel/air mixture, incorrect ignition timing and EGR system malfunctions.

9 Corrosion of the piston, in the form of small pits, indicates that coolant is leaking into the combustion chamber and/or the crankcase. Again, the cause must be corrected or the problem may persist in the rebuilt engine.

10 Measure the piston ring side clearance by laying a new piston ring in each ring groove and slipping a feeler gauge in beside it **(see illustration)**. Check the clearance at three or four locations around each groove. Be sure to use the correct ring for each groove - they are different. If the side clearance is greater than specified, new pistons will have to be used.

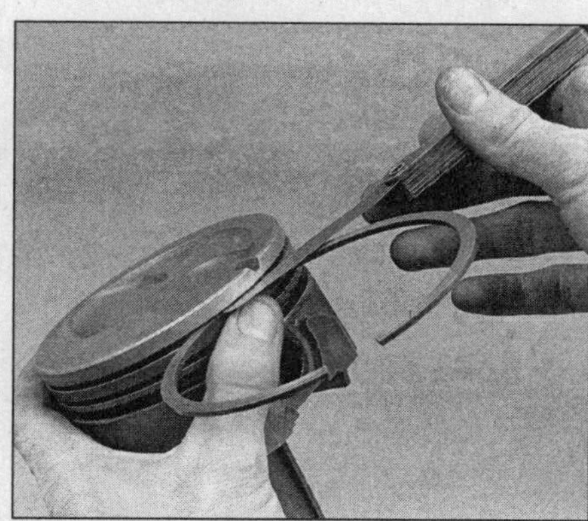

18.10 Check the ring side clearance with a feeler gauge at several points around the groove

11 On engines without moly-coated pistons, check the piston-to-bore clearance by measuring the bore (see Section 16) and the piston diameter. Make sure the pistons and bores are correctly matched. Measure the piston across the skirt, at a 90-degree angle to and in line with the piston pin **(see illustration)**. Subtract the piston diameter from the bore diameter to obtain the clearance. If it's greater than specified, the block will have to be rebored and new pistons and rings installed.

12 Check the piston-to-rod clearance by twisting the piston and rod in opposite directions. Any noticeable play indicates excessive wear, which must be corrected. The piston/connecting rod assemblies should be taken to an automotive machine shop to have the pistons and rods resized and new pins installed.

13 If the pistons must be removed from the connecting rods for any reason, they should be taken to an automotive machine shop. While they are there, have the connecting rods checked for bend and twist, since automotive machine shops have special equipment for this purpose. **Note:** *Unless new pistons and/or connecting rods must be installed, do not disassemble the pistons and connecting rods.*

18.11 On engines without moly-coated pistons, measure the piston diameter at a 90-degree angle to the piston pin and in line with it

14 Check the connecting rods for cracks and other damage. Temporarily remove the rod caps, lift out the old bearing inserts, wipe the rod and cap bearing surfaces clean and inspect them for nicks, gouges and scratches. After checking the rods, replace the old bearings, slip the caps into place and tighten the nuts finger tight. **Note:** *If the engine is being rebuilt because of a connecting rod knock, be sure to install new or remanufactured connecting rods.*

19 Crankshaft - inspection

Refer to illustration 19.6

1 Clean the crankshaft with solvent and dry it with compressed air (if available). Be sure to clean the oil holes with a stiff brush and flush them with solvent.

2 Check the main and connecting rod bearing journals for uneven wear, scoring, pits and cracks.

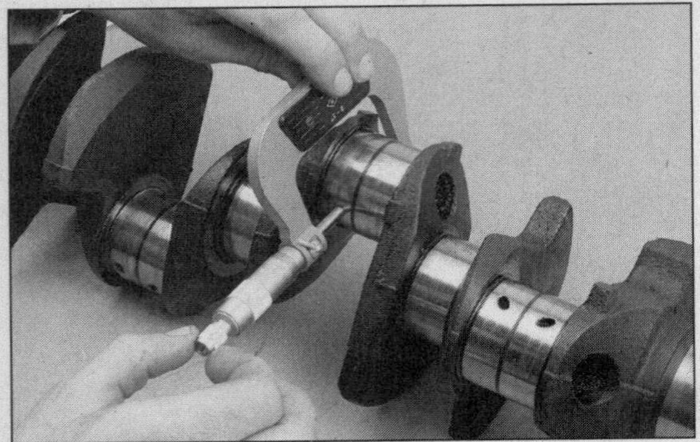

19.6 Measure at several points around each journal's circumference, then measure at each end - this will help you identify out-of-round and taper conditions

3 Rub a penny across each journal several times. If a journal picks up copper from the penny, it's too rough and must be reground.
4 Remove all burrs from the crankshaft oil holes with a stone, file or scraper.
5 Check the rest of the crankshaft for cracks and other damage. It should be magnafluxed to reveal hidden cracks - an automotive machine shop will handle the procedure.
6 Using a micrometer, measure the diameter of the main and connecting rod journals and compare the results to the specifications **(see illustration)**. By measuring the diameter at a number of points around each journal's circumference, you'll be able to determine whether or not the journal is out-of-round. Take the measurement at each end of the journal, near the crank throws, to determine if the journal is tapered.
7 If the crankshaft journals are damaged, tapered, out-of-round or worn beyond the limits given in the specifications, have the crankshaft reground by an automotive machine shop. Be sure to use the correct-size bearing inserts if the crankshaft is reconditioned.
8 Check the oil seal journals at each end of the crankshaft for wear and damage. If the seal has worn a groove in the journal, or if it's nicked or scratched, the new seal may leak when the engine is reassembled. In some cases, an automotive machine shop may be able to repair the journal by pressing on a thin sleeve. If repair isn't feasible, a new or different crankshaft should be installed.
9 On 4.7L V8 engines, inspect the reluctor ring at the rear of the crankshaft for damage, replacing it if necessary. Damage to this component can cause severe driveabilty problems.
10 Refer to Section 20 and examine the main and rod bearing inserts.

20 Main and connecting rod bearings - inspection

Refer to illustration 20.1

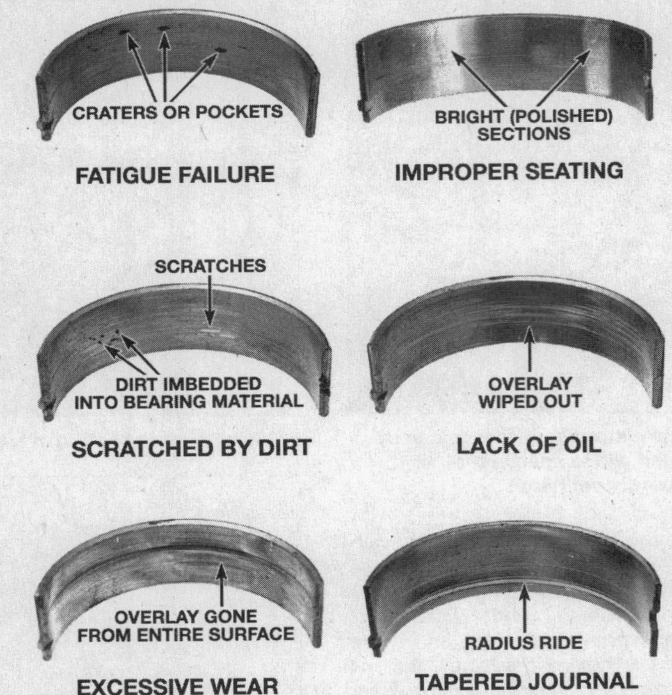

20.1 Before discarding the used bearings, examine them for indications of any possible problems with the crankshaft, noting the bearing location it came from - these are typical bearing failures

1 Even though the main and connecting rod bearings should be replaced with new ones during the engine overhaul, the old bearings should be retained for close examination, as they may reveal valuable information about the condition of the engine **(see illustration)**.
2 Bearing failure occurs because of lack of lubrication, the presence of dirt or other foreign particles, overloading the engine and corrosion. Regardless of the cause of bearing failure, it must be corrected before the engine is reassembled to prevent it from happening again.
3 When examining the bearings, remove them from the engine block, the main bearing caps, the connecting rods and the rod caps and lay them out on a clean surface in the same general position as their location in the engine. This will enable you to match any bearing problems with the corresponding crankshaft journal.
4 Dirt and other foreign particles get into the engine in a variety of ways. It may be left in the engine during assembly, or it may pass through filters or the PCV system. It may get into the oil, and from there into the bearings. Metal chips from machining operations and normal engine wear are often present. Abrasives are sometimes left in engine components after reconditioning, especially when parts are not thoroughly cleaned using the proper cleaning methods. Whatever the source, these foreign objects often end up embedded in the soft bearing material and are easily recognized. Large particles will not embed in the bearing and will score or gouge the bearing and journal. The best prevention for this cause of bearing failure is to clean all parts thoroughly and keep everything spotlessly clean during engine assembly. Frequent and regular engine oil and filter changes are also recommended.
5 Lack of lubrication (or lubrication breakdown) has a number of interrelated causes. Excessive heat (which thins the oil), overloading (which squeezes the oil from the bearing face) and oil leakage or throw-off (from excessive bearing clearances, worn oil pump or high engine speeds) all contribute to lubrication breakdown. Blocked oil passages, which usually are the result of misaligned oil holes in a bearing shell, will also oil starve a bearing and destroy it. When lack of lubrication is the cause of bearing failure, the bearing material is wiped or extruded from the steel backing of the bearing. Temperatures may increase to the point where the steel backing turns blue from overheating.
6 Driving habits can have a definite effect on bearing life. Low speed operation in too high a gear (lugging the engine) puts very high loads on bearings, which tends to squeeze out the oil film. These loads cause the bearings to flex, which produces fine cracks in the bearing face (fatigue failure). Eventually the bearing material will loosen in pieces and tear away from the steel backing. Short-trip driving leads to corrosion of bearings because insufficient engine heat is produced to drive off the condensed water and corrosive gases. These products collect in the engine oil, forming acid and sludge. As the oil is carried to the engine bearings, the acid attacks and corrodes the

bearing material.

7 Incorrect bearing installation during engine assembly will lead to bearing failure as well. Tight-fitting bearings leave insufficient bearing oil clearance and will result in oil starvation. Dirt or foreign particles trapped behind a bearing insert result in high spots on the bearing which lead to failure.

21 Engine overhaul - reassembly sequence

1 Before beginning engine reassembly, make sure you have all the necessary new parts, gaskets and seals as well as the following items on hand:

Common hand tools
3/8-inch and 1/2-inch-drive torque wrenches
Piston ring installation tool
Piston ring compressor
Vibration damper installation tool
Short lengths of rubber or plastic hose to fit over connecting rod bolts
Plastigage
Feeler gauges
A fine-tooth file
New engine oil
Engine assembly lube or moly-base grease
Gasket sealant
Thread locking compound

2 In order to save time and avoid problems, engine reassembly must be done in the following general order:

Six-cylinder engine and 5.2L and 5.9L V8 engines

New camshaft bearings (must be done by automotive machine shop)
Piston rings
Crankshaft and main bearings
Piston/connecting rod assemblies
Main bearing cap brace (six-cylinder engines)
Oil pump
Camshaft and lifters
Oil pan
Timing chain and sprockets
Cylinder head(s), pushrods and rocker arms
Timing cover
Crankshaft pulley
Intake and exhaust manifolds
Valve cover(s)
Flywheel/driveplate

4.7L V8 engine

Idler shaft
Piston rings
Crankshaft, main bearings and bedplate
Piston/connecting rod assemblies
Rear main oil seal
Oil pump
Oil pan gasket/windage tray
Oil pick up tube
Oil pan
Cylinder heads, hydraulic lash adjusters and camshafts
Timing chain guides
Primary and secondary timing chain assemblies
Rocker arms
Front timing cover and water pump
Crankshaft pulley
Intake and exhaust manifolds
Valve covers
Driveplate

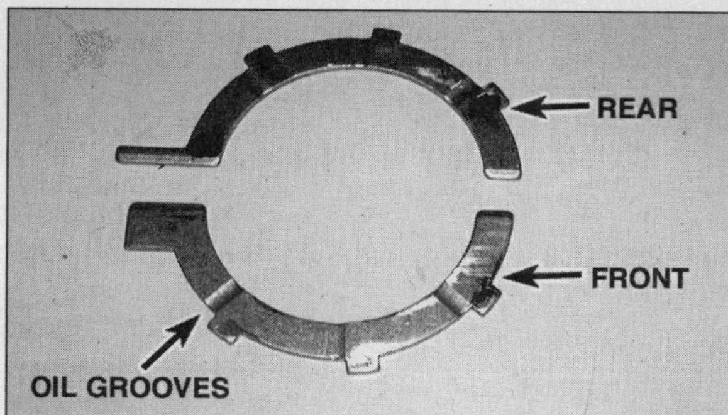

22.8a Thrust bearing identification - 4.7L V8 engine

22 Crankshaft - installation and main bearing oil clearance check

1 If you have determined that the engine overhaul will include having the crankshaft "turned" at a machine shop to a standard undersize such as 0.010-inch on the rod and main journals, bearing selection will be easy. You simply order a standard bearing set that is correct for a 0.010-inch-reground crankshaft. Even with a reground crank and new bearings, a clearance check is still required before final assembly (see Step 6).

2 On the factory engine assembly line, journals and bearings may vary slightly in size, and cranks and bearings are color-coded to match the right bearing with the right journal for a perfect fit. If you are using a standard-size crankshaft, you can match the correct new bearings to the crank by referring to the charts in Section 16. An automotive machine shop that specializes in engine rebuilding will be able to help you select the correct bearings.

3 The color code appears on the edge of the bearing insert.

4 The journal size codes are generally painted on the adjacent cheek toward the flanged end (rear) of the crankshaft, except for the rear main bearing journal, which is color marked on the rear crank flange on six-cylinder engines. On V8 engines, size codes are marked on the crankshaft rear flange with letters and numbers (see Section 16).

5 To obtain a select fit, upper and lower bearing inserts of different sizes may be used as a pair. For example, a standard insert is sometimes used in combination with a 0.001-inch undersize insert to reduce clearance by 0.0005-inch. **Caution:** *When using a standard crankshaft with factory bearings never use a pair of bearing inserts with a greater size difference than 0.001-inch. When replacing inserts, the odd-sized inserts must be either all on the top or all on the bottom.*

22.8b Insert the thrust bearing into the machined surface between the crankshaft and the upper bearing saddle, then rotate it down into the block until its flush with the parting line of the main bearing saddle - make sure the oil grooves on the thrust bearings face the crankshaft

Main bearing oil clearance check

Refer to illustrations 22.8a, 22.8b, 22.11, 22.12a, 22.12b and 22.15

6 Regardless of how the bearing sizes are determined, use the Plastigage clearance procedure below as a guide to ensure the proper size bearings are installed.

7 Clean the back sides of the new main bearing inserts and lay the bearing half with the oil groove in each main bearing saddle in the block. Lay the other bearing half from each bearing set in the corresponding main bearing cap or the corresponding saddle in the bedplate on 4.7L V8 engines. Make sure the tab on each bearing insert fits into the recess in the block or cap. Also, the oil holes in the block must line up with the oil holes in the bearing insert. **Caution:** *Do not hammer the bearings into place and don't nick or gouge the bearing faces. No lubrication should be used at this time.*

8 The thrust bearings on 4.7L V8 engines are a two piece design which must be installed in the machined surfaces of the

22.11 Lay the Plastigage strips (arrow) on the main bearing journals, parallel to the crankshaft centerline

upper bearing saddle in the engine block. There are front and rear thrust bearings so do not mix them up! Pry the crankshaft forward and install the front thrust bearing into the machined surface of the upper bearing saddle in the engine block **(see illustrations)**. Next pry the crankshaft toward the rear of the engine and install the rear thrust bearing into the machined surface of the upper bearing saddle in the engine block. **Note:** *Be sure to install the thrust bearings with the oil grooves facing the crankshaft.*

9 Clean the faces of the bearings in the block and the crankshaft main bearing journals with a clean, lint-free cloth. Check or clean the oil holes in the crankshaft, as any dirt here can go only one way - straight through the new bearings. Once you're certain the crankshaft is clean, carefully lay it in position in the main bearings.

10 Before the crankshaft can be permanently installed, the main bearing oil clearance must be checked.

11 Trim several pieces of the appropriate-size Plastigage (they must be slightly shorter than the width of the main bearings) and place one piece on each crankshaft main bearing journal, parallel with the journal axis **(see illustration)**.

12 Clean the faces of the bearings in the caps or the bedplate **(see illustrations)** and install the caps in their respective positions (don't mix them up) with the arrows pointing toward the front of the engine. Don't disturb the Plastigage. Apply a light coat of oil to the bolt threads and the under sides of the bolt heads, then install them.

13 Tighten the main bearing cap bolts, in three steps, to the torque listed in this Chapter's Specifications. Don't rotate the crankshaft at any time during this operation!

14 Remove the bolts and carefully lift off the main bearing caps or the bedplate on 4.7L V8 engines. Keep them in order. Don't disturb the Plastigage or rotate the crankshaft. If any of the main bearing caps are difficult to remove, tap them gently from side-to-side with a soft-face hammer to loosen them.

22.12a On 4.7L V8 engines, the bearings (arrows) are laid into the corresponding saddles in the bedplate, then the bedplate is flipped 180 degrees . . .

15 Compare the width of the crushed Plastigage on each journal to the scale printed on the Plastigage envelope to obtain the main bearing oil clearance **(see illustration)**. Check the Specifications to make sure it's correct. **Note:** *Make sure you are using the correct scale, as both metric and standard ones are included on the package.*

16 If the clearance is not as specified, the bearing inserts may be the wrong size (which means different ones will be required). Before deciding that different inserts are needed, make sure that no dirt or oil was between the bearing inserts and the caps or block when the clearance was measured. If the Plastigage is noticeably wider at one end than the other, the journal may be tapered.

17 Carefully scrape all traces of the Plastigage material off the main bearing journals and/or the bearing faces. Don't nick or scratch the bearing faces (a plastic credit card edge does the job).

Final crankshaft installation

Refer to illustrations 22.20a, 22.20b, 22.21, 22.23a, 22.23b, 22.23c and 22.24

18 Carefully lift the crankshaft out of the engine. Clean the bearing faces in the block, then apply a thin, uniform layer of engine assembly lube to each of the bearing surfaces.

19 Lubricate the crankshaft surfaces that contact the oil seals with clean engine oil.

20 On six-cylinder and 5.2L/5.9L V8 engines, install the upper half of the rear main seal into the block **(see illustrations)** with the yellow paint facing the rear of the engine (5.2L/5.9L V8 engines) and the seal lip facing

22.12b . . . and set over the crankshaft onto the dowels (arrow) on the engine block

22.15 Compare the width of the crushed Plastigage to the scale on the envelope, taking the measurement at the widest point of the Plastigage

22.20a On six-cylinder and 5.2L/5.9L engines, install the new neoprene upper seal half in the engine block . . .

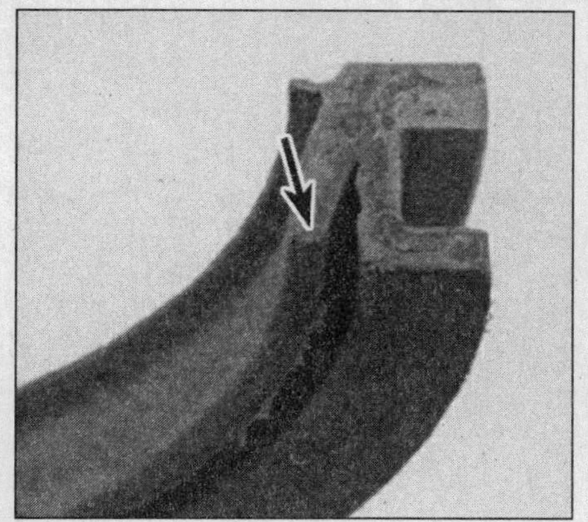

22.20b . . . with the seal lip (arrow) facing the front of the engine

22.21 Cylinder block-to-bedplate sealant location - 4.7L V8 engine

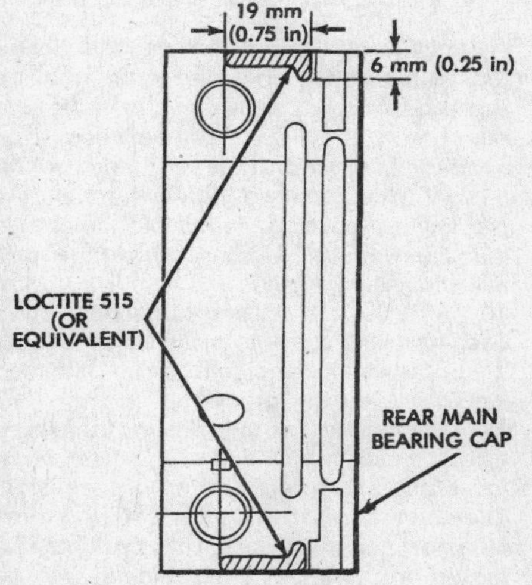

22.23a Apply Loctite 515 to the indicated area on the six-cylinder engine rear main cap

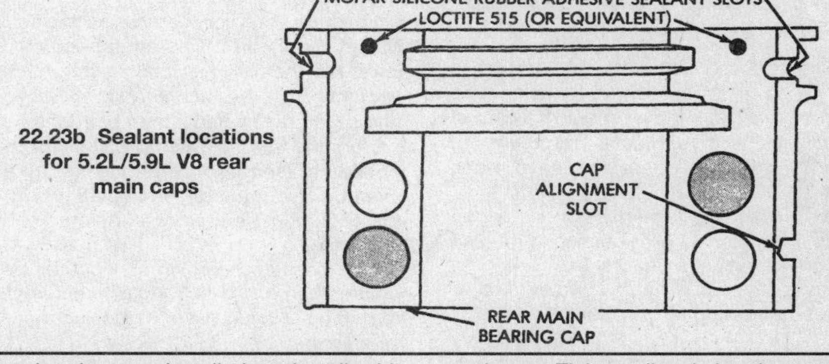

22.23b Sealant locations for 5.2L/5.9L V8 rear main caps

towards the front of the engine.

21 Make sure the crankshaft journals are clean, then lay the crankshaft back in place in the block. Clean the faces of the bearings in the caps or bedplate, then apply lubricant to them. On 4.7L V8 engines, make sure the thrust bearings are installed as described in Step 8, then apply a 1/8-inch bead of RTV sealant to the engine block **(see illustration)** and install the bedplate over the dowels on the block. On six-cylinder and 5.2L/5.9L V8 engines, install the caps in their respective positions with the arrows pointing toward the front of the engine.

22 Apply a light coat of oil to the bolt threads and the under sides of the bolt heads, then install them.

23 On six-cylinder and 5.2L/5.9L V8 engines, apply a dab of Loctite 515 on each side of the rear main cap, then install a new rear main oil seal lower half in the rear main cap **(see illustrations)** and install the rear main cap. Tighten all main bearing cap bolts, except the number three thrust bearing cap, to the torque listed in this Chapter's Specifications. Pry the crankshaft slightly back and forth in the block to seat the thrust bearings, then tighten the number 3 main cap to Specifications.

24 On 4.7L V8 engines, tighten the bedplate bolts in four steps, following the proper sequence to the torque listed in this Chapter's Specifications **(see illustration)**. **Note:** *Since RTV sealant is the only gasket material used between the bedplate and the engine block, it is critical to adhere to the sealant manufacturers instructions for application and drying times.*

22.23c Install the new neoprene lower seal half in the rear main cap

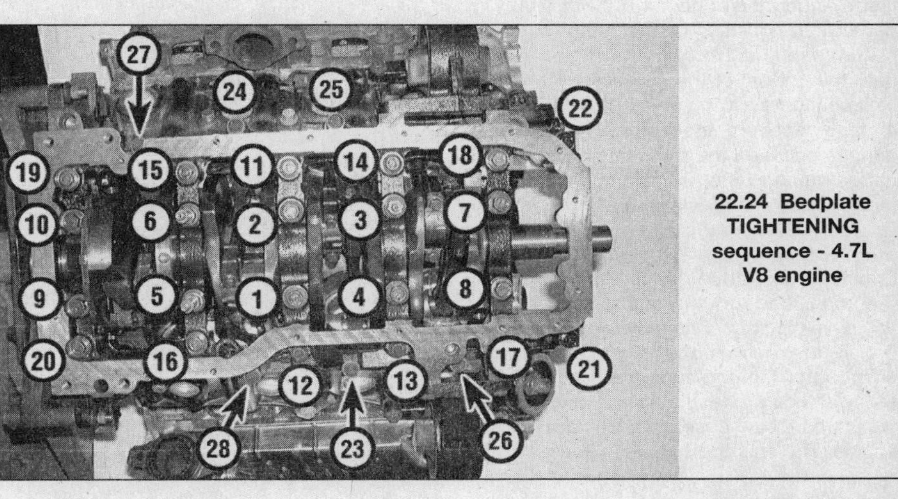

22.24 Bedplate TIGHTENING sequence - 4.7L V8 engine

2D

23.3 When checking piston ring end gap, the ring must be square in the cylinder bore - use a piston to push it down near the bottom of ring travel

25 Check the crankshaft endplay with a feeler gauge or a dial indicator as described in Section 14. The endplay should be correct if the crankshaft thrust faces aren't worn or damaged and new bearings have been installed.

26 On manual transmission-equipped models, install a new pilot bearing in the end of the crankshaft (see Chapter 8).

27 Rotate the crankshaft a number of times by hand to check for any obvious binding.

23 Piston rings - installation

Refer to illustrations 23.3, 23.4, 23.5, 23.9a, 23.9b, 23.12 and 23.13

1 Before installing the new piston rings, the ring end gaps must be checked. It's assumed that the piston ring side clearance has been checked and verified correct (see Section 18).

2 Lay out the piston/connecting rod assemblies and the new ring sets so the ring sets will be matched with the same piston and cylinder during the end gap measurement and engine assembly.

3 Insert the top (number one) ring into the first cylinder and square it up with the cylinder walls by pushing it in with the top of the piston **(see illustration)**. The ring should be near the bottom of the cylinder, at the lower limit of ring travel.

4 To measure the end gap, slip feeler gauges between the ends of the ring until a gauge equal to the gap width is found **(see illustration)**. The feeler gauge should slide between the ring ends with a slight amount of drag. Compare the measurement to the Specifications. If the gap is larger or smaller than specified, double-check to make sure you have the correct rings before proceeding.

5 If the gap is too small, it must be enlarged or the ring ends may come in contact with each other during engine operation, which can cause serious damage to the engine. The end gap can be increased by filing the ring ends very carefully with a fine file. Mount the file in a vise equipped with soft jaws, slip the ring over the file with the ends contacting the file face and slowly move the ring to remove material from the ends. When performing this operation, file only from the outside in **(see illustration)**. The ring ends must be filed squarely, and any tiny burrs on the end should be removed before final ring installation.

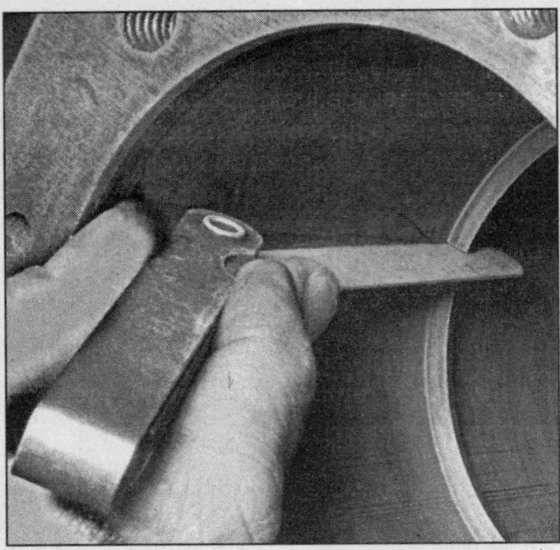
23.4 With the ring square in the cylinder, measure the end gap with feeler gauges

6 Excess end gap isn't critical unless it's greater than 0.040-inch. Again, double-check to make sure you have the correct rings for your engine.

7 Repeat the procedure for each ring that will be installed in the first cylinder and for each ring in the remaining cylinders. Remember to keep rings, pistons and cylinders matched up.

8 Once the ring end gaps have been checked/corrected, the rings can be installed on the pistons.

9 The oil control ring (lowest one on the piston) is usually installed first. It's normally composed of three separate components. Slip the spacer/expander into the groove **(see illustration)**. If an anti-rotation tang is used, make sure it's inserted into the drilled hole in the ring groove. Next, install the lower side rail. Don't use a piston ring installation tool on the oil ring side rails, as they may be damaged. Instead, place one end of the side rail into the groove between the spacer/expander and the ring land, hold it firmly in place and slide a finger around the piston while pushing the rail into the groove **(see illustration)**. Next, install the upper side rail in the same manner.

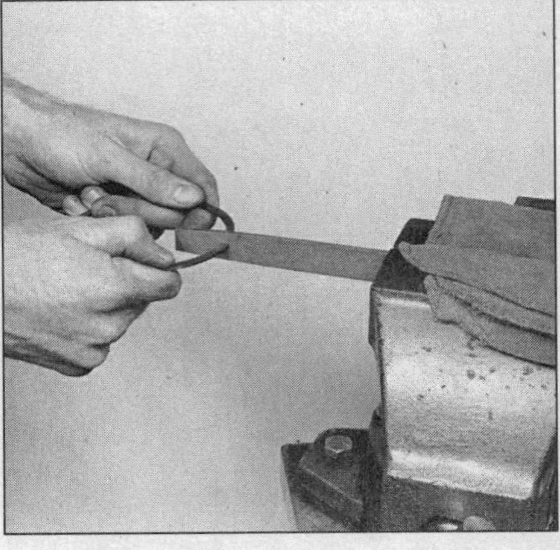
23.5 If the end gap is too small, clamp a fine file in a vise and file ring ends a little at a time and re-measure in the bore

10 After the three oil ring components have been installed, check to make sure that both the upper and lower side rails can be turned smoothly in the ring groove.

11 The number two (middle) ring is installed next. It's usually stamped with a mark which must face up, toward the top of the piston. The chamfer on the number two ring must face down on all engines, with two dots facing up on the six-cylinder engines and on V8's, the ID mark is a drill point, a stamped letter O, an oval depression or the word TOP. **Note:** *Always follow the instructions printed on the ring package or box - different manufacturers may require different approaches. Do not mix up the top and middle rings, as they have different cross sections.*

12 Use a piston ring installation tool and make sure the identification mark is facing

23.9a Installing the spacer/expander in the oil control ring groove

23.9b Slip the oil ring side rails in by hand - do not use a ring expander

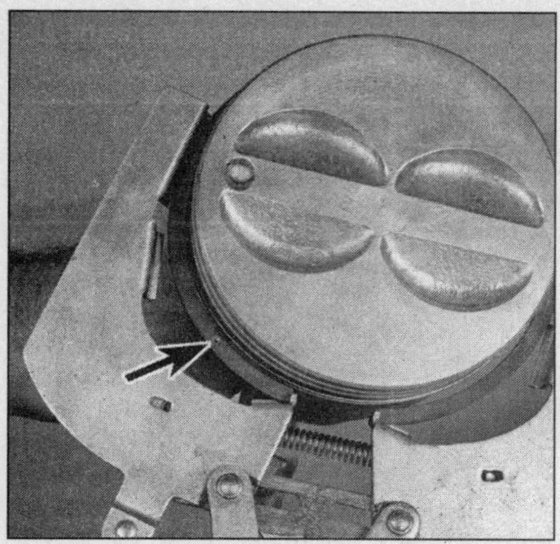

23.12 Install the compression rings using a ring expander tool like this one - note that the mark (arrow) on the ring must face UP

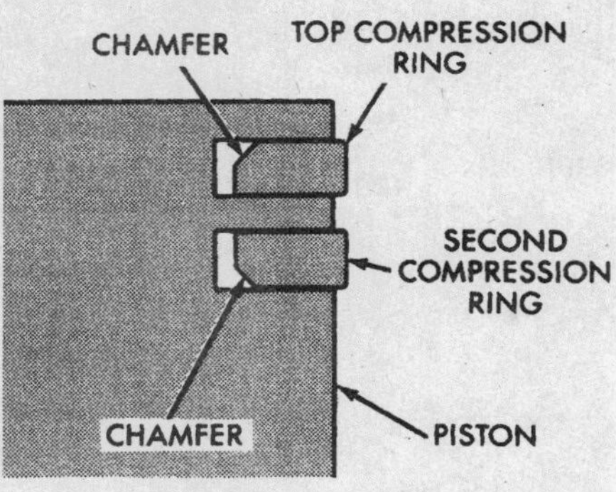

23.13 The chamfer on the second ring must face DOWN, while it must be UP on the top ring - don't mix the two up

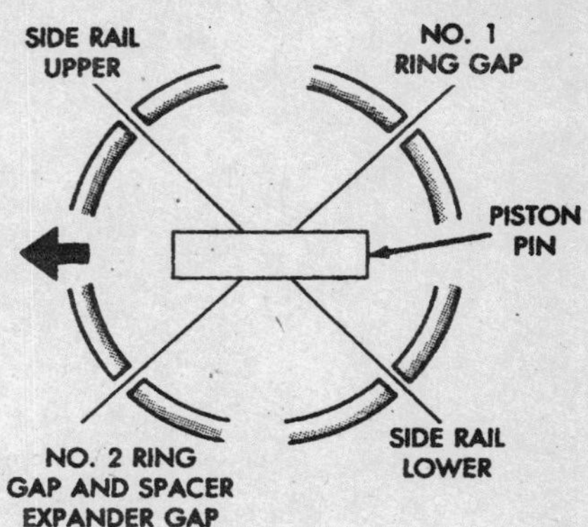

24.5 Position the piston ring gaps as shown here before installing the piston/connecting rod assemblies in the engine

the top of the piston, then slip the ring into the middle groove on the piston **(see illustration)**. Don't expand the ring any more than necessary to slide it over the piston.

13 Install the number one (top) ring in the same manner. Make sure the mark (dot) is facing up. Be careful not to confuse the number one and number two rings. The number one (top) ring must be installed with its chamfer UP while the second ring must have the chamfer DOWN **(see illustration)**. Generally, the second ring has **two** dots on the UP side, and the top ring has **one** dot on the UP side. Check the instructions with the ring set your are using.

14. Repeat the procedure for the remaining pistons and rings.

24 Pistons/connecting rods - installation and rod bearing oil clearance check

Refer to illustrations 24.5, 24.9, 24.11, 24.13 and 24.17

Caution: *The rod bolts on 4.7L V8 engines are a torque-to-yield design and must be replaced whenever they are loosened or removed.*

1 Before installing the piston/connecting rod assemblies, the cylinder walls must be perfectly clean, the top edge of each cylinder must be chamfered, and the crankshaft must be in place.

2 Remove the cap from the end of the number one connecting rod (refer to the marks made during removal). Remove the original bearing inserts and wipe the bearing surfaces of the connecting rod and cap with a clean, lint-free cloth. They must be kept spotlessly clean.

Connecting rod bearing oil clearance check

3 Clean the back side of the new upper bearing insert, then lay it in place in the connecting rod. Make sure the tab on the bearing fits into the recess in the rod. Don't hammer the bearing insert into place and be very careful not to nick or gouge the bearing face. Don't lubricate the bearing at this time.

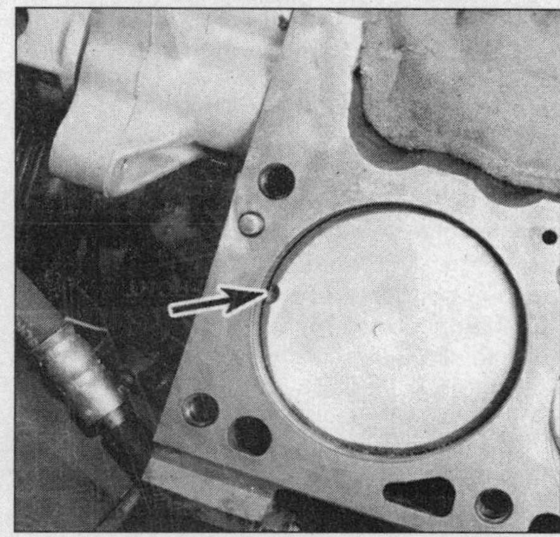

24.9 The notch (arrow) on the pistons must face forward

4 Clean the back side of the other bearing insert and install it in the rod cap. Again, make sure the tab on the bearing fits into the recess in the cap, and don't apply any lubricant. It's critically important that the mating surfaces of the bearing and connecting rod are perfectly clean and oil free when they're assembled.

5 Position the piston ring gaps at intervals around the piston **(see illustration)**.

6 Slip a section of plastic or rubber hose over each connecting rod cap bolt. On 4.7L V8 engines, use the connecting rod guide tools made during removal of the connecting rods.

7 Lubricate the piston and rings with clean engine oil and attach a piston ring compressor to the piston. Leave the skirt protruding about 1/4-inch to guide the piston into the cylinder. The rings must be compressed until they're flush with the piston.

8 Rotate the crankshaft until the number one connecting rod journal is at BDC (bottom dead center) and apply a coat of engine oil to the cylinder walls.

9 With the arrow or notch on top of the piston **(see illustration)** facing the front of the engine, gently insert the piston/connecting rod assembly into the number one cylinder bore and rest the bottom edge of the ring compressor on the engine block.

24.11 Drive the piston in gently with the end of a wooden hammer handle - go evenly and stop if you feel an obstruction

10 Tap the top edge of the ring compressor to make sure it's contacting the block around its entire circumference.

11 Gently tap on the top of the piston with the end of a wooden hammer handle **(see illustration)** while guiding the end of the connecting rod into place on the crankshaft journal. The piston rings may try to pop out of the ring compressor just before entering the cylinder bore, so keep some downward pressure on the ring compressor. Work slowly, and if any resistance is felt as the piston enters the cylinder, stop immediately. Find out what's hanging up and fix it before proceeding. Do not, for any reason, force the piston into the cylinder - you might break a ring and/or the piston.

12 Once the piston/connecting rod assembly is installed, the connecting rod bearing oil clearance must be checked before the rod cap is permanently bolted in place.

13 Cut a piece of the appropriate size Plastigage slightly shorter than the width of the

2D

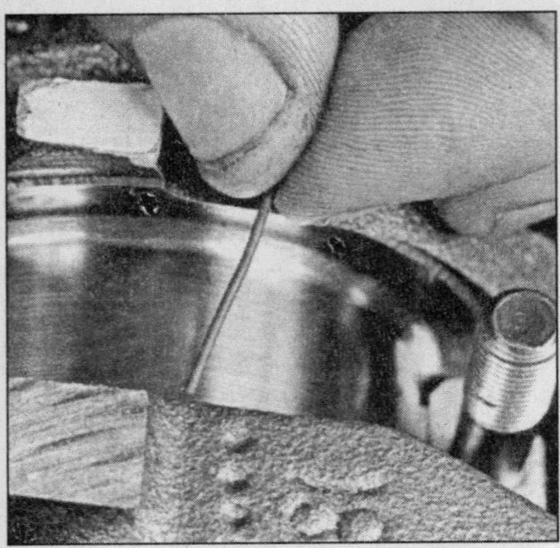

24.13 Lay the Plastigage strips on each rod bearing journal, parallel to the crankshaft centerline

connecting rod bearing and lay it in place on the number one connecting rod journal, parallel with the journal axis **(see illustration)**.

14 Clean the connecting rod cap bearing face, remove the protective hoses from the connecting rod bolts and install the rod cap. Make sure the mating mark on the cap is on the same side as the mark on the connecting rod.

15 Install the nuts (or bolts, on 4.7L V8 engines) and tighten them to the torque listed in this Chapter's Specifications, working up to it in three steps. If you're working on a 4.7L V8 engine, use a pair of the old bolts for checking the oil clearance - save the new bolts for final installation. **Note:** *Use a thin-wall socket to avoid erroneous torque readings that can result if the socket is wedged between the rod cap and nut. If the socket tends to wedge itself between the nut and the cap, lift up on it slightly until it no longer contacts the cap. Do not rotate the crankshaft at any time during this operation.*

16 Remove the nuts or bolts and detach the rod cap, being very careful not to disturb the Plastigage.

17 Compare the width of the crushed Plastigage to the scale printed on the Plastigage envelope to obtain the oil clearance **(see illustration)**. Compare it to the Specifications to make sure the clearance is correct.

18 If the clearance is not as specified, the bearing inserts may be the wrong size (which means different ones will be required). Before deciding that different inserts are needed, make sure that no dirt or oil was between the bearing inserts and the connecting rod or cap when the clearance was measured. Also, recheck the journal diameter. If the Plastigage was wider at one end than the other, the journal may be tapered, in which case the crankshaft needs to be machined.

Final connecting rod installation

19 Carefully scrape all traces of the Plastigage material off the rod journal and/or bearing face. Be very careful not to scratch the

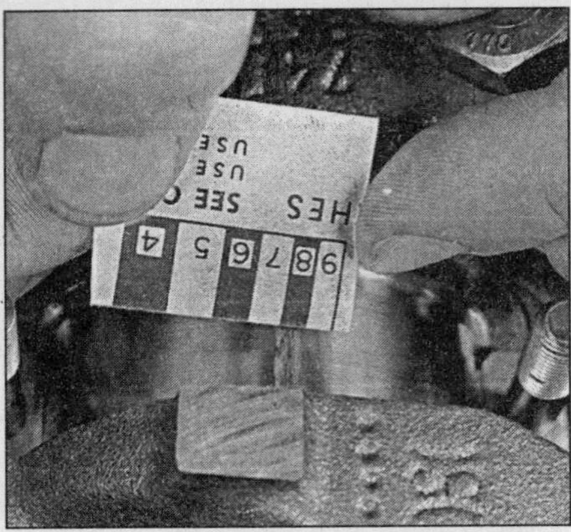

24.17 Measure the width of the crushed Plastigage to determine the rod bearing oil clearance and compare it to the Specifications

bearing - use your fingernail or the edge of a credit card.

20 Make sure the bearing faces are perfectly clean, then apply a uniform layer of clean moly-base grease or engine assembly lube to both of them. You'll have to push the piston into the cylinder to expose the face of the bearing insert in the connecting rod - be sure to slip the protective hoses over the rod bolts first.

21 Slide the connecting rod back into place on the journal, remove the protective hoses from the rod cap bolts, install the rod cap and tighten the nuts or bolts to the torque listed in this Chapter's Specifications. Again, work up to the torque in three steps.

22 Repeat the entire procedure for the remaining pistons/connecting rods.

23 The important points to remember are:

a) *Keep the back sides of the bearing inserts and the insides of the connecting rods and caps perfectly clean when assembling them.*
b) *Make sure you have the correct piston/rod assembly for each cylinder.*
c) *The notch or mark on the piston must face the front of the engine.*
d) *Stagger the ring end gaps* **(see illustration 24.5)**.
e) *Lubricate the cylinder walls with clean oil.*
f) *Lubricate the bearing faces when installing the rod caps after the oil clearance has been checked.*

24 After all the piston/connecting rod assemblies have been properly installed, rotate the crankshaft a number of times by hand to check for any obvious binding.

25 As a final step, the connecting rod endplay must be checked. Refer to Section 13 for this procedure.

26 Compare the measured endplay to the Specifications to make sure it's correct. If it was correct before disassembly and the original crankshaft and rods were reinstalled, it should still be right. If new rods or a new crankshaft were installed, the endplay may be inadequate. If so, the rods will have to be removed and taken to an automotive machine shop for resizing.

27 The balance of the engine assembly work is in attaching the cylinder heads, valvetrain, front cover, water pump, flywheel, intake and exhaust manifolds and accessories, all of which are covered in Chapter 2, Part A for six-cylinder engines, Part B for 5.2L/5.9L V8 engines or Part C for 4.7L V8 engines.

25 Initial start-up and break-in after overhaul

Warning: *Have a fire extinguisher handy when starting the engine for the first time.*

1 Once the engine has been installed in the vehicle, double-check the engine oil and coolant levels.

2 With the spark plugs out of the engine and the ignition and fuel systems disabled (see Sec-tion 3), crank the engine until oil pressure registers on the gauge or the light goes out.

3 Install the spark plugs, connect the plug wires and restore the ignition system functions (Section 3).

4 Start the engine. It may take a few moments for the fuel system to build up pressure, but the engine should start without a great deal of effort. **Note:** *If backfiring occurs through the throttle body, recheck the valve timing.*

5 After the engine starts, it should be allowed to warm up to normal operating temperature. Do not allow the engine to exceed a fast idle until the hydraulic lifters have bled down and become quiet again (usually about five minutes).

6 While the engine is warming up, make a thorough check for fuel, oil and coolant leaks. If a new camshaft and lifters have been installed during the overhaul, the engine should run at a fast idle for 15 minutes (keep an eye on the temperature gauge and don't allow the engine to overheat) to "break in" the cam and lifters.

7 Shut the engine off and recheck the engine oil and coolant levels.

8 Drive the vehicle to an area with minimum traffic, accelerate from 30 to 50 mph, then allow the vehicle to slow to 30 mph with the throttle closed. Repeat the procedure 10 or 12 times. This will load the piston rings and cause them to seat properly against the cylinder walls. Check again for oil and coolant leaks.

9 Drive the vehicle gently for the first 500 miles (no sustained high speeds) and keep a constant check on the oil level. It is not unusual for an engine to use oil during the break-in period.

10 At approximately 500 to 600 miles, change the oil and filter.

11 For the next few hundred miles, drive the vehicle normally. Do not pamper it or abuse it.

12 After 2000 miles, change the oil and filter again and consider the engine broken in.

Chapter 3
Cooling, heating and air conditioning systems

Contents

3

Specifications

General

Coolant capacity	See Chapter 1
Drivebelt tension	See Chapter 1
Radiator cap pressure rating	
1998 and earlier	14 to 18 psi
1999 and later	18 to 21 psi
Thermostat opening temperature	195° F

Torque specifications

	Ft-lbs
Fan blade-to-fan clutch bolts	18
Thermostat housing bolts	16
Water pump attaching bolts	
Six-cylinder engine	
1998 and earlier	22
1999 and later	17
V8 engines	
1999 and earlier	30
2000	40
Water pump pulley-bolts	20

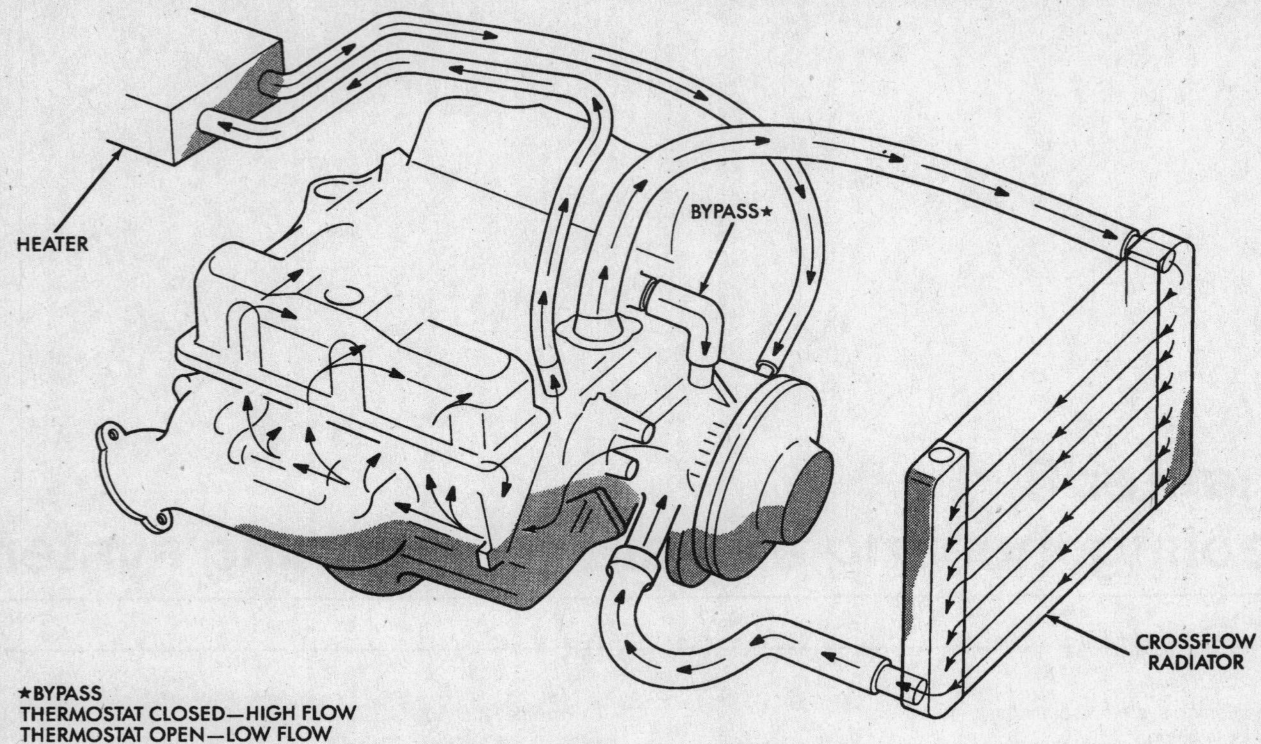

1.1a Typical 5.2L/5.9L V8 cooling system layout

1 General information

Engine cooling system

Refer to illustrations 1.1a, 1.1b and 1.1c

All vehicles covered by this manual employ a pressurized engine cooling system with thermostatically controlled coolant circulation **(see illustrations)**. An impeller-type water pump mounted on the front of the block pumps coolant through the engine. The coolant flows around each cylinder and toward the rear of the engine. Cast-in coolant passages direct coolant around the intake and exhaust ports, near the spark plug areas and in close proximity to the exhaust valve guides. The 4.7L V8 and six-cylinder cooling system uses a "reverse-rotation" design water pump.

A wax-pellet type thermostat is located in a housing near the front of the engine. During warm up, the closed thermostat prevents coolant from circulating through the radiator. As the engine nears normal operating temperature, the thermostat opens and allows hot coolant to travel through the radiator, where it's cooled before returning to the engine.

The cooling system is sealed by a pressure-type cap on the radiator, which raises the boiling point of the coolant and increases the cooling efficiency of the radiator. If the system pressure exceeds the cap pressure-relief value, the excess pressure in the system forces the spring-loaded valve inside the cap off its seat. This allows either excess pressure to escape to the atmosphere or coolant to escape through an overflow tube into a coolant recovery reservoir.

The coolant reservoir serves as both the point at which fresh coolant is added to the cooling system to maintain the proper fluid level and as a holding tank for overheated coolant.

Heating system

The heating system consists of a blower fan and heater core located in the heater/air conditioning unit under the dash, the hoses connecting the heater core to the engine cooling system and the heater/air conditioning control head on the instrument panel. Hot engine coolant is circulated through the heater core. When the heater mode is activated, a flap door opens to expose the heater box to the passenger compartment. A fan switch on the control head activates the blower motor, which forces air through the core, heating the air.

Air conditioning system

The air conditioning system consists of a condenser mounted in front of the radiator, an evaporator mounted adjacent to the heater core, a compressor mounted on the engine, an accumulator which stores liquid refrigerant and removes excess moisture, and the plumbing connecting all of the above components.

A blower fan forces the warmer air of the passenger compartment through the evaporator core (sort of a radiator-in-reverse), transferring the heat from the air to the refrigerant. The liquid refrigerant boils off into low pressure vapor, taking the heat with it when it leaves the evaporator.

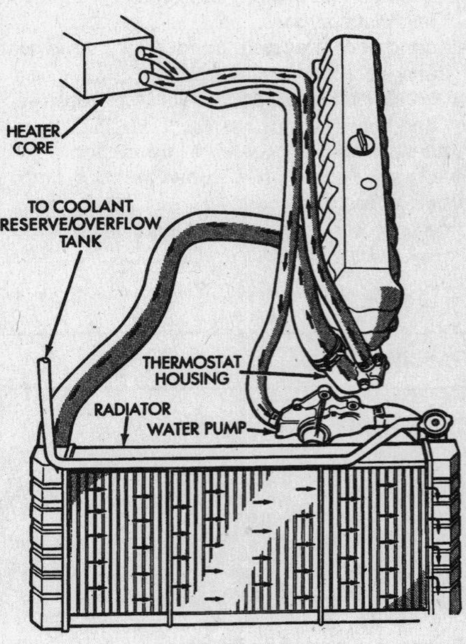

1.1b Typical six-cylinder cooling system layout

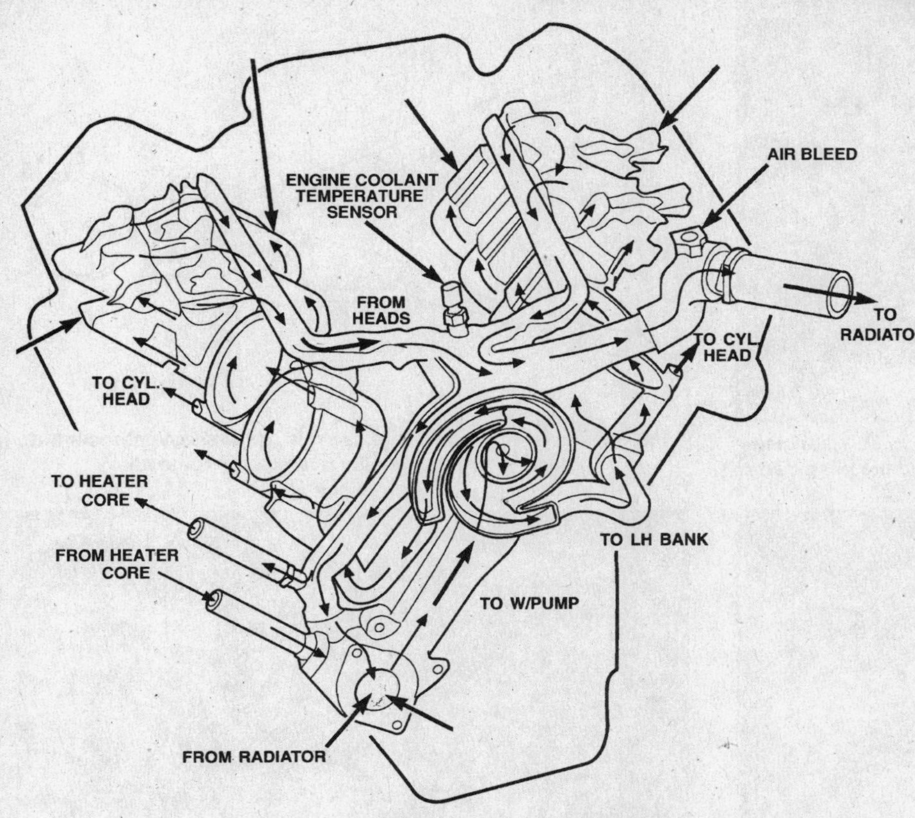

1.1c Typical 4.7L cooling system layout

2 Antifreeze/coolant - general information

Refer to illustration 2.6

Warning 1: *Use ethylene-glycol type antifreeze only. Do not use propylene-glycol type antifreeze solutions.*

Warning 2: *Do not allow antifreeze to come in contact with your skin or painted surfaces of the vehicle. Rinse off spills immediately with plenty of water. Antifreeze is highly toxic if ingested. Never leave antifreeze lying around in an open container or in puddles on the floor; children and pets are attracted by it's sweet smell and may drink it. Check with local authorities about disposing of used antifreeze. Many communities have collection centers which will see that antifreeze is disposed of safely. Never dump used anti-freeze on the ground or pour it into drains.*

Note: *Non-toxic antifreeze is now manufactured and available at local auto parts stores, but even these types should be disposed of properly.*

Keep antifreeze containers covered and repair leaks in your cooling system as soon as they are noticed.

The cooling system should be filled with a water/ethylene-glycol based antifreeze solution, which will prevent freezing down to at least -20°F, or lower if local climate requires it. It also provides protection against corrosion and increases the coolant boiling point.

The cooling system should be drained, flushed and refilled at the specified intervals (see Chapter 1). Old or contaminated antifreeze solutions are likely to cause damage and encourage the formation of rust and scale in the system. Use distilled water with the antifreeze.

Before adding antifreeze, check all hose connections, because antifreeze tends to leak through very minute openings. Engines don't normally consume coolant, so if the level goes down, find the cause and correct it.

The exact mixture of antifreeze-to-water which you should use depends on the relative weather conditions. The mixture should contain at least 50-percent antifreeze, but should never contain more than 70-percent antifreeze. Consult the mixture ratio chart on the antifreeze container before adding coolant. Hydrometers are available at most auto parts stores to test the coolant **(see illustration)**. Use antifreeze which meets the vehicle manufacturer's specifications.

2.6 An inexpensive hydrometer can be used to test the condition of your coolant

3 Thermostat - check and replacement

Warning: *Do not remove the radiator cap, drain the coolant or replace the thermostat until the engine has cooled completely.*

Note: *The following check applies to six-cylinder engines and 5.2L/5.9L V8 engines only.*

Check

1 Before assuming the thermostat is to blame for a cooling system problem, check the coolant level, drivebelt tension (Chapter 1) and temperature gauge (or light) operation.

2 If the engine seems to be taking a long time to warm up (based on heater output or temperature gauge operation), the thermostat is probably stuck open. Replace the thermostat with a new one.

3 If the engine runs hot, use your hand to check the temperature of the upper radiator hose. If the hose isn't hot, but the engine is, the thermostat is probably stuck closed, preventing the coolant inside the engine from escaping to the radiator. Replace the thermostat. **Caution:** *Don't drive the vehicle without a thermostat. The computer may stay in open loop and emissions and fuel economy will suffer.*

4 If the upper radiator hose is hot, it means that the coolant is flowing and the thermostat is open. Consult the *Troubleshooting* section at the front of this manual for cooling system diagnosis.

Replacement

Refer to illustrations 3.8a, 3.8b, 3.10a, 3.10b, 3.10c, 3.13, 3.14a and 3.14b

5 Disconnect the negative cable from the battery.

6 Drain the cooling system (see Chapter 1). If the coolant is relatively new or in good condition (see Chapter 1), save it and reuse it.

7 On six-cylinder and 5.2L/5.9L V8 engines, follow the upper radiator hose to the engine to locate the thermostat housing. On 4.7L V8 engines, follow the lower radiator hose to the engine to locate the thermostat housing.

3

3.8a The hose to the thermostat housing on 5.2L/5.9L V8 engines is in a tight spot - use pliers from above to squeeze the hose clamp

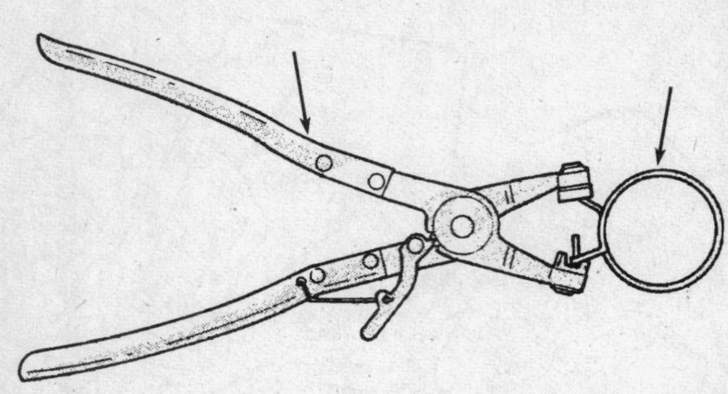

3.8b Special hose clamp pliers like this are available at most auto parts stores to simplify hose clamp removal

3.10a Thermostat housing locations - 5.2L/5.9L V8 engines

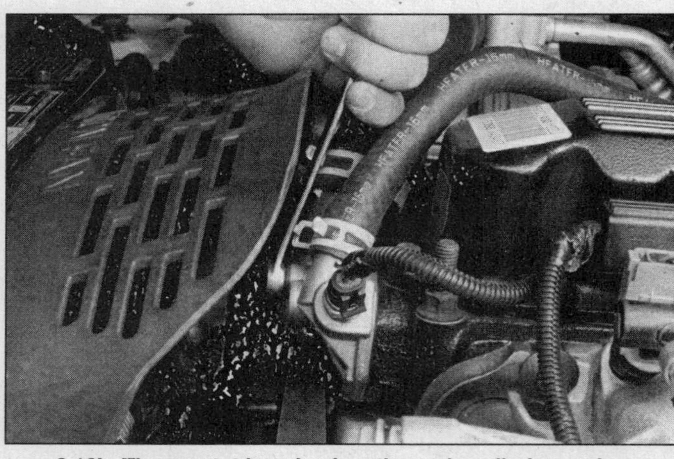

3.10b Thermostat housing location - six-cylinder engine

8 Squeeze the tabs on the hose clamp to loosen it from the hose(s), then reposition the clamp several inches back up the hose. Detach the hose(s) from the thermostat housing **(see illustration)**. **Note:** *Special hose clamps pliers are available at most auto parts stores* **(see illustration)**. If the hose is stuck, grasp it near the end with a pair of adjustable pliers and twist it to break the seal, then pull it off. If the hose is old or deteriorated, cut it off and install a new one.

9 If the outer surface of the thermostat housing that mates with the hose is deteriorated (corroded, pitted, etc.) it may be damaged further by hose removal. If it is, the thermostat housing will have to be replaced.

10 Remove the bolts and detach the thermostat housing from the engine **(see illustrations)**. If the cover is stuck, tap it with a soft-face hammer to jar it loose. Be prepared for some coolant to spill as the gasket seal is broken.

11 Note how the thermostat is installed and remove it from the engine.

12 On six-cylinder and 5.2L/5.9L V8 engines, stuff a rag into the engine opening, then remove all traces of old gasket material and sealant from the housing and cover with a gasket scraper. Remove the rag from the opening and clean the gasket mating surfaces with lacquer thinner or acetone. On 4.7L V8 engines, simply remove the rubber O-ring from around the thermostat.

13 On six-cylinder and 5.2L/5.9L V8 engines, install the new thermostat in the machined groove on the engine. Make sure the correct end faces out - the spring end is normally directed into the engine. Apply a thin, uniform layer of RTV sealant to both sides of the new gasket **(see illustration)** and position it over the thermostat on the engine.

14 On 4.7L V8 engines, install a new O-ring around the thermostat. Make sure to align the

3.10c Thermostat housing location - 4.7L V8 engine - view here is from below at the front of the engine

3.13 Before installing the gasket on six-cylinder and 5.2L/5.9L V8 engines, apply a thin, uniform layer of RTV sealant to both sides of the gasket

rubber tab on the inside of the O-ring groove with the notch on the thermostat **(see illustration)**. Then align the rubber tab on the outside of the O-ring with the notch on the thermostat housing and insert the thermostat and O-ring into the thermostat housing **(see illustration)**.

15 Install the thermostat housing and bolts onto the engine. Tighten the bolts to the torque listed in this Chapter's Specifications.

16 Reattach the hose(s) to the fitting(s) and tighten the hose clamp(s) securely.

17 Refill the cooling system (see Chapter 1).

18 Start the engine and allow it to reach normal operating temperature, then check for leaks and proper thermostat operation (as described in Steps 2 through 4).

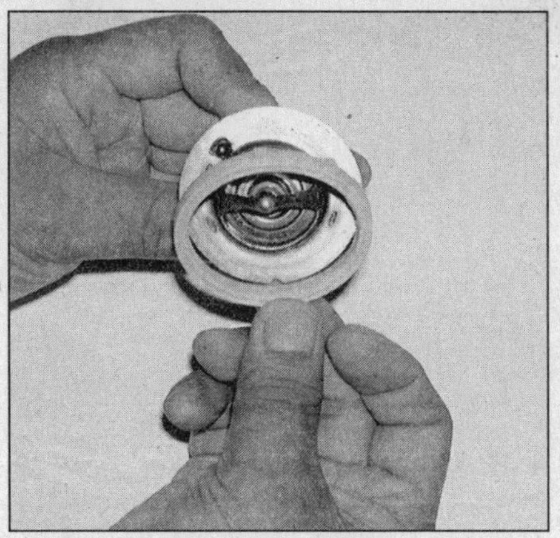

3.14a On 4.7L V8 engines, align the notch on the thermostat (arrow) with the rubber tab on the inner O- ring groove . . .

3.14b . . . then align the tab (lower arrow) on the outer diameter of the O-ring with the notch in the thermostat housing (upper arrow) and insert the thermostat into the thermostat housing

4 Engine cooling fan(s)- check and component replacement

Warning: *To avoid possible injury or damage, DO NOT operate the engine with a damaged fan. Do not attempt to repair fan blades - replace a damaged fan with a new one.*

Note: *Most of the models covered by this manual are equipped with a single mechanical clutch-type cooling fan only. However some engines are equipped with a "max" cooling package which consists of an electric cooling fan and a heavy duty radiator on earlier models or an electric cooling fan and a mechanical clutch type fan on 1999 and later models.*

Check

Mechanical cooling fan

1 Disconnect the negative battery cable and rock the fan back and forth by hand to check for excessive bearing play.

2 Visually inspect the clutch assembly for substantial fluid leakage. If problems are noted, replace the clutch assembly.

3 With the engine cold (and not running), turn the fan blades by hand. The fan should turn freely but never more than five turns.

4 With the engine completely warmed up, turn off the ignition switch and disconnect the negative battery cable from the battery. Turn the fan by hand. More drag should be evident. If the fan turns easily or it rotates the same number of turns it did when the clutch was cold, replace the fan clutch.

Electric cooling fan

1998 5.9L engine (with max cooling package)

Refer to illustrations 4.6a, 4.6b and 4.9

5 Check the 15 amp ignition relay fuse and the 60 amp maxi fuse in the power distribution center in the engine compartment. Refer to the label on the inside of the cover for exact fuse location. If a blown fuse is found it will be necessary to refer to the wiring diagrams at the end of this manual and repair the affected circuits for shorts to ground.

6 If the fuses are good, use a jumper wire to bridge the wiring harness connector terminals of the upper and lower hose fan switches **(see illustration)**. The cooling fan should respond as shown in the accompanying chart **(see illustration)**.

7 If the cooling fan now responds as described with the fan switch or switches bypassed, the electrical circuits to the fan are operating properly and the fan switch or switches are bad and should be replaced.

8 If the cooling fan did not respond as described, disconnect the cooling fan electrical connector. Using a fused jumper wire, apply battery voltage to the red and white and the dark green wire of the cooling fan. The cooling fan should operate in Low speed with battery voltage applied to the dark green wire and in High speed with battery voltage applied to both wires. If it doesn't respond, check the ground (black wire) connection to the motor. If the ground is OK, the cooling fan is bad and should be replaced.

9 If the cooling fan now responds with the jumper wires connected directly to the motor, the problem lies in the ignition relay, the cooling fan relay(s) or the wiring to these components. Refer to Chapter 12 to check the

3

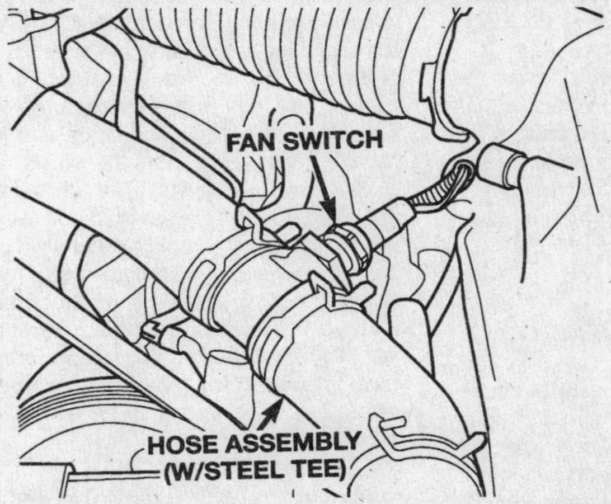

4.6a On 1998 5.9L engines, the upper and lower radiator hose fan switches control the cooling fan On/Off signal

COOLANT TEMP	A/C	UPPER HOSE SWITCH	LOWER HOSE SWITCH	FAN
-36° - 213°F	OFF	OPEN	OPEN	OFF
214° - 219°F	OFF	CLOSED	OPEN	ON/LOW
220°F AND UP	OFF	CLOSED	CLOSED	ON/HIGH
-36° - 213°F	ON	OPEN	OPEN	ON/LOW
214° - 219°F	ON	CLOSED	OPEN	ON/LOW
220°F AND UP	ON	CLOSED	CLOSED	ON/HIGH
WARNING: THE FAN CAN COME ON AT ANY TIME WHEN THE IGNITION KEY IS IN THE "RUN" POSITION.				

4.6b Cooling fan operating chart - 1998 5.9L engine

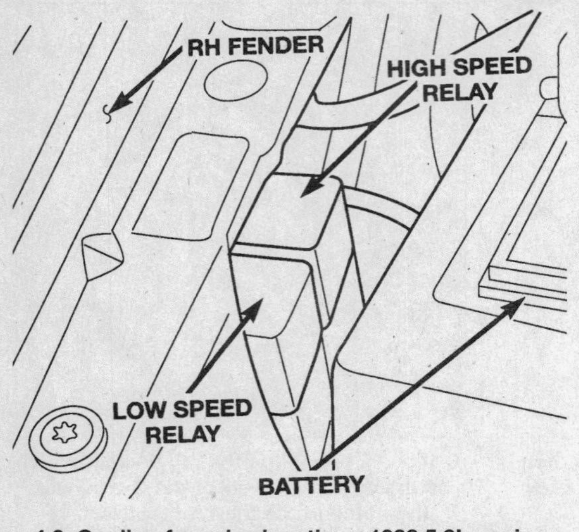

4.9 Cooling fan relay location - 1998 5.9L engine

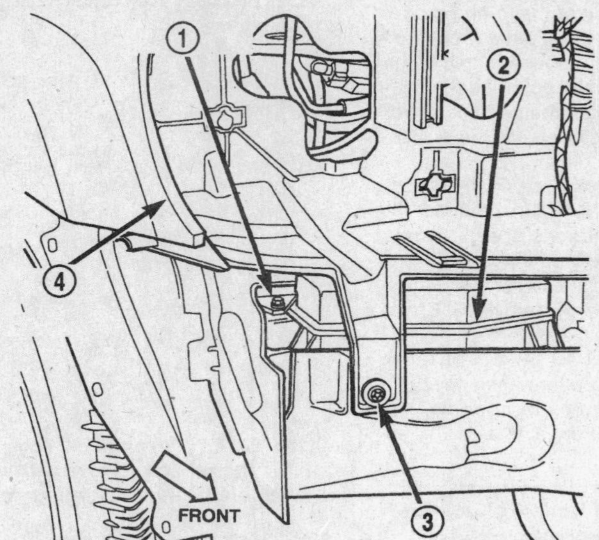

4.12 Cooling fan relay location - 1999 and later engines

1 *Radiator fan relay*
2 *Vacuum reservoir*
3 *Bolt*
4 *Right front fender*

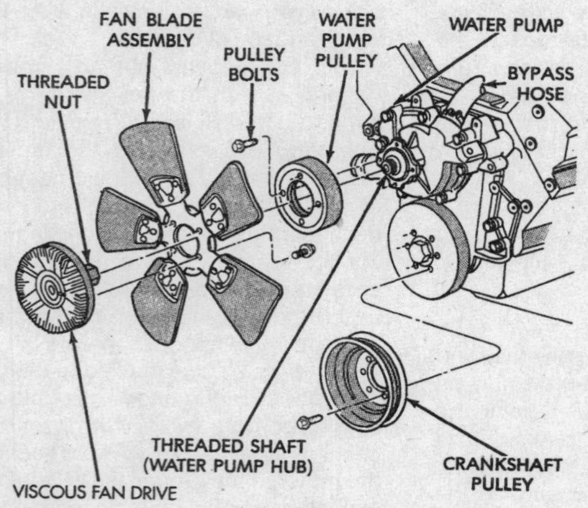

4.14a Engine cooling fan assembly details - 5.2L/5.9L V8 engine

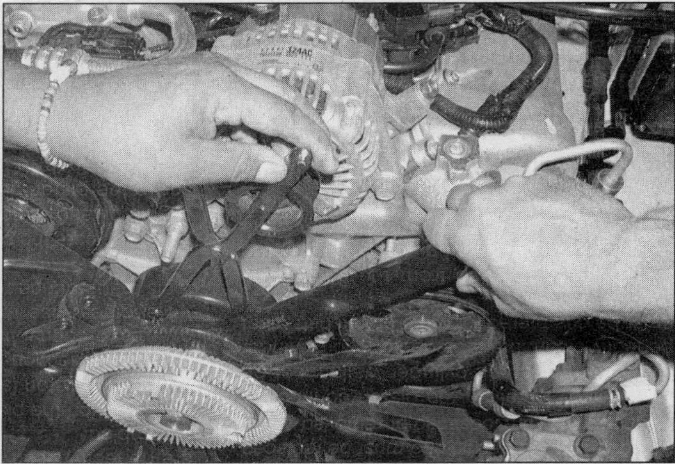

4.14b Use a special spanner wrench to hold the water pump pulley while the mechanical fan drive nut is loosened

relays and the wiring diagrams to check wiring for shorts or open circuits. **Note:** *The ignition relay is located in the power distribution center just behind the battery and the cooling fan relays are mounted to a bracket on the inner fenderwell to the right of the battery* **(see illustration)**.

1999 and later (with max cooling package)

Refer to illustration 4.12

10 Check the fuses in the interior compartment fuse block. Also check the 40 amp fuse in the power distribution center in the engine compartment. If any of the fuses are blown, it will be necessary to refer to the wiring diagrams at the end of this manual and repair the affected circuits for shorts to ground.

11 Disconnect the electrical connector from the cooling fan motor. Connect a fused jumper wire between one of the two terminals on the cooling fan connector and the positive battery terminal. Connect another jumper wire between the other terminal and ground. When the connections are made, the cooling fan should operate. Polarity may be critical, so, if the cooling fan doesn't operate, try swapping the jumper wires and test the operation again. If the cooling fan does not operate, it is faulty and should be replaced.

12 Because the electric cooling fan on the these models is directly controlled by the PCM and (pulse width modulated) cooling fan relay **(see illustration)**, all other tests on this system must be performed by a dealer service department or other qualified repair facility with a specialized scan tool.

Replacement

Refer to illustrations 4.14a, 4.14b, 4.16, 4.17 and 4.19

13 Disconnect the cable from the negative terminal of the battery.

14 If the vehicle is equipped with a mechanical type cooling fan, a special spanner wrench obtainable at most auto parts stores may be required to hold the water pump pulley while a large wrench is used to loosen the fan drive nut **(see illustrations)**. Sometimes it is possible to hold the water pump pulley by applying considerable hand pressure to the serpentine belt while the large nut is turned counterclockwise (right-hand threads), but it may require the tool if the fan drive nut is excessively tight. **Note:** *On 1998 and earlier six-cylinder engines, the fan attaches to the water pump with four studs/nuts; remove the nuts and pull the fan/clutch assembly off the studs.* In either case, carefully lower the fan and clutch assembly into the fan shroud. Be very careful not to damage the radiator fins while doing so.

15 If the vehicle is equipped with an electric cooling fan, make sure to disconnect the electrical connector from the fan motor.

16 Remove the fan shroud mounting bolts **(see illustration)**. Note that some of the mounting bolts can only be accessed from below.

17 Lift the shroud up until it clears the slots in the bottom bracket of the radiator, then lift the fan/clutch and shroud or the electric cooling fan and shroud assembly out of the engine compartment together **(see illustration). Note**

4.16 Remove the fan shroud mounting nuts/bolts

4.17 Lift the fan assembly and fan shroud out together

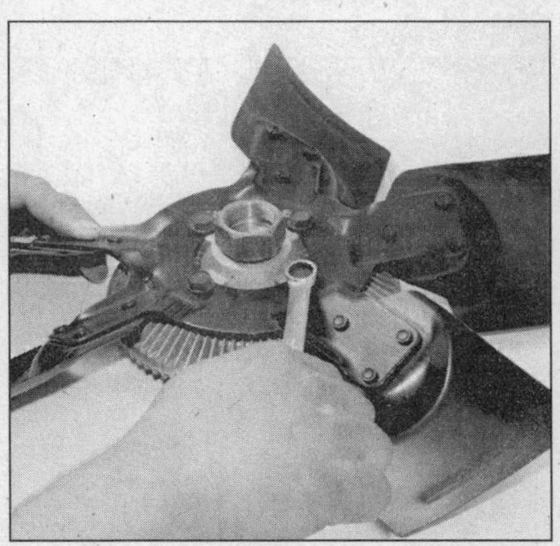

4.19 Remove the four bolts retaining the fan to the fan clutch - when the fan clutch is stored off the engine, place it with the radiator side DOWN

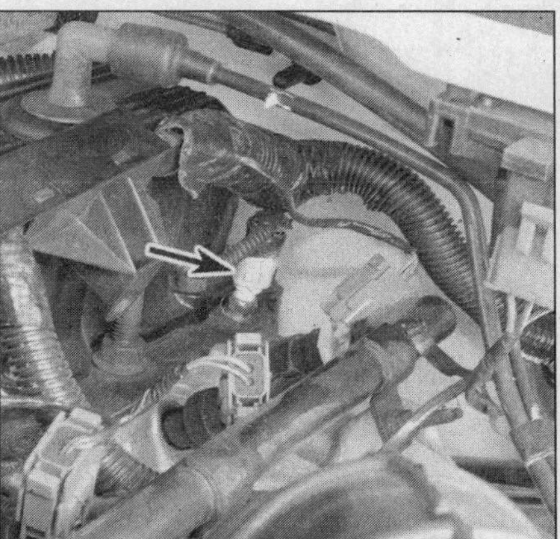

5.1a Coolant temperature sender location (arrow) - six-cylinder engine

5.1b Coolant temperature sender location (arrow) - V8 engine (the other temperature sensor at right is for the computer)

1: *If difficulty is encountered during the removal of the shroud, it will be necessary to drain the cooling system and remove the upper radiator hose to help facilitate removal of the shroud and fan.* **Note 2:** *On 1999 and later models, the electric cooling fan and the shroud are assembled as a unit and if defective must be replaced as a unit.*

18 Carefully inspect the fan blades for damage and defects. Replace it if necessary.

19 At this point, mechanical type fans may be unbolted from the clutch, if necessary **(see illustration)**. If the fan clutch is stored, position it with the radiator side facing down.

20 Installation is the reverse of removal. Be sure to tighten the fan and clutch mounting nuts evenly and securely.

5 Coolant temperature sending unit - check and replacement

Refer to illustrations 5.1a and 5.1b

Warning: *Wait until the engine is completely cool before beginning this procedure.*

Note: *This procedure applies to 1998 and earlier vehicles only. Refer to Chapter 6 for the check and replacement procedures on 1999 and later vehicles.*

1 The coolant temperature indicator system is composed of a light or temperature gauge mounted in the instrument panel and a coolant temperature sending unit mounted on the engine **(see illustrations)**. Some vehicles have more than one sensor unit, but only one is used for the indicator system, and it is usually a single-wire connector. Coolant temperature sensors for the computer usually have two wires.

Check

2 If an overheating indication occurs, check the coolant level in the system and then make sure the wiring between the light or gauge and the sending unit is secure and all fuses are intact.

3 With the ignition switch On and the starter motor cranking, the indicator light should glow (bulb check).

4 If the light does not glow, the bulb may be burned out, the ignition switch may be faulty or the circuit may be open. Test the circuit by grounding the wire to the sending unit with the ignition On (engine not running for safety). If the gauge deflects full scale or the light glows, replace the sending unit.

5 As soon as the engine starts, the light should go out and remain out unless the engine overheats. Failure of the light to go out may be due to a grounded wire between the light and the sending unit, a defective sending unit or a faulty ignition switch. Check the coolant to make sure it's the proper type. Plain water may have too low a boiling point to activate the sending unit.

Replacement

6 If the sending unit must be replaced, simply unscrew it from the engine and install the replacement. Use sealant on the threads. Make sure the engine is cool before removing the defective sending unit. There will be some coolant loss as the unit is removed, so be prepared to catch it. Check the level after the replacement has been installed.

6 Radiator and coolant reservoir - removal and installation

Coolant reservoir - removal and installation

Refer to illustration 6.2

Warning: *Wait until the engine is completely cool before beginning this procedure.*

1 Loosen the hose clamp and detach the hose from the reservoir. Also disconnect the

6.2 Remove the screw, nuts and hose (arrows) from the coolant reservoir and remove the reservoir from the engine compartment

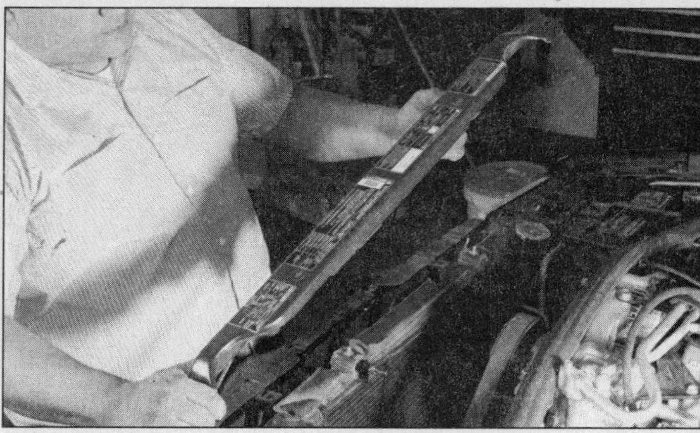

6.9 On early models, remove the radiator support with the hood latch attached and lay it aside

6.10 Radiator support panel mounting bolts - 1999 and later models

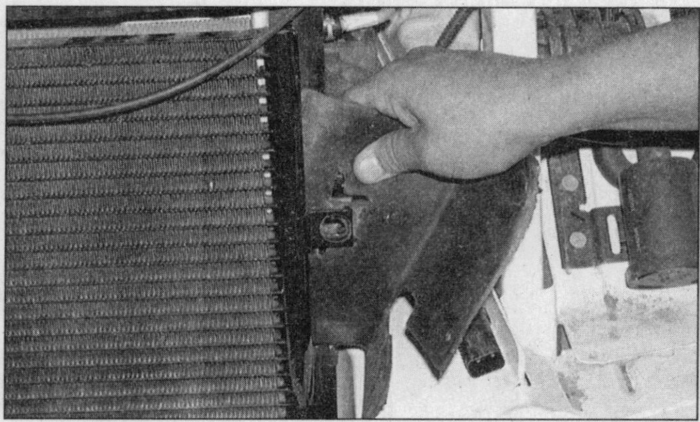

6.12 Detach the condenser mounting bolts (arrow) and remove the radiator side seals - 1999 and later models shown

electrical connector from the coolant level sensor if equipped.

2 Remove the screws/nuts and lift the reservoir from the vehicle **(see illustration)**.

3 Installation is the reverse of removal. Fill the reservoir to FULL mark with the proper type of coolant (see Chapter 1).

4 Disconnect the negative cable from the battery.

5 Drain the cooling system (see Chapter 1). If the coolant is relatively new or in good condition, save it and reuse it.

Radiator

Removal

Refer to illustrations 6.9, 6.10, 6.12 and 6.13

6 Loosen the hose clamps, then detach the radiator hoses from the fittings on the radiator. If they're stuck, grasp each hose near the end with a pair of adjustable pliers and twist it to break the seal, then pull it off - be careful not to distort the radiator fittings! If the hoses are old or deteriorated, cut them off and install new ones.

7 Disconnect the reservoir hose from the radiator filler neck.

8 Refer to Chapter 11 and remove the front grille.

9 On 1998 and earlier models, pry up two clips at each end holdi[illegible]he front hood seal to the body, then detach the radiator support panel mounting bolts and nuts and remove the support panel with the hood latch and cable still attached **(see illustration)**. Lay the support aside. Also remove the screws that attach the radiator fan shroud to the radiator and slide the shroud toward the engine.

10 On 1999 and later models, mark the position of the hood latch with a scribe or marker to aid in installation and remove the hood latch retaining bolts and position the hood latch aside. Detach the upper two bolts securing the radiator support panel to the body and remove radiator support panel **(see illustration)**. If the vehicle is equipped with a mechanical clutch-type cooling fan it will be necessary to refer to section 4 and remove the mechanical fan and electric cooling fan/shroud as an assembly. If the vehicle is only equipped with electric fan, the radiator and shroud can be removed together and separated after the radiator is removed from the vehicle, but make sure to disconnect the electrical connector from the cooling fan.

11 If the vehicle is equipped with an automatic transmission, disconnect the cooler lines from the rear of the radiator (see Chapter 7B). Use a drip pan to catch spilled fluid. Plug the lines and fittings.

12 If the vehicle is equipped with air conditioning, remove the bolts securing the A/C condenser to the radiator and remove the side air seals **(see illustration)**. If the vehicle is not equipped with A/C, simply remove the bolts securing the side seals to the radiator and detach the side seals to access the radiator mounting bolts.

13 Remove the radiator mounting bolts and carefully lift out the radiator **(see illustration)**. Don't spill coolant on the vehicle or scratch the paint. If the vehicle is equipped with an auxiliary transmission oil cooler, take special care not the tear the rubber seal on the right side of the radiator, as the oil cooler lines are routed through it.

Installation

Refer to illustration 6.15

14 With the radiator removed, it can be inspected for leaks and damage. If it needs repair, have a radiator shop or dealer service department perform the work as special techniques are required.

15 Whenever the radiator is removed from the vehicle, make note of the location of the various rubber seals (top, bottom and sides) and their location. They should be put back exactly as originally installed to ensure proper cooling **(see illustration)**.

16 Bugs and dirt can be removed from the front of the radiator with a garden hose, followed by compressed air and a soft brush.

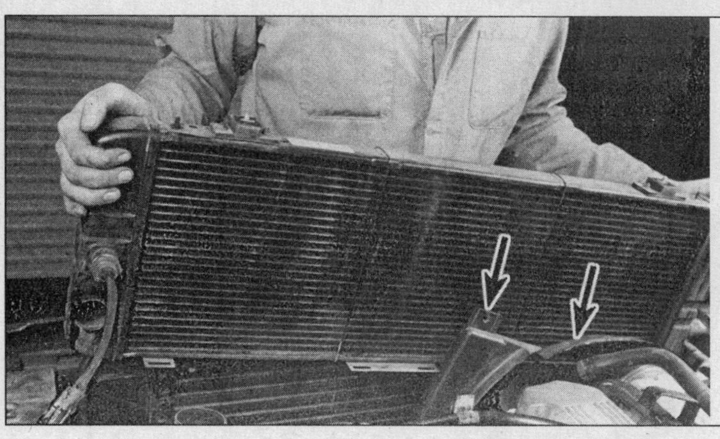

6.13 When lifting the radiator out, be careful not to damage the cooling fins with the fan and fan shroud (arrows)

Don't bend the cooling fins as this is done. When blowing out the core, direct the hose or air line only from the engine side out.

17 Check the radiator mounts for deterioration and make sure there's nothing in them when the radiator is installed.

18 Installation is the reverse of the removal procedure.

19 After installation, fill the cooling system with the proper mixture of antifreeze and water. Refer to Chapter 1 if necessary. **Caution:** *If you are working on a model with a 4.7L V8 engine, be sure to properly bleed the air from the cooling system* (see Chapter 1).

20 Start the engine and check for leaks. Allow the engine to reach normal operating temperature, indicated by the upper radiator hose becoming hot. Recheck the coolant level and add more if required.

21 If you're working on an automatic transmission-equipped vehicle, check and add fluid as needed.

7 Water pump - check

Refer to illustrations 7.3 and 7.4

1 A failure in the water pump can cause serious engine damage due to overheating.

2 There are several ways to check the operation of the water pump while it's installed on the engine. If the pump is defective, it should be replaced with a new or rebuilt unit.

3 Water pumps are equipped with weep or vent holes. If a failure occurs in the pump seal, coolant will leak from the hole. In most cases you'll need a flashlight to find the hole on the water pump from underneath to check for leaks **(see illustration)**.

4 If the water pump shaft bearings fail there may be a howling sound at the front of the engine while it's running. Shaft wear can be felt if the water pump pulley is rocked up and down **(see illustration)**. Don't mistake drivebelt slippage, which causes a squealing sound, for water pump bearing failure.

5 It is possible for a water pump to be bad, even if it doesn't howl or leak water. Sometimes the fins on the back of the impeller can corrode away until the pump is no longer effective. The only way to check for this is to remove the pump for examination.

8 Water pump - removal and installation

Refer to illustrations 8.6, 8.7a, 8.7b, 8.7c and 8.7d

Warning: *Wait until the engine is completely cool before beginning this procedure.*

Removal

1 Disconnect the negative cable from the battery.

2 Drain the cooling system (see Chapter 1). If the coolant is relatively new or in good condition, save it and reuse it.

3 Remove the cooling fan and shroud (see Section 4).

4 Remove the drivebelts (see Chapter 1) and the water pump pulley.

5 Loosen the clamps and detach the hoses from the water pump. If they're stuck, grasp each hose near the end with a pair of

3

7.3 The water pump weep hole (arrow) will drip coolant when the seal on the pump shaft fails

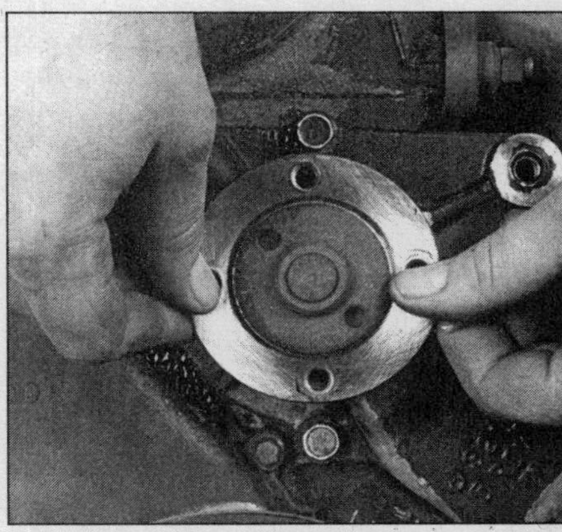

7.4 Grasp the water pump flange and try to rock the shaft back and forth to check for play

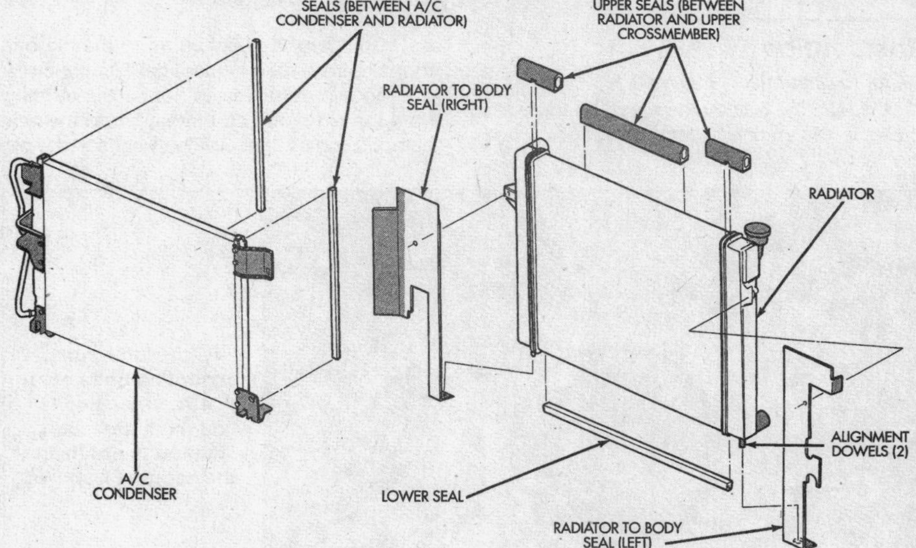

6.15 The rubber seals around the radiator and between the radiator and condenser must be installed in their original locations to retain proper cooling

8.6 On 4.0L six-cylinder engines, the power steering pump and bracket can be removed as a unit - the bracket is over the water pump

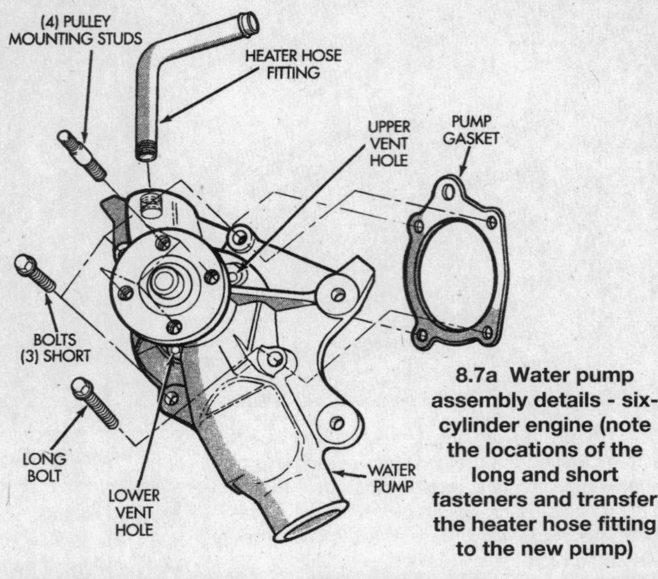

8.7a Water pump assembly details - six-cylinder engine (note the locations of the long and short fasteners and transfer the heater hose fitting to the new pump)

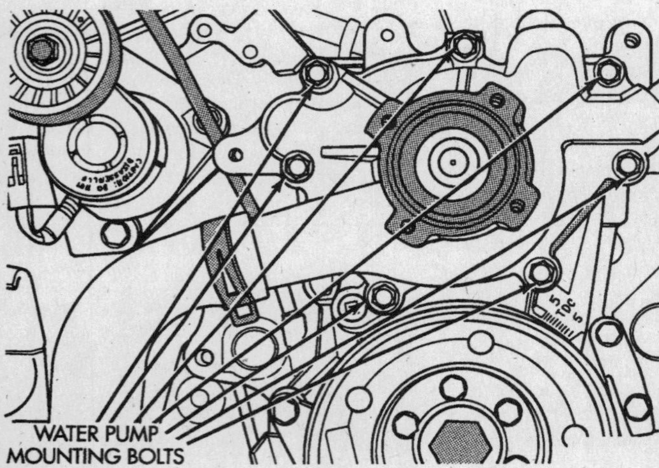

8.7b Water pump bolt locations - 5.2L/5.9L V8 engine

8.7c Reach behind the water pump with pliers to detach the bypass hose - V8 engine

adjustable pliers and twist it to break the seal, then pull it off. If the hoses are deteriorated, cut them off and install new ones. On 5.2L/5.9L V8 engines, unbolt the coolant return tube from the top of the engine and pull the end out of the water pump.

6 On six-cylinder models, remove the idler pulley and the power steering pump **(see illustration)**, don't disconnect the hoses and keep it upright to avoid spilling fluid.

7 Remove the bolts and detach the water pump from the engine. Note the locations of the various lengths and different types of bolts as they're removed to ensure correct installation **(see illustrations)**. On 5.2L/5.9L V8 models, reach above the water pump with pliers to release the hose clamp on the bypass hose **(see illustrations)**. **Note:** *On six-cylinder engines and 5.2L/5.9L V8 engines, be sure to check the condition of the bypass hose and replace it, if necessary (it's a good idea to replace it regardless of its apparent condition). It may be necessary to remove accessory drive brackets for access to the hose.*

Installation

Refer to illustrations 8.9 and 8.13

8 Clean the bolt threads and the threaded holes in the engine to remove corrosion and sealant.

9 Compare the new pump to the old one to make sure they're identical. On six-cylinder models, remove the heater hose fitting from the old pump and transfer it to the new pump, using sealant on the threads. If the old

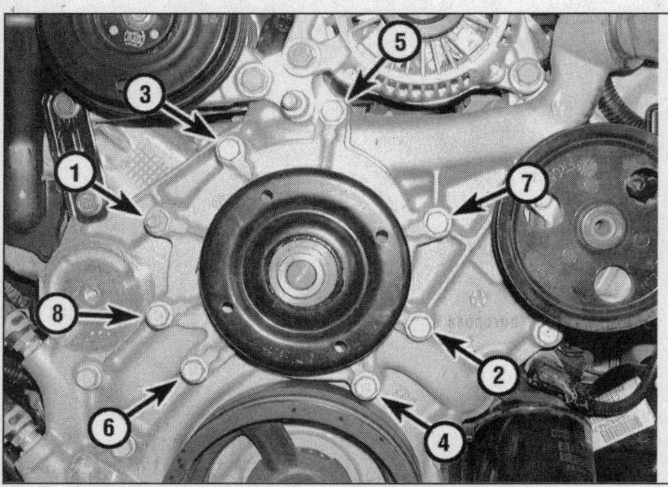

8.7d Water pump mounting bolts on a 4.7L V8 engine - during installation, tighten the bolts in the sequence shown

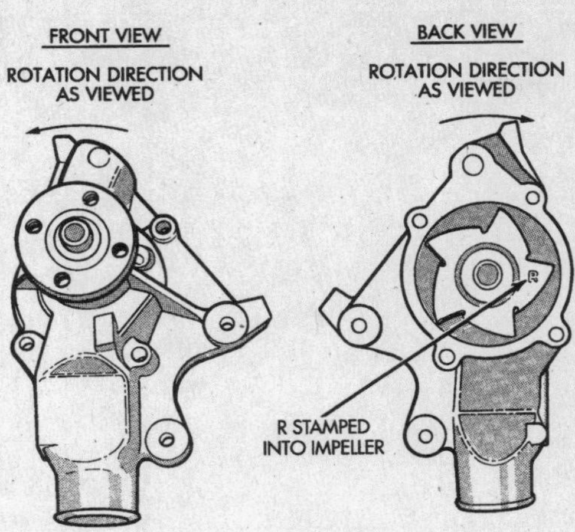

8.9 Six-cylinder replacement water pumps must be of reverse-rotation design - look for the R stamped on the impeller

8.13 When replacing the water pump, make sure the various-length bolts are in the correct locations

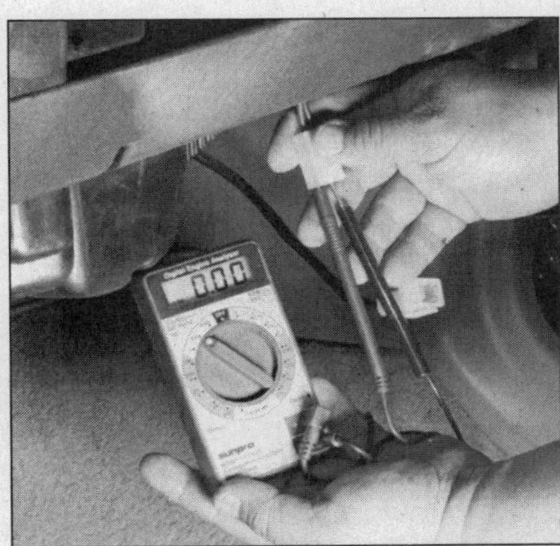

9.4 Disconnect the electrical connector to the blower motor and attach a voltmeter to the harness side of the connector

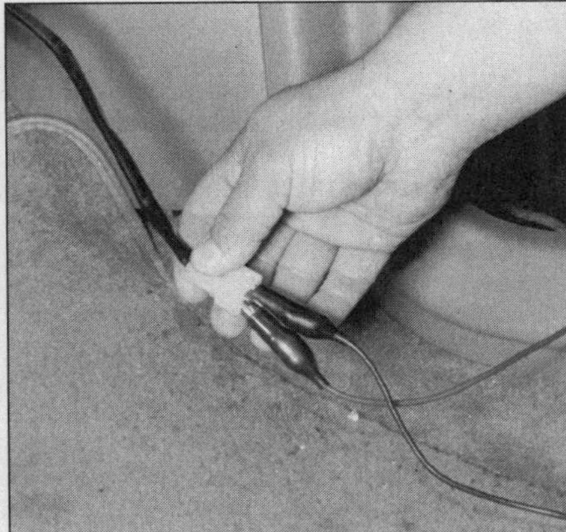

9.6 Supply a ground connection and fused battery voltage directly to the blower side of the connector to see if the blower operates

pump is being reused, check the impeller blades on the backside for corrosion. If any fins are missing or badly corroded, replace the pump with a new one. **Caution:** *Water pumps on 4.0L six-cylinder engines are of "reverse-rotation" design. They are marked on the impeller with an "R"* **(see illustration)**. *If you install a replacement pump, be sure it is marked this way, or overheating will result. (The water pump on 4.7L V8 engines is also a reverse-rotation pump, but is not marked.)*

10 Remove all traces of old gasket material from the engine with a gasket scraper.

11 Clean the engine and new water pump mating surfaces with lacquer thinner or acetone.

12 Apply a thin coat of RTV sealant to the engine side of the new gasket.

13 Apply a thin layer of RTV sealant to the gasket mating surface of the new pump, then carefully mate the gasket and the pump. Slip a couple of bolts through the pump mounting holes to hold the gasket in place **(see illustration)**.

14 Carefully attach the pump and gasket to the engine and thread the bolts into the holes finger tight.

15 Install the remaining bolts (if they also hold an accessory bracket in place, be sure to reposition the bracket at this time). Tighten them to the torque listed in this Chapter's Specifications in 1/4-turn increments. On 5.2L/5.9L V8 engines, install a new O-ring on the coolant return tube, connect the tube to the water pump and tighten the fasteners securely.

16 Reinstall all parts removed for access to the pump. Make sure the serpentine drivebelt is installed as originally routed (see Chapter 1) or overheating could result.

17 Refill the cooling system and check the drivebelt tension (see Chapter 1). Run the engine and check for leaks.

9 Heater and air conditioning blower motor - circuit check

Note: *The circuit checking procedure outlined below applies to manually operated heating and air conditioning systems only. Automatic heating and air conditioning systems on these models cannot be tested using conventional equipment. Due to the use of an integrated electronic control module that can only be tested with specialized equipment it will be necessary to take these vehicles to dealer service department or other qualified repair shop.*

Circuit check

Refer to illustrations 9.4, 9.6, 9.7a, 9.7b and 9.7c

1 Check the fuse and all connections in the circuit for looseness and corrosion. Make sure the battery is fully charged.

2 With the transmission in Park, the parking brake securely applied, turn the ignition switch to the On position (engine not running).

3 The blower motor is located under the glove compartment area of the dash, near the firewall. Bend back the clip on one of the blower motor mounting screws and release the wiring harness, pulling it down for easier disconnection of the connector.

4 Connect a voltmeter to the blower motor connector **(see illustration)**.

5 Move the blower switch through each of its positions and note the voltage readings. Changes in voltage indicates that the motor speeds will also vary as the switch is moved to the different positions. Slower speeds will deliver less voltage to the blower, and the HIGH position will bypass the resistor to supply a full 12 volts.

6 If there is voltage, but the blower motor does not operate, connect a jumper wire between the motor ground terminal and a good chassis ground. Connect a fused jumper wire between the battery positive terminal and the positive terminal on the motor side of the connector **(see illustration)**. If the motor now works, remove the jumper wire, and if the motor stops working when the ground wire is removed, check for bad ground and re-test. If the motor still doesn't work, the blower motor is probably faulty.

7 If there's no voltage at the blower motor, remove the resistor block connector and

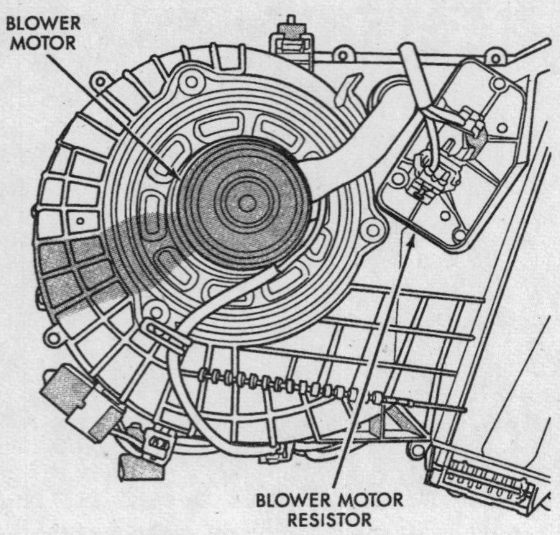

9.7a Blower motor resistor location

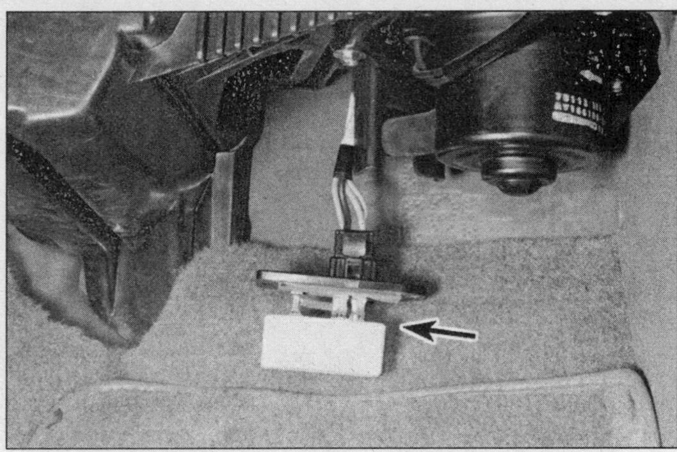
9.7b Remove the two screws and remove the blower motor resistor (arrow) from the case

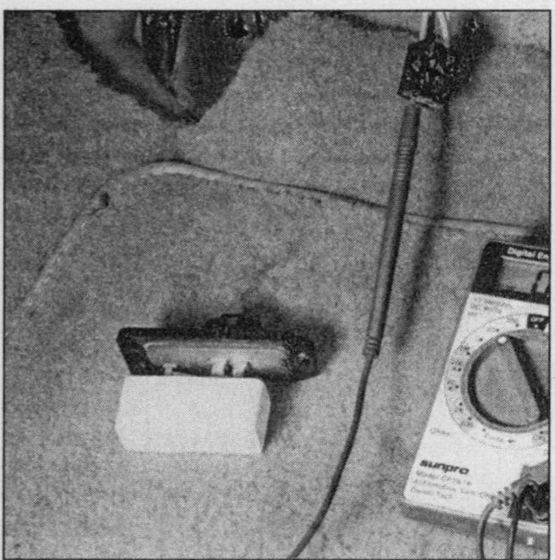
9.7c Check for voltage in the harness to the resistor, at each blower switch speed

check it for voltage **(see illustrations)**. If there's voltage at any of the connector terminals, check the resistor block for continuity and the wiring between the resistor block and the motor for an open or short.

8 If there's no voltage to the resistor, remove the heater control panel (see Section 11) and, with the ignition ON, check for voltage at the connector for the blower motor switch.

9 If there's no voltage at the blower switch, check the wiring between the fuse panel and the switch for an open or a short.

10 If there's voltage, connect one end of a jumper wire to the terminal of the switch connector with voltage. Connect the other end of the jumper to each of the terminals that supply voltage to the resistor block. If the motor now operates normally, replace the switch.

10 Blower motor - removal and installation

Refer to illustrations 10.3 and 10.4

Warning: *These models are equipped with airbags. The airbag is armed and can deploy (inflate) anytime the battery is connected. To prevent accidental deployment (and possible injury), disconnect the negative battery cable whenever working near airbag components. After the battery is disconnected, wait at least 2 minutes before beginning work (the system has a back-up capacitor that must fully discharge). For more information see Chapter 12.*

10.3 The blower motor is mounted under the instrument panel, near the glove box - disconnect the electrical connector, then remove the three blower motor mounting screws (arrows)

1 Disconnect the negative cable from the battery.

2 Disconnect the blower motor connector (see Section 9).

3 Remove the blower motor mounting screws and remove the motor **(see illustration)**.

4 Detach the fan retainer clip from the motor shaft **(see illustration)**.

5 Slip the fan off the old motor shaft, mount it on the new motor and install the retainer clip.

6 Place the blower motor in position, making sure the seal is in place and install the mounting screws.

7 Reconnect the electrical connector and battery cable. Test the motor operation.

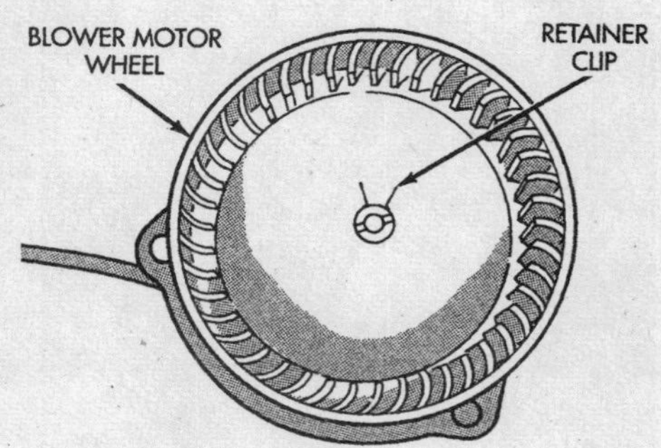

10.4 If the motor is being replaced, detach the fan clip, remove the fan and install it on the shaft of the new motor

11 Air conditioning and heater control assembly - removal and installation

Refer to illustrations 11.3, 11.4a and 11.4b

Warning: *These models are equipped with airbags. The airbag is armed and can deploy (inflate) anytime the battery is connected. To*

11.3 Remove the screws retaining the air conditioning and heater control panel

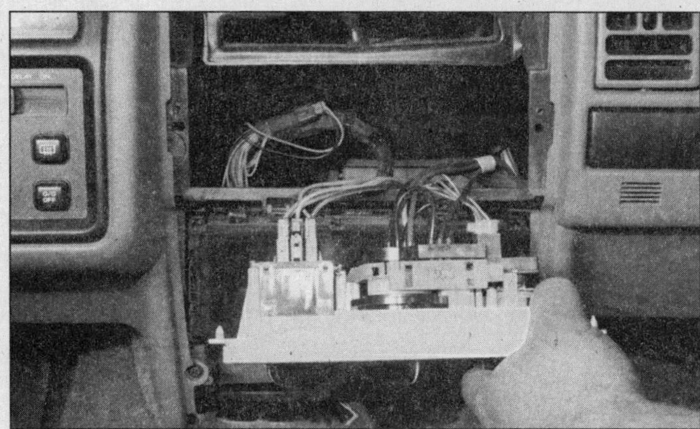

11.4a Pull the air conditioning and heater control panel out and disconnect the vacuum and electrical connectors

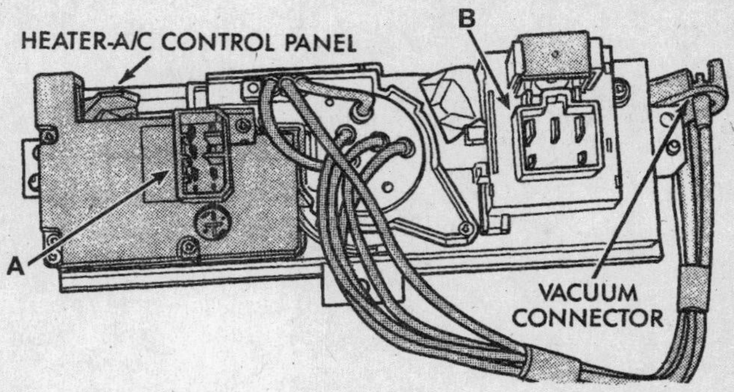

11.4b Air conditioning and heater control assembly connector details - 1998 and earlier

a) *Temperature control connection*
b) *Blower speed switch connection*

11.7 Use a hand-held vacuum pump to test the one-way check valve (arrow) in the vacuum line to the control panel

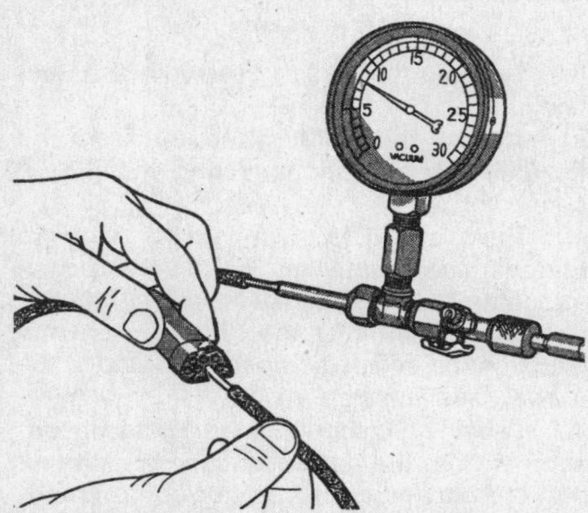

11.10 Apply 8 inches of vacuum to each of the seven ports in the vacuum harness connector

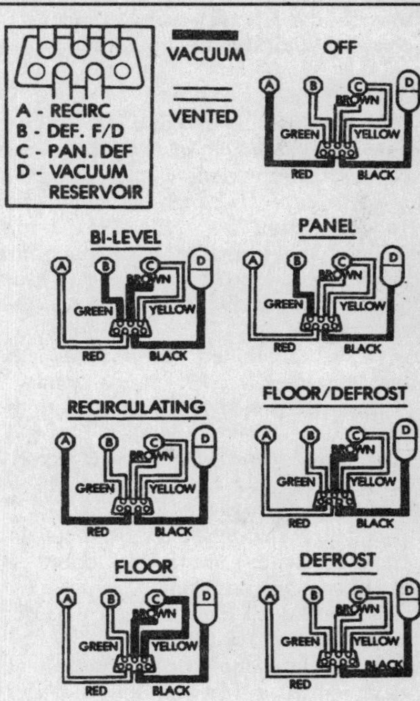

11.11 Vacuum circuit details

prevent accidental deployment (and possible injury), disconnect the negative battery cable whenever working near airbag components. After the battery is disconnected, wait at least 2 minutes before beginning work (the system has a back-up capacitor that must fully discharge). For more information see Chapter 12.

1 Disconnect the negative cable from the battery.

2 Refer to Chapter 11 and remove the center trim bezel(s).

3 Remove the four mounting screws from the air conditioning and heater control panel **(see illustration)**.

4 Pull the air conditioning and heater control panel forward and disconnect the electrical connectors and vacuum plug **(see illustrations)**.

5 To install the control, reverse the removal procedure.

Control panel vacuum tests

Refer to illustrations 11.7, 11.10 and 11.11

6 Vacuum is supplied to the climate control system by a line attached to the intake manifold. A check valve is used in this line.

7 With the engine OFF, pull the check valve (noting its direction) off the vacuum line from the intake manifold and apply 8 inches of vacuum with a hand-held pump on the engine side of the check valve **(see illustration)**. Vacuum should flow through.

8 With vacuum applied to the control panel side of the valve, vacuum should hold. If not, replace the check valve.

9 Reconnect the check valve in the vacuum line in the proper direction and plug the hose back onto the intake manifold.

10 With the control panel pulled out and the vacuum harness disconnected **(see illustration 11.4a)** apply 8 inches of vacuum in turn to each of the 7 ports on the connector **(see illustration)**.

11 If any of the circuits don't hold vacuum, follow the lines to their destination **(see illustration)** to isolate the source of the leak and repair it.

12 Heater core - removal and installation

Refer to illustrations 12.7a, 12.7b, 12.10, 12.14 and 12.16

Warning 1: *These models are equipped with airbags. The airbag is armed and can deploy (inflate) anytime the battery is connected. To prevent accidental deployment (and possible injury), disconnect the negative battery cable whenever working near airbag components. After the battery is disconnected, wait at least 2 minutes before beginning work (the system has a back-up capacitor that must fully discharge). For more information see Chapter 12.*

Warning 2: *The air conditioning system is under high pressure. Do not loosen any hose fittings or remove any components until after the system has been discharged. Air conditioning refrigerant should be properly discharged into an EPA-approved recovery/recycling unit at a dealer service department or an automotive air conditioning repair facility. Always wear eye protection when disconnecting air conditioning system fittings.*

Warning 3: *Wait until the engine is completely cool before beginning this procedure.*

12.7a On 1998 and earlier models, a spring-lock coupling tool like this is used to spread the garter-spring to allow disconnection of the air conditioning lines

12.7b On 1999 and later models, simply unbolt the refrigerant lines (arrows) from the firewall

Removal

1 **Note:** *Heater core removal on the Grand Cherokee is a difficult task for the home mechanic. It can be done with slow, careful attention to detail, but many fasteners and wiring connectors are difficult to get at behind the instrument panel. The entire instrument panel must be removed to allow the heater/air conditioning unit to be removed from the car.*

2 On air-conditioned models, have the system discharged (see **Warning** above).

3 Disconnect the negative cable from the battery.

4 Drain the cooling system (see Chapter 1) and disconnect the heater hoses at the heater core inlet and outlet.

5 Remove the console, lower the steering column and remove the instrument panel assembly (see Chapter 11).

6 Disconnect the blower motor electrical connector (see Section 10).

7 On air-conditioned models, disconnect the air conditioning lines from the evaporator at the firewall **(see illustrations)**. Special tools are required to disconnect the fittings on some models.

8 Remove the coolant reservoir (see Section 6).

9 Detach the PCM (computer) from the firewall, without disconnecting the 60-pin connector.

10 Pull the PCM away far enough to remove the heater/air conditioning unit attaching nuts (three behind the computer, one at the center of the firewall) from the studs on the engine compartment side of the firewall **(see illustration)**.

11 Label and disconnect the electrical connectors from the heater/air conditioning unit and the blower motor resistor/control module. Disconnect the vacuum hose at the vacuum motor.

12 Remove the evaporator drain tube.

13 Remove the defrost duct (above the heater/air conditioning unit).

14 Disconnect the rear floor heat duct **(see illustration)**.

15 Remove the nuts from the studs on the interior side of the firewall and pull the heater/air conditioning unit from the vehicle.

16 Remove the retaining screws **(see illustration)** and remove the heater core by pulling it straight up out of the housing.

Installation

17 Position the heater core in the housing and install the screws.

18 When installing the heater core, cement the seal in place to keep it from moving when the blower assembly is installed.

19 Reinstall the remaining components in the reverse order of removal. **Note:** *When installing the blower/evaporator housing, avoid pinching any wiring between the housing fresh air inlet and the instrument panel on the right side of the housing.*

20 Refill the cooling system (see Chapter 1).

21 Have the air conditioning system (if equipped) evacuated, recharged and leak-tested.

22 Start the engine and check for proper operation.

13 Air conditioning and heating system - check and maintenance

Air conditioning system

Refer to illustrations 13.1 and 13.5

Warning: *The air conditioning system is under high pressure. Do not loosen any hose fittings or remove any components until after the system has been discharged. Air conditioning refrigerant should be properly discharged into an EPA-approved recovery/recycling unit at a dealer service department or an automotive air conditioning repair facility. Always wear eye protection when disconnecting air conditioning system fittings.*

Caution 1: *All models covered by this manual use environmentally friendly R-134a. This refrigerant (and its appropriate refrigerant oils) are not compatible with R-12 refrigerant system components and must never be mixed or the components will be damaged.*

Caution 2: *When replacing entire components, additional refrigerant oil should be added equal to the amount that is removed with the component being replaced. Be sure to read the can before adding any oil to the system, to make sure it is compatible with the R-134a system.*

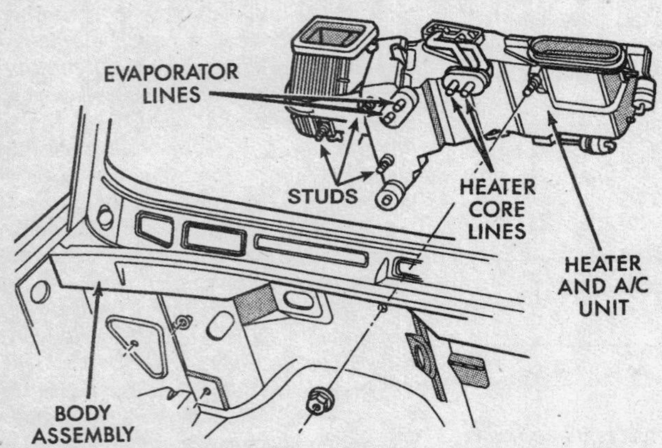

12.10 Remove the housing attaching nuts from the engine side of the firewall

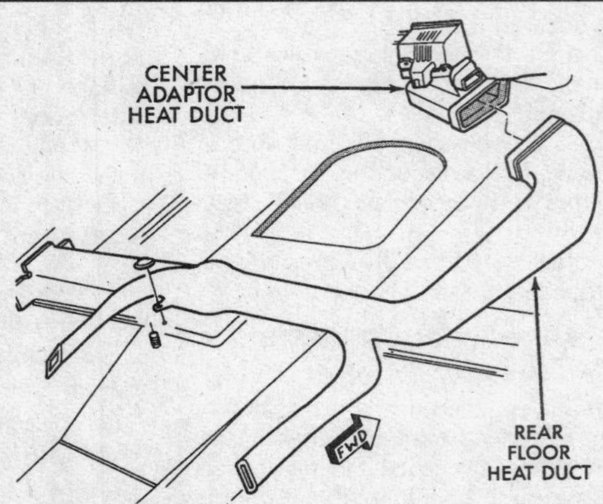

12.14 Disconnect the rear floor heat ducting

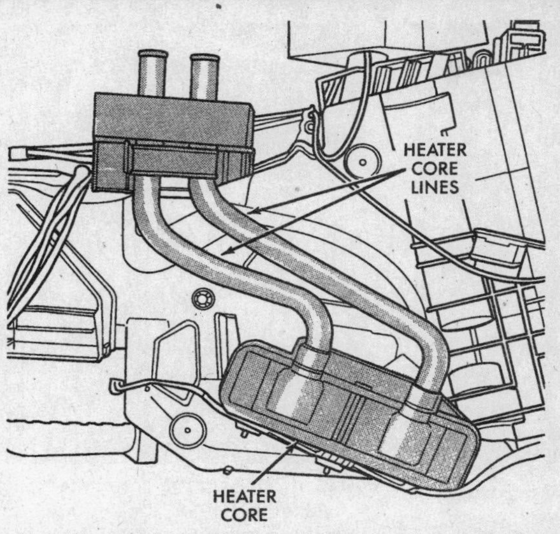

12.16 The heater core is removed from the top of the heater/air conditioning unit (1998 and earlier shown, 1999 and later similar)

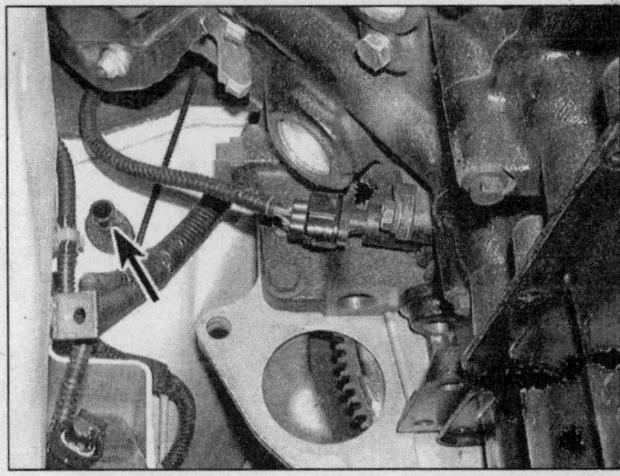

13.1 Check the evaporator housing drain tube (arrow) for blockage - the view here is from below the vehicle looking up at the firewall

1 The following maintenance checks should be performed on a regular basis to ensure that the air conditioning continues to operate at peak efficiency.

a) *Inspect the condition of the compressor drivebelt. If it is worn or deteriorated, replace it (see Chapter 1).*
b) *Check the drivebelt tension and, if necessary, adjust it (see Chapter 1).*
c) *Inspect the system hoses. Look for cracks, bubbles, hardening and deterioration. Inspect the hoses and all fittings for oil bubbles or seepage. If there is any evidence of wear, damage or leakage, replace the hose(s).*
d) *Inspect the condenser fins for leaves, bugs and any other foreign material that may have embedded itself in the fins. Use a "fin comb" or compressed air to remove debris from the condenser.*
e) *Make sure the system has the correct refrigerant charge.*
f) *If you hear water sloshing around in the dash area or have water dripping on the carpet, check the evaporator housing drain tube* **(see illustration)** *and insert a piece of wire into the opening to check for blockage.*

2 It's a good idea to operate the system for about 10 minutes at least once a month, particularly during the winter. Long-term non-use can cause hardening, and subsequent failure, of the seals.

3 Leaks in the air conditioning system are best spotted when the system is brought up to temperature and pressure, by running the engine with the air conditioning ON for five minutes. Shut the engine off and inspect the air conditioning hoses and connections. Traces of oil usually indicate refrigerant leaks.

4 Because of the complexity of the air conditioning system and the special equipment necessary to service it, in-depth troubleshooting and repairs are not included in this manual. However, simple checks and component replacement procedures are provided in this Chapter.

5 If the air conditioning system doesn't operate at all, check the fuse panel and the air conditioning relay, located in the relay box in the engine compartment **(see illustration)**.

6 The most common cause of poor cooling is simply a low system refrigerant charge. If a noticeable drop in cool air output occurs, the following quick check will help you determine if the refrigerant level is low. For more complete information on the air conditioning system, refer to the *Haynes Automotive Heating and Air Conditioning Manual*.

Checking the refrigerant charge

7 Warm the engine up to normal operating temperature.

8 Place the air conditioning temperature selector at the coldest setting and put the blower at the highest setting. Open the doors (to make sure the air conditioning system doesn't cycle off as soon as it cools the passenger compartment).

9 With the compressor engaged - the clutch will make an audible click and the center of the clutch will rotate - feel the evaporator inlet pipe between the fixed orifice tube and the evaporator with one hand while placing your other hand on the metal portion of the hose between the evaporator and the condenser.

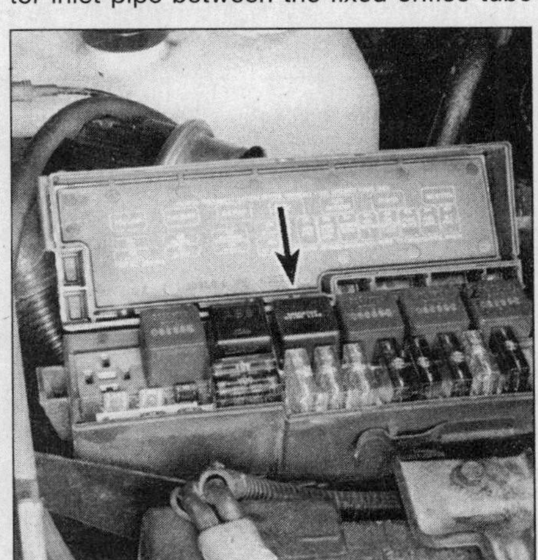

13.5 The air conditioning relay (arrow) is located in the power distribution center next to the battery

10 The pipe leading from the fixed orifice to the evaporator should be much colder than on the condenser side of the fixed orifice. If it isn't, the system charge is probably low. The earliest warning that a system is low on refrigerant is the air coming out of the ducts inside the vehicle. If the air isn't as cold as it used to be, the system probably needs a charge. Further inspection or testing of the system requires special tools and techniques and is beyond the scope of this manual.

11 If the inlet pipe has frost accumulation or feels cooler than the accumulator surface, the refrigerant charge is low.

Adding refrigerant

Refer to illustrations 13.12, 13.15a, 13.15b and 13.16

12 Buy an automotive charging kit at an auto parts store. A charging kit includes a 12-ounce can of refrigerant, a tap valve and a short section of hose that can be attached between the tap valve and the system low side service valve **(see illustration)**. Because

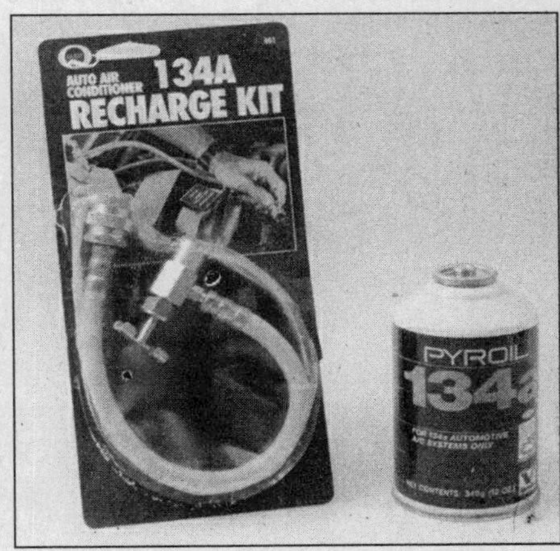

13.12 A basic charging kit for 134a systems is available at most auto parts stores - it must say 134a (not R-12) and so should the 12-ounce can of refrigerant

3

13.15a Add refrigerant to the system at the low-pressure port (1998 and earlier shown)

13.15b Location of the low-pressure service port - 1999 and later models

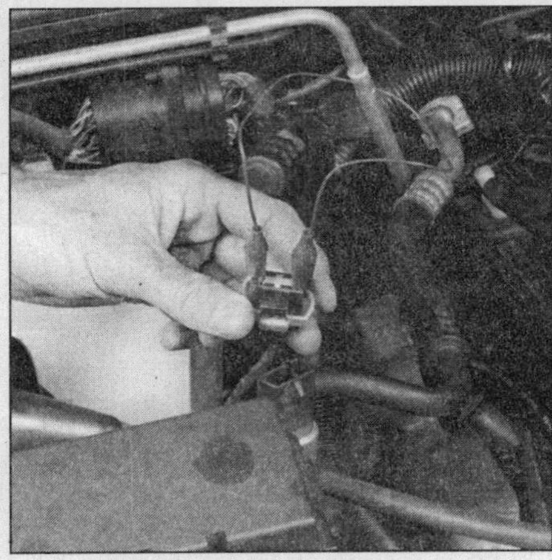

13.16 Unplug the connector from the cycling switch on the accumulator and jump the terminals with a jumper wire to keep the compressor clutch engaged

one can of refrigerant may not be sufficient to bring the system charge up to the proper level, it's a good idea to buy a couple of additional cans. Make sure that one of the cans contains red refrigerant dye. If the system is leaking, the red dye will leak out with the refrigerant and help you pinpoint the location of the leak. **Warning:** *Never add more than two cans of refrigerant to the system.*

13 Hook up the charging kit by following the manufacturer's instructions. **Warning:** *DO NOT hook the charging kit hose to the system high side!* The fittings on the charging kit are designed to fit **only** on the low side of the system.

14 Back off the valve handle on the charging kit and screw the kit onto the refrigerant can, making sure first that the O-ring or rubber seal inside the threaded portion of the kit is in place. **Warning:** *Wear protective eyewear when dealing with pressurized refrigerant cans.*

15 Remove the dust cap from the low-side charging port (by the coolant recovery tank) and attach the quick-connect fitting on the kit hose **(see illustrations)**.

16 Warm up the engine and turn on the air conditioner. Keep the charging kit hose away from the fan and other moving parts. The charging process requires the compressor to be running. If the clutch cycles off, remove the cycling switch connector from the accumulator **(see illustration)** and attach a jumper wire to the terminals on the harness side. This will keep the compressor engaged.

17 Turn the valve handle on the kit until the stem pierces the can, then back the handle out to release the refrigerant. You should be able to hear the rush of gas. Add refrigerant to the low side of the system until both the accumulator surface and the evaporator inlet pipe feel about the same temperature . Allow stabilization time between each addition.

18 If you have an accurate thermometer, you can place it in the center air conditioning duct inside the vehicle and keep track of the outlet air temperature. A charged system that is working properly should cool to approximately 40-degrees F. If the ambient (outside) air temperature is very high, say 110-degrees F, or if the relative humidity is high, the duct air temperature may be as high as 60- to 70-degrees F, but generally the air conditioning is 30 to 50-degrees F cooler than the ambient air.

19 When the can is empty, turn the valve handle to the closed position and release the connection from the low-side port. Replace the dust cap.

20 Remove the charging kit from the can and store the kit for future use with the piercing valve in the UP position, to prevent inadvertently piercing the can on the next use.

14 Automatic Temperature/Zone Control system - on-board diagnostics

Refer to illustrations 14.1 and 14.2

Note: *The diagnostic function of the ATC/AZC system is lost when the vehicle's battery is disconnected. If you are going to disconnect the battery, first check the diagnostic procedures below.*

1 Some Grand Cherokee models are equipped with automatic heating and air conditioning systems. 1998 and earlier models are equipped Automatic Temperature Control systems and the 1999 and later models are equipped with the Automatic Zone Control system. In either case, these systems include a digital display on the dash panel **(see illus-**

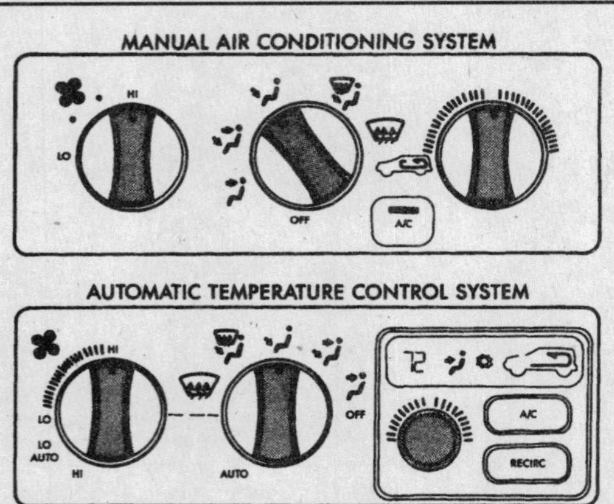

14.1 The control panel of the ATC system (bottom) compared to the standard, three-rotary-switch manual control panel (top)

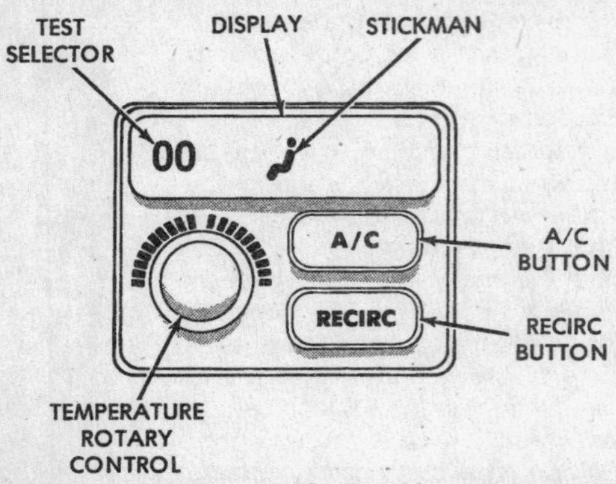

14.2 The upper left of the ATC display on 1998 and earlier vehicles normally indicates a two-digit temperature, but in the diagnostic mode shown here (no arrows showing on the stickman) the codes from 0 to 99 are displayed

tration). This system incorporates an on-board diagnostic function capable of troubleshooting input and output circuits and indicating both current and historical faults. Historical faults are those circuits which have had a problem in the past but are currently working. Many historical faults are the result of loose wiring connectors or connectors that have been disturbed by other repair work.

2 The same portion of the display that indicates the temperature **(see illustration)** acts as a test indicator to show fault codes.

3 If there is trouble in the ATC/AZC system, an "ER" message will display on the indicator once during an ignition cycle.

4 To access the fault codes, depress the AC and RECIRC buttons at the same time and hold them in while you rotate the temperature knob one click clockwise. **Note:** *On 1999 and later models rotate the left temperature control knob one click clockwise.* The complete display will light up (the segment test). If any of the lighted segments of the fluorescent vacuum tubes fail to illuminate, the ATC/AZC control panel is faulty and should be replaced.

5 After the segment test is performed release the buttons. On 1999 and later models, the display will clear momentarily and automatically begin displaying any stored trouble codes. If no trouble codes are stored a "0" will be displayed.

6 On 1998 and earlier models, "00" will show on the display (select test level). Now push either the AC or RECIRC buttons and the "stickman" figure will appear, indicating the Fault Code mode. Fault codes from 1 to 64 will appear if there are codes in memory. Record these codes and compare them to the chart to find the areas requiring further diagnosis..

7 If Fault Code 25 or 29 appears, the ATC/AZC control module itself must be replaced before any further testing is done. Current fault codes are cleared whenever the problem is solved or goes away. To clear historical faults, hold the AC or RECIRC button for 3 seconds. The faults have cleared when two horizontal bars appear in the display or they do not appear after several ON/OFF cycles of the ignition key.

8 The ATC/AZC diagnostic system is quite complicated and is capable of displaying data to check both input and output circuitry. However, if you can't find the problem through the basic Fault Code information (see diagnostic charts), we suggest that you take the vehicle to a dealer service department for further testing.

On-board diagnostic chart - 1998 and earlier

Automatic Temperature Control

Fault Code	Description	Problem Area
00	no faults	Normal operation
01	circuit open	Ambient Temperature Sensor
02	circuit open	In-vehicle Temperature Sensor
03	circuit open	Solar Sensor Input Circuit
04	circuit open	Front Panel Blower/Fan Control Input
05	circuit open	Front Panel Mode Control Input
06	circuit open	Blend Air Door Feedback Circuit
07	circuit open	Mode Door Feedback Circuit
08	feedback too high	Blower/Fan Feedback Circuit
09	circuit shorted	Ambient Temperature Sensor
10	circuit shorted	In-vehicle Temperature Sensor
11	circuit shorted	Solar Sensor Input Circuit
12	circuit shorted	Front Panel Blower/Fan Control Input
13	circuit shorted	Front Panel Mode Control Input
14	circuit shorted	Blend Air Door Feedback Circuit
15	circuit shorted	Mode Door Feedback Circuit
16	feedback too low	Blower/Fan Feedback Circuit
17	dimming input error	Pulse Width Dimming PWD Input
19	door not responding	Mode Door Feedback Circuit
20	door not responding	Blend Air Door Actuator Drive Circuit
21	door travel range too small	Mode Door Feedback Circuit
22	door travel range too large	Mode Door Feedback Circuit
23	door travel range too small	Blend Air Door Actuator Drive Circuit
24	door travel range too large	Blend Air Door Actuator Drive Circuit
25	calibration data error	Calibration and CPU Data
26	coolant temp message missing	Collision Detection C2D BUS Inputs
27	vehicle speed message missing	Collision Detection C2D BUS Inputs
28	engine RPM message missing	Collision Detection C2D BUS Inputs
29	CPU error	Calibration and CPU Data
33	circuit was open	Ambient Temperature Sensor
34	circuit was open	In-vehicle Temperature Sensor
35	circuit was open	Solar Sensor Input Circuit
36	circuit was open	Front Panel Blower/Fan Control Input
37	circuit was open	Front Panel Mode Control Input
38	circuit was open	Blend Air Door Feedback Circuit
39	circuit was open	Mode Door Feedback Circuit
40	feedback was too high	Blower/Fan Feedback Circuit
41	circuit was shorted	Ambient Temperature Sensor

Automatic Temperature Control (continued)

Fault Code	Description	Problem Area
42	circuit was shorted	In-vehicle Temperature Sensor
43	circuit was shorted	Solar Sensor Input Circuit
44	circuit was shorted	Front Panel Blower/Fan Control Input
45	circuit was shorted	Front Panel Mode Control Input
46	circuit was shorted	Blend Air Door Feedback Circuit
47	circuit was shorted	Mode Door Feedback Circuit
48	feedback was too low	Blower/Fan Feedback Circuit
49	dimming input was in error	Pulse Width Dimming PWD Input
51	door was not responding	Mode Door Feedback Circuit
52	door was not responding	Blend Air Door Actuator Drive Circuit
53	door range travel was too small	Mode Door Feedback Circuit
54	door range travel was too large	Mode Door Feedback Circuit
55	door range travel was too small	Blend Air Door Actuator Drive Circuit
56	door range travel was too large	Blend Air Door Actuator Drive Circuit
57	calibration data was in error	Calibration and CPU Data
58	coolant temp message was missing	Collision Detection C2D BUS Inputs
59	vehicle speed message was missing	Collision Detection C2D BUS Inputs
60	engine RPM message was missing	Collision Detection C2D BUS Inputs
61	CPU was in error	Calibration and CPU Data

AUTOMATIC ZONE CONTROL - 1999 AND LATER

FAULT CODES	CIRCUIT DESCRIPTION
01	IR thermister circuit open
02	IR thermister circuit shorted
03	Fan pot shorted
04	Fan pot open
05	Mode pot shorted
06	Mode pot open
07	IR sensor delta too large
08	Reserved
09	Reserved
10	Reserved
11	Engine air intake temperature Buss message missing
12	Country code Buss message missing
13	Mode motor not responding
14	AI (Recirc) motor not responding
15	Left temperature door not responding
16	Right temperature door not responding
17	Mode door travel range too small
18	Mode door travel range too large
19	AI (Recirc) door travel too small
20	AI (Recirc) door travel too large
21	Left temperature door travel too small
22	Left temperature door travel too large
23	Right temperature door travel too small
24	Right temperature door travel too large
25	Calibration check sum error
26	Engine coolant temp bus message missing
27	Vehicle speed bus message missing
28	Engine RPM bus message missing
29	OAT bus message missing
30	Display intensity bus message missing
31	VIN number bus message missing
32	Reserved

FAULT CODES	CIRCUIT DESCRIPTON
33	IR thermister circuit was open
34	IR thermister circuit was shorted
35	Fan pot was shorted
36	Fan pot was open
37	Mode pot was shorted
38	Mode pot was open
39	IR sensor delta was too large
40	Reserved
41	Reserved
42	Reserved
43	Engine air intake temperature Buss message missing
44	Country code Buss message missing
45	Mode motor was not responding
46	AI (Recirc) motor was not responding
47	Left temperature door was not responding
48	Right temperature door was not responding
49	Mode door travel range too small
50	Mode door travel range too large
51	AI (Recirc) door travel range too small
52	AI (Recirc) door travel range too large
53	Left temperature door travel too small
54	Left temperature door travel too large
55	Right temperature door travel too small
56	Right temperature door travel too large
57	Calibration check sum error
58	Engine coolant temp bus message missing
59	Vehicle speed bus message missing
60	Engine RPM bus message missing
61	OAT bus message missing
62	Display intensity bus message missing
63	VIN number bus message missing

15.3 Disconnect the clutch-cycling pressure switch from the accumulator

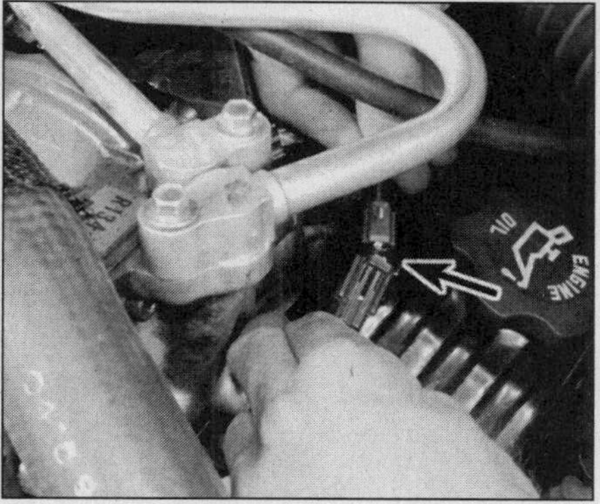
16.3 Disconnect the electrical connector (arrow) from the compressor - its a single-wire connector

16.5 The refrigerant lines on the compressor are retained by bolts, use a wrench to remove the lines, then plug the openings on the compressor and the lines

15 Air conditioning system accumulator - removal and installation

Refer to illustration 15.3

Warning: *The air conditioning system is under high pressure. Do not loosen any hose fittings or remove any components until after the system has been discharged. Air conditioning refrigerant should be properly discharged into an EPA-approved recovery/recycling unit at a dealer service department or an automotive air conditioning repair facility. Always wear eye protection when disconnecting air conditioning system fittings.*

1 Have the air conditioning system discharged (see **Warning** above).

2 Disconnect the negative cable from the battery.

3 Disconnect the electrical connector from the pressure switch near the top of the accumulator **(see illustration)**, which is located on the passenger side, next to the fenderwell.

4 Disconnect the refrigerant suction line from the compressor by unbolting it. **Note:** *The refrigerant lines do not detach from the accumulator - when the accumulator is replaced, the lines must be replaced along with it.*

5 Plug the open fittings to prevent entry of dirt and moisture.

6 Using a spring-lock coupling tool if necessary (see Section 12), disconnect the accumulator line where it goes through the firewall to the evaporator. Loosen the clamp bolt on the mounting bracket and lift the accumulator out with its attached lines.

7 If a new accumulator is being installed, remove the Schrader valve (under the pressure switch) from the old accumulator and pour the oil out into a measuring cup, noting the amount. Add fresh refrigerant oil (a type designated as compatible with refrigerant R-134a) to the new accumulator equal to the amount removed from the old unit, plus one ounce.

8 Installation is the reverse of removal. **Note:** *New R-134a compatible O-rings should be used in each spring-lock coupling during reassembly.*

9 Take the vehicle back to the shop that discharged it. Have the air conditioning system evacuated, charged and leak tested.

16 Air conditioning system compressor - removal and installation

Refer to illustrations 16.3, 16.5, 16.6a, 16.6b, 16.6c, 16.6d and 16.6e

Warning: *The air conditioning system is under high pressure. Do not loosen any hose fittings or remove any components until after the system has been discharged. Air conditioning refrigerant should be properly discharged into an EPA-approved recovery/recycling unit at a dealer service department or an automotive air conditioning repair facility. Always wear eye protection when disconnecting air conditioning system fittings.*

Note: *The accumulator* (see Section 15) *should be replaced whenever the compressor is replaced.*

1 Have the air conditioning system discharged (see **Warning** above).

2 Disconnect the negative cable from the battery.

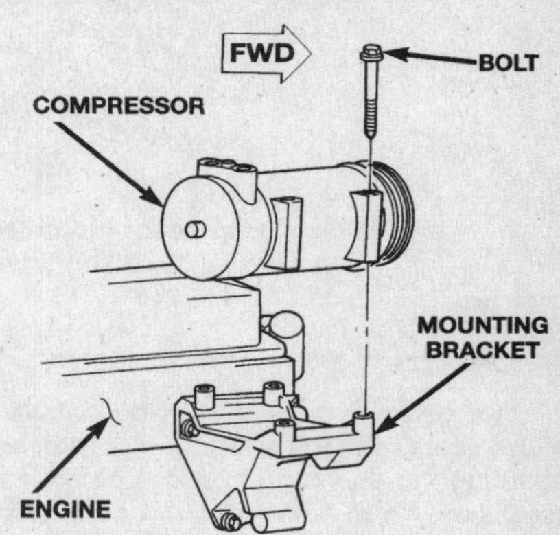

16.6a Compressor mounting details - 1998 and earlier six cylinder engine

3 Disconnect the compressor clutch electrical connector **(see illustration)**.

4 Remove the drivebelt (see Chapter 1).

5 Disconnect the refrigerant lines from the top rear of the compressor **(see illustration)**. Plug the open fittings to prevent entry of dirt and moisture.

6 Unbolt the compressor from the mounting brackets and lift it out of the vehicle **(see illustrations)**.

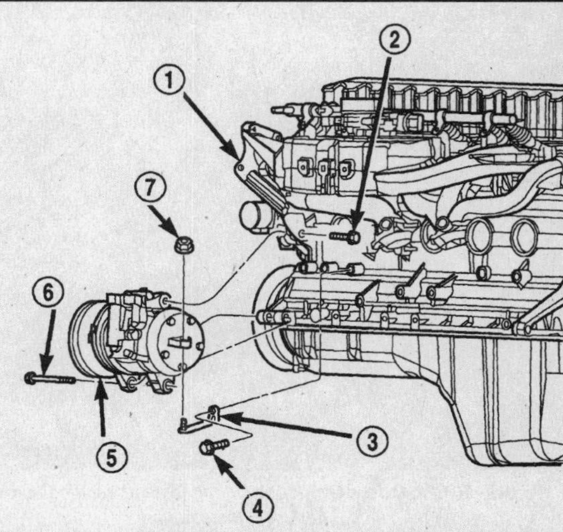

16.6b Compressor mounting details - 1999 and later six cylinder engine

1 *Power steering pump mounting bracket*
2 *Bolt*
3 *Brace*
4 *Bolt*
5 *A/C compressor*
6 *Bolt*
7 *Nut*

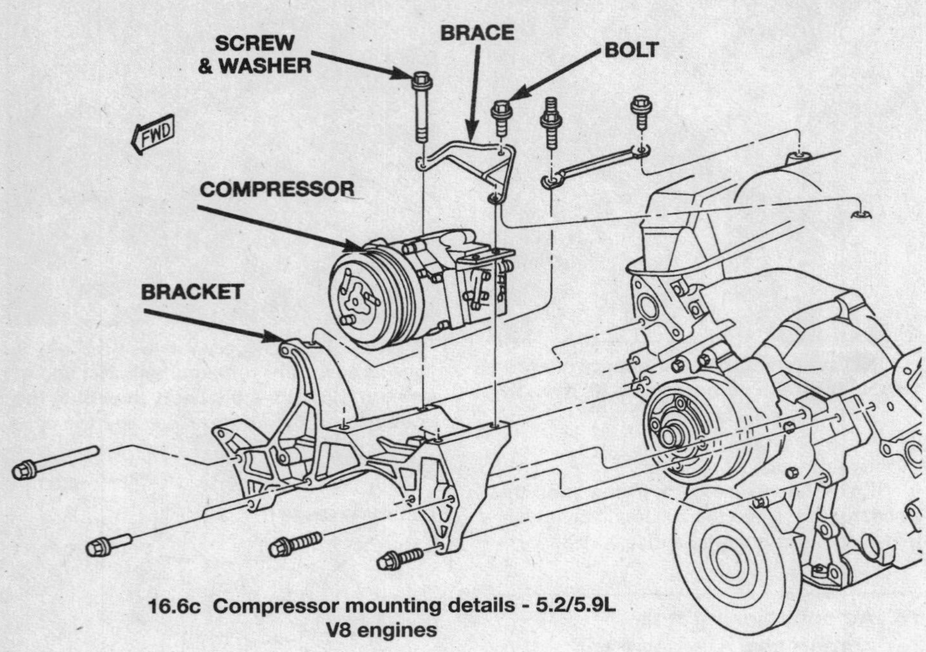

16.6c Compressor mounting details - 5.2/5.9L V8 engines

7 If a new compressor is being installed, follow the directions with the compressor regarding the draining of excess oil prior to installation. **Note:** *Any replacement compressor used must be designated as compatible with refrigerant R-134a.*

8 The clutch may have to be transferred from the original to the new compressor.

9 Installation is the reverse of removal. Replace all O-rings with new ones specifically made for air conditioning system use (and compatible with refrigerant R-134a) and lubricate them with refrigerant oil. Any refrigerant oil added must also be compatible with refrigerant R-134a.

10 Have the system evacuated, recharged and leak tested by the shop that discharged it.

17 Air conditioning system condenser - removal and installation

Refer to illustrations 17.5a, 17.5b and 17.6

Warning 1: *These models are equipped with airbags. The airbag is armed and can deploy (inflate) anytime the battery is connected. To prevent accidental deployment (and possible injury), disconnect the negative battery cable whenever working near airbag components. After the battery is disconnected, wait at least 2 minutes before beginning work (the system has a back-up capacitor that must fully discharge). For more information see Section 28 in Chapter 12.*

Warning 2: *The air conditioning system is under high pressure. Do not loosen any hose fittings or remove any components until after the system has been discharged. Air conditioning refrigerant should be properly discharged into an EPA-approved recovery/recycling unit at a dealer service department or an automotive air conditioning repair facility. Always wear eye protection when disconnecting air conditioning system fittings.*

Note: *The accumulator* (Section 15) *should be replaced whenever the condenser is replaced.*

1 Have the air conditioning system discharged (see **Warning** above).

2 Disconnect the negative battery cable from the battery.

3 Remove the radiator grille (see Chapter 11).

4 Remove the radiator (see Section 6).

5 Disconnect the refrigerant lines from the condenser. On later models the lines can simply be unbolted from the condenser. On early models remove the metal clips over the spring-lock connections **(see illustrations)**. Then clean the connections and apply the spring-lock tool. Wiggle and twist the pipes on either side of the connection and pull them apart.

6 Remove the mounting bolts from the condenser brackets **(see illustration)**.

7 Lift the condenser out of the vehicle and plug the lines to keep dirt and moisture out.

8 If the original condenser will be reinstalled, store it with the line fittings on top to prevent oil from draining out.

9 If a new condenser is being installed, pour one ounce of refrigerant oil into it prior to installation (an oil designated as compatible with refrigerant R-134a). **Note:** *New R-134a compatible O-rings should be used in each spring-lock coupling during reassembly.*

10 Reinstall the components in the reverse order of removal. Be sure the rubber pads are in place under the condenser. See Section 6 or proper location of sealing strips around the condenser and radiator.

16.6d Right side compressor mounting bolts (arrows) - 4.7L V8 engine

16.6e Left side compressor mounting bolt (arrow) - 4.7L V8 engine

17.5a These spring-lock connections must be disconnected to remove the condenser - first remove the metal clips (arrows)

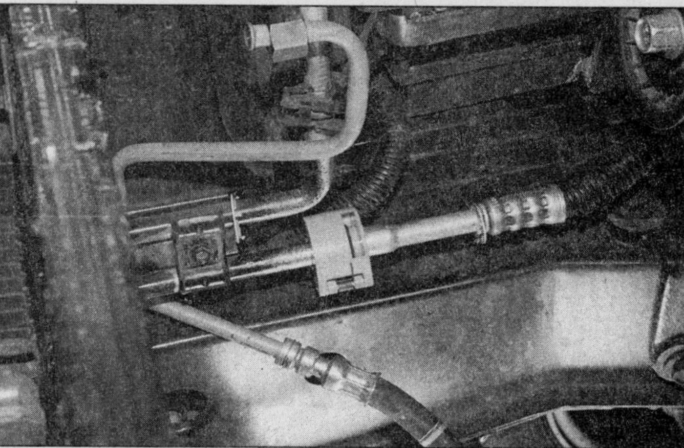
17.5b With a spring-lock connection tool in place, wiggle and twist the two lines before trying to separate the connection

11 Have the system evacuated, recharged and leak tested by the shop that discharged it.

18 Fixed orifice tube - removal and installation

General information

The fixed orifice tube assembly is the restriction between the high-pressure and low-pressure liquid refrigerant. It meters the flow of liquid refrigerant into the evaporator core.

Evaporator temperature is controlled by sensing the pressure inside the evaporator core with a pressure-operated electric switch. The switch controls compressor operation to keep the evaporator within the required pressure limits.

Note: *If, when the system is checked with the appropriate gauges, the high-pressure reads extremely high and low-pressure reads almost a vacuum, the fixed orifice tube is plugged and must be replaced.*

Replacement

Refer to illustration 18.3

Note: *On 1998 and earlier models, special spring-lock coupling tools are required to release the connectors used on the refrigerant lines throughout the air conditioning system.*

1 Disconnect the cable at the negative battery terminal.

2 Disconnect the liquid line from the condenser to the evaporator core at the firewall using a spring-lock coupling tool if necessary (see Section 5).

3 Disconnect the other end of the liquid line at the condenser, and remove the line from the vehicle. **Note:** *The fixed orifice tube is an integral part of this liquid line* **(see illustration)**, *and the line/orifice tube are replaced as a unit.*

4 Cap or plug the refrigerant openings immediately to prevent any dirt or excessive moisture from entering the system.

5 Reinstall the new line/orifice tube. **Note:** *New R-134a compatible O-rings should be used in each spring-lock coupling during reassembly.*

6 Have the system evacuated, recharged and leak-tested by a dealer service department or an air conditioning repair facility.

19 Air conditioning system evaporator - removal and installation

Refer to illustration 19.4

Warning 1: *These models are equipped with airbags. The airbag is armed and can deploy (inflate) anytime the battery is connected. To prevent accidental deployment (and possible injury), disconnect the negative battery cable whenever working near airbag components. After the battery is disconnected, wait at least 2 minutes before beginning work (the system has a back-up capacitor that must fully discharge). For more information see Chapter 12.*

Warning 2: *The air conditioning system is under high pressure. DO NOT loosen any fittings or remove any components until after the system has been discharged. Air conditioning refrigerant should be properly dis-*

17.6 Mounting bolts for the condenser are reached through the grille opening

18.3 The fixed orifice tube (arrow) is an integral part of the line to the evaporator (1998 and earlier shown, 1999 and later similar but mounted lower)

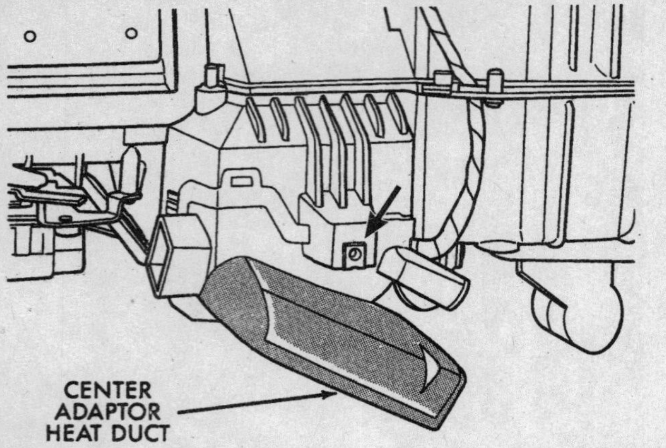

19.4 Remove the center adapter heat duct, allowing access to the last screw (arrow)

charged into an EPA-approved recovery/recycling unit at a dealer service department or an automotive air conditioning repair facility. Always wear eye protection when disconnecting air conditioning system fittings.

1 Disconnect the cable at the negative battery terminal.

2 Remove the heater/air conditioning unit from under the dash area (see Section 12).

3 Turn the heater/air conditioning unit upside down and remove the retaining screws retaining the two halves together.

4 Remove the center adapter heat duct **(see illustration)**, and remove the screw.

5 Turn the heater/air conditioning unit over, carefully, and remove the upper half of the housing.

6 Remove the evaporator from the lower housing.

7 Installation is the reverse of the removal procedure. If the evaporator was replaced with a new unit, add 2 ounces of refrigerant oil (of a type designated as compatible with refrigerant R-134a) to the system. **Note:** *New R-134a compatible O-rings should be used in each spring-lock coupling during reassembly.*

8 Have the system evacuated, recharged and leak tested by a dealer service department or an air conditioning repair facility.

Chapter 4
Fuel and exhaust systems

Contents

Specifications

General

Fuel pressure (engine running at idle speed)	
1993 and 1994	
With regulator vacuum hose attached	31 to 33 psi
With regulator vacuum hose disconnected	39 to 41 psi
1995 and 1996	37 to 41 psi
1997 and later	44.2 to 54.2 psi
Fuel injector resistance	11.3 to 15.7 ohms at 68-degrees F
Fuel level sending unit resistance	
1993 through 1995	
Empty	105 ohms
Full	5 ohms
1996 through 1998	
Empty	220 ohms
Full	20 ohms
1999 and later	
Empty	270 ohms
Full	20 ohms

Torque specifications

Throttle body mounting nuts/bolts	
4.0L I6	108 in-lbs
5.2L/5.9L V8	200 in-lbs
4.7L V8	105 in-lbs
Fuel rail mounting nuts/bolts	
4.0L I6	
1996 and earlier	20 ft-lbs
1997 and later	108 in-lbs
5.2L/5.9L V8	
1995 and earlier	20 ft-lbs
1996 and later	200 in-lbs
4.7L V8	100 in-lbs

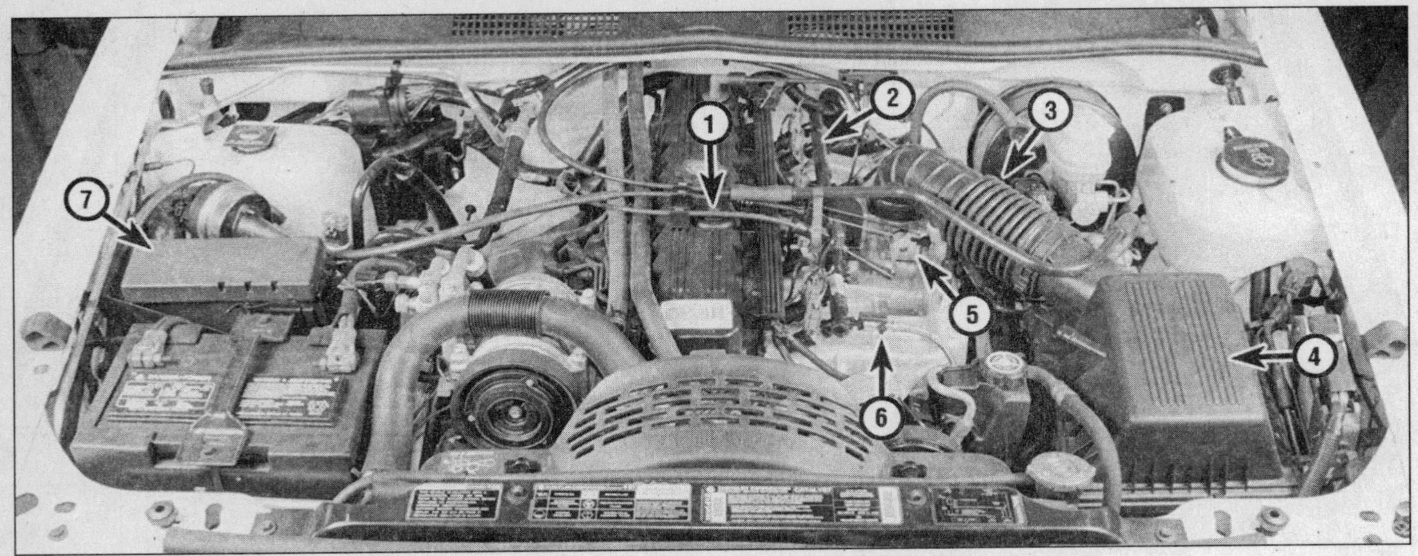

1.1a Typical fuel system components - 4.0L inline six-cylinder engine

1 *Accelerator cable*
2 *Fuel rail and injectors*
3 *Air intake duct*
4 *Air filter housing*
5 *Throttle body*
6 *Fuel supply line*
7 *Fuel pump relay (inside power distribution center)*

1 General information

Refer to illustrations 1.1a, 1.1b and 1.1c

Warning: *Gasoline is extremely flammable, so take extra precautions when you work on any part of the fuel system. Don't smoke or allow open flames or bare light bulbs near the work area, and don't work in a garage where a gas-type appliance (such as a water heater or a clothes dryer) is present. Since gasoline is carcinogenic, wear latex gloves when there's a possibility of being exposed to fuel, and, if you spill any fuel on your skin, rinse it off immediately with soap and water. Mop up any spills immediately and do not store fuel-soaked rags where they could ignite. The fuel system is under constant pressure, so, if any fuel lines are to be disconnected, the fuel pressure in the system must be relieved first. When you perform any kind of work on the fuel system, wear safety glasses and have a Class B type fire extinguisher on hand.*

All models covered by this manual are equipped with a sequential Multi Port Fuel Injection (MPFI) system **(see Illustrations)**. This system uses timed impulses to sequentially inject the fuel directly into the intake ports of each cylinder. The injectors are controlled by the Powertrain Control Module (PCM). The PCM monitors various engine parameters and delivers the exact amount of fuel, in the correct sequence, into the intake ports.

All models are equipped with an electric fuel pump, mounted in the fuel tank. It is necessary to remove the fuel tank for access to the fuel pump. The fuel level sending unit is an integral component of the fuel pump and it must be removed from the fuel tank in the same manner.

The exhaust system consists of exhaust manifolds, a catalytic converter, an exhaust pipe and a muffler. Each of these components is replaceable. For further information regarding the catalytic converter, refer to Chapter 6.

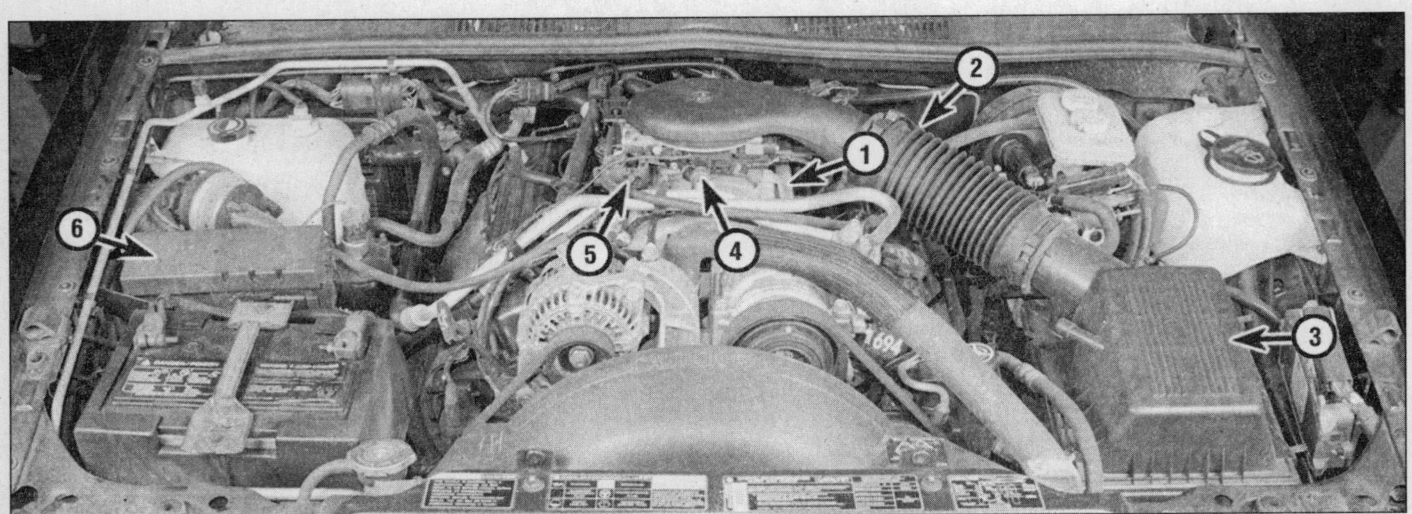

1.1b Typical fuel system components - 5.2L/5.9L V8 engine

1 *Fuel rail and injectors*
2 *Air intake duct*
3 *Air filter housing*
4 *Throttle body*
5 *Accelerator cable*
6 *Fuel pump relay (inside power distribution center)*

1.1c Typical fuel system components - 4.7L V8 engine

1 *Throttle body (under resonator)*
2 *Air intake duct/resonator*
3 *Fuel rail and injectors*
4 *Air filter housing*
5 *Fuel pump relay (inside power distribution center)*

2 Fuel pressure relief procedure

Warning: *See the* **Warning** *in Section 1.*

1 Remove the fuel filler cap (this will relieve any pressure that has built-up in the tank).

2 Remove the fuel pump relay from the power distribution center in the engine compartment (you can identify the relays by looking at the underside of the power distribution center cover).

3 Attempt to start the engine - the engine probably will not start or may start momentarily and immediately stall. Crank the engine for several seconds.

4 Turn the ignition key to the Off position. Install the fuel pump relay.

5 Disconnect the cable from the negative terminal of the battery before beginning work on the fuel system.

3 Fuel pump/fuel pressure - check

Warning: *See the* **Warning** *in Section 1.*

Preliminary check

Refer to illustration 3.3

1 If you suspect insufficient fuel delivery, first inspect all fuel lines to ensure that the problem is not simply a leak in a line.

2 Set the parking brake and have an assistant turn the ignition switch to the ON position while you listen to the fuel pump (inside the fuel tank). You should hear a "whirring" sound, lasting for a couple of seconds indicating the fuel pump is operating. If the fuel pump is operating, proceed to the pressure check.

3 If the fuel pump does not operate, disconnect the fuel pump electrical connector from the harness connector. Connect a voltmeter to the B+ (from fuel pump relay) and ground terminals of the harness connector **(see illustration)**. Have an assistant attempt to start the engine - battery voltage should be indicated on the meter. If power is not available at the fuel pump connector, check the fuel pump circuit, referring to Chapter 12 and the wiring diagrams. Check the related fuses, the fuel pump relay and the related wiring to ensure power is reaching the fuel pump connector. Check the ground circuit for continuity to a good chassis ground point.

4 If the power and ground circuits are good and the fuel pump does not operate when connected, replace the fuel pump (see Section 7).

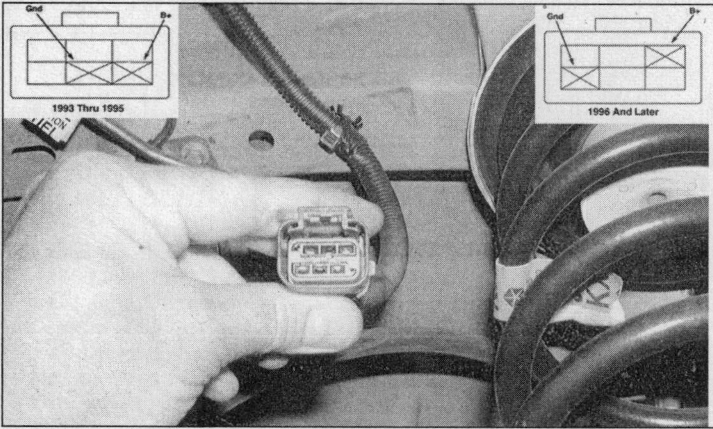

3.3 Check the battery power and ground circuits at the fuel pump connector

Pressure check

Refer to illustrations 3.6a, 3.6b, 3.6c and 3.6d

Note 1: *In order to perform the fuel pressure test, you will need a fuel pressure gauge capable of measuring high fuel pressure. The fuel gauge must be equipped with the proper fittings or adapters required to attach it to the fuel rail.*

Note 2: *In order to check the fuel pressure regulator on 1993 and 1994 models, you will need a special rubber adapter hose with the proper fittings or adapters required to attach it to the fuel rail and fuel return line. Do not pinch the vehicle's fuel lines shut or damage to the line may occur.*

5 Relieve the fuel pressure (see Section 2).

6 Remove the cap from the fuel pressure test port (if equipped) on the fuel rail and attach a fuel pressure gauge **(see illustrations)**. If the fuel rail is not equipped with a

3.6a Connect a fuel pressure gauge to the test port (4.0L six-cylinder engine shown)

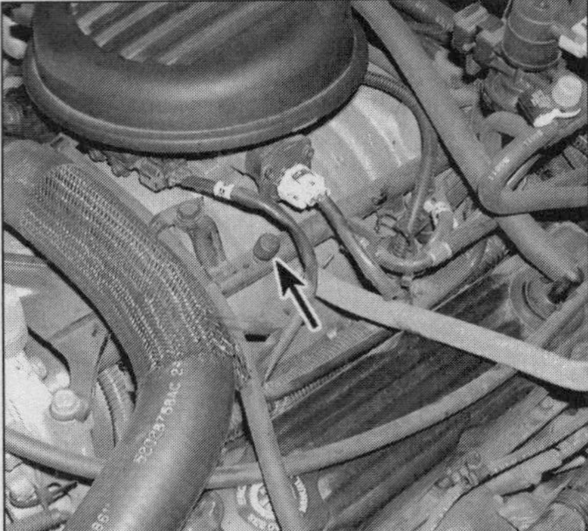
3.6b Fuel pressure test port location - 5.2L/5.9L V8 engines

3.6c Fuel pressure test port location - 4.7L V8 engine

test port, disconnect the fuel line from the fuel rail and connect a fuel pressure gauge using a special adapter **(see Illustration)**.

7 Start the engine and check the pressure on the gauge, comparing your reading with the pressure listed in this Chapter's Specifications.

8 On 1993 and 1994 models, disconnect and plug the vacuum hose from the fuel pressure regulator - the fuel pressure should increase approximately 8 to 10 psi. If the fuel pressure is not within specifications or the fuel pressure regulator doesn't respond as described, check the following:

a) *If the pressure is lower than specified, check for a restriction in the fuel filter or fuel line. If the fuel filter and fuel lines are good, slowly pinch the fuel return line shut. If the pressure rises above the maximum specified pressure, replace the fuel pressure regulator* (see Section 14). *If the pressure is still low, replace the fuel pump* (see Section 7).

b) *If the fuel pressure is higher than specified, check for vacuum at the fuel pressure regulator vacuum hose. If vacuum is not present, check the hose for a break or obstruction. If vacuum is present, check the fuel return line for restrictions. If no restrictions are found, replace the fuel pressure regulator* (see Section 14).

8 On 1995 and later models, if the fuel pressure is lower than specified, check the fuel lines and the fuel filter for restrictions. If no restriction is found, replace the fuel pump (see Section 7). If fuel pressure is higher than specified, replace the fuel pressure regulator (see Section 14).

4 Fuel lines and fittings - repair and replacement

Warning: *See the* **Warning** *in Section 1.*

1 Always relieve the fuel pressure before servicing fuel lines or fittings (see Section 2).

2 The fuel feed and vapor lines extend from the fuel tank to the engine compartment. The lines are secured to the underbody with clip and screw assemblies. These lines must be occasionally inspected for leaks, kinks and dents.

3 If evidence of dirt is found in the system or fuel filter during disassembly, the line should be disconnected and blown out. Check the fuel strainer on the fuel pump module for damage and deterioration.

Steel tubing

4 If replacement of a fuel line or emission line is called for, use tubes/hoses meeting the manufacturers specifications.

5 Don't use copper or aluminum tubing to replace steel tubing. These materials cannot withstand normal vehicle vibration.

6 Because fuel lines used on fuel-injected vehicles are under high pressure, they require special consideration.

7 Some fuel lines have threaded fittings with O-rings. Any time the fittings are loosened to service or replace components:

a) *Use a backup wrench while loosening and tightening the fittings.*

b) *Check all O-rings for cuts, cracks and deterioration. Replace any that appear hardened, worn or damaged.*

c) *If the lines are replaced, always use original equipment parts, or parts that meet the original equipment standards.*

Flexible hose

Refer to illustrations 4.11a, 4.11b, 4.11c, 4.11d and 4.11e

8 In the event of flexible fuel line damage it is necessary to replace the damaged lines with factory replacement parts. Others may

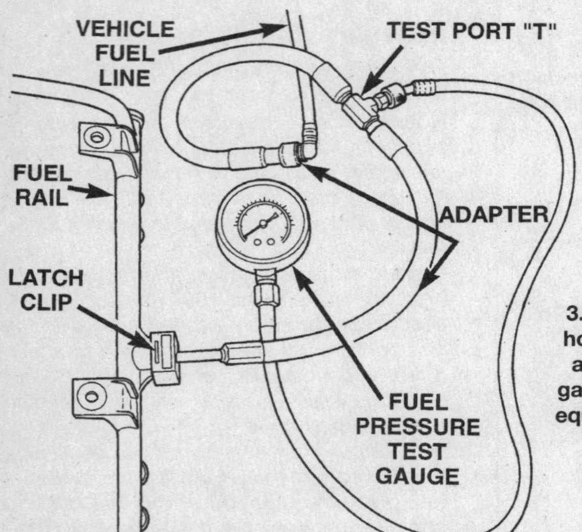

3.6d A special adapter hose will be required to attach a fuel pressure gauge if the fuel rail isn't equipped with a test port

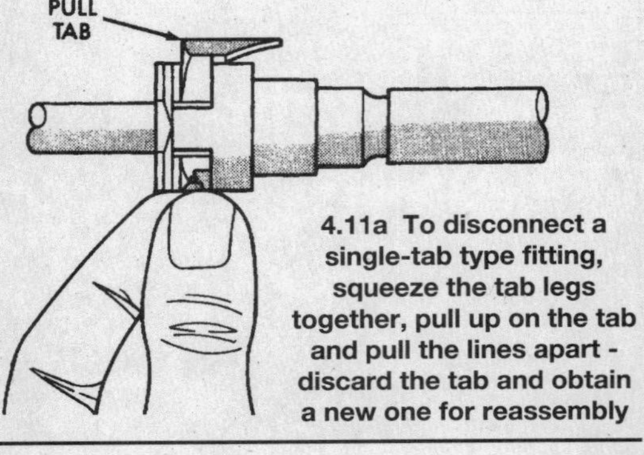

4.11a To disconnect a single-tab type fitting, squeeze the tab legs together, pull up on the tab and pull the lines apart - discard the tab and obtain a new one for reassembly

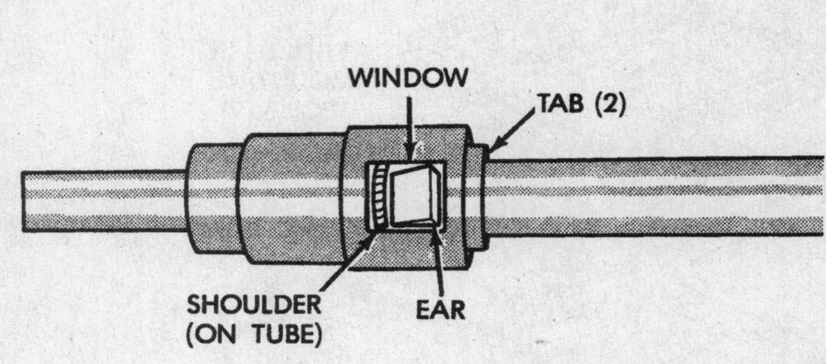

4.11b To disconnect a two-tab type fitting, squeeze the two tabs together and pull the lines apart

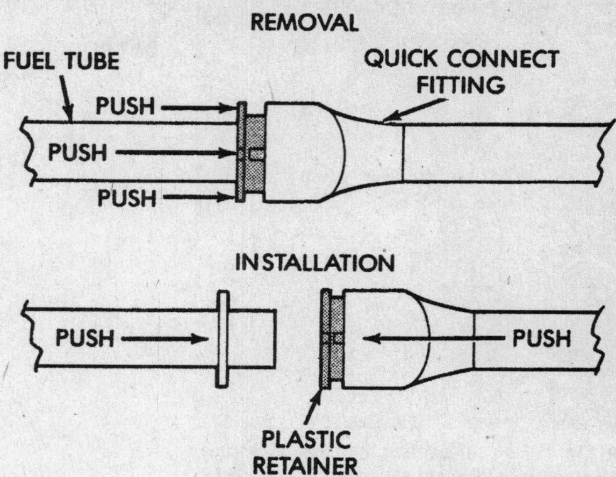

4.11c To disconnect a plastic ring type fitting, push the fuel lines together, push the retainer ring squarely into the fitting and pull the lines apart

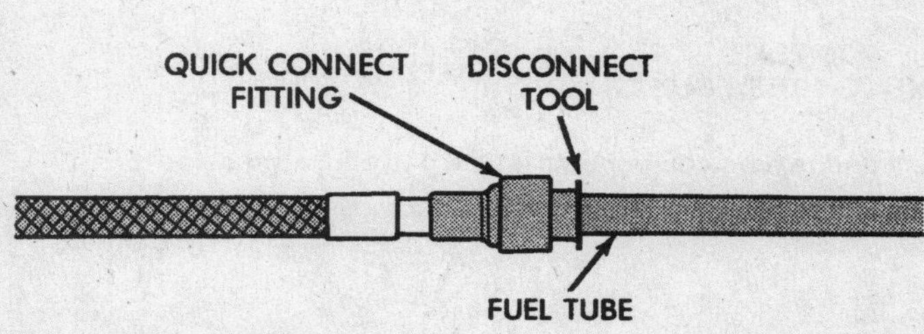

4.11d A special tool (available at most auto parts stores) is required to disconnect this type of metal fitting - remove the tethered clip, place the tool over the fuel line, insert it squarely into the fitting and pull the lines apart (the tool is not required to connect the lines)

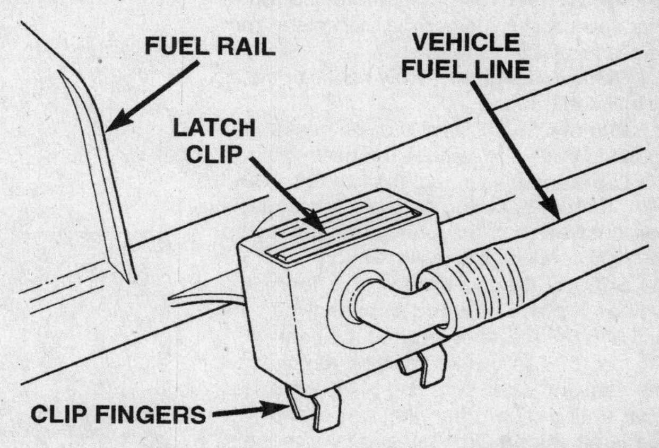

4.11e To disconnect a metal latch-clip type fitting, squeeze the clip fingers together, pull the clip straight up and pull the line off

fail from the high pressures of this system.

9 Relieve the fuel pressure before disconnecting the fitting (see Section 2).

10 Remove all fasteners attaching the lines to the vehicle body.

11 There are various methods of disconnecting the fittings, depending upon the type of quick-connect fitting installed on the fuel line **(see illustrations)**. Clean any debris from around the fitting. Disconnect the fitting and carefully remove the fuel line from the vehicle.

Caution: *The quick-connect fittings are not serviced separately. Do not attempt to repair these types of fuel lines in the event the fitting or line becomes damaged. Replace the entire fuel line as an assembly.*

12 Don't route fuel hose within four inches of any part of the exhaust system or within ten inches of the catalytic converter. Metal lines and rubber hoses must never be allowed to chafe against the frame. A minimum of 1/4-inch clearance must be maintained around a line or hose to prevent contact with the frame.

13 Installation is the reverse of removal with the following additions:

a) Clean the quick-connect fittings with a lint-free cloth and apply clean engine oil the fittings.

b) After connecting a quick-connect fitting, check the integrity of the connection by attempting to pull the lines apart.

c) Use new O-rings at the threaded fittings (if equipped).

d) Cycle the ignition key On and Off several times and check for leaks at the fitting, before starting the engine.

5 Fuel tank - removal and installation

Refer to illustrations 5.5, 5.6a, 5.6b, 5.9a, 5.9b, 5.9c, 5.9d, 5.11a, 5.11b, 5.11c, 5.12 and 5.13

Warning: *See the* **Warning** *in Section 1.*

1 Remove the fuel tank filler cap to relieve fuel tank pressure.

2 Relieve the fuel system pressure (see Section 2).

3 Detach the cable from the negative terminal of the battery.

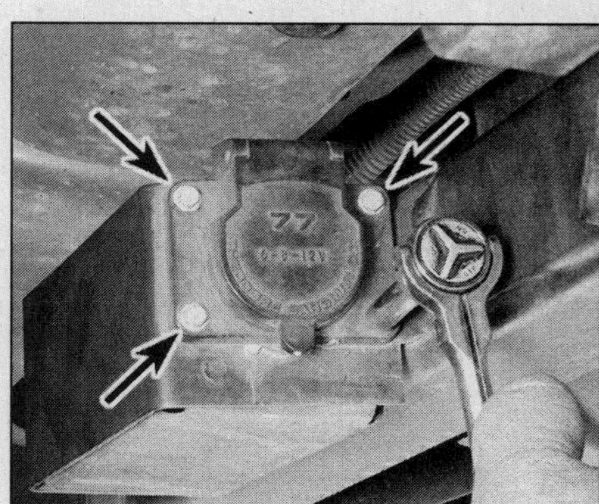

5.5 If the vehicle is equipped with a trailer package, remove the four bolts (arrows) that retain the service connector

4 Siphon the fuel into an approved gasoline container, using a siphoning kit (available at most auto parts stores). **Warning:** *Never start the siphoning action by mouth!*

5 If equipped, remove the electrical service connector mounted on the bumper **(see illustration)**.

4

5.6a On 1998 and earlier models, remove the bolts from the skid plate (be sure it is supported with a jack)

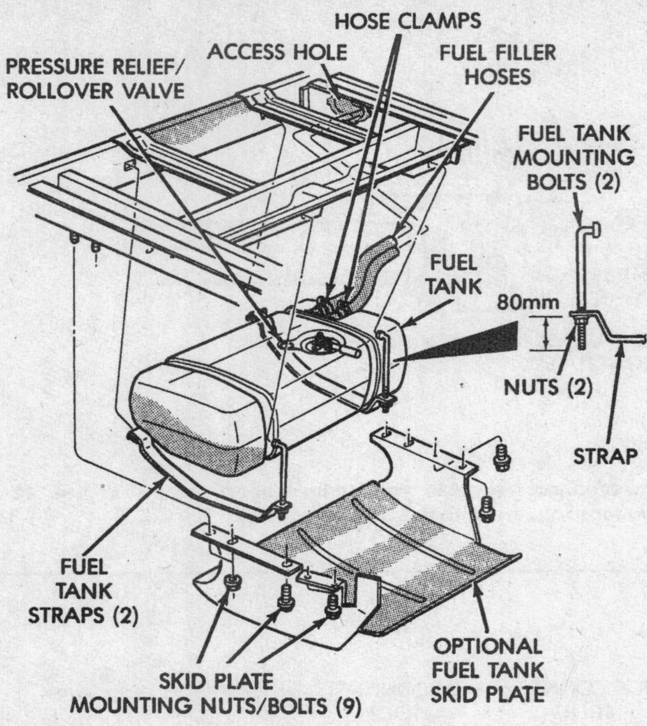

5.6b Fuel tank and related components - 1998 and earlier models

6 On 1998 and earlier models, support the skid plate with a floor jack (if equipped), remove the fuel tank skid plate mounting bolts and carefully lower the skid plate from the chassis **(see illustrations)**.

7 Remove the optional tow hooks or trailer hitch, if equipped.

8 Remove the exhaust tailpipe heat shield mounting bolts and remove the heat shield.

9 Label and disconnect the fuel hoses and any retainers securing them to the chassis **(see illustrations)**. Disconnect the fuel pump electrical connector **(see illustration)**.

10 Support the fuel tank with a floor jack. Position a piece of wood between the jack head and the fuel tank to protect the tank.

11 On 1998 and earlier models, remove fuel tank retaining strap nuts and pivot the straps down until they are hanging out of the way **(see illustration)**. On 1999 and later models, detach the clips at the rear bumper cover and remove the fuel tank retaining bolts **(see illustrations)**. **Note:** *On 1999 and later models, the fuel tank shield and fuel tank are removed as a unit.*

12 Remove the filler hose and the vapor return hose from the fuel tank **(see illustration)**

13 Remove the tank from the vehicle **(see illustration)**.

5.9a Pry open the fuel line retainer at the chassis

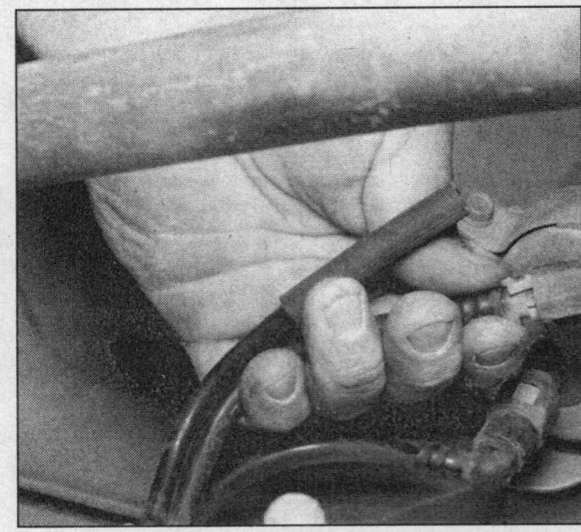
5.9b Depress the tabs and separate the fuel return line

5.9c Depress the tabs on the fuel filter line and separate the line

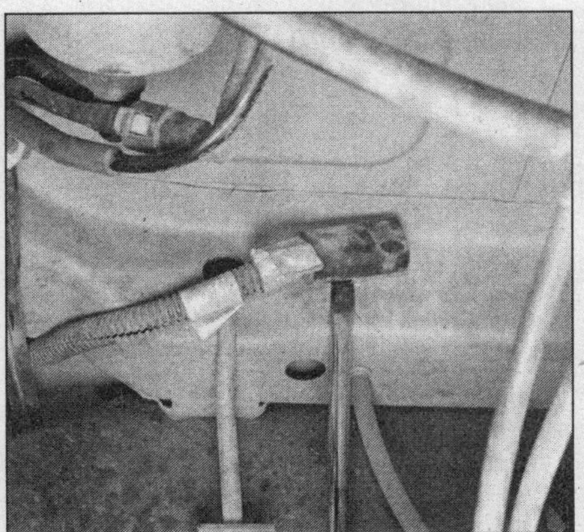
5.9d Disconnect the fuel pump electrical connector

5.11a On 1998 and earlier models, remove the fuel tank strap nuts

5.11b On 1999 and later models, detach the clips from the fuel tank shield at the rear bumper cover . . .

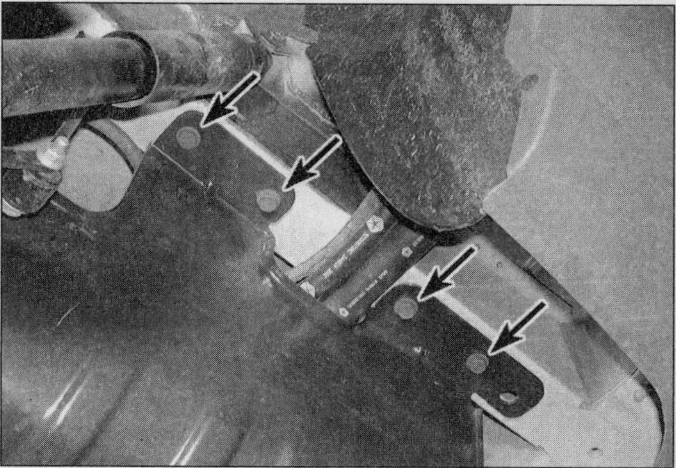

5.11c . . . and remove the fuel tank shield retaining bolts (left side shown, right side similar)

5.12 Remove the clamps that retain the fuel filler hose and the vapor return hose

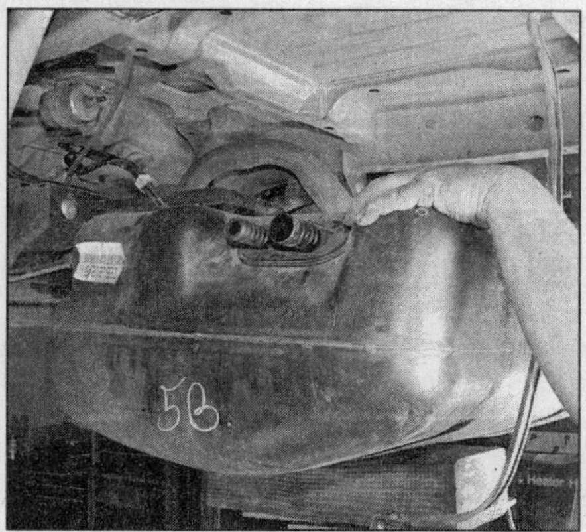

5.13 Slowly lower the fuel tank from the vehicle

14 On 1999 and later models, remove the nuts from the straps retaining the fuel tank to the shield and separate the fuel tank from the shield, if necessary.

15 Installation is the reverse of removal with the following additions:

a) *On 1998 and earlier models, tighten the fuel tank strap nuts until the nut is approximately 80 mm (3-5/32 inches) from the end of the mounting bolt* **(see illustration 4.6b).**
b) *On 1999 and later models, tighten the fuel tank mounting bolts to 60 ft-lbs.*

7.4 Before loosening the fuel pump module locknut, apply an alignment mark on the fuel pump module flange and fuel tank

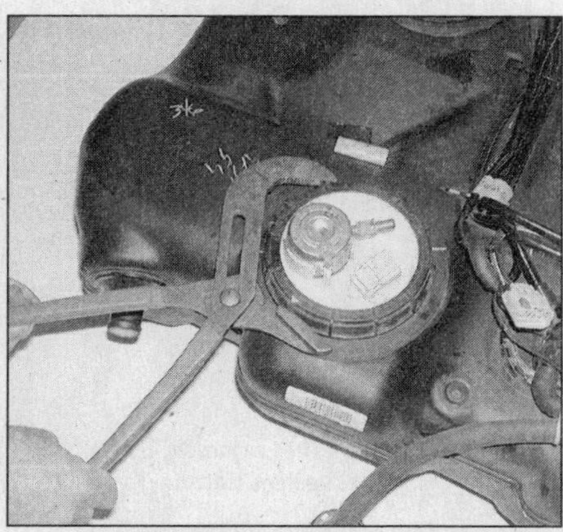

7.5 Loosen the fuel pump module locknut by turning it counterclockwise

6 Fuel tank cleaning and repair - general information

1 The fuel tank installed in the vehicles covered by this manual is not repairable. If the fuel tank becomes damaged, it must be replaced.

2 Cleaning the fuel tank (due to fuel contamination) should be performed by a professional with the proper training to carry out this critical and potentially dangerous work. Even after cleaning and flushing, explosive fumes may remain inside the fuel tank.

3 If the fuel tank is removed from the vehicle, it should not be placed in an area where sparks or open flames could ignite the fumes coming out of the tank. Be especially careful inside a garage where a gas-type appliance is located.

7 Fuel pump - removal and installation

Refer to illustrations 7.4, 7.5 and 7.6

Warning: *See the* **Warning** *in Section 1.*

1 Relieve the fuel system pressure (see Section 2).

2 Disconnect the cable from the negative battery terminal.

3 Remove the fuel tank from the vehicle (see Section 5). Disconnect the fuel lines and electrical connector from the module, as necessary.

4 Note the position of the fuel pump module in relation to the fuel tank. The fuel pump module must be installed in the original position or the fuel level float may contact the side of the tank. Apply an alignment mark if necessary **(see Illustration).**

5 Turn the locknut counterclockwise to remove it. If the locknut is difficult to turn, use a large pair of adjustable pliers or other appropriate tool to loosen it **(see Illustration).**

6 Remove the fuel pump module from the tank **(see illustration).** Angle the assembly

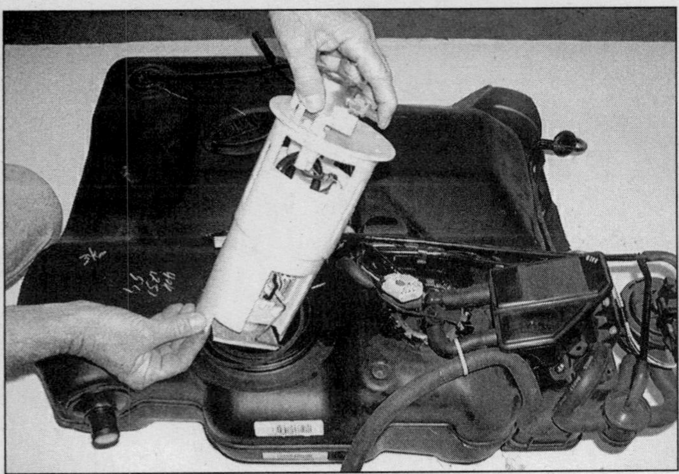

7.6 **Carefully remove the fuel pump module from the tank**

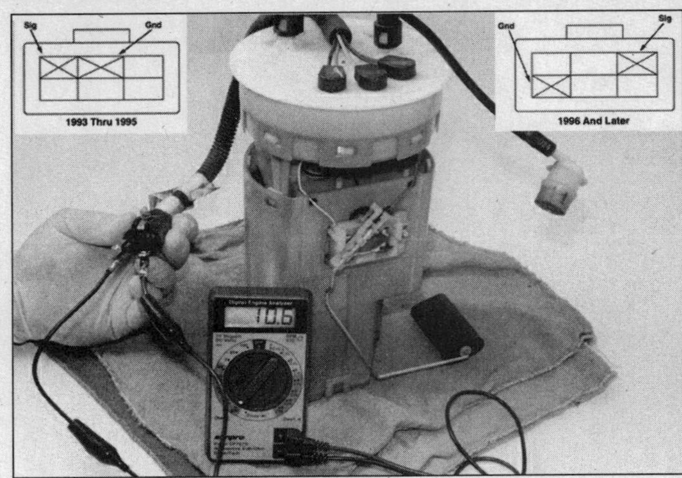

8.2 Connect an ohmmeter to the signal and ground terminals of the fuel pump module connector - measure the resistance of the fuel level sending unit with the float lowered (empty) and then with the float raised (full)

slightly to avoid damaging the fuel level sending unit float. **Warning:** *Some fuel may remain in the module reservoir and spill as the module is removed. Have several shop towels ready and a drain pan nearby to place the module in.*

7 The electric fuel pump is not serviced separately. In the event of failure, the complete assembly must be replaced. Transfer the fuel level sending unit and the fuel pressure regulator to the new unit as necessary (see Sections 8 and 14).

8 Installation is the reverse of removal with the following additions:

a) Replace the fuel pump module O-ring seal.

b) Install the module in the fuel tank, aligning the marks made in Step 4.

c) Press down on the module while tightening the locknut.

8 Fuel level sending unit - check and replacement

Warning: *See the* **Warning** *in Section 1.*

Check

Refer to illustration 8.2

1 Remove the fuel tank and the fuel pump module (see Sections 5 and 7).

2 Connect the probes of an ohmmeter to the fuel level sending unit signal and ground terminals of the fuel pump module electrical connector **(see illustration)**.

3 Position the float in the down (empty) position. Measure the resistance and compare it to the values listed in this Chapter's Specifications.

4 Move the float up to the full position. Measure the resistance and compare it to the values listed in this Chapter's Specifications.

5 If the fuel level sending unit resistance is incorrect or if the resistance does not change smoothly as the float travels from empty to full, replace the fuel level sending unit assembly.

Replacement

1996 and earlier models

6 On 1996 and earlier models, the fuel level sending unit is not serviced separately from the fuel pump module. If the fuel level sending unit requires replacement, replace the fuel pump module (see Section 7).

1997 and later models

Refer to illustration 8.8

7 Remove the fuel tank and the fuel pump module (see Sections 5 and 7).

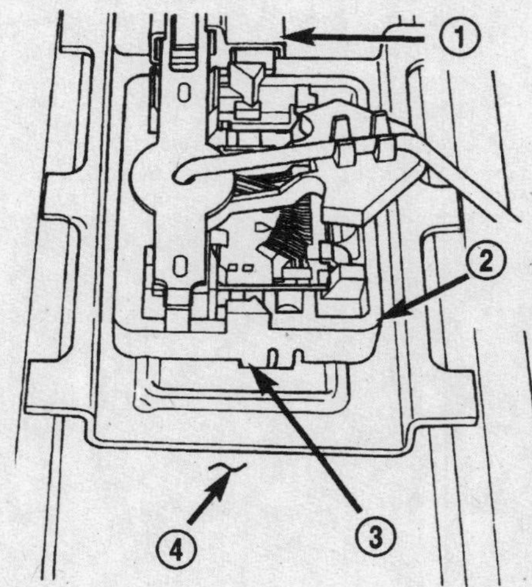

8.8 Fuel level sending unit installation details

1 Electrical connector
2 Fuel level sending unit
3 Lock tab
4 Fuel pump module

8 Disconnect the electrical connector from the fuel level sending unit inside the module **(see Illustration)**.

9 Using a small screwdriver or pick, pry the lock tab up and remove the fuel level sending unit from the module **(see illustration 8.8)**.

10 Position the sending unit onto the module and press it in until the lock tab snaps in place.

11 Connect the electrical connector to the fuel level sending unit and install the module in the fuel tank (see Section 7).

12 The remainder of installation is the reverse of removal.

9 Air filter housing - removal and installation

Refer to illustrations 9.1, 9.2a, 9.2b and 9.4

1 Loosen the clamp and detach the duct from the air filter housing cover **(see Illustration)**.

2 Working in the left front wheelwell,

9.1 Loosen the hose clamp (arrow) and detach the intake duct from the air cleaner cover

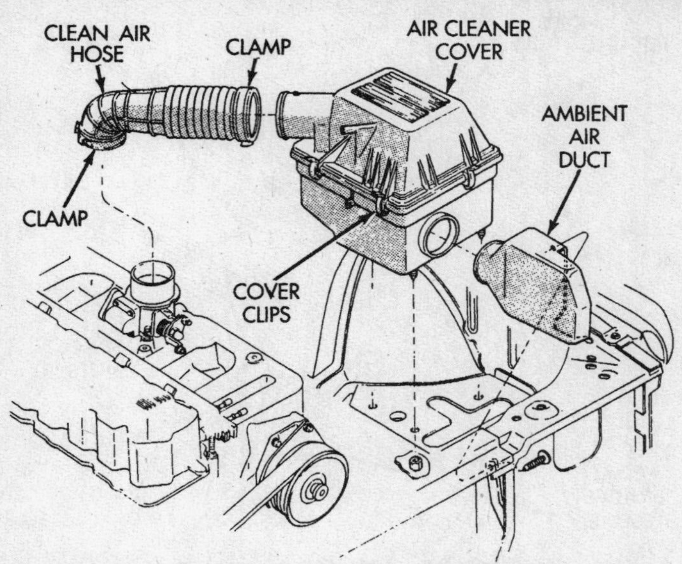

9.2a Typical 4.0L six-cylinder engine air filter assembly details

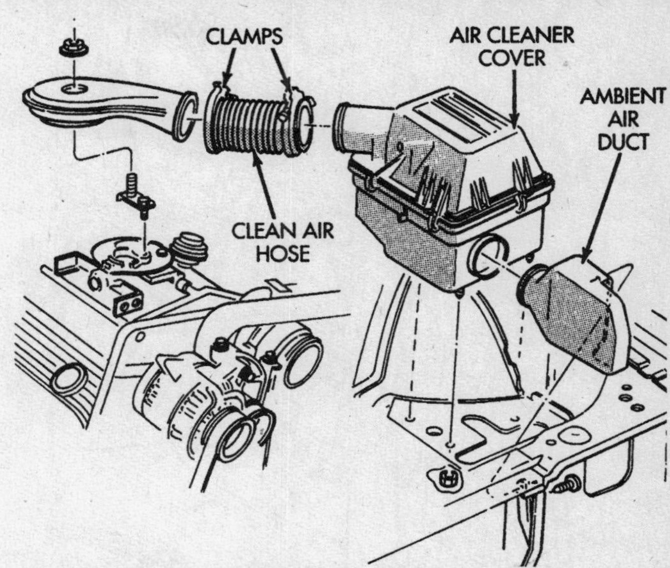

9.2b Typical 5.2L/5.9L V8 engine air filter assembly details

9.4 On 4.7L V8 models, the air intake duct resonator is fastened to the intake manifold with two screws beneath the resonator (arrows)

10.2 Rotate the throttle lever and detach the cable from the stud

remove the clips and pry back the inner fender lining (if necessary). Remove the air cleaner housing mounting nuts **(see illustrations)**.

3 Lift the housing up, detach the housing from the fresh air intake duct in the inner fender and remove the assembly from the engine compartment.

4 If necessary, loosen the hose clamp at the throttle body, remove the mounting nuts/screws (if equipped) and separate the air intake duct/resonator from the throttle body **(see illustration)**.

5 Installation is the reverse of removal.

10.3 On V8 models, use a flat bladed screwdriver to depress the locking tab and remove the cable from the bracket

10 Accelerator cable - replacement

Refer to illustrations 10.2, 10.3, 10.4 and 10.5

1 Remove the air intake duct/resonator from the throttle body.

2 Rotate the throttle lever and separate the cable end from the throttle lever stud **(see illustration)**.

3 On V8 models, push in on the tab to release the accelerator cable retainer from the bracket and slide the cable out of the bracket **(see Illustration)**.

4 On six-cylinder models, squeeze the tabs on the cable housing together and push the cable through the bracket **(see illustration)**.

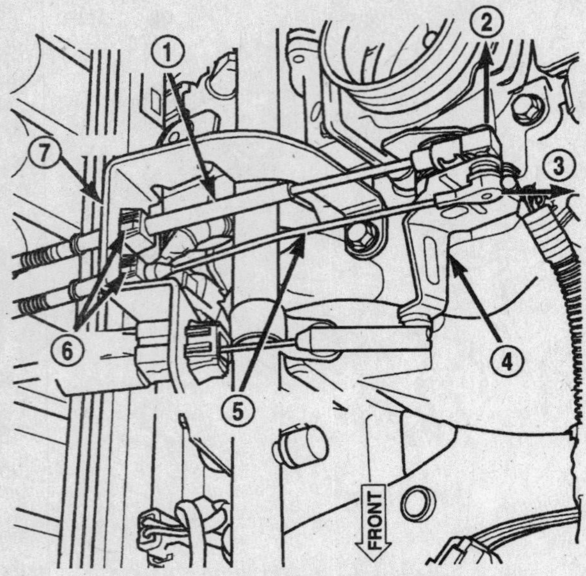

10.4 Accelerator cable installation details - 4.0L six-cylinder models

1 *Accelerator cable*
2 *Remove in direction of arrow*
3 *Remove in direction of arrow*
4 *Throttle lever*
5 *Cruise control cable*
6 *Release tabs*
7 *Bracket*

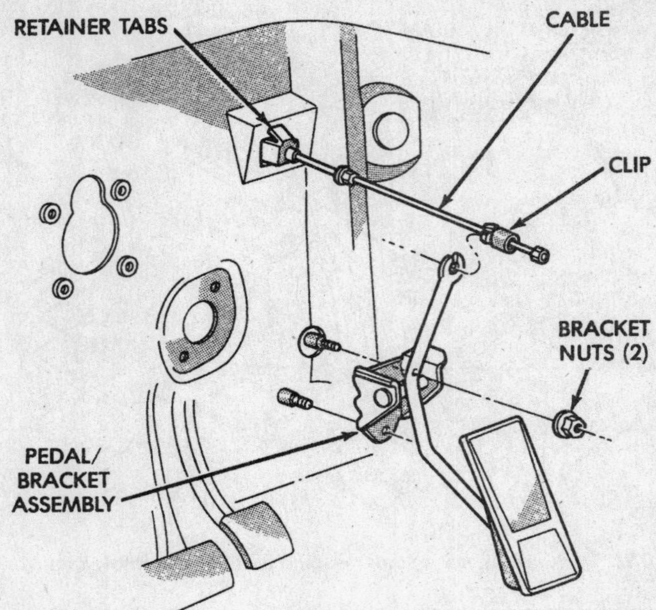

10.5 Pull the accelerator cable clip out of the pedal rod and slide the cable through the slot, then squeeze the housing retaining tabs together and push the cable assembly into the engine compartment

5 Working beneath the dash, detach the cable from the accelerator pedal **(see illustration)**.

6 Press in on the tabs of the cable retainer and push the cable through the firewall and into the engine compartment. Detach the cable from the routing clips and remove the cable from the engine compartment.

7 Installation is the reverse of removal.

11 Fuel injection system - general information

The sequential Multi Port Fuel Injection (MPFI) system consists of three sub-systems: air intake, emissions and engine control and fuel delivery. The system uses a Powertrain Control Module (PCM) along with the sensors (coolant temperature sensor, throttle position sensor, manifold absolute pressure sensor, oxygen sensor, etc.) to determine the proper air/fuel ratio under all operating conditions.

The fuel injection system and the emissions and engine control system are closely linked in function and design. For additional information, refer to Chapter 6.

Air intake system

The air intake system consists of the air filter, the air intake ducts, the throttle body, the idle control system, the air intake plenum and the intake manifold.

When the engine is idling, the air/fuel ratio is controlled by the idle air control system, which consists of the Powertrain Control Module (PCM) and the idle air control valve. The idle air control valve is controlled by the PCM and is opened and closed depending upon the running conditions of the engine (air conditioning system, power steering, cold and warm running etc.). This idle air control regulates the amount of airflow past the throttle plate and into the intake manifold, thus increasing or decreasing the engine idle speed. The PCM receives information from the sensors (vehicle speed, coolant temperature, air conditioning, power steering mode etc.) and adjusts the idle according to the demands of the engine and driver. Refer to Chapter 6 for information on the idle air control valve.

Emissions and engine control system

The emissions and engine control system is described in detail in Chapter 6.

Fuel delivery system

The fuel delivery system consists of these components: The fuel pump, the fuel filter, the fuel pressure regulator, the fuel rail and the fuel injectors.

The fuel pump is an electric-type located in the fuel tank. Fuel is drawn through a screen into the pump, flows through a one-way valve, passes through the main filter and is delivered to the fuel rail and injectors.

The fuel pressure regulator supplies regulated fuel pressure to the injectors. Four different types of fuel pressure regulators are used on the vehicles covered by this manual:

a) On 1993 and 1994 models, the fuel pressure regulator is a vacuum controlled regulator mounted on the fuel rail. The fuel pressure regulator supplies fuel at a slightly higher pressure than normal during periods of heavy demand (low vacuum). Excess pressure is bled off and fuel is routed back to the fuel tank through a fuel return line.

b) On 1995 and 1996 models, the fuel pressure regulator is incorporated into the fuel pump module inside the fuel tank. The pressure regulator supplies fuel to the fuel rail at a constant pressure. Excessive pressure is bled off directly into the fuel tank. If the fuel pressure regulator is defective, the entire fuel pump module must be replaced.

c) On 1997 and 1998 models, the fuel pressure regulator/fuel filter unit is mounted on top of the fuel pump module. The pressure regulator supplies fuel to the fuel rail at a constant pressure. Excessive pressure is bled off directly into the fuel tank. If the fuel pressure regulator is defective, the fuel pressure regulator/fuel filter may be replaced.

d) On 1999 and later models, a combination fuel filter/fuel pressure regulator is remotely mounted to the vehicle chassis near the fuel tank. Excessive pressure is bled off and fuel is returned to the fuel tank through a fuel return line. If the fuel pressure regulator is defective, the fuel pressure regulator/fuel filter may be replaced.

The injectors are solenoid-actuated pintle types consisting of a solenoid, plunger, needle valve and housing. When current is applied to the solenoid coil, the needle valve raises and pressurized fuel squirts out the nozzle. The injection quantity is determined by the length of time the valve is open (the length of time during which current is supplied to the solenoid coils). The fuel injector duration is controlled by the PCM.

The Automatic Shutdown (ASD) relay and the fuel pump relay are contained within the Power Distribution Center, which is located in the engine compartment. The ASD relay connects battery voltage to the fuel injectors and the ignition coil while the fuel pump relay connects battery voltage only to the fuel pump. If the PCM senses there is NO signal from the camshaft or crankshaft sensors (as with the engine not running or cranking), the PCM will de-energize both relays.

12 Fuel injection system - check

Refer to illustrations 12.7, 12.8 and 12.10

Note: *The following procedure is based on the assumption that the fuel pressure is adequate* (see Section 3).

1 Check all electrical connectors that are related to the system. Check the ground wire connections on the intake manifold for tightness. Loose connectors and poor grounds can cause many problems that resemble more serious malfunctions.

2 Check to see that the battery is fully charged, as the control unit and sensors depend on an accurate supply voltage in order to properly meter the fuel.

3 Check the air filter element - a dirty or partially blocked filter will severely impede performance and economy (see Chapter 1).

4 Check the related fuses. If a blown fuse is found, replace it and see if it blows again. If it does, search for a grounded wire in the harness.

5 Check the condition of all vacuum hoses connected to the intake manifold.

6 Remove the air intake duct from the throttle body and check for dirt, carbon or other residue build-up. If it's dirty, clean it with carburetor cleaner spray, a shop towel and a toothbrush, if necessary.

7 With the engine running, place an automotive stethoscope against each injector, one at a time, and listen for a clicking sound, indicating operation **(see illustration)**. If you don't have a stethoscope, place the tip of a screwdriver against the injector and listen through the handle.

8 Disconnect the injector electrical connectors and measure the resistance of each injector **(see illustration)**. Compare the measurements with the resistance values listed in this Chapter's Specifications.

9 Turn the ignition key On and check for battery voltage at the voltage supply wire terminal of one of the injector harness connectors. If battery voltage is not present, check the ASD relay and related wiring (see Chapter 12). **Note:** *The fuel injector voltage supply wire could be either dark green/black, dark green/orange or dark green/light green depending on model. The voltage supply wire is always the same color at each injector connector while the injector control wire color is different for each injector.*

12.7 Use a stethoscope to determine if the injectors are working properly - they should make a steady clicking sound that rises and falls with engine speed changes

10 Install an injector test light ("noid" light) into each injector electrical connector, one at a time **(see illustration)**. Crank the engine over. Confirm that the light flashes evenly on each connector. This tests the PCM control of the injectors. If the light does not flash, have the PCM checked at a dealer service department or other properly equipped repair facility.

11 The remainder of the engine control system checks can be found in Chapter 6.

13 Throttle body - removal and installation

Refer to illustrations 13.6a, 13.6b and 13.6c

Warning: *Wait until the engine is completely cool before beginning this procedure.*

1 Disconnect the cable from the negative battery terminal.

2 Remove the air intake duct/resonator from the throttle body.

3 Disconnect the electrical connectors from the throttle position sensor, idle air control valve and manifold absolute pressure sensor, as required.

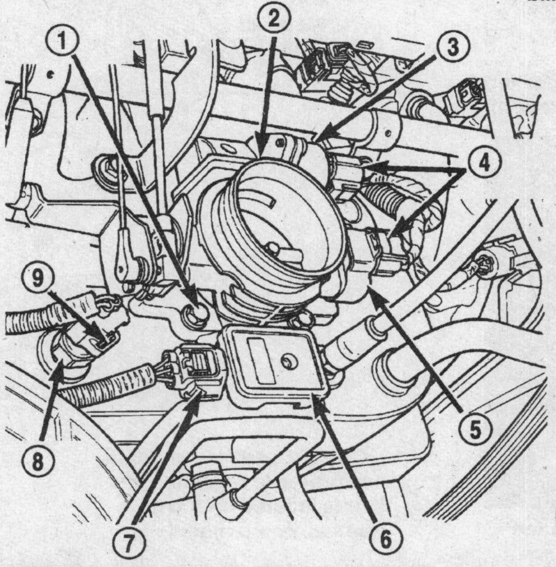

13.6a Throttle body and related components - 4.0L six-cylinder engine

1 *Mounting bolts*
2 *Throttle body*
3 *Idle air control valve*
4 *Electrical connector*
5 *Throttle position sensor*
6 *Manifold absolute pressure sensor*
7 *Electrical connector*
8 *Intake air temperature sensor*
9 *Electrical connector*

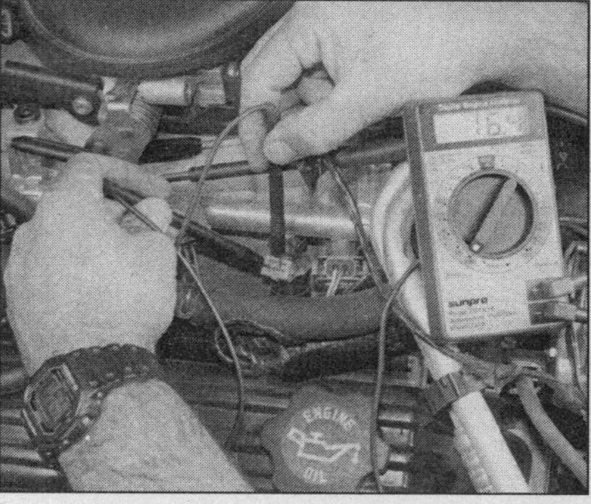

12.8 Measure the resistance of each injector across the two terminals of the injector

4 Label and detach all vacuum hoses from the throttle body.

5 Detach the accelerator cable (see Section 10) and, if equipped, the transmission throttle valve cable and cruise control cable.

6 Remove the mounting bolts/nuts and remove the throttle body and gasket **(see illustrations)**.

12.10 Install the "noid" light (available at most auto parts stores) into each injector electrical connector and confirm that it blinks when the engine is cranking

7 Remove all traces of old gasket material from the throttle body and intake manifold and install a new gasket.
8 Install the throttle body and tighten the bolts to the torque listed in this Chapter's Specifications.
9 The remainder of installation is the reverse of removal.

14 Fuel pressure regulator - replacement

Warning: *See the* **Warning** *in Section 1.*
Note: *Refer to Section 3 for fuel pressure regulator diagnosis.*

1993 and 1994 models

Refer to illustrations 14.3a and 14.3b

1 Relieve the fuel system pressure (see Section 2).
2 Detach the vacuum hose from the pressure regulator.
3 Remove the mounting bolt and the retaining clamp **(see illustrations)**.
4 Remove the pressure regulator from the fuel rail.
5 Installation is the reverse of removal. Be sure to use a new O-ring. Lubricate the O-ring with a light coat of clean engine oil before installation.
6 After installing the pressure regulator, cycle the ignition key On and Off several times and check for fuel leaks.

1995 and 1996 models

7 The fuel pressure regulator is an integral part of the fuel pump module and can't be replaced separately. If faulty, the entire fuel pump module must be replaced.

1997 and 1998 models

Refer to illustration 14.10

Note: *On 1997 and 1998 models, the fuel pressure regulator is part of the fuel filter/fuel pressure regulator unit located on top of the fuel pump module.*

8 Remove the fuel tank (see Section 4).
9 Thoroughly clean the area surrounding the fuel pump module.
10 Pry the retainer clamp from the tabs top of the fuel pump module **(see illustration)**. Discard the retainer clamp; it can't be reused.
11 Using two flat blade screwdrivers, carefully pry the fuel pressure regulator/fuel filter free from the top of the fuel pump module. Ensure that both upper and lower O-rings came out with the regulator.
12 Remove the gasket from the top of the fuel pump module and discard it.
13 Apply clean engine oil to the O-rings and ensure they are installed in the regulator grooves. **Warning:** *Don't install the O-rings in the fuel pump module as they will be damaged when the regulator is installed, resulting in a fuel leak.*
14 Install a new gasket onto the top of the fuel pump module and make sure it is properly seated.
15 Press the new fuel pressure regulator/fuel filter into the fuel pump module until it snaps into place (a positive click must be heard or felt to ensure proper engagement).
16 The arrow on top of the fuel pump module should be pointing towards the front of the vehicle (12 o'clock position). Rotate the fuel pressure regulator/fuel filter until the fuel supply fitting is pointing to the 11 o'clock position.

13.6b Throttle body mounting bolts (arrows) - 5.2L/5.9L V8 engines

13.6c Throttle body mounting bolts (arrows) - 4.7L V8 engine

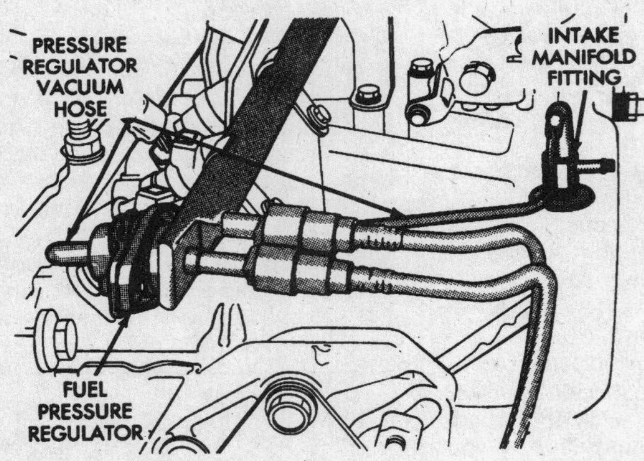

14.3a Fuel pressure regulator details - 1993 and 1994 4.0L six-cylinder engine

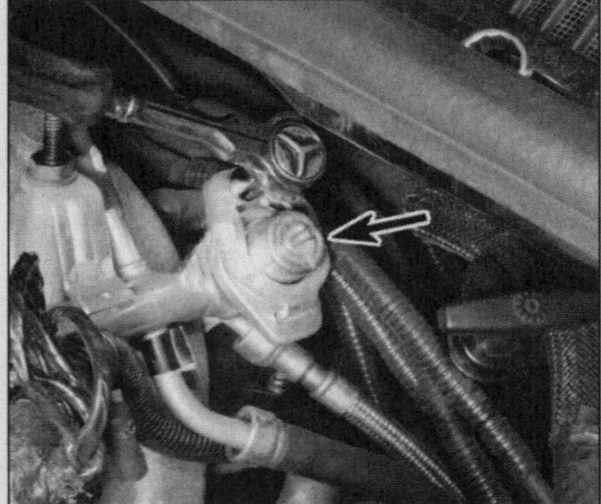

14.3b Fuel pressure regulator location - 1993 and 1994 5.2L/5.9L V8 engines

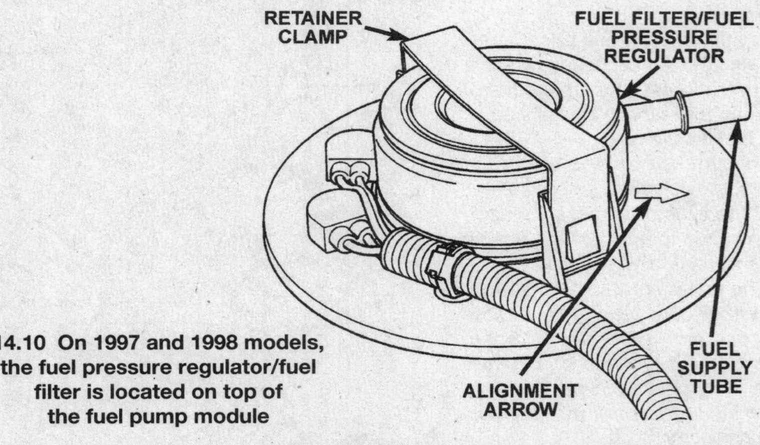

14.10 On 1997 and 1998 models, the fuel pressure regulator/fuel filter is located on top of the fuel pump module

17 Install a new retainer clamp over the top of the fuel pressure regulator/fuel filter and lock it into place onto the fuel pump module tabs. Ensure the clamp is locked in place on both sides.
18 Install the fuel tank (see Section 4).

1999 and later models

Refer to illustration 14.22

Note: *On 1999 and later models, the fuel pressure regulator is part of the inline fuel filter/fuel pressure regulator unit mounted to the vehicle chassis near the fuel tank.*

19 Relieve the fuel system pressure (see Section 2).
20 Raise the vehicle and support it securely on jackstands.
21 Disconnect the fuel lines from the fuel filter/fuel pressure regulator (see Section 4).
22 Remove the mounting bolts and remove the fuel filter/fuel pressure regulator **(see illustration)**.
24 Installation is the reverse of removal with the following additions:

- a) *Install a new O-ring on the fuel filter/fuel pressure regulator return line fitting.*
- b) *Clean the quick connect fittings with a lint free cloth and apply clean engine oil to the O-rings.*
- c) *After connecting the quick connect fittings, check the integrity of the fitting by attempting to pull the lines off.*
- d) *Cycle the ignition key On and Off several times and check for leaks at the fittings before starting the engine.*

15 Fuel rail and injectors - removal and installation

Warning: *See the* **Warning** *in Section 1.*
Note: *Refer to Section 12 for fuel injector diagnosis.*

Removal

Refer to illustrations 15.7, 15.8a, 15.8b, 15.8c, 15.9a, 15.9b, 15.9c, 15.10a and 15.10b

1 Relieve the fuel pressure (see Section 2).
2 Disconnect the cable from the negative battery terminal.
3 Remove the air intake duct/resonator. Clearly label and remove any vacuum hoses or electrical wiring that will interfere with the fuel rail removal. Detach any wiring harness retainers from the fuel rail.
4 On six-cylinder models, perform the following:

- a) *Disconnect the accelerator control cable, the transmission throttle valve cable and the cruise control cable (if equipped) from the throttle body.*
- b) *Remove the cable bracket mounting bolts and position the cables and bracket aside.*

5 On 5.2L/5.9L V8 models, perform the following:

- a) *Remove the throttle body* (see Section 13). *Position the control cables and bracket aside.*
- b) *Remove the air conditioning compressor-to-intake manifold bracket, if equipped.*
- c) *Disconnect the electrical connector from the intake manifold air temperature sensor.*
- d) *Remove the EVAP canister purge solenoid from the intake manifold.*

6 On 4.7L V8 models, perform the following:

- a) *Disconnect the electrical connectors from the alternator and position the wiring harness aside.*
- b) *Disconnect the electrical connectors from the throttle position sensor, idle air control valve, manifold absolute pressure sensor and intake air temperature sensor. Position the wiring harness aside.*
- c) *Remove the first three ignition coils from each cylinder bank* (see Chapter 5).

7 Disconnect the fuel injector electrical connectors and position the harness aside **(see illustration)**. **Note:** *Each connector should be numbered with the corresponding cylinder number. If the number tag is obscured or missing, renumber the connectors.*
8 Detach the fuel line(s) from the fuel rail **(see illustrations)**.

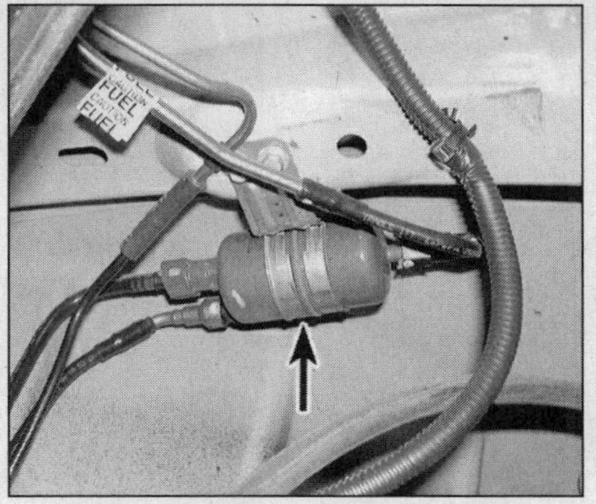

14.22 On 1999 and later models, the fuel pressure regulator/fuel filter is attached to the chassis near the fuel tank

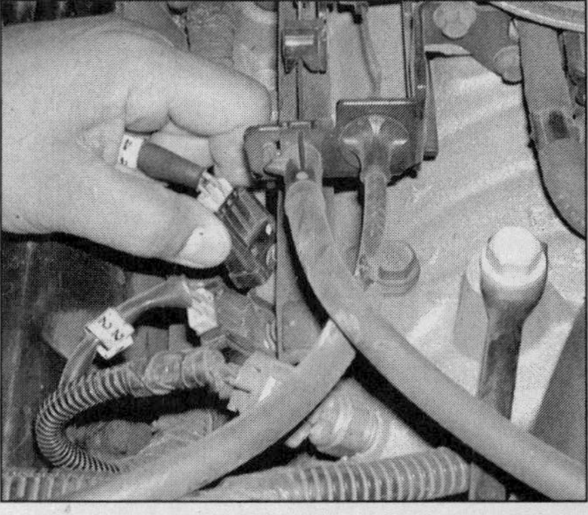

15.7 Disconnect the electrical connectors from the injectors - note the numbered tags on the wiring harness

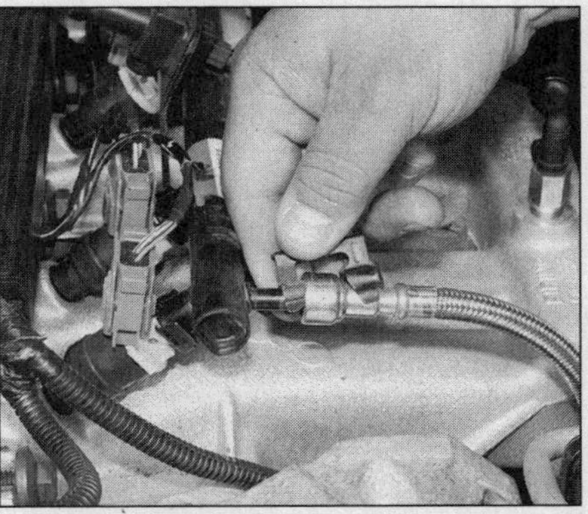

15.8a To disconnect the fuel line from the fuel rail, remove the safety clip from the quick-connect fitting . . .

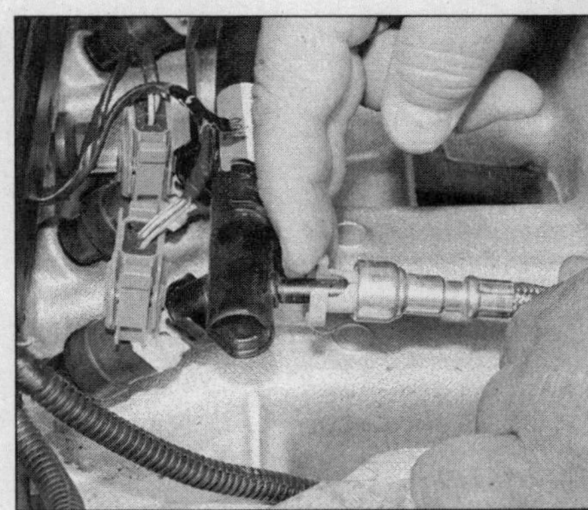

15.8b . . . insert the fuel line disconnect tool into the fitting and disconnect the fuel line from the fuel rail

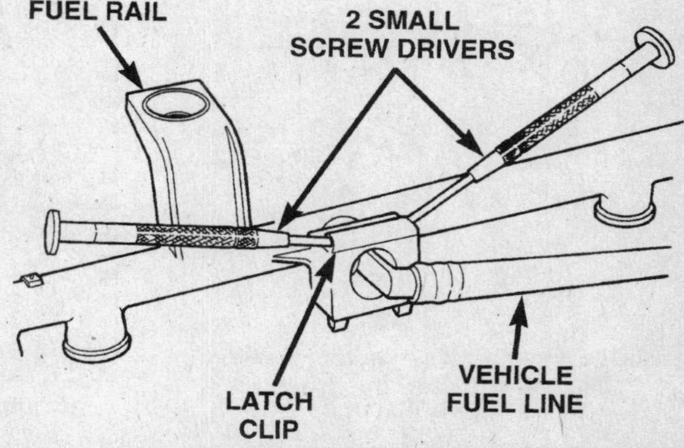

15.8c On some models the fuel line is secured to the fuel rail by a latch-clip type fitting - squeeze the legs of the clip together and pry the clip straight up

4

15.9a Fuel rail mounting bolts (arrows) - 4.0L six-cylinder engine

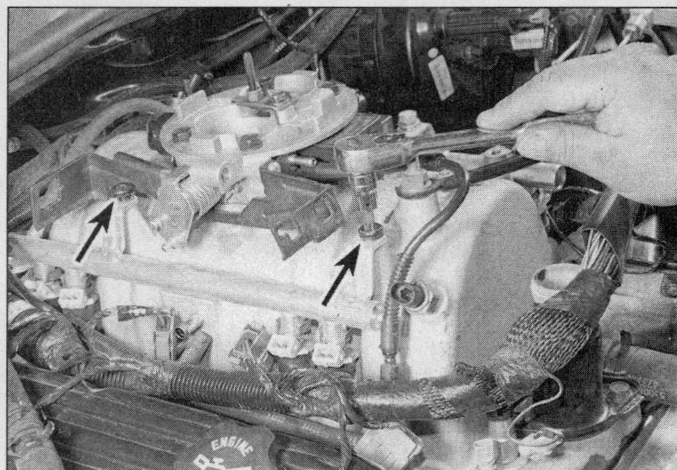
15.9b Fuel rail mounting bolts (arrows) - 5.2L/5.9L V8 engines (right side shown, left side similar)

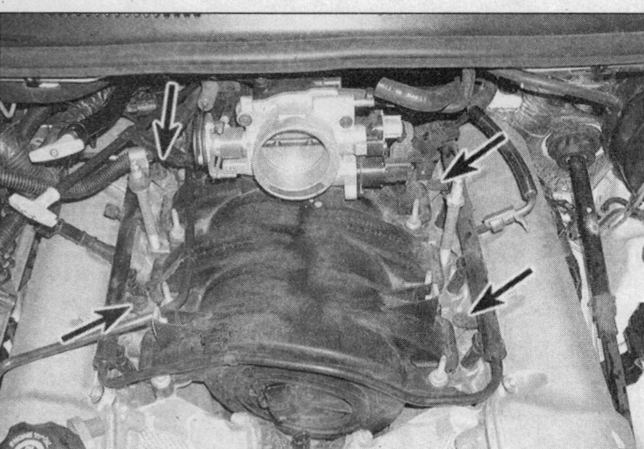
15.9c Fuel rail mounting bolts (arrows) - 4.7L V8 engine

9 Clean any debris from around the injectors. Remove the fuel rail mounting nuts/bolts **(see illustrations)**. Gently rock the fuel rail and injectors to loosen the injectors and remove the fuel rail and fuel injectors as an assembly. **Caution:** *Do not attempt to separate the left and right fuel rails on a V8 model. Both sides are serviced together as an assembly.*

10 Remove the injector(s) from the fuel rail, then remove and discard the O-rings **(see illustrations)**. **Note:** *Whether you're replacing an injector or a leaking O-ring, it's a good idea to remove all the injectors from the fuel rail and replace all the O-rings.*

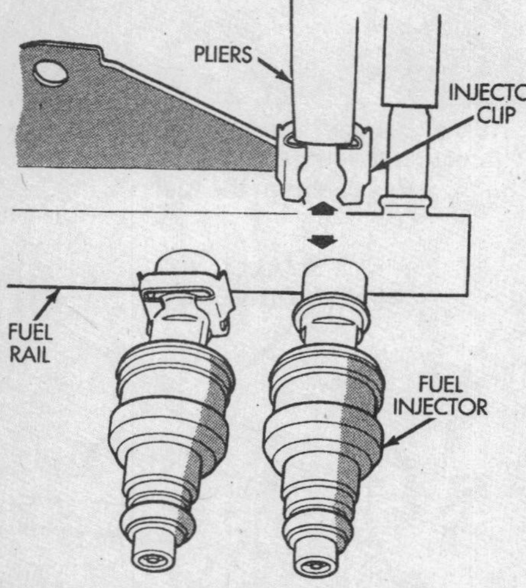

15.10a Remove the injector retaining clip and withdraw the injector from the fuel rail

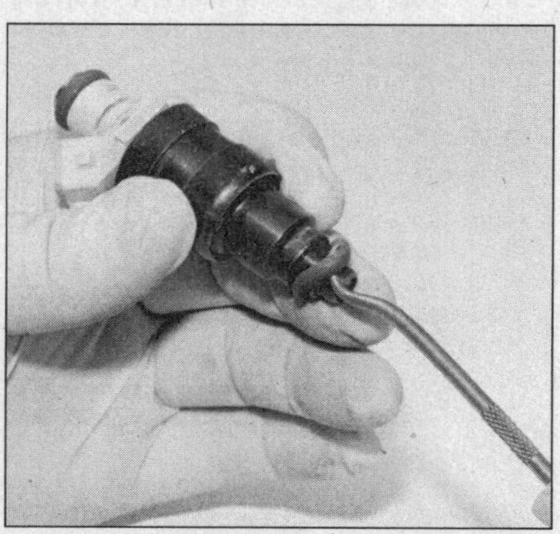
15.10b Carefully remove the O-rings from the injector

Installation

11 Coat the new seal rings (if equipped) with clean engine oil and slide them onto the injectors.

12 Coat the new O-rings with clean engine oil and install them on the injector(s), then insert each injector into its corresponding bore in the fuel rail. Install the injector retaining clip.

13 Install the injector and fuel rail assembly on the intake manifold. Make sure the injectors are fully seated, then tighten the fuel rail mounting nuts/bolts to the torque listed in this Chapter's Specifications.

14 Connect the fuel line(s) and make sure they are securely installed. If equipped with a latch-clip type fitting, make sure the latch-clip is properly seated.

15 Connect the electrical connectors to each injector, referring to the numbered tags.

16 The remainder of installation is the reverse of removal.

17 After the injector/fuel rail assembly installation is complete, turn the ignition switch to On, but don't operate the starter (this activates the fuel pump for about two seconds, which builds up fuel pressure in the fuel lines and the fuel rail). Repeat this about two or three times, then check the fuel lines, fuel rail and injectors for fuel leakage.

16 Exhaust system servicing - general information

Warning: *Inspection and repair of exhaust system components should be done only after enough time has elapsed after driving the vehicle to allow the system components to cool completely. Also, when working under the vehicle, make sure it is securely supported on jackstands.*

1 The exhaust system consists of the exhaust manifold(s), the catalytic converter, the muffler, the tailpipe and all connecting pipes, brackets, hangers and clamps. The

exhaust system is attached to the body with mounting brackets and rubber hangers. If any of the parts are improperly installed, excessive noise and vibration will be transmitted to the body.

Muffler and pipes

Refer to illustrations 16.2a and 16.2b

2 Conduct regular inspections of the exhaust system to keep it safe and quiet **(see Illustrations)**. Look for any damaged or bent parts, open seams, holes, loose connections, excessive corrosion or other defects which could allow exhaust fumes to enter the vehicle. Also check the catalytic converter when you inspect the exhaust system (see below). Deteriorated exhaust system components should not be repaired; they should be replaced with new parts.

3 If the exhaust system components are extremely corroded or rusted together, welding equipment will probably be required to remove them. The convenient way to accomplish this is to have a muffler repair shop remove the corroded sections with a cutting torch. If, however, you want to save money by doing it yourself (and you don't have a welding outfit with a cutting torch), simply cut off the old components with a hacksaw. If you have compressed air, special pneumatic cutting chisels can also be used. If you do decide to tackle the job at home, be sure to wear safety goggles to protect your eyes from metal chips and work gloves to protect your hands.

4 Here are some simple guidelines to follow when repairing the exhaust system:

a) Work from the back to the front when removing exhaust system components.
b) Apply penetrating oil to the exhaust system component fasteners to make them easier to remove.
c) Use new gaskets, hangers and clamps when installing exhaust systems components.
d) Apply anti-seize compound to the threads of all exhaust system fasteners during reassembly.
e) Be sure to allow sufficient clearance between newly installed parts and all points on the underbody to avoid overheating the floor pan and possibly damaging the interior carpet and insulation. Pay particularly close attention to the catalytic converter and heat shield.

Catalytic converter

Refer to illustration 16.6

Warning: *The converter gets very hot during operation. Make sure it has cooled down before you touch it.*

Note: *See Chapter 6 for additional information on the catalytic converter.*

5 Periodically inspect the heat shield for cracks, dents and loose or missing fasteners.

6 Inspect the converter for cracks or other damage **(see Illustration)**.

7 The catalytic converter is permanently welded into the exhaust system. If replacement is required, take the vehicle to a dealership service department or other qualified repair shop. Refer to Chapter 6 for additional information.

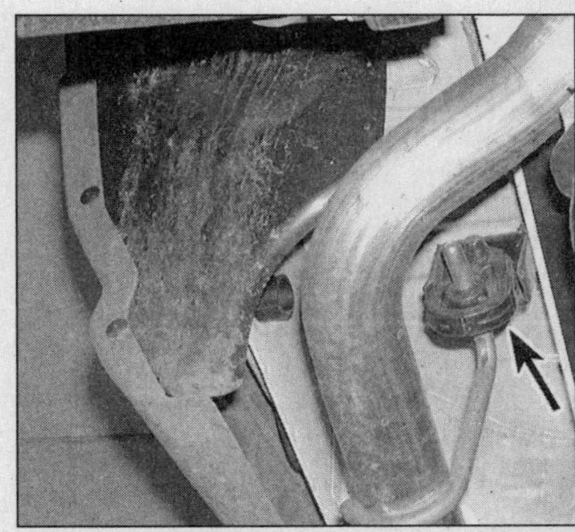

16.2a Inspect the exhaust system rubber hangers for damage

16.2b Inspect the exhaust pipe connections for leaks

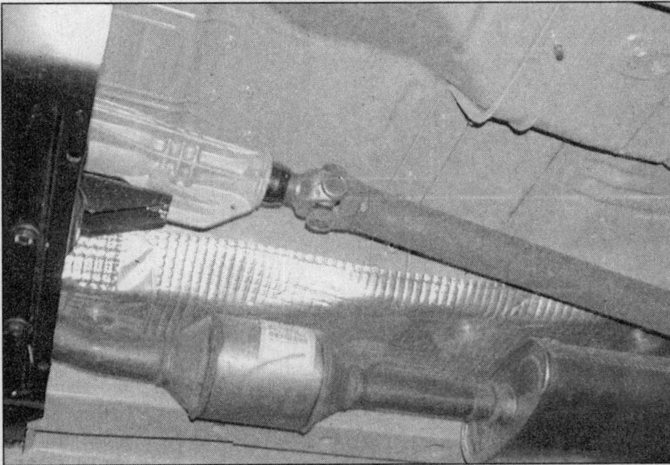

16.6 Inspect the catalytic converter and heat shield for damage

Notes

Chapter 5
Engine electrical systems

Contents

Specifications

General

Battery voltage
- Engine off: 12.0 to 12.6 volts
- Engine running: 13.5 to 15 volts

Firing order
- Six-cylinder engine: 1-5-3-6-2-4
- V8 engines: 1-8-4-3-6-5-7-2

Ignition system

Ignition coil resistance (at 70 to 80-degrees F)
- 1998 and earlier
  - Diamond
    - Primary resistance: 0.96 to 1.18 ohms
    - Secondary resistance: 11,300 to 15,300 ohms
  - Toyodenso
    - Primary resistance: 0.95 to 1.20 ohms
    - Secondary resistance: 11,300 to 13,300 ohms
- 1999 and 2000
  - Six-cylinder engine
    - Primary resistance: 0.71 to 0.88 ohms
    - Secondary resistance: Not available
  - V8 engine
    - Primary resistance: 0.6 to 0.9 ohms
    - Secondary resistance: 6,000 to 9,000 ohms

Spark plug/coil wire resistance
- Minimum: 3,000 ohms per foot
- Maximum: 12,000 ohms per foot

Torque specifications

	Ft-lbs
Distributor hold-down bolt	17
Alternator mounting bolts	
Six-cylinder engine	41
5.2L/5.9L V8 engines	30
4.7L V8 engine	
Vertical bolt	40
Horizontal bolts	
Short bolt	55
Long bolt	40
Starter mounting bolt/nut	
Six-cylinder engine	
1995 and earlier	33
1996 and later	
Lower bolt	30
Upper bolt	40
5.2L/5.9L V8 engines	
1995 and earlier	
Lower bolt	50
Upper nut	20
1996 and later	50
4.7L V8 engine	40

1 General information

Refer to illustration 1.1

The engine electrical systems include all ignition, charging and starting components **(see illustration)**. Because of their engine-related functions, these components are discussed separately from chassis electrical devices such as the lights, the instruments, etc. (which are included in Chapter 12).

Always observe the following precautions when working on the electrical systems:

a) *Be extremely careful when servicing engine electrical components. They are easily damaged if checked, connected or handled improperly.*

b) *Never leave the ignition switch on for long periods of time with the engine off.*

c) *Don't disconnect the battery cables while the engine is running.*

d) *Maintain correct polarity when connecting a battery cable from another vehicle during jump starting.*

e) *Always disconnect the negative cable first and hook it up last or the battery may be shorted by the tool being used to loosen the cable clamps.*

It's also a good idea to review the safety-related information regarding the engine electrical systems located in the *Safety First* section near the front of this manual before beginning any operation included in this Chapter.

2 Battery - emergency jump starting

Refer to the *Booster battery (jump) starting* procedure at the front of this manual.

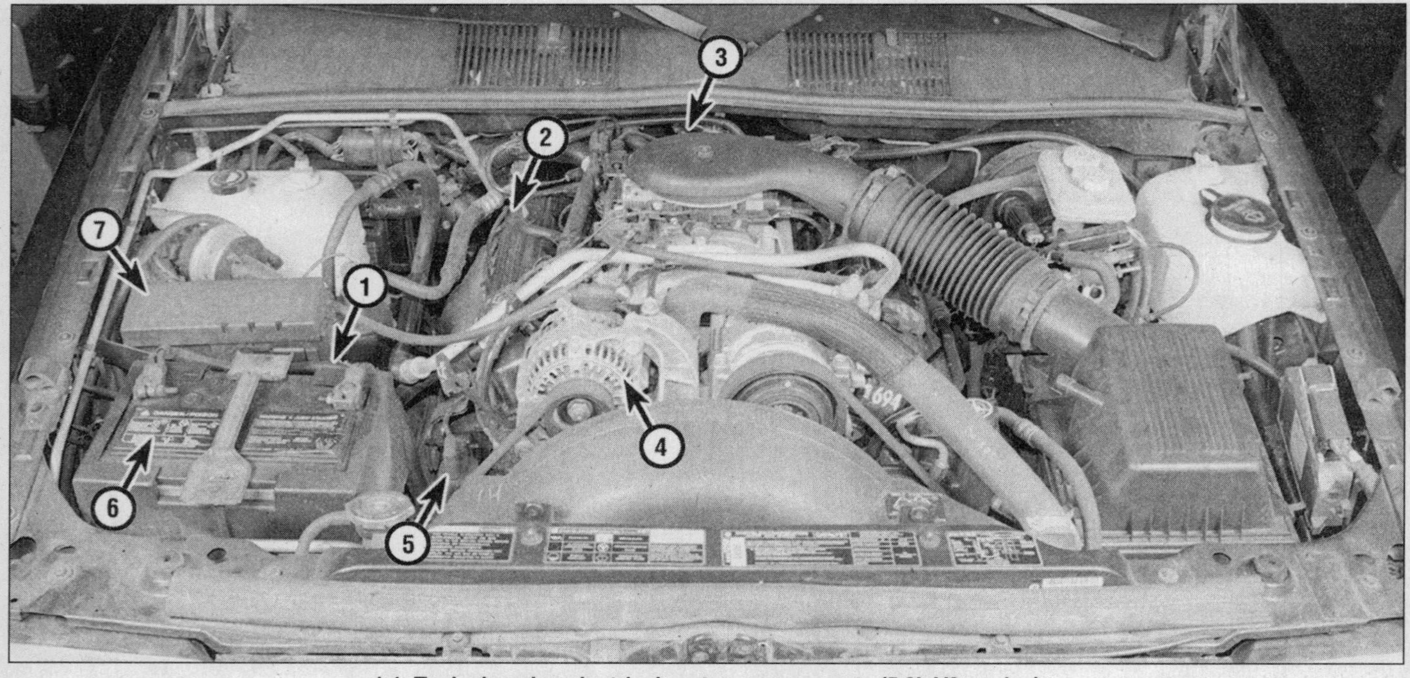

1.1 Typical engine electrical system components (5.2L V8 engine)

1 *Battery cable*
2 *Spark plug wi*
3 *Distributor*
4 *Alternator*
5 *Ignition coil*
6 *Battery*
7 *Power distribution center*

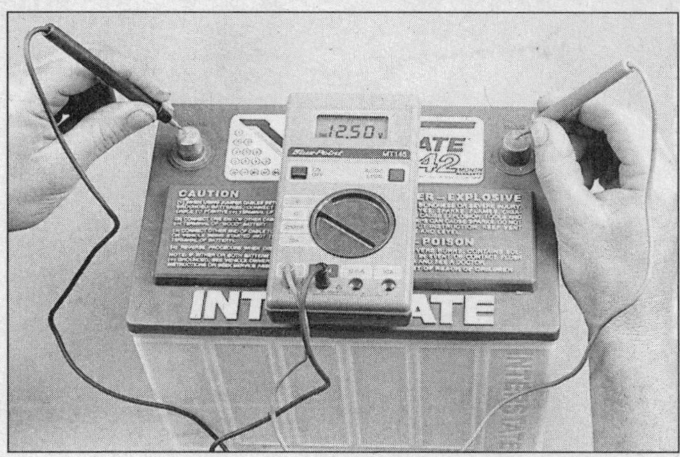

3.2 To test the open circuit voltage of the battery, connect a voltmeter to the battery - a fully charged battery should measure at least 12.4 volts (depending on outside air temperature)

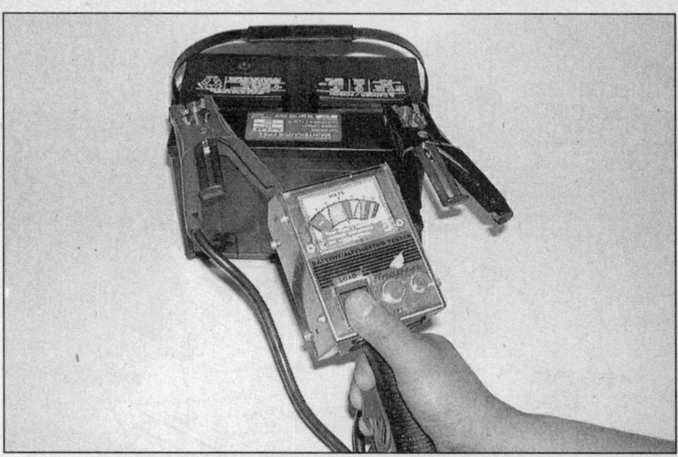

3.3 Connect a battery load tester to the battery and check the battery condition under load following to the tool manufacturers instructions

3 Battery - check and replacement

Warning: *Hydrogen gas is produced by the battery, so keep open flames and lighted cigarettes away from it at all times. Always wear eye protection when working around a battery. Rinse off spilled electrolyte immediately with large amounts of water.*

Caution: *Always disconnect the negative cable first and hook it up last or the battery may be shorted by the tool being used to loosen the cable clamps.*

Check

Refer to illustrations 3.2 and 3.3

1 Disconnect the negative battery cable, then the positive cable from the battery.

2 Check the battery state of charge. Visually inspect the indicator eye on the top of the battery. If the indicator eye is clear, charge the battery as described in Chapter 1. Next perform an open voltage circuit test using a voltmeter **(see illustration)**. **Note:** *The battery's surface charge must be removed before accurate voltage measurements can be made. Turn on the high beams for ten seconds, then turn them off and let the vehicle stand for two minutes.* With the engine and all accessories off, touch the negative probe of the voltmeter to the negative terminal of the battery and the positive probe to the positive terminal of the battery. The battery voltage should be 12.4 volts or more. If the battery is less than the specified voltage, charge the battery before proceeding to the next test. Do not proceed with the battery load test unless the battery charge is correct.

3 Perform a battery load test. An accurate check of the battery condition can only be performed with a battery load tester (available at most auto parts stores). This test evaluates the ability of the battery to operate the starter and other accessories during periods of heavy amperage draw (load). The tool utilizes a carbon pile to increase the load demand (amperage draw) on the battery. Install a special battery load testing tool onto the terminals **(see illustration)**. Load test the battery according to the tool manufacturer's instructions. Typically a load of 50-percent of the cold cranking amperage rating is applied during the test. The cold cranking amperage rating can usually be found on the battery label. Maintain the load on the battery for a maximum of 15 seconds. The battery voltage should not drop below 9.6 volts during the test. If the battery condition is weak or defective, the tool will indicate this condition immediately. **Note:** *Cold temperatures will cause the voltage readings to drop slightly. Follow the chart given in the tool manufacturer's instructions to compensate for cold climates. Minimum load voltage for freezing temperatures (32-degrees F) should be approximately 9.1 volts.*

Replacement

Refer to illustrations 3.5a, 3.5b, 3.8a and 3.8b

4 Disconnect the negative battery cable, then the positive cable from the battery.

5 Remove the battery hold-down clamp bolt. On 1998 and earlier models remove the thermo-guard **(see Illustrations)**. On 1999 and 2000 models, the thermo-guard is removed with the battery.

6 Lift out the battery. Be careful - it's heavy. On 1999 and 2000 models, remove

3.5a Remove the bolts (arrows) and detach the hold-down clamp from the battery tray

3.5b On 1998 and earlier models, remove the thermo-guard from the battery

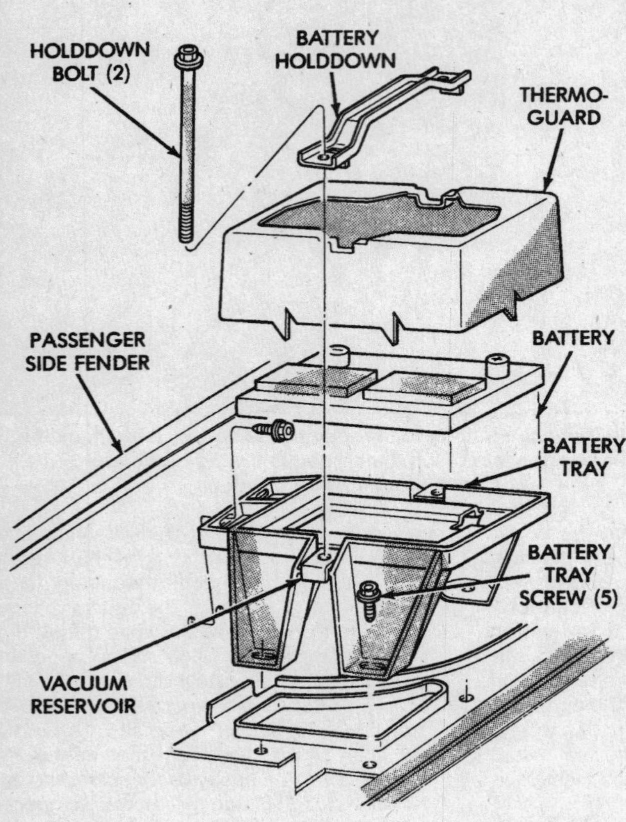

3.8a Battery and battery tray installation details - 1998 and earlier models

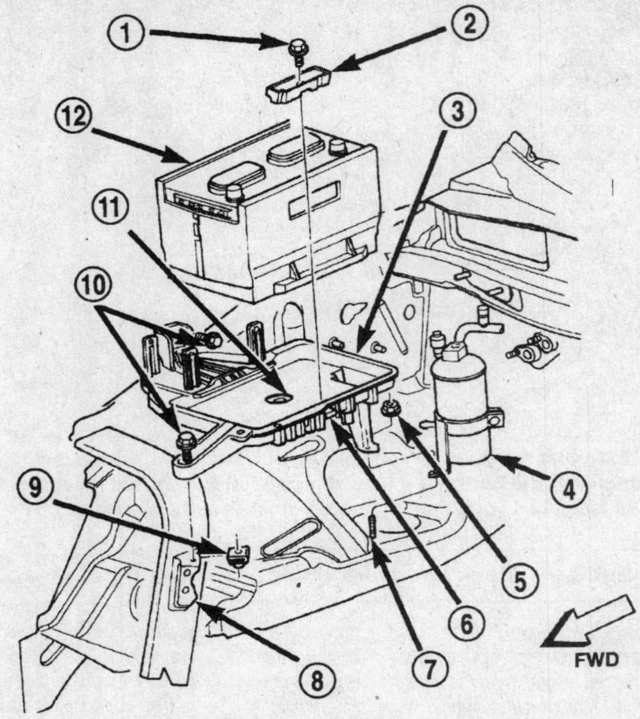

3.8b Battery and battery tray installation details - 1999 and 2000 models

1 Screw
2 Hold-down bracket
3 Battery tray
4 Accumulator
5 Nut
6 U-nut
7 Stud
8 Radiator support bracket
9 U-nut
10 Screw
11 Battery temperature sensor
12 Battery and thermoguard

the thermo-guard. **Note:** *Battery straps and handlers are available at most auto parts stores for a reasonable price. They make it easier to remove and carry the battery.*

7 While the battery is out, inspect the area underneath the tray for corrosion.

8 If corrosion has leaked down past the battery tray, remove the screws/nuts and remove the battery tray **(see illustrations)**. On 1999 and 2000 models, detach the power distribution center from the battery tray and disconnect the electrical connector from the battery temperature sensor.

9 Use baking soda to clean the deposits from the metal to prevent further oxidation, then spray the area with a rust-inhibiting paint.

10 If you are replacing the battery, make sure you get one that's identical, with the same dimensions, amperage rating, cold cranking rating, etc.

11 Installation is the reverse of removal.

4 Battery cables - check and replacement

Refer to illustration 4.4

1 Periodically inspect the entire length of each battery cable for damage, cracked or burned insulation and corrosion. Poor battery cable connections can cause starting problems and decreased engine performance.

2 Check the cable-to-terminal connections at the ends of the cables for cracks, loose wire strands and corrosion. The presence of white, fluffy deposits under the insulation at the cable terminal connection is a sign that the cable is corroded and should be replaced. Check the terminals for distortion, missing mounting bolts and corrosion.

3 When removing the cables, always disconnect the negative cable first and hook it up last or the battery may be shorted by the tool used to loosen the cable clamps. Even if only the positive cable is being replaced, be sure to disconnect the negative cable from the battery first (see Chapter 1 for further information regarding battery cable removal).

4 Disconnect the old cables from the battery, then trace each of them to their opposite ends and detach them from the starter solenoid, power distribution center and ground terminals, as necessary. Note the routing of each cable to ensure correct installation. Cut the electrical tape and remove the protective sheathing from the battery cables **(see illustration)**.

5 If you are replacing either or both of the battery cables, take them with you when buying new cables. It is vitally important that you replace the cables with identical parts. Cables have characteristics that make them easy to identify: positive cables are usually red and larger in cross-section; ground cables are usually black and smaller in cross-section.

6 Clean the threads of the solenoid or

4.4 The battery cables are wrapped in a protective sheathing (arrow) - if necessary, cut the electrical tape and remove the sheathing

ground connection with a wire brush to remove rust and corrosion. Apply a light coat of battery terminal corrosion inhibitor or petroleum jelly to the threads to prevent future corrosion.

7 Attach the cable to the solenoid or ground connection and tighten the mounting nut/bolt securely.

8 Before connecting a new cable to the battery, make sure that it reaches the battery post without having to be stretched.

9 Connect the positive cable first, followed by the negative cable.

5 Ignition system - general information

1998 and earlier models

1998 and earlier models are equipped with a computer controlled electronic distributor-type ignition system. The ignition system consists of the spark plugs, spark plug wires, distributor, ignition coil, camshaft position sensor, crankshaft position sensor and the Powertrain Control Module (PCM). The PCM controls the ignition timing, spark and advance characteristics for the engine. The ignition timing is not adjustable, therefore, changing the position of the distributor will not affect the timing in any way.

The distributor is gear-driven from the camshaft. The distributor contains the rotor, camshaft position sensor and pulse ring. The distributor is not equipped with centrifugal or vacuum advance mechanisms. The camshaft position sensor is a hall-effect device triggered by the pulse ring connected to the distributor shaft. The signal generated by the camshaft position sensor is used by the computer for cylinder identification and fuel synchronization. The camshaft position sensor may be removed from the distributor while in the vehicle. Refer to Chapter 6 for additional information on the camshaft position sensor. **Note:** *The camshaft position sensor may be referred to as a sync signal generator or a switch plate assembly by dealership or parts store personnel.*

1999 and 2000 models

1999 and 2000 models are equipped with a distributorless ignition system. The ignition system consists of the battery, ignition coils, spark plugs, camshaft position sensor, crankshaft position sensor and the Powertrain Control Module (PCM). The PCM controls the ignition timing and spark advance characteristics for the engine. The ignition timing is not adjustable.

The 4.0L six-cylinder ignition system uses a "waste spark" method of spark distribution. The system utilizes three ignition coils in a coil rail assembly. Each coil is paired with its opposing cylinder in the firing order (1-6, 5-2, 3-4) so one cylinder under compression fires simultaneously with its opposing cylinder, where the piston is on the exhaust stroke. Since the cylinder on the exhaust stroke requires very little of the available voltage to fire its plug, most of the voltage is used to fire the plug of the cylinder on the compression stroke. Conventional ignition coils have one end of the secondary winding connected to the engine ground. On a waste spark system, neither end of the secondary winding is grounded - instead, one end of the coil's secondary winding is directly attached to the spark plug and the other end is attached to the spark plug of the companion cylinder.

The 4.7L V8 uses a "coil-over-plug" system. The system utilizes eight individual coils located above each spark plug and connected directly to the spark plug. The PCM controls the operation of the ignition coils, firing each coil in sequence.

All models

The PCM controls the ignition system by opening and closing the ignition coil ground circuit. The computerized ignition system provides complete control of the ignition timing by determining the optimum timing in response to engine speed, coolant temperature, throttle position and vacuum pressure in the intake manifold. These parameters are relayed to the PCM by the camshaft position sensor, crankshaft position sensor, throttle position sensor, coolant temperature sensor and manifold absolute pressure sensor. The PCM and the crankshaft position sensor are very important components of the ignition system. The ignition system will not operate and the engine will not start if the PCM or the crankshaft position sensor are defective. Refer to Chapter 6 for additional information on the various sensors.

6 Ignition system - check

Warning 1: *Because of the high voltage generated by the ignition system, extreme care should be taken whenever an operation is performed involving ignition components. This not only includes the ignition coil, but related components and test equipment.*

Warning 2: *The following procedure requires the engine to be cranked during testing. Make sure the meter leads, loose clothing, long hair, etc. are away from the moving parts of the engine (drivebelt, cooling fan, etc.) before cranking the engine.*

1 Before proceeding with the ignition system, check the following items:

- a) *Make sure the battery cable clamps, where they connect to the battery, are clean and tight.*
- b) *Test the condition of the battery* (see Section 3). *If it does not pass all the tests, replace it with a new battery.*
- c) *Check the ignition system wiring and connections for tightness, damage, corrosion or any other signs of a bad connection.*
- d) *Check the related fuses inside the power distribution center* (see Chapter 12). *If they're burned, determine the cause and repair the circuit.*

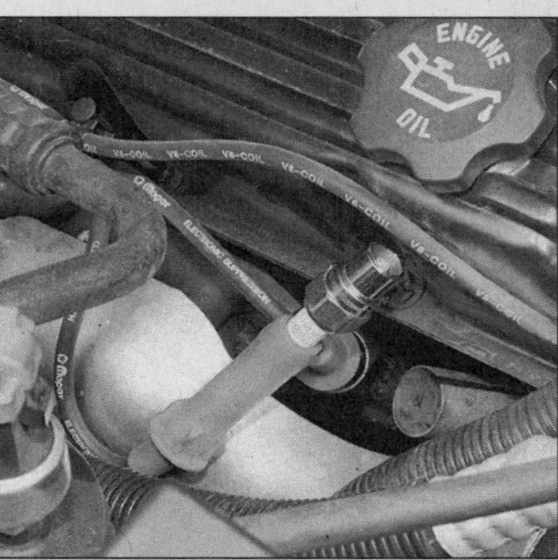

6.4 To use a calibrated ignition tester, simply disconnect a spark plug wire, connect it to the tester, clip the tester to a convenient ground and crank the engine over - if there's enough power to fire the plug, bright blue sparks will be visible between the electrode tip and the tester body (weak sparks or intermittent sparks are the same as no sparks)

2 If the engine turns over but won't start or has a severe misfire, make sure there is sufficient secondary ignition voltage to fire the spark plugs according to model as follows.

1998 and earlier models

Refer to illustrations 6.4 and 6.9

3 Disable the fuel system by removing the fuel pump relay from the power distribution center (see Chapter 12).

4 Disconnect a spark plug wire from one of the spark plugs and attach a calibrated ignition system tester (available at most auto parts stores) to the spark plug boot. Connect the clip on the tester to a bolt or metal bracket on the engine **(see illustration)**. Crank the engine and watch the end of the tester to see if a bright blue, well-defined spark occurs (weak spark or intermittent spark is the same as no spark).

5 If spark occurs, sufficient voltage is reaching the plug to fire it. Repeat the check at the remaining spark plug wires to verify that the distributor cap, rotor, spark plug wires, ignition coil and control systems are functioning properly. If the ignition system is operating properly the problem lies elsewhere; i.e. a mechanical or fuel system problem. However, the spark plugs may be fouled, so remove and check them as described in Chapter 1.

6 If no spark occurs at one or more wires, remove the suspected spark plug wire from the ignition coil and check the terminals at both ends for damage. Connect an ohmmeter to the ends of the spark plug wire and check the wire for an open or high resistance. If the spark plug wire resistance is greater

6.9 **Disconnect the ignition coil electrical connector and check for battery voltage at the dark green/black or dark green/orange wire terminal with the ignition On**

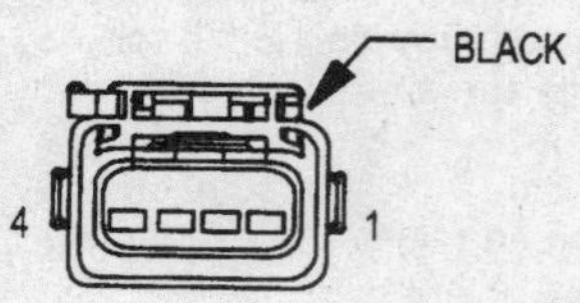

CAV	CIRCUIT	FUNCTION
1	K91 16TN/RD	COIL DRIVER NO. 1
2	A142 16DG/OR	AUTOMATIC SHUT DOWN RELAY OUTPUT
3	K92 16TN/PK	COIL DRIVER NO. 2
4	K93 16TN/OR	COIL DRIVER NO. 3

6.17 **Ignition coil rail electrical connector terminal identification (harness side) - 1999 and 2000 4.0L six-cylinder engine**

than the value listed in this Chapter's Specifications, replace the wire.

7 Disconnect the ignition coil wire from the distributor cap, connect the calibrated ignition tester to the coil wire, crank the engine and check for spark **(see illustration 6.4)**. If adequate sparks occur at the coil wire, remove the distributor cap and check the distributor cap and rotor as described in Chapter 1. Replace the defective parts as necessary. Crank the engine while watching the distributor rotor. If the rotor does not turn, the distributor drive gear is stripped or the distributor shaft, timing chain/gears or camshaft are broken or damaged.

8 If no spark occurs at the coil wire, use an ohmmeter to check the resistance of the coil wire. Compare your measurement with the value listed in this Chapter's specifications. Replace the coil wire if defective.

9 Disconnect the electrical connector from the ignition coil. Turn the ignition key On and check for battery voltage at the dark green/black or dark green/orange wire terminal of the harness connector **(see illustration)**. If battery voltage is not available at the ignition coil, check the auto shut-down relay (see Chapter 12) and the circuits from the power distribution center to the coil (don't forget to check the fuses first). Also check the auto shut-down relay control circuit from the relay to the PCM for continuity.

10 Using an ohmmeter, check the primary and secondary resistance of the ignition coil (see Section 7). Replace the ignition coil if defective.

11 If the previous checks are correct, check the trigger signal from the Powertrain Control Module. Attach the leads of a test light to the two terminals of the ignition coil connector. Crank the engine. The test light should blink with the engine cranking if a trigger signal is present. If a trigger signal is not present, check the related circuits for continuity. If the circuits are good, have the PCM checked by a dealer service department or other qualified repair shop. **Note:** *Refer to the wiring diagrams at the end of Chapter 12 for wire color identification for testing and additional information on the circuits.*

1999 and 2000 models

Six-cylinder models

Refer to illustration 6.17

Caution: *Because of the ignition system design on six-cylinder models, a special set of spark plug wires must be obtained or fabricated before an ignition system check can be performed. A test set can be fabricated from bulk spark plug wire and terminals available at most auto parts stores. As an alternative, individual jumper wires can be used to ground the spark plug terminals not being tested. Whichever method you chose, make sure all the spark plug terminals are either connected to a spark plug or grounded to the engine block before cranking the engine or damage to the ignition coils may result.*

12 Disable the fuel system by removing the fuel pump relay from the power distribution center (see Chapter 12).

13 Remove the bolts retaining the ignition coil rail (see Section 7). Carefully detach the coil rail from the spark plugs, turn the coil rail over and place it in a secure location. Do not disconnect the electrical connector from the ignition coil rail.

14 Attach a calibrated ignition system tester (available at most auto parts stores) to one of the spark plug boots. Using a heavy gauge jumper wire, connect the clip on the tester to a bolt or metal bracket on the engine. Connect the remaining ignition coil spark plug terminals to the spark plugs or ground them to the engine. Crank the engine and watch the end of the tester to see if a bright blue, well-defined spark occurs (weak spark or intermittent spark is the same as no spark).

15 If spark occurs, sufficient voltage is reaching the plug to fire it (repeat the check at the remaining spark plug terminals to verify that the ignition coils are good). If the ignition system is operating properly the problem lies elsewhere; i.e. a mechanical or fuel system problem. However, the spark plugs may be fouled, so remove and check them as described in Chapter 1.

16 If no spark occurs at one or more of the coils, remove the ignition coil rail and measure the primary resistance of the ignition coils (see Section 7). Check the spring inside each spark plug boot for damage. Replace the ignition coil rail if defective.

17 If no spark occurs at all the coils, disconnect the electrical connector from the coil rail and check for battery voltage to the ignition coil rail from the auto shut-down relay and power distribution center. Attach a 12 volt test light or voltmeter to a good engine ground point and check for power at the dark green/orange wire terminal of the ignition coil rail harness connector **(see illustration)**. Battery voltage should be available with the ignition key On. If there is no battery voltage present at the coil rail connector, check the auto shut-down relay (see Chapter 12) and the circuits between the power distribution center and ignition coil rail (don't forget to check the fuses). Also check the auto shut-down relay control circuit from the relay to the PCM for continuity. **Note:** *Refer to the wiring diagrams at the end of Chapter 12 for wire color identification for testing and additional information on the circuits.*

18 If the previous checks are correct, check for a trigger signal from the Powertrain Control Module. Attach a test light to the dark green/orange wire terminal and one of the other terminals of the coil rail harness connector. Crank the engine. The test light should blink with the engine cranking if a trigger signal is present. Check for a trigger signal between the dark green/orange wire terminal and each of the other terminals in the connector. If a trigger signal is present at the coil rail connector, the Powertrain Control Module and the crankshaft position sensor are functioning properly. If a trigger signal is not present at the coil terminals check the crankshaft position sensor (see Chapter 6). If

6.20 To use a calibrated ignition tester on a 1999 or 2000 V8 model, remove an ignition coil and connect the tester to the spark plug boot, clip the tester to a convenient ground and crank the engine over - if there's enough power to fire the plug, bright blue sparks will be visible between the electrode tip and the tester body (weak sparks or intermittent sparks are the same as no sparks)

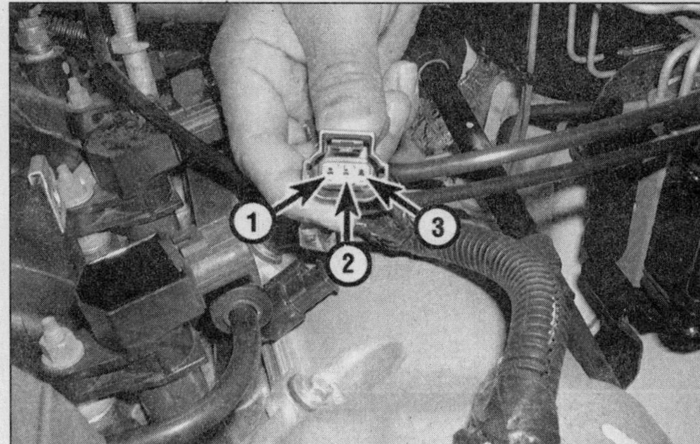

6.23 Ignition coil electrical connector terminal identification - 1999 and 2000 V8 models

1 *Coil driver*
2 *B+ (from auto shut-down relay)*
3 *Not used*

the crankshaft position sensor is good, have the PCM checked by a dealer service department or other qualified repair shop.

V8 models

Refer to illustrations 6.20 and 6.23

19 Disable the fuel system by removing the fuel pump relay from the power distribution center (see Chapter 12).

20 If the engine turns over but won't start, make sure there is sufficient secondary ignition voltage to fire the spark plug. Remove an ignition coil (see Section 7) and attach a calibrated ignition system tester (available at most auto parts stores) to the spark plug boot (be sure to reconnect the electrical connector to the coil). Connect the clip on the tester to a bolt or metal bracket on the engine **(see illustration)**. Crank the engine and watch the end of the tester to see if a bright blue, well-defined spark occurs (weak spark or intermittent spark is the same as no spark).

21 If spark occurs, sufficient voltage is reaching the plug to fire it (repeat the check at the remaining ignition coils to verify that the ignition coils are good). If the ignition system is operating properly the problem lies elsewhere; i.e. a mechanical or fuel system problem. However, the plugs themselves may be fouled, so remove and check them as described in Chapter 1.

22 If no spark occurs at one or more coil, remove the coil and check the spring inside the spark plug boot for damage. Using an ohmmeter, measure the primary and secondary resistance of the coil (see Section 7). Replace the coil if defective.

23 If no spark occurs at all the coils, check for battery voltage to the ignition coils. Attach a 12 volt test light or voltmeter to a good engine ground point. Disconnect each coil electrical connector and check for power at the dark green/orange wire terminal **(see illustration)**. Battery voltage should be available with the ignition key On. If there is no battery voltage present, check the auto shut-down relay (see Chapter 12) and the circuits between the power distribution center and ignition coil (don't forget to check the fuses). Also check the auto shut-down relay control circuit from the relay to the PCM for continuity. **Note:** *Refer to the wiring diagrams at the end of Chapter 12 for wire color identification for testing and additional information on the circuits.*

24 If battery voltage is available to the ignition coil, attach a test light to terminals no. 1 and 2 of an ignition coil harness connector **(see illustration 6.23)**. Crank the engine. The test light should blink with the engine cranking if a trigger signal is present. If a trigger signal is present at the coil, the Powertrain Control Module, the camshaft position sensor and the crankshaft position sensor are functioning properly.

25 If a trigger signal is not present at the ignition coil, refer to Chapter 6 and check the camshaft position sensor and crankshaft position sensor. If the camshaft position sensor and crankshaft position sensor are good, have the PCM checked by a dealer service department or other qualified repair shop.

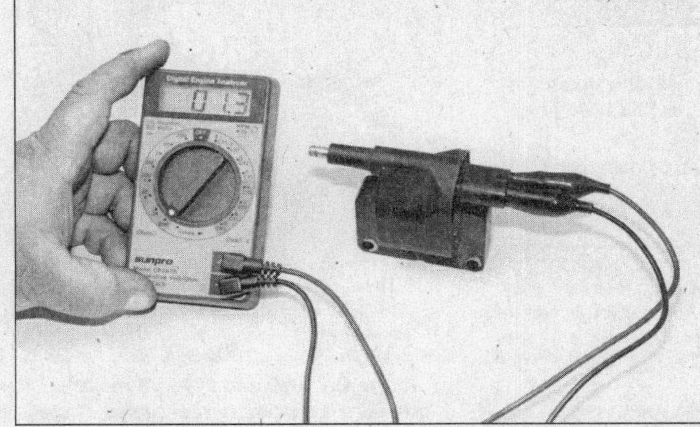

7.3 Connect an ohmmeter to the two primary wiring connector terminals and check the coil primary resistance (1998 and earlier models)

7 Ignition coil - check and replacement

1998 and earlier models

Check

Refer to illustrations 7.3 and 7.4

1 Remove the ignition coil (see Steps 6 and 7).

2 Clean the outer case and check it for cracks and other damage. Clean the coil primary terminals and check the coil tower terminal for corrosion. Clean it with a wire brush if any corrosion is found. It is essential for proper operation of the ignition system that all coil terminals and wires be kept clean and dry.

3 Check the coil primary resistance by attaching the leads of an ohmmeter to the two terminals of the connector **(see illustration)**.

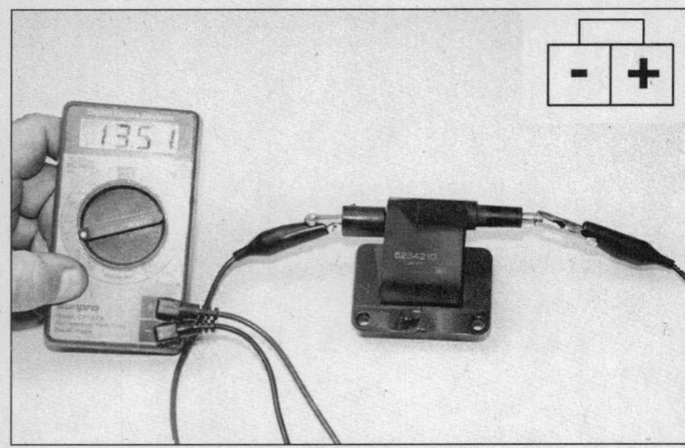

7.4 Connect an ohmmeter to the positive primary wiring connector terminal and the high-tension terminal and check the coil secondary resistance (1998 and earlier models)

7.7 Remove the ignition coil mounting bolts and detach the coil from the bracket (1998 and earlier V8 model shown)

7.11a Ignition coil terminal identification - 1999 and 2000 4.0L six-cylinder engine

1 *Coil driver no. 1*
2 *B+ (from auto shut-down relay)*
3 *Coil driver no. 2*
4 *Coil driver no. 3*

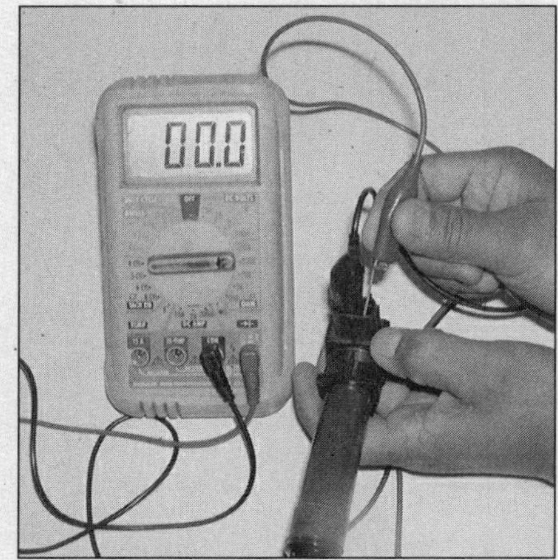

7.11b On 1999 and 2000 V8 models, measure the primary resistance of the ignition coil across the two terminals of the coil connector

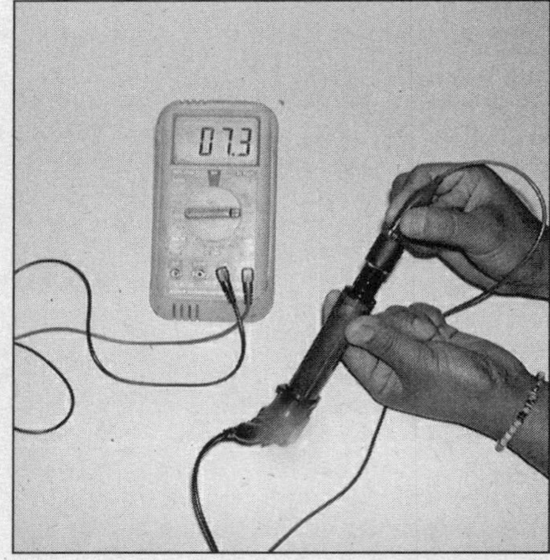

7.12 To measure the secondary resistance on 1999 and 2000 V8 models, connect an ohmmeter to one of the primary terminals and the spark plug terminal

4 Check the coil secondary resistance by connecting one of the ohmmeter leads to the positive terminal of the connector and the other ohmmeter lead to the high-tension terminal **(see illustration)**.

5 Compare your measurements to the resistance values listed in this Chapter's Specifications. If either measured resistance value is not as specified, replace the ignition coil.

Replacement

Refer to illustration 7.7

6 Disconnect the primary wiring electrical connector and remove the high-tension lead from the coil.

7 Remove the ignition coil mounting bolts and detach it from the coil bracket **(see Illustration)**.

8 Installation is the reverse of removal.

1999 and 2000 models

Check

Refer to illustrations 7.11a, 7.11b and 7.12

9 Remove the ignition coil (see Replacement procedure).

10 Clean the outer case and check it for cracks and other damage. Clean the coil primary terminals and check the spring and spark plug boot for damage. If either the spring or spark plug boot is damaged, the coil must be replaced.

11 Check the coil primary resistance according to engine type as follows **(see illustrations)**:

a) *On six-cylinder models, connect an ohmmeter to terminals no. 2 and 1; 2 and 3; 2 and 4; of the coil rail connector. Each measurement is the primary resistance of one of the three coils in the coil rail assembly.*
b) *On V8 models, connect an ohmmeter to the two terminals of the ignition coil connector.*

12 On V8 models, check the coil secondary resistance by connecting the ohmmeter to either of the primary terminals of the connector and the high-tension terminal **(see illustration)**. **Note:** *It is not possible to measure the secondary resistance on a six-cylinder model coil.*

13 Compare your measurements to the resistance values listed in this Chapter's Specifications. If either measured resistance value is not as specified, replace the ignition coil. **Note:** *On six-cylinder models, the entire coil rail assembly must be replaced if one coil is defective.*

Replacement

Refer to illustrations 7.14, 7.16 and 7.23

Six-cylinder models

Note: *The ignition coil rail assembly consists of the ignition coils, spark plug boots, springs, electrical connector and internal wiring. None of the components are serviced separately. The entire assembly is removed and replaced as a unit.*

14 Remove the coil rail mounting bolts **(see illustration)**.

15 Carefully work the spark plug boots off the spark plugs by alternately prying at one end of the coil rail, then the other. If difficulty is encountered, twist the spark plug boots with a spark plug boot removal tool to loosen the seal.

16 Position the coil rail to access the electrical connector. Disconnect the electrical

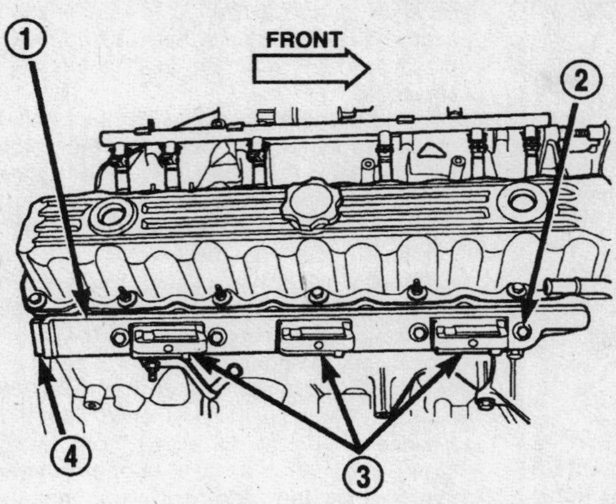

7.14 Ignition coil rail installation details - 1999 and 2000 4.0L six-cylinder models

1 Coil rail
2 Mounting bolts (1 of 4)
3 Ignition coils
4 Electrical connector

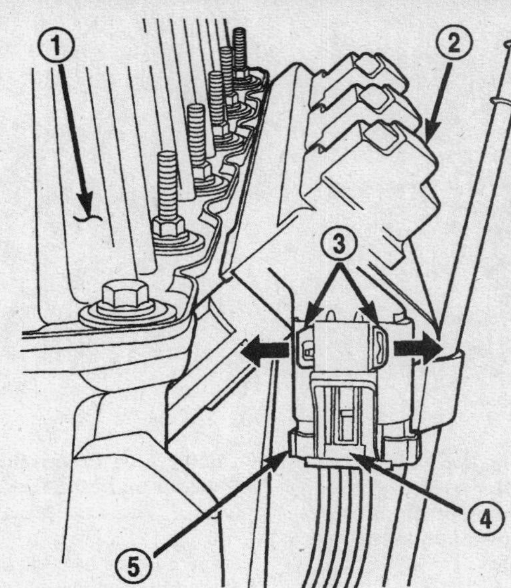

7.16 Ignition coil rail electrical connector details - 1999 and 2000 4.0L six-cylinder models

1 Valve cover
2 Coil rail
3 Slide tab
4 Release lock
5 Electrical connector

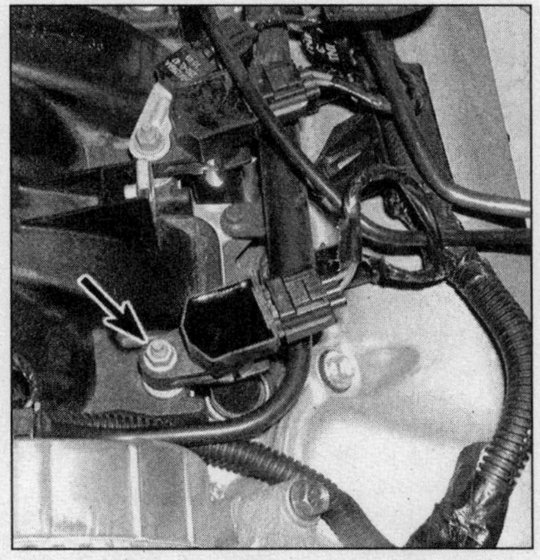

7.23 Ignition coil mounting stud/nut (arrow) - 1999 and 2000 V8 models

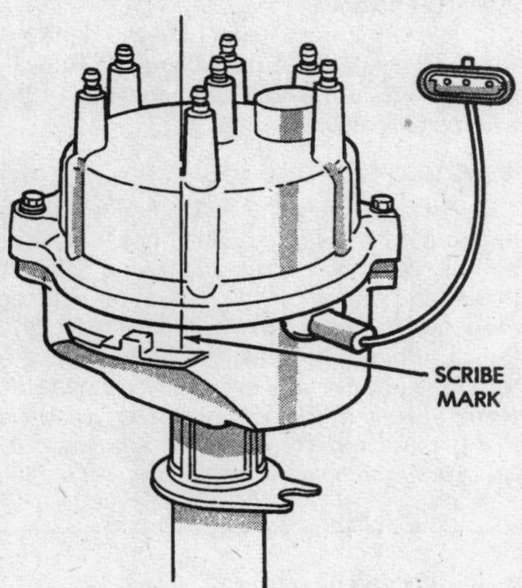

8.3 Locate the number one spark plug wire terminal and mark the location on the distributor body

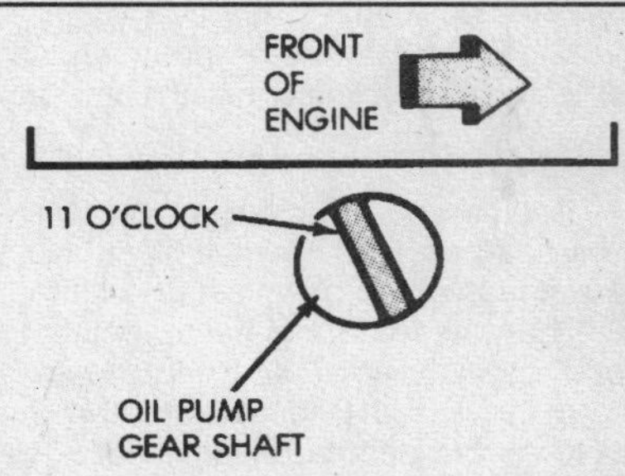

8.8 Before installing the distributor, position the slot in the oil pump drive gear as shown

5

connector from the coil rail by first pushing the slide tab up, then press the release lock and pull the connector off **(see illustration)**.

17 Remove the coil rail assembly.

18 When installing the coil rail, position all the spark plug boots over the spark plugs and press the coil rail down until the boots are completely seated.

19 Install the mounting bolts and tighten them in stages until secure.

20 Connect the electrical connector and slide the tab down until locked in place.

V8 models

21 Remove the air intake duct/resonator, if necessary.

22 Disconnect the electrical connector from the coil.

23 Remove the nut from the ignition coil mounting stud **(see Illustration)**. Carefully pull the coil up with a twisting motion.

24 Installation is the reverse of removal.

8 Distributor - removal and installation

Six-cylinder engine

Removal

Refer to illustration 8.3

1 Position the engine with number one cylinder at TDC on the compression stroke (see Chapter 2A).

2 Disconnect the cable from the negative battery terminal.

3 On 1993 and 1994 models, locate the number one spark plug wire terminal on the distributor cap and apply an alignment mark on the distributor body below the number one spark plug terminal **(see illustration)**.

4 Disconnect the coil wire from the distributor cap. Loosen the distributor cap mounting screws, remove the cap from the distributor and position the cap (with the spark plug wires attached) aside.

5 Disconnect the electrical connector from the distributor.

6 Remove the distributor hold-down bolt and clamp. Pull the distributor straight up and out of the engine block. Remove and discard the gasket.

Installation

Refer to illustrations 8.8, 8.10a, 8.10b, 8.13, 8.14a and 8.14b

7 If the crankshaft has been moved while the distributor is out, the number one piston must be repositioned at TDC. This can be done by removing the number one cylinder spark plug and feeling for compression pressure at the number one spark plug hole as the crankshaft is rotated. Once compression is felt, continue rotating the crankshaft until the mark on the crankshaft damper is aligned with the zero or TDC mark on the timing indicator.

8 Check the position of the slot in the oil pump drive. The slot should be pointing at approximately the eleven o'clock position **(see Illustration)**. If it isn't, insert a large flat-

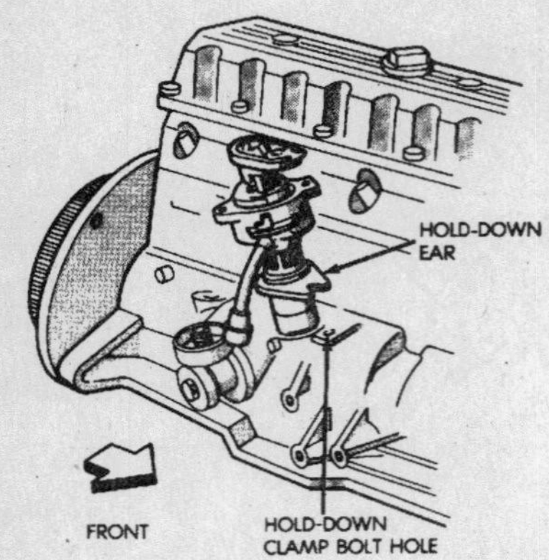

8.10a When installing the distributor, position the distributor over the hold-down bolt hole in the engine block, align the slot in the hold-down flange with the bolt hole . . .

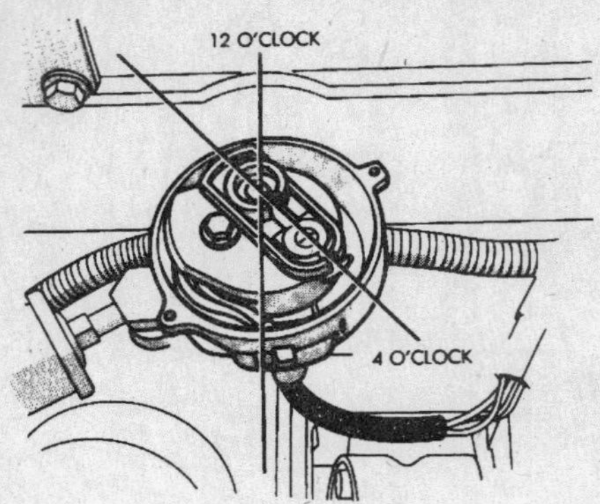

8.10b . . . and turn the rotor to point the rotor tip at the four o'clock position

blade screwdriver into the slot and rotate it to the proper position.

9 Clean the distributor mounting surface on the engine block and install a new gasket.

1993 and 1994 models

10 Position the distributor over the hold-down bolt hole in the engine block and align the slot in the hold-down flange with the bolt hole. Turn the rotor until the tip points to the four o'clock position **(see illustrations)**.

11 Insert the distributor into the engine block. As the distributor gear engages the rotor will turn due to the helical cut gears. When the distributor is fully seated, the rotor should be pointing at the alignment mark made in Step 3. If the rotor isn't aligned with the mark when the distributor is fully seated and the hold-down bolt hole is centered in the slot in the hold-down flange, remove the distributor and repeat the procedure. **Note:** *It may be necessary to rotate the rotor slightly to engage the distributor shaft with the slot in the oil pump drive.*

12 Install the hold-down bolt and tighten it to the torque listed in this Chapter's Specifications. The remainder of installation is the reverse of removal.

1995 and later models

13 Remove the camshaft position sensor and rotate the distributor shaft until the alignment hole in the plastic ring aligns with the hole in the bottom of the distributor housing **(see Illustration)**. Insert a 3/16-inch drift punch through the holes to lock the distributor in the number one cylinder firing position. **Note:** *Several alignment holes are provided in the plastic ring for different applications. Make sure the hole for the six-cylinder engine (as shown in the illustration) is the one you're using. On new replacement distributors, a plastic pin installed by the manufacturer should be in place.*

14 Install the rotor on the distributor shaft. Insert the distributor into the engine block with the slot in the distributor base positioned at one o'clock. Lower the distributor into position, as the distributor gear engages the camshaft gear the distributor will rotate clockwise. When fully seated, the mounting bolt hole should be centered in the slot **(see illustrations)**. If necessary, rotate the distributor slightly to engage the oil pump.

15 When the distributor is correctly installed; the engine is positioned with number one cylinder at TDC on the compression stroke, the rotor is pointing at approximately five o'clock, the alignment pin is installed through the alignment holes and the mounting bolt hole is centered in the base slot. If all the conditions are correct, install the mounting clamp and bolt and tighten the bolt to the torque listed in this Chapter's Specifications.

16 Remove the alignment tool from the distributor. If a new distributor was installed, pull the plastic pin from the bottom of the housing. Install the camshaft position sensor and the rotor.

17 The remainder of installation is the reverse of removal.

V8 engines

Removal

Refer to illustrations 8.22, 8.23 and 8.24

18 Position the engine with number one cylinder at TDC on the compression stroke (see Chapter 2B).

19 Disconnect the cable from the negative battery terminal. Remove the air intake duct/resonator.

20 Disconnect the electrical connector from the distributor.

21 Disconnect the coil wire from the distributor cap. Loosen the distributor cap mount-

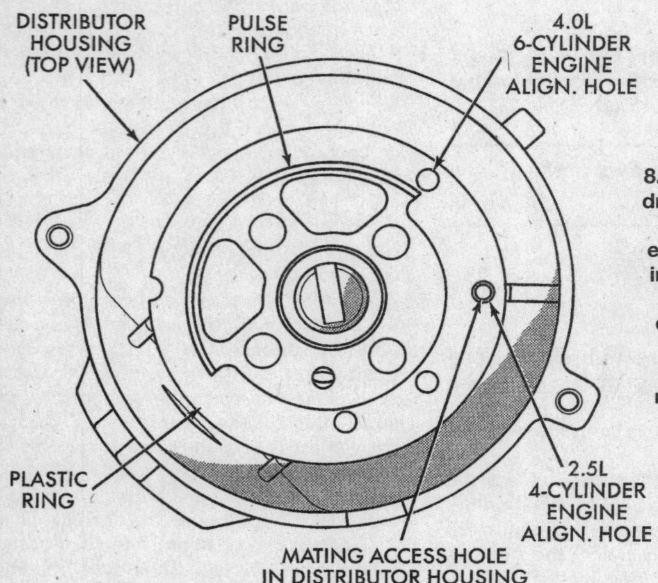

8.13 Insert a 3/16-inch drift punch through the 4.0L six-cylinder engine alignment hole in the plastic ring and into the hole in the distributor housing - this locks the distributor in the number one cylinder firing position

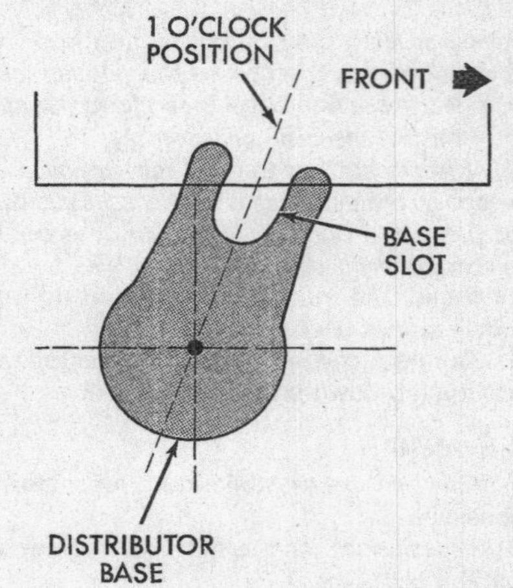

8.14a Insert the distributor into the engine block with the base slot at one o'clock . . .

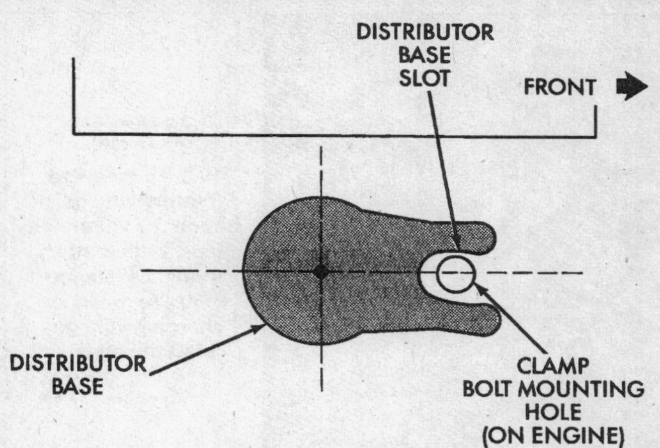

8.14b . . . as the camshaft gear is engaged, the distributor will rotate clockwise (due to the helical cut gears) - when the distributor is seated the mounting bolt hole should be centered in the base slot

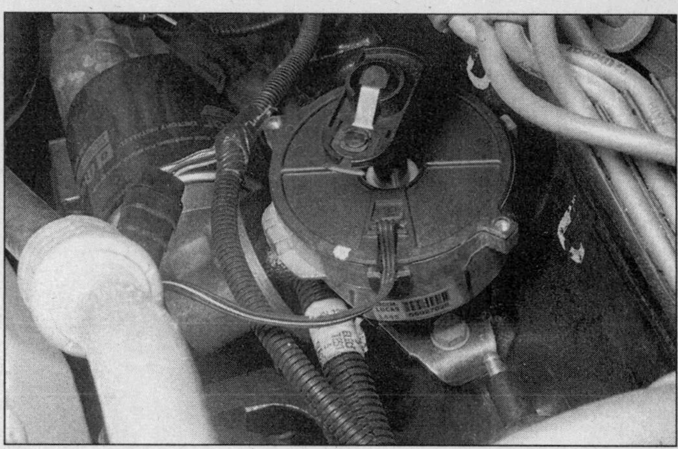

8.22 With number one cylinder at TDC on the compression stroke, the rotor should be pointing at the camshaft position sensor alignment mark - apply a dab of white paint on the mark for better visibility

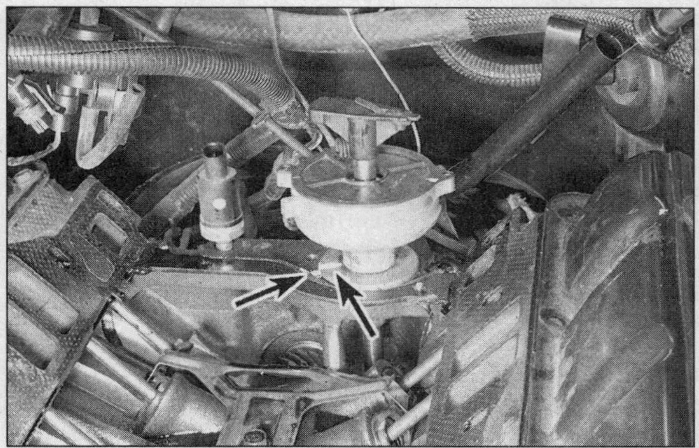

8.23 Mark the position of the distributor body to the engine (arrows) (intake manifold removed for clarity)

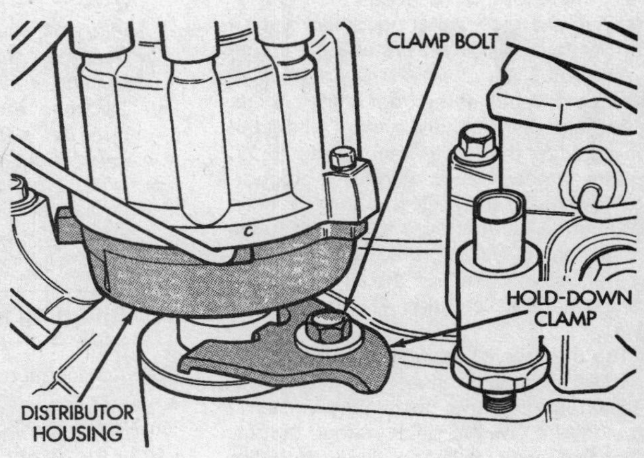

8.24 Remove the distributor hold-down clamp and bolt

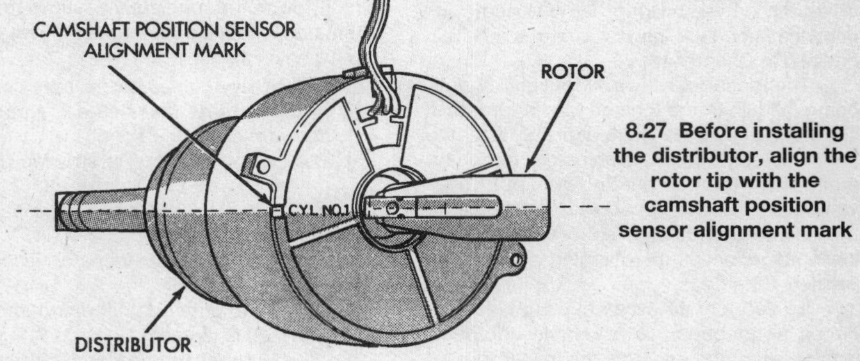

8.27 Before installing the distributor, align the rotor tip with the camshaft position sensor alignment mark

ing screws, remove the cap from the distributor and position the cap (with the spark plug wires attached) aside.

22 The distributor rotor should be pointing at the camshaft position sensor alignment mark, if necessary apply a paint mark on the edge of the distributor body directly below the rotor tip and in line with it **(see Illustration)**.

23 Mark the position of the distributor base to the engine to ensure the distributor can be re-installed in exactly the same position as originally installed **(see Illustration)**.

24 Remove the distributor hold-down bolt and pull out the distributor **(see illustration)**. Remove and discard the O-ring.

Installation

Refer to illustration 8.27

25 If the crankshaft has been moved while the distributor is out, the number one piston must be repositioned at TDC. This can be done by feeling for compression pressure at the number one spark plug hole as the crankshaft is rotated. Once compression is felt, continue rotating the crankshaft until the mark on the crankshaft damper is aligned with the zero or TDC mark on the timing indicator.

26 Install a new O-ring on the distributor housing.

27 Turn the rotor until it aligns with the camshaft position sensor alignment mark **(see Illustration)**. **Note:** *If the engine is being reassembled (as in an engine overhaul), make sure the distributor drive gear is properly installed* (see Chapter 2B).

28 Insert the distributor into the engine block and align the marks made in Step 23. The distributor must be installed the same

5

position it was in before removal. Make sure the rotor remains aligned with the camshaft position sensor alignment mark.

29 Install the distributor hold-down clamp and tighten the bolt to the torque listed in this Chapter's Specifications.

30 The remainder of installation is the reverse of removal. **Note:** *If the alignment marks are lost or driveability symptoms appear after installing the distributor, take the vehicle to a dealer service department or other properly equipped repair facility and have the fuel synchronization reset with special equipment.*

9 Charging system - general information and precautions

The charging system includes the alternator, a charge indicator light, the battery, the Powertrain Control Module (PCM), a fusible link and the wiring between all the components. The charging system supplies electrical power for the ignition system, the lights, the radio, etc. The alternator is driven by a serpentine drivebelt at the front of the engine.

The alternator control system within the PCM regulates the DC current output of the alternator in accordance with driving conditions. Depending upon electric load, vehicle speed, battery temperature and accessories (air conditioning system, radio, cruise control etc.), the system will adjust the amount of DC current generated, creating less load on the engine.

The purpose of the voltage regulator is to limit the alternator voltage output to a preset value. This prevents power surges, circuit overloads, etc., during peak voltage output. The voltage regulator is contained within the PCM and in the event of failure, the PCM must be replaced.

The charging system doesn't ordinarily require periodic maintenance. However, the drivebelt, battery, wiring harness and connections should be inspected at the intervals outlined in Chapter 1.

The dashboard warning light should come ON when the ignition key is turned to ON, but it should go off immediately after the engine is started. If it remains on, there is a malfunction in the charging system. Some vehicles are also equipped with a voltmeter. If the voltmeter indicates abnormally high or low voltage, check the charging system (see Section 10).

Be very careful when making electrical circuit connections to a vehicle equipped with an alternator and note the following:

a) *When reconnecting wires to the alternator from the battery, be sure to note the polarity.*
b) *Before using arc welding equipment to repair any part of the vehicle, disconnect the wires from the alternator and the battery terminals.*
c) *Never start the engine with a battery charger connected.*
d) *Always disconnect both battery cables before using a battery charger.*
e) *The alternator is turned by an engine drivebelt which could cause serious injury if your hands, hair or clothes become entangled in it with the engine running.*
f) *Because the alternator is connected directly to the battery, it could arc or cause a fire if overloaded or shorted out.*
g) *Wrap a plastic bag over the alternator and secure it with rubber bands before steam-cleaning the engine.*

10.3 To measure battery voltage, attach the voltmeter leads to the battery terminals (engine OFF) - to measure charging voltage, start the engine

10 Charging system - check

Refer to illustration 10.3

Note: *These vehicles are equipped with an On-Board Diagnostic (OBD) system that is useful for detecting charging system problems. Refer to Chapter 6 for the list of diagnostic codes and procedures for obtaining the codes.*

1 If a malfunction occurs in the charging circuit, do not immediately assume that the alternator is causing the problem. First check the following items:

a) *The battery cables where they connect to the battery. Make sure the connections are clean and tight.*
b) *The battery electrolyte specific gravity (by observing the charge indicator on the battery). If it is low, charge the battery.*
c) *Check the external alternator wiring and connections.*
d) *Check the drivebelt condition and tension* (see Chapter 1).
e) *Check the alternator mounting bolts for tightness.*
f) *Run the engine and check the alternator for abnormal noise.*

2 Using a voltmeter, check the battery voltage with the engine off. It should be approximately 12.4 to 12.6 volts with a fully charged battery.

3 Start the engine and check the battery voltage again. It should now be greater than the voltage recorded in Step 2, but not more than 15 volts **(see illustration)**.

4 If the indicated voltage reading is less or more than the specified charging voltage, have the charging system checked at a dealer service department or other properly equipped repair facility. The voltage regulator on these models is contained within the PCM and it cannot be adjusted, removed or tampered with in any way.

11 Alternator - removal and installation

Refer to illustrations 11.2a, 11.2b, 11.4a, 11.4b, 11.4c and 11.4d

1 Disconnect the cable from the negative battery terminal.

2 Disconnect the output wire and the field terminals from the alternator **(see illustrations)**.

3 Loosen the tension on the drivebelt, then remove the serpentine drivebelt (see Chapter 1).

4 Remove the mounting nuts/bolts and separate the alternator from the engine **(see illustrations)**.

5 If you are replacing the alternator, take the old one with you when purchasing a replacement unit. Make sure the new/rebuilt unit looks identical to the old alternator. Look at the terminals - they should be the same in number, size and location as the terminals on the old alternator. Finally, look at the identification numbers - they will be stamped into the housing or printed on a tag attached to the housing. Make sure the numbers are the same on both alternators.

6 Many new/rebuilt alternators do not have a pulley installed, so you may have to switch the pulley from the old unit to the new/rebuilt one. When buying an alternator, find out the shop's policy regarding pulleys; some shops will perform this service free of charge.

7 Installation is the reverse of removal. Tighten the mounting bolts to the torque listed in this Chapter's Specifications.

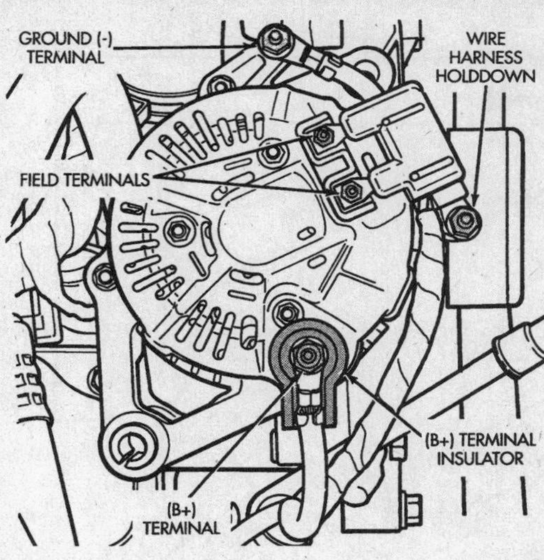

11.2a Alternator electrical connections - 1998 and earlier models

11.2b Disconnect the alternator electrical connections - 1999 and 2000 models

A B+ terminal (output) B Field terminals

11.4a Remove the alternator mounting nut/bolts (arrows) and remove the alternator from the bracket (5.2L/5.9L V8 model shown)

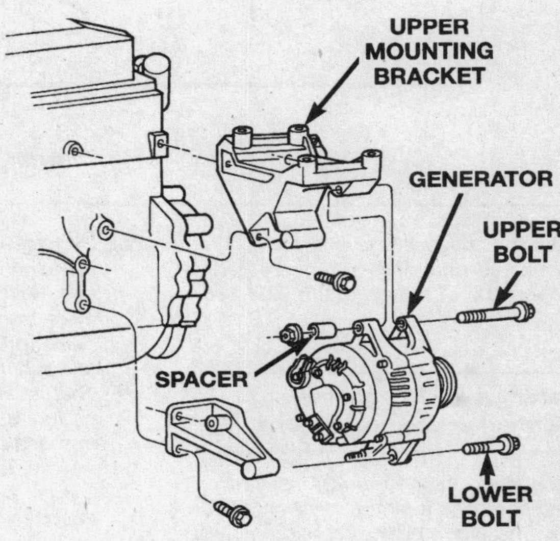

11.4b Alternator installation details - 1998 and earlier 4.0L six-cylinder models

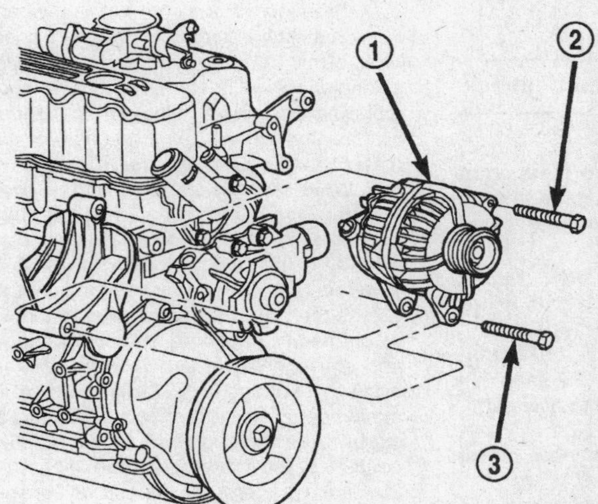

11.4c Alternator installation details - 1999 and 2000 4.0L six-cylinder models

1 Alternator 2 Upper bolt 3 Lower bolt

11.4d Alternator mounting bolts (arrows) - 1999 and 2000 4.7L V8 models

5

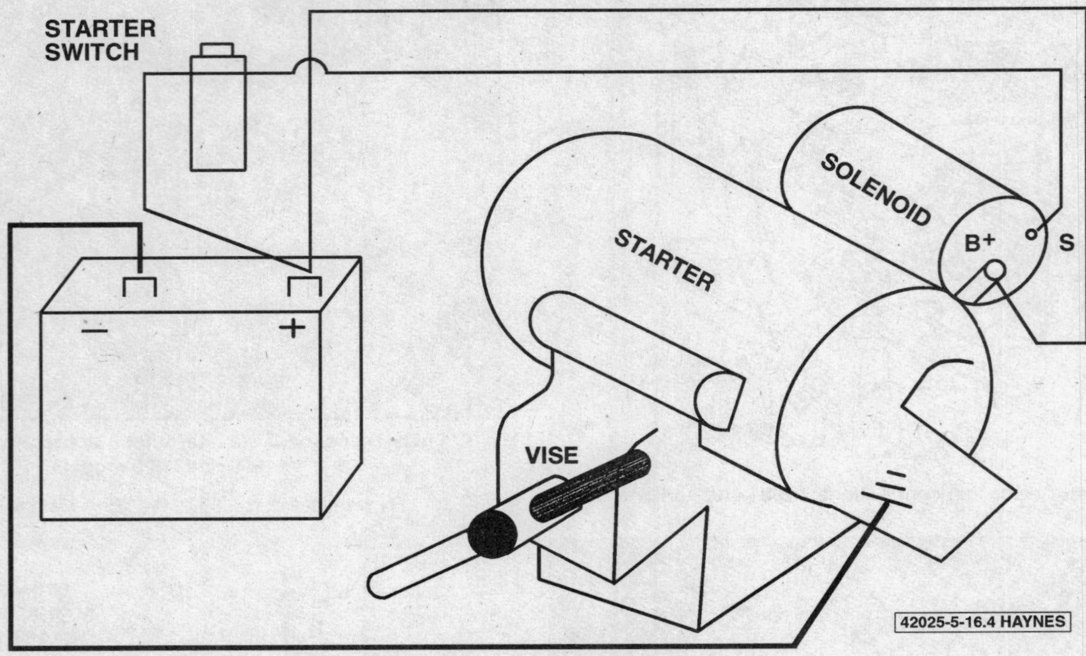

13.4 Starter motor bench testing details

8 Install the drivebelt (see Chapter 1).

9 Check the charging voltage to verify proper operation of the alternator (see Section 10).

12 Starting system - general information and precautions

The starter motor assembly installed on these engines uses a planetary gear reduction drive. This type starter motor assembly is serviced as a complete unit. If any component of the starter motor fails, including the solenoid, the entire assembly must be replaced.

The sole function of the starting system is to turn over the engine quickly enough to allow it to start. The starting system consists of the battery, starter relay, starter motor assembly and the wiring connecting the components.

The starter motor assembly is installed on the lower part of the engine, next to the transmission bellhousing.

When the ignition key is turned to the START position, the starter solenoid is actuated through the starter control circuit which includes a starter relay located in the power distribution center. The starter solenoid then connects the battery to the starter motor. The battery supplies the electrical energy to the starter motor, which does the actual work of cranking the engine.

Always observe the following precautions when working on the starting system:

a) *Excessive cranking of the starter motor can overheat it and cause serious damage. Never operate the starter motor for more than 15 seconds at a time without pausing to allow it to cool for at least two minutes.*
b) *The starter is connected directly to the battery and could arc or cause a fire if mishandled, overloaded or shorted.*
c) *Always detach the cable from the negative terminal of the battery before working on the starting system.*

13 Starter motor and circuit - check

Refer to illustration 13.4

1 If a malfunction occurs in the starting circuit, do not immediately assume that the starter is causing the problem. First, check the following items:

a) *Make sure the battery cable clamps, where they connect to the battery, are clean and tight.*
b) *Check the condition of the battery cables* (see Section 4). *Replace any defective battery cables with new parts.*
c) *Test the condition of the battery* (see Section 3). *If it does not pass all the tests, replace it with a new battery.*
d) *Check the starter motor wiring and connections.*
e) *Check the starter motor mounting bolts for tightness.*
f) *Check the related fuses in the engine compartment fuse box* (see Chapter 12). *If they're blown, determine the cause and repair the circuit.*
g) *Check the ignition switch circuit for correct operation* (see Chapter 12).
h) *Check the starter relay* (see Chapter 12).
i) *Check the operation of the clutch safety switch* (see Chapter 8) *or the Park/Neutral position switch* (see Chapter 7B). *These systems must operate correctly to provide battery voltage to the starter solenoid.*

2 If the starter does not activate when the ignition switch is turned to the start position, check for battery voltage to the starter solenoid. This will determine if the solenoid is receiving the correct voltage from the ignition switch. Connect a 12-volt test light or a voltmeter to the starter solenoid positive terminal. While an assistant turns the ignition switch to the start position, observe the test light or voltmeter. The test light should shine brightly or battery voltage should be indicated on the voltmeter. If voltage is not available to the starter solenoid, refer to the wiring diagrams in Chapter 12 and check the fuses and starter relay in series with the starting system. If voltage is available but there is no movement from the starter motor, remove the starter from the engine (see Section 14) and bench test the starter (see Step 4).

3 If the starter turns over slowly, check the starter cranking voltage and the current draw from the battery. This test must be performed with the starter assembly on the engine. Crank the engine over (for 10 seconds or less) and observe the battery voltage. It

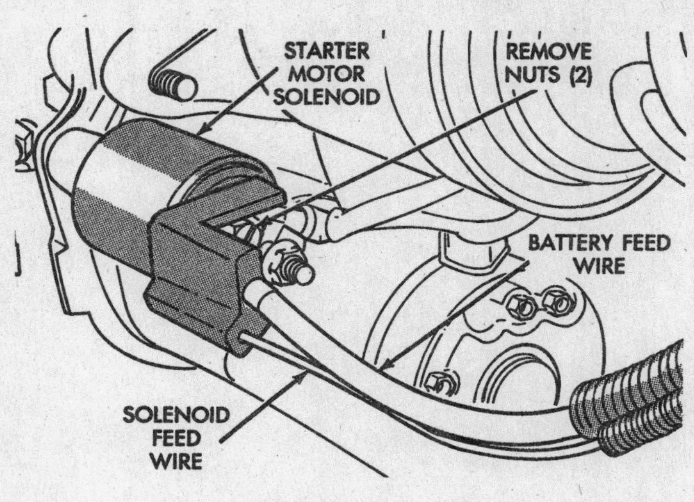

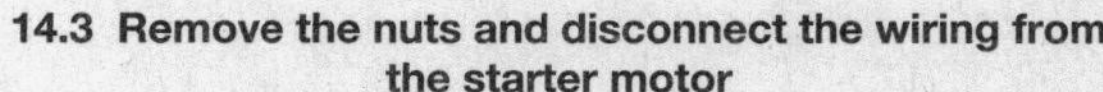
14.3 Remove the nuts and disconnect the wiring from the starter motor

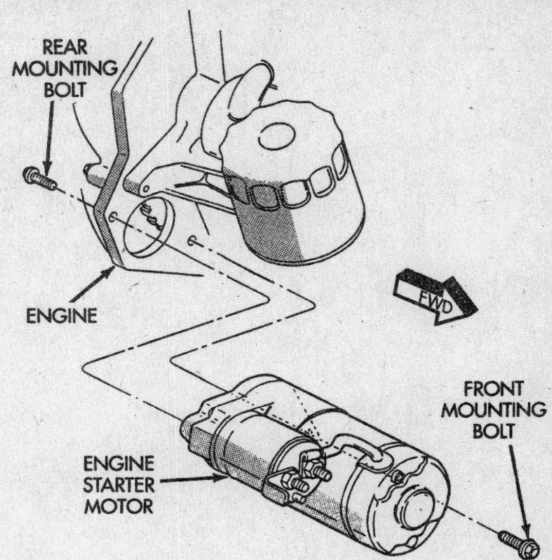

14.4 Starter motor installation details - typical

should not drop below 8.5 volts. Also, observe the current draw using an amp meter. Typically a starter should not exceed 300 amps (slightly less for the six-cylinder engine). If the starter motor amperage draw is excessive, have it tested by a dealer service department or other qualified repair shop. There are several conditions that may affect the starter cranking potential. The battery must be in good condition and the battery cold-cranking rating must not be under-rated for the particular application. Be sure to check the battery specifications carefully. The battery terminals and cables must be clean and not corroded. Also, in cases of extreme cold temperatures, make sure the battery and/or engine block is warmed before performing the tests.

4 If the starter is receiving voltage but does not activate, remove and check the starter motor assembly on the bench. Most likely the solenoid is defective. In some rare cases, the engine may be seized so be sure to try and rotate the crankshaft pulley (see Chapter 2A, 2B or 2C) before proceeding. With the starter assembly mounted in a vise on the bench, install one jumper cable from the positive terminal of a test battery to the B+ terminal on the starter. Install another jumper cable from the negative terminal of the battery to the body of the starter **(see illustration)**. Install a starter switch and apply battery voltage to the solenoid S terminal (for 10 seconds or less) and observe the solenoid plunger, shift lever and overrunning clutch extend and rotate the pinion drive. If the pinion drive extends but does not rotate, the solenoid is operating but the starter motor is defective. If there is no movement but the solenoid clicks, the solenoid and/or the starter motor is defective. If the solenoid plunger extends and rotates the pinion drive, the starter assembly is operating properly.

14 Starter motor - removal and installation

Refer to illustrations 14.3 and 14.4

1 Detach the cable from the negative terminal of the battery.

2 On 4WD models, remove the front axle skid plate (if equipped).

3 Disconnect the wires from the terminals on the starter motor solenoid **(see illustration)**. Disconnect any clips securing the wiring to the starter. **Note:** *It may be easier to disconnect the battery cable from the starter after removing the mounting bolts and lowering the starter. Be sure to support the starter, do not allow it to hang by the wiring.*

4 Remove the mounting bolts and detach the starter **(see illustration)**.

5 Installation is the reverse of removal.

Notes

Chapter 6
Emissions and engine control systems

Contents

1 General information

Refer to illustrations 1.1a, 1.1b, 1.1c, 1.3a, 1.3b, 1.6a and 1.6b

To prevent pollution of the atmosphere from incompletely burned and evaporating gases, and to maintain good driveability and fuel economy, a number of emission control systems are incorporated **(see illustrations)**. They include the:

Catalytic converter
Crankcase ventilation system
Electronic engine control system
Evaporative emissions control system
Exhaust Gas Recirculation (EGR) system

All of these systems are linked, directly or indirectly, to the emission control system.

1.1a Typical emission and engine control system components - 4.0L six-cylinder models

1. *Crankcase ventilation system orifice*
2. *Manifold Absolute Pressure (MAP) sensor (1995 and earlier)*
3. *Intake Air Temperature (IAT) sensor (on intake manifold)*
4. *EVAP purge solenoid*
5. *Throttle body (with throttle position sensor, idle air control valve and on 1996 and later models, manifold absolute pressure sensor)*
6. *Engine coolant temperature sensor (on thermostat housing)*
7. *Powertrain Control Module (PCM)*
8. *Camshaft position sensor (inside distributor)*

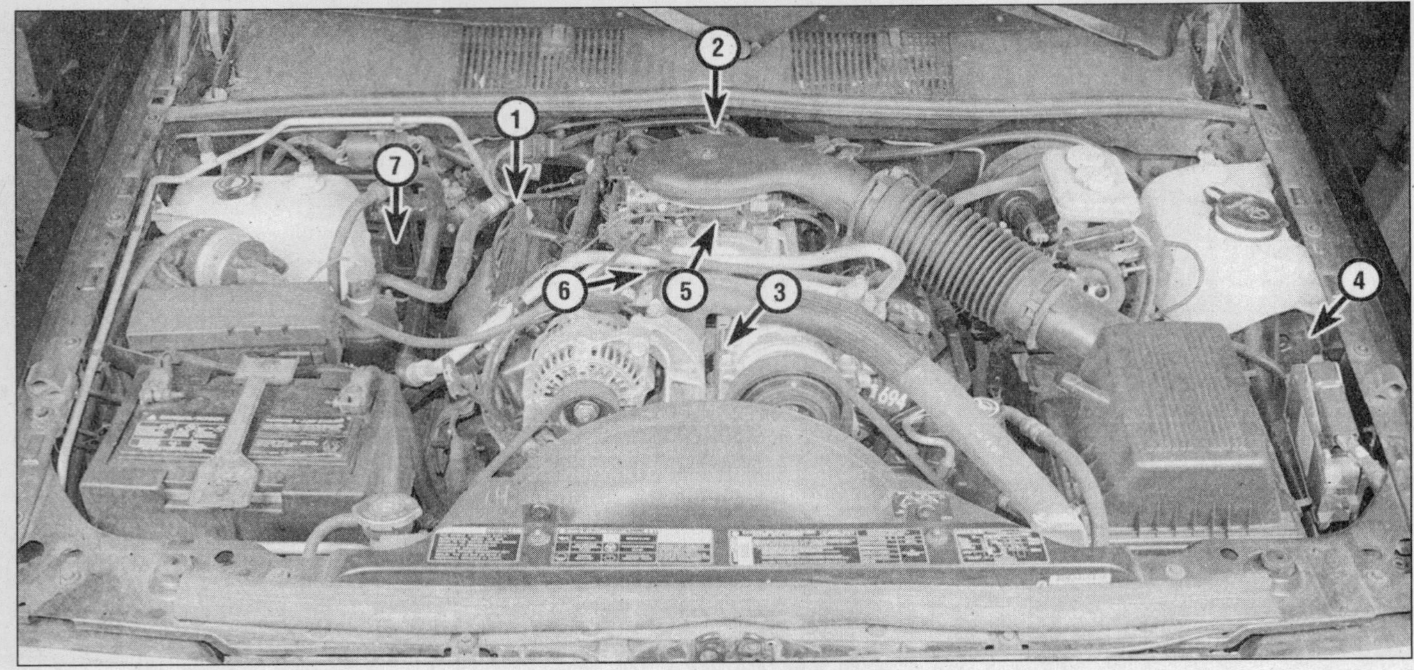

1.1b Typical emission and engine control system components - 5.2L/5.9L V8 models

1. *PCV valve*
2. *Camshaft position sensor (inside distributor)*
3. *Engine coolant temperature sensor*
4. *EVAP purge solenoid*
5. *Throttle body (with throttle position sensor, idle air control valve and manifold absolute pressure sensor)*
6. *Intake Air Temperature (IAT) sensor*
7. *Powertrain Control Module (PCM)*

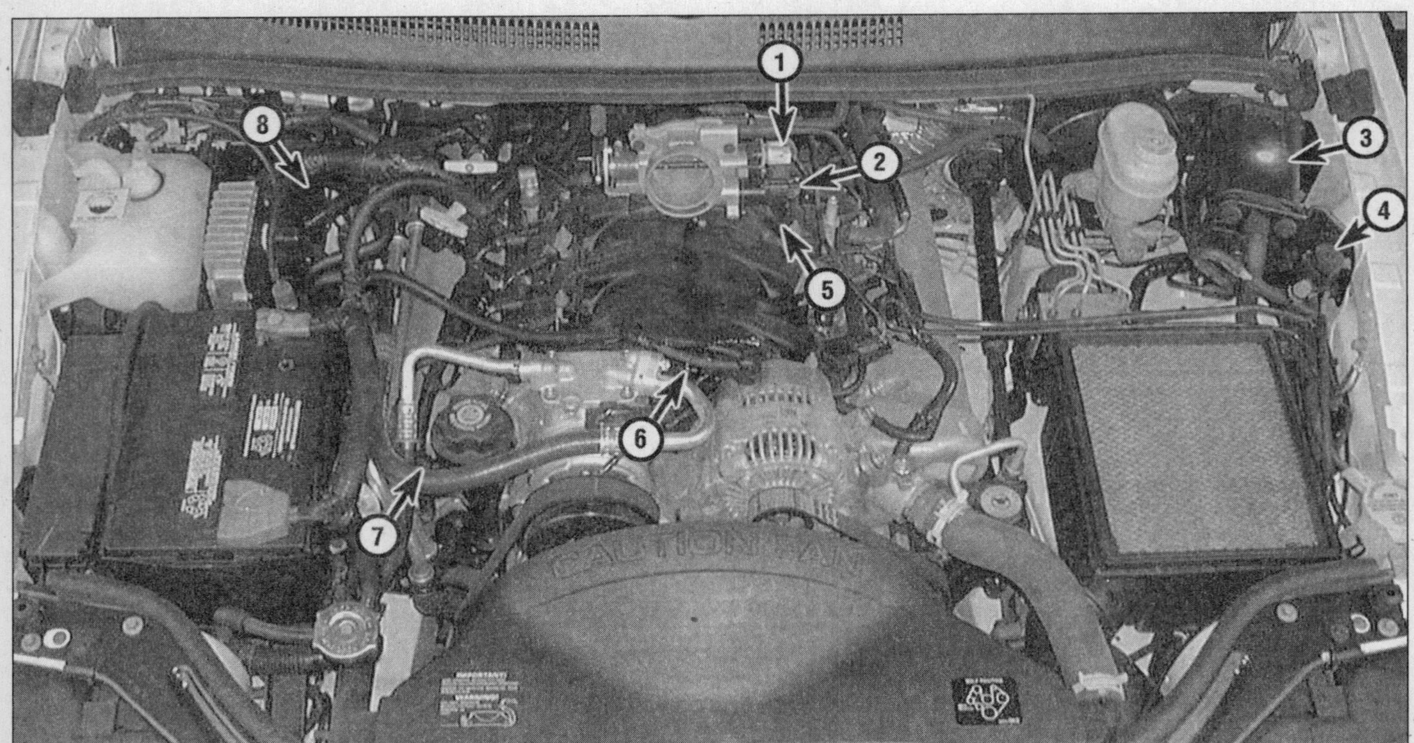

1.1c Typical emission and engine control system components - 4.7L V8 models

1. *Throttle Position Sensor (TPS)*
2. *Idle Air Control (IAC) valve*
3. *EVAP canister*
4. *EVAP purge solenoid*
5. *Intake Air Temperature (IAT) sensor*
6. *Manifold Absolute Pressure (MAP) sensor*
7. *PCV valve*
8. *Powertrain Control Module (PCM)*

The Sections in this Chapter include general descriptions, checking procedures within the scope of the home mechanic and component replacement procedures (when possible) for each of the systems listed above. Before assuming that an emissions control system is malfunctioning, check the fuel and ignition systems carefully. The diagnosis of some emission control devices requires specialized tools, equipment and training. If checking and servicing become too difficult or if a procedure is beyond your ability, consult a dealer service department or other qualified repair shop. Remember, the most frequent cause of emissions problems is simply a loose or broken vacuum hose or wire, so always check the hose and wiring connections first **(see illustrations)**.

This doesn't mean, however, that emission control systems are particularly difficult to maintain and repair. You can quickly and easily perform many checks and do most of the regular maintenance at home with common tune-up and hand tools. **Note:** *Because of a federally mandated warranty which covers the emission control system components, check with your dealer about warranty coverage before working on any emissions-related systems. Once the warranty has expired, you may wish to perform some of the component checks and/or replacement procedures in this Chapter to save money.*

Pay close attention to any special precautions outlined in this Chapter. It should be noted that the illustrations of the various systems may not exactly match the system installed on the vehicle you're working on because of changes made by the manufacturer during production or from year-to-year.

A Vehicle Emissions Control Information (VECI) label is located in the engine compartment **(see illustrations)**. This label contains important emissions specifications and adjustment information, as well as a vacuum hose schematic with emissions components identified. When servicing the engine or emissions systems, the VECI label in your particular vehicle should always be checked for up-to-date information.

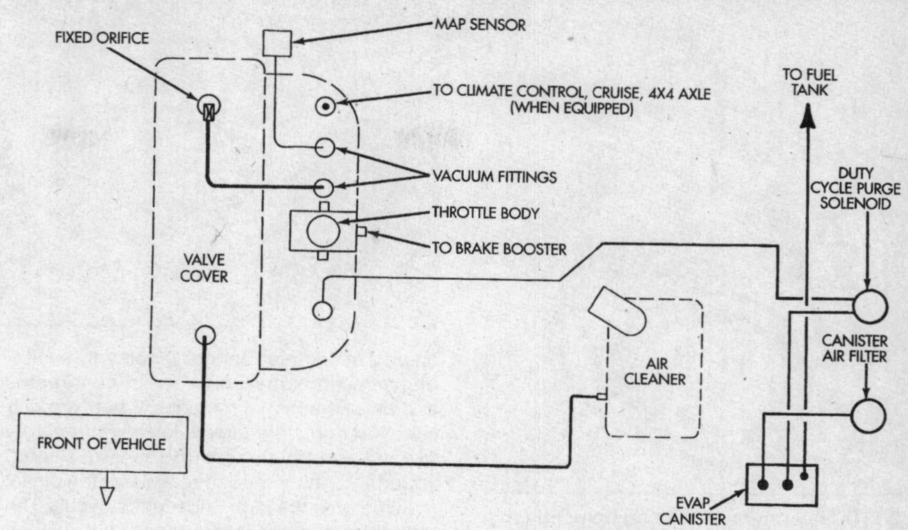

1.3a Typical vacuum schematic - 4.0L six-cylinder engine

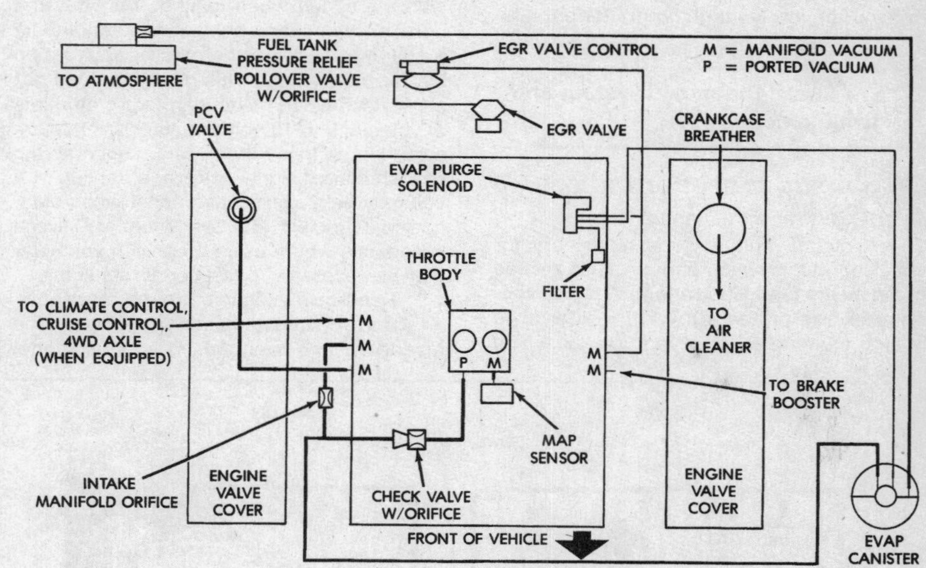

1.3b Typical vacuum schematic - 5.2L/5.9L V8 engines

1.6a The Vehicle Emission Control Information (VECI) label, located in the engine compartment, contains information on the emission devices installed on your vehicle

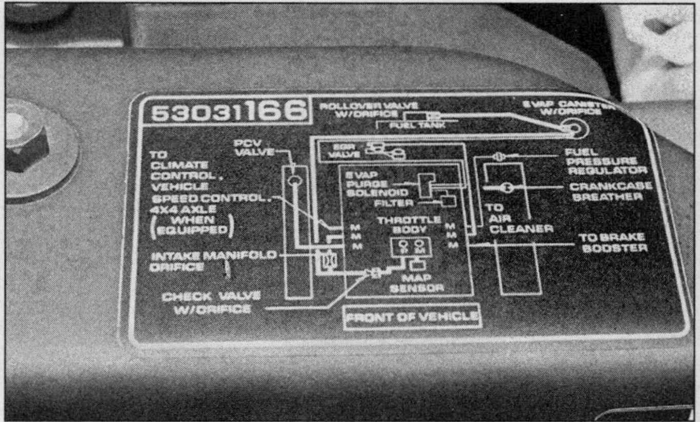

1.6b A vacuum line routing schematic is located in the engine compartment adjacent to the VECI label

6

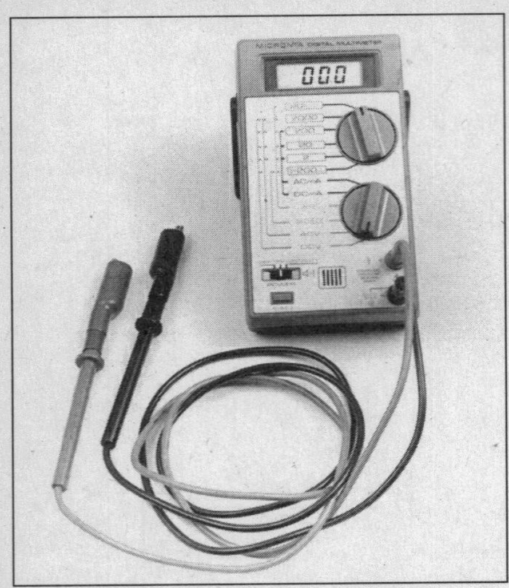

2.1 Digital multimeters can be used for testing all types of circuits; because of their high impedance, they are much more accurate than analog meters for measuring low-voltage computer circuits

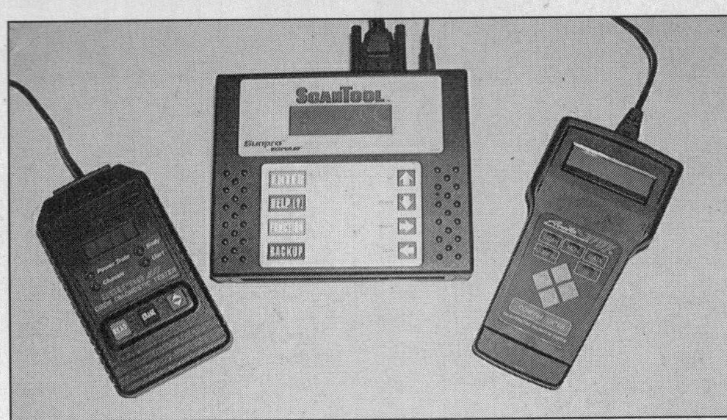

2.2 Scanners like the Actron OBD-II diagnostic tester, the Actron Scantool and the AutoXray XP240 are powerful diagnostic aids - programmed with comprehensive diagnostic information, they can tell you just about anything you want to know about your engine management system

2 On Board Diagnosis system and trouble codes

Diagnostic tool information

Refer to illustrations 2.1 and 2.2

1 A digital multimeter is necessary for checking fuel injection and emission related components **(see illustration)**. A digital volt-ohmmeter is preferred over the older style analog multimeter for several reasons. The analog multimeter cannot display the volts-ohms or amps measurement in hundredths and thousandths increments. When working with electronic circuits which are often very low voltage, this accurate reading is most important. Another good reason for the digital multimeter is the high impedance circuit. The digital multimeter is equipped with a high resistance internal circuitry (10 million ohms). Because a voltmeter is hooked up in parallel with the circuit when testing, it is vital that none of the voltage being measured should be allowed to travel the parallel path set up by the meter itself. This dilemma does not show itself when measuring larger amounts of voltage (9 to 12 volt circuits) but if you are measuring a low voltage circuit such as the oxygen sensor signal voltage, a fraction of a volt may be a significant amount when diagnosing a problem. However, there are several exceptions where using an analog voltmeter may be necessary to test certain sensors.

2 Hand-held scanners are the most powerful and versatile tools for analyzing engine management systems used on later model vehicles **(see illustration)**. Each brand scan tool must be examined carefully to match the year, make and model of the vehicle you are working on. Often interchangeable cartridges are available to access the particular manufacturer (Ford, GM, Chrysler, etc.). Some manufacturers will specify by continent (Asia, Europe, USA, etc.).

3 With the arrival of the federally mandated emission control system (OBD-II), a specially designed scanner has been developed. Several tool manufacturers have released OBD-II scan tools for the home mechanic. Ask the parts salesman at a local auto parts store for additional information concerning dates and costs.

4 Other types of code readers may be available for some models at parts stores. These tools simplify the procedure for extracting codes from the engine management computer by simply "plugging in" to the diagnostic connector on the vehicle wiring harness.

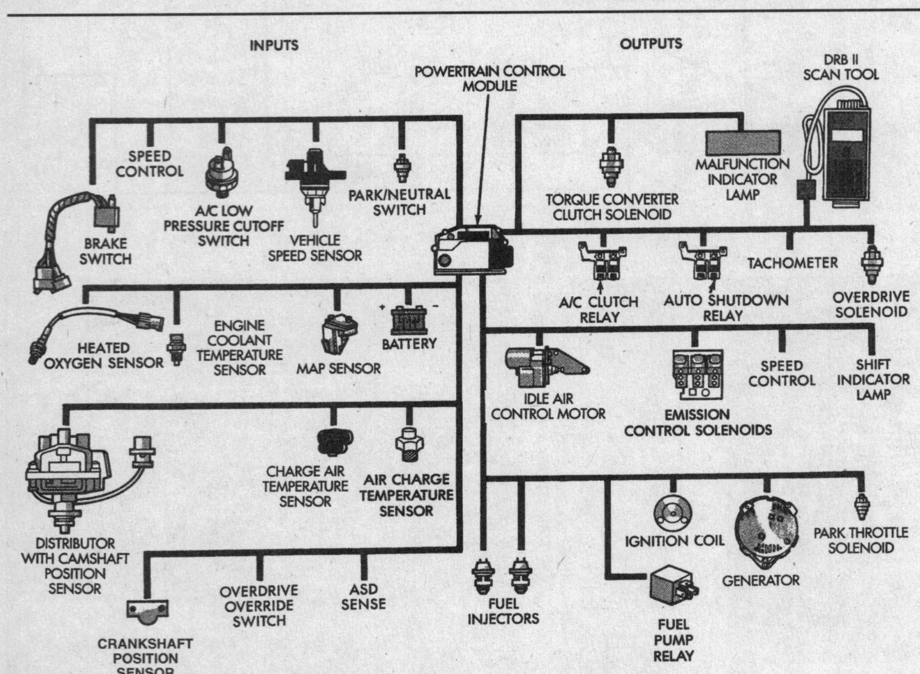

2.5 Typical emissions and engine control system components

On Board Diagnostic system general description

Refer to illustration 2.5

5 All models described in this manual are equipped with an On Board Diagnostic system. The systems consist of an onboard computer, known as the Powertrain Control Module (PCM), and information sensors, which monitor various functions of the engine and send data to the PCM **(see illustration)**. 1996 and later models are equipped with the second generation On Board Diagnostic system (OBD-II).

6 Based on the data and the information programmed into the computer's memory, the PCM generates output signals to control various engine functions via control relays, solenoids and other output actuators. The PCM is specifically calibrated to optimize the emissions, fuel economy and driveability of the vehicle.

7 Because of a federally mandated warranty which covers the emissions system components and because any owner-induced damage to the PCM, the sensors and/or the control devices may void the warranty, it isn't a good idea to attempt diagnosis or replacement of the PCM at home while the vehicle is under warranty. Take the vehicle to a dealer service department if the PCM or a system component malfunctions.

Information sensors

8 **Battery temperature sensor** - 1996 and later models are equipped with a battery temperature sensor. The battery temperature sensor senses the temperature of the battery. The PCM uses this information in controlling the voltage output of the alternator.

9 **Camshaft position sensor** - The camshaft position sensor provides information on camshaft position. The PCM uses this information, along with the crankshaft position sensor information, to control ignition timing and fuel injection synchronization.

10 **Crankshaft position sensor** - The crankshaft position sensor senses crankshaft position (TDC) during each engine revolution. The PCM uses this information to control ignition timing and fuel injection synchronization.

11 **Engine coolant temperature sensor** - The engine coolant temperature sensor senses engine coolant temperature. The PCM uses this information to control the fuel delivery and ignition timing.

12 **Intake air temperature sensor** - The intake air temperature senses the temperature of the air entering the intake manifold. The PCM uses this information to control fuel delivery.

13 **Manifold absolute pressure sensor** - The manifold absolute pressure monitors intake manifold pressure and ambient barometric pressure. The PCM uses this input signal to determine engine load and adjust fuel delivery accordingly.

14 **Oxygen sensor** - The oxygen sensors generate a voltage signal that varies with the difference between the oxygen content of the exhaust and the oxygen in the surrounding air. The PCM uses this information to determine if the fuel system is running rich or lean.

15 **Throttle position sensor** - The throttle position sensor senses throttle movement and position. This signal enables the PCM to determine when the throttle is closed, in a cruise position, or wide open. The PCM uses this information to control fuel delivery and ignition timing.

16 **Transmission pressure and temperature sensors** - The transmission pressure and temperature sensor signals are used by the PCM to control shift operation (four-speed electronic automatic transmission only).

17 **Vehicle speed sensor** - The vehicle speed sensor provides information to the PCM to indicate vehicle speed. 1998 and earlier models are equipped with a vehicle speed sensor. On 1999 and 2000 models, the PCM receives vehicle speed information from the antilock brake system rear wheel speed sensor.

18 **Miscellaneous PCM inputs** - In addition to the various sensors, the PCM monitors various switches and circuits to determine vehicle operating conditions. The switches and circuits include:

a) Air conditioning system
b) Auto shutdown relay
c) Battery voltage
d) Brake On/Off switch
e) Cruise control system
f) Engine oil pressure
g) EVAP leak detection pump operation
h) Fuel level
i) Ignition switch
j) Overdrive switch
k) Park/neutral position switch
l) Sensor signal and ground circuits

Output actuators

19 **Air conditioning clutch relay** - The PCM controls the operation of the air conditioning compressor clutch with the air conditioning clutch relay.

20 **Automatic shutdown relay** - The automatic shutdown relay supplies battery power to the fuel injectors, ignition coil and oxygen sensor heaters. The PCM controls the operation of the automatic shutdown relay.

21 **Check Engine light** - The PCM will illuminate the Check Engine light if a malfunction in the emissions system or electronic engine control system occurs.

22 **Cruise control vacuum and vent solenoids** - The cruise control system operation is controlled by the PCM.

23 **Engine cooling fan relay** - The auxiliary electric engine cooling fan, on 1999 and 2000 models, is controlled by the PCM according to information received from the engine coolant temperature sensor.

24 **EVAP canister purge solenoid** - The evaporative emission canister purge solenoid is a solenoid valve, operated by the PCM to purge the fuel vapor canister and route fuel vapor to the intake manifold for combustion.

25 **EVAP leak detection pump** - Some models are equipped with a self-diagnostic EVAP leak detection system. The system checks the integrity of the EVAP system when the engine is started cold. The PCM controls the operation of the EVAP leak detection pump.

26 **Exhaust Gas Recirculation (EGR) control solenoid valve** - The PCM controls the amount of EGR flow with the EGR control solenoid valve (1993 through 1995 5.2L models only).

27 **Fuel injectors** - The PCM opens the fuel injectors individually in firing order sequence. The PCM also controls the time the injector is held open (pulse width). The pulse width of the injector (measured in milliseconds) determines the amount of fuel delivered. For more information on the fuel delivery system and the fuel injectors, including injector replacement, refer to Chapter 4.

28 **Fuel pump relay** - The fuel pump relay is activated by the PCM with the ignition switch in the Start or Run position. When the ignition switch is turned on, the relay is activated to supply initial line pressure to the system. Refer to Chapter 12 or your owner's manual for more information on relay location. For more information on fuel pump check and replacement, refer to Chapter 4.

29 **Idle air control valve** - The idle air control valve controls the amount of air allowed to bypass the throttle plate when the throttle valve is closed or at idle position. The more air allowed to bypass the throttle plate, the higher the idle speed. The idle air control valve opening and the resulting idle speed is controlled by the PCM.

30 **Ignition coil** - The PCM controls the ignition coil by grounding the primary circuit to the ignition coil which generates high voltage in the secondary circuit, thus sending spark from the ignition coil to the distributor or spark plugs. The PCM controls ignition timing depending on engine operation conditions. Refer to Chapter 5 for more information on the ignition coil.

31 **Tachometer** - The PCM operates the tachometer from the information received from the crankshaft position sensor.

32 **Transmission shift control solenoids** - The PCM receives input signals from various sensors and switches such as the vehicle speed sensor, transmission temperature sensor, throttle position sensor and manifold absolute pressure sensor to determine shifting points, required line pressure and torque converter lock-up operations of the transmission (four-speed electronic automatic transmission only).

33 **Voltage regulator** - The PCM controls the charging system voltage by grounding the alternator field driver circuit depending on electrical load requirements. The voltage regulator circuitry is fully contained within the PCM. Any failure in the voltage regulator circuitry requires replacement of the PCM.

Obtaining diagnostic trouble codes

Refer to illustration 2.36

Note: *The diagnostic trouble codes on all models can be extracted from the Powertrain Control Module (PCM) using a specialized scan tool. On 1997 and earlier models, the PCM can be placed in the diagnostic mode and the trouble codes flashed on the Check Engine light. On 1998 and later models, have the vehicle diagnosed by a dealer service department or other qualified automotive repair facility if the proper scan tool is not available.*

34 The PCM will illuminate the CHECK ENGINE light (also known as the Malfunction Indicator Lamp) on the dash if it recognizes a fault in the system. The light will remain illuminated until the problem is repaired and the code is cleared or the PCM does not detect any malfunction for several consecutive drive cycles.

35 The diagnostic codes for the On Board Diagnostic (OBD) system can be extracted from the PCM using a scan tool on all models. On 1997 and earlier models, the system can be accessed to flash the trouble codes on the CHECK ENGINE light. To extract the diagnostic trouble codes on 1997 and earlier models using this method, proceed as follows:

a) *Without starting the engine, turn the ignition key ON, OFF, ON, OFF and finally ON. The CHECK ENGINE light on the dash will begin to flash.*

b) *If any trouble codes are stored in the PCM memory, the CHECK ENGINE light will flash the number of the first digit, pause and flash the number of the second digit. For example: Code 23, air temperature sensor circuit would be indicated by two flashes, pause, then three flashes. A long pause will appear between individual codes if more than one code is present. Carefully observe the flashes and record the exact code number(s) onto paper.*

c) *After the stored codes have been indicated (or if everything in the self diagnosis system is functioning properly), the CHECK ENGINE light will flash a Code 55. On 1996 and 1997 models, refer to the "Check Engine Light Flash Code" column of the trouble code chart for trouble code identification.*

d) *If the ignition key is turned OFF during the code extraction process and possibly turned back ON, the self diagnostic system will automatically invalidate the procedure. Restart the procedure to extract the codes.* **Note:** *The self diagnostic system cannot be accessed with the engine running.*

36 The preferred code extraction method requires a special scan tool that is programmed to interface with the OBD system by plugging into the diagnostic connector

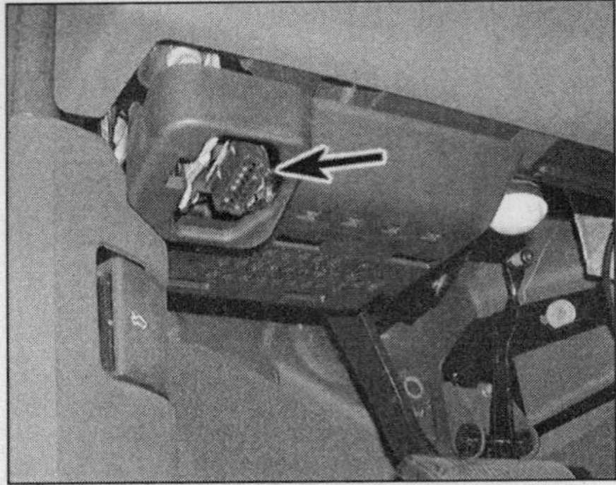

2.36 The diagnostic connector is typically located under the instrument panel

(see illustration). When used, the scan tool has the ability to diagnose in-depth driveability problems and it allows freeze frame data to be retrieved from the PCM stored memory. Freeze frame data is an OBD II PCM feature that records all related sensor and actuator activity on the PCM data stream whenever an engine control or emissions fault is detected and a trouble code is set. This ability to look at the circuit conditions and values when the malfunction occurs provides a valuable tool when trying to diagnose intermittent driveability problems. If the tool is not available and intermittent driveability problems exist, have the vehicle checked at a dealer service department or other qualified repair shop. **Note:** *On 1996 and later models, the scan tool trouble code designation is referred to as a P0 or P1 code. Refer to the "Generic Scan Tool Code" column of the trouble code chart for the code identification.*

Clearing diagnostic trouble codes

37 After the system has been repaired, the codes must be cleared from the PCM memory using a scan tool. **Caution:** *Do not disconnect the battery from the vehicle in an attempt to clear the codes. If necessary, have the codes cleared by a dealer service department or other qualified repair facility.*

38 Always clear the codes from the PCM before starting the engine after a new electronic emission control component is installed onto the engine. The PCM stores the operating parameters of each sensor. The PCM may set a trouble code if a new sensor is allowed to operate before the parameters from the old sensor have been erased.

Diagnostic trouble code identification

39 The accompanying list of diagnostic trouble codes is a compilation of all the codes that may be encountered **(see the following tables)**. Not all codes pertain to all models and not all codes will illuminate the Check Engine light when set. The codes listed under the "Check Engine Light Flash Code" column are codes that may be displayed by the Check Engine light on 1997 and earlier models (see Step 35). 1998 and later models require a scan tool to access the diagnostic trouble codes.

Diagnostic Trouble Code Identification Chart - 1995 and earlier models

Trouble code	Trouble code description
Code 11	No distributor reference signal detected during engine cranking. Check the circuit between the distributor and the PCM.
Code 12	Problem with the battery connection. Direct battery input to PCM disconnected within the last 50 ignition key cycles.
Code 13	Indicates a problem with the manifold absolute pressure sensor vacuum source.
Code 14	Manifold absolute pressure sensor voltage too low or too high
Code 15	A problem with the vehicle speed sensor signal. No distance/speed sensor signal detected during road load conditions.
Code 16	Loss of battery voltage.
Code 17	Engine is cold too long. Engine coolant temperature remains below normal operating temperatures during operation (check the thermostat).
Code 21	Problem with oxygen sensor signal circuit. Sensor voltage to computer not fluctuating.
Code 22	Coolant sensor voltage too high or too low. Test coolant temperature sensor.
Code 23	Indicates that the intake air temperature sensor input is below the minimum acceptable voltage or sensor input is above the maximum acceptable voltage.
Code 24	Throttle position sensor voltage high or low. Test the throttle position sensor.
Code 25	Idle air control valve circuits. A shorted condition is detected in one or more of the idle air control valve circuits.
Code 27	One of the injector control circuit output drivers does not respond properly to the control signal. Check the circuits.
Code 31	Problem with the EVAP canister purge solenoid circuit.
Code 32	An open or shorted condition detected in the EGR solenoid circuit Possible air/fuel ratio imbalance not detected during diagnosis
Code 33	Air conditioning clutch relay circuit. An open or shorted condition detected in the air conditioning clutch relay circuit.
Code 34	Open or shorted condition detected in the cruise control vacuum or vent solenoid circuits.
Code 37	Torque converter clutch solenoid circuit. An open or shorted condition detected in the torque converter part throttle unlock solenoid circuit (automatic transmissions models only).
Code 41	Problem with the charging system. Occurs when battery voltage from the auto shutdown relay is below 11.75-volts.

Trouble code	Trouble code description
Code 42	Auto shutdown relay control circuit indicates an open or shorted circuit condition.
Code 44	Battery temperature sensor volts malfunction. Problem with the battery temperature voltage circuit in the PCM.
Code 45	Overdrive solenoid. Problem detected in the overdrive solenoid circuit
Code 46	Charging system voltage too high. PCM indicates that the battery voltage is not properly regulated.
Code 47	Charging system voltage too low. Battery voltage sense input below target charging voltage during engine operation and no significant change in voltage detected during active test of alternator output.
Code 51	Oxygen sensor signal input indicates lean fuel/air ratio condition during engine operation.
Code 52	Oxygen sensor signal input indicates rich fuel/air ratio condition during engine operation.
Code 53	Internal PCM failure detected.
Code 54	No camshaft position sensor signal from distributor. Problem with the distributor synchronization circuit.
Code 55	Completion of fault code display on CHECK ENGINE lamp. This is an End of message code.
Code 62	Unsuccessful attempt to update service reminder mileage in the PCM memory.
Code 63	Controller failure. PCM memory write denied. Check the PCM.

Diagnostic Trouble Code Identification Chart - 1996 and later models

Check Engine Light Flash Code	Generic Scan Tool Code	DRB Scan Tool Display	Description of Diagnostic Trouble Code
12			Battery power to PCM was disconnected
54	P0340	No Cam Signal at PCM	No camshaft signal detected during engine cranking.
53	P0601	Internal Controller Failure	PCM Internal fault condition detected.
47	P0162	Charging System Voltage Too Low	Battery voltage sense input below target charging during engine operation. Also, no significant change detected in battery voltage during active test of generator output circuit.
46	P1594	Charging System Voltage Too High	Battery voltage sense input above target charging voltage during engine operation.
42	P1388	Auto Shutdown Relay Control Circuit	An open or shorted condition detected in the auto shutdown relay circuit.
41	P0622	Generator Field Not Switching Properly	An open or shorted condition detected in the generator field control circuit.
37	P0743	Torque Converter Clutch Soleniod/Trans Relay Circuits	An open or shorted condition detected in the torque converter part throttle unlock solenoid control circuit (3 speed auto RH trans. only).
35	P1491	Rad Fan Control Relay Circuit	An open or shorted condition detected in the low speed radiator fan relay control circuit.
34	P1595	Speed Control Solenoid Circuits	An open or shorted condition detected in the Speed Control vacuum or vent solenoid circuits.
33	P0645	A/C Clutch Relay Circuit	An open or shorted condition detected in the A/C clutch relay circuit.
31	P0443	EVAP Purge Solenoid Circuit	An open or shorted condition detected in the duty cycle purge solenoid circuit.
27	P0203 or	Injector #3 Control Circuit	Injector #3 output driver does not respond properly to the control signal.
	P0202 or	Injector #2 Control Circuit	Injector #2 output driver does not respond properly to the control signal.
	P0201	Injector #1 Control Circuit	Injector #1 output driver does not respond properly to the control signal.
25	P0505	Idle Air Control Motor Circuits	A shorted or open condition detected in one or more of the idle air control motor circuits.
24	P0122 or	Throttle Position Sensor Voltage Low	Throttle position sensor input below the minimum acceptable voltage
	P0123	Throttle Position Sensor Voltage High	Throttle position sensor input above the maximum acceptable voltage.

Diagnostic Trouble Code Identification Chart - 1996 and later models

Check Engine Light Flash Code	Generic Scan Tool Code	DRB Scan Tool Display	Description of Diagnostic Trouble Code
22	P0117	ECT Sensor Voltage Too Low	Engine coolant temperature sensor input below minimum acceptable voltage.
	or		
	P0118	ECT Sensor Voltage Too High	Engine coolant temperature sensor input above maximum acceptable voltage.
17	1281	Engine Is Cold Too Long	Engine did not reach operating temperature within acceptable limits.
	P0107	MAP Sensor Voltage Too Low	MAP sensor input below minimum acceptable voltage.
14	or		
	P0108	MAP Sensor Voltage Too High	MAP sensor input above maximum acceptable voltage.
13	P1297	No Change in MAP From Start to Run	No difference recognized between the engine MAP reading and the barometric (atmospheric) pressure reading from start-up.
11	P0320	No Crank Reference Signal at PCM	No crank reference signal detected during engine cranking.
	P0351	Ignition Coil #1 Primary Circuit	Peak primary circuit current not achieved with maximum dwell time.
42	P1389	No ASD Relay Output Voltage at PCM	An Open condition Detected In The ASD Relay Output Circuit.
63	P1696	PCM Failure EEPROM Write Denied	Unsuccessful attempt to write to an EEPROM location by the PCM.
37	P0753	Trans 3-4 Shift Sol/Trans Relay Circuits	Current state of output port for the solenoid is different from expected state.
	P0112	Intake Air Temp Sensor Voltage Low	Intake air temperature sensor input below the maximum acceptable voltage.
23	or		
	P0113	Intake Air Temp Sensor Voltage High	Intake air temperature sensor input above the minimum acceptable voltage.
27	P0204	Injector #4 Control Circuit	Injector #4 output driver does not respond properly to the control signal.
21	P0132	Left Upstream O2S Shorted to Voltage	Oxygen sensor input voltage maintained above the normal operating range.
	P0154	O2 2/1 Signal Inactive	No signal at O2 2/1 sensor
	P0152	O2 2/1 Shorted High	Oxygen sensor input voltage sustained above the normal operating range.
53	PO600	PCM Failure SPI Communications	PCM internal fault condition detected
	P0205	Injector #5 Control Circuit	Injector #5 output driver does not respond properly to the control signal.
27	or		
	P0206	Injector #6 Control Circuit	Injector #6 output driver does not respond properly to the control signal.
45	P0712	Trans Temp Sensor Voltage Too Low	Voltage less than 1.55 volts.
	or		

Check Engine Light Flash Code	Generic Scan Tool Code	DRB Scan Tool Display	Description of Diagnostic Trouble Code
45	P0713	Trans Temp Sensor Voltage Too High	Voltage greater than 3.76 volts.
27	P0207	Injector #7 Control Circuit	Injector #7 output driver does not respond properly to the control signal.
	or		
	P0208	Injector #8 Control Circuit	Injector #8 output driver does not respond properly to the control signal.
77	P1683	SPD CTRL PWR RLY; or S/C 12V Driver CKT	Malfuntion detected with power feed to speed control servo solenoids
34	P1596	MUX S/C Switch High	Speed control switch input above the maximum acceptable voltage.
	or		
	P1597	MUX S/C Switch Low	Speed control switch input below the minimum acceptable voltage.
42	P1282	Fuel Pump Relay Control Circuit	An open or shorted condition detected in the fuel pump relay control circuit.
21	P0133 or P0152	O2 1/1 Slow Response	Oxygen sensor response slower than minimum required switching frequency.
	or		
	P0135	O2 1/1 Heater Circuit	Upstream oxygen sensor heating element circuit malfunction
	P0139	O2 1/1 Slow Response	Oxygen sensor response slower than minimum required switching frequency.
	P0141	O2 1/2 Heater Circuit	Oxygen sensor heating element circuit malfunction.
43	P0300	Multiple Cylinder Mis-fire	Misfire detected in multiple cylinders.
	or		
	P0301	Cylinder #1 Mis-fire	Misfire detected in cylinder #1.
	or		
	P0302	Cylinder #2 Mis-fire	Misfire detected in cylinder #2.
	or		
	P0303	Cylinder #3 Mis-fire	Misfire detected in cylinder #3.
	or		
	P0304	Cylinder #4 Mis-fire	Misfire detected in cylinder #4.
72	P0420	Catalyst 1/1 Effic	Catalyst efficiency below required level.
31	P0441	Incorrect Purge Flow	Insufficient or excessive vapor flow detected during evaporative emission system operation.
37	P1899	P/N Switch Stuck in Park or in Gear	Incorrect input state detected for the Park/Neutral switch, auto. trans. only.
	P0551	Pwr Steering Sw Perf	Power steering high pressure seen at high speed.
52	P0172	Left Bank or Fuel System Rich	A rich air/fuel mixture has been indicated by an abnormally lean correction factor.
51	P0171	Right Rear (or just) Fuel System Lean	A lean air/fuel mixture has been indicated by an abnormally rich correction factor.

Diagnostic Trouble Code Identification Chart - 1996 and later models

Check Engine Light Flash Code	Generic Scan Tool Code	DRB Scan Tool Display	Description of Diagnostic Trouble Code
	P0175	Fuel System 2/1 Rich	A rich air/fuel mixture has been indicated by an abnormally lean correction factor.
	P0174	Fuel System 2/1 Lean	A lean air/fuel mixture has been indicated by an abnormally lean correction factor.
	P0153	O2 2/1 Slow Response	Oxygen sensor response slower than minimum required switching frequency.
	P0159	O2 2/1 Slow Response	Oxygen sensor response slower than minimum required switching frequency.
	P0155	O2 2/1 Heater circuit	Oxygen sensor heater element malfunction.
	P0161	O2 2/1 Heater circuit	Oxygen sensor heater element malfunction.
21	P0138	Left Bank Downstream or Downstream and Pre-Catalyst O2S Shorted to Voltage	Oxygen sensor input voltage maintained above the normal operating range.
	P0158	O2 2/2 Shorted High	Oxygen sensor input voltage maintained above the normal operating range.
17	P0125	Closed Loop Temp Not Reached	Engine does not reach 20°F within 5 minutes with a vehicle speed signal.
24	P0121	TPS Voltage Does Not Agree With MAP	TPS signal does not correlate to MAP sensor
14	P1296	No 5 Volts To MAP Sensor	5 Volt output to MAP sensor open.
25	P1294	Target Idle Not Reached	Actual idle speed does not equal target idle speed.
37	P1756 or P1757	Governor Pressure Not Equal to Target @ 15-20 PSI Governor Pressure Above 3 PSI In Gear With 0 MPH	Governor sensor input not between 10 and 25 psi when requested. Governor pressure greater than 3 psi when requested to be 0 psi.
37	P0740	Torq Conv Clu, No RPM Drop At Lockup	Relationship between engine speed and vehicle speed indicates no torque converter clutch engagement (auto. trans. only).
42	P0462 or P0463 or P0460	Fuel Level Sending Unit Volts Too Low Fuel Level Sending Unit Volts Too High Fuel Level Unit No Change Over Miles	Open circuit between PCM and fuel gauge sending unit. Circuit shorted to voltage between PCM and fuel gauge sending unit. No movement of fuel level sender detected.
44	P1493 or P1492	Ambient/Batt Temp Sen VoltsToo Low Ambient/Batt Temp Sensor VoltsToo High	Battery temperature sensor input voltage below an acceptable range. Battery temperature sensor input voltage above an acceptable range.

Check Engine Light Flash Code	Generic Scan Tool Code	DRB Scan Tool Display	Description of Diagnostic Trouble Code
21	P0131 or P0137	Left Bank and Upstream O2S Shorted to Ground Downstream, Left Bank Downstream and Pre-Catalyst O2S Shorted to Ground	O2 sensor voltage too low, tested after cold start. O2 sensor voltage too low, tested after cold start.
11	P1391	Intermittent Loss of CMP or CKP	Intermittent loss of either camshaft or crankshaft position sensor
31	P0442 or P0455	Evap Leak Monitor Small Leak Detected Evap Leak Monitor Large Leak Detected	A small leak has been detected by the leak detection monitor The leak detection monitor is unable to pressurize Evap system, indicating a large leak.
45	P0711	Trans Temp Sensor, No Rise After Start	Sump temp did not rise more than 16°F within 10 minutes when starting temp is below 40°F or sump temp is above 260°F with coolant below 100°F.
37	P0783	3-4 Shift Sol, No RPM Drop @ 3-4 Shift	The ratio of engine rpm/output shaft speed did not change beyond on the minimum required.
15	P0720	Low Ouput Spd Sensor RPM Above 15 mph	Output shaft speed is less than 60 rpm with vehicle speed above 15 mph.
45	P1764 or P1763 or P1762	Governor Pessure Sensor Volts Too Low Governor Pressure Sensor Volts Too HI Governor Press Sen Offset Volts Too Lo or High	Voltage less than .10 volts. Voltage greater than 4.89 volts. Sensor input greater or less than calibration for 3 consecutive Neutral/Park occurances.
37	P0748	Governor Pressure Sol Control/Trans Relay Circuits	Current state of solenoid output port is different than expected.
37	P1765	Trans 12 Volt Supply Relay Ctrl Circuit	Current state of solenoid output port is different than expeted.
43	P0305 or P0306 or P0307 or P0308	Cylinder #5 Mis-fire Cylinder #6 Mis-fire Cylinder #7 Mis-fire Cylinder #8 Mis-fire	Misfire detected in cylinder #5. Misfire detected in cylinder #6. Misfire detected in cylinder #7. Misfire detected in cylinder #8.
	P0432	Catalyst 2/1 EFFIC	Catalyst 2/1 efficiency below required level
	P0151	O2 2/1 Voltage Low	Oxygen sensor input voltage maintained below normal operating range.

Diagnostic Trouble Code Identification Chart - 1996 and later models

Check Engine Light Flash Code	Generic Scan Tool Code	DRB Scan Tool Display	Description of Diagnostic Trouble Code
	P0157	O2 2/1 Voltage Low	Oxygen sensor input voltage maintained below normal operating range.
31	P1495 or P1494	Leak Detection Pump Solenoid Circuit Leak detection pump SW or mechanical fault	Leak detection pump solenoid circuit fault (open or short) Leak detection pump switch does not respond to input.
11	P1398	Mis-fire Adaptive Numerator at Limit	CKP sensor target windows have too much variation
31	P1486	Evap leak monitor pinched hose found	Plug or pinch detected between purge solenoid and fuel tank
45	P0751	O/D Switch Pressed (LO) More Than 5 Min	Overdrive Off switch input too low for more than 5 minutes.
	P0147	O2 1/3 Heater Circuit	Oxygen sensor heater element malfunction.
21	P0133 or P1195	Cat Mon slow O2 1/1	A slow switching oxygen sensor has been detected in bank 1/1 during catalyst monitor test.
	P0153 or P1196	Cat Mon slow O2 2/1	A slow switching oxygen sensor has been detected in bank 2/1 during catalyst monitor test.
	P0129 or P1197	Cat Mon slow O2 1/2	A slow switching oxygen sensor has been detected in bank 1/2 during catalyst monitor test.
15		No vehicle speed sensor signal	No Vehicle speed sensor signal detected during driving conditions.

3.4 Loosen the bolt (arrow) and disconnect the electrical connectors from the PCM

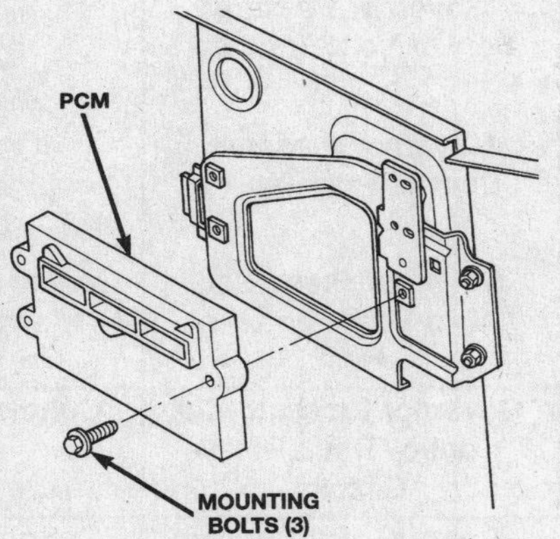

3.5a Remove the PCM mounting bolts - 1998 and earlier models

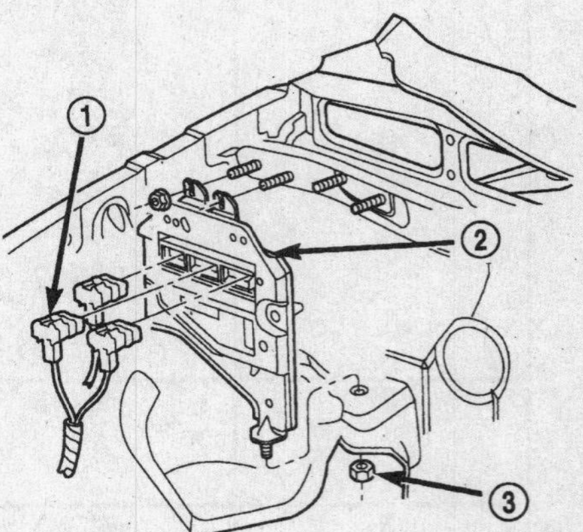

3.5b PCM installation details - 1999 and later models

1	Electrical connectors	2	Bracket
		3	Bracket nuts

3 Powertrain Control Module - removal and installation

Refer to illustrations 3.4, 3.5a and 3.5b

Caution: *Avoid static electricity damage to the Powertrain Control Module (PCM) by grounding yourself to the body of the vehicle before touching the PCM and using a special anti-static pad to store the PCM on, once it is removed.*

Note 1: *Anytime the PCM is replaced with a new unit on 1998 and later models, the PCM must be reprogrammed by a dealership service department with special equipment. The following procedure pertains to removal and installation of the original PCM only. If, on 1998 and later models, the PCM must be replaced with a new unit, take the vehicle to a dealership service department.*

Note 2: *Anytime the battery is disconnected, stored operating parameters may be lost from the PCM causing the engine to run rough for a period of time while the PCM relearns the information.*

1 Disconnect the negative battery cable from the battery terminal.

2 On 1999 and 2000 models, remove the Transmission Control Module, if equipped (see Chapter 7B).

3 Remove the coolant reserve overflow tank (see Chapter 3).

4 Remove the wiring harness retaining bolt for the PCM electrical connector and disconnect the connectors from the PCM **(see illustration)**.

5 Remove the PCM mounting bolts/nuts and remove the assembly from the vehicle **(see illustrations)**.

6 On 1999 and later models, remove the PCM from the mounting bracket.

7 Installation is the reverse of removal.

4.2a Throttle body and related components - 4.0L six-cylinder engine

1. *Throttle body*
2. *Idle air control valve*
3. *Throttle position sensor*
4. *Manifold absolute pressure sensor*

4.2b Throttle position sensor location - 4.7L V8 models

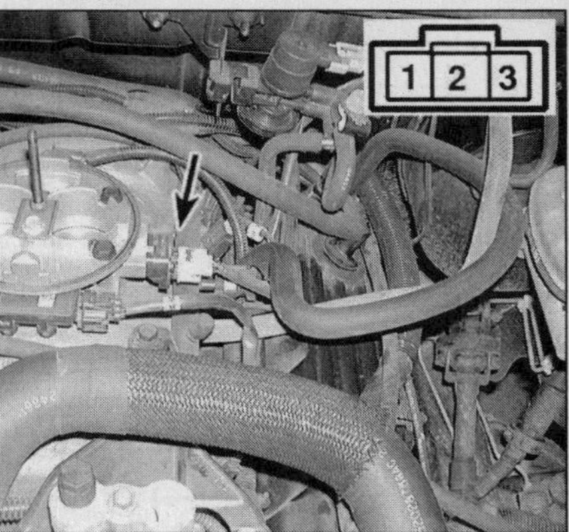

4.3 Throttle position sensor location (5.2L/5.9L V8 models) and connector terminal identification (all)

4 Throttle Position Sensor (TPS) - check and replacement

1 The throttle position sensor is located on the throttle body at the end of the throttle shaft. By monitoring the output voltage from the throttle position sensor, the PCM can determine fuel delivery based on throttle valve angle (driver demand). A broken or loose throttle position sensor can cause intermittent bursts of fuel from the injectors and an unstable idle because the PCM thinks the throttle is moving.

Check

Refer to illustrations 4.2a, 4.2b, 4.3 and 4.4

Note: *Performing the following test will set a diagnostic trouble code and illuminate the Check Engine light. Clear the diagnostic trouble code after performing the tests and making the necessary repairs* (see Section 2).

2 Locate the throttle position sensor on the throttle body **(see illustrations)**.

3 Before checking the throttle position sensor, check the voltage supply and ground circuits from the PCM. Disconnect the electrical connector from the throttle position sensor. On all except 4.7L V8 models, connect the positive lead of a voltmeter to terminal number three and the negative lead to terminal number one of the harness connector **(see illustration)**. On 4.7L V8 models, connect the positive lead of a voltmeter to terminal number one and the negative lead to terminal number three of the harness connector Turn the ignition key On - the voltage should read approximately 5.0 volts. If the voltage is incorrect, check the wiring from the throttle position sensor to the PCM. If the circuits are good, have the PCM checked at a dealer service department or other properly equipped repair facility.

4 To check the throttle position sensor operation, reconnect the connector to the throttle position sensor and using a suitable probe, backprobe the center wire terminal of the throttle position sensor connector **(see illustration)**. **Note:** *Refer to Chapter 12 for additional information on how to backprobe a connector.* Connect the positive lead of a voltmeter to the probe and the negative lead to a good engine ground point. Turn the ignition key On - with the throttle fully closed the voltage should read between 0.20 and 0.90 volts. Gradually open the throttle - the voltage should increase smoothly to approximately 4.5 volts at wide-open throttle. If the test results are incorrect, replace the throttle position sensor.

4.4 To check the throttle position sensor, backprobe the center wire terminal of the throttle position sensor and connect a voltmeter

Replacement

Refer to illustrations 4.7 and 4.8

5 Remove the air intake duct/resonator.

6 Disconnect the electrical connector from the throttle position sensor.

7 Remove the throttle position sensor mounting screws and remove the throttle position sensor from the throttle body **(see illustration)**.

8 When installing the throttle position sensor, be sure to align the socket locating tangs on the throttle position sensor with the throttle shaft in the throttle body **(see illustration)**. Install the throttle position sensor and slightly rotate it against the throttle shaft to apply a slight amount of tension on the throttle position sensor. If necessary remove the sensor

4.7 Remove the throttle position sensor mounting screws (arrows)

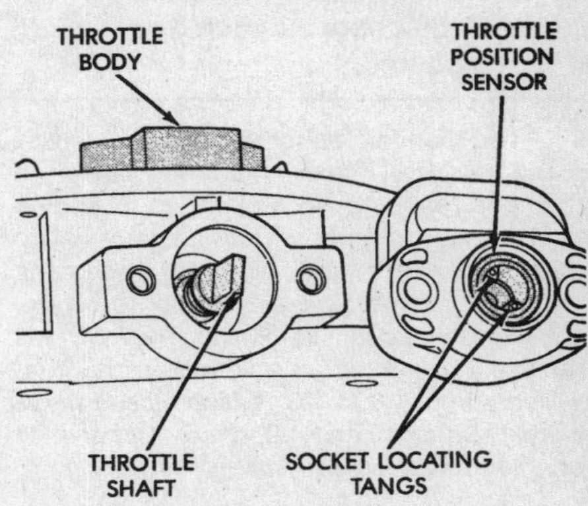

4.8 When installing the throttle position sensor, align the socket locating tangs on the throttle position sensor with the throttle shaft

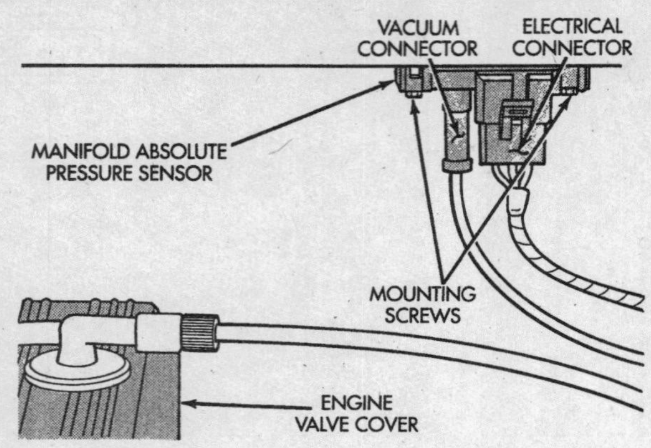

5.2a Manifold absolute pressure (MAP) sensor location - 1995 and earlier 4.0L six-cylinder models

5.2b Manifold absolute pressure sensor location - 4.7L V8 models

5.2c Manifold absolute pressure sensor location (5.2L/5.9L V8 models) and connector terminal identification (all)

5.4 To check the manifold absolute pressure sensor, backprobe the center wire terminal of the manifold absolute pressure sensor and connect a voltmeter

and install it with the socket locating tangs on the other side of the throttle shaft.

9 The remainder of installation is the reverse of removal.

5 Manifold Absolute Pressure (MAP) sensor - check and replacement

1 The manifold absolute pressure sensor monitors the intake manifold pressure changes resulting from changes in engine load and speed and converts the information into a voltage output. The PCM uses the manifold absolute pressure sensor to control fuel delivery and ignition timing. The PCM will receive information as a voltage signal that will vary from 1.0 to 1.5 volts at closed throttle (high vacuum) and 4.0 to 4.5 volts at wide open throttle (low vacuum).

Check

Refer to illustrations 5.2a, 5.2b, 5.2c and 5.4

Note: *Performing the following test will set a diagnostic trouble code and illuminate the Check Engine light. Clear the diagnostic trouble code after performing the tests and making the necessary repairs* (see Section 2).

2 On 1995 and earlier six-cylinder models, the manifold absolute pressure sensor is located under the cowl panel **(see illustration)**. On 1996 and later six-cylinder models and all 5.2L/5.9L V8 models, the manifold absolute pressure sensor is located on the throttle body. On 4.7L V8 models, the manifold absolute pressure sensor is located at the front of the intake manifold **(see illustration)**. On six-cylinder models and 5.2L/5.9L V8 models, check the vacuum hose connecting the manifold absolute pressure sensor to the throttle body for damage and a snug fit. On all models, check the terminals in the connector and the wires leading to the sensor for looseness and breaks. Repair as required.

3 Before checking the manifold absolute pressure sensor, check the voltage supply and ground circuits from the PCM. Disconnect the electrical connector from the manifold absolute pressure sensor. On all except 4.7L V8 models, connect the positive lead of a voltmeter to terminal number three and the negative lead to terminal number one of the harness connector **(see illustration 5.2c)**. On 4.7L V8 models, connect the positive lead of a voltmeter to terminal number one and the negative lead to terminal number three of the harness connector. Turn the ignition key On - the voltage should read approximately 5.0 volts. If the voltage is incorrect, check the wiring from the manifold absolute pressure sensor to the PCM. If the circuits are good, have the PCM checked at a dealer service department or other properly equipped repair facility.

4 To check the manifold absolute pressure sensor operation, reconnect the connector to the manifold absolute pressure sensor and using a suitable probe, backprobe the center wire terminal of the manifold absolute pressure sensor connector **(see illustration)**. **Note:** *Refer to Chapter 12 for additional information on how to backprobe a connector.* Connect the positive lead of a voltmeter to the probe and the negative lead to a good engine ground point. Turn the ignition key On - with the engine not running, the voltage should read 4.0 to 5.0 volts. Start the engine and allow it to idle - the voltage should decrease to approximately 1.5 to 2.0 volts. If the test results are incorrect, replace the manifold absolute pressure sensor.

5.6 Remove the manifold absolute pressure sensor mounting screws (arrows)

6.2a Intake air temperature sensor location - 4.0L six-cylinder models

6.2b Intake air temperature sensor location - 5.2L/5.9L V8 models

6.2c Intake air temperature sensor location - 4.7L V8 models

TEMPERATURE		RESISTANCE (OHMS)	
C	F	MIN	MAX
-40	-40	291,490	381,710
-20	-4	85,850	108,390
-10	14	49,250	61,430
0	32	29,330	35,990
10	50	17,990	21,810
20	68	11,370	13,610
25	77	9,120	10,880
30	86	7,370	8,750
40	104	4,900	5,750
50	122	3,330	3,880
60	140	2,310	2,670
70	158	1,630	1,870
80	176	1,170	1,340
90	194	860	970
100	212	640	720
110	230	480	540
120	248	370	410

6.3 Intake air temperature sensor and engine coolant temperature sensor approximate temperature vs. resistance values

Replacement

Refer to illustration 5.6

5 Remove the air intake duct/resonator.

6 Disconnect the electrical connector from the manifold absolute pressure sensor and remove the manifold absolute pressure sensor as follows, according to model **(see illustration)**:

a) *On 1995 and earlier six-cylinder models, disconnect the vacuum hose from the sensor, remove the screws retaining the manifold absolute pressure sensor to the cowl panel and remove the sensor.*

b) *On 1996 and later six-cylinder models and all 5.2L/5.9L V8 models, remove the screws retaining the manifold absolute pressure sensor to the throttle body. Remove the manifold absolute pressure sensor and disconnect the hose from the throttle body.*

c) *On 4.7L V8 models, remove the screws retaining the manifold absolute pressure sensor to the intake manifold and withdraw the sensor from the intake manifold. Inspect the O-ring for damage and replace it, if necessary.*

7 Installation is the reverse of removal.

6 Intake Air Temperature (IAT) sensor - check and replacement

1 The intake air temperature sensor is a thermistor (a resistor which varies the value of its resistance in accordance with temperature changes). The change in the resistance values will directly affect the voltage signal from the sensor to the PCM. As the sensor temperature INCREASES, the resistance values will DECREASE. As the sensor temperature DECREASES, the resistance values will INCREASE.

Check

Refer to illustrations 6.2a, 6.2b, 6.2c and 6.3

Note: *Performing the following test will set a diagnostic trouble code and illuminate the Check Engine light. Clear the diagnostic trouble code after performing the tests and making the necessary repairs* (see Section 2).

2 The intake air temperature sensor is located in the intake manifold **(see illustrations)**. Before checking the intake air temperature sensor, check the voltage supply and ground circuits from the PCM. Disconnect the electrical connector from the intake air temperature sensor and connect a voltmeter to the two terminals of the harness connector. Turn the ignition key On - the voltage should read approximately 5.0 volts. If the voltage is incorrect, check the wiring from the intake air temperature sensor to the PCM. If the circuits are good, have the PCM checked at a dealer service department or other properly equipped repair facility.

3 With the ignition switch OFF, disconnect the electrical connector from the intake air temperature sensor. Using an ohmmeter, measure the resistance between the two terminals on the sensor while it is completely cold (50 to 80-degrees F). Reconnect the electrical connector to the sensor, start the engine and warm it up until it reaches operating temperature (180 to 200-degrees F), then turn off the engine, disconnect the connector and check the resistance again. Compare your measurements to the resistance chart **(see illustration)**. If the sensor resistance test results are incorrect, replace the intake air temperature sensor. **Note:** *A more accurate check may be performed by removing the sensor and suspending the tip of the sensor in a container of water. Heat the water on the stove while you monitor the resistance of the sensor.*

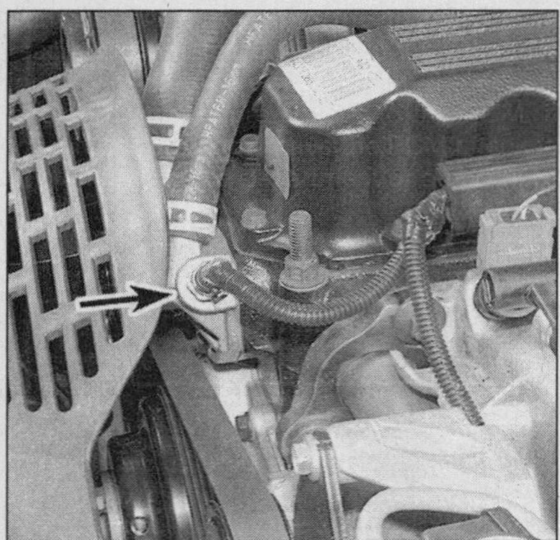
7.2a Engine coolant temperature sensor location - 4.0L six-cylinder models

7.2b Engine coolant temperature sensor location - 5.2L/5.9L V8 models

7.2c Engine coolant temperature sensor location - 4.7L V8 models

Replacement

4 Disconnect the electrical connector from the intake air temperature sensor.

5 Unscrew the sensor from the intake manifold and remove the intake air temperature sensor.

6 Installation is the reverse of removal.

7 Engine coolant temperature sensor - check and replacement

1 The engine coolant temperature sensor is a thermistor (a resistor which varies the value of its resistance in accordance with temperature changes). The change in the resistance values will directly affect the voltage signal from the sensor to the PCM. As the sensor temperature INCREASES, the resistance values will DECREASE. As the sensor temperature DECREASES, the resistance values will INCREASE.

Check

Refer to illustrations 7.2a, 7.2b and 7.2c

Note: *Performing the following test will set a diagnostic trouble code and illuminate the Check Engine light. Clear the diagnostic trouble code after performing the tests and making the necessary repairs* (see Section 2).

2 The engine coolant temperature sensor is located in the thermostat housing on six-cylinder models, in the intake manifold, next to the thermostat housing on 5.2L/5.9L V8 models and at the front of the engine block on 4.7L V8 models **(see illustrations)**. Before checking the engine coolant temperature sensor, check the voltage supply and ground circuits from the PCM. Disconnect the electrical connector from the engine coolant temperature sensor and connect a voltmeter to the two terminals of the harness connector. Turn the ignition key On - the voltage should read approximately 5.0 volts. If the voltage is incorrect, check the wiring from the engine coolant temperature sensor to the PCM. If the circuits are good, have the PCM checked at a dealer service department or other properly equipped repair facility.

3 With the ignition switch OFF, disconnect the electrical connector from the engine coolant temperature sensor. Using an ohmmeter, measure the resistance between the two terminals on the sensor while it is completely cold (50 to 80-degrees F). Reconnect the electrical connector to the sensor, start the engine and warm it up until it reaches operating temperature (180 to 200-degrees F), disconnect the connector and check the resistance again. Compare your measurements to the resistance chart **(see illustration 6.3)**. If the sensor resistance test results are incorrect, replace the engine coolant temperature sensor. **Note:** *A more accurate check may be performed by removing the sensor and suspending the tip of the sensor in a container of water. Heat the water on the stove while you monitor the resistance of the sensor.*

Replacement

Warning: *Wait until the engine is completely cool before beginning this procedure.*

4 Partially drain the cooling system (see Chapter 1).

5 Disconnect the electrical connector from the sensor and carefully unscrew the sensor. **Note:** *On 5.2L/5.9L V8 models, use a long hooked tool to disconnect the electrical connector and a deep socket and long extension to unscrew the sensor.*

6 Before installing the new sensor, wrap the threads with Teflon sealing tape to prevent leakage and thread corrosion.

7 Installation is the reverse of removal. Refill the cooling system.

8 Crankshaft position sensor - check and replacement

1 The crankshaft position sensor determines the timing for the fuel injection and ignition on each cylinder. It also detects engine RPM. The crankshaft position sensor is a Hall-Effect device triggered by notches in the flywheel/driveplate (all except 4.7L V8 engines). On 4.7L V8 models the crankshaft position sensor is triggered by a reluctor wheel at the rear of the crankshaft. On six-cylinder models, the sensor is mounted on the transmission bellhousing at the left rear of the engine. On 5.2L/5.9L V8 models, the crankshaft sensor is mounted on the engine block behind the right cylinder head. On 4.7L models, the crankshaft position sensor is mounted on the engine block above the starter motor. The engine will not operate if the PCM does not receive a crankshaft position sensor input.

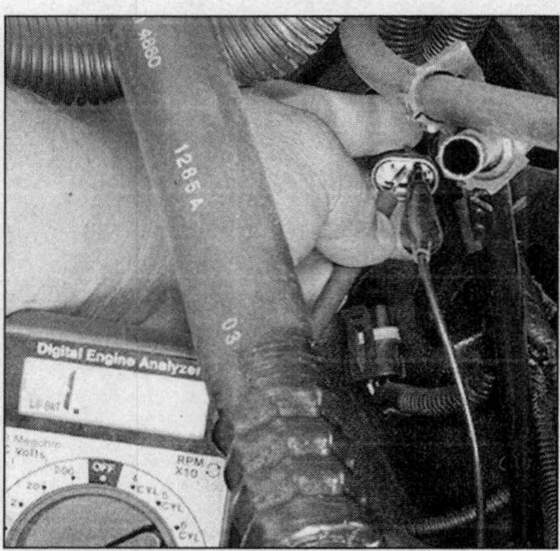

8.2 Disconnect the electrical connector from the crankshaft position sensor and check the power and ground circuits on the harness side

Check

Refer to illustrations 8.2 and 8.3

Note 1: *Performing the following test will set a diagnostic trouble code and illuminate the Check Engine light. Clear the diagnostic trouble code after performing the tests and making the necessary repairs* (see Section 2).

Note 2: *An analog (non-digital) voltmeter is required to check the crankshaft position sensor operation.*

2 Before checking the crankshaft position sensor, check the voltage supply and ground circuits from the PCM. On all except 4.7L V8 models, follow the wiring harness up from the

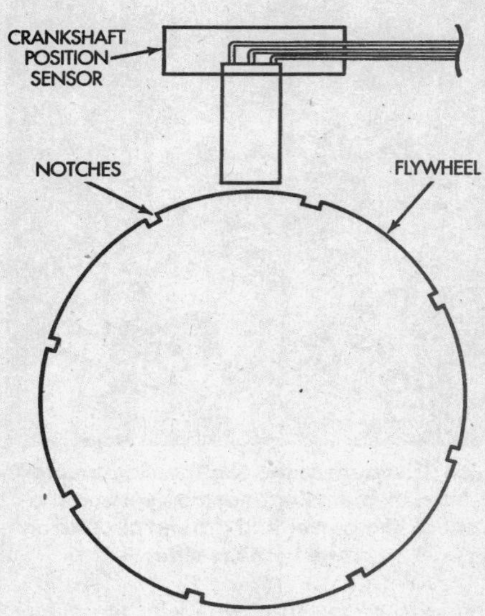

8.3 The crankshaft position sensor detects notches in the flywheel/driveplate (V8 engine shown)

crankshaft position sensor to locate the harness connector **(see illustrations 8.5 and 8.12)**. Disconnect the electrical connector and connect the positive lead of a voltmeter to white/black (1998 and earlier) or orange (1999 and later) wire terminal and the negative lead to the black/light blue terminal of the harness connector **(see illustration)**. Turn the ignition key On - the voltage should read approximately 8.0 volts (1995 and earlier) or 5.0 volts (1996 and later). If the voltage is incorrect, check the wiring from the crankshaft position sensor to the PCM. If the circuits are good, have the PCM checked at a dealer service department or other properly equipped repair facility. **Note:** *On 4.7L V8 models, the crankshaft position sensor and electrical connector may be difficult, if not impossible, to access without removing the starter.*

3 To check the crankshaft position sensor operation, reconnect the connector to the crankshaft position sensor and using a suitable probe, backprobe the orange/light green (1998 and earlier) or gray/black (1999 and later) wire terminal of the crankshaft position sensor connector **Note:** *Refer to Chapter 12 for additional information on how to backprobe a connector.* Connect the positive lead of an analog voltmeter to the probe and the negative lead to a good engine ground point. Turn the ignition key On. Rotate the engine slowly with a breaker bar and socket attached to the crankshaft pulley center bolt while watching the meter. The voltage should fluctuate from approximately 5.0 volts to 0.0 volts as the notches in the flywheel/driveplate pass the sensor **(see illustration)**. If the test results are incorrect, replace the crankshaft position sensor. **Note:** *Rotate the engine slowly through at least one complete revolution. The notches are equally spaced on V8 models. On six-cylinder models, they are 120-degrees apart. Removing the spark plugs from the engine will make the crankshaft much easier to turn.*

Replacement

Six-cylinder models

Refer to illustrations 8.5 and 8.6

4 Disconnect the crankshaft sensor wiring harness connector.

5 Remove the sensor mounting bolt(s) and remove the sensor **(see illustration)**.

6 On models with an automatic transmission, the sensor is retained by a single bolt and the sensor bolt hole is slotted. If installing the original sensor, obtain an original-type spacer from a dealership parts department and apply it to the tip of the sensor to properly set the sensor depth **(see illustration)**. A new sensor should be equipped with the paper spacer in place. Install the sensor, lightly press the spacer against the driveplate and tighten the bolt to 60 in-lbs. **Caution:** *Damage to the sensor will occur at startup if the sensor-to-driveplate clearance is not correct.*

7 On manual transmission models, use only the original bolts to mount the sensor

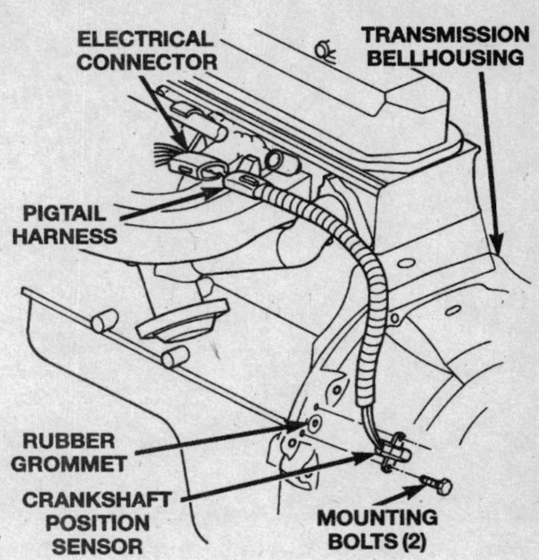

8.5 Crankshaft position sensor installation details - 4.0L six-cylinder models

(they are machined to correctly space the sensor to the flywheel). Tighten the bolts to 14 ft-lbs.

8 Connect the crankshaft sensor wiring harness connector.

5.2L/5.9L V8 models

Refer to illustration 8.12

9 Remove the spark plug wires from the right bank spark plugs. Mark the correct cylinder location of each wire as it is removed. Detach the wires from the mounting stud at the rear of the valve cover and position the wires out of the way.

10 Remove the right exhaust manifold heat shield (see Chapter 2B).

11 Remove the EGR valve and tube (see Section 15). Remove the oil pressure sending unit for access to the tube nut, if necessary.

12 Disconnect the crankshaft sensor wiring harness connector, remove the mounting bolts and remove the sensor **(see illustration)**.

13 Installation is the reverse of removal. Tighten the crankshaft position sensor mounting bolts to 70 in-lbs.

6

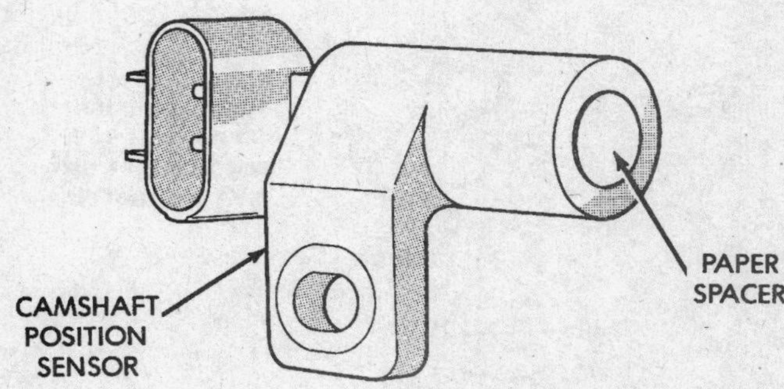

8.6 On six-cylinder models with an automatic transmission, make sure the paper spacer is in place before installing the crankshaft position sensor

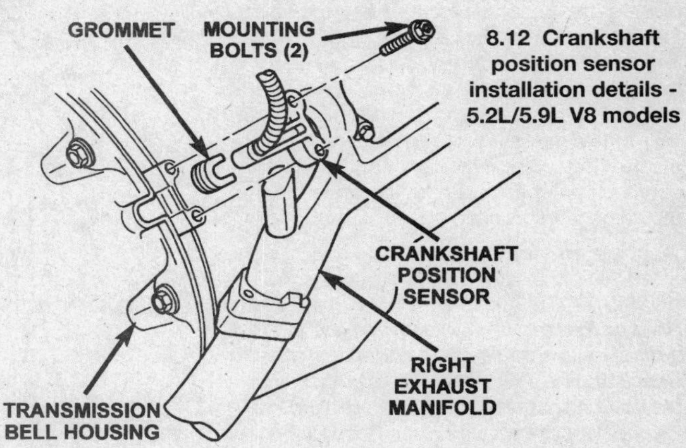

8.12 Crankshaft position sensor installation details - 5.2L/5.9L V8 models

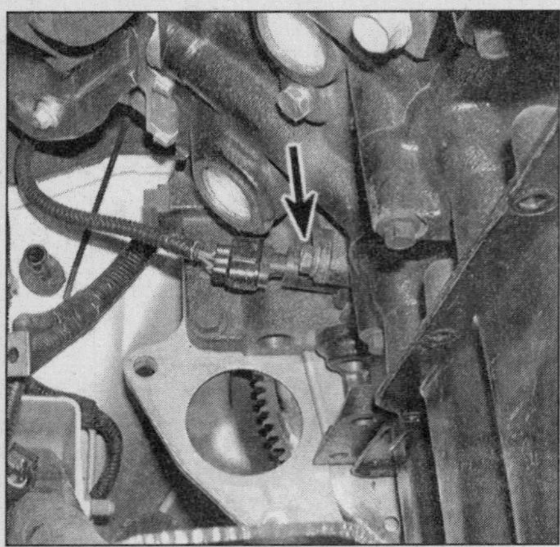

8.17 On 4.7L V8 models, remove the bolt (arrow) and withdraw the crankshaft position sensor from the engine block

4.7L V8 models

Refer to illustration 8.17

14 Disconnect the cable from the negative battery terminal.

15 Remove the starter motor (see Chapter 5).

16 Disconnect the electrical connector from the crankshaft position sensor.

17 Remove the mounting bolt and carefully withdraw the sensor from the engine block **(see illustration)**.

18 Inspect the O-ring and replace it if necessary.

19 Apply a small amount of engine oil on the O-ring and press the sensor in the engine block until fully seated. Install the mounting bolt and tighten it to 21 ft-lbs.

20 The remainder of installation is the reverse of removal.

9 Camshaft position sensor - check and replacement

1 The camshaft position sensor, in conjunction with the crankshaft position sensor, determines the timing for the fuel injection on each cylinder. The camshaft position sensor is a Hall-Effect device. On 1998 and earlier models, the sensor is located in the distributor and is triggered by a pulse ring. On 1999 and 2000 six-cylinder models, the sensor is located on top of the oil pump drive and is triggered by a pulse ring. On 1999 and 2000 V8 models, the sensor is located at the front of the right-bank cylinder head and is triggered by notches in a reluctor wheel behind the camshaft sprocket.

Check

Refer to illustration 9.4

Note 1: *Performing the following test will set a diagnostic trouble code and illuminate the Check Engine light. Clear the diagnostic trouble code after performing the tests and making the necessary repairs* (see Section 2).

Note 2: *An analog (non-digital) voltmeter is required to check the camshaft position sensor operation.*

2 On 5.2L/5.9L V8 models, remove the air intake duct.

3 Disconnect the electrical connector from the ignition coil, ignition coil rail or each individual coil, as applicable. Remove the fuel pump relay from the power distribution center. **Warning:** *Failure to disable the fuel and ignition system before performing this procedure could result in vehicle damage or personal injury.*

4 Before checking the camshaft position sensor, check the voltage supply and ground circuits from the PCM. Disconnect the electrical connector from the camshaft position sensor and connect the positive lead of a voltmeter to the white/black (1998 and earlier) or orange (1999 and later) wire terminal and the negative lead to the black/light blue wire terminal of the harness connector **(see illustration)**. Turn the ignition key On - the voltage should read approximately 8.0 volts (1995 and earlier) or 5.0 volts (1996 and later). If the voltage is incorrect, check the wiring from the camshaft position sensor to the PCM. If the circuits are good, have the PCM checked at a dealer service department or other properly equipped repair facility.

5 To check the camshaft position sensor operation, reconnect the connector to the camshaft position sensor and using a suitable probe, backprobe the gray/black (1998 and earlier) or tan/yellow (1999 and later) wire terminal of the camshaft position sensor connector. **Note:** *Refer to Chapter 12 for additional information on how to backprobe a connector.* Connect the positive lead of an analog voltmeter to the probe and the negative lead to a good engine ground point. Turn the ignition key to Start - with the engine cranking, the voltage should fluctuate from approximately 5.0 volts to 0.0 volts. If the test results are incorrect, replace the camshaft position sensor. **Warning:** *Make sure that the meter leads, loose clothing, long hair, etc. are away from the moving parts of the engine (drivebelt, cooling fan, etc.) before cranking the engine over.*

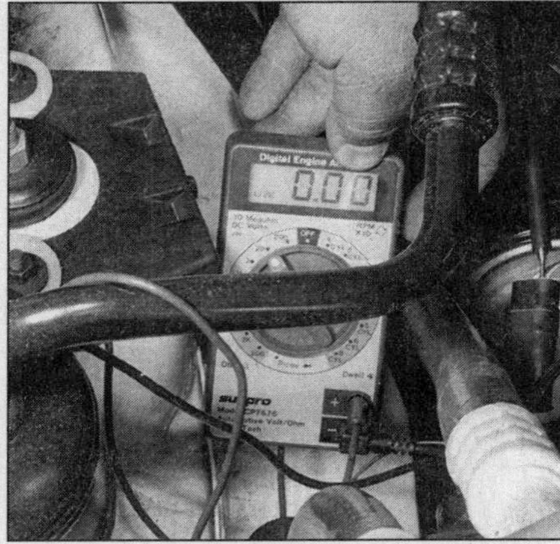

9.4 Disconnect the electrical connector from the camshaft position sensor and check the power and ground circuits on the harness side

Replacement

1993 and 1994 six-cylinder models

Refer to illustration 9.9

6 Disconnect the negative battery cable from the battery terminal. **Caution:** *If the stereo in your vehicle is equipped with an anti-theft system, make sure you have the*

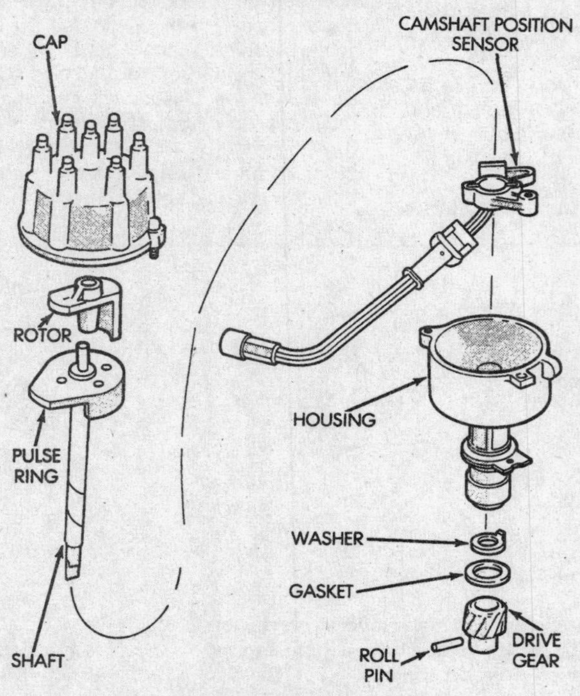

9.9 Distributor components - 1993 and 1994 4.0L six-cylinder models

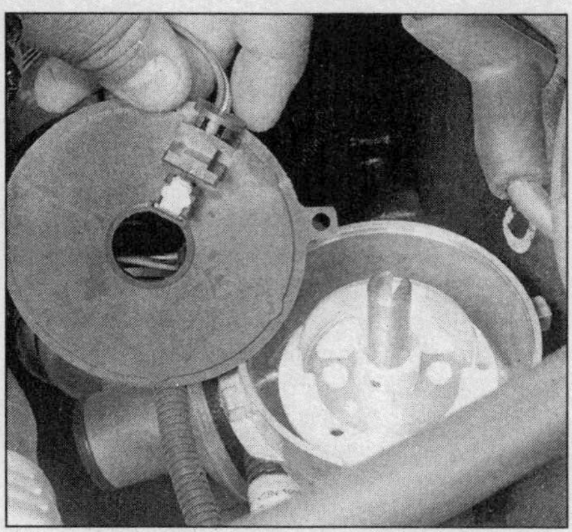

9.18 Carefully remove the camshaft position sensor from the distributor - 1995 through 1998 six-cylinder models and all 5.2L/5.9L V8 models

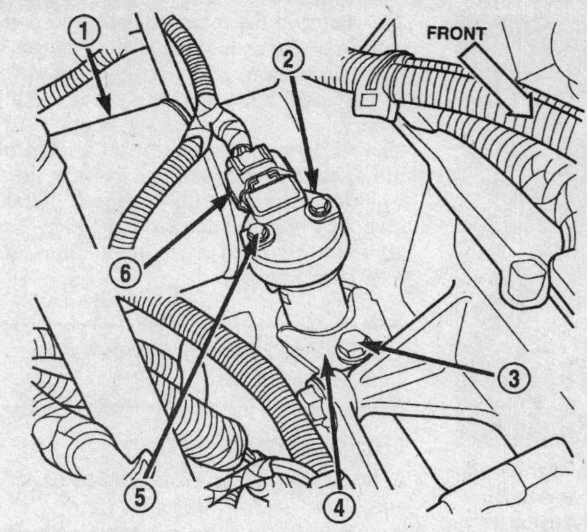

9.21 Camshaft position sensor installation details - 1999 and later 4.0L six-cylinder models

1 *Oil filter*
2 *Camshaft position sensor*
3 *Clamp bolt*
4 *Oil pump drive clamp*
5 *Camshaft position sensor mounting bolts*
6 *Electrical connector*

correct activation code before disconnecting the battery.

7 Remove the distributor (see Chapter 5).

8 Mark the position of the gear and the shaft in relation to the roll pin. This will be necessary for reassembly.

9 Use a small punch and remove the roll pin from the distributor shaft **(see illustration)**.

10 Remove the gear and thrust washer from the distributor.

11 Slide the distributor shaft out of the distributor housing.

12 Remove the camshaft position sensor mounting screw and the arm.

13 Slide the wire harness grommet out of the slot in the distributor housing and lift the sensor out.

14 Installation is the reverse of removal.

1995 through 1998 six-cylinder models and all 5.2L/5.9L V8 models

Refer to illustration 9.18

15 On V8 models, remove the air intake duct.

16 Remove the distributor cap and rotor from the distributor (see Chapter 1).

17 Disconnect the electrical connector from the camshaft position sensor.

18 Remove the camshaft position sensor from the distributor **(see illustration)**.

19 Installation is the reverse of removal. Align the tab on the camshaft position sensor with the notch in the distributor housing.

1999 and later models

Six-cylinder models

Refer to illustrations 9.21 and 9.23

Note: *The camshaft position sensor is mounted on top of the oil pump drive assembly. It is possible to replace the camshaft position sensor without removing the oil pump drive assembly. If the oil pump drive assembly must be removed, the mounting flange must be precisely marked in relation to the engine block or it will be necessary to reset the camshaft position sensor with a scan tool.*

20 Disconnect the electrical connector from the camshaft position sensor.

21 If removing the sensor only, remove the two mounting bolts and remove the sensor from the oil pump drive **(see illustration)**. Install the new sensor, tighten the bolts securely and connect the electrical connector.

22 If removing the oil pump drive assembly, position number one cylinder at TDC on the compression stroke (see Chapter 2A).

23 Remove the sensor (see Step 21). Insert an appropriate size alignment pin (such as a drill bit) through the holes in the sensor base and pulse ring **(see illustration)**. If the holes do not align, rotate the crankshaft until they do.

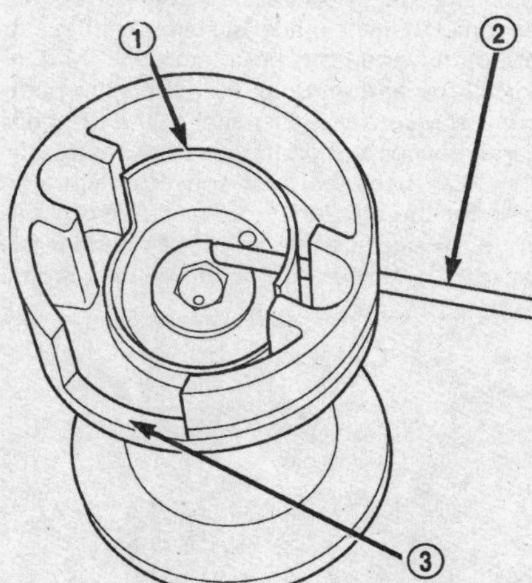

9.23 Before removing the camshaft position sensor on a 1999 or later 4.0L six-cylinder model, position the engine with number one cylinder at TDC and install an appropriate size alignment pin through the hole in the base and pulse ring

1 *Pulse ring*
2 *Alignment pin*
3 *Oil pump driveshaft assembly*

24 Mark the position of the oil pump drive mounting flange base to the engine to ensure the oil pump drive can be re-installed in exactly the same position as originally installed.

25 Remove the oil pump drive hold-down bolt and withdraw the oil pump drive assembly. Remove and discard the O-ring.

26 If the crankshaft has been moved while the oil pump drive is out, the number one piston must be repositioned at TDC. This can be done by feeling for compression pressure at the number one spark plug hole as the crankshaft is rotated. Once compression is felt, continue rotating the crankshaft until the mark on the crankshaft damper is aligned with the zero or TDC mark on the timing indicator (see Chapter 2A).

27 Turn the pulse ring until the hole in the pulse ring aligns with the hole in the housing and install an alignment pin through the holes **(see illustration 9.23)**. Install a new O-ring on the oil pump drive assembly.

28 Insert the oil pump drive into the engine block and align the marks made in Step 24. Because of the helical-cut drive gear, the assembly will rotate clockwise as the gears engage. Be sure to compensate by starting the installation with the drive assembly positioned counterclockwise from the desired finished position. Installing the oil pump drive assembly is very similar to installing a distributor, see Chapter 5 for additional information, if necessary. The oil pump drive must be installed in the same position as originally installed before removal. When properly installed, a line drawn through the center of the camshaft position sensor and electrical connector should be parallel with the centerline of the engine. **Note:** *If the alignment marks are lost or poor driveability symptoms appear after installing the oil pump drive, take the vehicle to a dealer service department or other properly equipped repair facility and have the fuel synchronization reset with a scan tool.*

29 Install the oil pump drive hold-down clamp and tighten the bolt to 17 ft-lbs.

9.33 On 4.7L V8 models, remove the bolt (arrow) and withdraw the camshaft position sensor from the right cylinder head (seen here from below)

30 The remainder of installation is the reverse of removal.

V8 models

Refer to illustration 9.33

31 Raise the vehicle and support it securely on jackstands.

32 Disconnect the electrical connector from the camshaft position sensor.

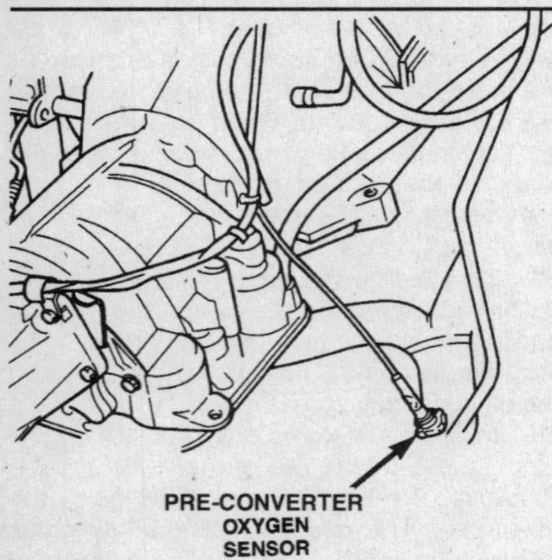

10.3a The pre-converter oxygen sensor is located in the exhaust pipe between the exhaust manifold and the catalytic converter

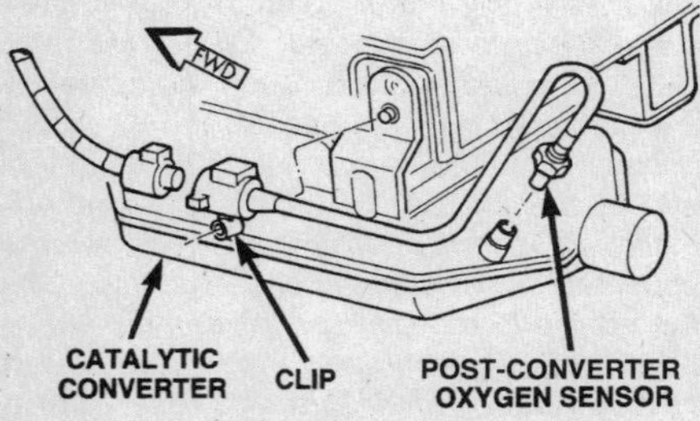

10.3b On 1996 through 1998 models, the post-converter oxygen sensor is located in the catalytic converter outlet pipe

33 Remove the mounting bolt and carefully withdraw the sensor from the cylinder head **(see illustration)**.

34 Inspect the O-ring and replace it if necessary.

35 Apply a small amount of engine oil on the O-ring and press the sensor into the cylinder head until fully seated. Install the mounting bolt and tighten it to 106 in-lbs.

36 The remainder of installation is the reverse of removal.

10 Oxygen sensor - check and replacement

General description

Refer to illustrations 10.3a, 10.3b, 10.3c and 10.3d

1 The oxygen in the exhaust reacts with the elements inside the oxygen sensor to produce a voltage output that varies from 0.1 volt (high oxygen, lean mixture) to 0.9 volt (low oxygen, rich mixture). The pre-converter oxygen sensor (mounted in the exhaust system before the catalytic converter) provides a feedback signal to the PCM that indicates the amount of leftover oxygen in the exhaust. The PCM monitors this variable voltage continuously to determine the required fuel injector pulse width and to control the engine air/fuel ratio. A mixture ratio of 14.7 parts air to 1 part fuel is the ideal ratio for minimum exhaust emissions, as well as the best combination of fuel economy and engine performance. Based on oxygen sensor signals, the PCM tries to maintain this air/fuel ratio of 14.7:1 at all times.

2 The post-converter oxygen sensor (mounted in the exhaust system after the catalytic converter) has no effect on PCM control of the air/fuel ratio. However, the post-converter sensor is identical to the pre-converter sensor and operates in the same way. The PCM uses the post-converter signal to monitor the efficiency of the catalytic converter. A post-converter oxygen sensor will produce a slower fluctuating voltage signal that reflects the lower oxygen content in the post-catalyst exhaust.

3 Oxygen sensor configuration varies depending on model, and is as follows **(see illustrations)**:

a) 1993 through 1995 models are equipped with a single pre-converter oxygen sensor.

b) 1996 through 1998 models are equipped with one pre-converter oxygen sensor and one post-converter oxygen sensor.

c) 1999 and later Federal emissions models are equipped with one pre-converter oxygen sensor and one post-converter oxygen sensor.

d) 1999 and later California emissions models are equipped with four oxygen sensors; two pre-converter oxygen sensors and two post-converter oxygen sensors.

4 An oxygen sensor produces no voltage when it is below its normal operating temperature of about 600-degrees F. During this warm-up period, the PCM operates in an open-loop fuel control mode. It does not use the oxygen sensor signal as a feedback indication of residual oxygen in the exhaust. Instead, the PCM controls fuel metering based on the inputs of other sensors and its own programs.

5 Proper operation of an oxygen sensor depends on four conditions:

a) ***Electrical*** *- The low voltages generated by the sensor require good, clean connections which should be checked whenever a sensor problem is suspected or indicated.*

b) ***Outside air supply*** *- The sensor needs air circulation to the internal portion of the sensor. Whenever the sensor is installed, make sure the air passages are not restricted.*

c) ***Proper operating temperature*** *- The PCM will not react to the sensor signal until the sensor reaches approximately 600-degrees F. This factor must be considered when evaluating the performance of the sensor.*

10.3c 1999 and later California emissions models are equipped with four oxygen sensors - two pre-converter oxygen sensors . . .

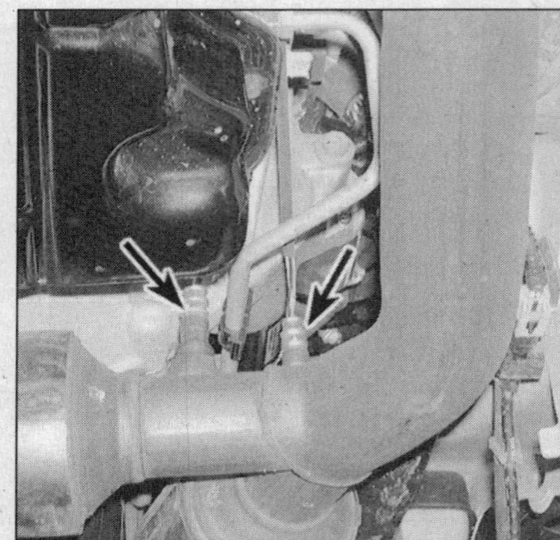

10.3d . . . and two post-converter sensors (arrows)

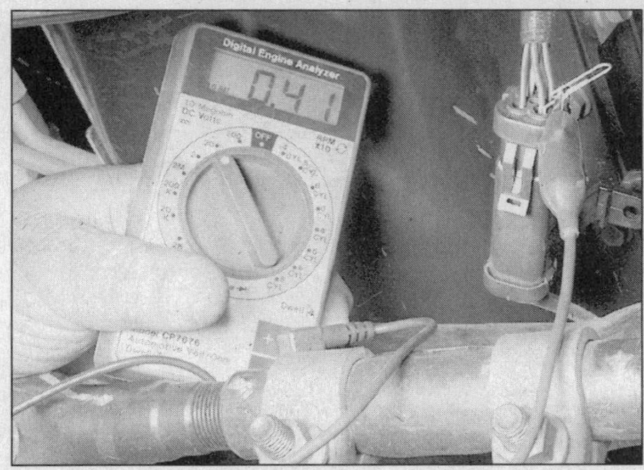

10.7a Backprobe the oxygen sensor electrical connector signal wire and monitor the sensor output voltage

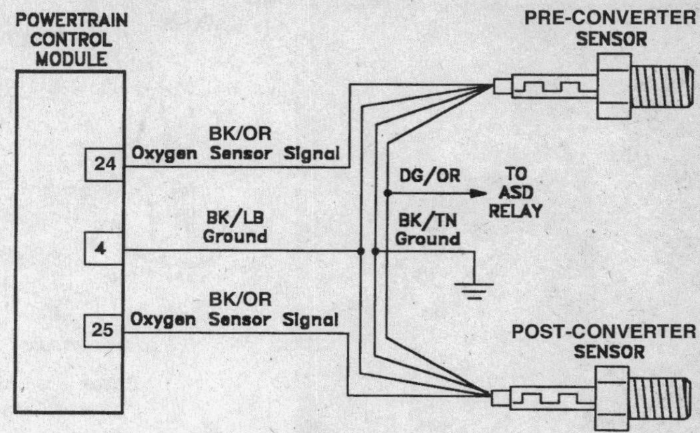

10.7b Typical oxygen sensor circuit - refer to the wiring diagrams to determine the exact wire colors for testing on your particular model

d) ***Unleaded fuel** - Unleaded fuel is essential for proper operation of the sensor.*

6 The PCM can detect several different oxygen sensor problems and set diagnostic trouble codes to indicate the specific fault (see Section 2). When an oxygen sensor fault occurs, the PCM will disregard the oxygen sensor signal voltage and revert to open-loop fuel control as described previously.

Check

Refer to illustrations 10.7a, 10.7b and 10.9

Caution: *The oxygen sensor is very sensitive to excessive circuit loads and circuit damage of any kind. For safest testing, disconnect the oxygen sensor connector, install jumper wires between the two connectors and connect your voltmeter to the jumper wires. If jumper wires aren't available, carefully backprobe the wires in the connector shell with suitable probes (such as T-pins). Do not puncture the oxygen sensor wires or try to backprobe the sensor itself. Use only a digital voltmeter to test an oxygen sensor.*

Note 1: *On some models the wire connecter for the oxygen sensor is located at the back of the engine block. It may be necessary to lower the transmission* (see Chapter 7) *or remove some components to disconnect it.*

Note 2: *Performing the following test will set a diagnostic trouble code and illuminate the Check Engine light. Clear the diagnostic trouble code after performing the tests and making the necessary repairs* (see Section 2).

7 Turn the ignition ON but do not start the engine. Connect your voltmeter negative (-) lead to a good ground and the positive (+) lead to the oxygen sensor signal wire at the oxygen sensor connector **(see illustrations)**. The meter should read approximately 400 to 450 millivolts (0.40 to 0.45 volt). If it doesn't, trace and repair the circuit from the sensor to the PCM. **Note:** *Refer to the wiring diagrams at the end of Chapter 12 to identify the signal wire by the color code.*

8 Start the engine and let it warm up to normal operating temperature; again check the oxygen sensor signal voltage.

1995 AND EARLIER

1996 AND LATER

10.9 Oxygen sensor terminal identification (looking into the face of the harness connector)

	1995 and earlier		*1996 and later*
1	*Sensor ground*	*1*	*B+ (from ASD relay)*
2	*Sensor signal*	*2*	*Heater ground*
3	*Heater ground*	*3*	*Sensor ground*
4	*B+ (from fuel pump relay)*	*4*	*Sensor signal*

a) *Voltage from a pre-converter sensor should range from 100 to 900 millivolts (0.1 to 0.9 volt) and switch actively between high and low readings.*

b) *Voltage from a post-converter sensor should also read between 100 to 900 millivolts (0.1 to 0.9 volt) but it should not switch actively. The post-converter oxygen sensor voltage may stay toward the center of its range (about 400 millivolts) or stay for relatively longer periods of time at the upper or lower limits of the range.*

9 Check the battery voltage supply and ground circuits to the oxygen sensor heater. Disconnect the electrical connector and connect the voltmeter negative (-) lead to the heater ground terminal and the positive (+) lead to the B+ terminal of the sensor connector **(see illustration)**. With the ignition ON, the meter should read more than 10 volts. Battery voltage is supplied to the sensors through the fuel pump relay (1995 and earlier) or ASD relay (1996 and later) for only about three seconds when the engine is not running. Have an assistant turn the ignition ON while you read the voltmeter. Refer to the wiring diagrams in Chapter 12 for more information on the circuits and relays.

10 Check the resistance of the oxygen sensor heater. With the connector disconnected, connect an ohmmeter to the two heater terminals of the connector (oxygen sensor side). The oxygen sensor pigtail is generally not color coded, but the heater wires are usually the white wires. The oxygen sensor heater resistance should be 4.0 to 7.0 ohms. If an open circuit or excessive resistance is indicated, replace the oxygen sensor. **Note:** *If the tests indicate that a sensor is good, and not the cause of a driveability problem or diagnostic trouble code, check the wiring harness and connectors between the sensor and the PCM for an open or short circuit. If no problems are found, have the vehicle checked by a dealer service department or other qualified repair shop.*

Replacement

Refer to illustration 10.14

11 Observe the following guidelines when replacing an oxygen sensor:

a) *The sensor has a permanently attached pigtail and electrical connector which should not be removed from the sensor. Damage or removal of the pigtail or electrical connector can harm operation of the sensor.*

10.14 A special slotted socket, allowing clearance for the wiring harness, may be required for oxygen sensor removal (the tool is available at most auto parts stores)

- b) *Keep grease, dirt and other contaminants away from the electrical connector and the louvered end of the sensor.*
- c) *Do not use cleaning solvents of any kind on the oxygen sensor.*
- d) *Do not drop or roughly handle the sensor.*

12 Raise the vehicle and place it securely on jackstands.

13 Disconnect the electrical connector from the sensor.

14 Using a suitable wrench or specialized oxygen sensor socket, unscrew the sensor from the exhaust pipe **(see illustration)**.

15 Anti-seize compound must be used on the threads of the sensor to aid future removal. The threads of most new sensors will be coated with this compound. If not, be sure to apply anti-seize compound before installing the sensor.

16 Install the sensor and tighten it securely.

17 Reconnect the electrical connector to the sensor and lower the vehicle.

11 Battery temperature sensor - check and replacement

1 The battery temperature sensor provides the PCM with information on battery temperature. The PCM uses this information to regulate the charging system voltage. The PCM increases the charging system voltage at colder temperatures and decreases the charging system voltage at warmer temperatures. The battery temperature sensor is used on 1996 and later models only.

Check

Refer to illustration 11.2

2 Locate the sensor harness from under the battery and follow the lead to the electrical connector **(see illustration)**. Disconnect the connector.

3 Using an ohmmeter, measure the resistance of the sensor across the two terminals of

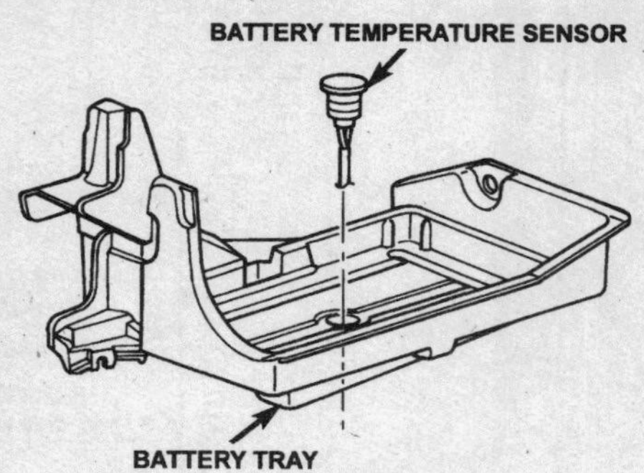

11.2 Battery temperature sensor installation details

the sensor connector. At 75 to 80 degrees F the sensor resistance should be 9,000 to 11,000 ohms.

4 If the resistance is not as specified, replace the sensor.

Replacement

5 Remove the battery (see Chapter 5).

6 Disconnect the electrical connector from the sensor harness.

7 Pry the sensor from the battery tray and pull the harness and connector through the hole.

8 Installation is the reverse of removal.

12 Vehicle Speed Sensor (VSS) - check and replacement

1 The Vehicle Speed Sensor (VSS) is a variable-reluctance sensor mounted on the transmission or transfer case and driven by a speedometer pinion gear. It produces a pulsating voltage, the frequency of which is proportional to vehicle speed. The PCM uses the sensor input signal for several different engine and transmission control functions. The VSS signal also drives the speedometer on the instrument panel. A defective VSS can cause various driveability and transmission problems. 1998 and earlier models are equipped with a vehicle speed sensor. On 1999 and later models, the PCM receives vehicle speed information from the antilock brake system rear wheel speed sensor.

Check

Refer to illustration 12.3

Note: *Performing the following test will set a diagnostic trouble code and illuminate the Check Engine light. Clear the diagnostic trouble code after performing the tests and making the necessary repairs* (see Section 2).

2 Raise the vehicle and support it securely on jackstands.

3 Before checking the VSS, check the voltage supply and ground circuits from the PCM. Disconnect the electrical connector from the VSS and connect the positive lead of a voltmeter to terminal number three and the negative lead to terminal number two of

12.3 Vehicle speed sensor location and connector terminal identification

1 Sensor signal
2 Ground
3 5-volt supply

the harness connector **(see illustration)**. Turn the ignition key On - the voltage should read approximately 5.0 volts. If the voltage is incorrect, check the wiring from the VSS to the PCM. If the circuits are good, have the PCM checked at a dealer service department or other properly equipped repair facility.

4 To check the VSS operation, reconnect the connector to the VSS and using a suitable probe, backprobe terminal number one of the VSS connector. Connect the positive lead of a voltmeter to the probe and the negative lead to a good chassis ground point. Turn the ignition key On.

5 Have an assistant hold one of the rear tires (or block the tire so it will not turn). Rotate the other tire by hand while watching the voltmeter, the voltage should fluctuate between zero and five volts as the speedometer pinion turns.

6 If the test results are incorrect, replace the vehicle speed sensor.

Replacement

Refer to illustration 12.9

7 Raise the vehicle and support it securely on jackstands.

8 Disconnect the electrical connector from the VSS.

9 Remove the bolt and remove the VSS from the speedometer pinion adapter **(see illustration)**. It isn't necessary to remove the complete adapter assembly from the transmission.

10 Inspect the O-ring on the sensor and replace it if damaged. If you are installing a new sensor, use a new O-ring.

11 Installation is the reverse of removal.

13 Idle Air Control (IAC) valve - check and replacement

1 The idle speed is controlled by the idle air control valve, located on the throttle body. The idle air control valve regulates the air

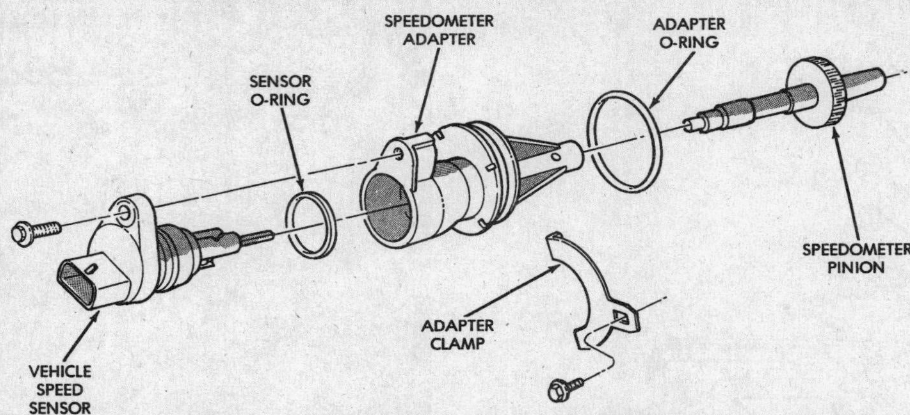

12.9 Vehicle speed sensor, adapter and speedometer pinion installation details (1998 and earlier models only)

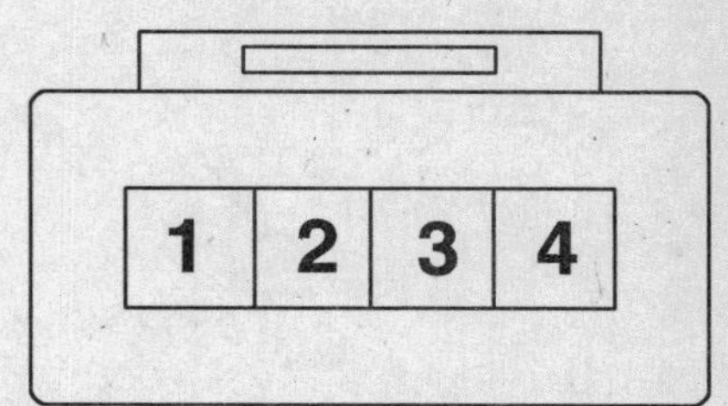

13.4 Idle air control valve terminal identification for testing

bypassing the throttle plate by moving the pintle in or out of the air passage. The idle air control valve is controlled by the PCM, adjusting the idle speed depending upon the running conditions of the engine (air conditioning system, power steering, cold and warm running etc.).

Check

Refer to illustration 13.4

Note: *Performing the following test will set a diagnostic trouble code and illuminate the Check Engine light. Clear the diagnostic trouble code after performing the tests and making the necessary repairs* (see Section 2).

2 A scan tool is required for complete testing of the idle air control valve and circuits. However, there are several tests the home mechanic can perform on the idle air control system to verify operation but they are limited and are useful only in the case of definite idle air control valve failure.

3 When the engine is started cold, the idle air control valve should vary the idle as the engine begins to warm-up. Allow the engine to warm-up, then disconnect the electrical connector from the idle air control valve and listen carefully for a change in the idle. Connect the idle air control valve electrical connector and place a load on the engine by placing the transmission in gear (automatic), turning the air conditioning on and/or operating the power steering. The idle should remain steady or increase slightly. If the engine stumbles or stalls, or if there are obvious signs that the idle air control valve is not working, stop the engine and continue testing.

4 Disconnect the electrical connector from the idle air control valve. Using an ohmmeter, measure the resistance across terminals 1 and 3 (1998 and earlier) or 1 and 4 (1999 and later) of the idle air control valve **(see illustration)**. Then measure the resistance across terminals 2 and 4 (1998 and earlier) or 2 and 3 (1999 and later) - the resistance should be equal on each measurement. If one or both checks indicate an open or short circuit, replace the idle air control valve.

5 If the idle air control valve is good, have the PCM diagnosed by a dealer service department or other qualified repair shop.

Replacement

Refer to illustrations 13.8a, 13.8b and 13.8c

6 Remove the air intake duct/resonator from the throttle body.

7 Disconnect the electrical connector from the idle air control valve.

8 Remove the two mounting screws and withdraw the idle air control valve from the throttle body **(see illustrations)**.

9 Installation is the reverse of removal. Be sure to install a new O-ring.

14 Crankcase ventilation system

Refer to illustrations 14.2, 14.3a, 14.3b and 14.3c

1 The crankcase ventilation system reduces hydrocarbon emissions by scavenging crankcase vapors. It does this by circulating fresh air from the air cleaner through the crankcase, where it mixes with blow-by gases. The gases are drawn by intake manifold vacuum into the intake manifold, where they mix with the incoming air/fuel mixture and are consumed during normal combustion.

2 The six-cylinder engine is equipped with a fixed orifice crankcase ventilation system. This system operates the same as a conventional Positive Crankcase Ventilation system but it does not use the vacuum control valve. Instead, a molded vacuum tube connects manifold vacuum to the valve cover. The vacuum tube contains a fixed orifice of a cali-

13.8a Idle air control valve location - 4.0L six-cylinder models

13.8b Idle air control valve location - 5.2L/5.9L V8 models

13.8c Idle air control valve location - 4.7L V8 models

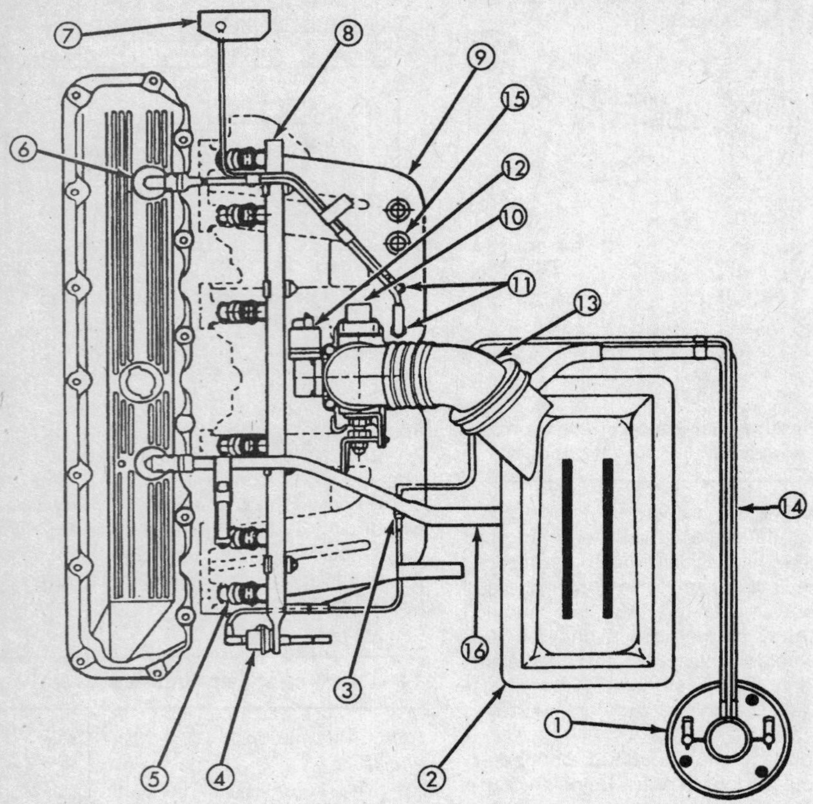

14.2 Fixed orifice crankcase ventilation system - typical 4.0L six-cylinder model

1. *EVAP canister*
2. *Air cleaner*
3. *Vacuum fitting*
4. *Fuel pressure regulator*
5. *Fuel injectors*
6. *Fixed orifice*
7. *Manifold absolute pressure sensor*
8. *Fuel rail*
9. *Intake manifold*
10. *Throttle position sensor*
11. *Manifold fittings*
12. *Idle air control valve*
13. *Fresh air hose*
14. *Vacuum tube assembly*
15. *Intake air temperature sensor*

brated size **(see illustration)**. It meters the amount of crankcase vapors drawn out of the engine. A fresh air supply hose from the air intake duct is connected to the valve cover. When the engine is running, fresh air enters the engine and mixes with the crankcase vapors. Manifold vacuum draws the vapor/air mixture through the fixed orifice and into the intake manifold.

14.3b On 4.7L V8 models, the PCV valve is located in the oil filler tube

3 V8 models are equipped with a Positive Crankcase Ventilation (PCV) system. The main component of this system is the PCV valve **(see illustrations)**. Fresh air flows from the air intake duct through a vent tube into the crankcase. Crankcase vapors are drawn from the crankcase by the PCV valve **(see illustration)**. To maintain idle quality and good driveability, the PCV valve restricts the flow when the intake manifold vacuum is high. When intake manifold vacuum is lower, maximum vapor flow is allowed through the valve.

14.3a On 5.2L/5.9L V8 models, the PCV valve is located in the right valve cover

4 Checking and replacement of the PCV valve is covered in Chapter 1.

15 Exhaust Gas Recirculation (EGR) system

Note: *If the EGR valve control solenoid becomes disconnected, damaged or the electrical signal is lost, the EGR valve will be open at all times. This will cause poor performance, rough idle and driveability problems during warm-up, idle and some driving conditions.*

General description

Refer to illustration 15.2

1 The EGR system is used on the 1993 through 1995 5.2L V8 models only. The EGR system reduces oxides of nitrogen by recirculating exhaust gas through the EGR valve passage and intake manifold into the combustion chambers.

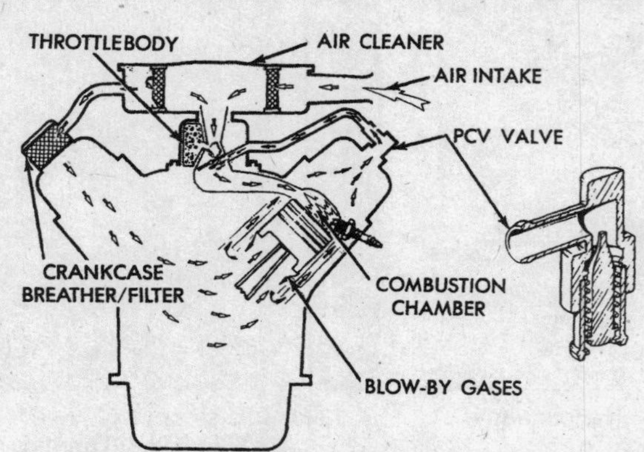

14.3c Gas flow in a typical PCV system

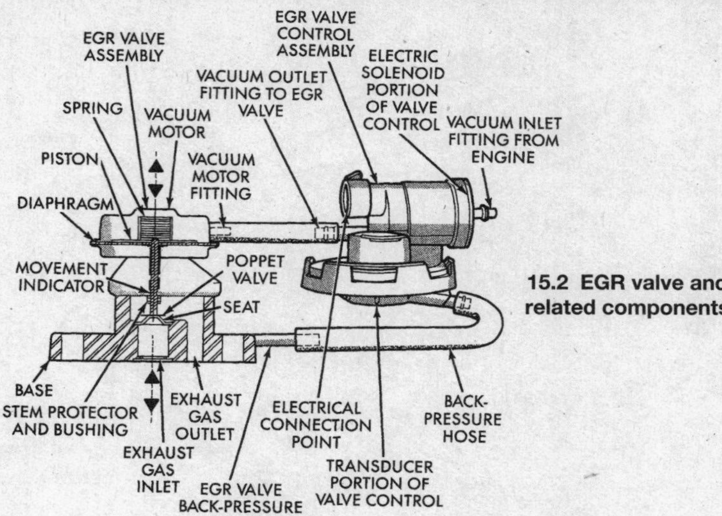

15.2 EGR valve and related components

15.4 Check for battery power to the EGR control solenoid with the ignition key ON

2 The EGR system consists of the EGR valve, EGR backpressure transducer, EGR control solenoid and the Powertrain Control Module (PCM) **(see illustration)**. When the ground is removed from the EGR control solenoid, vacuum is applied to the EGR valve and exhaust gas is allowed to flow into the intake manifold. The PCM controls the ground signal to the EGR control solenoid, producing the ideal EGR flow for the various engine operating conditions. The EGR backpressure transducer monitors exhaust backpressure. If exhaust backpressure is not strong enough for the desired EGR flow, the transducer bleeds off the vacuum signal, preventing EGR operation.

Check

Refer to illustrations 15.4, 15.5 and 15.9

3 Check the condition of the EGR system vacuum hoses. Check for leaks, cracks, kinks or hardening of the rubber hoses. Make sure all the hoses are intact before proceeding with the EGR check. Check the vacuum schematic for the correct EGR system hose routing. Reroute the hoses if necessary.

4 Disconnect the electrical connector from the EGR control solenoid and connect a voltmeter to the two terminals of the electrical connector. Turn the ignition key On and verify the EGR control solenoid is receiving battery voltage **(see illustration)**. If battery voltage is not present, check the circuit from the power distribution center to the EGR control solenoid (don't forget to check the fuses first). Check the circuit from the EGR control solenoid to the PCM. If the circuits are good, have the PCM diagnosed by a dealer service department or other qualified repair shop.

5 Detach the vacuum hose from the EGR valve and attach a hand-held vacuum pump directly to the EGR valve **(see illustration)**. Start the engine and apply 15 in-Hg of vacuum to the EGR valve. With the engine at idle speed, the idle speed should drop considerably or even stall as vacuum is applied.

6 If the engine speed does not change, a faulty EGR valve or plugged passages in the intake manifold are indicated. Watch the stem on the EGR valve for movement as you apply vacuum to the EGR valve. If the stem doesn't move or if the EGR valve fails to hold vacuum, replace the EGR valve. If the valve opens and the engine does not run rough or stall, remove the EGR valve and check the passages in the EGR valve and intake manifold. Check the EGR tube for damage or restrictions. **Note:** *If the EGR valve is severely plugged with carbon deposits, do not attempt to scrape them out. Replace the unit.*

7 Turn the engine off and remove the air intake duct from the throttle body. Disconnect the backpressure transducer vacuum hose from the base of the EGR valve. Using an air nozzle with a rubber tip, apply approximately 50 psi of compressed air to the metal fitting at the base of the EGR valve. Open the throttle plate and listen for the air being emitted into the intake manifold. If you hear air being emitted into the intake manifold, replace the EGR valve, the valve is leaking.

8 Disconnect the vacuum hose from the transducer fitting at the bottom of the EGR valve control solenoid. Attach a hand-held vacuum pump to the fitting and apply 10 in-Hg of vacuum to the transducer. If the transducer fails to hold vacuum, replace the EGR control solenoid/transducer, the diaphragm is leaking. If the transducer holds vacuum, remove the vacuum pump and reconnect the vacuum hose.

9 Disconnect the vacuum hose from the EGR control solenoid inlet fitting and install a vacuum gauge to the hose. Start the engine and verify that the control valve is receiving full manifold vacuum **(see illustration)**. If vacuum is not present, repair the vacuum line to the vacuum source. If vacuum is present, remove the vacuum gauge and reconnect the vacuum hose.

10 Disconnect the vacuum hose from the EGR control solenoid outlet fitting (to the EGR valve). Connect the vacuum gauge to the outlet fitting. Disconnect the electrical connector from the EGR control solenoid. Start the engine and increase the engine

15.5 Apply vacuum directly to the EGR valve with the engine running - the engine should run rough or stall

15.9 Check for manifold vacuum at the EGR control solenoid vacuum supply hose with the engine running

6

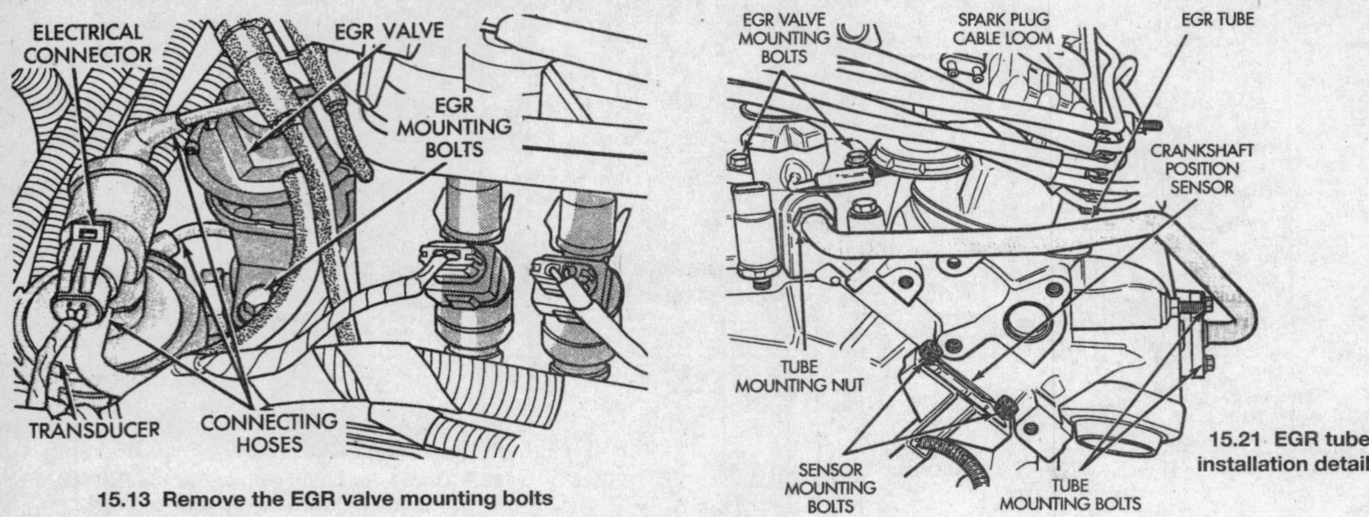

15.13 Remove the EGR valve mounting bolts

15.21 EGR tube installation details

speed to approximately 2000 rpm. Vacuum should be indicated at the gauge, if not, replace the EGR control solenoid. **Note:** *Sufficient exhaust backpressure must be present to allow full manifold vacuum through the control solenoid valve. If necessary, restrict the tailpipe opening to increase backpressure. Do not restrict the exhaust any longer than necessary or the engine may overheat.*

11 If the previous checks fail to reveal the problem, have the control system diagnosed by a dealer service department or other qualified repair shop.

Component replacement

Refer to illustrations 15.13 and 15.21

EGR valve and control solenoid

Note: *The EGR valve, EGR control solenoid, backpressure transducer and connecting hoses are replaced as a unit.*

12 Disconnect the vacuum hose and the electrical connector from the EGR valve control solenoid.

13 Remove the two EGR valve mounting bolts and remove the EGR valve/control solenoid assembly **(see illustration)**.

14 Clean the mating surfaces of the EGR valve and intake manifold.

15 Install the EGR valve assembly, using a new gasket. Tighten the bolts securely.

16 Connect the vacuum hose and electrical connector to the EGR control solenoid.

EGR tube

17 Disconnect the spark plug wires from the right-side cylinder bank spark plugs. Detach the wire retainers from the valve cover and position the spark plug wires aside.

18 Remove the heat shield from the right-side exhaust manifold.

19 Remove the EGR valve.

20 Remove the engine oil pressure sending unit.

21 Loosen the EGR tube nut from the intake manifold fitting. Remove the EGR tube mounting bolts from the exhaust manifold and remove the tube **(see illustration)**.

22 Discard the old gaskets from the EGR valve and EGR tube and clean the gasket mating surfaces.

23 Install the EGR tube with a new gasket at the exhaust manifold. Loosely install the mounting bolts at the exhaust manifold and connect the nut to the intake manifold fitting. Tighten the nut at the intake manifold, then tighten the mounting bolts at the exhaust manifold.

24 The remainder of installation is the reverse of removal with the following additions:

a) *Install the EGR valve with a new gasket.*
b) *Apply thread sealing compound to the threads of the oil pressure sending unit.*

16 Evaporative emissions control (EVAP) system

General description

1 The fuel evaporative emissions control (EVAP) system absorbs fuel vapors from the fuel tank and, during engine operation, releases them into the engine intake system where they mix with the incoming air/fuel mixture.

2 The evaporative system employs a canister filled with activated charcoal to absorb fuel vapors.

3 The fuel tank filler cap is fitted with a two-way valve as a safety device. The valve vents fuel vapors to the atmosphere if the evaporative control system fails. After passing through the vapor control valve, fuel vapor is carried by vent hoses to the charcoal canister. The activated charcoal in the canister absorbs and stores these vapors.

4 Another fuel cut-off valve (fuel tank rollover valve), mounted on the fuel tank, regulates fuel vapor flow from the fuel tank to the charcoal canister, based on the pressure or vacuum caused by temperature changes. The fuel tank rollover valve prevents fuel flow to the canister in the event the vehicle rolls over in an accident.

5 On 1995 and earlier six-cylinder Federal emissions models, the system is strictly a vacuum operated (non-electric) system. When engine vacuum is applied to a purge orifice on top of the canister, fresh air is allowed to flow through the canister, purging the vapors through a hose connected to the air cleaner housing. On all other models, the system is an electronic system controlled by the PCM.

6 When the engine is running and warmed to a pre-set temperature, a purge control solenoid allows a purge control diaphragm valve in the charcoal canister to be opened by intake manifold vacuum. Fuel vapors from the canister are then drawn through the purge control diaphragm valve by intake manifold vacuum. The duty cycle of the EVAP purge control solenoid regulates the rate of flow of the fuel vapors from the canister to the throttle body. The PCM controls the purge control solenoid. During cold running conditions and hot start time delay, the PCM does not energize the solenoid. After the engine has warmed up to the correct operating temperatures, the PCM purges the vapors into the throttle body according to the running conditions of the engine. The PCM will cycle (On then Off) the purge control solenoid about 5 to 10 times per second. The flow rate will be controlled by the pulse width or length of time the solenoid is allowed to be energized.

7 Some 1996 and later models are equipped with a leak detection monitor system. The system is a self-diagnostic system designed to detect a leak in the EVAP system. Each time the engine is started cold, the PCM energizes the leak detection pump. The pump pressurizes the EVAP system then shuts off. The PCM is able to detect a leak if the pump continues to run, unable to pressurize the system. If a leak is detected, the PCM will trigger a diagnostic trouble code (see Section 2).

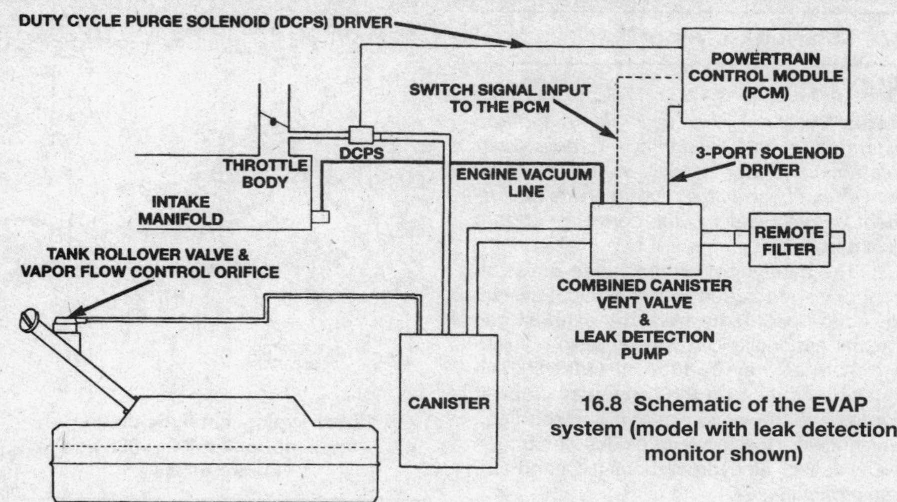

16.8 Schematic of the EVAP system (model with leak detection monitor shown)

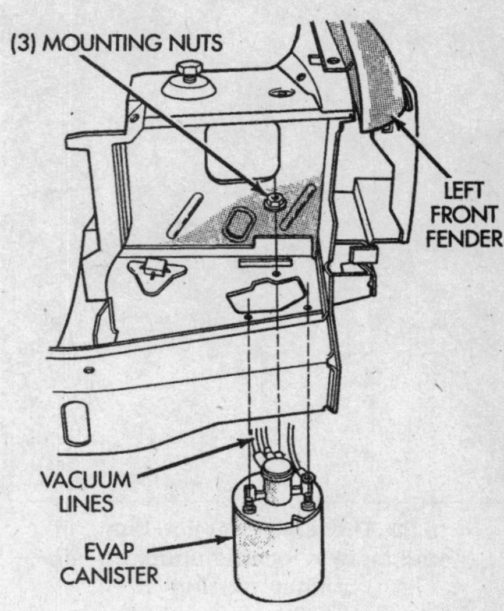

16.11 On 1998 and earlier models, the EVAP canister is mounted at the front of the left front fender

Check

Refer to illustration 16.8

Note: *The evaporative control system, like all emission control systems, is protected by a federally-mandated warranty (5 years or 50,000 miles at the time this manual was written). The EVAP system probably won't fail during the service life of the vehicle; however, if it does, the hoses or charcoal canister are usually to blame.*

8 Always check the hoses first. A disconnected, damaged or missing hose is the most likely cause of a malfunctioning EVAP system. Refer to the Vacuum Hose Routing Diagram (attached to the radiator support) to determine whether the hoses are correctly routed and attached. Repair any damaged hoses or replace any missing hoses as necessary **(see illustration)**.

9 Check the related fuses and wiring to the purge control solenoid. Refer to the wiring diagrams at the end of Chapter 12, if necessary. The purge control solenoid valve is normally closed - no vapors will pass through the ports. When the PCM energizes the solenoid (by completing the circuit to ground), the valve opens and vapors flow through.

10 A scan tool is required to thoroughly check the system. If the above checks fail to identify the problem area, have the system diagnosed by a dealer service department or other qualified repair shop.

Component replacement

EVAP canister

Refer to illustrations 16.11 and 16.15

1998 and earlier models

11 The EVAP canister is located at the front of the left fender, under the headlight **(see illustration)**.

12 Remove the radiator grille and the front bumper cover (see Chapter 11).

13 Label and remove the hoses from the canister. Remove the mounting nuts and remove the canister.

14 Installation is the reverse of removal.

1999 and later models

15 The EVAP canister is located in the engine compartment mounted to the left inner fender **(see illustration)**.

16 Remove the purge control solenoid. Disconnect the hoses from the canister.

17 Remove the canister front mounting bracket bolts and withdraw the canister from the rear mounting bracket.

18 Installation is the reverse of removal.

Purge control solenoid

Refer to illustrations 16.19a and 16.19b

19 On 1998 and earlier models (except 1995 and earlier 5.2L V8 models), the purge control solenoid is mounted to the left inner fender **(see illustration)**. On 1995 and earlier 5.2L V8 models, the purge control solenoid is mounted on the intake manifold **(see illustration)**. On 1999 and later models, the purge control solenoid is mounted to the EVAP canister **(see illustration 16.15)**.

20 Disconnect the electrical connector.

16.15 EVAP canister and purge control solenoid locations - 1999 and later models

1 *EVAP canister*
2 *Purge control solenoid*

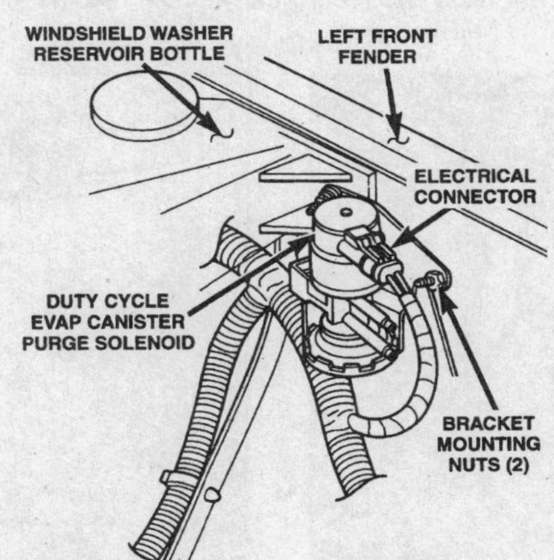

16.19a On 1998 and earlier models (except 1995 and earlier 5.2L V8 models), the purge control solenoid is mounted on the left front inner fender

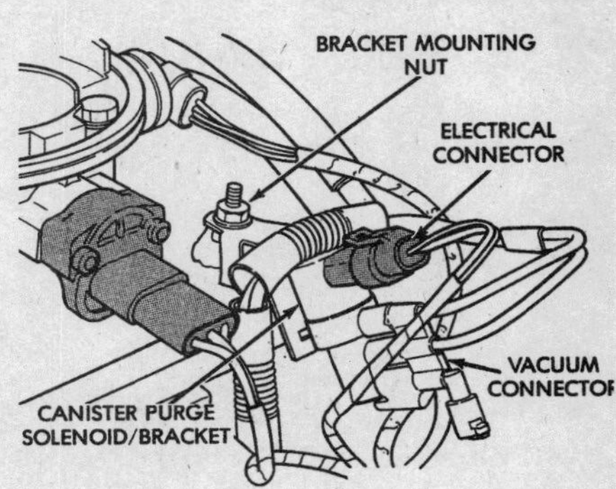

16.19b On 1995 and earlier 5.2L V8 models, the purge control solenoid is located on the left side of the intake manifold

16.23 The leak detection pump (if equipped) is located under the air filter housing

Label and remove the hoses from the purge control solenoid.

21 Remove the mounting bolts and remove the purge control solenoid and bracket.

22 Installation is the reverse of removal.

Leak detection pump

Refer to illustration 16.23

23 The leak detection pump (if equipped) is located under the air filter housing **(see illustration)**.

24 Remove the air filter housing (see Chapter 4).

25 On 1998 and earlier models, remove the purge control solenoid.

26 Disconnect the electrical connector from the leak detection pump.

27 Carefully label and remove the hoses from the leak detection pump.

28 Remove the mounting bolts and remove the leak detection pump.

29 Installation is the reverse of removal. **Note:** *The vapor and vacuum lines must be securely connected to the leak detection pump or a diagnostic trouble code may be set.*

17 Catalytic converter

Refer to illustrations 17.3a, 17.3b and 17.3c

Note: *Because of a federally mandated extended warranty which covers emissions-related components such as the catalytic converter, check with a dealer service department before replacing the converter at your own expense.*

1 The catalytic converter is an emission control device added to the exhaust system to reduce pollutants from the exhaust gas stream. A monolithic converter design is used with a three-way (reduction) catalyst. The catalytic coating on the three-way catalyst contains platinum, palladium and rhodium, which lowers the levels of oxides of nitrogen (NOx) as well as hydrocarbons (HC) and carbon monoxide (CO).

2 The test equipment for a catalytic converter is expensive and highly sophisticated. If you suspect that the converter on your vehicle is malfunctioning, take it to a dealer or authorized emissions inspection facility for diagnosis and repair.

3 Whenever the vehicle is raised for servicing of underbody components, check the converter for leaks, corrosion, dents and other damage **(see illustrations)**. Check the welds/flange bolts that attach the front and rear ends of the converter to the exhaust system. If damage is discovered, the converter should be replaced.

4 Although catalytic converters don't break too often, they can become plugged. The easiest way to check for a restricted converter is to use a vacuum gauge to diagnose the effect of a blocked exhaust on intake vacuum.

a) *Connect a vacuum gauge to an intake manifold vacuum source.*
b) *Warm the engine to operating temperature, place the transmission in park (automatic) or neutral (manual) and apply the parking brake.*
c) *Note and record the vacuum reading at idle.*

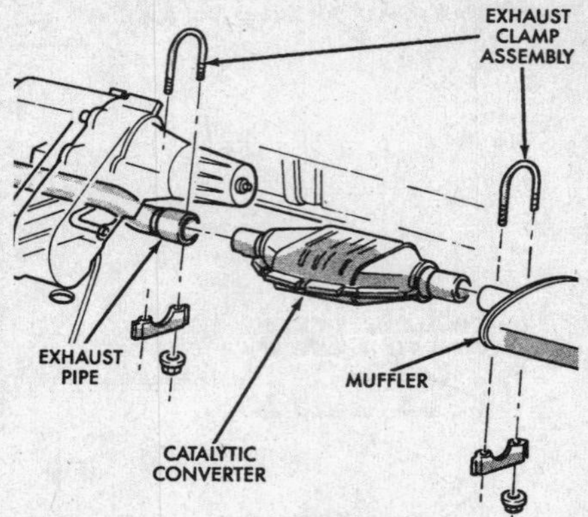

17.3a Typical catalytic converter installation details - 1998 and earlier models

d) *Open the throttle until the engine speed is about 2000 rpm.*
e) *Release the throttle quickly and record the vacuum reading.*
f) *Perform the test three more times, recording the reading after each test.*
g) *If the reading after the fourth test is more than one in-Hg lower than the reading recorded at idle, the catalytic converter, muffler or exhaust pipes may be plugged or restricted.*

5 If the catalytic converter requires replacement on 1998 and earlier models, remove the exhaust pipe clamps at the inlet and outlet pipes. Heat the inlet and outlet pipes until they are cherry red and twist the converter back-and-forth to loosen the pipes. Detach the exhaust hangers and separate the muffler from the converter, then remove the converter from the header pipe.

6 On 1999 and later models, the catalytic converter(s) are permanently welded into the exhaust system. If replacement is required, take the vehicle to a dealership service department or other qualified repair shop. Refer to the exhaust system servicing section in Chapter 4 for additional information.

17.3b On 1999 and later California emissions models, two mini-catalytic converters are located in the header pipe near the exhaust manifolds . . .

17.3c . . . and the main catalytic converter is located ahead of the muffler - the catalytic converters are designed to last the lifetime of the vehicle

Chapter 7 Part A
Manual transmission

Contents

Specifications

General

Transmission lubricant type	See Chapter 1

Torque specifications

	Ft-lbs
Clutch housing-to-engine bolts	45
Crossmember-to-frame bolts	30
Rear mount-to-transmission bolts	33
Rear mount-to-crossmember bracket through bolt	45

1 General information

1993 through 1995 models were available with either an automatic transmission or a five-speed manual transmission. The manual transmission option was dropped after the 1995 model year; all 1996 and later models are equipped with an automatic transmission only. All information on the manual transmission is included in this Part of Chapter 7. Information on the automatic transmission can be found in Part B of this Chapter. Vehicles with four-wheel-drive also use a transfer case. Information on the transfer case is in Part C of this Chapter.

All models with a manual transmission use an AX-15 five-speed with a fifth-gear overdrive. Because of the complexity, unavailability of replacement parts and special tools needed, internal repair is not recommended. The information in this Chapter is limited to general information and removal and installation of the transmission.

Depending on the expense involved in having a faulty transmission overhauled, it may be a good idea to replace the unit with either a new or rebuilt unit. Your local dealer or transmission shop should be able to supply you with information concerning cost, availability and exchange policy. Regardless of how you decide to remedy a transmission problem, you can still save a lot of money by removing and installing the unit yourself.

2 Manual transmission - removal and installation

Removal

Refer to illustrations 2.12 and 2.16

1 Disconnect the negative cable from the battery.
2 Raise the vehicle and support it securely on jackstands.
3 Remove the skid plate.
4 Remove the driveshaft(s) (see Chapter 8). Use a plastic bag to cover the end of the transmission to prevent fluid loss and contamination.
5 Disconnect the transfer case shift linkage from the shift lever or range lever, if equipped (see Chapter 7, Part C).
6 Unplug the electrical connector from the vehicle speed sensor.
7 Remove the harness wires from the clips on the transmission case.
8 Detach the transmission and transfer case vent hoses.
9 Unplug the wires at the transfer case electrical switch, if equipped.
10 Support the engine. This can be done from above with an engine hoist, or by placing a jack (with a block of wood as an insulator) under the engine oil pan. The engine should remain supported at all times while the transmission is out of the vehicle.
11 Support the transmission with a jack - preferably a special jack made for this purpose. Safety chains will help steady the transmission on the jack.

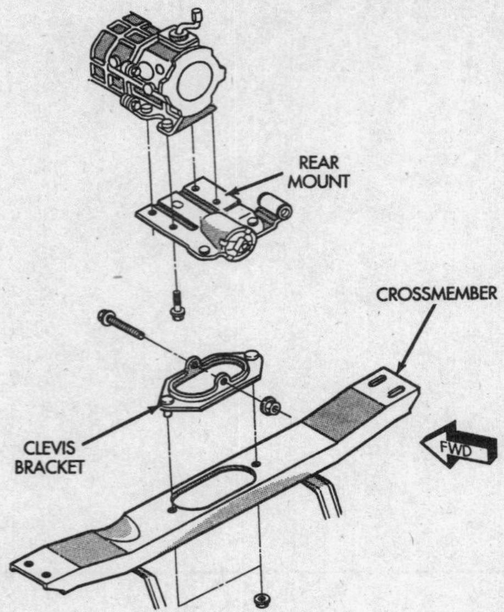

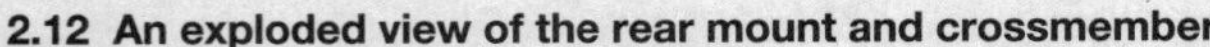

2.12 An exploded view of the rear mount and crossmember

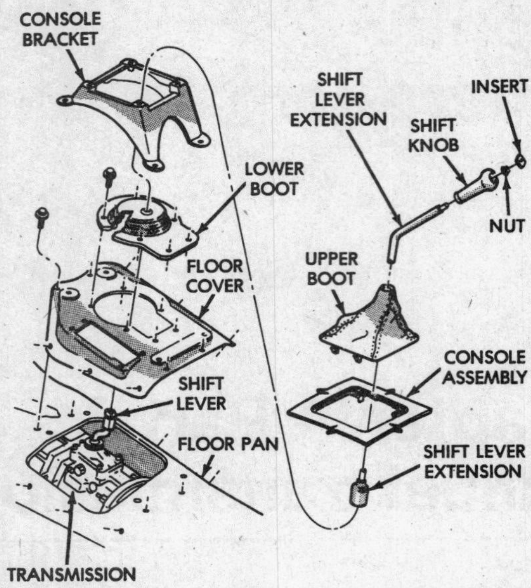

2.16 An exploded view of the shift lever assembly

12 Remove the rear mount-to-crossmember bracket through bolt. Raise the transmission slightly, remove the crossmember-to-frame nuts/bolts and remove the crossmember **(see illustration)**.

13 Remove the rear mount-to-transmission bolts, detach the rear mount from the exhaust pipe hanger and remove the rear mount.

14 Remove the transfer case from the transmission, if equipped (see Chapter 7, Part C).

15 Lower the transmission enough to provide access to the shift lever.

16 Reach up and around the transmission case and unseat the shift lever dust boot from the transmission shift tower **(see illustration)**. Move the boot up on the shift lever for access to the shift lever retainer.

17 To disengage the shift lever from the transmission:

a) *Reach up and around the transmission case and press the shift lever retainer down with your fingers.*
b) *Turn the retainer counterclockwise to release it.*
c) *Lift the lever and retainer out of the shift tower.* **Note:** *It isn't necessary to remove the shift lever from the floorpan boot. Simply leave the lever in place for later installation.*

18 Disconnect and remove the crankshaft position sensor (see Chapter 6). Don't misplace the crankshaft position sensor mounting bolts; the sensor must be reinstalled with the original bolts.

19 Detach the clutch slave cylinder from the clutch housing (see Chapter 8) and move the cylinder aside for working clearance. DO NOT disconnect the clutch hydraulic fluid line from the slave cylinder. If you disconnect the line from the slave cylinder, the system cannot be bled. The clutch hydraulic release system (master cylinder/reservoir, fluid line and slave cylinder) is a one-piece assembly that must not be disassembled. **Caution:** *Be very careful not to kink the clutch fluid hydraulic fluid line.*

20 Remove the clutch housing-to-engine bolts.

21 Make a final check that all electrical connectors are unplugged and all hoses and lines are disconnected, then move the transmission and jack toward the rear of the vehicle until the transmission input shaft is clear of the clutch housing.

22 Once the input shaft is clear, lower the transmission and remove it from under the vehicle. **Caution:** *Do not depress the clutch pedal while the transmission is out of the vehicle.*

23 Remove and inspect the clutch components, if necessary (see Chapter 8). In most cases, new clutch components should be routinely installed if the transmission is removed.

Installation

24 If removed, install the clutch components (see Chapter 8).

25 With the transmission secured to the jack as on removal, raise the transmission into position behind the engine and then carefully slide it forward, engaging the input shaft with the clutch plate hub. Do not use excessive force to install the transmission - if the input shaft does not slide into place, readjust the angle of the transmission so it is level and/or turn the input shaft so the splines engage properly with the clutch.

26 Install the clutch housing-to-engine bolts. Tighten the bolts to the torque listed in this Chapter's Specifications.

27 Install the crossmember and transmission support. Tighten all nuts and bolts to the torque listed in this Chapter's Specifications.

28 Remove the jacks supporting the transmission and the engine.

29 Install the various items removed previously in the reverse order of removal. Refer to Chapter 7 Part C for installation of the transfer case, if applicable, and to Chapter 8 for the installation of the driveshaft.

30 Make a final check that all electrical connectors are plugged in.

31 Make sure that the transmission lubricant is filled to the proper level (see Chapter 1).

32 Remove the jackstands and lower the vehicle.

33 On 4WD models, connect the transfer case shift linkage (see Chapter 7 Part C).

34 Connect the negative battery cable.

35 Road test the vehicle for proper operation and check for leakage.

3 Manual transmission overhaul - general information

Overhauling a manual transmission is a difficult job for the do-it-yourselfer. It involves the disassembly and reassembly of many small parts. Numerous clearances must be precisely measured and, if necessary, changed with select fit spacers and snap-rings. Therefore, if transmission problems arise, overhaul should be left to a transmission repair shop. However, you can remove and install the transmission yourself. Rebuilt transmissions may be available; check with your dealer parts department and auto parts stores. A rebuilt unit is usually considerably less expensive than having your old transmission overhauled.

Chapter 7 Part B
Automatic transmission

Contents

Specifications

General

Torque converter bolt length (1995 and earlier models only)	
42RE/44RE	
On 9.5-inch, 3-lug converter	0.46 inch
On 10.75-inch, 4-lug converter	0.44 inch
42RH/46RH	
On 9.5-inch, 3-lug converter	0.46 inch
On 9.5-inch, 4-lug converter	0.52 inch
On 10.0-inch, 4-lug converter	0.52 inch
On 10.75-inch, 4-lug converter	0.44 inch

Torque specifications

	Ft-lbs (unless otherwise indicated)
Transmission-to-engine bolts	
1993 models	
AW-4 **(refer to illustration 12.22a)**	
"A" bolts	37 to 47
"B" bolts/nuts	159 to 203 in-lbs
1994 and 1995 models	
42RE/44RE **(refer to illustration 12.22b)**	
"A" bolts	37 to 47
"B" bolts	22 to 26
"C" bolts	96 to 180 in-lbs
"D" bolts	35 to 40
42RH/46RH **(refer to illustration 12.22c)**	
"A" bolts	30 to 50 in-lbs
"B" bolts	20 to 40
"C" bolts	35 to 65
"D" bolts	80 to 120 in-lbs
1996 and later models	
46RE	30
All others	50
Driveplate-to-torque converter bolts	
AW-4	22
42RE/44RE	
With 9.5-inch, 3-lug converter	40
With 10.75-inch, 4-lug converter	23
42RH/46RH	
With 9.5-inch, 3-lug converter	40
With 9.5-inch, 4-lug converter	55
With 10.0-inch, 4-lug converter	55
With 10.75-inch, 4-lug converter	23
46RE	23
45RFE	23

Torque specifications	Ft-lbs (unless otherwise indicated)
Neutral start switch	
AW-4 (1993 only)	
Adjusting bolt	108 in-lbs
Attaching nut	61 in-lbs
All others	25
Transmission speed sensor bolt (1993 AW-4 only)	65 in-lbs
Vehicle speed sensor/speedometer adapter	
Sensor retaining screw	15 to 27 in-lbs
Adapter clamp screw	90 to 110 in-lbs

1 General information and identification

General information

All vehicles covered in this manual come equipped with a five-speed manual transmission or an automatic transmission. All information on the automatic transmission is included in this Part of Chapter 7. Information on the manual transmission can be found in Part A of this Chapter. Vehicles with four-wheel-drive also use a transfer case. Information on the transfer case is in Part C of this Chapter.

Due to the complexity of the automatic transmissions covered in this manual and the need for specialized equipment to perform most service operations, this Chapter contains only general diagnosis, routine maintenance, adjustment and removal and installation procedures.

If the transmission requires major repair work, it should be left to a dealer service department or an automotive or transmission repair shop. You can, however, remove and install the transmission yourself and save the expense, even if the repair work is done by a transmission shop.

Identification

Several different electronically controlled automatic transmissions are used in the vehicles covered by this manual. The AW-4 is used on 1993 models only. The 42RE, 42RH, 44RE, 46RH, 46RE transmissions used in later models are all of similar design (although different from the AW-4). The 42RE and 42RH are used on six-cylinder models, while the 44RE, 46RH and 46RE are used on 5.2L/5.9L V8 models.

In 1999 the 45RFE was introduced on models with the 4.7L V8 engine. This electronically controlled transmission is a multi-speed automatic which uses an additional planetary gearset that allows the transmission control module to choose between two second gears, which results in smoother passing gear operation.

2 Diagnosis - general

Note: *Automatic transmission malfunctions may be caused by five general conditions: poor engine performance, improper adjustments, hydraulic malfunctions, mechanical malfunctions or malfunctions in the computer or its signal network. Diagnosis of these problems should always begin with a check of the easily repaired items: fluid level and condition* (Chapter 1), *shift linkage adjustment and throttle linkage adjustment. Next, perform a road test to determine if the problem has been corrected or if more diagnosis is necessary. If the problem persists after the preliminary tests and corrections are completed, additional diagnosis should be done by a dealer service department or transmission repair shop. Refer to the* Troubleshooting *Section at the front of this manual for information on symptoms of transmission problems.*

Preliminary checks

1 Drive the vehicle to warm the transmission to normal operating temperature.

2 Check the fluid level as described in Chapter 1:

a) *If the fluid level is unusually low, add enough fluid to bring the level within the designated area of the dipstick, then check for external leaks (see below).*
b) *If the fluid level is abnormally high, drain off the excess, then check the drained fluid for contamination by coolant. The presence of engine coolant in the automatic transmission fluid indicates that a failure has occurred in the internal radiator walls that separate the coolant from the transmission fluid* (see Chapter 3).
c) *If the fluid is foaming, drain it and refill the transmission, then check for coolant in the fluid or a high fluid level.*

3 Check the engine idle speed. **Note:** *If the engine is malfunctioning, do not proceed with the preliminary checks until it has been repaired and runs normally.*

4 Check the throttle valve linkage for freedom of movement. Adjust it if necessary (Section 5). **Note:** *The throttle valve linkage may function properly when the engine is shut off and cold, but it may malfunction once the engine is hot. Check it cold and at normal engine operating temperature.*

5 Inspect the shift control linkage (see Section 4). Make sure it's properly adjusted and the linkage operates smoothly.

Fluid leak diagnosis

6 Most fluid leaks are easy to locate visually. Repair usually consists of replacing a seal or gasket. If a leak is difficult to find, the following procedure may help.

7 Identify the fluid. Make sure it's transmission fluid and not engine oil or brake fluid (automatic transmission fluid is a deep red color).

8 Try to pinpoint the source of the leak. Drive the vehicle several miles, then park it over a large sheet of cardboard. After a minute or two, you should be able to locate the leak by determining the source of the fluid dripping onto the cardboard.

9 Make a careful visual inspection of the suspected component and the area immediately around it. Pay particular attention to gasket mating surfaces. A mirror is often helpful for finding leaks in areas that are hard to see.

10 If the leak still cannot be found, clean the suspected area thoroughly with a degreaser or solvent, then dry it.

11 Drive the vehicle several miles at normal operating temperature and varying speeds. After driving the vehicle, visually inspect the suspected component again.

12 Once the leak has been located, the cause must be determined before it can be properly repaired. If a gasket is replaced but the sealing flange is bent, the new gasket will not stop the leak. The bent flange must be straightened.

13 Before attempting to repair a leak, check to make sure the following conditions are corrected or they may cause another leak. **Note:** *Some of the following conditions cannot be fixed without highly specialized tools and expertise. Such problems must be referred to a transmission repair shop or a dealer service department.*

Gasket leaks

14 Check the pan periodically. Make sure the bolts are tight, no bolts are missing, the gasket is in good condition and the pan is flat (dents in the pan may indicate damage to the valve body inside).

15 If the pan gasket is leaking, the fluid level or the fluid pressure may be too high, the vent may be plugged, the pan bolts may be too tight, the pan sealing flange may be warped, the sealing surface of the transmission housing may be damaged, the gasket may be damaged or the transmission casting may be cracked or porous. If sealant instead of gasket material has been used to form a seal between the pan and the transmission housing, it may be the wrong sealant.

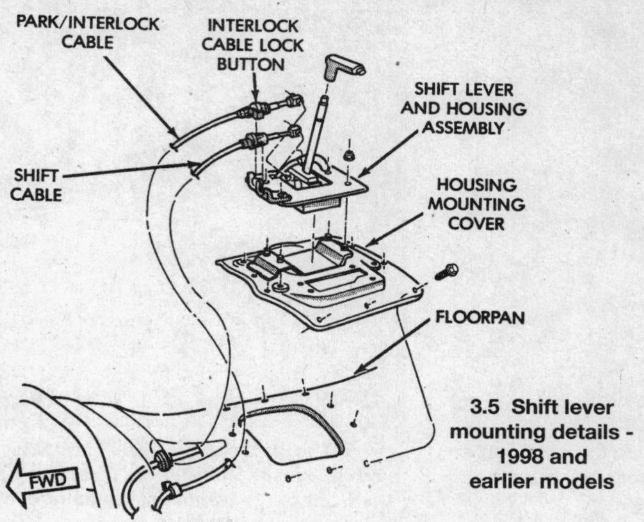

3.5 Shift lever mounting details - 1998 and earlier models

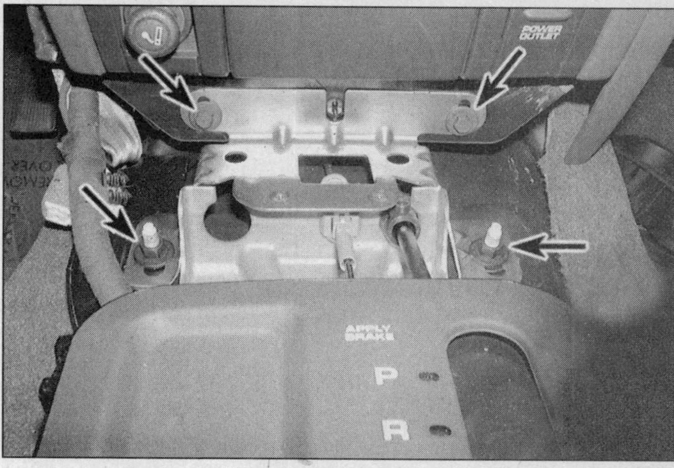

3.12 The reinforcement plate at the center of the instrument panel is secured by two bolts and two nuts (arrows) - 1999 and later models

Seal leaks

16 If a transmission seal is leaking, the fluid level or pressure may be too high, the vent may be plugged, the seal bore may be damaged, the seal itself may be damaged or improperly installed, the surface of the shaft protruding through the seal may be damaged or a loose bearing may be causing excessive shaft movement.

17 Make sure the dipstick tube seal is in good condition and the tube is properly seated. Periodically check the area around the speedometer gear or sensor for leakage. If transmission fluid is evident, check the O-ring for damage.

Case leaks

18 If the case itself appears to be leaking, the casting is porous and will have to be repaired or replaced.

19 Make sure the oil cooler hose fittings are tight and in good condition.

Fluid comes out the vent pipe or fill tube

20 If this condition occurs, the transmission is overfilled, there is coolant in the fluid, the case is porous, the dipstick is incorrect, the vent is plugged or the drain back holes are plugged.

3 Shift lever - removal and installation

Warning: *These models are equipped with airbags. Always disable the airbag system before working in the vicinity of any airbag system component to avoid the possibility of accidental deployment of the airbag(s), which could cause personal injury* (see Chapter 12).

1998 and earlier models

Refer to illustration 3.5

1 Disconnect the cable from the negative terminal of the battery.

2 Remove the handle from the transmission shift lever. Grasp the handle and pull up sharply to remove the handle from the lever.

3 Remove the center console (see Chapter 11).

4 Remove the console bracket retaining screws and the bracket (see Chapter 11).

5 Remove the nuts which attach the shift lever assembly to the floor pan mounting cover **(see illustration)**.

6 Lift the shift lever assembly up for access to the cables.

7 Disengage the shift and interlock cables (see Sections 4 and 7).

8 Remove the shift lever assembly.

9 Installation is the reverse of removal. If necessary, adjust the shift cable (see Section 4), transfer case shift cable (if equipped; see Chapter 7C) and the Park interlock cable (see Section 7).

1999 and later models

Refer to illustrations 3.12 and 3.17

10 Disconnect the cable from the negative terminal of the battery.

11 Remove the center console (see Chapter 11).

12 Remove the reinforcement plate at the front of the shifter base **(see illustration)**.

13 Detach the shift cable from the pin on the shift lever and from the shifter base (see Section 4).

14 Detach the Park interlock cable from the shift lever and from the shifter base (see Section 7).

15 If you are working on a 4WD model, detach the transfer case shift cable from the transfer case lever and from the shifter base (see Chapter 7C).

16 Unplug the electrical connector at the rear of the shifter base.

17 Unscrew the mounting nuts and remove the shifter assembly **(see illustration)**.

18 Installation is the reverse of removal. If necessary, adjust the shift cable (see Section 4), transfer case shift cable (if equipped; see Chapter 7C) and the Park interlock cable (see Section 7).

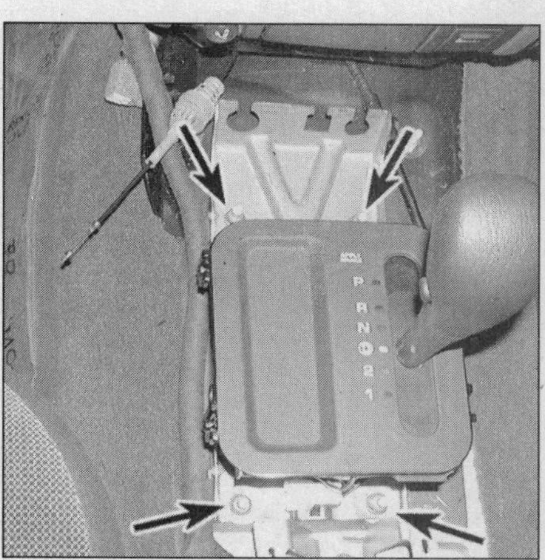

3.17 The shifter assembly is secured by four nuts (arrows) - 1999 and later models

4 Shift cable - check, replacement and adjustment

Check

1 Firmly apply the parking brake and try to momentarily operate the starter in each shift lever position. The starter should operate in Park and Neutral only. If the starter operates in any position other than Park or Neutral, adjust the shift linkage (see below). If, after adjustment, the starter still operates in positions other than Park or Neutral, the Park/Neutral Position (PNP) switch is defective or in need of adjustment (see Section 6).

Replacement

Refer to illustrations 4.5, 4.6, 4.7, 4.9a, 4.9b and 4.10

Warning: *These models are equipped with airbags. Always disable the airbag system before working in the vicinity of any airbag system component to avoid the possibility of accidental deployment of the airbag(s), which could cause personal injury* (see Chapter 12).

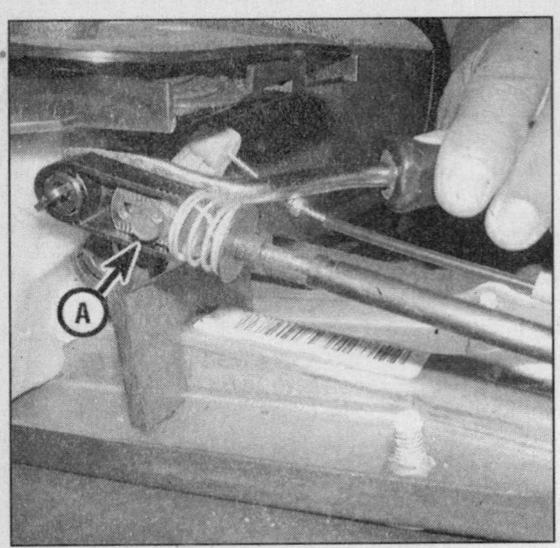

4.5 Prying the shift cable off the pin at the shift lever; (A) indicates the cable adjuster bolt

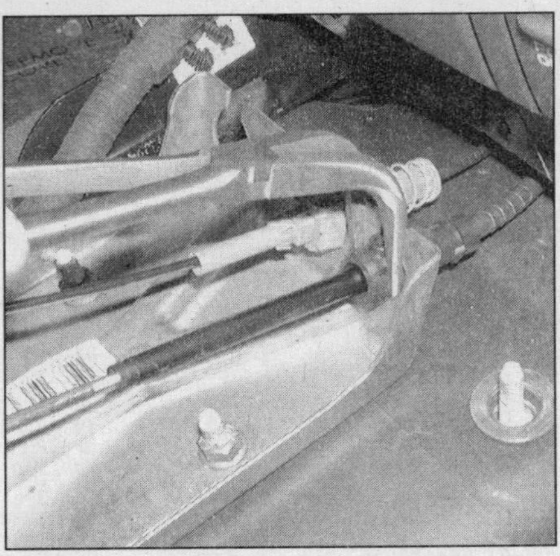
4.6 Squeeze the tangs on the shift cable retainer and detach the cable from the shifter base

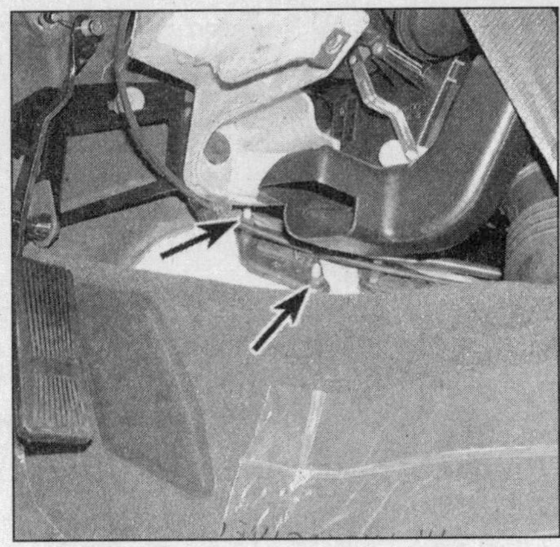
4.7 The shift cable on 1999 and later models passes through the floor pan near the front of the transmission tunnel - it's secured by two nuts

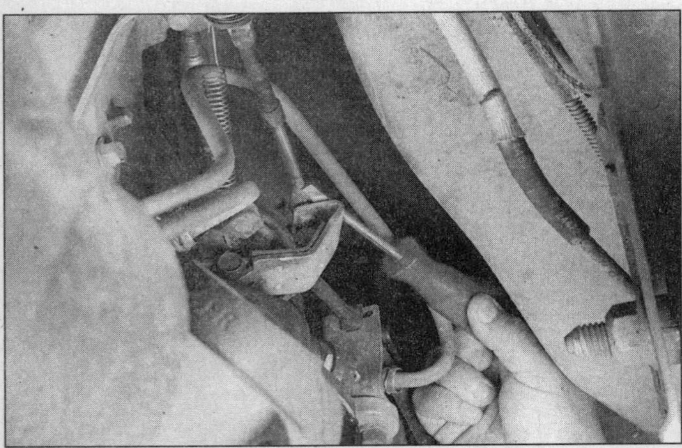
4.9a The shift cable is connected to the transmission shift lever with a ball-and-socket type connector; to disconnect it, pry the cable end loose from the small pin on the lever

4.9b Detaching the shift cable from the manual lever on a 45RFE transmission (1999 and later models with a 4.7L V8 engine)

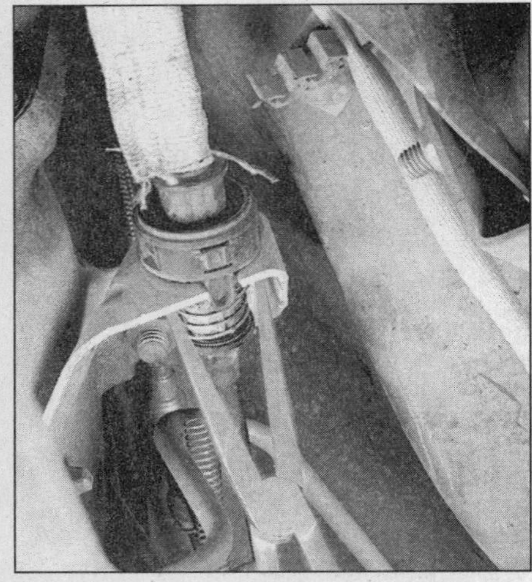
4.10 To detach the shift cable from the bracket at the transmission, pinch the housing tabs together with a pair of pliers, then pull the cable and housing through the bracket

2 Disconnect the cable from the negative terminal of the battery.

3 Remove the center console (see Chapter 11). On 1999 and later models, also remove the trim panel from the left side of the dash, under the steering column.

4 If you're working on a 1998 or earlier model, unbolt the shifter assembly from the floor pan (see Section 3).

5 Pry the cable end off the pin on the shift lever **(see illustration)**.

6 Detach the cable from the shifter base **(see illustration)**. **Note:** *If you're working on a 1999 or later model, it will be necessary to remove the reinforcement plate from the center of the instrument panel* **(see illustration 3.12)**.

7 If you're working on a 1999 or later model, remove the nuts securing the cable to the floor pan **(see illustration)**.

8 Raise the vehicle and support it securely on jackstands.

9 Detach the cable end from the shift lever on the transmission **(see illustrations)**.

10 Detach the cable from the bracket at the transmission **(see illustration)**.

11 Push the cable through the floor pan and remove it from the vehicle (on 1998 and earlier models you'll have to dislodge the grommet from the floor pan).

12 Installation is the reverse of removal. Adjust the cable (see the next Step).

Adjustment

13 Place the shift lever in Park. Raise the vehicle and support it securely on jackstands.

14 If you're working on a 1998 or earlier model, release the cable housing from the shift cable bracket **(see illustration 4.10)**, then unsnap the shift cable adjuster clamp to unlock the cable.

15 Be sure the shift lever on the transmission is all the way to the rear, in the last detent. This is the Park position.

16 Make sure the park lock is engaged by trying to rotate the driveshaft. The driveshaft will not rotate if the park lock is functioning properly.

17 On 1998 and earlier models, snap the shift cable housing into the bracket, then press down on the adjuster clamp until it

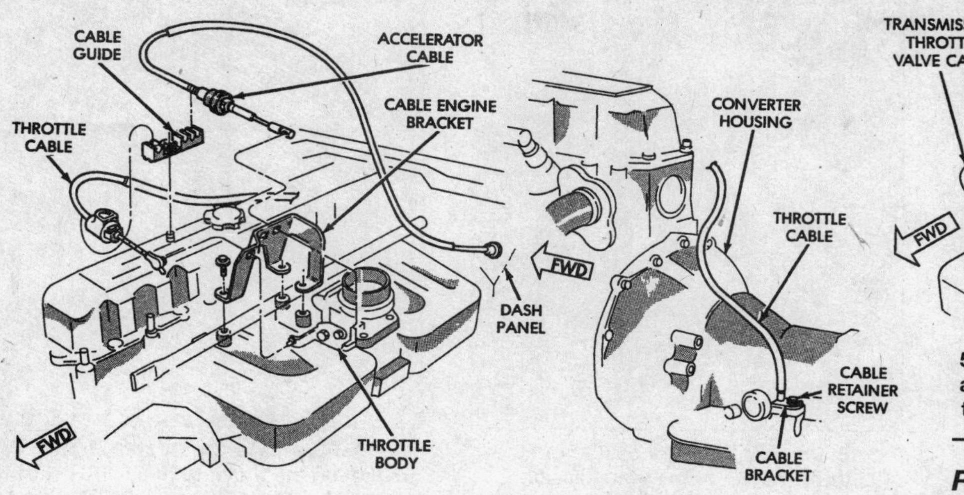

5.3a Details of the throttle valve cable assembly used on the AW-4 transmission (1993 models only)

snaps in place to lock the shift cable.

18 On 1999 and later models, loosen the cable adjuster screw at the shift lever **(see illustration 4.5)**. Recheck to make sure the shift lever at the transmission is in the Park position and the gearshift lever inside the vehicle is also in the Park position. Now tighten the adjustment screw.

19 Lower the vehicle.

20 With the parking brake firmly applied, make sure the engine starts only in Park and Neutral. If the linkage appears to be adjusted properly, but the starter operates in positions other than Park or Neutral, check the Park/Neutral Position (PNP) switch (see Section 6).

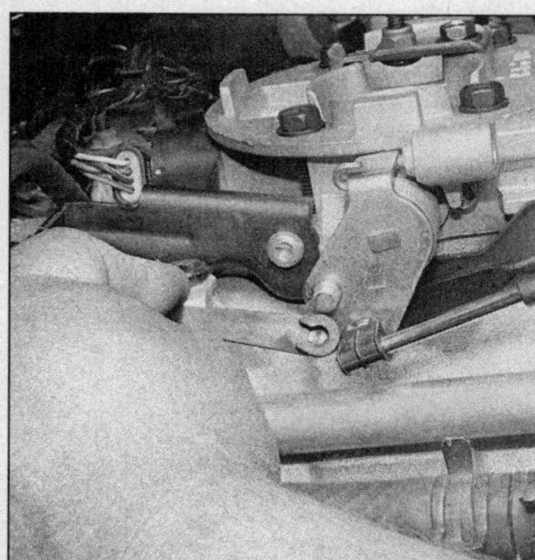

5.3c To disconnect the throttle valve cable from the throttle linkage on a 42RH or 46RH transmission, slide this cable end forward (the cable is turned around so you can see the groove in the side of the cable which normally faces toward the throttle body; this groove engages the pin on the throttle lever)

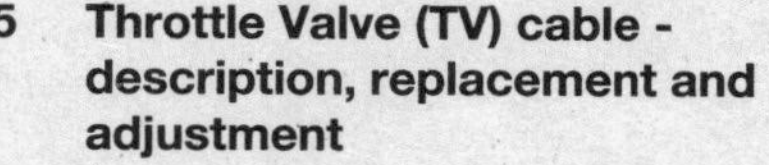

5 Throttle Valve (TV) cable - description, replacement and adjustment

Note: *This procedure does not apply to the 45RFE transmission.*

Description

1 The transmission throttle valve (TV) is operated by a cam on the throttle lever, which in turn is operated by an adjustable cable. The cable is attached to an arm mounted on the throttle lever shaft at the transmission. A lock button at the engine end of the cable is provided for cable adjustment. A correctly adjusted throttle valve cable will synchronize the throttle lever on the transmission with the throttle shaft lever on the throttle body so that the throttle lever on the transmission neither moves ahead of nor lags behind the lever on the throttle body.

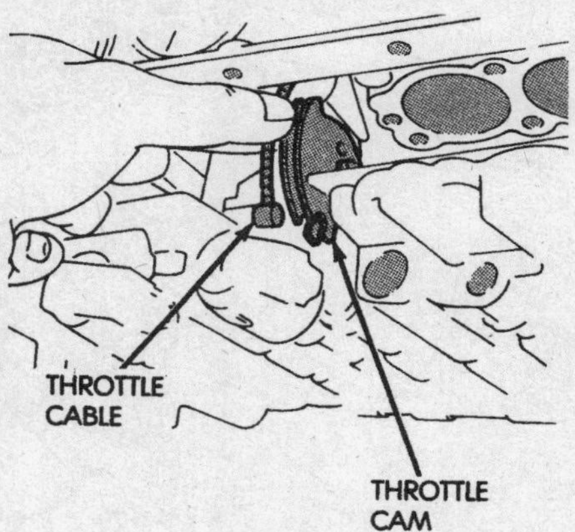

5.7 Disengage the TV cable from the throttle valve cam (1993 AW-4 only)

5.3b Details of the throttle valve cable assembly on the 42RE/44RE and 46RE transmissions (1994 and later models)

Replacement

Refer to illustrations 5.3a, 5.3b and 5.3c

2 Remove the air cleaner (see Chapter 4).

3 Disconnect the cable from the throttle linkage on the throttle body **(see illustrations)**.

4 Compress the cable mounting ears and remove the cable from the engine bracket.

5 Raise the vehicle and support it securely on jackstands.

AW-4 (1993 models only)

Refer to illustrations 5.7 and 5.8

6 Remove the transmission oil pan (see Chapter 1).

7 Disengage the cable from the throttle valve cam **(see illustration)**.

8 Remove the cable bracket bolt and remove the cable and bracket from the case **(see illustration)**.

9 Remove and discard the cable seal.

10 Lubricate and install a new seal on the cable.

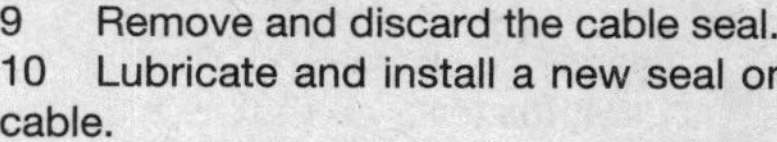

11 Insert the cable in the transmission case.

12 Attach the cable to the throttle cam.

13 Install the cable bracket on the case and tighten the bracket bolt securely.

14 Install the pan (see Chapter 1).

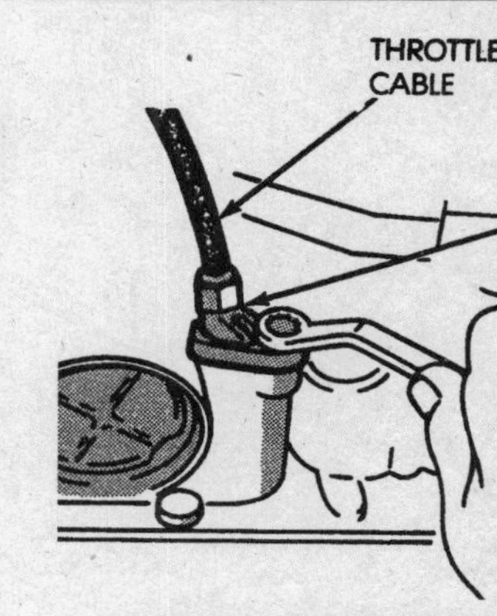

5.8 Remove the cable bracket bolt and remove the cable and bracket from the case (1993 AW-4 only)

42RE/44RE, 46RE and 42RH/46RH (1994 and later models)

Refer to illustrations 5.15 and 5.16

15 Disengage the cable from the stud on the throttle lever at the transmission **(see illustration)**.

16 Detach the cable from its bracket on the transmission **(see illustration)**.

17 Reattach the new cable to the bracket.

18 Engage the cable with the stud on the throttle lever.

All models

19 Remove the jackstands and lower the vehicle.

20 Connect the new cable to the engine bracket and throttle linkage (see Chapter 4).

21 Install the air cleaner (see Chapter 4).

22 Refill the transmission with the specified transmission fluid (see Chapter 1).

23 Adjust the cable (see below).

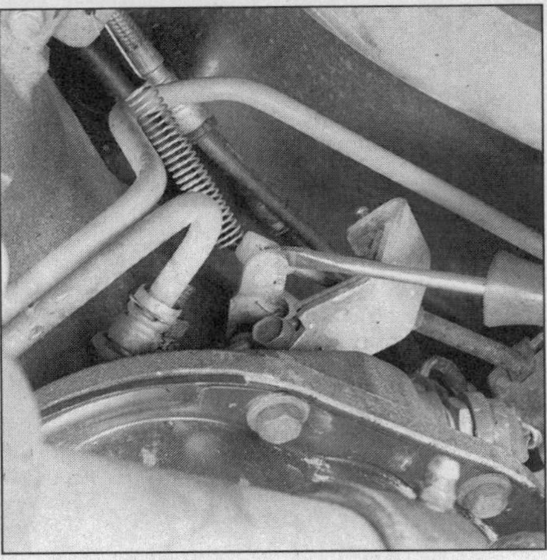

5.15 Disengage the TV cable from the throttle lever at the transmission (46RH shown; 42RE/44RE, 46RE and 42RH similar)

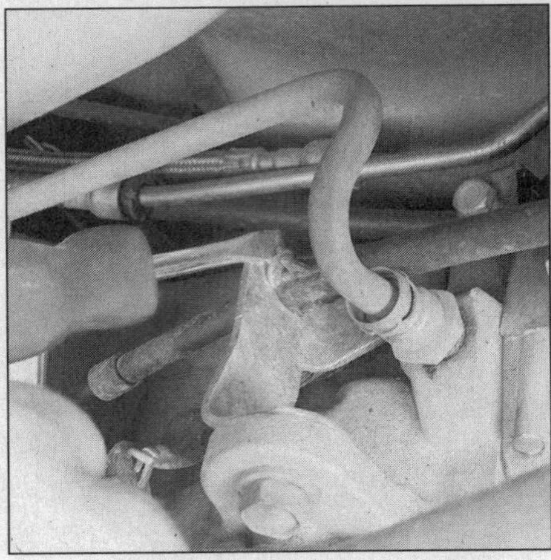

5.16 Disengage the TV cable from its bracket on the transmission (46RH shown; 42RE/44RE, 46RE and 42RH similar)

Adjustment

AW-4 (1993 models only)

Refer to illustrations 5.24 and 5.25

24 Shift the transmission into Park and make sure the ignition switch is in the Off position. Depress the cable release button **(see illustration)**, then push the cable conduit back into the sheath as far as it will go.

25 Rotate the throttle lever on the throttle body to the wide open throttle position **(see illustration)**. The cable will ratchet to the correct adjustment point as the lever is rotated. Release the throttle lever.

42RE/44RE, 46RE and 42RH/46RH (1994 and later models)

Refer to illustrations 5.35a and 5.35b

26 Remove the air cleaner (see Chapter 4).

27 Verify that the lever on the throttle body is at its curb idle (closed) position and that the transmission throttle lever is also at idle (fully forward) position.

28 Disengage the cable from the attachment stud on the throttle body lever **(see illustrations 5.3b and 5.3c)**.

29 Compare the position of the cable end to the attachment stud on the throttle body lever:

- a) *The cable end and the attachment stud should be aligned (or centered on one another) to within 1/32-inch in either direction.*
- b) *If the cable end and attachment stud are misaligned (off center), the cable must be adjusted (see below).*

30 Reattach the cable end to the attachment stud.

31 With the help of an assistant, note the movement of the transmission throttle lever and the lever on the throttle body.

- a) *If both levers move simultaneously from idle to half-throttle and back to idle position, the adjustment is correct.*
- b) *If the transmission throttle lever moves ahead of or lags behind the throttle body lever, the cable must be adjusted.*
- c) *If the throttle body lever prevents the transmission lever from returning to the closed position, the cable must be adjusted.*

32 Disconnect the cable end from the attachment stud on the throttle lever at the throttle body **(see illustrations 5.3b and 5.3c)**.

33 Raise the vehicle and support it securely on jackstands.

34 Verify that the transmission throttle lever at the transmission is in the fully closed position. Then verify that the lever on the throttle body is at its curb idle position.

35 Depress the cable lock button to release the cable **(see illustrations)**. The lock button only has to move about 5/64-inch to release the cable in the adjuster head.

36 Center the cable end on the attachment stud to within 1/32-inch and release the lock button **(see illustration 5.35a)**.

37 Check the cable adjustment. Make sure the transmission throttle lever and the lever

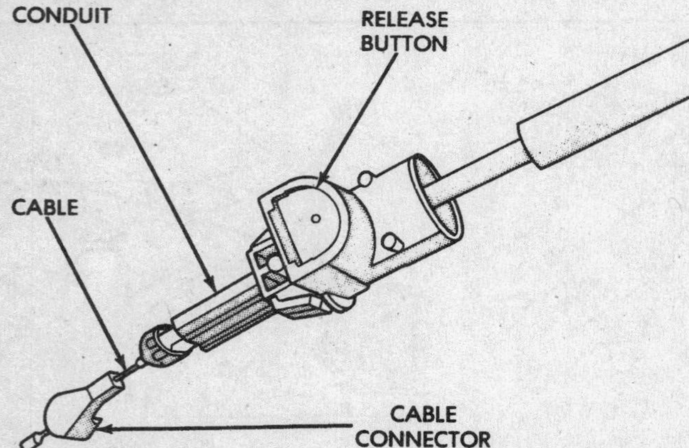

5.24 Depress the cable release button, then push the cable conduit back into the sheath as far as it will go (AW-4 only)

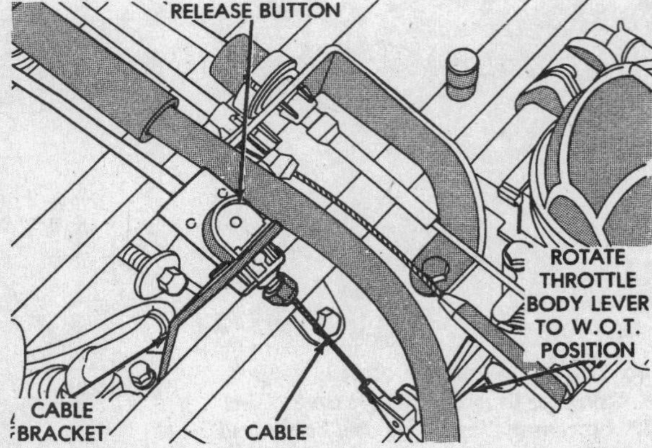

5.25 Rotate the throttle lever on the throttle body to the wide open throttle position, let the cable ratchet to the correct adjustment point as the lever is rotated, then release the throttle lever (AW-4)

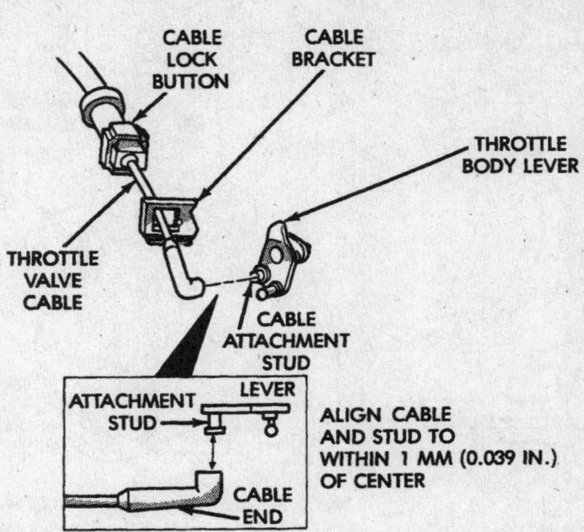

5.35a Installation details of the throttle valve cable at the throttle body lever (1994 42RE/44RE and 42RH/46RH)

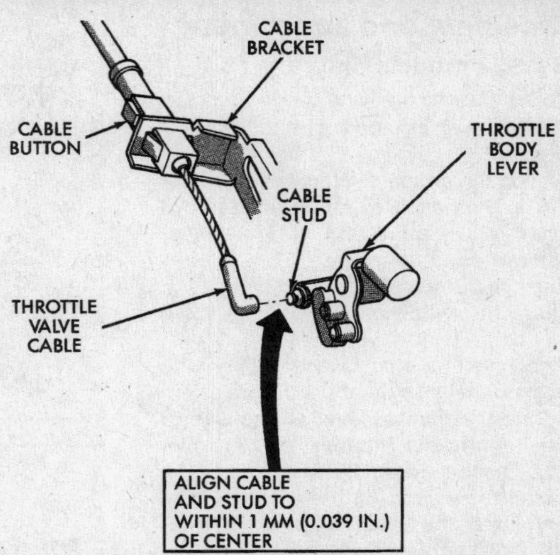

5.35b Installation details of the throttle valve cable at the throttle body lever (1995 and later 42/44RE and 46RH)

on the throttle body move simultaneously as described in Step 31.

6 Park/Neutral Position (PNP) switch - check, replacement and adjustment

Note: *This procedure does not apply to the 45RFE transmission used on 1999 and later models equipped with the 4.7L V8 engine. The PNP switch on the 45RFE transmission is referred to as the Transmission Range (TR) sensor and is mounted above the valve body. Due to the sensitive nature of the valve body and its importance in the operation of the transmission, replacement of this sensor on the 45RFE transmission should be left to a transmission specialist.*

Check

1 Raise the front of the vehicle and support it securely on jackstands.

AW-4 (1993 models only)

Refer to illustration 6.2

2 Unplug the switch electrical connector and compare the continuity of the indicated terminals to the continuity specified in the accompanying table **(see illustration)**. If continuity is not as specified, try adjusting the switch (see below), then retest it. If continuity is still not as specified, replace the switch (see below).

3 If the switch is okay, plug in the connector and lower the vehicle. If it isn't, proceed to Step 8.

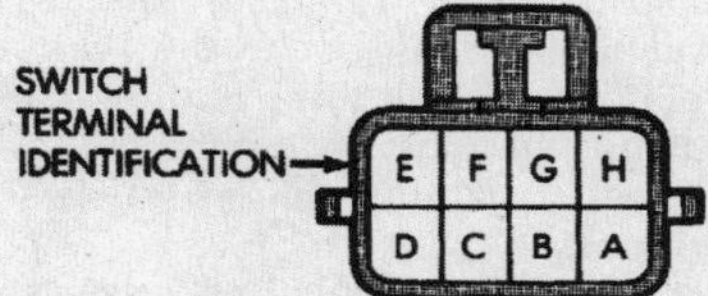

	B	C	A	E	G	H
P	O	O				
R			O	O		
N	O	O				
D						
3			O	—	O	
1-2			O	—	—	O

6.2 Park/Neutral Position switch electrical connector terminal guide (above) and continuity table (below) (1993 AW-4 transmission only)

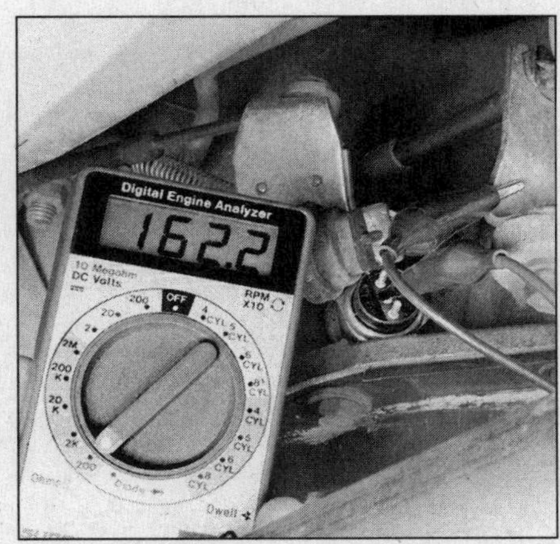

6.5 Check continuity between the center terminal and the transmission case; there should be continuity only when the transmission is in Park or Neutral (42RE/44RE and 42RH/46RH)

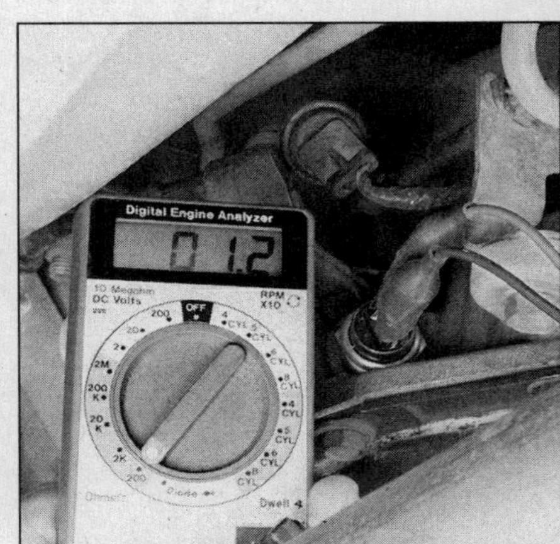

6.6 Shift the transmission into Reverse and check continuity at the switch outer terminals; there should be continuity only when the transmission is in Reverse, but there should be no continuity between the outer terminals and the transmission case

42/44RE, 46RE and 42RH/46RH (1994 and later models)

Refer to illustrations 6.5 and 6.6

4 Verify that the shift cable is correctly adjusted (see Section 4). **Note:** *The switch will not operate properly if the cable isn't correctly adjusted.*

5 Unplug the switch electrical connector and test for continuity between the center terminal and transmission case **(see illustration)**. There should be continuity only when the transmission is in Park or Neutral.

6 Shift the transmission into Reverse and test for continuity at the switch outer terminals **(see illustration)**. There should be continuity only when the transmission is in Reverse, but there should be no continuity between the outer terminals and the case.

7 If the switch is okay, plug in the connector and lower the vehicle. If it isn't, proceed to Step 20.

7B

Replacement and adjustment

AW-4 (1993 models only)

Refer to illustrations 6.10 and 6.17

8 Raise the vehicle and support it securely on jackstands.

9 Unplug the switch electrical connector.

10 Pry up the washer lock tabs and remove the switch retaining nut and tabbed washer **(see illustration)**.

11 Remove the switch adjusting bolt.

12 Slide the switch off the manual valve shaft.

13 Disconnect the shift linkage rod from the shift lever on the left side of the transmission.

14 Rotate the manual lever all the way to the rear, then rotate the lever forward two detent positions to Neutral.

15 Install the switch on the manual valve shaft and install the switch adjusting bolt finger tight. Don't tighten the bolt yet.

16 Install the tabbed washer on the manual valve shaft and install the switch retaining nut. Tighten the nut to the torque listed in this Chapter's Specifications, but do not bend down the washer lock tabs yet.

17 Make sure the transmission is in Neutral and rotate the switch until the neutral standard line on the housing lines up with the vertical groove in the manual valve shaft **(see illustration)**.

18 Tighten the adjusting bolt to the torque listed in this Chapter's Specifications and bend the washer lock tabs over the switch retaining nut.

19 Plug in the electrical connector, connect the shift linkage, lower the vehicle and check the switch operation to make sure the starter operates in Park or Neutral only. Be sure to firmly apply the parking brake when making this check.

42RE/44RE, 46RE and 42RH/46RH (1994 and later models)

Refer to illustration 6.23

20 Raise the vehicle and support it securely on jackstands.

21 Position a drain pan underneath the switch.

22 Unplug the switch electrical connector and remove the switch from the transmission.

23 Move the shift lever to the Park and Neutral positions and verify that the switch operating lever fingers are centered in the switch opening in the transmission case **(see illustration)**.

24 Install a new seal on the new switch and install the switch in the transmission case. Tighten the switch to the torque listed in this Chapter's Specifications.

25 Plug in the switch connector, lower the vehicle and top off the transmission with the specified fluid (see Chapter 1).

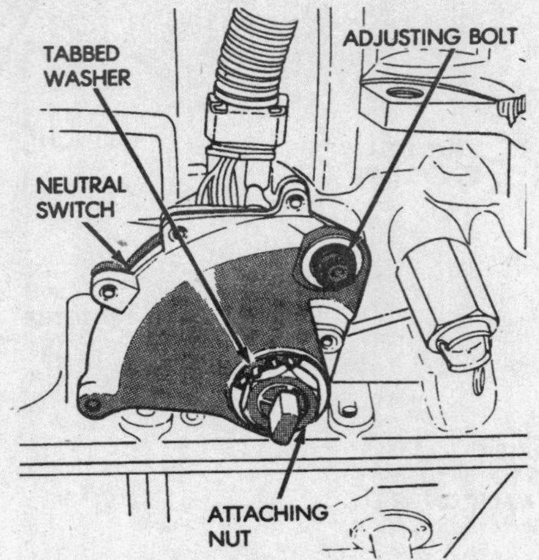

6.10 Pry the washer lock tabs on the tabbed washer and remove the switch retaining nut and the tabbed washer (1993 AW-4 transmission only)

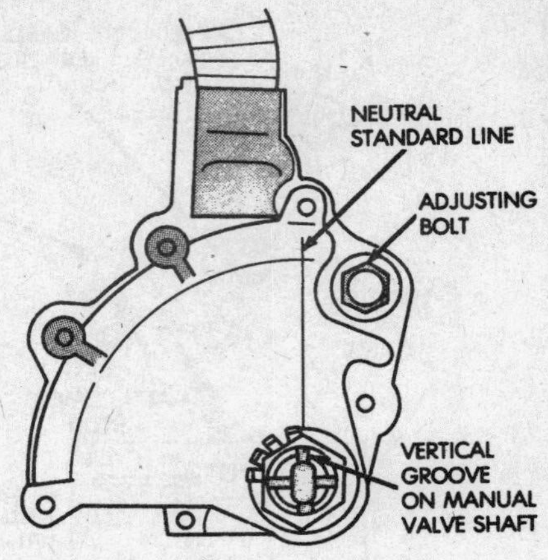

6.17 With the transmission in Neutral, rotate the switch until the neutral standard line on the switch housing lines up with the vertical groove in the manual valve shaft (1993 AW-4 transmission only)

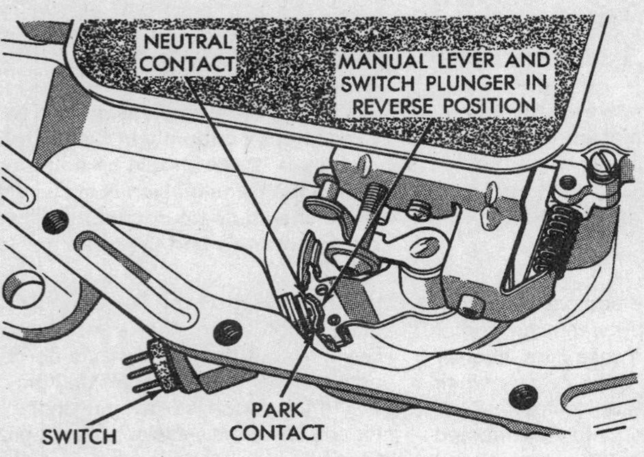

6.23 Move the shift lever to the Park and Neutral positions and verify that the switch operating fingers are centered in the switch opening in the transmission case

7 Park/Lock system - description, adjustment and component replacement

Warning: *These models are equipped with airbags. Always disable the airbag system before working in the vicinity of any airbag system component to avoid the possibility of accidental deployment of the airbag(s), which could cause personal injury* (see Chapter 12).

Description

1 The Park/Lock system prevents the shift lever from being moved out of Park unless the brake pedal is depressed simultaneously. It also prevents the ignition key from being removed from the ignition switch unless the shift lever is in the Park position. When the ignition key is turned to the Accessory or Run position, a solenoid in-line with the interlock cable is energized, locking the shift lever in Park; when the brake pedal is depressed, the solenoid is de-energized, unlocking the shift lever so that it can be moved into some other gear.

Park/Lock cable

Check and adjustment

Refer to illustrations 7.5a and 7.5b

2 With the ignition key turned to the On position, verify that the shift lever moves through all gear positions.

3 While moving the shift lever through all gear positions, verify that the ignition key can't be turned to the Lock position.

4 Verify that the ignition key can be removed when it's in the Lock position and the shift lever is in the Park position.

5 If the Park/Lock cable fails any of the above tests, adjust it as follows.

- *a) Put the shift lever in Park.*
- *b) Turn the ignition key to Lock.*
- *c) Remove the console (see Chapter 11).*
- *e) Pry up the locking tab on the cable adjuster, then pull the cable forward and release it* **(see illustrations)**. *The adjuster is spring loaded, so it will automatically remove the slack from the cable.* **Note:** *On 1999 and later models it will be necessary to remove the reinforcement plate at the front of the shifter base.*
- *f) Push in on the locking tab until it clicks back into place.*

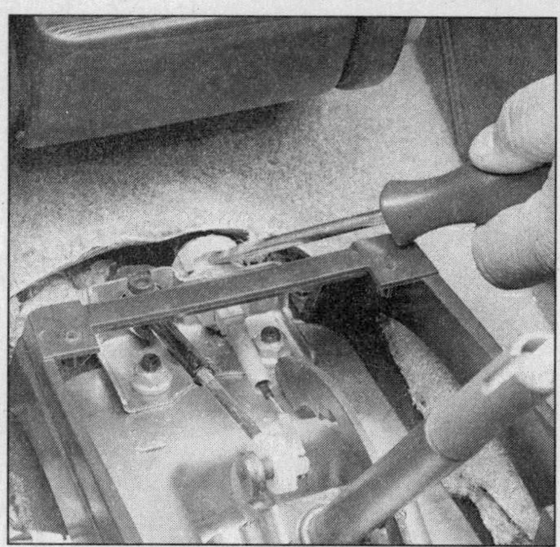

7.5a Pull the cable lock button up to release the cable (1998 or earlier model shown; on 1999 and later models, the shift cable and Park/Lock cable are reversed)

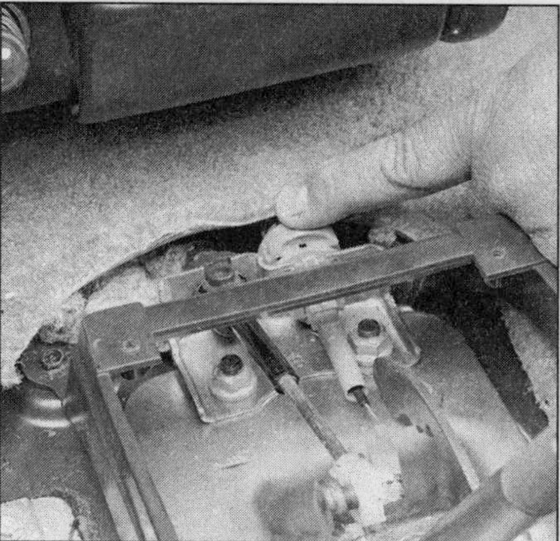

7.5b Pull the cable forward, then release it and press down the cable lock button until it snaps into place

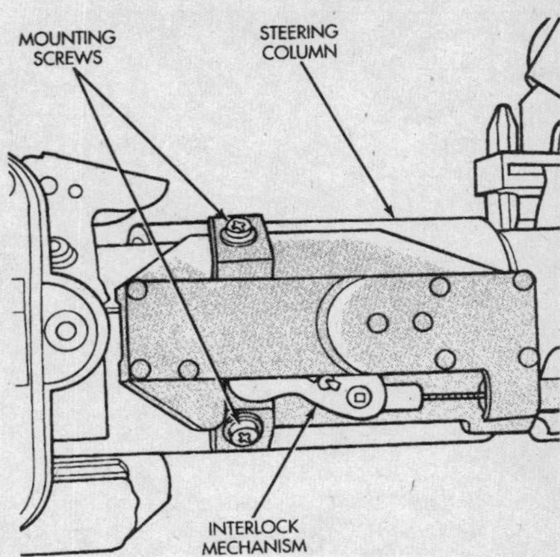

7.10 On 1998 and earlier models the park interlock mechanism is retained to the steering column by two screws

7.11 To detach the Park/Lock cable from the lock cylinder housing on a 1999 or later model, depress the plastic locking tang and pull the cable end out of the housing

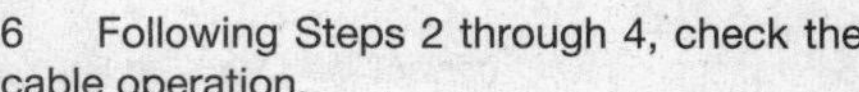

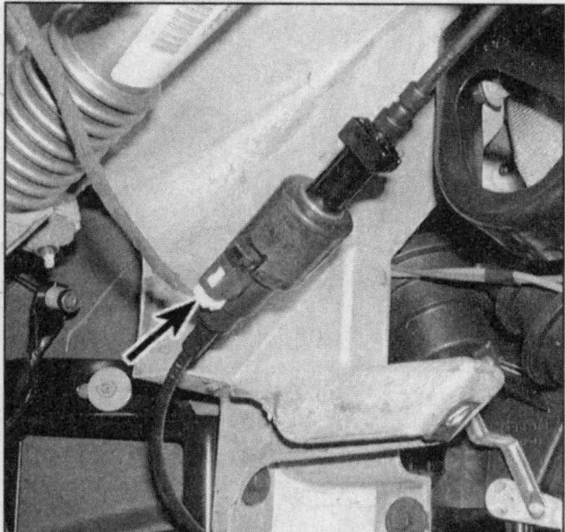

7.12 Unplug the electrical connector from the interlock solenoid on the cable

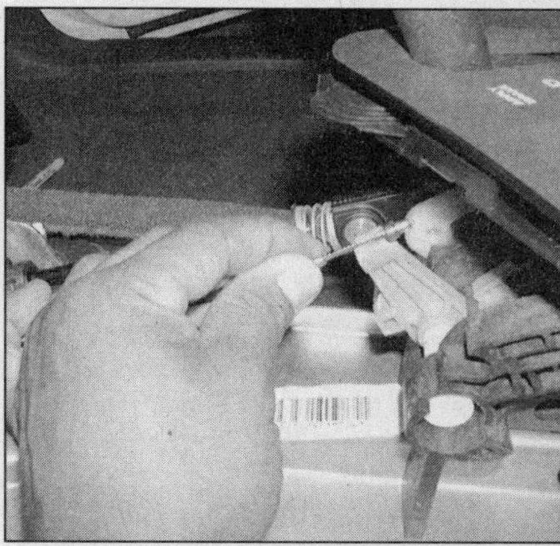

7.13 Pull the end of the cable back and pass it through the slot in the lever

6 Following Steps 2 through 4, check the cable operation.

Replacement

Refer to illustrations 7.10, 7.11, 7.12, 7.13 and 7.14

7 Disconnect the cable from the negative terminal of the battery.

8 Remove the console, the left-side under-dash panel and steering column knee bolster panel (see Chapter 11).

9 Put the shift lever in the Park position.

10 If you're working on a 1998 or earlier model, remove the two screws that attach the interlock mechanism to the steering column **(see illustration)**, then detach the mechanism.

11 If you're working on a 1999 or later model, depress the locking tang and slide the end of the interlock cable out of the lock cylinder housing **(see illustration)**.

12 Unplug the electrical connector from the solenoid on the cable **(see illustration)**.

13 Detach the shifter end of the cable from the Park/Lock lever **(see illustration)**.

14 Depress the tabs on the sides of the cable adjuster **(see illustration)** and disengage the adjuster from the cable bracket.

15 Remove any cable clips and remove the Park/Lock cable.

16 Installation is the reverse of removal. Be sure to adjust the cable as described earlier in this Section.

17 Check the operation of the Park/Lock cable.

Brake/Transmission Shift Interlock (BTSI) solenoid

Check

18 Turn the ignition key to the Run position and, without depressing the brake pedal, attempt to move the shift lever out of Park; you shouldn't be able to depress the button on the shift lever. If you can depress the button, either the solenoid is not receiving voltage, is not grounded, or it is defective.

a) *Check the fuses.*
b) *Check the adjustment of the shift cable and the Park/Neutral Position switch.*
c) *Check for power to the BTSI solenoid. If voltage is not available, repair the circuit between the solenoid and the fuse block. If voltage is available, check the ground circuit for continuity. If continuity is present, replace the BTSI solenoid.*

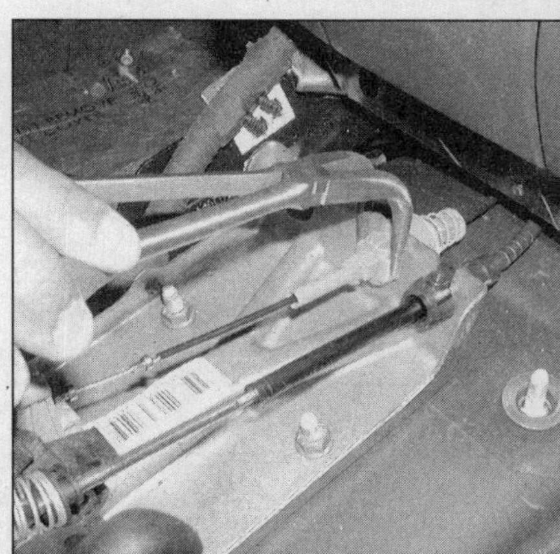

7.14 Squeeze the tangs on the sides of the cable adjuster, then detach the cable from the shifter base

7B

8.2 To check the transmission mount, insert a large screwdriver or prybar between the extension housing and crossmember (2WD models) or between the transfer case and crossmember (4WD models) and pry the transmission up, looking for cracks in the rubber

19 Now depress the brake pedal - you should be able to push the button on the shifter in and move the lever out of Park. If you can't, the solenoid isn't de-activating.

a) *Make sure the shift cable is properly adjusted.*
b) *Unplug the electrical connector from the solenoid and check for continuity to ground on the harness terminal that does not have power to it. There should be continuity when the brake pedal is at rest and no continuity (or high resistance) when the pedal is depressed at least 1/2-inch.*
c) *If the ground path from the solenoid is cut when the brake pedal is depressed, but the solenoid still doesn't release, it is stuck; replace it.*

Replacement

20 The solenoid is an integral part of the Park/Lock cable. Follow the cable replacement procedure described earlier in this Section to replace the solenoid/cable assembly.

8 Transmission mount - check and replacement

Refer to illustrations 8.2 and 8.3

1 Raise the vehicle and support it securely on jackstands.

2 Working underneath the vehicle, insert a large screwdriver or prybar into the space between the transmission extension housing and the crossmember (2WD models) or the transfer case and the crossmember (4WD models) **(see illustration)**. Try to pry the transmission or transfer case up slightly. There should be very little movement between the center of the mount and the rubber insulator. If any cracks are noticed, replace the mount.

3 To replace the mount, remove the bolts

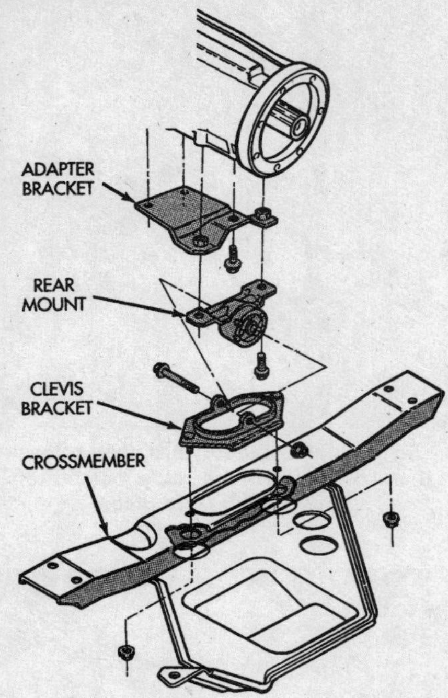

8.3 An exploded view of a typical transmission rear mount assembly

that attach the mount to the crossmember and to the transmission **(see illustration)**.

4 Raise the transmission or transfer case slightly with a jack and remove the insulator, noting which holes are used in the crossmember for proper alignment during installation.

5 Installation is the reverse of the removal procedure. Be sure to tighten the nuts/bolts securely.

9 Oil seal replacement

1 Oil leaks frequently occur due to wear of the extension housing or transfer case (4WD models) oil seal and/or the speed sensor/speedometer adapter O-rings. Replacement of these seals and O-rings is relatively easy, since they can be replaced without removing the transmission or transfer case (4WD models) from the vehicle.

9.5 Using a screwdriver or seal removal tool, carefully pry the oil seal out of the rear of the transmission or transfer case (but don't damage the splines on the transmission output shaft)

Extension housing/transfer case seal

Refer to illustrations 9.5, 9.6 and 9.7

2 The extension housing oil seal is located at the extreme rear of the transmission or transfer case, where the driveshaft is attached. If you suspect leakage at the seal, raise the vehicle and support it securely on jackstands. If the seal is leaking, transmission lubricant will be built up on the front of the driveshaft and may be dripping from the rear of the transmission or transfer case.

3 Remove the driveshaft (see Chapter 8).

4 Using a soft-faced hammer, carefully tap the dust shield (if equipped) to the rear and remove it from the transmission. Be careful not to distort it.

5 Using an oil seal removal tool, carefully pry the oil seal out of the rear of the transmission or transfer case **(see illustration)**. Do not damage the splines on the transmission output shaft.

6 If the oil seal cannot be removed with a screwdriver or seal removal tool, a slide hammer and a special oil seal puller tool (available at some auto parts stores) will be required **(see illustration)**.

7 Using a large seal driver or a very large deep socket as a drift, install the new oil seal

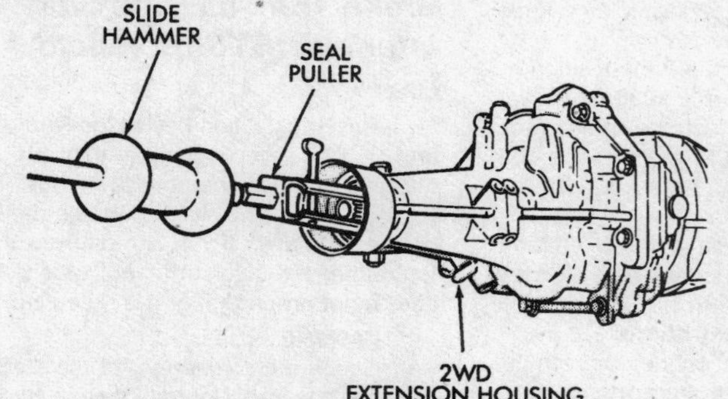

9.6 If the oil seal cannot be removed with a screwdriver or pry bar, you'll need a special oil seal removal tool like this (available at some auto parts stores)

9.7 Using a seal driver or a very large deep socket as a drift, install the new oil seal; drive it into the bore squarely and make sure it's completely seated

(see illustration). Drive it into the bore squarely and make sure it's completely seated.

8 Reinstall the dust shield by carefully tapping it into place. Lubricate the splines of the transmission output shaft and the outside of the driveshaft sleeve yoke with chassis grease, then install the driveshaft. Be careful not to damage the lip of the new seal.

Vehicle speed sensor/speedometer adapter O-rings (1998 and earlier models)

Refer to illustrations 9.9, 9.17a, 9.17b, 9.17c and 9.24

Note: *1999 and later models are not equipped with a vehicle speed sensor.*

9 The vehicle speed sensor/speedometer adapter assembly is located on the side of the extension housing on 2WD models or on the back of the transfer case on 4WD models **(see illustration)**. Look for transmission fluid around the cable housing to determine if the seal and O-ring are leaking.

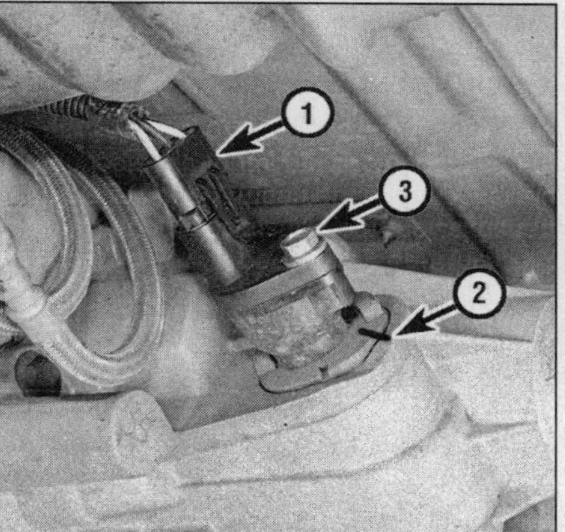

9.9 The vehicle speed sensor/speedometer adapter assembly is located on the back of the transfer case (4WD models - shown here) or on the left side of the extension housing on 2WD models. To remove the assembly, disconnect the electrical connector (1), mark the relation of the assembly to the housing (2), then remove the bolt (3)

10 Unplug the electrical connector from the speed sensor.

11 If you're going to reinstall the same speedometer adapter, mark the relationship of the adapter to the extension housing or transfer case; if you're going to install a new adapter, you'll have to index it as described below during installation.

12 Remove the sensor retaining bolt and adapter clamp.

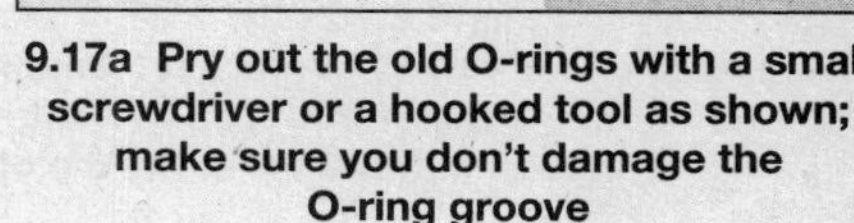

9.17a Pry out the old O-rings with a small screwdriver or a hooked tool as shown; make sure you don't damage the O-ring groove

13 Remove the vehicle speed sensor and speedometer adapter.

14 Inspect the terminal pins in the speed sensor. If the pins are dirty, clean them with electrical spray cleaner.

15 Remove the old O-rings from the speed sensor and from the adapter.

16 Thoroughly clean the adapter flange and adapter mounting surface in the housing. These surfaces must be clean in order to align the speedometer adapter properly.

17 Install new O-rings on the sensor and adapter **(see illustrations)**.

18 Lubricate the sensor and adapter O-rings with transmission fluid.

19 Install the vehicle speed sensor in the speedometer adapter. Tighten the retaining screw to the torque listed in this Chapter's Specifications.

20 Install the speedometer pinion in the adapter.

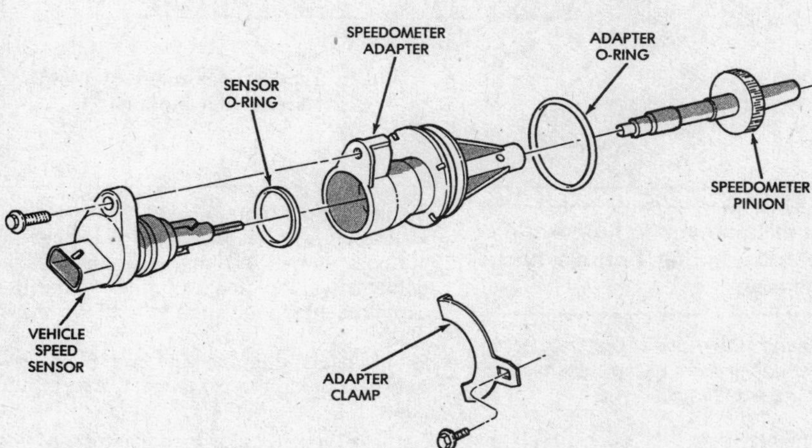

9.17b An exploded view of a typical transmission-mounted vehicle speed sensor/speedometer adapter assembly

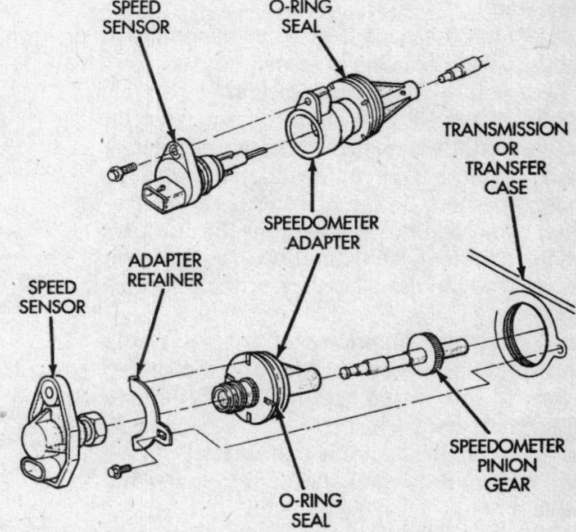

9.17c An exploded view of two typical transfer case-mounted vehicle speed sensor/speedometer adapter assemblies

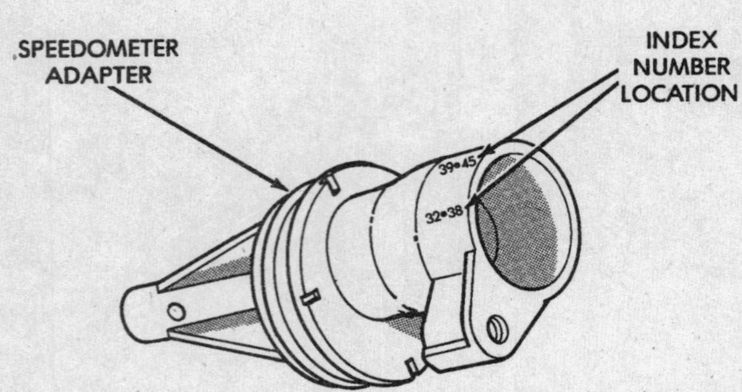

9.24 The index numbers on the side of the speedometer adapter correspond to various pinions used on different models; during installation, the adapter must be oriented so that the correct range of numbers is at the 6 o'clock position

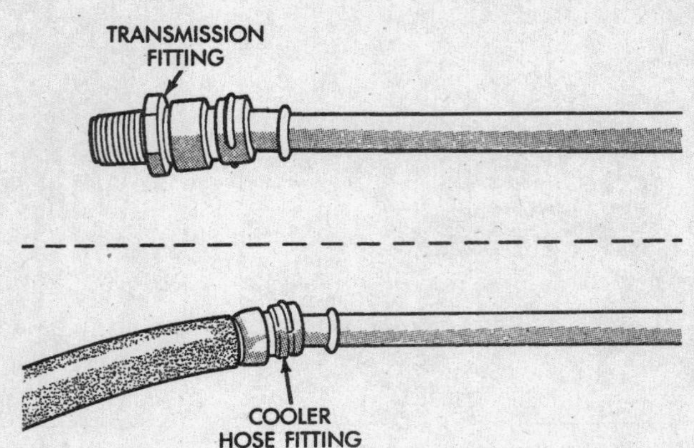

10.1 Typical transmission-to-cooler line fitting (top) and cooler line-to-cooler hose fitting (below)

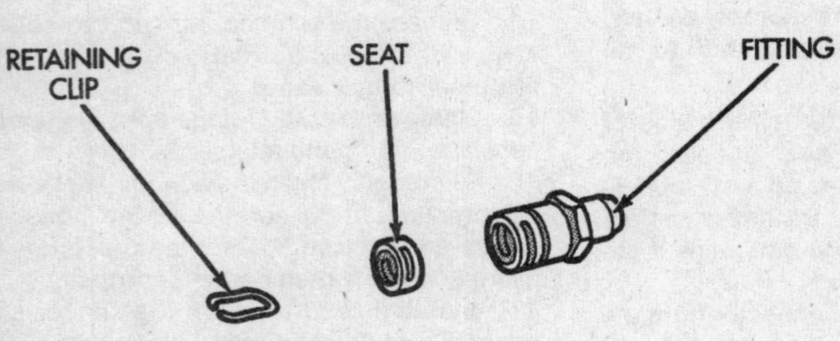

10.2a Type 1 quick-connect fitting (all 1993 models and some 1994 models)

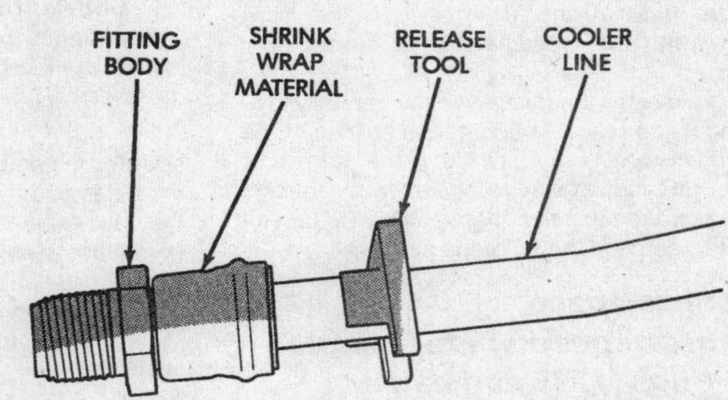

10.2b Type 2 quick-connect fitting (some 1994 and 1995 models)

21 If you're using a new adapter, count the number of teeth on the speedometer pinion before installing the sensor/adapter assembly.

22 Lubricate the pinion teeth with transmission fluid.

23 Install the sensor/adapter assembly in the extension housing or transfer case.

24 If you're using the old adapter, simply align the marks you made on the adapter flange and the extension housing or transfer case. If you're using a new adapter, note the index numbers on the adapter **(see illustration)** and rotate the adapter until the numbers on the adapter which correspond to the number of teeth on the pinion are at the 6 o'clock position.

25 Install the speedometer adapter clamp and retaining screw and tighten the clamp screw to the torque listed in this Chapter's Specifications.

26 Plug in the electrical connector.

27 Remove the jackstands and lower the vehicle.

28 Test drive the vehicle, make sure the speedometer works, then check for leaks from the extension housing or transfer case at the sensor/adapter assembly.

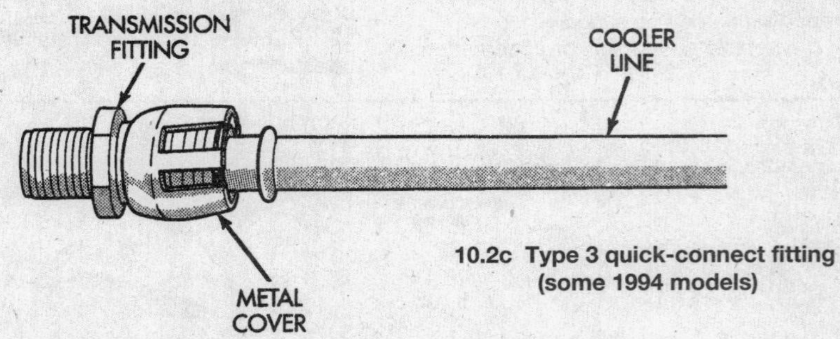

10.2c Type 3 quick-connect fitting (some 1994 models)

10 Transmission cooler hoses and lines - description, removal and installation

Note: *1998 and later models use threaded fittings on the cooler lines at the transmission, not quick-connect fittings.*

Description

Refer to illustrations 10.1, 10.2a, 10.2b, 10.2c and 10.4

1 The transmission cooler hoses and lines on 1997 and earlier models are attached with quick-connect fittings **(see illustration)**. A flange on the cooler line serves as the sealing surface. A wire retaining clip secures the cooler line in the fitting with this flange. The clip fits behind the flange to hold the line in place.

2 Depending on the year and model, any one of four different fitting styles may be used:

a) *On Type 1 fittings* **(see illustration)**, *the retaining clip is exposed.*
b) *On Type 2 fittings* **(see illustration)**, *the retaining clip and fitting body are encased in a shrink-wrap material.*

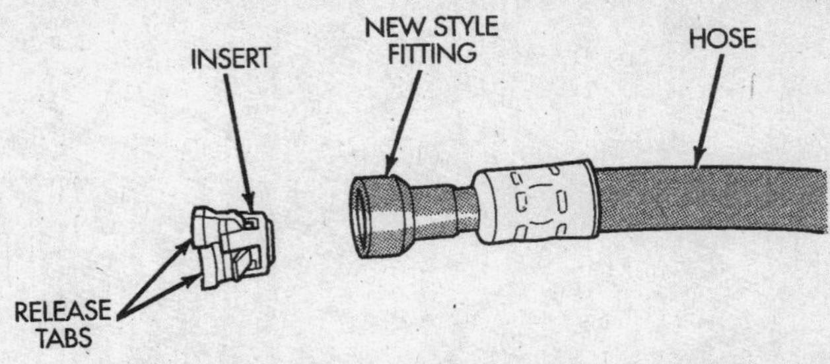

10.4 New-style fitting (late 1995 through 1997 models)

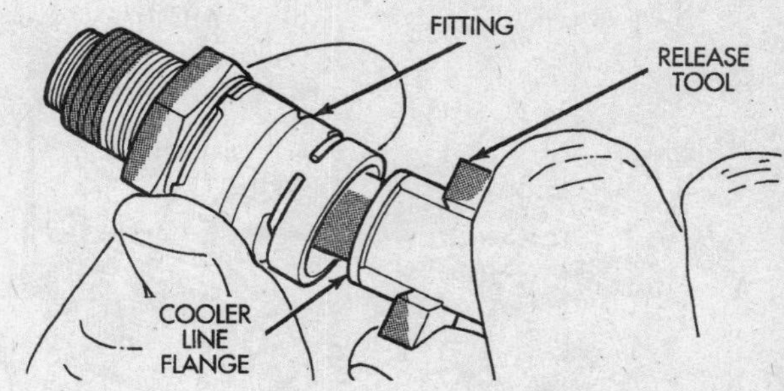

10.7 Slide the plastic release tool into the fitting until the tool bottoms against the flange on the cooler line, then push and turn the tool to spread the retaining clip and pull the cooler line out of the fitting (Type 2 fitting shown)

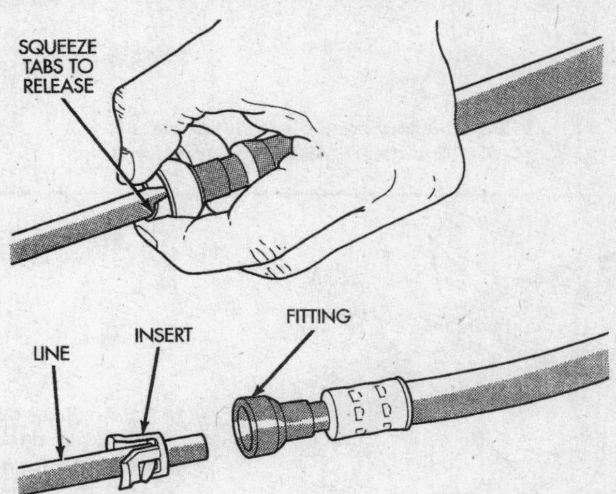

10.8 To release a new-style fitting, simply compress the release tabs with finger pressure (note that the fitting insert remains on the cooler line after release)

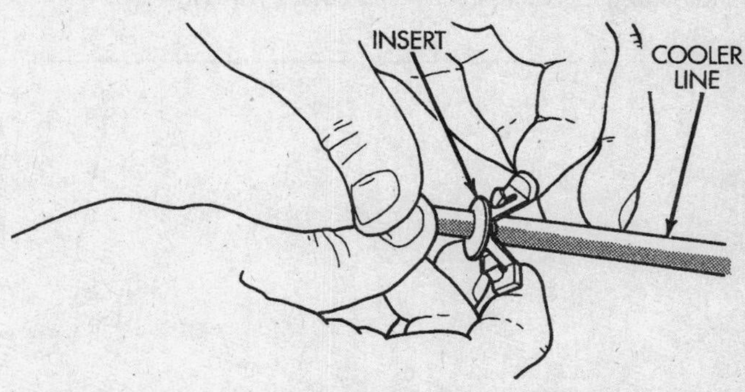

10.9 If a new-style fitting insert is damaged, simply spread the release tabs far enough to release the insert and slide it off the cooler line - replace the insert with a new one

c) *On Type 3 fittings* **(see illustration)**, *the retaining clip is covered by a metal sleeve crimped onto the body.*

d) *The new-style fitting used on later 1995 models has built-in release tabs.*

3 A special release tool is needed to disconnect most types of fittings:

a) *On 1993 models, a special release tool (available at most auto parts stores) must be used to release the Type 1 fittings.*

b) *On 1994 models with Type 2 or Type 3 fittings, the release tool is clipped directly to one of the cooler lines. The tool spreads the wire retainer clip to allow the cooler line to release. This tool can also be used to disconnect Type 1 fittings, which are still used on some 1994 models.*

c) *Type 2 fittings are also used on early 1995 models, so the same tool works on them.*

4 A new-style fitting, which is used on late 1995 through 1997 models, requires no release tool. This fitting has a plastic insert with built-in release tabs **(see illustration)**.

5 The transmission cooler hoses, lines and fittings cannot be repaired. Damaged fittings or cooler hoses or lines must be replaced as a single assembly.

Removal and installation

6 If the fittings and/or the hoses or lines are covered in dirt, mud or grease, clean them off with a cleaner/degreaser before disconnecting anything. The plastic release tool won't fit into the retainer clip if the fitting is filled with foreign matter.

All fittings except new-style

Refer to illustration 10.7

7 Slide the plastic release tool into the fitting until the tool bottoms against the flange on the cooler line **(see illustration)**, then push and turn the tool to spread the retaining clip and pull the cooler line out of the fitting.

New-style fittings (late 1995 through 1997 models)

Refer to illustrations 10.8 and 10.9

8 To release the new-style fittings, simply compress the release tabs with finger pressure **(see illustration)**. Note that the fitting insert remains on the cooler line after release.

9 It's not necessary to remove the insert from the cooler line unless the insert is damaged. If the fitting insert is damaged, simply spread the release tabs far enough to release the insert and slide it off the cooler line **(see illustration)**.

All fittings

10 Cover the open ends of the cooler lines and fittings to prevent dirty from entering the transmission.

11 Inspect the fittings. If a fitting or retaining clip attached to a line is damaged, the complete fitting must be replaced as an assembly (but you can still use the old metal line). Be sure to coat the threads with a non-hardening thread sealant. If a fitting or retaining clip wedged onto a hose is damaged, the hose and fittings must be replaced as a single assembly.

12 Wipe off the fitting and hose or line with a clean, dry shop rag.

13 Insert the cooler line into the fitting and push in the line until the retaining clip secures the line. You'll hear and feel a snapping or clicking sound when the retaining clip seats behind the cooler line flange.

14 Try to pull the cooler hoses and lines apart at the fittings to verify that they're properly secured. **Caution:** *On fittings which utilize a wire retaining clip, the clip must secure the cooler line in the fitting. If the clip is deformed or distorted, normal fluid pressure could unseat the cooler line, resulting in loss of fluid and damage to the transmission. Make SURE all cooler lines are properly*

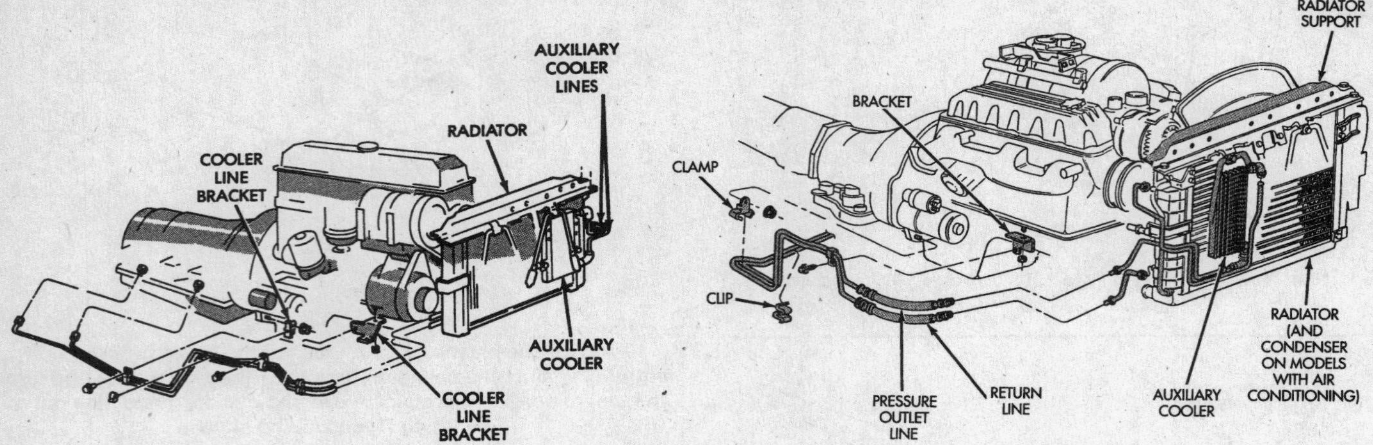

11.6a Auxiliary cooler installation details (1993 models)

11.6b Auxiliary cooler installation details (42/44RE and early 46RH transmissions)

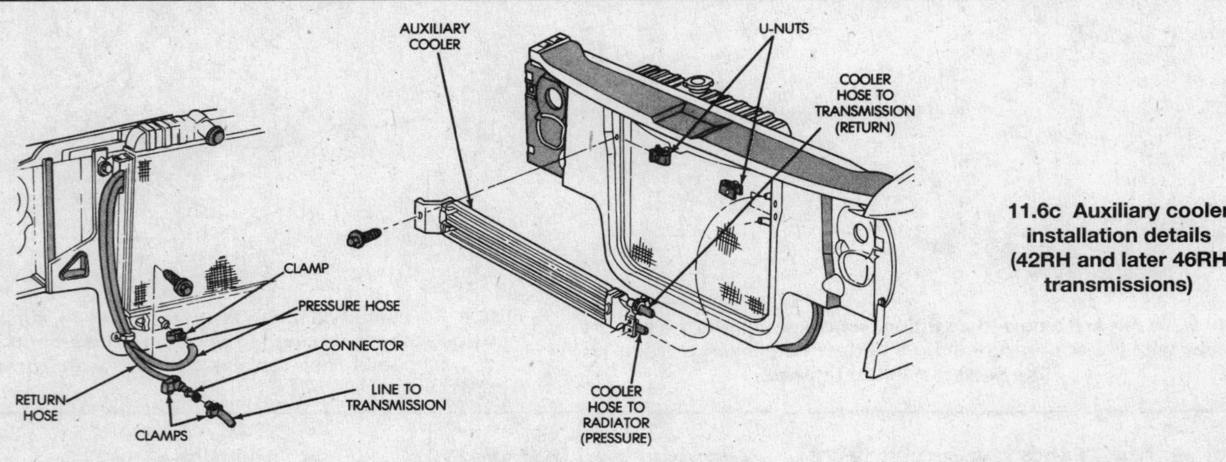

11.6c Auxiliary cooler installation details (42RH and later 46RH transmissions)

engaged by their retaining clips as described above.

15 Check the transmission fluid and add as necessary to bring it to the appropriate level (see Chapter 1).

16 Test drive the vehicle and check for leaks.

11 Auxiliary fluid cooler - removal and installation

Refer to illustrations 11.6a, 11.6b and 11.6c

1 Remove the grille (see Chapter 11).

2 Remove the bumper fascia (see Chapter 11). Also remove the grill opening reinforcement.

3 Label the cooler lines to assure proper reassembly.

4 Position a drain pan underneath the cooler hose clamps.

5 Loosen the cooler hose clamps and disconnect the clamps.

6 Remove all fasteners which attach the cooler to the radiator and support **(see illustrations)**.

7 Remove the auxiliary cooler.

8 Inspect the cooler hoses for cracks and tears. Cut off any worn hose material from the hose ends. Replace the hoses if they're cracked or torn anywhere besides the ends.

9 Installation is the reverse of removal. Be sure to tighten all cooler fasteners and cooler hose clamps securely.

10 Check the transmission fluid level and add as necessary to bring it to the appropriate level (see Chapter 1).

12 Automatic transmission - removal and installation

Removal

Refer to illustrations 12.2, 12.5a, 12.5b, 12.5c, 12.15a, 12.15b, 12.15c, 12.19, 12.20, 12.22a, 12.22b and 12.22c

Note 1: *Some transmission oil cooler lines have quick-connect fittings that require a special tool. Make sure you obtain the proper tool before beginning this procedure* (see Section 10).

Note 2: *If you are replacing a 45RFE transmission, a DRB scan tool must be used to perform a "Quick Learn" procedure, which recalibrates the transmission and Transmission Control Module (TCM) to work in the most efficient manner. This is a procedure*

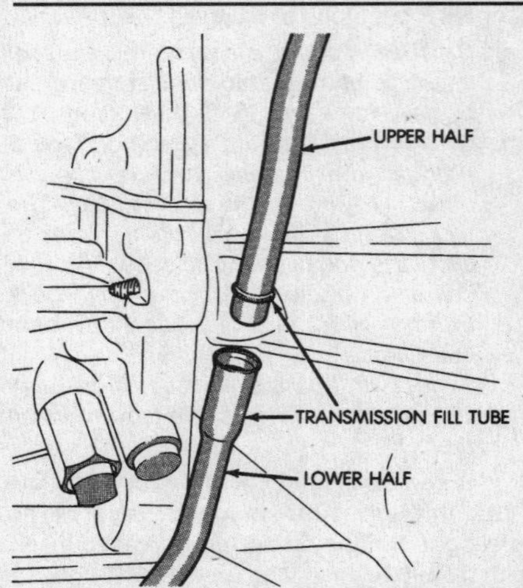

12.2 On 1993 models with an AW-4 transmission, remove the upper half of the transmission fill tube

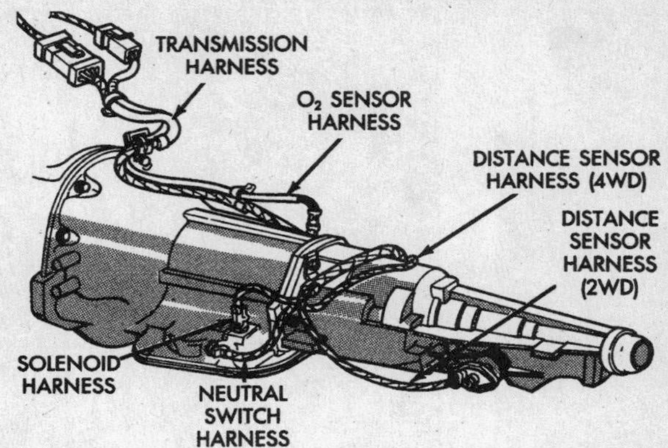

12.5a Unplug all electrical connectors, unclip the wire harnesses from their retaining clips on the transmission and/or transfer case, and set the harnesses aside; it's a good idea to label all connectors with colored electrical tape to simplify reassembly

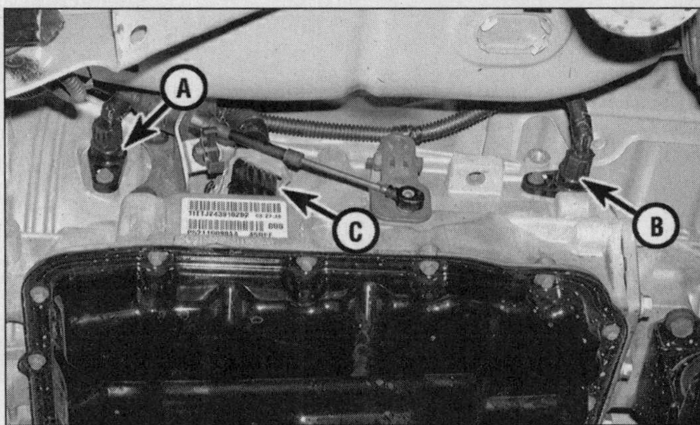

12.5b On the 45FRE transmission (1999 and later models with a 4.7L V8 engine), disconnect the electrical connectors from the input speed sensor (A), the output speed sensor (B), the solenoid and pressure switch assembly (C) . . .

which must be performed by a dealer service department or other repair shop equipped with the required special tool.

1 Disconnect the negative cable from the battery.

2 On 1993 models with an AW-4 transmission, remove the upper half of the transmission fill tube **(see illustration)**.

3 Raise the vehicle and support it securely on jackstands.

4 Remove the skid plate, if equipped.

5 Unplug all electrical connectors **(see illustrations)**, unclip the wiring harnesses from their retaining clips on the transmission and/or transfer case, and set the harnesses aside. It's a good idea to label all connectors with colored electrical tape to simplify reassembly. The following harnesses are used on various transmissions:

a) *AW-4: Vehicle speed sensor, neutral start switch and shift and lock-up solenoid wires.*
b) *42RE/44RE, 46RE: Vehicle speed sensor, transmission solenoid, neutral start switch and lock-up solenoid wires; output shaft speed sensor wires (4WD only).*
c) *42RH/46RH: Vehicle speed sensor, neutral start switch and lock-up solenoid wires.*
d) *45RFE: Input speed sensor, output speed sensor, solenoid and pressure switch assembly harness, and the line pressure sensor electrical connector.*

6 Clearly label, then detach, all vacuum lines to the transmission and/or transfer case.

7 Remove any exhaust system components that are in the way (see Chapter 4).

8 Drain the transmission fluid (see Chapter 1).

9 Reinstall the transmission pan.

10 Disconnect the transmission cooler lines from the transmission. Plug the cooler lines to prevent leakage and to protect the transmission from contamination.

11 Disconnect the shift cable from the transmission (see Section 4), and secure it aside.

12 Disconnect the Throttle Valve (TV) cable (if equipped) from the transmission (see Section 5), move it aside and secure it to the pan.

13 On 4WD models, disconnect the transfer case shift linkage or cable at the transfer case range lever (see Chapter 7 Part C). Then remove the linkage bracket bolts and remove the linkage and bracket from the transfer case. Move the linkage aside for clearance.

12.5c . . . and the line pressure sensor

14 Support the transmission with a floor jack and, if available, a transmission jack adapter. Safety chains will help steady the transmission on the jack.

15 Remove the bolts/nuts which attach the rear insulator to the rear crossmember **(see illustration)**. Raise the transmission and trans-

7B

12.15a To detach the transmission from the rear crossmember, remove the large center bolt (arrow)

12.15b Raise the transmission and transfer case (if equipped) slightly, remove the three crossmember-to-frame bolts (arrows) on the left side . . .

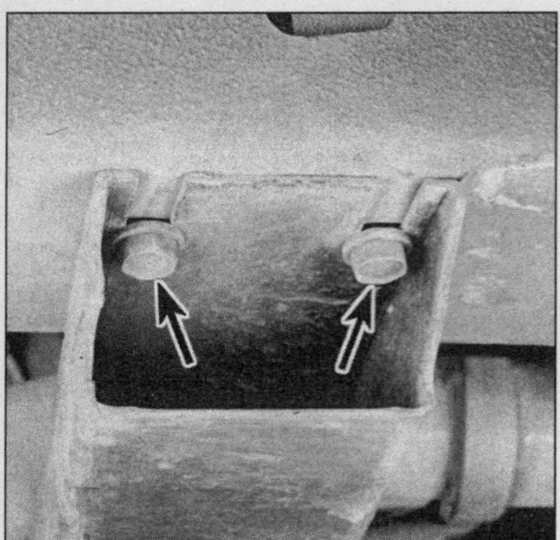

12.15c . . . remove the two bolts (arrows) on the right side, then detach the crossmember

12.19 Mark the relationship of the torque converter to the driveplate with white paint so they can correctly aligned on reassembly - this is critical for engine balancing!

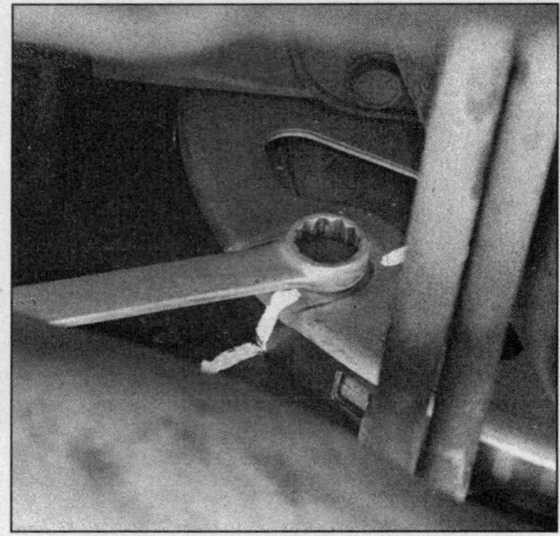

12.20 To remove the torque converter-to-driveplate bolts, turn the crankshaft for access to each bolt; turn the crankshaft in a clockwise direction only (as viewed from the front)

fer case (if equipped) slightly, then remove the crossmember-to-frame bolts **(see illustrations)** and detach the crossmember.

16 On 4WD models, remove the transfer case (see Chapter 7 Part C). **Note:** *If you are not planning to replace the transmission, but are removing it in order to gain access to other components such as the torque converter, it's not really necessary to remove the transfer case. However, the transmission and transfer case are awkward and heavy when removed and installed as a single assembly; they're much easier to maneuver off and on as separate units.* **Warning:** *If you decide to leave the transfer case attached to the transmission, be sure to use safety chains to help stabilize the transmission and transfer case assembly and to prevent it from falling off the jack head, which could cause serious damage to the transmission and/or transfer case and serious bodily injury to you.*

17 Remove the torque converter access cover from the transmission. Some models use three bolts, some use four.

18 Remove the starter motor (see Chapter 5).

19 Mark the relationship of the torque converter to the driveplate with white paint so they can be installed in the same position **(see illustration)**.

20 Remove the torque converter-to-driveplate bolts **(see illustration)**. Turn the crankshaft for access to each bolt. Turn the crankshaft in a clockwise direction only (as viewed from the front).

21 Support the engine with a floor jack or an engine hoist (see Chapter 2).

22 Remove the lower transmission-to-engine bolts **(see illustrations)**.

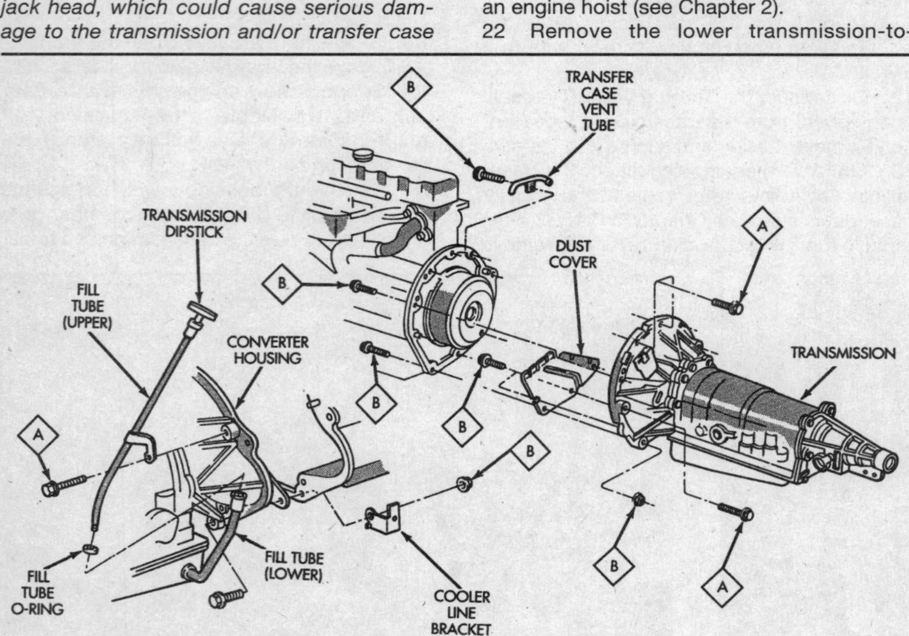

12.22a Transmission fastener torque guide (AW-4) (refer to the Specifications for torque values)

23 Lower the transmission (and, on 4WD models, the transfer case, if still attached) slightly and remove the upper transmission-to-engine bolts.

24 On all except 4.7L V8 models, unplug and remove the crankshaft position sensor (see Chapter 6). **Caution:** *The crankshaft position sensor can be damaged if the transmission is removed (or installed) with the sensor installed.*

25 Remove the transmission dipstick tube. Remove and discard the dipstick tube O-ring.

26 Move the transmission to the rear to disengage it from the engine block dowel pins and make sure the torque converter is detached from the driveplate. Secure the torque converter to the transmission so it won't fall out during removal.

27 Lower the transmission and move it out from under the vehicle.

Installation

Refer to illustration 12.28

28 Prior to installation, make sure the torque converter hub is securely engaged in the pump. The distance from the bellhousing mating surface to the machined converter lugs must be 1/2-inch **(see illustration)**. If it isn't, rotate the converter while pushing in until the converter "clunks" into place (then recheck the measurement). **Note:** *If you removed the converter from a 45RFE transmission, make sure the O-ring on the converter hub is in good condition.*

29 With the transmission (and, on 4WD models, the transfer case, if you left it attached) secured to the jack, raise it into position. Be sure to keep it level so the torque converter does not slide forward. Connect the transmission fluid cooler lines.

30 Turn the torque converter to line up its bolt holes with the holes in the driveplate. The white paint mark on the torque converter and the driveplate made in Step 19 must line up.

31 Move the transmission (and on 4WD

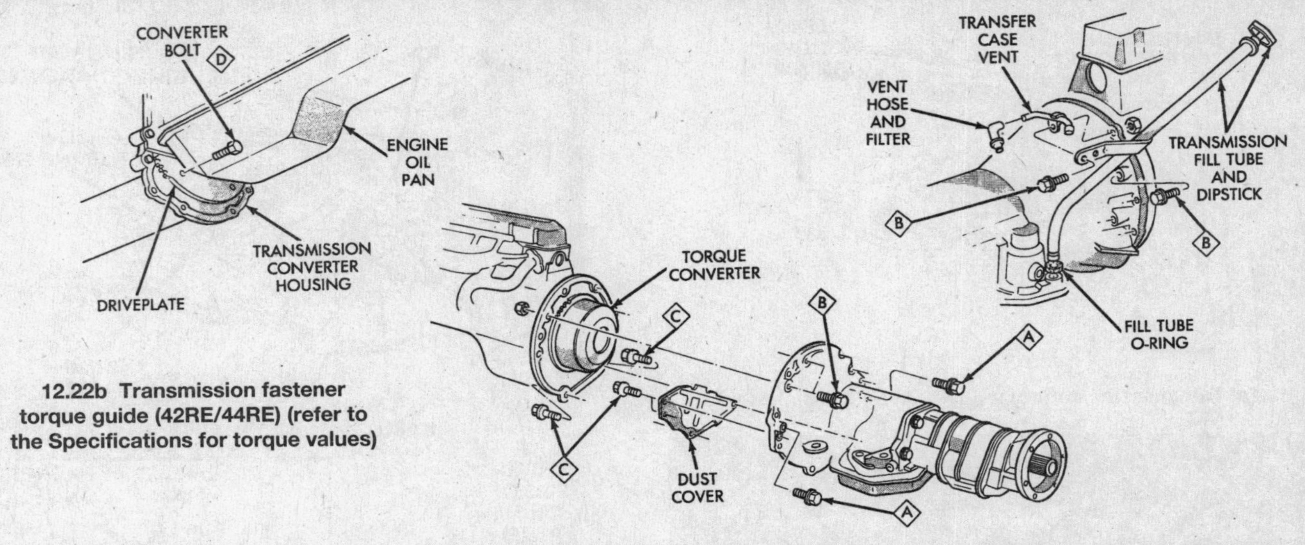

12.22b Transmission fastener torque guide (42RE/44RE) (refer to the Specifications for torque values)

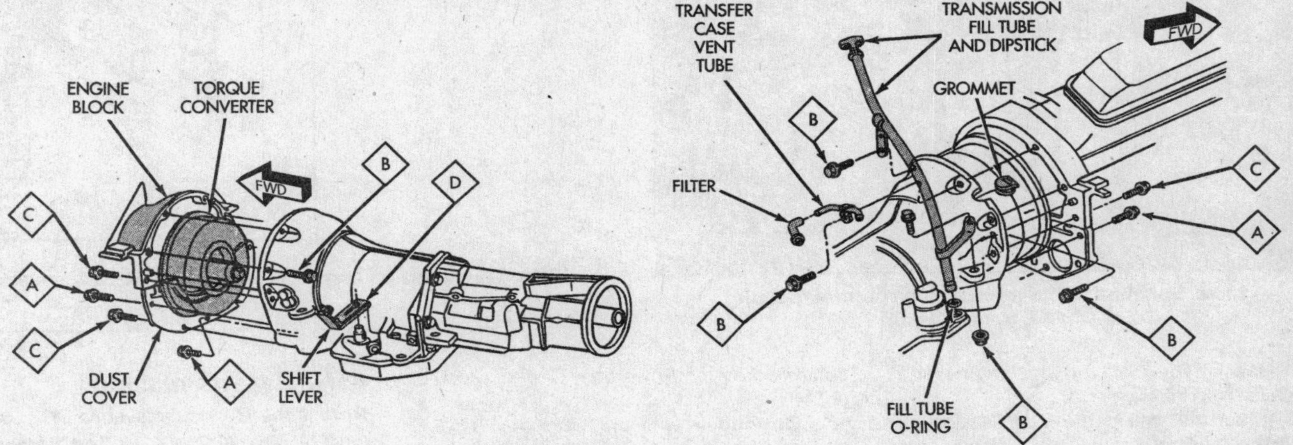

12.22c Transmission fastener torque guide (42RH/46RH) (refer to the Specifications for torque values) - 1995 and earlier models

models, the transfer case, if already attached) forward carefully until the dowel pins and the torque converter are engaged.

32 Install the transmission housing-to-engine bolts. Tighten the fasteners to the torque values listed in this Chapter's Specifications. On 1995 and earlier models, refer to the accompanying bolt torque guide **(see illustration 12.22a, 12.22b or 12.22c)**.

33 Install the torque converter-to-driveplate bolts. Tighten the bolts to the torque listed in this Chapter's Specifications. Install the torque converter access cover and tighten the bolts securely. **Note (for 1995 and earlier models only):** *If you're using new torque converter bolts, measure the length of the bolts (from the underside of the bolt head to the end of the bolt threads) and compare your measurements to the dimensions listed in this Chapter's Specifications to ensure that you're using the right bolt length for the converter in your vehicle. Using a bolt that's too long will damage the modulated clutch surfaces in the converter.*

34 Attach the rear mount-to-crossmember and crossmember-to-frame bolts. Tighten the bolts securely.

35 Remove the jacks supporting the transmission (and on 4WD models, the transfer case, if already attached).

36 Install the dipstick tube.

37 If you removed the transfer case prior to transmission removal, install it now (see Chapter 7 Part C).

38 Install the starter motor (see Chapter 5).

39 Reattach all vacuum hose(s), plug in all electrical connectors and reattach all wire harness clips.

40 Connect the shift cable (see Section 4) and the TV cable (see Section 5). On 4WD models, connect the transfer case shift linkage (see Chapter 7 Part C).

41 Install the driveshaft(s) (see Chapter 8).

42 Install any exhaust system components that were removed or disconnected.

43 Remove the jackstands and lower the vehicle.

44 Fill the transmission with the specified

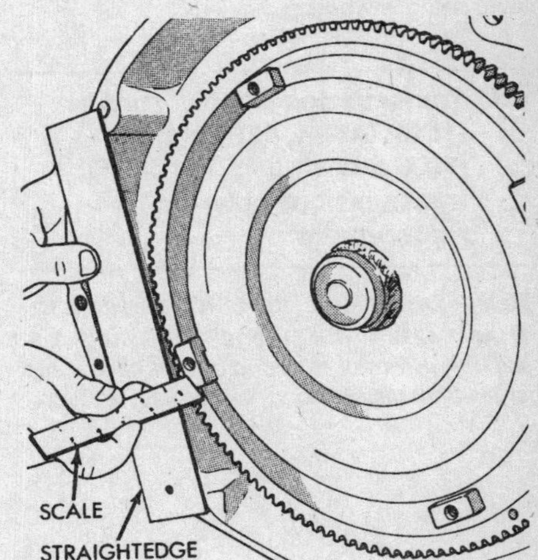

12.28 Before installing the transmission, make sure the distance from the bellhousing mating surface to the converter lugs is 1/2-inch

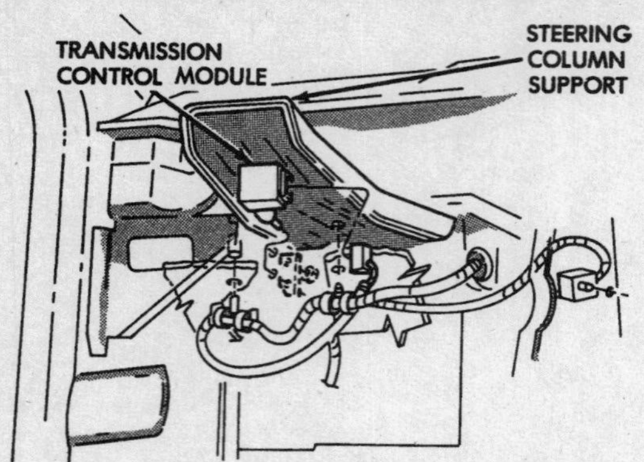

13.1a Transmission control module (AW-4 transmission)

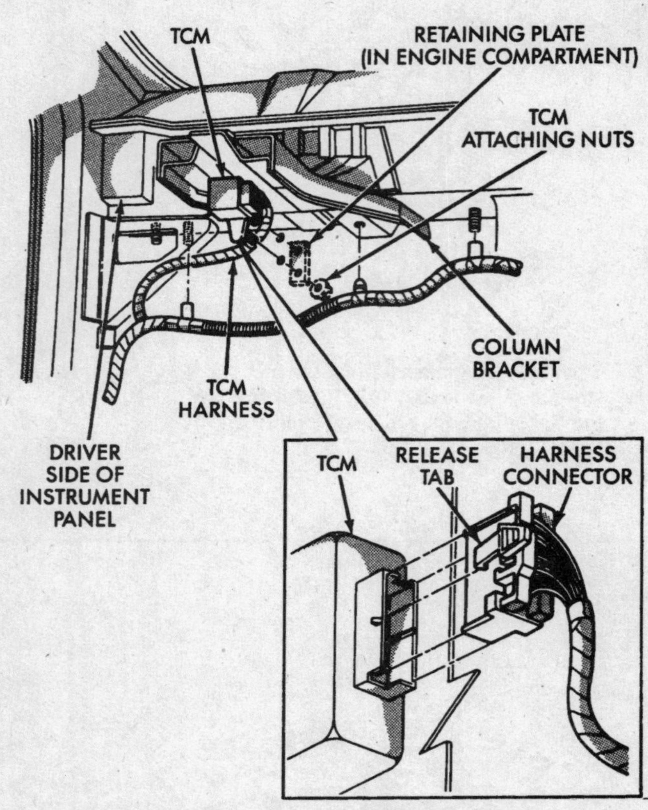

13.1b Transmission control module (42/44RE transmission)

13.1c Location of the transmission control module (45RFE transmission)

fluid (see Chapter 1), run the engine and check for fluid leaks.

45 Adjust the shift cable (see Section 4), the TV cable, if equipped (see Section 5) and, on 4WD models, the transfer case shift linkage or cable (see Chapter 7 Part C).

46 Test drive the vehicle and make sure everything is working properly.

13 Transmission Control Module (TCM) (AW-4, 42RE/44RE [1995 and earlier] and 45RFE transmissions only) - replacement

Refer to illustrations 13.1a, 13.1b and 13.1c

Note: *42RH, 46RH and 1996 and later 42RE/44RE transmissions do not have a separate transmission control module; shifting is controlled by the Powertrain Control Module (PCM).*

1 On models with an AW-4 or 42/44RE transmission, the transmission control module is located underneath the driver's side of the dashboard, to the left of the steering column **(see illustrations)**. On models with the 45RFE transmission, it's mounted to a bracket on the left-side inner fender panel **(see illustration)**.

AW-4 and 42/44RE transmissions

2 Disconnect the cable from the negative terminal of the battery.

3 Unplug the electrical connector.

4 Remove the module mounting screws and/or nuts.

5 Remove the module from the cowl panel.

6 Installation is the reverse of removal.

45RFE transmission

Note: *If the TCM is replaced, a DRB scan tool must be used to perform a "Quick Learn" procedure, which recalibrates the transmission and TCM to work in the most efficient manner. This is a procedure which must be performed by a dealer service department or other repair shop equipped with the required special tool.*

7 Disconnect the cable from the negative terminal of the battery.

8 Remove the two bolts holding the TCM bracket to the inner fender panel, and the nut securing it to the firewall.

9 Lift up the TCM and mounting bracket, then unplug the electrical connector.

10 The TCM can now be unbolted from its bracket, if necessary.

11 Installation is the reverse of removal.

Chapter 7 Part C
Transfer case

Contents

Specifications

Torque specifications

	Ft-lbs (unless otherwise indicated)
Shift linkage trunnion lock bolt	120 in-lbs
Transfer case-to-transmission nuts	
5/16-inch stud	26
3/8-inch stud	35

1 General information

The transfer case is a device which transmits power from the transmission to the front and rear driveshafts. The models covered by this manual may be equipped with any one of the following transfer cases. The transfer cases were originally manufactured by New Process (NP) which later became New Venture (NV).

a) *The NP/NV231 is a part-time transfer case with a low-range reduction gear system. The low-range reduction system provides increased low-speed torque capability.*
b) *The NP/NV242 is a full- and part-time transfer case with four operating ranges. It provides both 2WD and full-time 4WD, and has a locking mechanism for undifferentiated 4WD in high and low ranges.*
c) *The NP/NV249 is a full-time 4WD transfer case with two operating ranges and a neutral position. Its two operating ranges are 4-high and 4-low (4-low is for extra pulling power in off-road situations). Engine torque is distributed to the front and rear axles through a differential and viscous coupling in the transfer case. The NV249 is also equipped with low-range gear reduction (4-low) for increased low-speed and off-road torque capability.*
d) *The NV247 is a 4WD on-demand transfer case with two operating ranges and a neutral position. Its two operating ranges are 4-high and 4-low. Under normal conditions, with the transfer case in 4-high, the majority of the torque is applied to the rear wheels. When rear wheel slip is sensed, torque is transferred to the front axle, allowing the axle with the best traction to receive the majority of the torque. The system is totally mechanical (non-electronic) and requires no driver input to operate. 4-low is engaged for extra pulling power in off-road situations.*

2 Shift linkage adjustment (1998 and earlier models)

Refer to illustrations 2.3a, 2.3b and 2.3c

1 Shift the transfer case into the Neutral position.

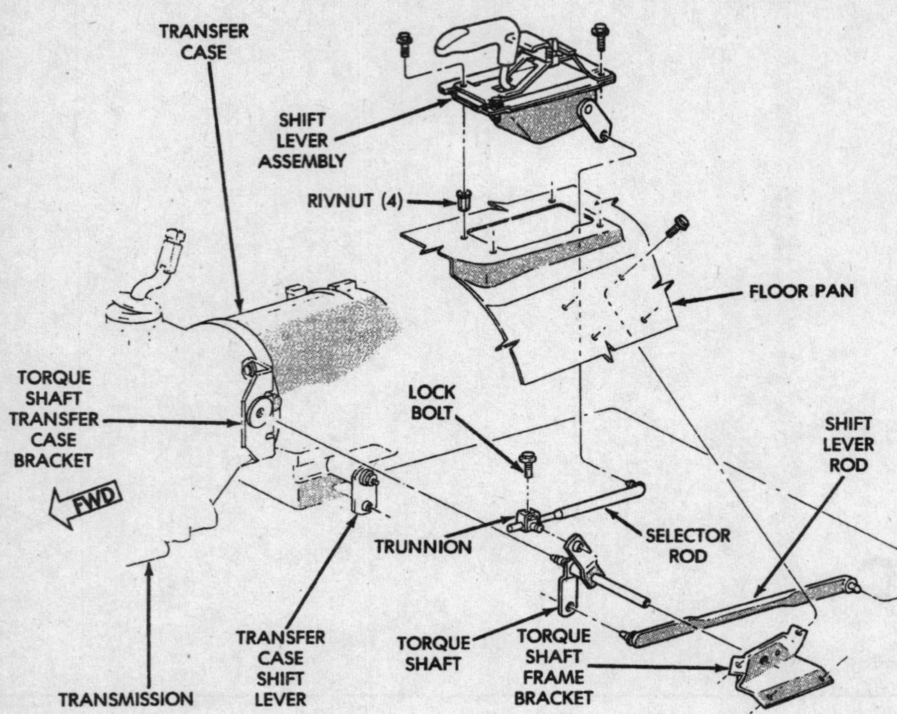

2.3a Typical shift linkage installation details (models equipped with a manual transmission)

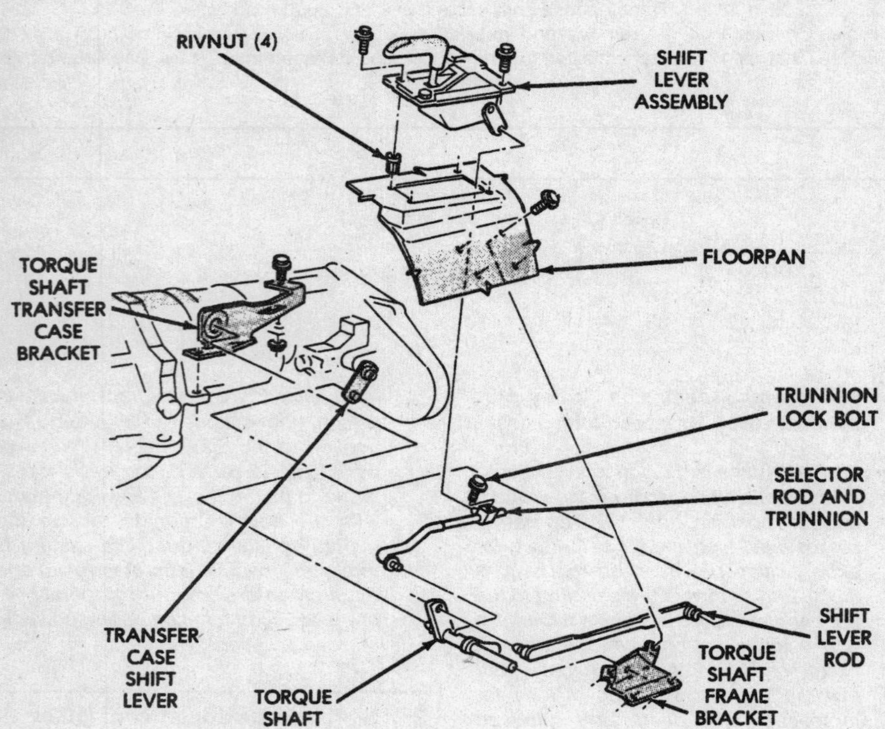

2.3b Typical shift linkage installation details (models equipped with an automatic transmission)

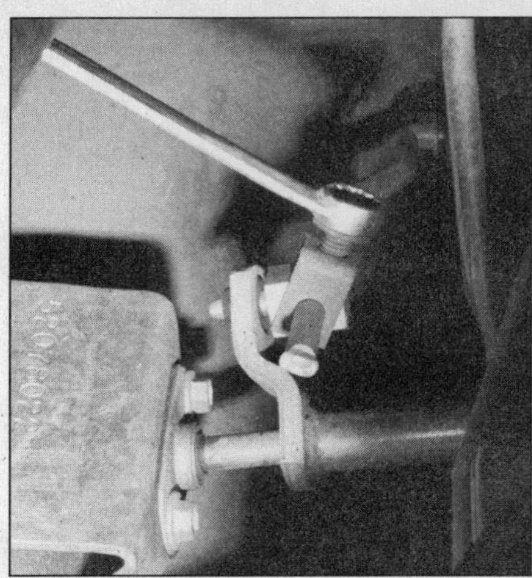

2.3c Loosen the lock bolt in the selector rod trunnion (typical)

2 Raise the vehicle and support it securely on jackstands.

3 Loosen the lock bolt in the selector rod trunnion **(see illustrations)**.

4 Verify that the transfer case shift lever is in the Neutral position and the transfer case is actually in neutral.

5 Move the trunnion forward or rearward on the selector rod as necessary, then tighten the trunnion lock bolt to the torque listed in this Chapter's Specifications.

6 Lower the vehicle and check the transfer case shifting action. Test drive the vehicle and verify that the transfer case is fully engaged in the 2H, 4H and 4L positions. If it isn't fully engaged in all three positions, readjust it.

3 Transfer case shift cable (1999 and 2000 models) - removal and installation

Refer to illustrations 3.3, 3.5 and 3.6

Warning: *These models are equipped with airbags. Always disable the airbag system before working in the vicinity of any airbag system component to avoid the possibility of accidental deployment of the airbag(s), which could cause personal injury* (see Chapter 12).

1 Apply the parking brake, shift the transmission into the Park position and shift the transfer case into the Neutral position.

2 Raise the vehicle and support it securely on jackstands.

3 Disconnect the shift cable from the shift lever at the transfer case. Remove the retaining clip and detach the cable from the support bracket **(see illustration)**.

4 Remove the center console (see Chapter 11).

5 Disconnect the cable from the shift lever. Remove the clip and detach the cable from the shift lever base **(see illustration)**.

6 Remove the seal plate mounting nuts from the floor pan studs and pull the shift cable through the floor pan and into the passenger compartment **(see illustration)**.

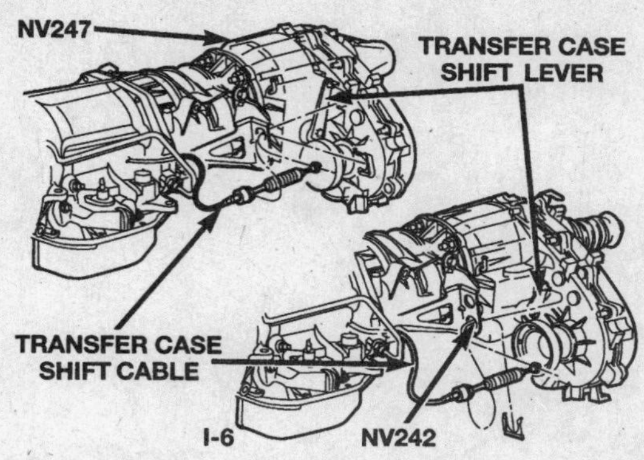

3.3 Transfer case shift cable installation details - at transfer case

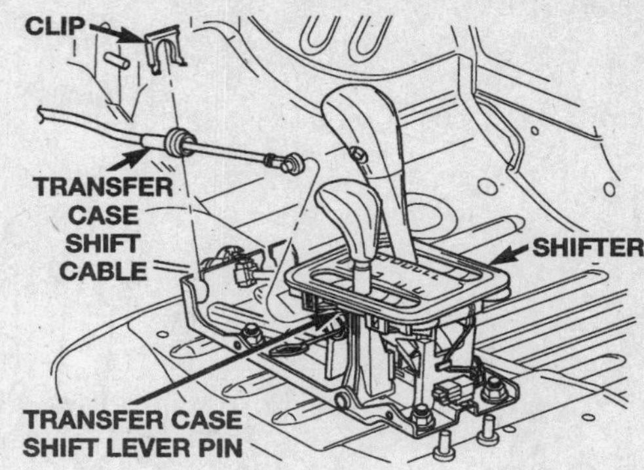

3.5 Transfer case shift cable installation details - at shift lever

7 Installation is the reverse of removal. Check the transfer case operation in all operating ranges.

4 Transfer case - removal and installation

Removal

Refer to illustrations 4.9 and 4.11

1 Disconnect the negative cable from the battery.

2 Raise the vehicle and support it securely on jackstands.

3 Drain the transfer case lubricant (see Chapter 1).

4 Unplug the electrical connector from the vehicle speed sensor, if equipped. Remove the vehicle speed sensor/speedometer adapter assembly to protect it from damage during transfer case removal (see Chapter 7B).

5 Disconnect the shift lever rod or shift cable from the transfer case shift lever (see Section 2 or 3).

6 Detach all vacuum/vent lines and wire harness connectors from the transfer case.

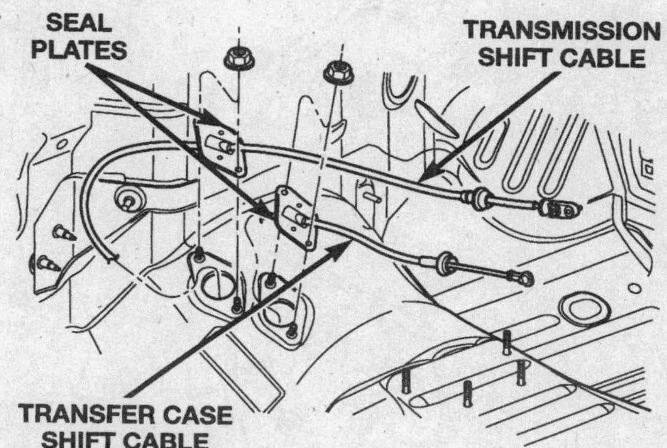

3.6 Remove the transfer case shift cable seal plate nuts from the mounting studs

7 Remove the exhaust system components as necessary for clearance (see Chapter 4).

8 Remove the driveshafts (see Chapter 8).

9 Support the transmission with a transmission jack and remove the skid plate (if equipped) and the rear crossmember **(see illustration)**.

10 Support the transfer case with a jack - preferably a special jack made for this purpose. Safety chains will help steady the transfer case on the jack.

11 Remove the transfer case-to-transmission nuts from the mounting studs **(see illustration)**.

12 Make a final check that all wires and hoses have been disconnected from the transfer case and then move the transfer case

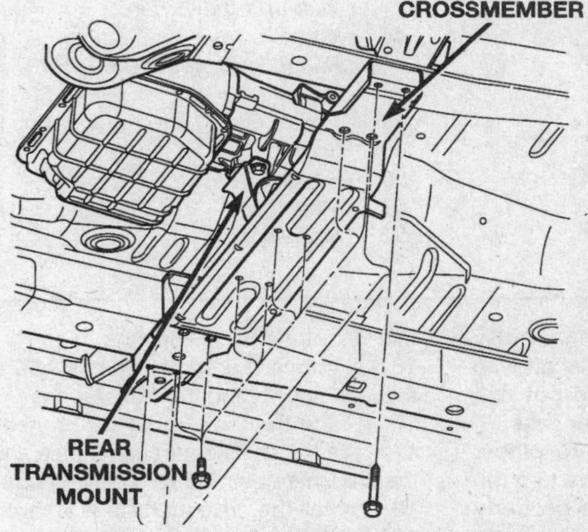

4.9 Rear crossmember installation details (typical)

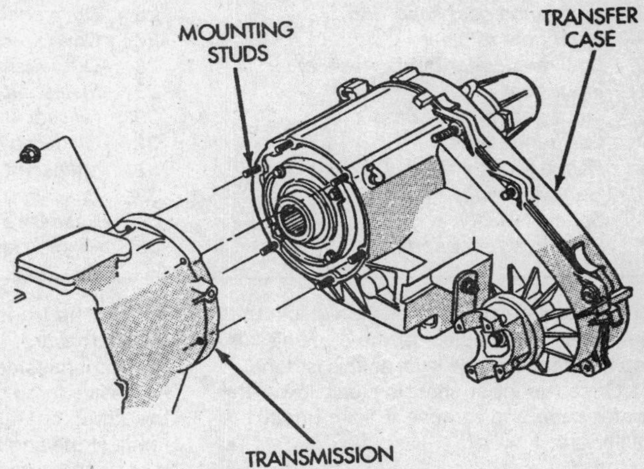

4.11 Remove the nuts from the transfer case mounting studs

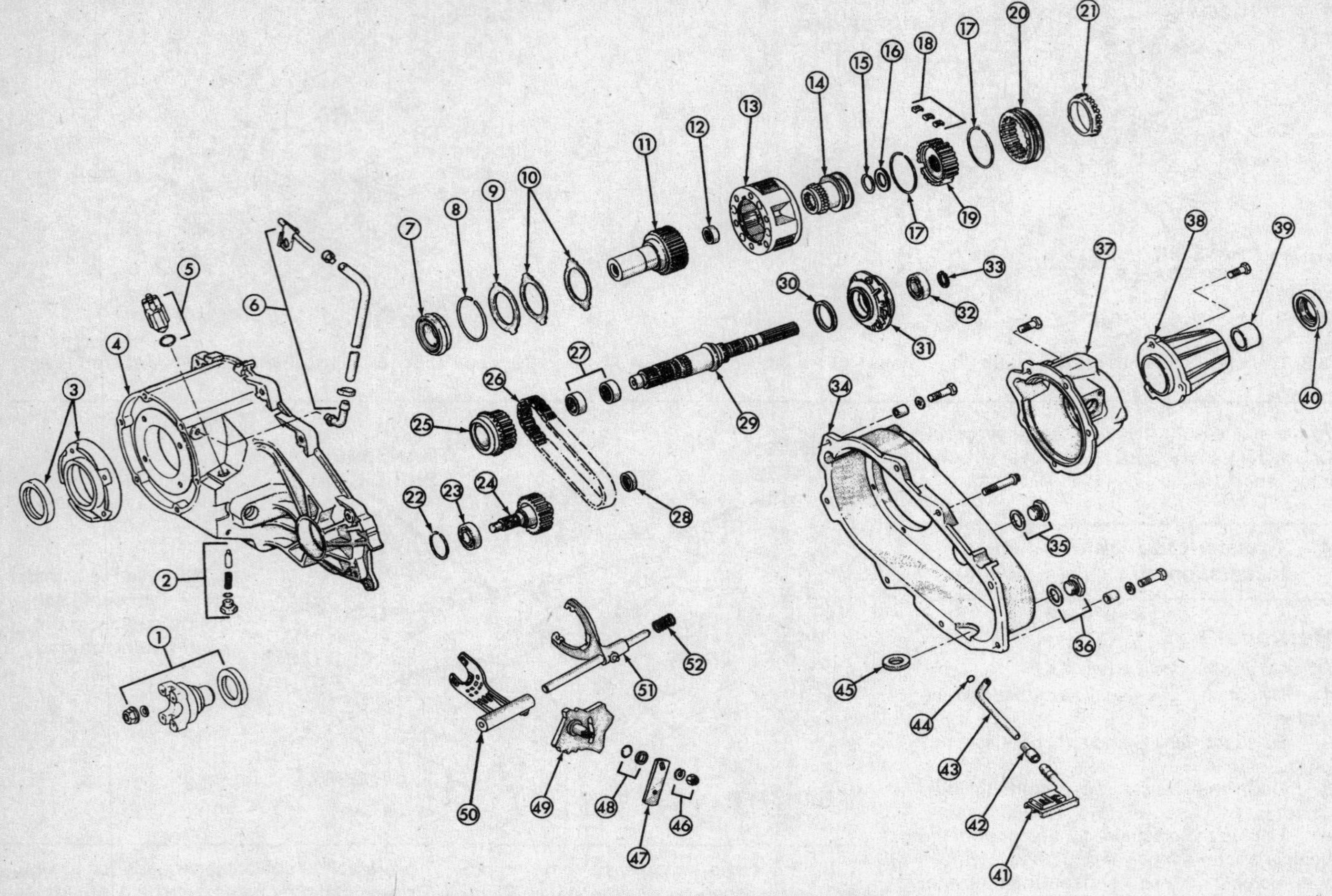

5.4a An exploded view of the New Process 231 transfer case

1 *Front yoke nut, seal washer, yoke and oil seal*
2 *Shift detent plug, spring and pin*
3 *Front retainer and seal*
4 *Front case*
5 *Vacuum switch and seal*
6 *Vent assembly*
7 *Input gear bearing and snap-ring*
8 *Low range gear snap-ring*
9 *Input gear retainer*
10 *Low range gear thrust washers*
11 *Input gear*
12 *Input gear pilot bearing*
13 *Low range gear*
14 *Range fork shift hub*
15 *Hub snap-ring*
16 *Spacer washer*
17 *Synchronizer springs*
18 *Synchronizer struts*
19 *Hub*
20 *Sleeve*
21 *Stop-ring*
22 *Front bearing snap-ring*
23 *Front bearing output shaft*
24 *Front output shaft*
25 *Drive sprocket*
26 *Drive chain*
27 *Drive sprocket bearings*
28 *Output shaft rear bearing*
29 *Mainshaft*
30 *Oil seal*
31 *Oil pump assembly*
32 *Mainshaft rear bearing*
33 *Snap-ring*
34 *Rear case*
35 *Fill plug and gasket*
36 *Drain plug and gasket*
37 *Rear retainer*
38 *Extension housing*
39 *Bushing*
40 *Oil seal*
41 *Oil pick-up screen*
42 *Tube connector*
43 *Oil pick-up tube*
44 *Pick-up tube O-ring*
45 *Magnet*
46 *Range lever nut and washer*
47 *Range lever*
48 *Sector O-ring and retainer bushing*
49 *Sector*
50 *Range fork*
51 *Mode fork*
52 *Mode spring*

and jack toward the rear of the vehicle until the transfer case is clear of the transmission. Keep the transfer case level as this is done.

13 Once the input shaft is clear, lower the transfer case and remove it from under the vehicle.

Installation

14 With the transfer case secured to the jack as on removal, raise it into position behind the transmission and then carefully slide it forward, engaging the input shaft with the transmission output shaft. Do not use excessive force to install the transfer case - if the input shaft does not slide into place, readjust the angle so it is level and/or turn the input shaft so the splines engage properly with the transmission.

15 Install the transfer case-to-transmission nuts. Tighten the nuts to the torque listed in this Chapter's Specifications.

16 Remove the safety chains and remove the jack supporting the transfer case.

17 Place the jack under the transmission, raise the transmission/transfer case and install the crossmember and skid plate (if equipped).

18 Install the driveshafts (see Chapter 8).

19 Install the exhaust system (see Chapter 4).

20 Reattach all vacuum/vent lines.

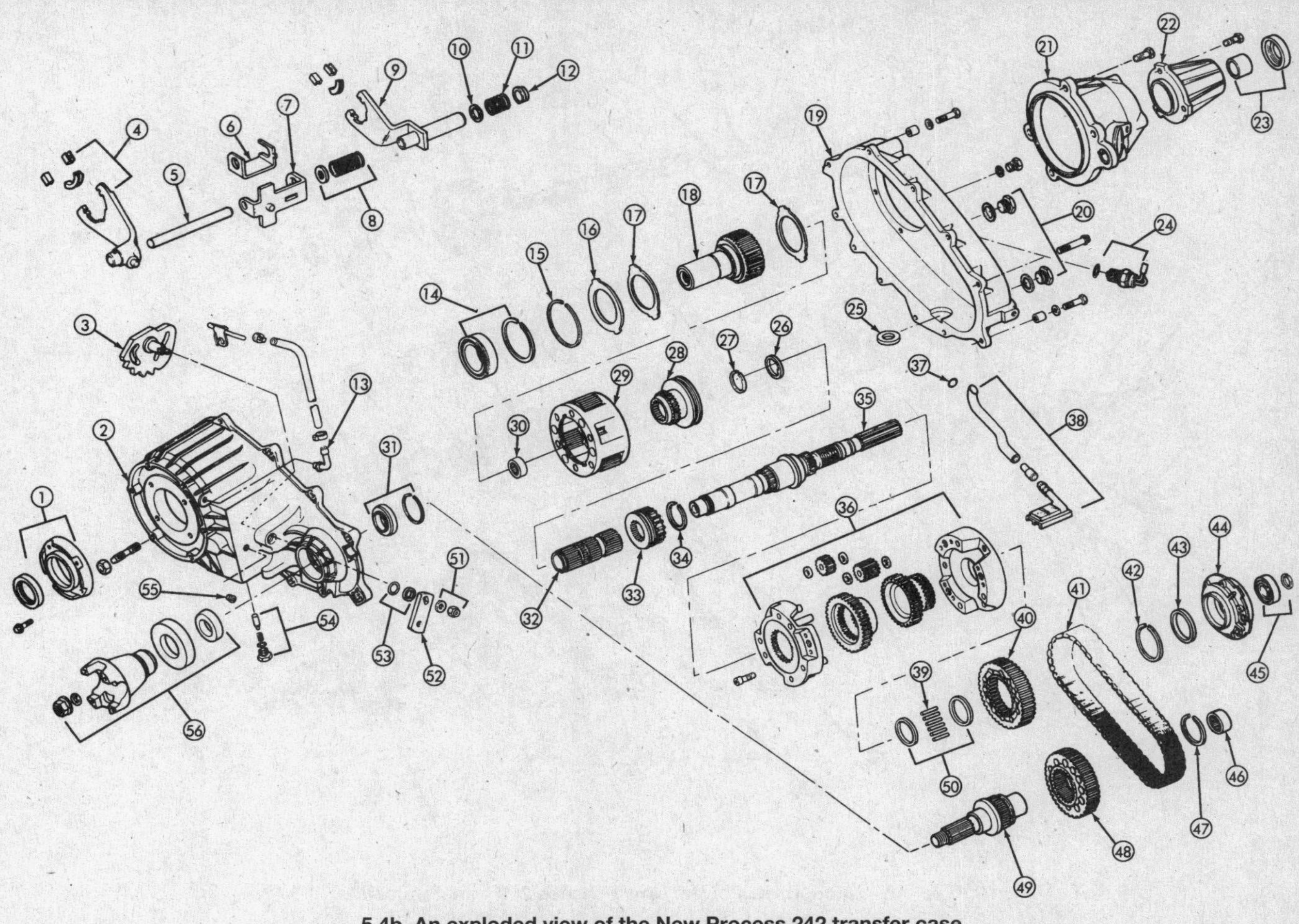

5.4b An exploded view of the New Process 242 transfer case

1. Front bearing retainer and seal
2. Front case
3. Shift sector
4. Low range fork and inserts
5. Shift rail
6. Shift bracket
7. Slider bracket
8. Bushing and spring
9. Mode fork and inserts
10. Bushing
11. Fork spring
12. Bushing
13. Vent tube assembly
14. Input gear bearing and snap-ring
15. Low range gear snap-ring
16. Retainer, low range gear
17. Thrust washer, low range gear
18. Input gear
19. Rear case
20. Drain/fill plugs
21. Rear bearing retainer
22. Extension housing
23. Bushing and oil seal
24. Vacuum switch
25. Magnet
26. Thrust ring
27. Snap-ring
28. Shift sleeve
29. Low range gear
30. Pilot bushing (input gear/mainshaft)
31. Front output shaft front bearing and snap-ring
32. Intermediate clutch shaft
33. Shift sleeve
34. Snap-ring
35. Mainshaft
36. Differential assembly
37. Oil pump tube O-ring
38. Oil pump pick-up tube and screen
39. Mainshaft bearing rollers
40. Drive sprocket
41. Drive chain
42. Snap-ring
43. Oil pump seal
44. Oil pump
45. Rear bearing and snap-ring
46. Front output shaft rear bearing
47. Snap-ring
48. Driven sprocket
49. Front output shaft
50. Mainshaft bearing spacers
51. Shift lever washer and nut
52. Shift lever
53. Sector O-ring and seal
54. Detent pin, spring and plug
55. Seal plug
56. Front yoke nut, seal washer, yoke, slinger and oil seal

21 Reconnect the shift lever rod or shift cable to the transfer case shift lever and on 1998 and earlier models, adjust the shift linkage (see Section 2).

22 Install the vehicle speed sensor/speedometer adapter assembly (if equipped) and plug in the electrical connector (see Chapter 7 Part B).

23 Check the lubricant level in the transfer case and refill as necessary (see Chapter 1). If the vehicle is equipped with a manual transmission, check the lubricant level for the manual transmission, as well (see Chapter 1).

24 Remove the jackstands and lower the vehicle.

25 Connect the negative battery cable.

26 Road test the vehicle for proper operation and check for leakage.

5 Transfer case overhaul - general information

Refer to illustrations 5.4a, 5.4b and 5.4c

Overhauling a transfer case is a difficult job for the do-it-yourselfer. It involves the disassembly and reassembly of many small parts. Numerous clearances must be precisely measured and, if necessary, changed with select fit spacers and snap-rings. As a result, if transfer case problems arise, it can be removed and installed by a competent do-it-yourselfer, but overhaul should be left to a transmission repair shop. Rebuilt transfer cases may be available - check with your dealer parts department and auto parts stores. At any rate, the time and money involved in an overhaul is almost sure to

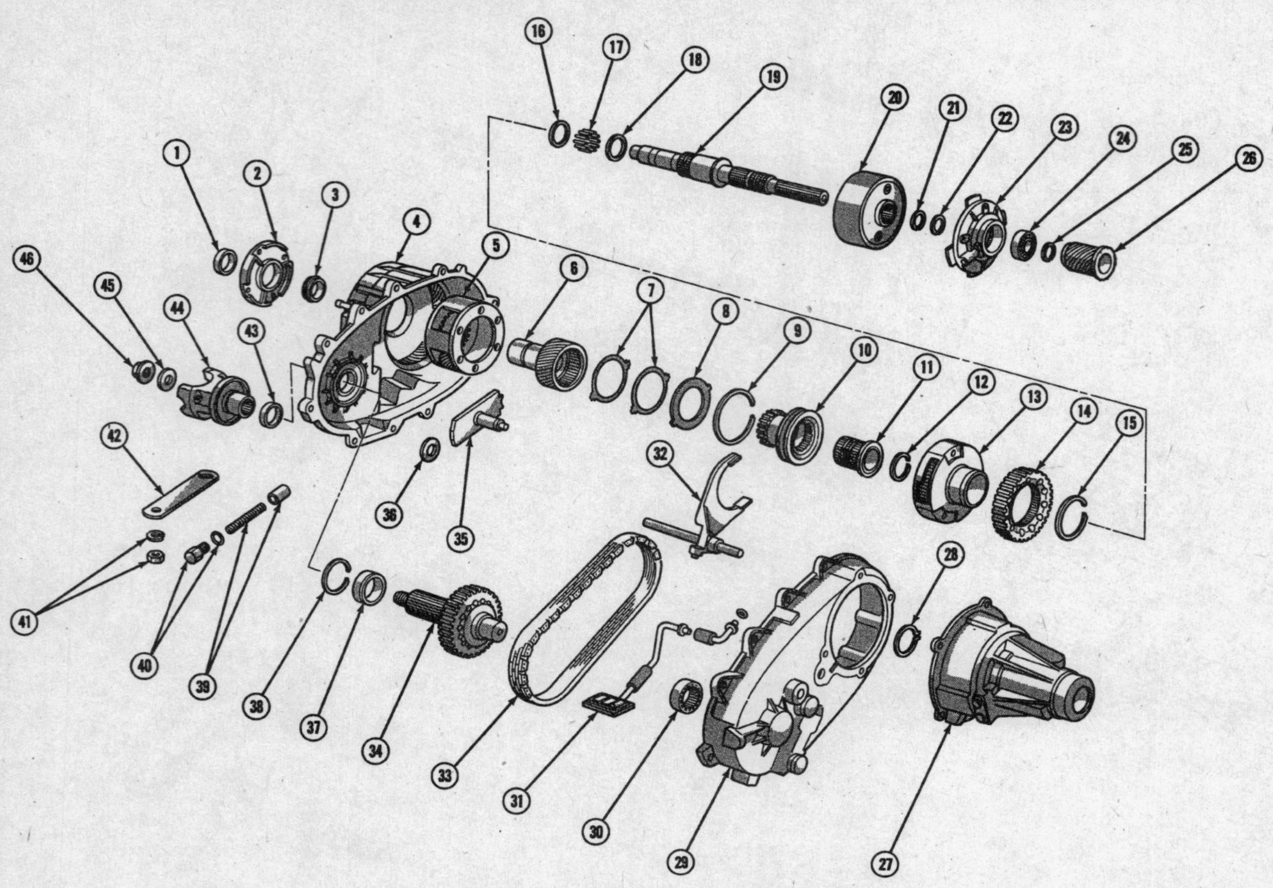

5.4c An exploded view of the New Process 249 transfer case

1 Oil seal
2 Front bearing retainer
3 Mainshaft front bearing and locating ring
4 Front case (includes low range annulus gear and shift rail bushing)
5 Low range gear
6 Input gear
7 Tabbed thrust washer
8 Retainer plate
9 Input gear snap ring
10 Clutch sleeve
11 Clutch shaft
12 Differential snap ring
13 Differential assembly
14 Mainshaft drive gear
15 Drive gear snap ring
16 Mainshaft bearing spacer
17 Mainshaft needle bearings
18 Mainshaft bearing spacer
19 Mainshaft
20 Viscous coupling
21 Viscous coupling snap ring
22 Oil pump locating snap ring
23 Oil pump
24 Mainshaft rear bearing
25 Rear bearing snap ring
26 Speedometer drive gear
27 Rear retainer assembly (includes retainer cap, oil seal, bushing, access cover and gasket)
28 Rear bearing locating ring
29 Front case
30 Front output shaft rear bearing
31 Oil pick-up tube assembly (includes connecting hoses, pick-up screen, tubes and O-ring)
32 Range fork and shift rail assembly (includes range fork pads)
33 Drive chain
34 Front output shaft
35 Shift sector
36 Case magnet
37 Front output shaft front bearing
38 Bearing snap ring
39 Detent plunger and spring
40 Detent plug and O-ring
41 Range lever locknut and washer
42 Range lever
43 Oil seal (front output shaft front bearing)
44 Output shaft yoke
45 Yoke seal washer
46 Yoke nut

exceed the cost of a rebuilt unit.

Nevertheless, it's not impossible for an inexperienced mechanic to rebuild a transfer case if the special tools are available and the job is done in a deliberate step-by-step manner so nothing is overlooked.

The tools necessary for an overhaul include internal and external snap-ring pliers, a bearing puller, a slide hammer, a set of pin punches, a dial indicator and possibly a hydraulic press. In addition, a large, sturdy workbench and a vise or transmission stand will be required.

During disassembly of the transfer case, make careful notes of how each piece comes off, where it fits in relation to other pieces and what holds it in place. Exploded views are included **(see illustrations)** to show where the parts go - but actually noting how they are installed when you remove the parts will make it much easier to get the transfer case back together.

Before taking the transfer case apart for repair, it will help if you have some idea what area of the transfer case is malfunctioning. Certain problems can be closely tied to specific areas in the transfer case, which can make component examination and replacement easier. Refer to the *Troubleshooting* section at the front of this manual for information regarding possible sources of trouble.

Chapter 8
Clutch and drivetrain

Contents

Specifications

Fluid type	See Chapter 1
Type 3 driveshaft CV joint length	5-39/64 inches

Torque specifications

Ft-lbs (unless otherwise indicated)

Clutch

Clutch master cylinder/reservoir	
Master cylinder mounting nuts	200 to 300 in-lbs
Reservoir mounting screws	30 to 50 in-lbs
Clutch slave cylinder mounting nuts	200 to 300 in-lbs
Pressure plate-to-flywheel bolts	
1993 and 1994	
5/16-inch bolts	17
3/8-inch bolts	30
1995	38

Drivetrain

Differential cover bolts	See Chapter 1
Driveshaft	
Front driveshaft	
U-joint-to-pinion yoke strap bolts (1993 and 1994 models only)	14
CV joint-to-pinion flange bolts	
1995 through 1998	30
1999 and later	23.5
U-joint/CV joint-to-transfer case bolts	
1993 through 1998	20
1999 and later	23.5
Rear driveshaft	
U-joint-to-rear axle yoke strap bolts	168 in-lbs
Front axleshaft or stub axle hub nut	175
Front hub and bearing assembly-to-steering knuckle bolts	75
Pinion mate shaft lock bolt	168 in-lbs
Rear axle retaining plate nuts (1999 and later models)	45

1 General information

The information in this Chapter deals with the components from the rear of the engine to the drive wheels, except for the transmission and transfer case, which are dealt with in the previous Chapter. For the purposes of this Chapter, these components are grouped into three categories: clutch, driveshaft and axles. Separate Sections within this Chapter offer general descriptions and checking procedures for components in each of the three groups.

Since nearly all the procedures covered in this Chapter involve working under the vehicle, make sure it's securely supported on sturdy jackstands or on a hoist where the vehicle can be easily raised and lowered.

2 Clutch - description and check

Refer to illustration 2.1

1 All vehicles with a manual transmission use a single dry plate, diaphragm spring type clutch **(see illustration)**. The clutch disc has a splined hub which allows it to slide along the splines of the transmission input shaft. The clutch and pressure plate are held in contact by spring pressure exerted by the diaphragm in the pressure plate.

2 The clutch release system is operated by hydraulic pressure. The hydraulic release system consists of the clutch pedal, a master cylinder and fluid reservoir, the hydraulic line, a slave cylinder, and a release lever mounted inside the clutch housing.

3 When pressure is applied to the clutch pedal to release the clutch, hydraulic pressure is exerted against the release bearing through the release lever. The bearing pushes against the fingers of the diaphragm spring of the pressure plate assembly, which in turn releases the clutch plate.

4 Terminology can be a problem when discussing the clutch components because common names are in some cases different from those used by the manufacturer. For example, the driven plate is also called the clutch plate or disc, the clutch release bearing is sometimes called a throwout bearing, the release cylinder is sometimes called the operating or slave cylinder.

5 Other than replacement of components with obvious damage, some preliminary checks should be performed to diagnose clutch problems.

a) *The first check should be of the fluid level in the clutch master cylinder. If the fluid level is low, add fluid as necessary and inspect the hydraulic system for leaks.*

b) *To check "clutch spin down time," run the engine at normal idle speed with the transmission in Neutral (clutch pedal up - engaged). Disengage the clutch (pedal down), wait several seconds and shift the transmission into Reverse. No grinding noise should be heard. A grinding noise would most likely indicate a problem in the pressure plate or the clutch disc (assuming the transmission is in good condition).*

c) *To check for complete clutch release, run the engine (with the parking brake applied to prevent movement) and hold the clutch pedal approximately 1/2-inch from the floor. Shift the transmission between 1st gear and Reverse several times. If the shift is rough, component failure is indicated. Check the release cylinder pushrod travel. With the clutch pedal depressed completely, the release cylinder pushrod should extend substantially. If it doesn't, check the fluid level in the clutch master cylinder.*

d) *Visually inspect the pivot bushing at the top of the clutch pedal to make sure there is no binding or excessive play.*

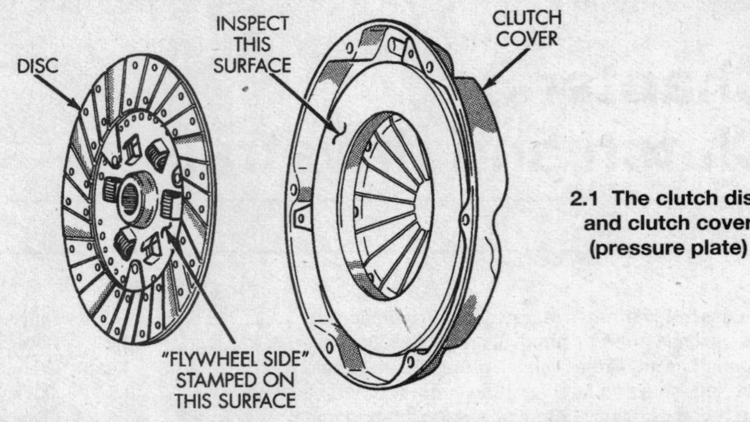

2.1 The clutch disc and clutch cover (pressure plate)

3 Clutch hydraulic release system - removal and installation

Refer to illustrations 3.2 and 3.6

Note: *The clutch hydraulic release system is serviced as an assembly. If the system is leaking or has air in it, replace the entire assembly.*

1 Raise the vehicle and support it securely on jackstands.

2 Remove the nuts attaching the slave cylinder to the clutch housing **(see illustration)**.

3 Remove the slave cylinder and fluid line retaining clip from the housing.

4 Disengage the clutch hydraulic fluid line from the retaining clips on the body.

5 Lower the vehicle.

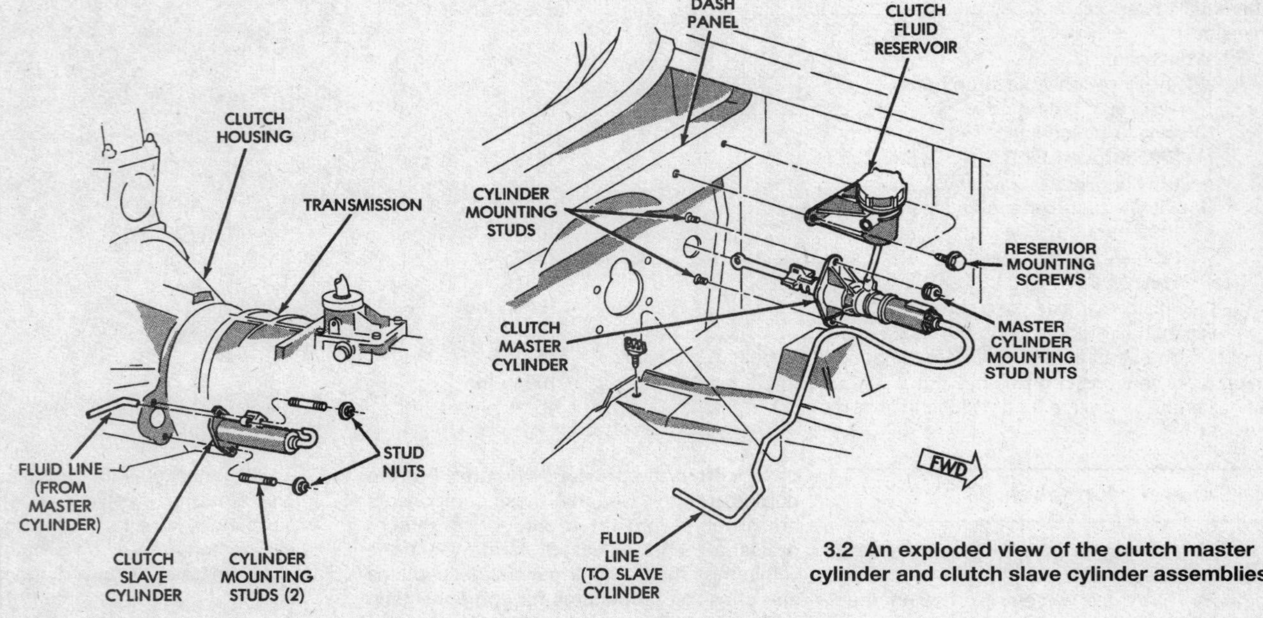

3.2 An exploded view of the clutch master cylinder and clutch slave cylinder assemblies

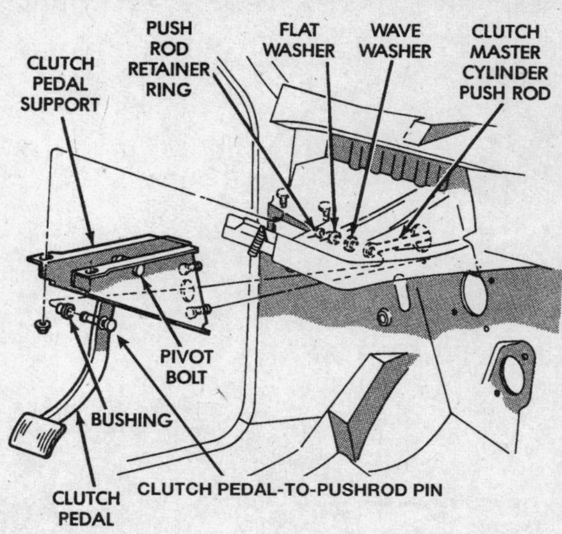

3.6 Installation details of the clutch pushrod-to-clutch pedal connection

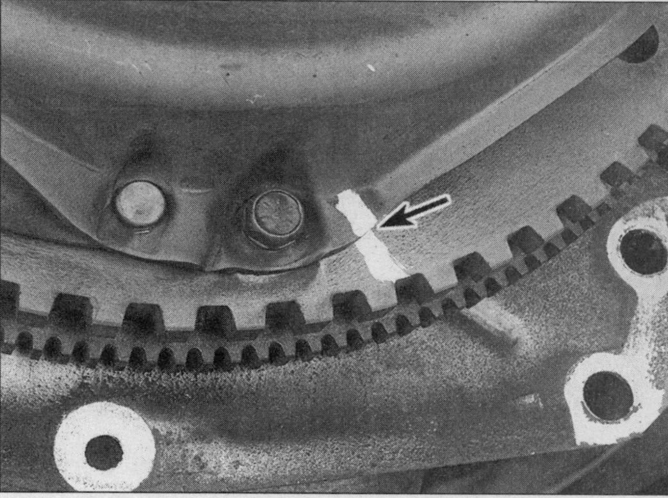

4.5 Mark the relationship of the pressure plate to the flywheel (arrow) (in case you are going to re-use the same pressure plate)

6 Remove the retainer ring, flat washer and wave washer that attach the clutch master cylinder pushrod to the clutch pedal **(see illustration)**.

7 Unplug the electrical connector for the clutch pedal position switch wires, then slide the clutch master cylinder pushrod off the clutch pedal-to-pushrod pin. Inspect the condition of the pin bushing. If it's worn, replace it.

8 To avoid spillage, verify that the cap on the master cylinder reservoir is tight. Remove the reservoir mounting screws and the master cylinder mounting stud nuts that attach the reservoir and master cylinder to the firewall.

9 Remove the entire assembly. Do not attempt to disconnect the clutch fluid hydraulic line from the master cylinder or the slave cylinder.

10 Installation is the reverse of removal. Be sure to tighten all fasteners to the torque values listed in this Chapter's Specifications.

4 Clutch components - removal, inspection and installation

Warning: *Dust produced by clutch wear and deposited on clutch components is hazardous to your health. DO NOT blow it out with compressed air and DO NOT inhale it. DO NOT use gasoline or petroleum-based solvents to remove the dust. Brake system cleaner should be used to flush the dust into a drain pan. After the clutch components are wiped clean with a rag, dispose of the contaminated rags and cleaner in a covered, marked container.*

Removal

Refer to illustration 4.5

1 Access to the clutch components is normally accomplished by removing the transmission, leaving the engine in the vehicle. If, of course, the engine is being removed for major overhaul, then check the clutch for wear and replace worn components as necessary. However, the relatively low cost of the clutch components compared to the time and trouble spent gaining access to them warrants their replacement anytime the engine or transmission is removed, unless they are new or in near perfect condition. The following procedures are based on the assumption the engine will stay in place.

2 Disconnect the clutch slave cylinder from the clutch housing (see Section 3). **Caution:** *Do not disconnect the clutch hydraulic fluid line from the slave cylinder. The hydraulic release system cannot be bled if air is allowed to enter the system.*

3 Remove the transmission from the vehicle (see Chapter 7, Part A). Support the engine while the transmission is out. Preferably, an engine hoist should be used to support it from above. However, if a jack is used underneath the engine, make sure a piece of wood is positioned between the jack and oil pan to spread the load. **Caution:** *The pickup for the oil pump is very close to the bottom of the oil pan. If the pan is bent or distorted in any way, engine oil starvation could occur.*

4 To support the clutch disc during removal, install a clutch alignment tool through the clutch disc hub.

5 Carefully inspect the flywheel and pressure plate for indexing marks. The marks are usually an X, an O or a white letter. If they cannot be found, scribe marks yourself so the pressure plate and the flywheel will be in the same alignment during installation **(see illustration)**.

6 Turning each bolt a little at a time, loosen the pressure plate-to-flywheel bolts. Work in a criss-cross pattern until all spring pressure is relieved. Then hold the pressure plate securely and completely remove the bolts, followed by the pressure plate and clutch disc.

Inspection

Refer to illustrations 4.8, 4.9 and 4.10

7 Ordinarily, when a problem occurs in the clutch, it can be attributed to wear of the clutch driven plate assembly (clutch disc). However, all components should be inspected at this time.

8 Inspect the flywheel for cracks, heat checking, grooves and other obvious defects **(see illustration)**. If the imperfections are slight, a machine shop can machine the surface flat and smooth, which is highly recommended regardless of the surface appearance. Refer to Chapter 2 for the flywheel removal and installation procedure.

4.8 Check the surface of the flywheel for cracks, hot spots (dark colored areas) and other obvious defects; resurfacing by a machine shop will correct minor defects - the surface on this flywheel is in fairly good condition; however, resurfacing is always a good idea

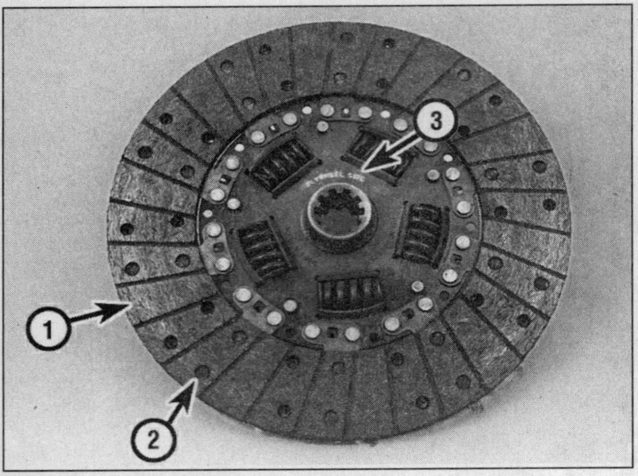

4.9 The clutch disc

1) ***Lining*** *- will wear down in use*
2) ***Rivets*** *- secure the lining and will damage the pressure plate or flywheel surface if allowed to contact it*
3) ***Marks*** *- "Flywheel side" or something similar*

4.14 Center the clutch disc in the pressure plate with an alignment tool before the bolts are tightened

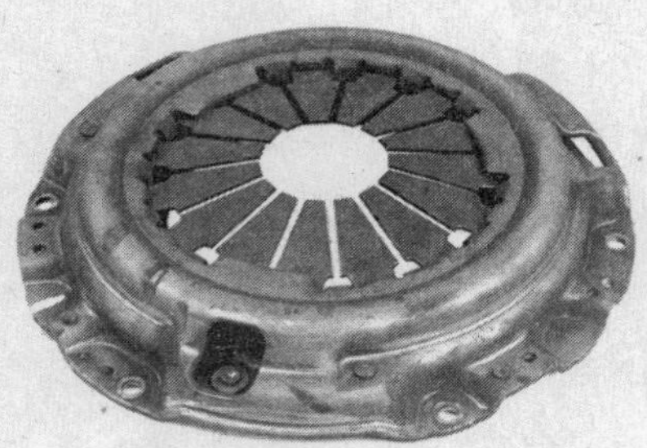

NORMAL FINGER WEAR

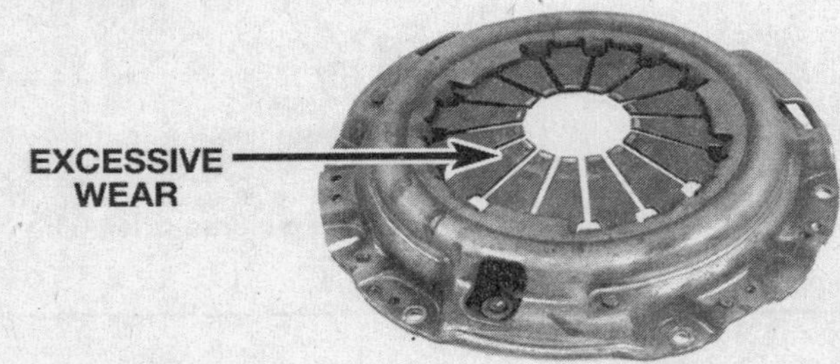

EXCESSIVE FINGER WEAR

BROKEN OR BENT FINGERS

4.10 Replace the pressure plate if excessive wear or damage is noted

9 Inspect the lining on the clutch disc. There should be at least 1/16-inch of lining above the rivet heads. Check for loose rivets, distortion, cracks, broken springs and other obvious damage **(see illustration)**. As mentioned above, ordinarily the clutch disc is routinely replaced, so if in doubt about the condition, replace it with a new one.

10 Check the machined surfaces and the diaphragm spring fingers of the pressure plate **(see illustration)**. If the surface is grooved or otherwise damaged, replace the pressure plate. Also check for obvious damage, distortion, cracking, etc. Light glazing can be removed with sandpaper or emery cloth. If a new pressure plate is required, new and factory-rebuilt units are available.

11 Check the condition of the release bearing (see Section 5).

12 Inspect the pilot bearing (see Section 6).

Installation

Refer to illustration 4.14

13 Before installation, clean the flywheel and pressure plate machined surfaces with brake system cleaner, lacquer thinner or acetone. It's important that no oil or grease is on these surfaces or the lining of the clutch disc. Handle the parts only with clean hands.

14 Position the clutch disc and pressure plate against the flywheel with the clutch held in place with an alignment tool **(see illustration)**. Make sure it's installed properly (most replacement clutch plates will be marked "flywheel side" or something similar - if not marked, install the clutch disc with the damper springs toward the transmission).

15 Tighten the pressure plate-to-flywheel bolts only finger tight, working around the pressure plate.

16 Center the clutch disc by ensuring the alignment tool extends through the splined hub and into the pilot bearing in the crankshaft. Wiggle the tool up, down or side-to-side as needed to bottom the tool in the pilot bearing. Tighten the pressure plate-to-flywheel bolts a little at a time, working in a criss-cross pattern to prevent distorting the cover. After all of the bolts are snug, tighten them to the torque listed in this Chapter's Specifications. Remove the alignment tool.

17 Using high-temperature grease, lubricate the inner groove of the release bearing (see Section 5). Also place grease on the release lever contact areas and the transmission input shaft bearing retainer.

18 Install the clutch release bearing (see Section 5).

19 Install the transmission (see Chapter 7, Part A).

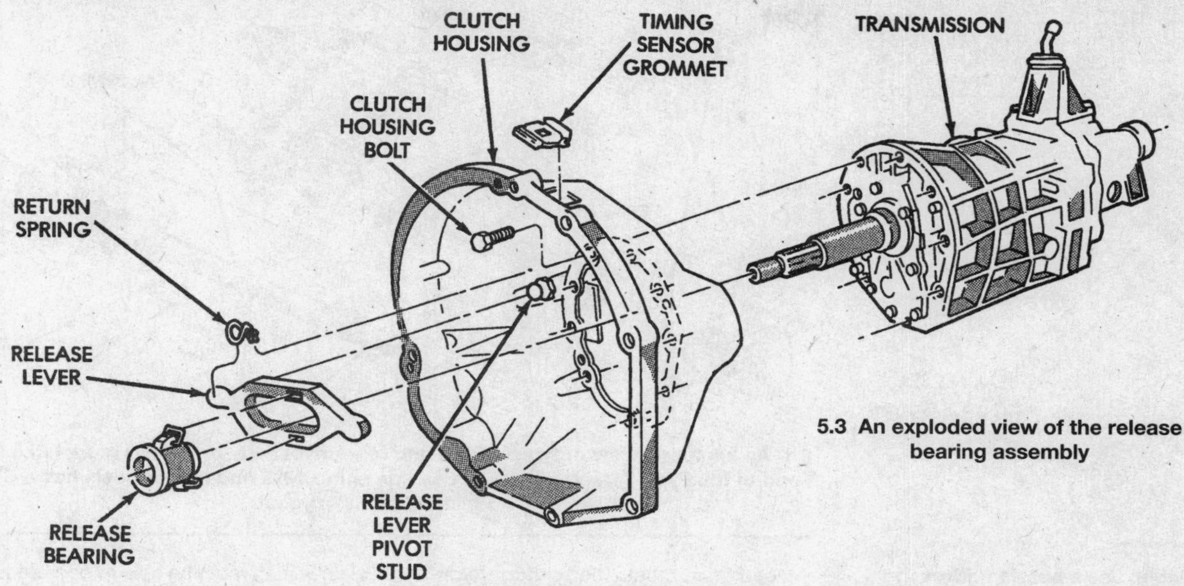

5.3 An exploded view of the release bearing assembly

20 Install the slave cylinder (see Section 3).
21 Remove the jackstands and lower the vehicle.
22 Test drive the vehicle to verify that the transmission and clutch assembly are working properly.

5 Clutch release bearing - removal, inspection and installation

Warning: *Dust produced by clutch wear and deposited on clutch components is hazardous to your health. DO NOT blow it out with compressed air and DO NOT inhale it. DO NOT use gasoline or petroleum-based solvents to remove the dust. Brake system cleaner should be used to flush the dust into a drain pan. After the clutch components are wiped clean with a rag, dispose of the contaminated rags and cleaner in a covered, marked container.*

Removal

Refer to illustration 5.3

1 Unbolt the clutch slave cylinder from the clutch housing (see Section 3) and pull it out of its recess. Hang it out of the way with a piece of wire but be careful not to kink the clutch hydraulic fluid line. **Caution:** *Do not disconnect the clutch hydraulic fluid line. The clutch release system is a one-piece assembly and cannot be bled.*

2 Remove the transmission (see Chapter 7, Part A).

3 Disengage the release lever from the return spring and the pivot stud **(see illustration)**, slide the bearing off the transmission input shaft bearing retainer and separate the bearing from the lever.

Inspection

4 Hold the center portion of the bearing stationary and rotate the outer portion while applying pressure. If the bearing doesn't turn smoothly or if it's noisy, replace it with a new one. Wipe the bearing with a clean rag and inspect it for damage and wear. Don't immerse the bearing in solvent - it is sealed for life and immersion would ruin it. Inspect the bearing for fluid leakage. If any leakage is evident, replace the entire assembly.

5 Inspect the release lever and the pivot stud. Make sure the pivot stud is tight and in good condition. Make sure the lever isn't distorted or worn. Inspect the release lever return spring. If it's bent or damaged, replace it.

Installation

6 Installation is basically the reverse of removal. Be sure to lightly lubricate the crankshaft pilot bearing, the transmission input shaft splines, the transmission bearing retainer sliding surfaces, the release lever and the release lever pivot stud with high-temperature grease.

7 Install the transmission (see Chapter 7).

8 Reattach the clutch slave cylinder (see Section 3).

6.5 A small slide-hammer puller is handy for removing the pilot bearing

6 Pilot bearing - inspection and replacement

Refer to illustrations 6.5 and 6.6

1 The clutch pilot bearing is pressed into the rear of the crankshaft. It is greased at the factory and does not require additional lubrication. Its primary purpose is to support the front of the transmission input shaft. The pilot bearing should be inspected whenever the clutch components are removed from the engine. Due to its inaccessibility, if you are in doubt as to its condition, replace it with a new one. **Note:** *If the engine has been removed from the vehicle, disregard the following steps which do not apply.*

2 Remove the transmission (refer to Chapter 7 Part A).

3 Remove the clutch components (see Section 4).

4 Inspect for any excessive wear, scoring, lack of grease, dryness or obvious damage. If any of these conditions are noted, the bearing should be replaced. A flashlight will be helpful to direct light into the recess.

5 Removal can be accomplished with a

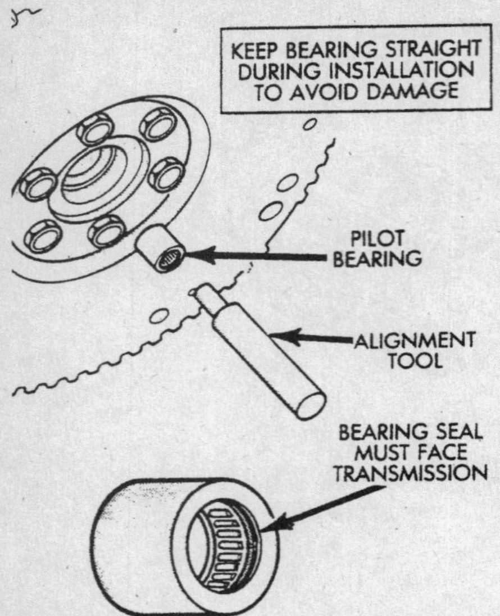

6.6 If available, use a bearing driver to install the pilot bearing; note that the seal in the bearing must face the transmission

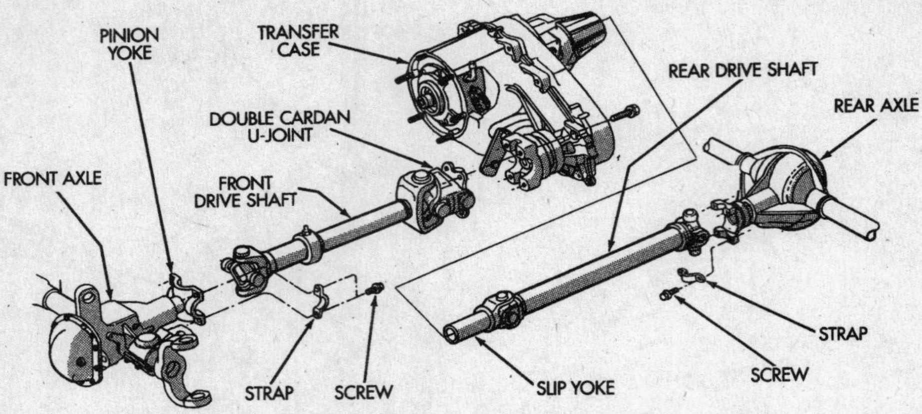

8.1 An exploded view of typical front and rear driveshafts on a 4WD model (the forward end of the Type 3 front driveshaft used on some 1995 and later models has a CV joint instead of a slip yoke)

slide hammer fitted with a puller attachment **(see illustration)**, which are available at most auto parts stores or equipment rental yards.

6 To install the new bearing, lightly lubricate the outside surface with multi-purpose grease, then drive it into the recess with a hammer and bearing/bushing driver **(see illustration)**. Make sure the bearing seal faces toward the transmission. If you don't have a bearing driver, carefully tap it into place with a hammer and a socket. **Caution:** *Be careful not to let the bearing become cocked in the bore.*

7 Install the clutch components, transmission and all other components removed previously, tightening all fasteners properly.

7 Clutch pedal position switch - check and replacement

Check

1 The clutch pedal position switch, which is part of the starter relay circuit, closes the starter relay circuit only when the clutch pedal is fully depressed.

2 To test the switch, verify that the engine will not crank over when the clutch pedal is in the released position, and that it does crank over with the pedal depressed.

3 If the engine starts without depressing the clutch pedal, replace the switch. The continuity of the switch can also be checked with an ohmmeter; there should be no continuity through the switch with the pedal at rest, and continuity with the pedal depressed.

Replacement

4 The switch is an integral part of the clutch master cylinder pushrod and can't be serviced separately. If the switch must be replaced, so must the clutch hydraulic release system (see Section 3).

8 Driveshafts, differentials and axles - general information

Driveshafts

Refer to illustrations 8.1, 8.3a, 8.3b, 8.5, 8.6a, 8.6b and 8.6c

1 The driveshaft **(see illustration)** transmits power from the transmission (2WD models) or the transfer case (4WD models) to the rear axle; on 4WD models, this driveshaft is referred to as the rear driveshaft. On 4WD models, another driveshaft also transmits power from the transfer case to the front axle; this driveshaft is referred to as the front driveshaft.

2 As the vehicle travels over irregular surfaces, the axles "float" up and down, so the driveshaft must be able to operate while constantly changing its angle relative to the transmission/transfer case and the axle. This is made possible by universal joints (U-joints) at each end of the driveshaft. As the angle between the transmission/transfer case and the axle changes, so does the *length* between them, so the driveshaft must also be able to change its length. The necessary change in length is made possible by a sliding yoke, which is also referred to as a sleeve yoke or slip joint.

3 On 1993 models, the front driveshaft uses the 1310 series U-joint and the rear driveshaft uses the 2950 series U-joint. Two different types of U-joint systems - single-cardan and double-cardan **(see illustrations)** - are used.

4 On 1994 models, the front and rear driveshafts both use the 1310 series U-joint. Again, two different types of U-joints - single-cardan and double-cardan - are used.

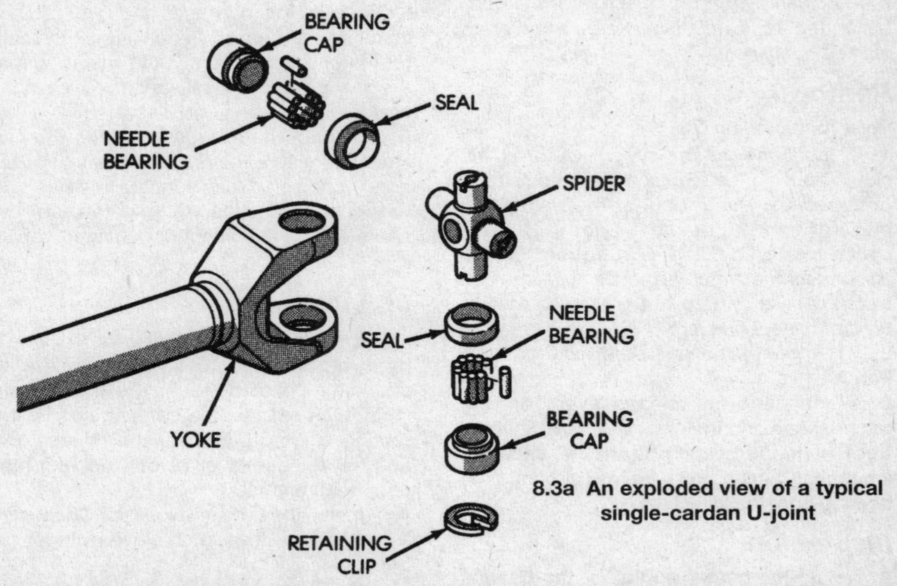

8.3a An exploded view of a typical single-cardan U-joint

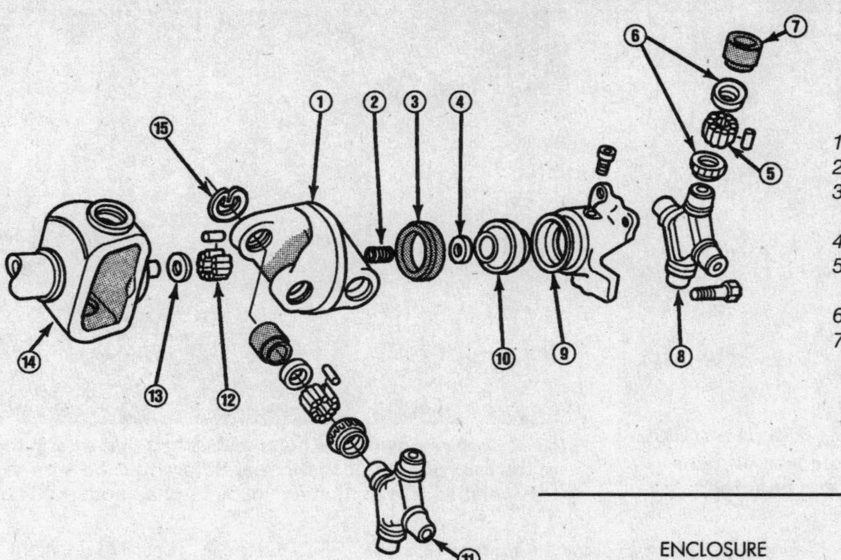

8.3b An exploded view of a typical double-cardan U-joint

1 Link yoke
2 Socket spring
3 Socket ball retainer
4 Thrust washer
5 Needle bearings
6 Seal
7 Bearing cap
8 Rear spider
9 Socket yoke
10 Socket ball
11 Front spider
12 Needle bearings
13 Thrust washer
14 Driveshaft yoke
15 Retaining clip

5 On 1995 and later models, three types of U-joints - single-cardan and double-cardan U-joints, and constant velocity (CV) joints **(see illustration)** - are used. The CV-joint is not repairable; if the joint or its boot is worn or damaged, the driveshaft must be replaced.

6 Three types of front driveshafts are used on 1995 and later 4WD models. Type 1 and Type 2 driveshafts **(see illustrations)** both use a single-cardan joint at the axle end and a double-cardan joint at the transfer case end. The only difference between the two is the means used to protect the slip yoke from dirt and dust. The Type 1 uses a dust cap and seal to protect the yoke; the Type 2 uses a rubber dust boot. The Type 3 driveshaft **(see illustration)** has no slip yoke. It uses a double-cardan joint at the transfer case and a CV joint at the front axle end. The CV joint contracts and expands, so no slip yoke is needed. A splined shaft on the CV joint allows the overall driveshaft length to be adjusted for optimal joint travel. However, the overall driveshaft length has already been preset during manufacturing. Do NOT attempt to adjust the length of the driveshaft.

7 The U-joints installed by the factory require no lubrication. The slip yoke on the

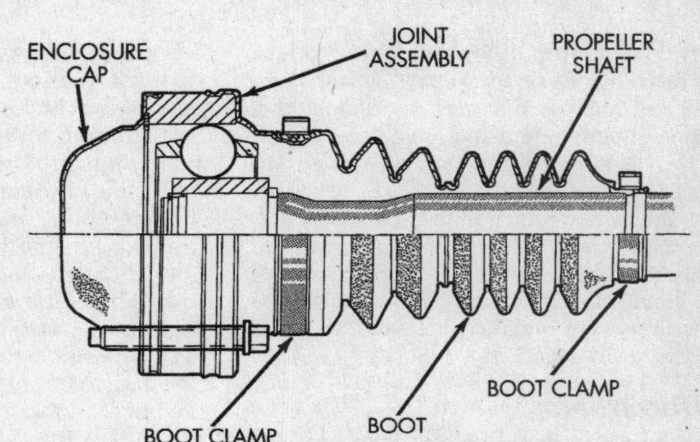

8.5 A cutaway view of the CV joint used on the Type 3 front driveshaft

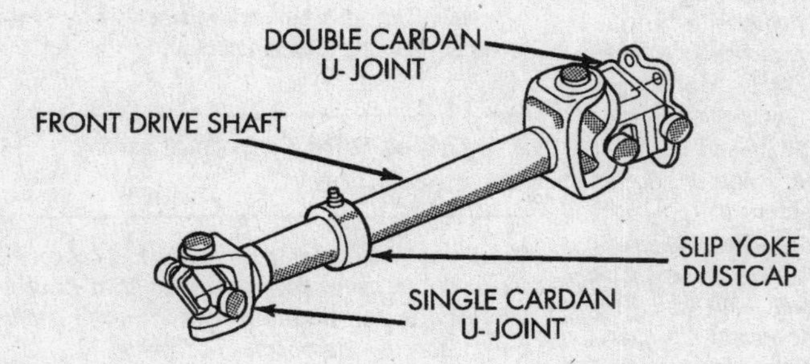

8.6a A Type 1 front driveshaft uses a dust cap for the slip yoke

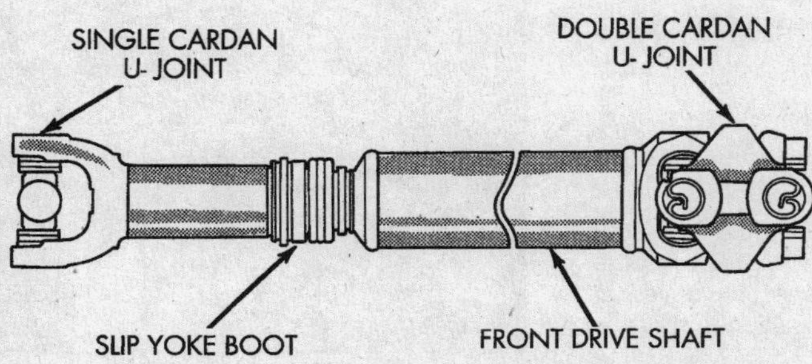

8.6b A Type 2 front driveshaft uses a dust boot for the slip yoke

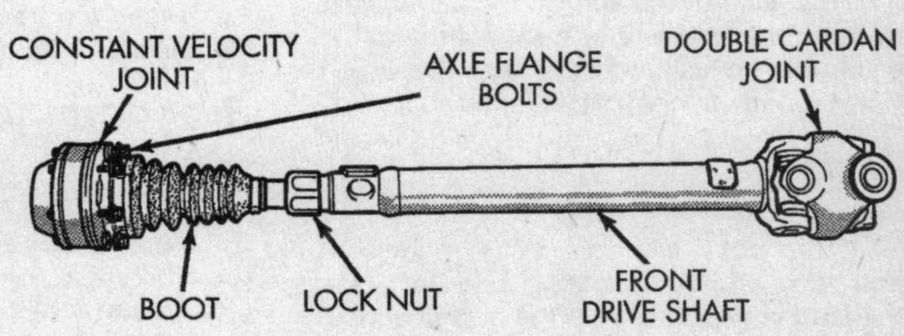

8.6c A Type 3 front driveshaft uses a constant velocity (CV) joint instead of a slip yoke at its forward end

8.11 This tag on the differential housing cover contains information such as the part number and gear ratio

9.3 Before removing the bolts and straps that attach the rear end of the rear driveshaft to the rear differential, be sure to mark the relationship of the two yokes to one another as shown

front driveshaft of 1993 and 1994 models, and the slip yoke on the Type 1 driveshaft on 1995 and later models, must be lubricated at routine maintenance intervals (see Chapter 1). The slip yoke is equipped with a zerk-type fitting for this purpose.

8 The driveshafts are finely balanced during production. Whenever they are removed or disassembled, they must be reassembled and installed in the exact manner and positions they were originally installed, to avoid excessive vibration.

Differentials

9 The differential is located inside a cast-iron housing which is an integral part of the axle assembly. A stamped steel cover bolted to the open rear side of the housing allows inspection and servicing of the differential. The differential itself is a hypoid-gear design with the centerline of the pinion below the centerline of the ring gear. Besides the standard differential, an optional limited-slip unit is also available. If one wheel spins and loses traction, the differential transfers torque from that wheel to the other wheel. Two different limited slip units are available - Trac-Lock and, on 1999 and later models, Vari-Loc. The Trac-Loc unit uses two clutch packs to generate this transfer of torque. The Vari-Loc unit uses a gerotor pump and two clutch packs to accomplish the torque transfer, and is much more efficient.

Because of the special tools and skills needed to adjust and/or overhaul the differential, this work should be left to a dealer service department or qualified repair shop.

Axles

Refer to illustration 8.11

10 The front axle on 4WD models is equipped with semi-floating axleshafts, i.e. the load is supported by the hub bearings. The axleshafts are retained by nuts at the hub bearings. Hub bearings are serviced as an assembly, i.e. they cannot be rebuilt. The outer ends of the axleshafts are equipped with either a universal joint or a constant velocity (CV) joint to allow the front wheels to turn left and right. The axle is equipped with an anti-lock brake system (ABS). The sensors are attached to the knuckle assemblies; the tone rings are pressed onto the axleshafts near the hub flanges. You'll find the part number and gear ratio listed on a tag attached to the housing cover. Build-date identification codes are stamped on the backside of the axleshaft tube. Always refer to these numbers when replacing the axle.

11 The rear axleshafts are housed inside the rear axle assembly, which is virtually identical to the front axle assembly. Again, you'll find the part number and gear ratio listed on a tag attached to the differential housing cover and the build-date identification codes stamped on the backside of the axleshaft tube **(see illustration)**. Always refer to these numbers when replacing the axle.

9 Driveshafts - removal and installation

Caution: *Whenever a driveshaft is removed, the manufacturer recommends that new U-joint straps and bolts be used during installation. Use only the correct hardware.*

1 Raise the vehicle and support it securely on jackstands. Chock the wheels at the opposite end of the vehicle being raised to prevent the vehicle from rolling.

2 Place the transmission in Neutral with the parking brake off.

Rear driveshaft

Removal

Refer to illustrations 9.3 and 9.4

3 Mark the relationship of the driveshaft to the differential yoke, in line with each other **(see illustration)** to ensure that the driveshaft is reinstalled in the same position to preserve the balance.

4 Remove the rear universal joint bolts and straps **(see illustration)**. Turn the driveshaft (or tires) as necessary to bring the bolts into the most accessible position.

5 Tape the bearing caps to the spider to prevent the caps from coming off during removal.

6 Lower the rear of the driveshaft and then slide the front out of the transmission or transfer case.

7 To prevent loss of fluid and protect against contamination while the driveshaft is out, wrap a plastic bag over the transmission or transfer case housing and hold it in place with a rubber band.

Installation

8 Remove the plastic bag from the transmission or transfer case and wipe the area clean. Inspect the oil seal carefully. If it's damaged or worn, replace it (see Chapter 7, Part B).

9 Slide the front of the driveshaft into the transmission (2WD models) or transfer case (4WD models).

10 Raise the rear of the driveshaft into position, checking to be sure the marks are in alignment. If not, turn the rear wheels to match the pinion flange and the driveshaft.

11 Remove the tape securing the bearing

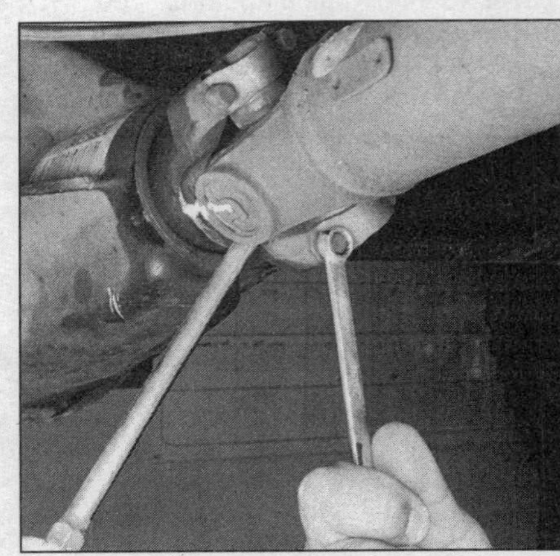

9.4 To prevent the driveshaft from turning when loosening the flange bolts, insert a screwdriver through the yoke as shown

9.12a Before disconnecting the U-joint from the front axle differential pinion yoke on 1993 and 1994 models, mark the relationship of the U-joint joint to the yoke (shown); before disconnecting the CV joint from the front axle differential pinion yoke on 1995 and later models, mark the relationship of the CV joint to the yoke (not shown)

caps and install new straps and bolts. Tighten the bolts to the torque listed in this Chapter's Specifications.

Front driveshaft (4WD models)

Removal

Refer to illustrations 9.12a and 9.12b

12 Using white paint, chalk or a scribe, mark the relationship of the front U-joint (1993 and 1994 models) or the CV joint (1995 and later models with a Type 3 driveshaft) to the front differential pinion yoke **(see illustration)**. Also mark the relationship of the rear U-joint flange to the transfer case companion flange **(see illustration)**.

13 Unbolt the front end of the driveshaft from the pinion yoke and the rear end from the transfer case companion flange **(see illustration 9.4)** and remove the driveshaft from the vehicle. On 1993 and 1994 models, tape the bearing caps to the spider to prevent the caps from falling off.

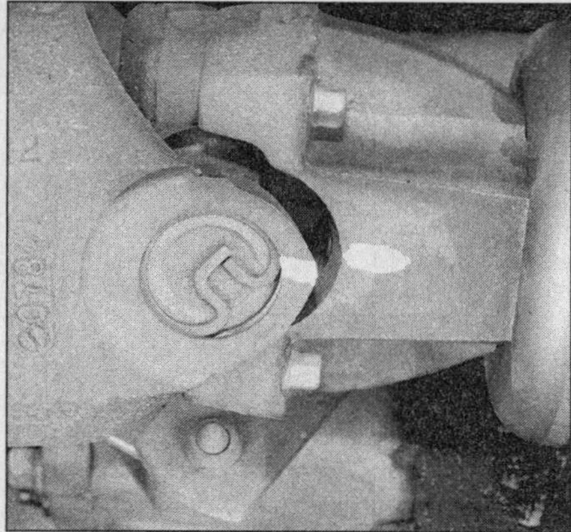

9.12b Before disconnecting the U-joint from the transfer case companion flange, mark the relationship of the U-joint to the flange

Installation

14 Installation is the reverse of removal. If you're installing a new Type 3 driveshaft on a 1995 and later model, refer to the measurement and adjustment procedure below. Be sure to tighten all fasteners to the torque listed in this Chapter's Specifications.

15 Remove the jackstands and lower the vehicle.

Type 3 driveshaft adjustment (some 1995 and later models)

Refer to illustration 9.16

Caution: *If you're installing a **new** Type 3 driveshaft, it must be measured and adjusted before the vehicle is driven. Don't try to adjust a driveshaft that has been in use.*

16 Measure the distance from the CV joint cup to the end of the CV joint boot **(see illustration). Note:** *This measurement must be taken with the new driveshaft installed and the vehicle at curb height.* Compare your measurement to the dimension listed in this Chapter's Specifications.

17 If necessary, adjust the driveshaft by loosening the locknut and moving one end of the shaft in or out until the distance from the CV joint cup to the end of the CV joint boot is correct, then tighten the locknut to the torque listed in this Chapter's Specifications.

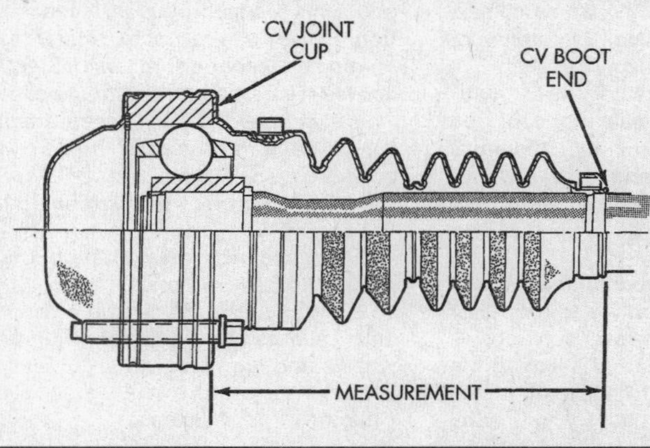

9.16 If you're installing a new Type 3 driveshaft on a 1995 or later model, this dimension must be measured and adjusted before driving the vehicle

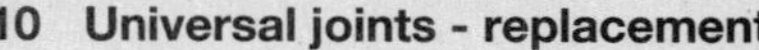

10 Universal joints - replacement

Refer to illustrations 10.2, 10.4 and 10.9

Note: *A press or large vise will be required for this procedure. It may be a good idea to take the driveshaft to a repair or machine shop where the universal joints can be replaced for you, normally at a reasonable charge.*

1 Remove the driveshaft as outlined in the previous Section.

Single cardan U-joint

2 Using a small pair of pliers, remove the snap-rings from the spider **(see illustration).**

3 Supporting the driveshaft, place it in position on either an arbor press or on a workbench equipped with a vise.

4 Place a piece of pipe or a large socket with the same inside diameter over one of the bearing caps. Position a socket which is of slightly smaller diameter than the cap on the opposite bearing cap **(see illustration)** and use the vise or press to force the cap out (inside the pipe or large socket), stopping just before it comes completely out of the yoke. Use the vise or large pliers to work the cap the rest of the way out.

10.2 A pair of needle-nose pliers can be used to remove the universal joint snap-rings

10.4 To press the universal joint out of the driveshaft yoke, set it up in a vise with the small socket pushing the joint and bearing cap into the large socket

5 Transfer the sockets to the other side and press the opposite bearing cap out in the same manner.

6 Pack the new universal joint bearings with grease. Ordinarily, specific instructions for lubrication will be included with the universal joint servicing kit and should be followed carefully.

7 Position the spider in the yoke and partially install one bearing cap in the yoke. If the replacement spider is equipped with a grease fitting, be sure it's offset in the proper direction (toward the driveshaft).

8 Start the spider into the bearing cap and then partially install the other cap. Align the spider and press the bearing caps into position, being careful not to damage the dust seals.

9 Install the snap-rings. If difficulty is encountered in seating the snap-rings, strike the driveshaft yoke sharply with a hammer. This will spring the yoke ears slightly and allow the snap-rings to seat in the groove **(see illustration)**.

10 Install the grease fitting and fill the joint with grease. Be careful not to overfill the joint, as this could blow out the grease seals.

11 Install the driveshaft. Tighten the flange bolts to the specified torque.

Double cardan U-joint

12 Use the above procedure, but note that it will have to be repeated because the double cardan joint is made up of two single cardan joints. **Note:** *The double cardan joint must be replaced as an assembly - don't attempt to replace half of it, even if only one cross-and-roller assembly is worn.*

Constant velocity joint (1995 and later models with a Type 3 driveshaft)

13 CV joints on Type 3 driveshafts cannot be replaced separately. If the joint or its boot is damaged, replace the entire driveshaft assembly (see Section 9). Note that this driveshaft must also be measured and, if necessary, adjusted before driving the vehicle (the measurement and adjustment procedure is also in Section 9).

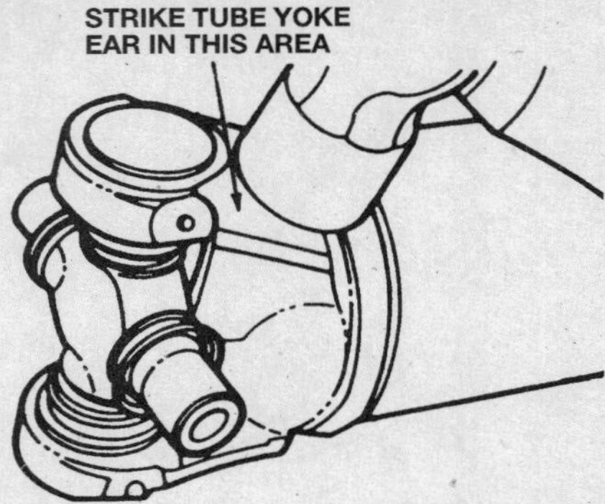

10.9 If the snap-ring will not seat in the groove, strike the yoke with a hammer. This will relieve tension that has set up in the yoke, and slightly spring the yoke ears. This should also be done if the joint feels tight when assembled

11.6a Mark the relationship of the pinion shaft to the pinion shaft yoke

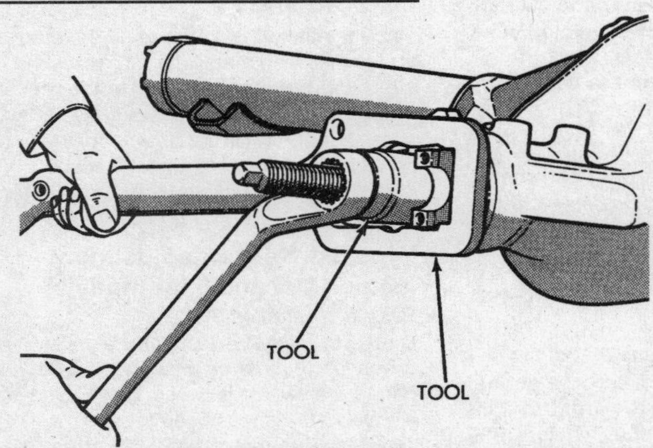

11.6b Hold the pinion shaft yoke stationary while removing the nut

11 Pinion oil seal - replacement

Refer to illustrations 11.6a and 11.6b

1 A pinion shaft oil seal failure results in the leakage of differential gear lubricant past the seal and onto the driveshaft yoke or flange. The seal is replaceable without removing or disassembling the differential.

2 Loosen the wheel lug nuts, raise the vehicle and support it securely on jackstands. Chock the wheels at the opposite end to prevent the vehicle from rolling.

3 Remove the wheels and brake drums or brake calipers and discs (see Chapter 9).

4 Disconnect the driveshaft from the pinion shaft yoke (see Section 9).

5 Rotate the yoke at least ten times, then, using an inch-pound torque wrench, measure and record the torque required to turn the pinion.

6 Mark the relationship of the pinion shaft yoke to the pinion shaft **(see illustration)**. Remove the pinion shaft nut while preventing the yoke from turning, using a special tool (available at most auto parts stores) or a chain wrench **(see illustration)**.

7 Remove the yoke from the shaft, using a puller if necessary.

8 Using a large screwdriver or seal removal tool, carefully pry the pinion oil seal out of the differential housing. Be careful not to damage the splines of the pinion shaft.

9 Lubricate the new seal lip with multi-purpose grease or differential lubricant and carefully place it in position in the differential housing. Using a hammer and a seal driver or a large socket with a diameter slightly smaller than that of the new seal, drive the seal into place.

10 Clean the sealing lip contact surface of the pinion shaft yoke. Apply a thin coat of multi-purpose grease to the seal contact surface and the shaft splines and, aligning the matchmarks made prior to removal, install the yoke onto the shaft. If necessary, use the old nut to pull the yoke into place. **Caution:** *The manufacturer recommends replacing the nut with a new one whenever it has been removed.*

11 Install the washer and a new pinion shaft nut. Tighten the nut just enough to eliminate all endplay in the pinion shaft.

12 Turn the pinion shaft yoke several times to seat the bearings.

13 Using an inch-pound torque wrench, measure the torque required to turn the pinion shaft. Tighten the nut in small increments until the torque recorded in Step 5, plus an additional five inch-pounds, is reached. **Caution:** *Be very careful not to exceed the torque readings described in this Step, which applies the proper preload to the pinion bearings. If the preload is exceeded, the differential will have to be disassembled and a new collapsible spacer installed. DO NOT loosen the pinion nut once the tightening process has begun.*

14 Connect the driveshaft to the pinion shaft yoke (see Section 9).

15 Install the brake drums or brake discs and calipers (see Chapter 9). Install the wheels and lug nuts. Lower the vehicle and tighten the lug nuts to the torque listed in the Chapter 1 Specifications.

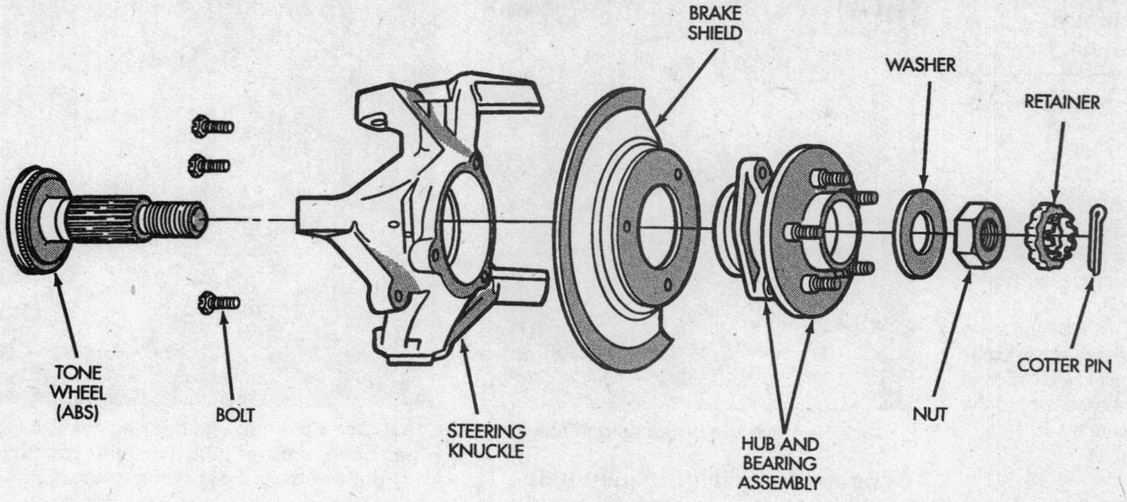

12.1a An exploded view of the front hub and bearing assembly and related components

12.1b Before raising the vehicle, remove the cotter pin and nut lock, then loosen the axle hub nut

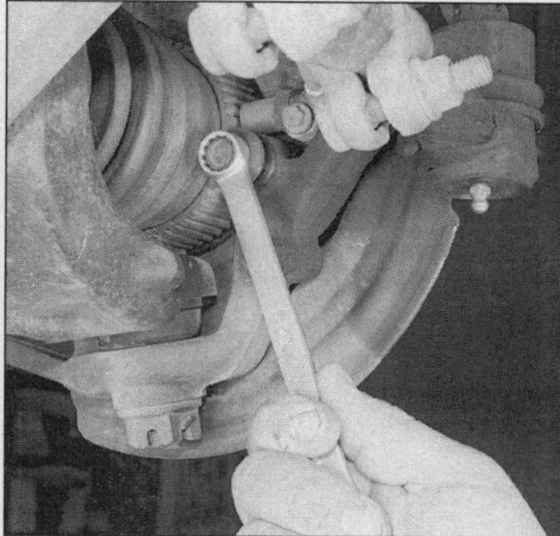

12.4a Remove the front bolt (arrow) . . .

12.4b . . . and the two rear bolts (arrows) that secure the hub assembly to the steering knuckle (you'll need a 12-point socket)

16 Test drive the vehicle and check around the pinion shaft yoke for evidence of leakage.

12 Hub and bearing assembly (front) - removal and installation

Note: *The hub and bearing assembly cannot be disassembled. If either the hub or bearing is damaged or worn, the entire assembly must be replaced.*

Removal

Refer to illustrations 12.1a, 12.1b, 12.4a, 12.4b and 12.4c

1 Pry the hubcap from the wheel, remove the cotter pin and nut lock then loosen the axle hub nut **(see illustrations)**.

2 Loosen the front wheel lug nuts, raise the front of the vehicle and support it securely on jackstands. Remove the wheel. Also remove the ABS wheel speed sensor (see Chapter 9).

3 Remove the disc brake caliper and disc (see Chapter 9).

4 Remove the hub nut. If you're working on 2WD model, pull the stub axle through the hub and remove it, being careful not to damage the teeth on the ABS tone wheel. Unbolt the hub assembly from the steering knuckle and tap it out of the knuckle bore **(see illustrations)**. **Note:** *If the axle or stub axle sticks in the hub splines, push it out of the hub with a puller.*

Installation

5 Using sandpaper or emery cloth, clean the opening in the steering knuckle to remove any rust or dirt that may be present. Lubricate the axleshaft splines with wheel bearing grease. Install the hub assembly onto the axleshaft (4WD models). Tighten the hub-to-knuckle bolts to the torque listed in this Chapter's Specifications. If you're working on a 2WD model, lubricate the splines of the stub axle and install it through the back side of the hub/bearing assembly.

6 Install the washer and hub nut, tightening the nut securely. Install the ABS wheel speed sensor, brake disc and caliper (see Chapter 9), mount the wheel temporarily and lower the vehicle.

7 Tighten the hub nut to the torque listed in this Chapter's Specifications. Now raise the vehicle again, remove the wheel and install the nut lock and a new cotter pin then reinstall the wheel. Tighten the lug nuts to the torque listed in the Chapter 1 Specifications. Install the hubcap.

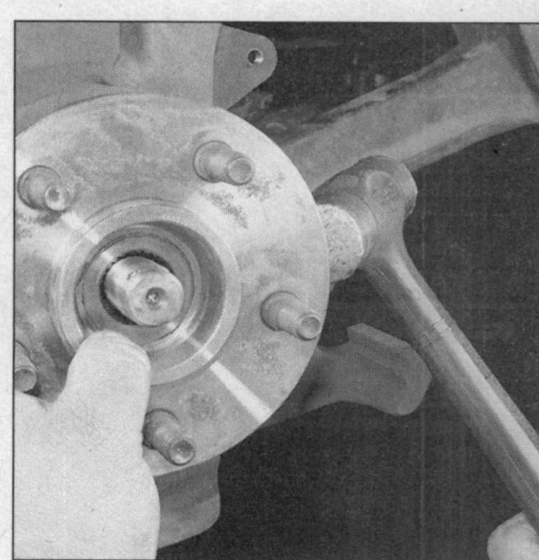

12.4c Carefully tap the hub assembly out of the steering knuckle

13 Axleshaft (front) - removal, overhaul and installation

Note: *This procedure applies to 4WD models only.*

Removal

Refer to illustrations 13.1 and 13.2

1 Following the procedure described in Section 12, remove the front axle hub and bearing assembly. Remove the disc brake splash shield **(see illustration)**.

2 Slide the axle straight out of the axle housing **(see illustration)**. Be extremely careful not to damage the axleshaft seal, which is located at the inner end of the tube, where the axleshaft is splined into the differential.

13.1 Remove the brake splash shield

13.2 Once the hub and bearing assembly has been removed, the axleshaft can be pulled straight out of the housing

Overhaul

Single-cardan U-joint type

Refer to illustrations 13.3a and 13.3b

3 Follow the U-joint replacement procedure in Section 10, but note that the snap-rings are inboard of the yoke ears and fit into grooves in the bearing caps **(see illustration)**. They are removed by driving them out with a screwdriver **(see illustration)**.

Constant Velocity (CV) joint type

Refer to illustrations 13.5a, 13.5b, 13.6a, 13.6b, 13.7, 13.8, 13.9a, 13.9b, 13.12, 13.13 and 13.18

Note: *The constant velocity outer front axle joint is not repairable - if it becomes worn, noisy or breaks, the entire joint must be replaced. The joint can and should, however, be disassembled, cleaned and repacked with grease in the event of a boot failure.*

4 Cut both boot retaining clamps and discard them. Slide the boot back on the axleshaft.

5 Mount the axleshaft in a vise. The jaws of the vise must be lined with wood or rags to avoid damage to the axle. Using a hammer and a brass punch positioned on the inner

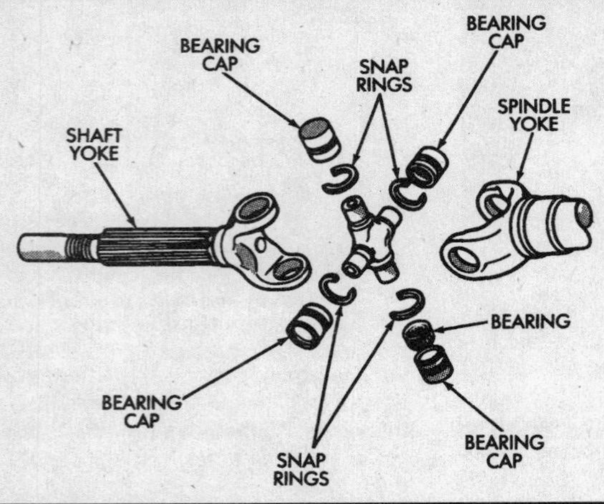

13.3a An exploded view of the U-joint assembly for the front axleshaft

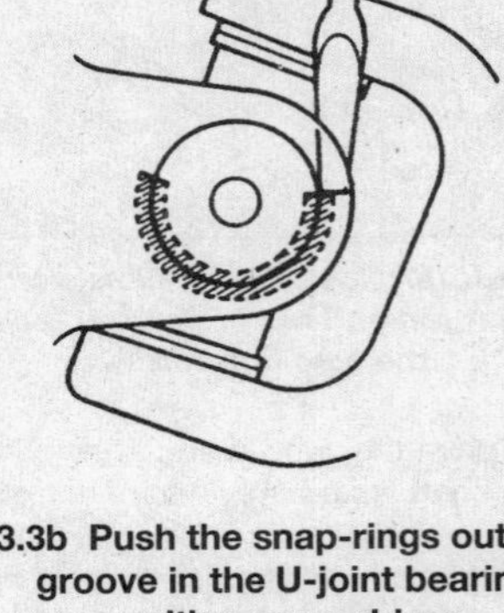

13.3b Push the snap-rings out of the groove in the U-joint bearings with a screwdriver

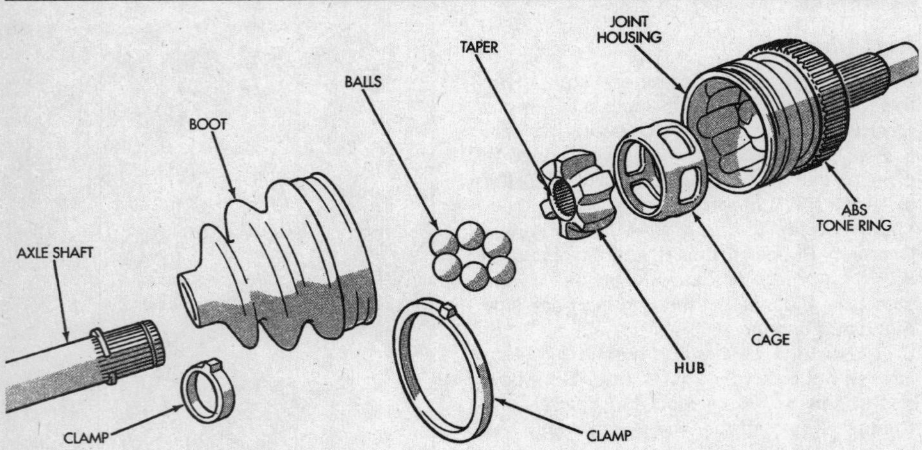

13.5a An exploded view of the double offset (CV) joint

13.5b Dislodge the CV joint assembly from the axleshaft with a brass punch and hammer (but don't allow the joint to fall!)

13.6a Tilt the inner race far enough to allow ball removal - a brass punch can be used if the inner race is difficult to move

13.6b If necessary, pry the ball bearings out with a screwdriver

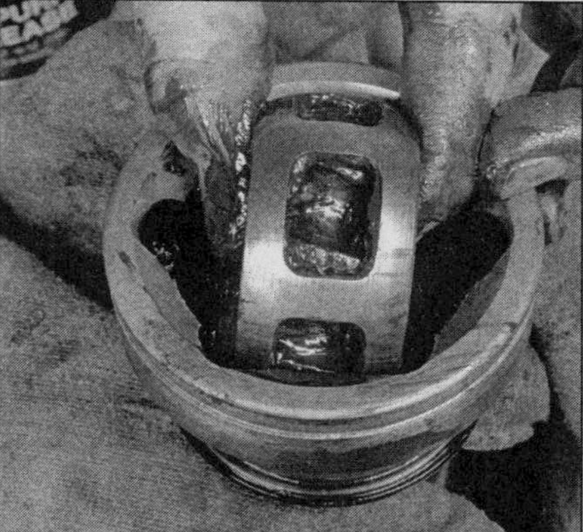

13.7 Tilt the inner race and cage 90-degrees, then align the windows in the cage with the lands and rotate the inner race up and out of the outer race

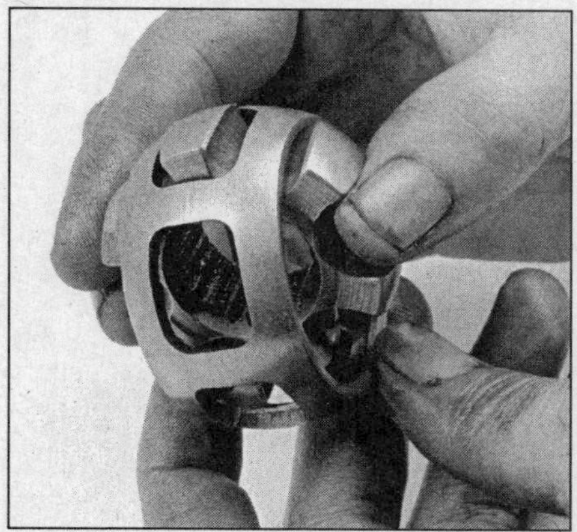

13.8 Align the inner race lands with the cage windows and rotate the inner race out of the cage

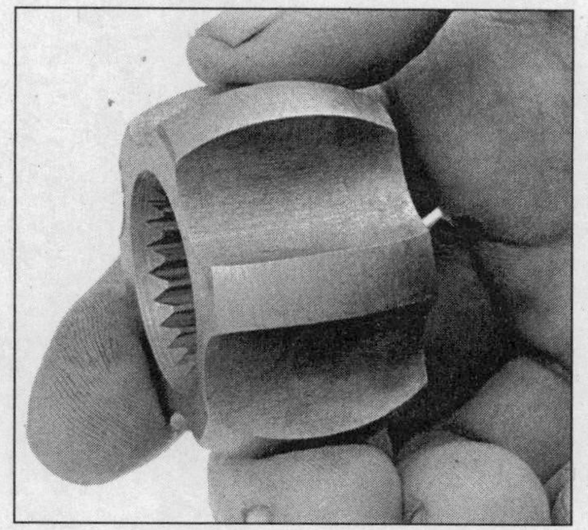

13.9a Check the inner race lands and grooves for pitting and score marks

13.9b Check the cage for cracks, pitting and score marks (shiny spots are normal and don't affect operation)

race, knock the joint off the axleshaft **(see illustrations)**.

6 Remove the axleshaft from the vise. Place the joint assembly in the vise, with the stub axle pointing down. Again, wood or rags must line the vise jaws to prevent marring the stub axle surface. Tap the inner race with the brass punch to angle it far enough to allow a ball bearing to be removed **(see illustration)**. Repeat this until all the balls are removed. If the balls are stuck, a screwdriver can be used to pry them from the cage **(see illustration)**.

7 With all of the balls removed, tilt the inner race and cage assembly 90-degrees, align the cage windows with the outer race (joint housing) lands and remove the assembly from the outer race **(see illustration)**.

8 Align the inner race lands with the cage windows and rotate the inner race out of the cage **(see illustration)**.

9 Wash all of the components in solvent. Inspect the cage and races for pitting, score marks, cracks and other signs of wear and damage **(see illustration)**. Shiny, polished spots are normal and don't affect CV joint operation **(see illustration)**.

10 Install the inner race in the cage by reversing the technique described in Step 8.

11 Install the inner race and cage assembly in the outer race by reversing the removal method used in Step 7. The small diameter of the cage must face out and the stopping groove in the inner race must face in.

12 Press the balls into the cage windows **(see illustration)**.

13 Pack the CV joint assembly with half of the lubricant supplied in the boot/joint kit through the inner splined hole. Force the grease into the bearing by inserting a wooden dowel through the splined hole and pushing it to the bottom of the joint **(see illustration)**. Repeat this procedure until the joint is thoroughly packed (the other half of the grease is

13.12 Align the cage windows and the inner and outer race grooves, then tilt the cage and inner race to insert the balls

13.13 Apply grease through the splined hole, then insert a wooden dowel into the hole and push down - the dowel will force grease into the joint

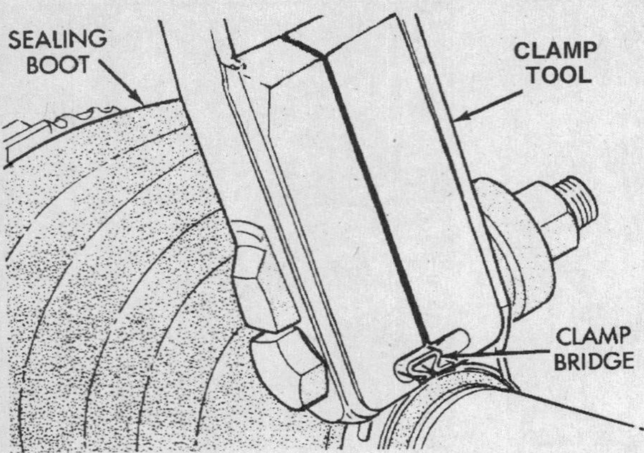

13.18 A tool like this, available at most auto parts stores, is required for tightening the boot clamps

15.3 Remove the lock bolt securing the pinion mate shaft

to be placed in the boot). **Caution:** *Use CV joint grease only.*

14 Wrap the axleshaft splines with tape to avoid damaging the boot. Place the small boot clamp in the groove in the small end of the boot, then slide the boot onto the axleshaft. Remove the tape and smear the remainder of the grease into the boot.

15 Install the replacement retaining ring and spacer ring on the shaft.

16 Slide the CV joint onto the shaft until the inner race contacts the inner retaining ring.

17 Install the large boot clamp in its groove in the boot and slide the boot over the outer race (joint housing).

18 Tighten both boot clamps **(see illustration)**.

Installation

19 Installation is the reverse of the removal procedure.

14 Axle assembly (front) - removal and installation

Note: *This procedure applies to both 4WD and 2WD vehicles. If you are working on a 2WD vehicle, ignore references to the driveshaft and differential.*

Removal

1 Loosen the front wheel lug nuts, raise the front of the vehicle and support it securely on jackstands positioned under the frame rails. Remove the front wheels.

2 Unbolt the front brake calipers and hang them out of the way with pieces of wire - don't let the calipers hang by the brake hose (see Chapter 9).

3 Remove the brake pads, anchor plates and brake discs (see Chapter 9).

4 Mark the relationship of the front driveshaft to the front differential pinion shaft yoke, then disconnect the driveshaft from the yoke (see Section 9).

5 Disconnect the stabilizer bar links from the brackets on the axle (see Chapter 10).

6 Disconnect the tie-rod and drag link from the steering knuckle arms (see Chapter 10). Position them out of the way and hang them with pieces of wire from the underbody.

7 If you're working on a 4WD model, place a floor jack under each axle tube. On 2WD models, position a floor jack in the center of the axle.

8 Unbolt the lower ends of the shock absorbers from the axle.

9 Remove the steering dampener (see Chapter 10).

10 Remove the ABS wheel speed sensors (if equipped) (see Chapter 9).

11 Slowly lower the assembly to the ground and remove the coil springs.

12 Unbolt the upper and lower suspension arms from the axle (see Chapter 10).

13 Remove the axle assembly out from under the vehicle.

Installation

14 Installation is the reverse of the removal procedure. When raising the axle into position, make sure the coil springs seat properly. Be sure to use new U-joint straps and bolts. Tighten all fasteners to the specified torque values. **Note:** *Don't tighten the control arm-to-axle housing fasteners until the vehicle is sitting at normal ride height.*

15.4 Carefully remove the pinion mate shaft from the differential case (don't turn the differential case after the shaft has been removed, or the pinion gears may fall out)

15.5 Push the axle flange in, then remove the C-lock from the inner end of the axleshaft

15 Axleshaft, seal and bearing (rear) - removal and installation

1998 and earlier models

Removal

Refer to illustrations 15.3, 15.4, 15.5, 15.7a and 15.7b

1 Loosen the wheel lug nuts, raise the rear of the vehicle and support it securely on jackstands. Remove the wheel and the brake drum or brake disc (see Chapter 9).

2 Remove the cover from the differential carrier and allow the oil to drain into a container (see Chapter 1).

3 Remove the lock bolt that secures the differential pinion mate shaft **(see illustration)**.

4 Remove the pinion mate shaft **(see illustration)**. **Caution:** *Do not rotate the differential with the shaft removed or the pinion gears may be dislodged.*

5 Push the outer (flanged) end of the axle-

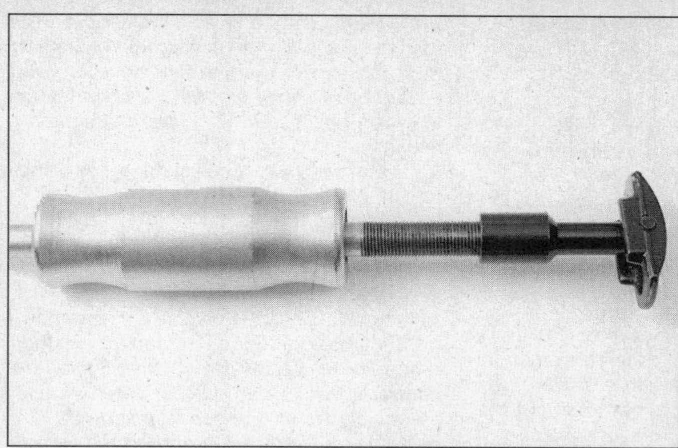
15.7a A typical slide hammer and axleshaft bearing removal attachment

15.7b Using the slide hammer and puller attachment to remove the axleshaft bearing

shaft in and remove the C-lock from the inner end of the shaft **(see illustration)**. Withdraw the axleshaft.

6 If you are just replacing the oil seal, pry the seal out of the axle housing with a seal removal tool or a prybar (if you are replacing the bearing, skip to the next step - the seal will come out as you remove the bearing).

7 A bearing puller which grips the bearing from behind will be required for bearing removal **(see illustration)**. Attach a slide hammer to the puller and extract the bearing from the axle housing **(see illustration)**.

Installation

8 Clean out the bearing recess and drive in the new bearing with a bearing installer or a large socket positioned against the outer bearing race. Make sure the bearing is tapped in to the full depth of the recess and the numbers on the bearing are visible from the outer end of the axle housing.

9 Apply high-temperature grease to the oil seal recess and tap the new seal evenly into place with a hammer and seal installation tool or a large socket so the lips are facing in and the metal face is visible from the end of the axle housing. When correctly installed, the face of the oil seal should be flush with the end of the axle housing.

10 Lubricate the seal lip and the bore in the axle bearing with clean differential lubricant.

11 Install the axleshaft and C-lock, then insert the pinion mate shaft. Apply a non-hardening thread-locking compound to the threads of the lock bolt, then tighten the lock bolt to the torque listed in this Chapter's Specifications.

12 Clean off all traces of sealant or old gasket material from the differential cover and rear axle housing, then apply a bead of RTV sealant to the cover. Install the cover and bolts, tightening the bolts in a criss-cross pattern to the torque listed in the Chapter 1 Specifications.

13 Refill the differential with the correct grade of lubricant (see Chapter 1).

14 The remainder of installation is the reverse of removal. Install the wheel and lug nuts, lower the vehicle and tighten the lug nuts to the torque listed in the Chapter 1 Specifications.

1999 and later models

Refer to illustrations 15.17 and 15.18

Caution: *The manufacturer recommends replacing the axle retaining plate nuts and bolts (on models so equipped) with new ones whenever they are removed.*

Removal

15 Loosen the wheel lug nuts, raise the rear of the vehicle and support it securely on jackstands. Chock the front wheels to prevent the vehicle from rolling. Remove the wheel.

16 Remove the brake caliper and brake disc (see Chapter 9). Don't disconnect the brake hose from the caliper. Hang the caliper with a piece of wire - don't let it hang by the hose. Also remove the ABS wheel speed sensor (so it doesn't get damaged).

17 Remove the nuts securing the axle retaining plate to the axle housing **(see illustration)**. **Note:** *Some 1999 models use bolts and nuts, instead of studs and nuts as shown in the photo. On models with bolts and nuts, the nuts can be accessed through the hole in the axle flange.*

18 Connect a slide hammer and adapter to

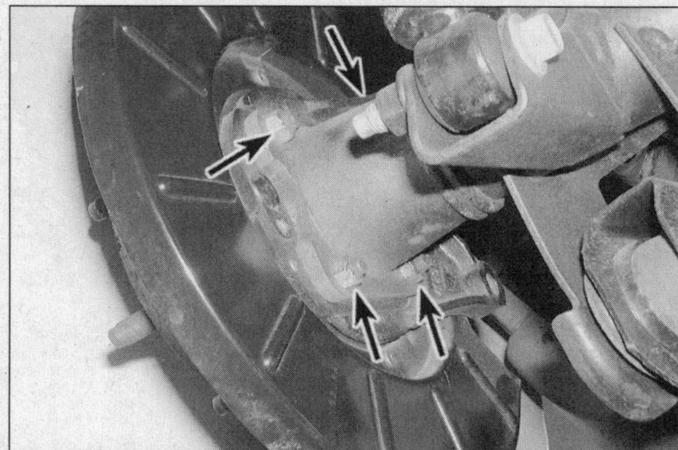
15.17 On 1999 and later models, the rear axle is retained by four nuts (arrows) (see the note in text regarding some 1999 models)

15.18 After removing the axle retaining plate fasteners, pull the axle from the housing with a slide hammer and adapter connected to the axle flange

the axle flange and pull the axle from the housing **(see illustration)**.

19 If the bearing or seal must be replaced, take the assembly to an automotive machine shop to have the old bearing and seal removed and new ones installed.

Installation

20 Wipe the bearing bore in the axle housing clean. Pack the bearing with wheel bearing grease and apply a thin coat of grease to the outer surface of the bearing.

21 Guide the axleshaft straight into the axle housing. Rotate the shaft slightly to engage the splines on the shaft with the splines in the differential side gear.

22 Install new axle retaining plate nuts (and bolts, on some 1999 models) and tighten them to the torque listed in this Chapter's Specifications.

23 The remainder of installation is the reverse of removal. Tighten the caliper mounting bracket and caliper bolts to the torque listed in the Chapter 9 Specifications. Check the differential lubricant level and add some, if necessary.

24 Install the wheel and lug nuts. Lower the vehicle and tighten the lug nuts to the torque listed in the Chapter 1 Specifications.

16 Axle assembly (rear) - removal and installation

1 Loosen the rear wheel lug nuts, raise the vehicle and support it securely on jackstands placed underneath the frame. Remove the wheels.

2 Support the rear axle assembly with a floor jack placed underneath the differential.

3 Detach the shock absorbers from the axle and compress them to get them out of the way (see Chapter 10).

4 Disconnect the driveshaft from the differential pinion shaft yoke and hang the rear of the driveshaft from the underbody with a piece of wire (see Section 9).

5 Unbolt the stabilizer bar from the stabilizer bar links, if so equipped (see Chapter 10).

6 Disconnect the parking brake cables from the equalizer (see Chapter 9).

7 Disconnect the flexible brake hose from the fitting where it meets the rigid brake line. Plug the end of the line and hose or wrap plastic bags tightly around the open fittings to prevent excessive fluid loss and contamination.

8 Remove the rear coil springs (see Chapter 10).

9 Detach the control arms from the assembly (see Chapter 10).

10 Lower the jack and move the axle assembly out from under the vehicle.

11 Installation is the reverse of the removal procedure. Tighten all fasteners to their proper torque values. **Note:** *Don't tighten the control arm-to-axle housing fasteners until the vehicle is sitting at normal ride height (this can be simulated by supporting the vehicle with jackstands placed under the rear axle tubes).*

12 Bleed the brake system (it will probably only be necessary to bleed the rear brakes if you plugged the brake line right after it was disconnected).

13 Tighten the wheel lug nuts to the torque listed in the Chapter 1 Specifications.

Chapter 9 Brakes

Contents

Specifications

General

Brake fluid type .... See Chapter 1

Disc brakes

Disc minimum thickness .... Stamped or cast into disc
Maximum allowable disc thickness variation
  Front disc .... No more than 0.0005 inch point to point
  Rear disc .... No more than 0.001 inch point to point
Disc runout (maximum)
  Front
    1998 and earlier .... 0.005 inch
    1999 .... 0.003 inch
    2000 .... 0.002 inch
  Rear
    1998 and earlier .... 0.005 inch
    1999 and later .... 0.003 inch
Brake pad minimum thickness .... See Chapter 1

Drum brakes

Drum maximum diameter .... Stamped or cast into drum
Minimum brake lining thickness .... See Chapter 1

9

Torque specifications

	Ft-lbs (unless otherwise indicated)
Power brake booster mounting nuts	
1996 and earlier	30
1997 and later	29
Master cylinder mounting nuts	
1996 and earlier	120 to 216 in-lbs
1997 and 1998	155 in-lbs
1999 and later	18
Caliper mounting bolts (front and rear) - 1998 and earlier models	84 to 180 in-lbs
Caliper slide pins (front and rear) - 1999 and later models	21 to 30
Caliper mounting bracket bolts (front and rear) - 1999 and later models only	66 to 85
Brake hose-to-caliper banjo bolt (front and rear)	23
Wheel cylinder mounting bolts	90 in-lbs
Wheel speed sensor mounting bolt	
Front	
1998 and earlier	34 to 50 in-lbs
1999 and later	106 to 124
Rear	106 to 124
Wheel lug nuts	See Chapter 1

1 General information

The vehicles covered by this manual are equipped with hydraulically operated front and rear brake systems. The front brakes are ventilated discs; the rear brakes are either drums (all 1993 and some 1994 models) or solid discs (some 1994 and all 1995 and later models). Both the front and rear brakes are self adjusting. The front and rear disc brakes automatically compensate for pad wear, while the rear drum brakes incorporate an adjustment mechanism which is activated as the brakes are applied when the vehicle is driven in reverse.

Hydraulic circuits

The hydraulic system consists of two separate circuits; if a leak or failure occurs in one hydraulic circuit, the other circuit will remain operative. The master cylinder has separate reservoirs for each circuit. A visual warning of circuit failure or air in the system is given by a warning light activated by displacement of the piston in the pressure differential switch portion of the combination valve from its normal "in balance" position.

Master cylinder/reservoir assembly

A special ABS master cylinder is used on all vehicles. The ABS unit is similar to a conventional master cylinder: When the brake pedal is applied, a pushrod connecting the brake pedal to the power brake booster moves forward. This forward motion is amplified by the power brake booster unit and transmitted to the primary and secondary pistons inside the master cylinder, which generates hydraulic pressure inside the brake system. Internally, the master cylinder has one-way valves in the center of the pistons so the ABS unit can pull brake fluid from the master cylinder reservoir without damaging the piston seals.

Because of its unique design, the master cylinder is not rebuildable. If it malfunctions, you must replace it. The master cylinder used on 1993 and 1994 models has two filler caps; the unit used on 1995 and later models has a single filler cap. The two units are not interchangeable. The reservoir, reservoir grommets and connecting hoses are the only replaceable components.

Combination valve

The combination valve, used on 1998 and earlier models, performs several functions. A metering valve limits pressure to the front brakes until a predetermined front-brake pressure is reached and the rear brakes have been activated.

A fixed-rate proportioning valve meters outlet pressure to the rear brakes once a predetermined rear input pressure has been reached, preventing premature rear-wheel lock-up during heavy braking.

A brake pressure differential switch continuously compares the front and rear brake pressures. If a failure occurs somewhere in the system, a red warning light on the instrument cluster comes on. Once a failure has occurred, the warning light remains on until the system has been repaired.

Power brake booster

A dual-diaphragm, vacuum-operated power brake booster, mounted between the firewall and the master cylinder, utilizes engine manifold vacuum and atmospheric pressure to provide assistance to the hydraulic brake system.

A pedal travel sensor for the Anti-lock Brake System (ABS) is installed in the power brake booster on 1993 and 1994 models. This sensor is not used on 1995 and later models, so the booster for these models is slightly different than earlier units. The two types of boosters are not interchangeable.

Parking brake system

The parking brake system consists of a lever between the front seats, the cables connecting this lever to the rear brakes, an equalizer for balancing the tension of the two rear cables, and the parking brake shoes. On all 1993 models and 1994 models with rear drums, the parking brake system utilizes the brake shoes themselves. On 1994 models with rear discs, and on all 1995 and later models, small parking brake shoes are housed inside the rear disc hubs (the hubs serve as the parking brake drums).

Service

After completing any operation involving disassembly of any part of the brake system, always test drive the vehicle to check for proper braking performance before resuming normal driving. When testing the brakes, perform the tests on a clean, dry flat surface. Conditions other than these can lead to inaccurate test results.

Test the brakes at various speeds with both light and heavy pedal pressure. The vehicle should stop evenly without pulling to one side or the other. Avoid locking the brakes because this slides the tires and diminishes braking efficiency and control of the vehicle.

Tires, vehicle load and front-end alignment are factors which also affect braking performance.

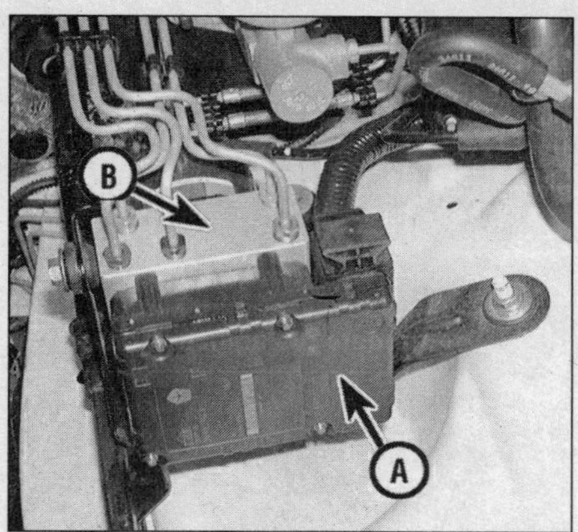

2.2 The ABS ECU (A) on later models is attached to the Hydraulic Control Unit (B) (on earlier models it's bolted to the left inner fender panel)

2 Anti-lock Brake System (ABS) - general information

The Anti-lock Brake System (ABS), which is standard on all models covered by this manual, is designed to maintain steering control, directional stability and optimal deceleration under severe braking conditions, regardless of the road surface. It does so by monitoring the rotational speed of each wheel and controlling the brake line pressure to each wheel during braking. This prevents the wheels from locking-up, which means the vehicle is easier to control during panic stops.

Components

Electronic Control Unit (ECU)

Refer to illustration 2.2

The Electronic Control Unit (ECU), which is located on the left (driver's side) inner fender panel in the engine compartment (and, on later models, attached to the hydraulic control unit), is the "brain" of the ABS system **(see illustration)**. The ECU has two microprocessors which receive identical sensor signals that are processed and compared simultaneously. A self-check program illuminates an amber warning light on the instrument cluster when a system fault is detected. Fault codes are stored in a diagnostic program memory, but cannot be accessed without a DRB scan tool. ABS system faults remain in memory until they're cleared, or until the vehicle has been started about 50 times. Stored faults are NOT erased simply by disconnecting the battery.

Hydraulic control unit (HCU)

The hydraulic control unit (HCU) consists of a valve body and a pump/motor assembly **(see illustration 2.2)**. The valve body contains three electrically-operated solenoid valves - one for each of the front wheels and one for both rear wheels - that modulate brake hydraulic pressure when the brakes are applied. During braking, these solenoid valves are cycled on and off rapidly and continuously by the ECU, modulating brake hydraulic pressure as necessary to control wheel lock-up.

The pump/motor assembly supplies the extra fluid needed during anti-lock braking. The pump is connected to the master cylinder reservoir by supply and return hoses, and is operated by an integral DC electric motor which is controlled by the ECU.

Wheel speed sensors

Refer to illustrations 2.6 and 2.7

Wheel speed sensors are located at each wheel and generate small electrical pulsations when the gear-type tone rings are turning, sending a signal to the ECU, indicating wheel rotational speed.

The front wheel speed sensors are mounted on the steering knuckles in close relationship to the tone rings, which are pressed onto the front axleshafts (4WD models) or integral with the front stub axle (2WD models) **(see illustration)**.

The rear sensors bolt to the brake backing plates **(see illustration)**. The tone rings are pressed onto the rear axleshafts.

Diagnosis and repair

If a dashboard warning light comes on and stays on while the vehicle is in operation, the ABS system requires attention. Although a DRB scan tool is necessary to properly diagnose the ABS system (and is required to access any trouble codes that may have been stored), you can perform a few preliminary checks before taking the vehicle to a dealer who is equipped with this tester.

a) Check the brake fluid level in the master cylinder reservoir.
b) Verify that the electrical connector for the ECU is plugged in.
c) Check the electrical connector at the pump motor.
d) Check the ABS fuses (see Chapter 12).
e) Follow the wiring harness to each wheel and check that all connections are secure and that the wiring is not damaged.

If the above preliminary checks do not rectify the problem, the vehicle should be diagnosed by a dealer service department or other repair facility equipped with the necessary scan tool. Due to the rather complex nature of this system and the high operating pressures involved, all actual repair work must be done by a dealer service department or other qualified repair shop.

3 Disc brake pads - replacement

Refer to illustrations 3.5a, 3.5b, 3.6a through 3.6p, or 3.7a through 3.7h

Warning: *Disc brake pads must be replaced on both front wheels or rear wheels at the same time - never replace the pads on only one wheel. Also, the dust created by the brake system is harmful to your health. Never blow it out with compressed air and don't inhale any of it. An approved filtering mask should be worn when working on the brakes. Do not, under any circumstances, use petroleum-based solvents to clean brake parts. Use brake system cleaner only!*

Note: *This procedure applies to both the front and the rear disc brakes.*

1 Remove the cap(s) from the brake fluid reservoir.

2 Loosen the wheel lug nuts, raise the front of the vehicle and support it securely on jackstands.

3 Remove the front or rear wheels. Work on one brake assembly at a time, using the assembled brake for reference if necessary.

4 Inspect the brake disc carefully as outlined in Section 5. If machining is necessary, follow the information in that Section to remove the disc.

5 Push the piston back into the bore to provide room for the new brake pads. A C-clamp can be used to accomplish this on 1998 and earlier front calipers and all rear calipers **(see illustration)**. On 1999 and later front calipers, use a screwdriver to pry the caliper outward, which will depress the piston **(see illustration)**. As the piston is depressed to the bottom of the caliper bore, the fluid in

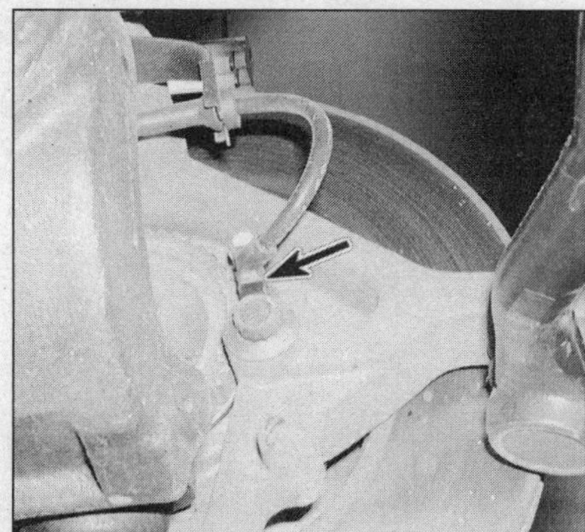

2.6 The front wheel speed sensors are bolted to the steering knuckle

2.7 The rear wheel speed sensors are bolted to the brake backing plate

3.5a A large C-clamp can be used to push the piston back into the caliper bore on 1998 and earlier front calipers and all rear calipers

3.5b On 1999 and later front calipers, insert a screwdriver through the caliper window and pry the caliper outward to depress the piston

3.6a Before removing the caliper, wash off all traces of brake dust with brake system cleaner

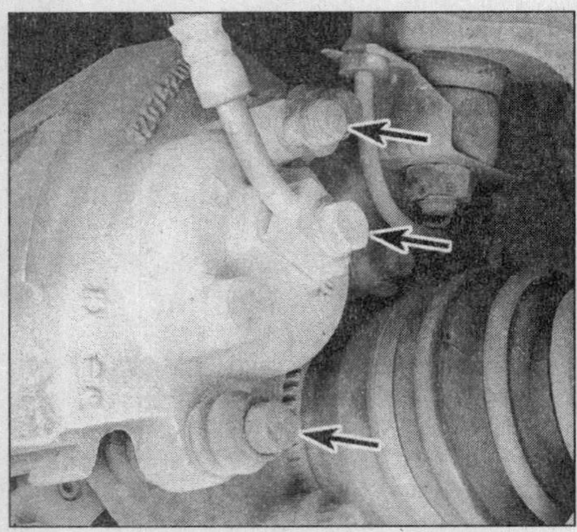
3.6b To remove the caliper, remove these two mounting bolts (upper and lower arrows); don't remove the brake hose-to-caliper banjo bolt (center arrow) unless you intend to replace the caliper or hose

3.6c Remove the caliper; this is a good time to check for fluid leakage around the caliper piston boot, which would indicate the need for replacement

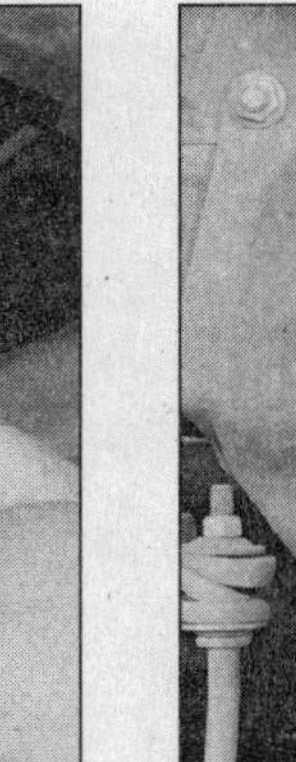
3.6d Whenever you have to let go of the caliper, hang it from the coil spring with a piece of wire - DON'T let it hang by the brake hose!

the master cylinder will rise. Make sure it doesn't overflow. If necessary, siphon off some of the fluid. **Warning:** *Brake fluid is poisonous - never siphon it by mouth. Use a suction gun or old poultry baster. If a baster is used, never again use it for the preparation of food.* **Caution:** *Brake fluid will damage paint. If any fluid is spilled, wash it off immediately with plenty of clean, cold water.*

6 If you're working on a 1998 or earlier model, follow the accompanying illustration sequence, beginning with **illustration 3.6a**, for the actual pad replacement procedure. Be sure to stay in order and read the caption under each illustration.

7 If you're working on a 1999 or later model, follow the accompanying illustrations, beginning with **illustration 3.7a**, for the actual pad replacement procedure. Be sure to stay in order and read the caption under each illustration.

3.6e To remove the outer brake pad from the caliper, push up on the ears at either end of the pad, disengage the pad retaining clips from the caliper and remove the pad

3.6f To remove the inner brake pad, pull on the pad to disengage the retaining clips from the bore of the piston

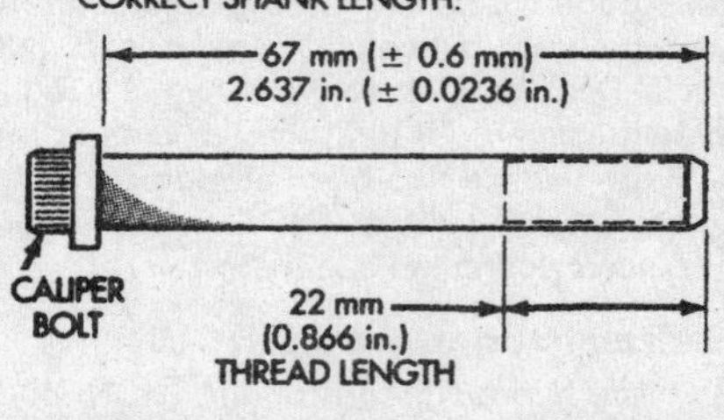

3.6g If you're working on a 1994 through 1998 model, measure the length of the caliper mounting bolts; if they aren't within specifications, replace them (if they have stretched, they can contact the inner brake pad, causing the brake to drag)

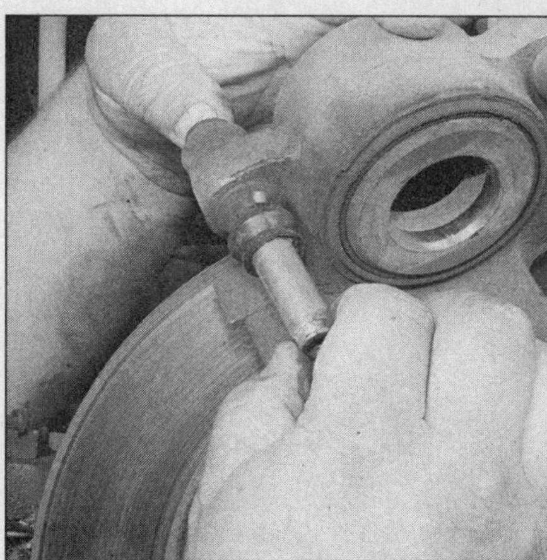

3.6h Pull out both pin bushings and inspect them for corrosion and wear; if either pin is damaged, replace it

3.6i If either rubber dust boot for the pins and bushings is cracked, torn or dried out, replace it

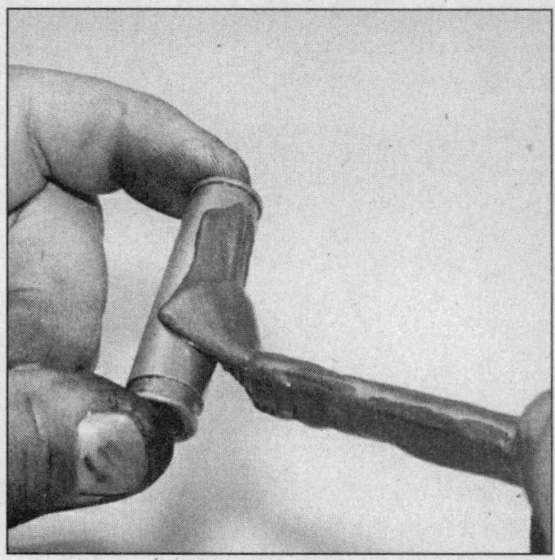

3.6j Coat the bushings with high temperature grease

3.6k Install the bushings

3.6l Before installing the new brake pads, depress the piston all the way into the caliper to facilitate installation of the new pads over the disc

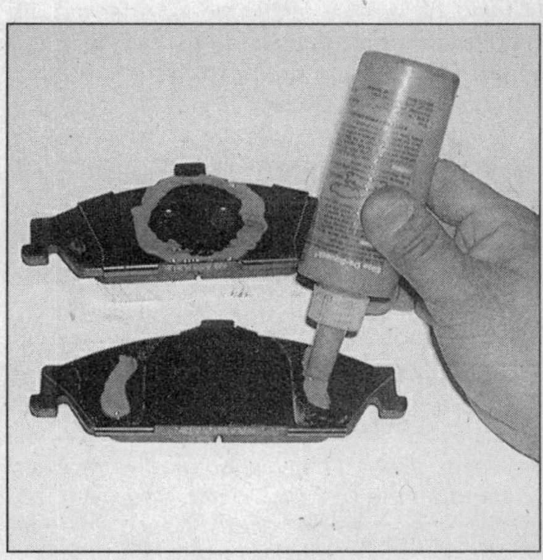

3.6m Apply anti-squeal compound to the backs of the new pads (follow the instructions on the container)

3.6n Install the new inner brake pad by pushing the retaining clips into the piston; make sure the clips are fully engaged (the pad should be seated flat against the piston)

3.6o Install the new outer brake pad by pushing down firmly on the pad with your thumbs as shown; make sure the pad is properly seated and the retaining clips fully engaged

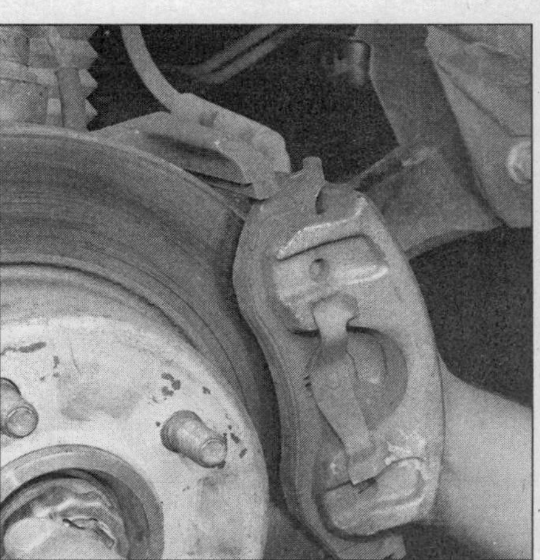

3.6p Install the caliper and pads over the disc, install the caliper bolts and tighten them to the torque listed in this Chapter's Specifications

3.7a Wash the brake with brake system cleaner (see illustration 3.6a), then pry off the caliper support spring

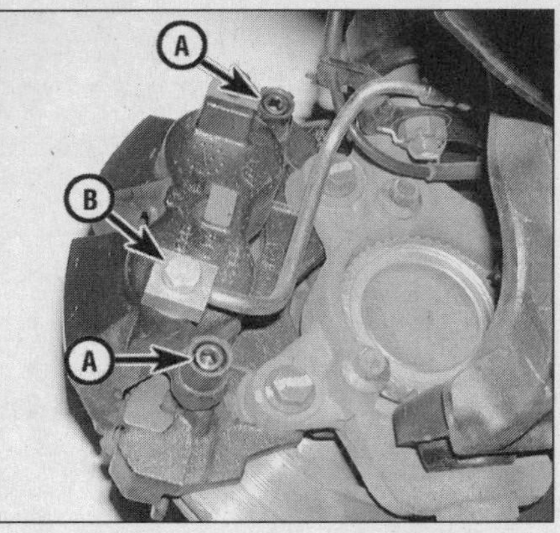

3.7b Remove the caps from the caliper slide pins (A), then unscrew the slide pins. The banjo bolt (B) shouldn't be removed unless you're replacing the caliper or the hose

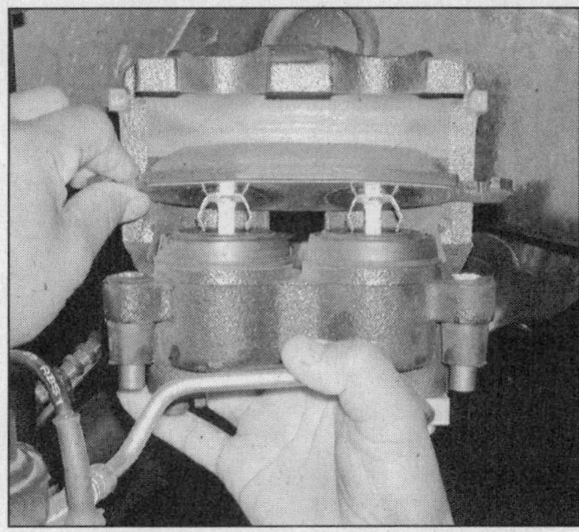

3.7c Lift off the caliper and pull on the inner pad to disengage the retaining clip(s) from the bore(s) of the piston(s). After removing the pad, hang the caliper with a length of wire (see illustration 3.6d)

8 When reinstalling the caliper, be sure to tighten the mounting bolts to the torque listed in this Chapter's Specifications. After the job has been completed, firmly depress the brake pedal a few times to bring the pads into contact with the disc.

9 Check for fluid leakage and make sure the brakes operate normally before driving in traffic.

4 Disc brake caliper - removal and installation

Warning: *Dust created by the brake system is harmful to your health. Never blow it out with compressed air and don't inhale any of it. An approved filtering mask should be worn when working on the brakes. Do not, under any circumstances, use petroleum-based solvents to clean brake parts. Use brake system cleaner only!*

Note: *If replacement is indicated (usually because of fluid leakage) explore all options before beginning the job. New and factory rebuilt calipers are available on an exchange basis, which makes this job quite easy. Always replace the calipers in pairs (on the front or rear) - never replace just one of them.*

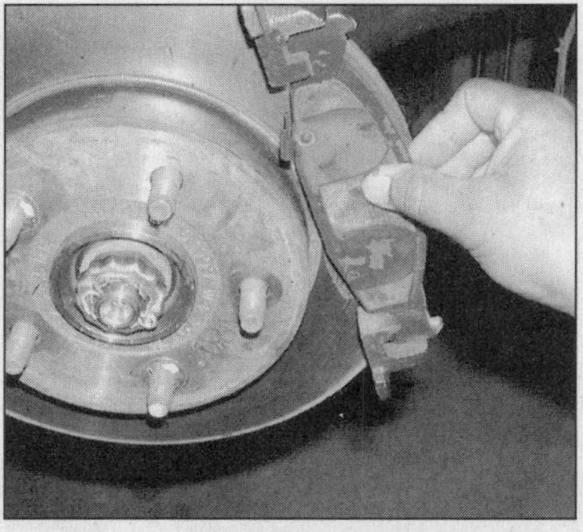

3.7d Remove the outer brake pad from the caliper mounting bracket

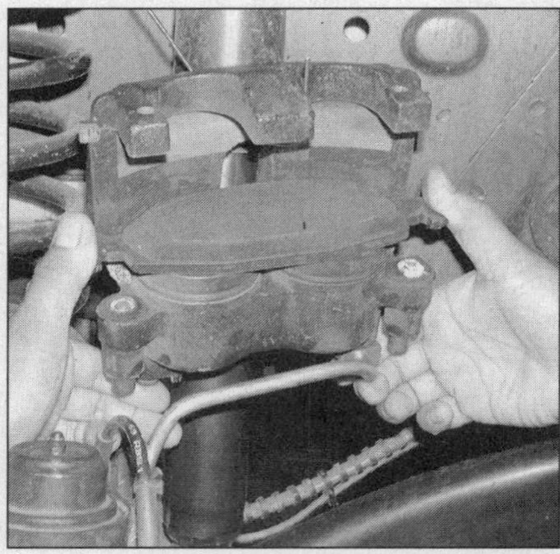

3.7e Apply anti-squeal compound to the backs of the new pads (see illustration 3.6m), then install the new inner pad by pushing the retaining clip(s) straight into the piston(s); make sure the clips are fully engaged - the pad should be seated flat against the piston(s)

Removal

1 Remove the cover from the brake fluid reservoir, siphon off two-thirds of the fluid

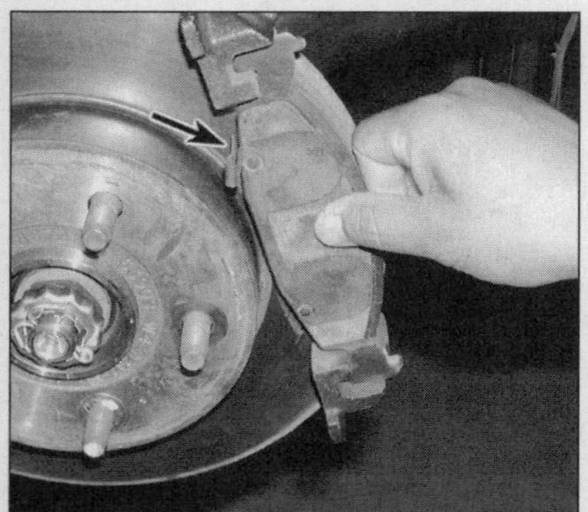

3.7f Install the outer pad in the caliper mounting bracket - the wear indicator (arrow) must be positioned at the top (front pads only)

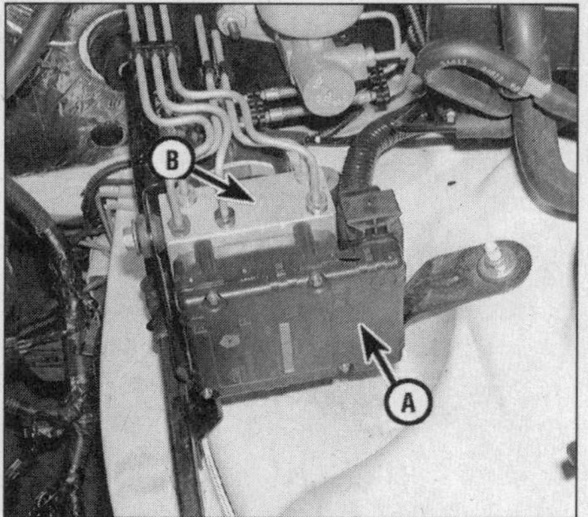

3.7g Install the caliper onto the mounting bracket, then lubricate the slide pins with high-temperature grease. Install the slide pins and tighten them to the torque listed in this Chapter's Specifications

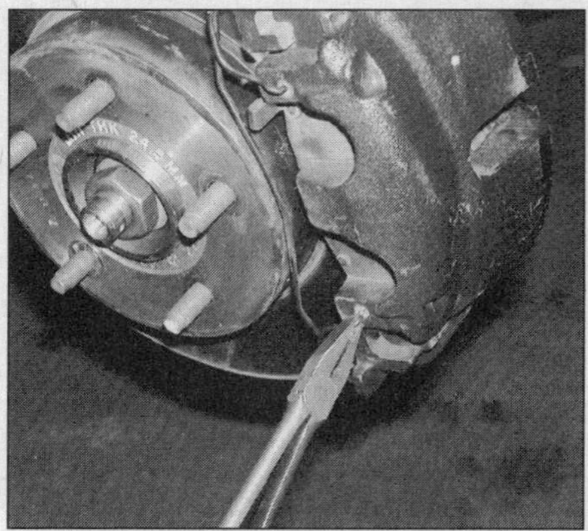

3.7h Install the caliper support spring, making sure it seats properly

5.4a Use a dial indicator to check disc runout - if the reading exceeds the specified allowable runout limit, the disc will have to be machined or replaced

5.4b Using a swirling motion, remove the glaze from the disc surface with sandpaper or emery cloth

into a container and discard it. **Warning:** *Brake fluid is poisonous - never siphon it by mouth. Use a suction gun or old poultry baster. If a baster is used, never again use it for the preparation of food.* **Caution:** *Brake fluid will damage paint. If any fluid is spilled, wash it off immediately with plenty of clean, cold water.*

2 Loosen the wheel lug nuts, raise the front of the vehicle and support it securely on jackstands. Remove the front wheels.

3 Push the piston back into the caliper bore **(see illustration 3.5a or 3.5b)**.

4 To disconnect the brake hose from the caliper, remove the brake hose-to-caliper banjo bolt **(see illustration 3.6b [1998 and earlier models] or 3.7b [1999 and later models])**. **Note:** *Do not disconnect the brake hose from the caliper if you are only removing the caliper.* Have a rag handy to catch spilled fluid and immediately wrap a plastic bag tightly around the end of the hose to prevent fluid loss and contamination.

5 Unscrew the two caliper mounting bolts (1998 and earlier models) or slide pins (1999 and later models) and detach the caliper (see Section 3). If you're removing the caliper for access to other components, hang it with a piece of wire from the coil spring - DON'T let it hang by the brake hose **(see illustration 3.6d)**.

Installation

6 Inspect the caliper mounting bolts or slide pins for excessive corrosion, replacing them if necessary. If you're working on a 1994 through 1998 model, measure the length of the bolts and replace them if they have stretched **(see illustration 3.6g)**.

7 Clean and lubricate the caliper bushings (1998 and earlier models) or slide pins **(see illustrations 3.6h through 3.6k [1998 and earlier models] or 3.7g [1999 and later models])**.

8 Install the caliper, tightening the mounting bolts or slide pins to the torque listed in this Chapter's Specifications.

9 Install the brake hose and banjo bolt. Use new copper washers, then tighten the bolt to the torque listed in this Chapter's Specifications.

10 Bleed the brakes (see Section 11).

11 Install the wheels and lower the vehicle. Tighten the wheel lug nuts to the torque listed in the Chapter 1 Specifications.

12 After the job has been completed, firmly depress the brake pedal a few times to bring the pads into contact with the disc.

13 Check brake operation before driving the vehicle in traffic.

5 Brake disc - inspection, removal and installation

Inspection

Refer to illustrations 5.4a, 5.4b, 5.5a, 5.5b and 5.5c

1 Loosen the wheel lug nuts, raise the vehicle and support it securely on jackstands. Remove the wheel and install the lug nuts to hold the disc in place. **Note:** *If the lug nuts don't contact the disc when screwed on all the way, install washers under them.* If you're working on a rear disc brake, release the parking brake.

2 Remove the brake caliper (see Section 3). It isn't necessary to disconnect the brake hose. After removing the caliper mounting pins, suspend the caliper out of the way with a piece of wire **(see illustration 3.6d)**. Don't let the caliper hang by the hose and don't stretch or twist the hose.

3 Visually check the disc surface for score marks and other damage. Light scratches and shallow grooves are normal after use and may not always be detrimental to brake operation, but deep score marks - over 0.015-inch (0.38 mm) - require disc removal and refinishing by an automotive machine shop. Be sure to check both sides of the disc. If pulsating has been noticed during application of the brakes, suspect disc runout. Be sure to check the wheel bearings to make sure they're properly adjusted.

4 To check disc runout, place a dial indicator at a point about 1/2-inch from the outer edge of the disc **(see illustration)**. Set the indicator to zero and turn the disc. The indicator reading should not exceed the specified allowable runout limit. If it does, the disc should be refinished by an automotive machine shop. **Note:** *Professionals recommend resurfacing of brake discs regardless of the dial indicator reading (to produce a smooth, flat surface that will eliminate brake pedal pulsations and other undesirable symptoms related to questionable discs). At the very least, if you elect not to have the discs resurfaced, deglaze them with sandpaper or emery cloth (use a swirling motion to ensure a nondirectional finish)* **(see illustration)**.

5 The disc must not be machined to a thickness less than the specified minimum refinish thickness. The minimum (or discard) thickness is cast into the inside of the disc **(see illustrations)**. The disc thickness can be checked with a micrometer **(see illustration)**.

Removal

Refer to illustrations 5.6, 5.7a and 5.7b

6 Remove lug nuts you installed to hold the disc in place. If the disc is held in place by pressed-metal retainers, cut them off with

5.5a The minimum thickness is cast into the inside of the disc on early models (typical)

5.5b On later models the minimum thickness is cast into the outside of the disc (typical)

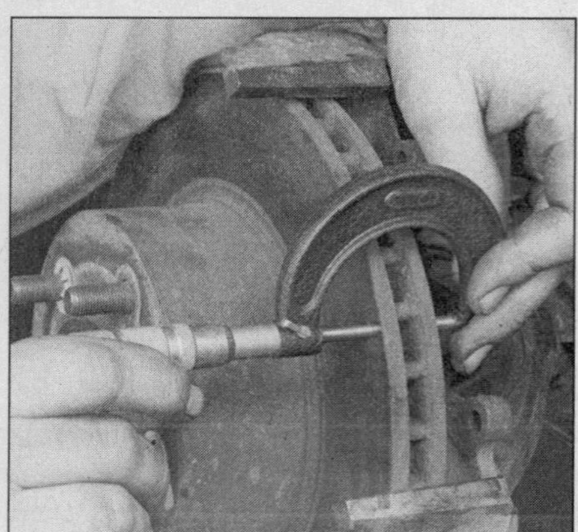

5.5c Use a micrometer to measure disc thickness at several points

5.6 If the disc is held on by pressed-metal retainers, cut them off and discard them (it isn't necessary to install new ones)

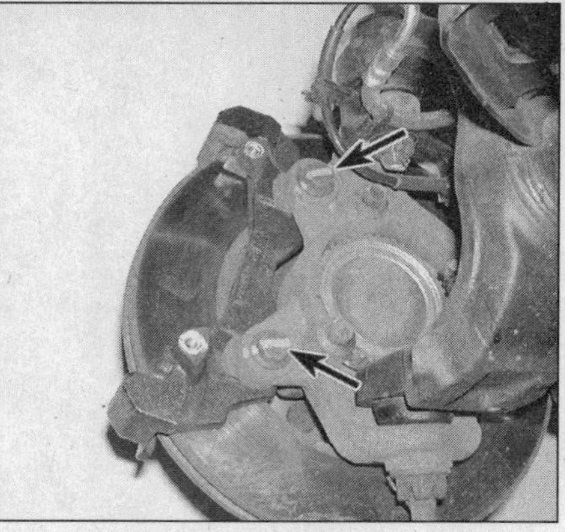

5.7a If you're working on a 1999 or later model, the caliper mounting bracket will have to be removed to allow removal of the disc - here are the front caliper mounting bracket-to-steering knuckle bolts . . .

5.7b . . . and here are the rear caliper mounting bracket-to-axle housing bolts

diagonal cutters and discard them **(see illustration)**.

7 If you're working on a 1999 or later model, remove the caliper mounting bracket **(see illustrations)**.

8 Slide the disc off the hub.

Installation

9 Place the disc over the wheel studs on the hub. If you're working on a 1999 or later model, install the caliper mounting bracket, tightening the bolts to the torque listed in this Chapter's Specifications.

10 Install the caliper and brake pads (see Section 3). Tighten the caliper mounting bolts (1998 and earlier models) or slide pins (1999 and later models) to the torque listed in this Chapter's Specifications.

11 Install the wheel, then lower the vehicle to the ground. Tighten the wheel lug nuts to the torque listed in the Chapter 1 Specifications.

12 Depress the brake pedal a few times to bring the brake pads into contact with the disc. Bleeding of the system will not be necessary unless the brake hose was disconnected from the caliper. Check the operation of the brakes carefully before placing the vehicle into normal service.

6 Drum brake shoes - replacement

Refer to illustrations 6.4a through 6.4x, 6.5 and 6.7

Warning: *Drum brake shoes must be replaced on both wheels at the same time - never replace the shoes on only one wheel. Also, the dust created by the brake system is harmful to your health. Never blow it out with compressed air and don't inhale any of it. An approved filtering mask should be worn when working on the brakes. Do not, under any circumstances, use petroleum-based solvents to clean brake parts. Use brake system cleaner only!*

Caution: *Whenever the brake shoes are replaced, the retracting and hold-down springs should also be replaced. Due to the continuous heating/cooling cycle that the springs are subjected to, they lose their tension over a period of time and may allow the shoes to drag on the drum and wear at a much faster rate than normal.*

1 Loosen the wheel lug nuts, raise the rear of the vehicle and support it securely on jackstands. Block the front wheels to keep the vehicle from rolling.

2 Release the parking brake.

3 Remove the wheel and brake drum (see Chapter 1). **Note:** *All four rear brake shoes must be replaced at the same time, but to avoid mixing up parts, work on only one brake assembly at a time.*

4 Follow the accompanying illustrations **(6.4a through 6.4x)** for the replacement of the brake shoes. Be sure to stay in order and read the caption under each illustration. **Note:** *If the brake drum cannot be easily pulled off, make sure the parking brake is*

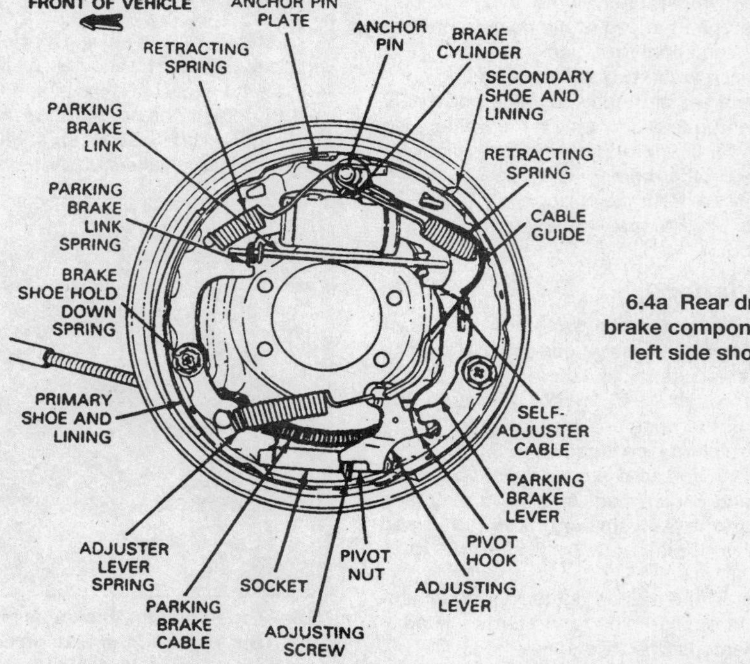

6.4a Rear drum brake components - left side shown

6.4b Before removing any internal drum brake components, wash them off with brake cleaner and allow them to dry - position a drain pan under the brake to catch the residue - DO NOT USE COMPRESSED AIR TO BLOW THE BRAKE DUST FROM THE PARTS!

6.4c Pull out on the adjuster lever and turn the star wheel to retract the brake shoes

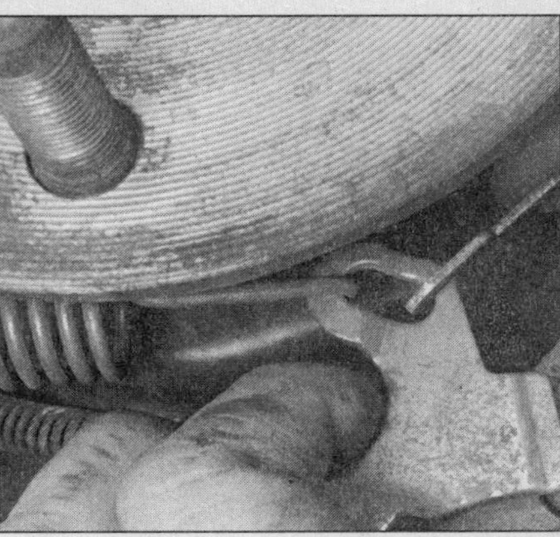

6.4d Pull back on the self-adjuster cable and push the adjusting lever toward the rear, unhooking it from the secondary brake shoe

completely released, then apply some penetrating oil at the hub-to-drum joint. Allow the oil to soak in and try to pull the drum off. If the drum still cannot be pulled off, the brake shoes will have to be retracted. This is accomplished by first removing the plug from the backing plate. With the plug removed, pull the lever off the adjusting star wheel with one narrow screwdriver while turning the adjusting wheel with another narrow screwdriver, moving the shoes away from the drum **(see illustration 6.7)**. *The drum should now come off.*

5 Before reinstalling the drum it should be checked for cracks, score marks, deep scratches and hard spots, which will appear as small discolored areas. If the hard spots cannot be removed with fine emery cloth or if

6.4e Remove the primary and secondary shoe retracting springs - the spring removal tool shown here can be purchased at most auto parts stores and greatly simplifies this step

6.4f Remove the self-adjuster cable and anchor pin plate from the anchor pin

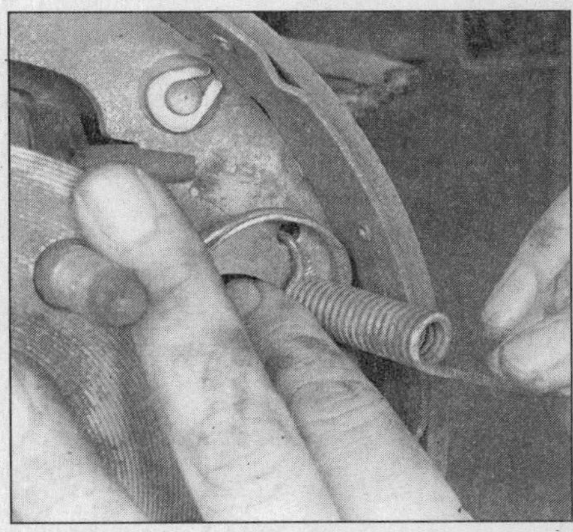

6.4g Remove the secondary shoe retracting spring and cable guide from the secondary shoe

6.4h Remove the primary shoe hold-down spring and pin . . .

6.4i . . . then lift the primary shoe and adjusting screw from the backing plate

6.4j Remove the parking brake link

6.4k Remove the secondary shoe hold-down spring and pin then lift the shoe from the backing plate - pry the parking brake lever retaining clip off of the pivot pin . . .

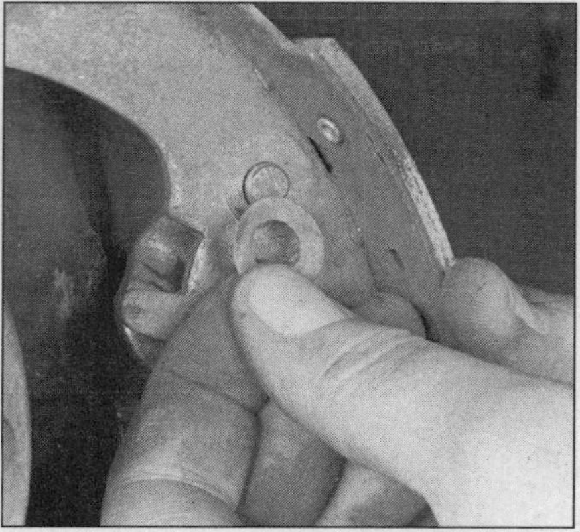

6.4l . . . then separate the lever from the secondary shoe - be careful not to lose the spring washer

6.4m Lubricate the brake shoe contact areas (arrows) with high temperature grease

6.4n Attach the new shoe to the parking brake lever, install the spring washer and retaining clip on the pivot pin, then crimp the clip closed with a pair of pliers

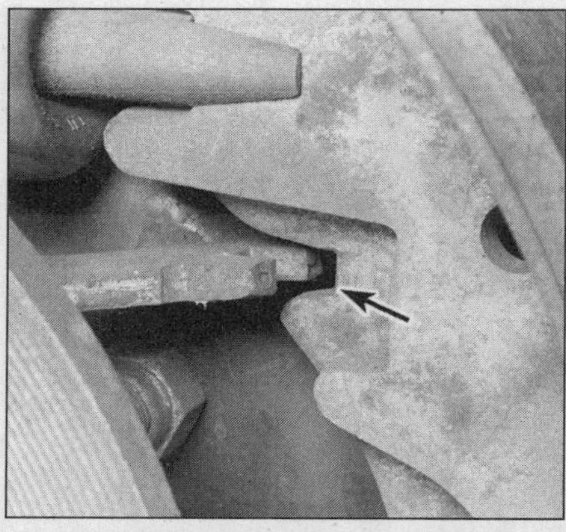

6.4o Install the secondary shoe and hold-down spring to the backing plate, then position the end of the parking brake link into the notch (arrow)

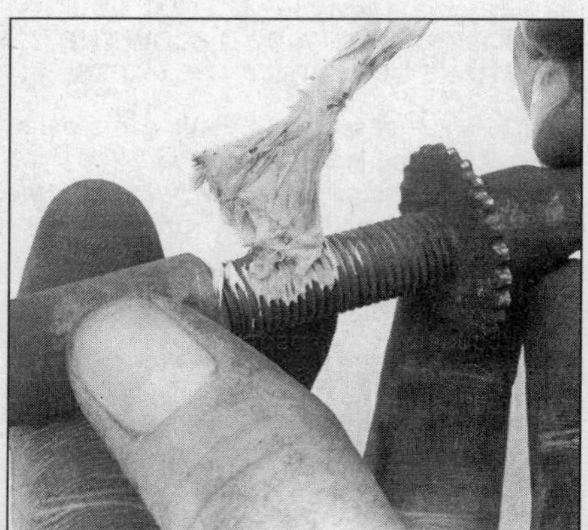

6.4p Lubricate the adjusting screw threads and socket with high-temperature grease

any of the other conditions listed above exist, the drum must be taken to an automotive machine shop to have it resurfaced. **Note:** *Professionals recommend resurfacing the drums whenever a brake job is done. Resurfacing will eliminate the possibility of out-of-round drums. If the drums are worn so much that they can't be resurfaced without exceeding the maximum allowable diameter (stamped into the drum)* **(see illustration)**, *then new ones will be required. At the very least, if you elect not to have the drums resurfaced, remove the glazing from the surface with sandpaper or emery cloth using a swirling motion.*

6.4q Place the primary shoe against the backing plate, then install the hold-down spring - make sure the parking brake strut and wheel cylinder pushrods engage in the brake shoe slots

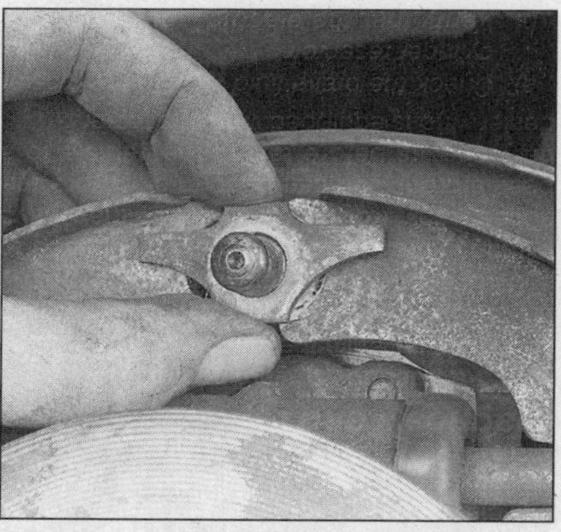

6.4r Install the anchor pin plate . . .

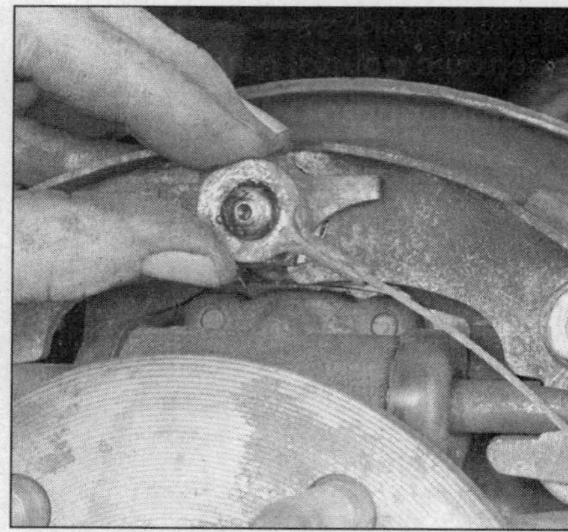

6.4s . . . and the self-adjuster cable

6.4t Hook the end of the secondary shoe retractor spring through the cable guide and into the hole in the shoe, then stretch the spring over the anchor pin

6.4u Install the primary shoe retractor spring - the tool shown here is available at most auto parts stores and makes this step much easier and safer

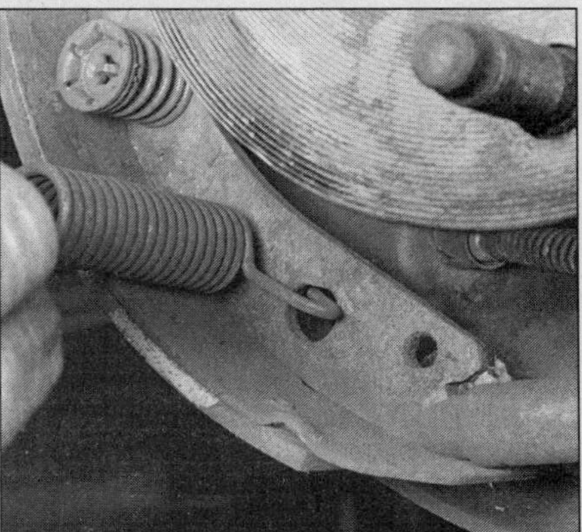

6.4v Hook the adjuster lever spring into the hole at the bottom of the primary shoe

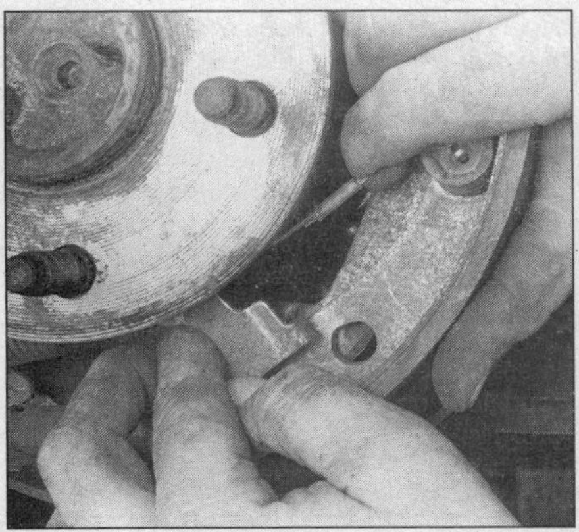

6.4w Hook the adjuster lever spring and cable into the adjuster lever and pull the cable down and to the rear, inserting the hook on the lever into the hole in the secondary shoe

6.4x Wiggle the assembly to ensure the shoes are centered on the backing plate

6.5 The maximum allowable drum diameter is cast into the drum (typical)

6 Once the new shoes are in place, install the drums on the axle flanges. Remove the rubber plugs from the brake backing plates.

7 Insert a narrow screwdriver or brake adjusting tool through the adjustment hole and turn the star wheel until the brakes drag slightly as the drum is turned **(see illustration)**.

8 Turn the star wheel in the opposite direction until the shoes don't drag and the drum turns freely.

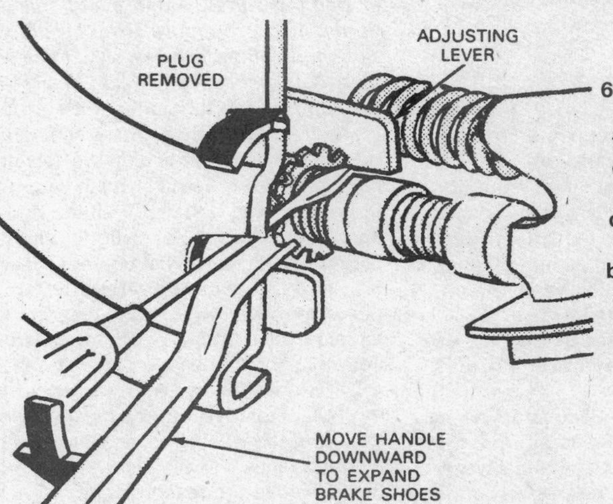

6.7 Using a screwdriver or brake adjuster tool, turn the adjuster wheel in the direction shown until the shoes drag on the brake drum; then, insert a small screwdriver through the backing plate to move the adjuster lever away from the adjuster wheel and turn the wheel in the opposite direction until the drum turns freely

9 Repeat the adjustment on the opposite wheel and install the backing plate plugs.

10 Mount the wheel, install the lug nuts, then lower the vehicle. Tighten the lug nuts to the torque listed in the Chapter 1 Specifications.

11 Make a number of forward and reverse stops to allow the brakes to further adjust themselves.

12 Check brake operation before driving the vehicle in traffic.

7 Wheel cylinder - removal and installation

Note: *If replacement is indicated (usually because of fluid leakage or sticky operation), it is recommended that new wheel cylinders be installed. Never replace only one wheel cylinder - always replace both of them at the same time.*

Removal

Refer to illustration 7.4

1 Raise the rear of the vehicle and support it securely on jackstands. Block the front wheels to keep the vehicle from rolling.

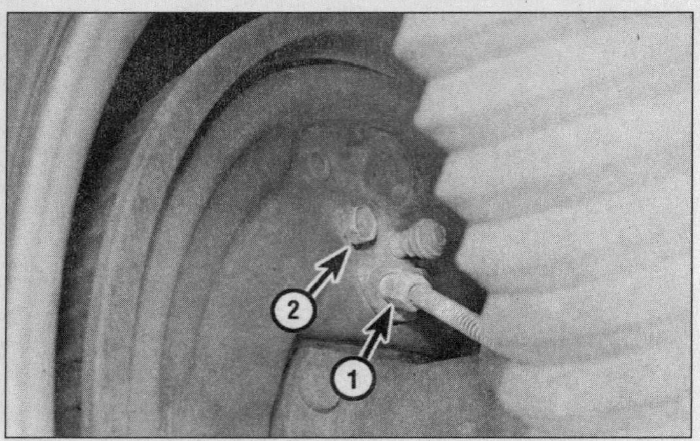

7.4 Completely loosen the brake line fitting (1), then remove the two wheel cylinder mounting bolts (2) (the front bolt isn't visible in this photo)

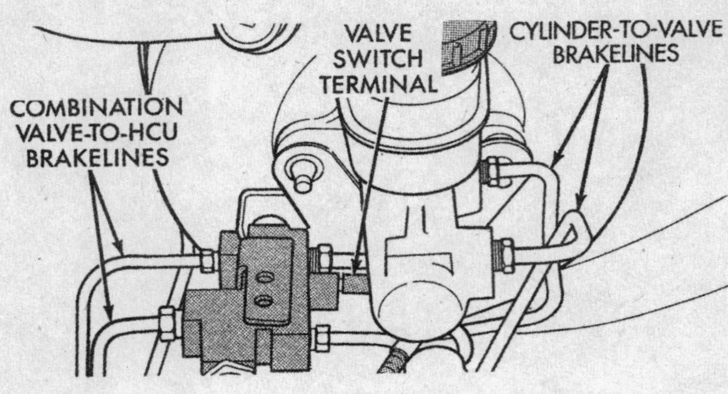

8.8 The combination valve is located next to the master cylinder (1998 and earlier models only; on later models the valve's functions are handled by the ABS HCU)

2 Remove the brake shoe assembly (see Section 6).

3 Remove all dirt and foreign material from around the wheel cylinder.

4 Completely loosen the brake line fitting **(see illustration)**. Don't pull the brake line away from the wheel cylinder.

5 Remove the wheel cylinder mounting bolts **(see illustration 7.4)**.

6 Detach the wheel cylinder from the brake backing plate. Immediately plug the brake line to prevent fluid loss and contamination. **Note:** *If the brake shoe linings are contaminated with brake fluid, install new brake shoes.*

Installation

7 Place the wheel cylinder in position and, while it's still loose, connect the brake line to it, being careful not to cross-thread the fitting.

8 Install the bolts and tighten them to the torque listed in this Chapter's Specifications. Tighten the brake line fitting securely. Install the brake shoe assembly.

9 Bleed the brakes (see Section 11).

10 Check brake operation before driving the vehicle in traffic.

8 Combination valve - removal and installation

Note: *This procedure applies to 1998 and earlier models only.*

Check

Metering valve

1 To check the metering valve, have an assistant apply and release the brake pedal while you watch the valve stem (the small black button on the front of the combination valve). If the valve is operating correctly, the stem will extend slightly when the brakes are applied and retract when the brakes are released. If the valve does not protrude slightly each time the brake pedal is applied, replace the combination valve.

Pressure differential switch

2 Raise the vehicle and place it securely on jackstands.

3 Connect a bleeder hose to a rear wheel cylinder and immerse the other end of the hose in a container partially filled with brake fluid. Loosen the bleeder screw slightly.

4 To check the pressure differential switch, have your helper sit inside the vehicle, apply the brake pedal, and watch the red brake warning light.

5 If the warning light comes on, the switch is operating correctly. Bleed the rear wheel cylinder to ensure no air was let into the system.

6 If the warning light doesn't come on, check the circuit fuse, the bulb and the wiring (see Chapter 12). (One quick way to eliminate the fuse and bulb as the problem is to apply the parking brake; if the parking brake switch turns on the light, the fuse and bulb are okay). Repair the circuit or replace parts as necessary and retest the warning light switch. Don't forget to bleed the rear wheel cylinder to ensure no air was let into the system.

7 If, after making the necessary repairs and/or part replacements, the warning light still doesn't come on, the switch is faulty. Replace the combination valve.

Replacement

Refer to illustration 8.8

8 Unplug the electrical connector from the combination valve **(see illustration)**.

9 Disconnect the brake lines from the combination valve, using a flare nut wrench if available. Plug the ends of the lines to prevent loss of brake fluid and the entry of dirt.

10 Remove the combination valve mounting bolt and detach the valve.

11 Installation is the reverse of removal. Be sure to tighten the brake line fitting nuts securely.

12 Bleed the system (see Section 11) when you're done. **Warning:** *To bleed air from the hydraulic control unit, the manufacturer states that the system must be bled using the conventional method (as described in Section 11), then with the use of a DRB scan tool to purge any air from the HCU. After that has been done the system must be bled using the conventional method once again. This tool is very expensive and generally available only to dealership service departments. If, after bleeding the system you do not have a firm brake pedal, or if the ABS light on the instrument panel does not go off, or if you have any doubts whatsoever about the effectiveness of the brake system, have it towed to a dealer service department or other repair shop equipped with the necessary tools for bleeding the system.*

9 Master cylinder - removal, installation and reservoir/grommet replacement

Removal

Refer to illustration 9.3

1 Place rags under the brake line fittings and prepare caps or plastic bags to cover the ends of the lines once they're disconnected. **Caution:** *Brake fluid will damage paint. Cover all painted surfaces and avoid spilling fluid during this procedure. Brake fluid can be siphoned out of the reservoir using a squeeze bulb, but wear safety goggles.*

2 Loosen the tube nuts at the ends of the brake lines where they enter the master cylinder. To prevent rounding off the flats on these nuts, a flare-nut wrench, which wraps around the nut, should be used. Pull the brake lines away from the master cylinder slightly and plug the ends to prevent contamination.

3 If you're working on a 1998 or earlier model, remove the nuts and detach the combination valve bracket from the master cylinder mounting studs **(see illustration)**.

4 Remove the two master cylinder mounting nuts. Remove the master cylinder from the booster, taking care not to kink the hydraulic lines.

5 Remove the reservoir cap, then discard any fluid remaining in the reservoir.

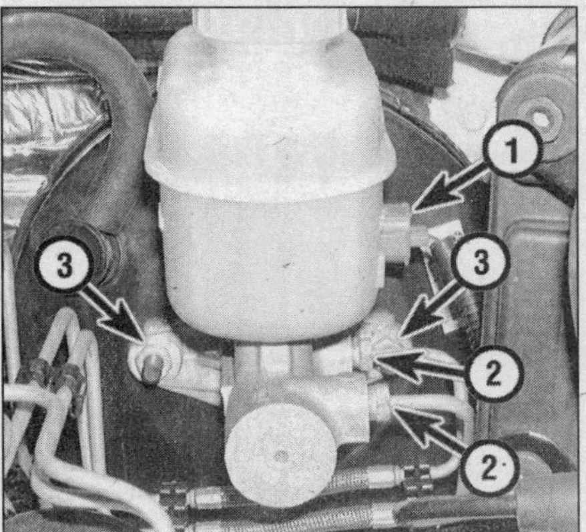

9.3 Master cylinder mounting details

1 *Electrical connector for fluid level sensor*
2 *Brake line fittings*
3 *Mounting nuts*

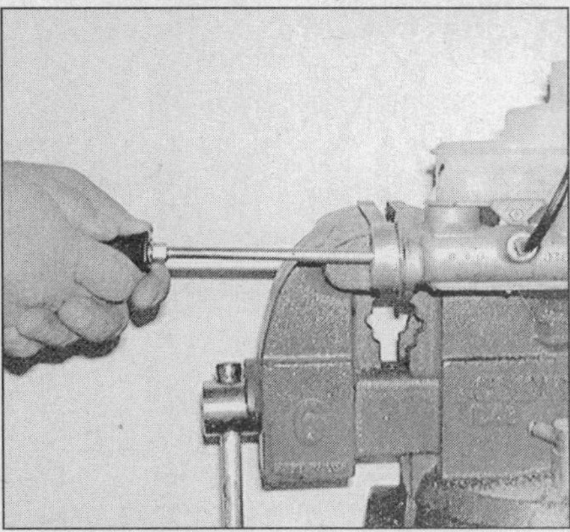

9.7 The best way to bleed air from the master cylinder before installing it on the vehicle is with a pair of bleeder tubes that direct brake fluid into the reservoir during bleeding

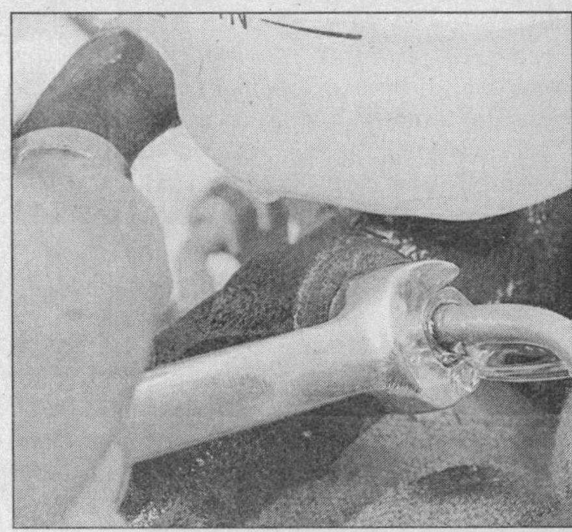

9.15 Have an assistant depress the brake pedal and hold it down, then loosen the fitting nut, allowing air and brake fluid to escape; repeat this procedure on both fittings until the fluid is clear of air bubbles

Installation

Refer to illustrations 9.7 and 9.15

6 Bench bleed the new master cylinder before installing it. Mount the master cylinder in a vise, with the jaws of the vise clamping on the mounting flange.

7 Attach a pair of master cylinder bleeder tubes to the outlet ports of the master cylinder **(see illustration)**.

8 Fill the reservoir with brake fluid of the recommended type (see Chapter 1).

9 Slowly push the pistons into the master cylinder (a large Phillips screwdriver can be used for this) - air will be expelled from the pressure chambers and into the reservoir. Because the tubes are submerged in fluid, air can't be drawn back into the master cylinder when you release the pistons.

10 Repeat the procedure until no more air bubbles are present.

11 Remove the bleed tubes, one at a time, and install plugs in the open ports to prevent fluid leakage and air from entering. Install the reservoir cap.

12 Install the master cylinder over the studs on the power brake booster and tighten the attaching nuts only finger tight at this time.

13 Thread the brake line fittings into the master cylinder. Since the master cylinder is still a bit loose, it can be moved slightly in order for the fittings to thread in easily. Do not strip the threads as the fittings are tightened.

14 Tighten the mounting nuts to the torque listed in this Chapter's Specifications, then the brake line fittings. If you're working on a 1998 or earlier model, attach the combination valve bracket and install the nuts, tightening them to the same torque as the master cylinder mounting nuts.

15 Fill the master cylinder reservoir with fluid, then bleed the master cylinder. **Caution:** *Have plenty of rags on hand to catch the fluid - brake fluid will ruin painted surfaces. After the bleeding procedure is completed, rinse the area under the master cylinder with clean water.* To bleed the cylinder on the vehicle, have an assistant depress the brake pedal and hold the pedal to the floor. Loosen the fitting nut to allow air and fluid to escape **(see illustration)**. Repeat this procedure on both fittings at the master cylinder until the fluid is clear of air bubbles, then bleed the lines on both sides of the combination valve (1998 and earlier models) and at the HCU inlet lines. **Warning:** *To bleed air from the hydraulic control unit, the manufacturer states that the system must be bled using the conventional method (as described in Section 11), then with the use of a DRB scan tool to purge any air from the HCU. After that has been done the system must be bled using the conventional method once again. This tool is very expensive and generally available only to dealership service departments. If, after bleeding the system you do not have a firm brake pedal, or if the ABS light on the instrument panel does not go off, or if you have any doubts whatsoever about the effectiveness of the brake system, have it towed to a dealer service department or other repair shop equipped with the necessary tools for bleeding the system.*

16 Test the operation of the brake system carefully before placing the vehicle into normal service. **Warning:** *Do not operate the vehicle if you are in doubt about the effectiveness of the brake system.*

Reservoir/grommet replacement

Refer to illustrations 9.19, 9.20 and 9.22

Note: *The brake fluid reservoir can be replaced separately from the master cylinder body if it becomes damaged. If there is leakage between the reservoir and the master cylinder body, the grommets in the master cylinder can be replaced.*

17 Remove as much fluid as possible from the reservoir with a suction gun, large syringe or a poultry baster. **Warning:** *If a poultry baster is used, never again use it for the preparation of food.*

18 Place rags under the master cylinder to absorb any fluid that may spill out once the reservoir is detached from the master cylinder. **Caution:** *Brake fluid will damage paint. Cover all body parts and be careful not to spill fluid during this procedure.*

19 If you're working on a 1998 or earlier model, use a hammer and punch to drive out the roll pins that retain the reservoir **(see illustration)**.

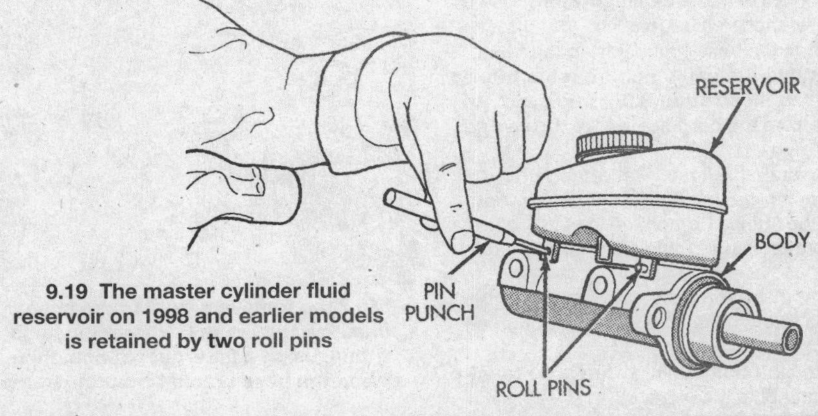

9.19 The master cylinder fluid reservoir on 1998 and earlier models is retained by two roll pins

9

20 If you're working on a 1999 or later model, use the tool provided with the new reservoir and insert it under the reservoir to dislodge the retaining tabs **(see illustration)**. If you are replacing the grommets and not the reservoir (and don't have the special tool), use a couple of small screwdrivers to dislodge the tabs.

21 Pull the reservoir out of the master cylinder body.

22 If you are simply replacing the grommets, remove them from the master cylinder and install new ones **(see illustration)**.

23 Lubricate the reservoir O-rings with clean brake fluid, then press the reservoir straight down into place on the master cylinder body.

24 If you're working on a 1998 or earlier model, install new roll pins.

25 If you're working on a 1999 or later model, make sure the retaining tabs on the bottom of the reservoir are completely engaged.

26 Refill the reservoir with the recommended brake fluid (see Chapter 1) and check for leaks.

27 Bleed the master cylinder **(see illustration 9.15)**.

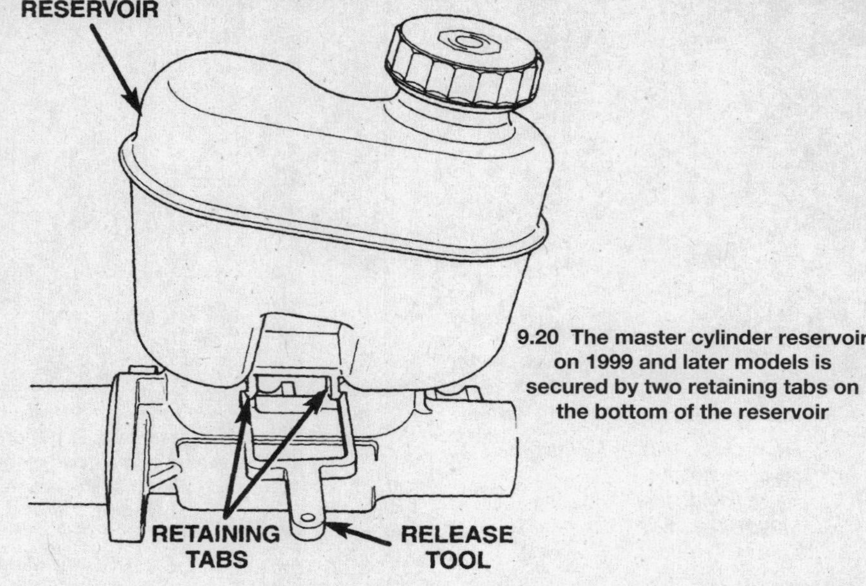

9.20 The master cylinder reservoir on 1999 and later models is secured by two retaining tabs on the bottom of the reservoir

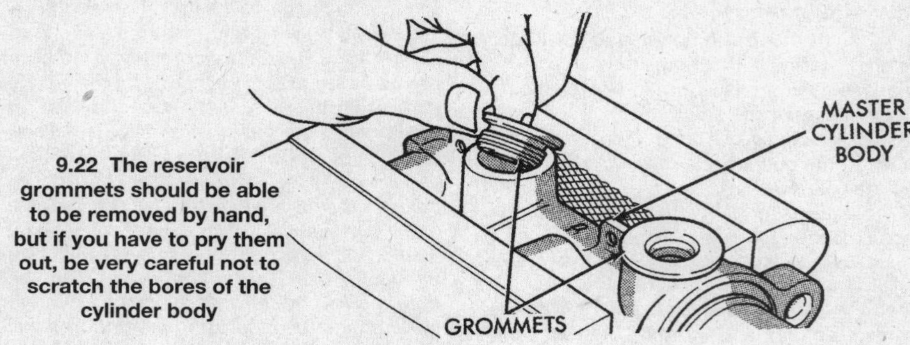

9.22 The reservoir grommets should be able to be removed by hand, but if you have to pry them out, be very careful not to scratch the bores of the cylinder body

10 Brake hoses and lines - inspection and replacement

Inspection

1 About every six months, with the vehicle raised and supported securely on jackstands, the flexible hoses that connect the steel brake lines with the front and rear brake assemblies should be inspected for cracks, chafing of the outer cover, leaks, blisters and other damage. These are important and vulnerable parts of the brake system and inspection should be complete. A light and mirror will be helpful for a thorough check. If a hose exhibits any of the above conditions, replace it with a new one.

Replacement

Front brake hose

Refer to illustration 10.2

2 Disconnect the brake line from the hose fitting, being careful not to bend the frame bracket or brake line **(see illustration)**. Use a flare nut wrench, if available.

3 Unbolt the hose bracket from the frame.

4 Remove the banjo bolt from the brake caliper **(see illustration 3.6b)** and separate the hose from the caliper. Discard the sealing washers.

5 To install the hose, first attach it to the caliper, using new sealing washers on both sides of the fitting. Tighten the inlet fitting bolt to the torque listed in this Chapter's Specifications.

6 Without twisting the hose, connect the brake line to the hose fitting, but don't tighten it yet.

7 Install the bolt retaining the hose to the frame, tightening it securely.

8 Tighten the brake line fitting securely.

9 When the brake hose installation is complete, there should be no kinks in the hose. Make sure the hose doesn't contact any part of the suspension. Check this by turning the wheels to the extreme left and right positions. If the hose makes contact, remove it and correct the installation as necessary. Bleed the system (see Section 11).

Rear brake hose

Refer to illustrations 10.11 and 10.12

10 Using a back-up wrench, disconnect the hose at the frame bracket, being careful not

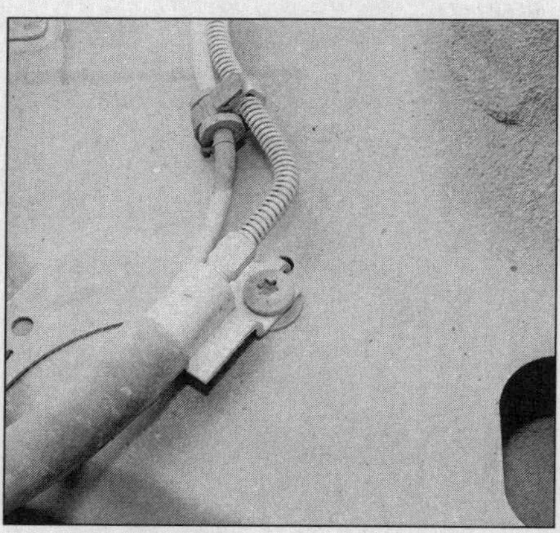

10.2 To remove a front brake hose, unscrew the brake line from the hose fitting, using a flare-nut wrench; then unbolt the hose bracket from the frame

10.11 To disconnect a rear brake hose from its metal line, unscrew the threaded fitting, then remove the U-clip

10.12 To disconnect the rear brake hose from the axle, unscrew the brake line fittings (A) from the junction block, then unscrew the vent hose bolt (B) and remove the hose

11.10 When bleeding the brakes, a hose is connected to the bleeder valve at the caliper or wheel cylinder and then submerged in brake fluid. Air will be seen as bubbles in the tube and container. All air must be expelled before moving to the next wheel

to bend the bracket or steel lines.

11 Remove the U-clip with a pair of pliers **(see illustration)** and separate the female fitting from the bracket.

12 Disconnect the two hydraulic lines at the junction block, then unbolt and remove the hose **(see illustration)**.

13 Bolt the junction block to the axle housing and connect the lines, tightening them securely. Without twisting the hose, install the female end of the hose in the frame bracket.

14 Install the U-clip retaining the female end to the bracket.

15 Using a back-up wrench, attach the steel line fittings to the female fittings. Again, be careful not to bend the bracket or steel line.

16 Make sure the hose installation did not loosen the frame bracket. Tighten the bracket if necessary.

17 Fill the master cylinder reservoir and bleed the system (see Section 11).

Metal brake lines

18 When replacing brake lines be sure to use the correct parts. Don't use copper tubing for any brake system components. Purchase steel brake lines from a dealer or auto parts store.

19 Prefabricated brake line, with the tube ends already flared and fittings installed, is available at auto parts stores and dealers. These lines can be bent to the proper shapes using a tubing bender.

20 When installing the new line, make sure it's securely supported in the brackets and has plenty of clearance between moving or hot components.

21 After installation, check the master cylinder fluid level and add fluid as necessary. Bleed the brake system as outlined in Section 11 and test the brakes carefully before driving the vehicle in traffic.

11 Brake system bleeding

Refer to illustration 11.10

Warning 1: *The following procedure is a manual bleeding procedure. This is the only bleeding procedure which can be performed at home without special tools. However, if air has found its way into the hydraulic control unit, the entire system must be bled manually, then with a DRB scan tool, then manually a second time. If the brake pedal feels "spongy" even after bleeding the brakes, or the ABS light on the instrument panel does not go off, or if you have any doubts whatsoever about the effectiveness of the brake system, have the vehicle towed to a dealer service department or other repair shop equipped with the necessary tools for bleeding the system.*

Warning 2: *Wear eye protection when bleeding the brake system. If the fluid comes in contact with your eyes, immediately rinse them with water and seek medical attention.*

Note: *Bleeding the hydraulic system is necessary to remove any air that manages to find its way into the system when it's been opened during removal and installation of a hydraulic component.*

1 It will be necessary to bleed the complete system if air has entered the system due to low fluid level, or if the brake lines have been disconnected at the master cylinder.

2 If a brake line was disconnected only at a wheel, then only that caliper or wheel cylinder must be bled.

3 If a brake line is disconnected at a fitting located between the master cylinder and any of the brakes, that part of the system served by the disconnected line must be bled. The following procedure describes bleeding the entire system, however.

4 Remove any residual vacuum from the brake power booster by applying the brake several times with the engine off.

5 Remove the cap(s) from the master cylinder reservoir and fill the reservoir with brake fluid. Reinstall the cap(s). **Note:** *Check the fluid level often during the bleeding operation and add fluid as necessary to prevent the fluid level from falling low enough to allow air bubbles into the master cylinder.*

6 Have an assistant on hand, as well as a supply of new brake fluid, a clear container partially filled with clean brake fluid, a length of clear tubing to fit over the bleeder valve and a wrench to open and close the bleeder valve.

7 Begin the bleeding process by bleeding the master cylinder **(see illustration 9.15)**.

8 Using the same technique, bleed the lines at the combination valve (1998 and earlier models only) and where the lines enter the ABS HCU.

9 Moving to the right rear wheel, loosen the bleeder valve slightly, then tighten it to a point where it is snug but can still be loosened quickly and easily.

10 Place one end of the hose over the bleeder valve and submerge the other end in brake fluid in the container **(see illustration)**.

11 Have the assistant push the brake pedal slowly to the floor, then hold the pedal firmly depressed.

12 While the pedal is held depressed, open the bleeder valve just enough to allow a flow of fluid to leave the valve. Watch for air bubbles to exit the submerged end of the tube. When the fluid flow slows after a couple of seconds, close the valve and have your assistant release the pedal.

13 Repeat Steps 11 and 12 until no more air is seen leaving the tube, then tighten the bleeder valve and proceed to the left rear wheel, the right front wheel and the left front wheel, in that order, and perform the same procedure. Be sure to check the fluid in the master cylinder reservoir frequently.

14 Never use old brake fluid. It contains moisture which can boil, rendering the brakes inoperative.

15 Refill the master cylinder with fluid at the end of the operation.

16 Check the operation of the brakes. The pedal should feel solid when depressed, with no sponginess. If necessary, repeat the entire process. **Warning:** *If, after bleeding the system you do not have a firm brake pedal, or if the ABS light on the instrument panel does not go off, or if you have any doubts whatsoever about the effectiveness of the brake system, have it towed to a dealer service department or other repair shop equipped with the necessary tools for bleeding the system.*

12 Power brake booster - check, removal and installation

Operating check

1 Depress the pedal and start the engine. If the pedal goes down slightly, operation is normal.

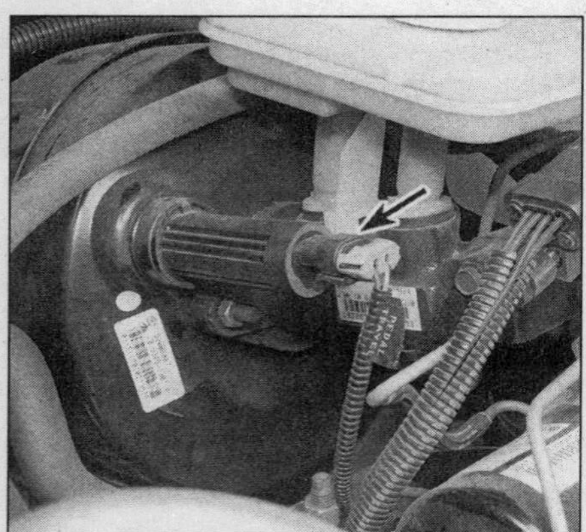

12.5 On 1993 and 1994 models, unplug the electrical connector for the pedal travel sensor

2 Depress the brake pedal several times with the engine running and make sure that there is no change in the pedal reserve distance.

Airtightness check

3 Start the engine and turn it off after one or two minutes. Depress the brake pedal several times slowly. If the pedal goes down farther the first time but gradually rises after the second or third depression, the booster is airtight.

4 Depress the brake pedal while the engine is running, then stop the engine with the pedal depressed. If there is no change in the pedal reserve travel after holding the pedal for 30 seconds, the booster is airtight.

Removal

Refer to illustrations 12.5 and 12.9

5 If you're working on a 1993 or 1994 model, unplug the electrical connector for the pedal travel sensor **(see illustration)**.

6 Unbolt the combination valve bracket (if equipped) and the brake master cylinder from the booster (see Section 8) and carefully pull it forward. Don't disconnect the brake lines, but be careful not to kink the lines. Also, it may be necessary to unbolt the ABS HCU and, on some models, the evaporative emissions control canister, to provide enough room to get the booster out.

7 Detach the vacuum hose from the booster.

8 Working inside the vehicle, remove the under-dash panel and locate the rear (brake pedal) end of the booster pushrod. This end of the pushrod is connected to a small pin on the brake pedal arm.

9 Remove the pushrod retaining clip **(see illustration)** and slip the pushrod off the pin. Unplug the brake light switch electrical connector. **Note:** *Some models have a flat washer and a wave washer behind the clip - don't lose them.*

10 Remove the four nuts holding the brake booster to the firewall. You may need a light to see these, as they are up under the dash area **(see illustration 12.9)**.

11 Slide the booster straight out from the firewall until the studs clear the holes and pull the booster, brackets and gaskets from the engine compartment area.

Installation

12 Installation is the reverse of removal. Be sure to tighten the booster mounting nuts to the torque listed in this Chapter's Specifications.

13 Brake light switch - check, replacement and adjustment

Refer to illustrations 13.1a and 13.1b

1 The brake light switch is mounted on a flange on the pedal support bracket. The adjustable switch used on 1993 models is secured in the bracket with a retainer clip **(see illustration)**; the self-adjusting switch used on 1994 and later models is secured in the bracket by means of an integral retainer on the switch body **(see illustration)**.

Check

1993 models

2 Verify that the brake lights come on when the brake pedal is depressed.

3 If the brake lights do not come on when the pedal is depressed, check the fuse and the brake light bulbs (see Chapter 12), then check the switch again.

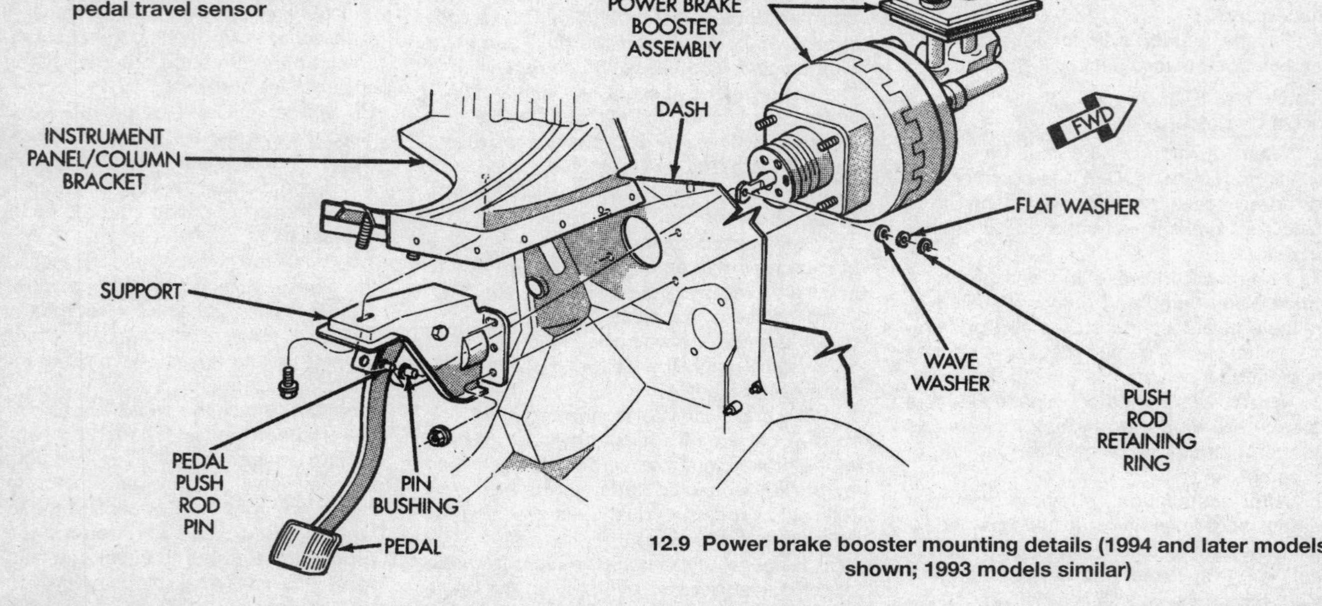

12.9 Power brake booster mounting details (1994 and later models shown; 1993 models similar)

4 If the brake lights still don't come on when the brake pedal is depressed, try adjusting the switch (see below), then check it again.

5 If the brake lights still fail to come on even after the switch has been adjusted, verify that the switch is getting voltage (see the wiring diagrams at the end of Chapter 12).

6 If the switch is getting voltage, verify that there is voltage at the terminal that feeds the wire to the brake lights when the pedal is depressed.

7 If the switch doesn't work as described, replace it. If the switch is operating properly but the brake lights are inoperative, locate the open circuit condition between the switch and the brake lights.

1994 and later models

Refer to illustration 13.9

8 First, do all the external checks described above for 1993 models (fuse, brake light bulbs, voltage to the switch, voltage between the switch and the brake lights, etc.).

9 If the switch still doesn't function properly, remove the switch, unplug the electrical connector from the switch, reinstall the switch (see below) and, usihg an ohmmeter or a continuity tester, check continuity between terminal pins 5 and 6 **(see illustration)** as follows:

a) *Pull the plunger all the way out to its fully extended position.*
b) *Attach the ohmmeter test leads to pins 5 and 6 and note the ohmmeter reading.*

10 If there is continuity between these pins, proceed to the next test. If there isn't, replace the switch.

11 Using an ohmmeter or a continuity tester, check continuity between terminals pins 1 and 2, and between pins 3 and 4, as follows:

a) *Push the switch plunger in to its fully retracted position.*
b) *Attach the ohmmeter test leads to pins 1 and 2 and note the ohmmeter reading.*
c) *If there is continuity between these pins, the switch is okay. If there isn't, replace the switch (it's open or shorted).*
d) *Repeat this test for terminal pins 3 and 4.*

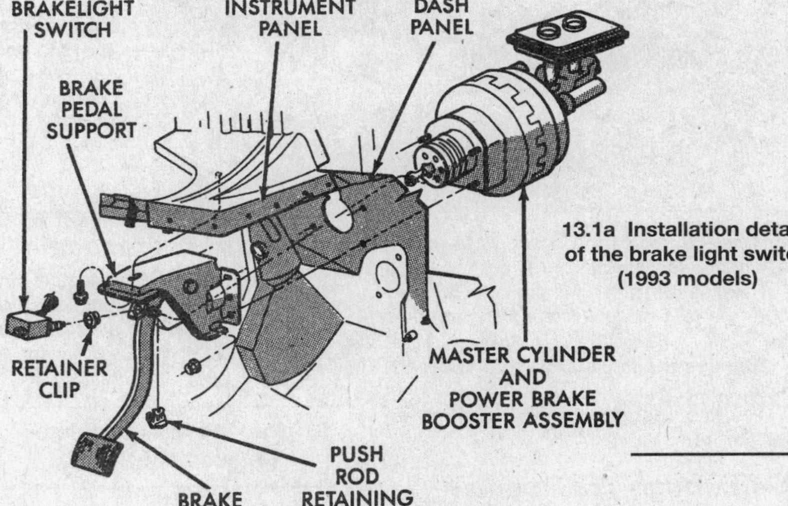

13.1a Installation details of the brake light switch (1993 models)

Replacement

12 Remove the steering column cover and the lower trim panel from the dash (see Chapter 11).

1993 models

13 Unplug the switch electrical connector.

14 Thread the switch out of the retainer, or rock it up and down and pull it to the rear, out of its retainer.

15 Thread the new switch into place until it's fully seated in the retainer and the plunger is touching the brake pedal.

16 Plug in the electrical connector.

17 Adjust the switch (see Step 29).

1994 and later models

18 Press the brake pedal down to its fully-applied position.

19 Rotate the switch about 30-degrees in a counterclockwise direction to unlock the switch retainer, then pull the switch to the rear and out of its bracket.

20 Unplug the switch electrical connector and remove the switch.

21 Pull the switch plunger all the way out to its fully extended position.

22 Plug in the electrical connector.

23 Press and hold the brake pedal in its fully-applied position.

24 Align the tab on the switch with the notch in the switch bracket, then insert the switch in the bracket and turn it clockwise about 30 degrees to lock it into place.

25 Release the brake pedal, then pull the pedal all the way to the rear. This sets the plunger to its correct position as the pedal pushes the plunger into the switch body. The switch should make a ratcheting (clicking) sound as it self-adjusts.

All models

26 Install the under-dash trim panel.

28 Depress the brake pedal and verify that the switch is operating properly.

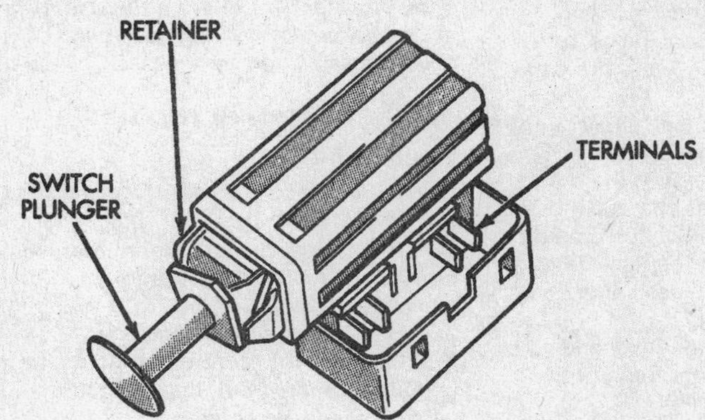

13.1b Brake light switch (1994 and later models)

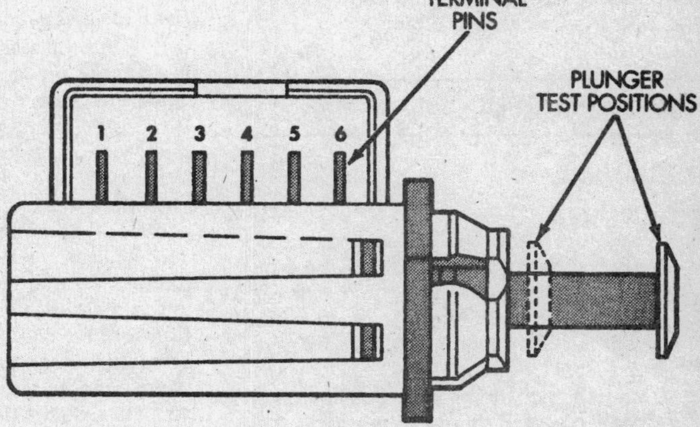

13.9 Terminal pin identification guide (1994 and later models)

1) *Terminal pins 1 and 2 - brake sensor circuit*
2) *Terminal pins 5 and 6 - brake light circuit*
3) *Terminals pins 3 and 4 - speed control circuit*

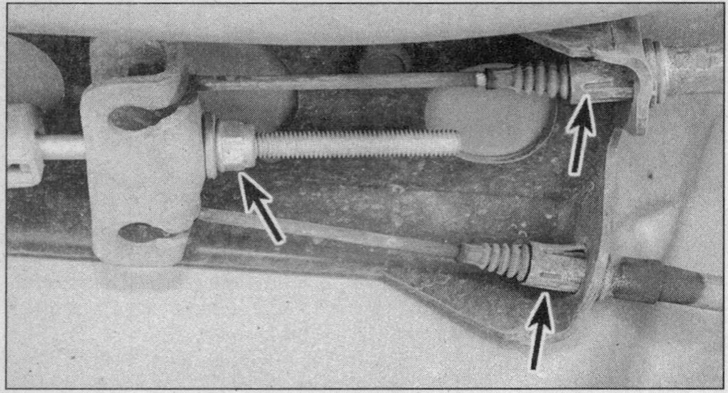

14.3 On 1993 models, mark the position of the adjusting nut (left arrow) on the threaded end of the cable tensioner, then tighten the nut about 1/2-inch farther down the tensioner; if you're replacing a rear cable, compress its retainer (right arrows) with a small worm-drive type hose clamp

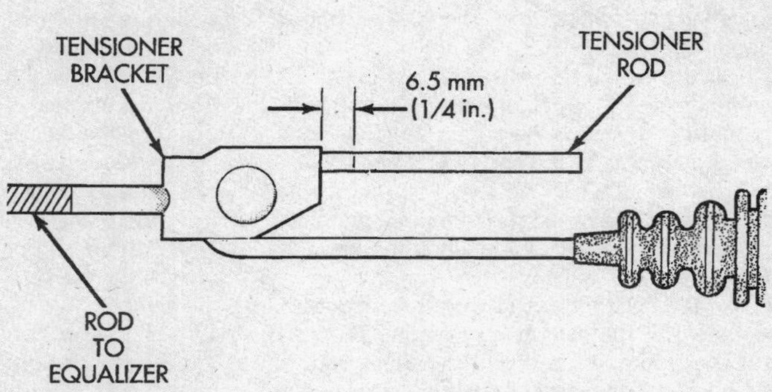

14.10 On 1994 through 1998 models, mark the tensioner rod about 1/4-inch from the tensioner bracket

Adjustment (1993 models only)

29 Move the brake pedal forward by hand and note the operation of the switch plunger (you'll need a flashlight and a small mechanic's mirror for this operation). The plunger should still be fully extended at the point at which all pedal freeplay is taken up and brake application begins. The clearance between the plunger and the pedal at this point should be about 1/16 to 1/8-inch.

30 If the switch plunger-to-pedal clearance is okay, and the brake lights are operating properly, no adjustment is required.

31 If the switch plunger is not fully extended, i.e. clearance between the plunger and the brake pedal is insufficient, adjust the switch position as follows:

- a) *Grasp the brake pedal and pull it to the rear as far as possible.*
- b) *The pedal should contact the switch barrel, push it to the rear in its retaining clip and stop at the correct position.*
- c) *Check the plunger-to-pedal clearance again and make sure it's now correct. If not, repeat this procedure.*

32 Depress the brake pedal and verify that the switch is operating properly.

14 Parking brake - adjustment

1993 models

Refer to illustration 14.3

1 Raise the rear of the vehicle and support it securely on jackstands. Block the front wheels to prevent the vehicle from rolling.

2 Verify that the drums or discs rotate freely without drag, then fully apply the parking brake lever.

3 Mark the position of the adjusting nut on the threaded end of the cable tensioner **(see illustration)**.

4 Tighten the adjusting nut about 1/2-inch farther down the threaded end of the cable tensioner. If there aren't enough threads left for proper adjustment, replace the cable tensioner (see Section 15). Do not try to modify or re-use the tensioner.

15.4a To detach the front cable from its bracket, compress the retainer tangs (arrow) (1998 and earlier models)

5 Release the parking brake lever and verify that the rear wheels rotate freely, without drag.

6 Remove the jackstands and lower the vehicle.

7 Test the operation of the parking brake on an incline.

1994 through 1998 models

Refer to illustration 14.10

8 Raise the rear of the vehicle and support it securely on jackstands. Block the front wheels to prevent the vehicle from rolling.

9 Verify that the drums or discs rotate freely without drag, then fully apply the parking brake lever.

10 Mark the tensioner rod **(see illustration)** about 1/4-inch from the tensioner bracket.

11 Tighten the adjusting nut at the equalizer until the mark on the tensioner rod moves into alignment with the tensioner bracket. **Caution:** *Do NOT loosen or tighten the tensioner adjusting nut for any reason after completing this adjustment.*

12 Release the parking brake lever and verify that the rear wheels rotate freely without any drag.

13 Remove the jackstands and lower the vehicle.

14 Test the operation of the parking brake on an incline.

1999 and later models

15 Cable tension is adjusted automatically on these models. If the parking brake fails to hold the vehicle on an incline, adjust the parking brake shoes (see Section 16, Step 7).

15 Parking brake cables - replacement

Warning: *These models are equipped with airbags. Always disable the airbag system before working in the vicinity of any airbag system component to avoid the possibility of accidental deployment of the airbag(s), which could cause personal injury* (see Chapter 12).

1998 and earlier models

Front cable

Refer to illustrations 15.4a, 15.4b, 15.5 and 15.6

1 Disconnect the cable from the negative terminal of the battery. Remove the center console (see Chapter 11).

2 Release the parking brake.

3 Remove the front cable adjusting nut **(see illustration 14.3)**, then detach the front cable from the equalizer.

4 Disengage the front cable retainer **(see illustration)**, then remove the insert from the floor pan **(see illustration)**.

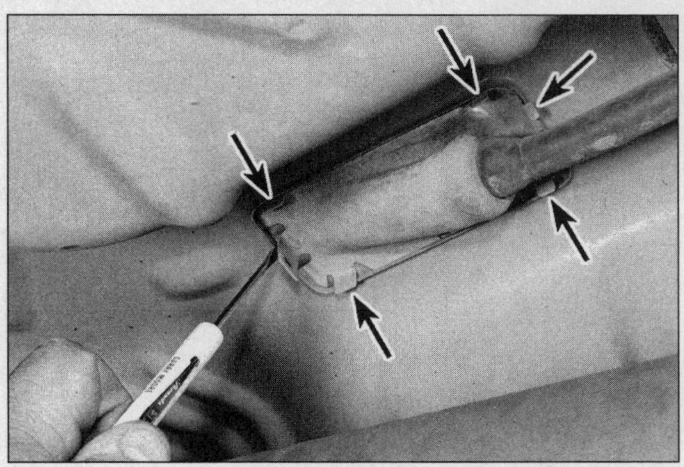

15.4b To remove the cable insert from the floor pan, pop these retainer tabs (arrows) loose with a small screwdriver (1998 and earlier models)

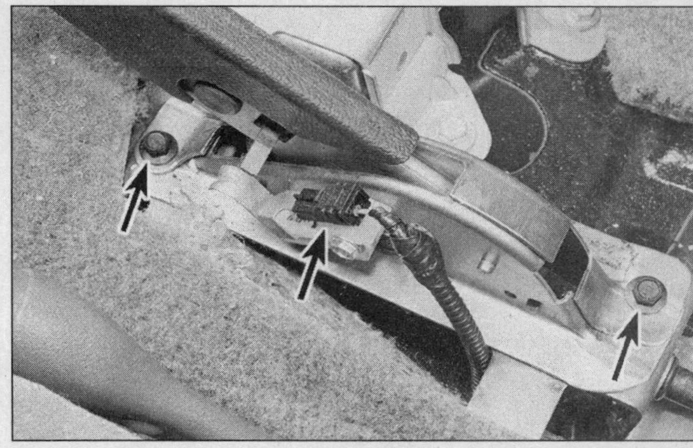

15.5 To remove the parking brake lever, unplug the parking brake switch electrical connector (center arrow) and remove the parking brake lever retaining bolts (left and right arrows) (1998 and earlier models)

15.6 To remove the airbag control module, remove the four mounting bolts (arrows), lift the module up and unplug the two electrical connectors (not visible in this photo) (1998 and earlier models)

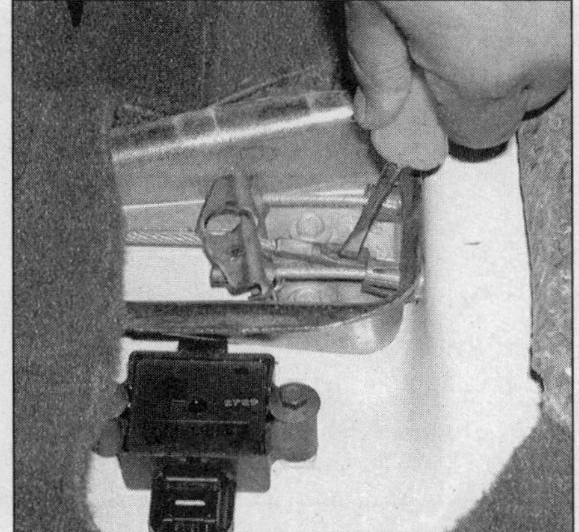

15.22a Using a screwdriver inserted through the cable eyelet, pry the front cable back . . .

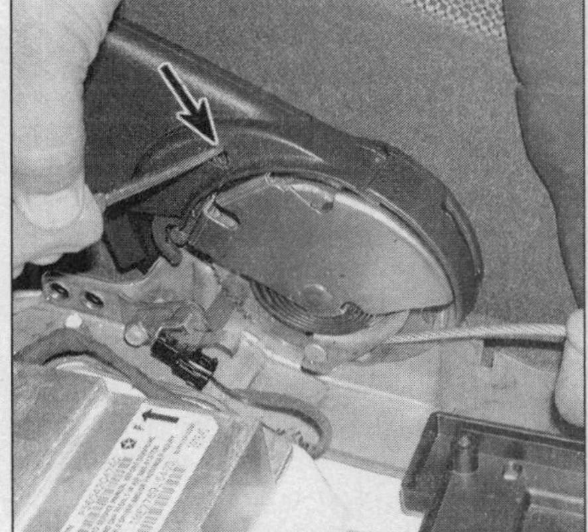

15.22b . . . then, using another screwdriver, pry the lockout spring down (arrow); this will lock the tensioner mechanism and provide slack in the cables

5 From inside the vehicle, unplug the electrical connectors from the parking brake switch **(see illustration)** and the airbag control module (there are two connectors at the front of the module).

6 Remove the airbag control module mounting bolts **(see illustration)**, remove the module and set it aside for access to the parking brake lever.

7 Remove the parking brake lever mounting bolts, remove the lever assembly and flip it upside down for access to the parking brake cable.

8 Disconnect the parking brake cable from the lever and pull it through the floor pan.

9 Slide the floor pan insert off the old cable and slide it onto the new cable.

10 Installation is the reverse of removal.

11 Be sure to adjust the parking brake cable (see Section 14) when you're done.

Rear cable

12 Raise the vehicle and place it securely on jackstands. Block the front wheels to prevent the vehicle from rolling. On models with rear drum brakes, remove the rear wheels.

13 Disengage the cable from the equalizer and from the bracket just behind the equalizer **(see illustration 14.3)**.

14 Disengage the cable from all body and chassis retainer clips.

15 On models with rear drum brakes, remove the brake drums, remove the secondary brake shoe (see Section 6) and disconnect the cable from the parking brake lever. Then compress the cable retainer tangs and remove the cable from the brake backing plate.

16 On models with rear disc brakes, disengage the cable eyelet from the actuating lever, then compress the retainer which secures the cable to the bracket attached to the caliper.

17 Installation is the reverse of the removal procedure.

18 Be sure to adjust the parking brake as described in Section 14.

1999 and later models

Front cable

Refer to illustrations 15.22a, 15.22b, 15.23, 15.24, 15.25 and 15.26

19 Disconnect the cable from the negative terminal of the battery. **Caution:** *If the stereo in your vehicle is equipped with an anti-theft system, make sure you have the correct activation code before disconnecting the battery.* Remove the center console (see Chapter 11).

20 Release the parking brake.

21 Remove the rear seat cushion and the front seats (see Chapter 11). Remove the sill trim plates and peel up the carpet for access to the parking brake cables.

22 Pry the front cable back, using a screwdriver inserted through the eyelet **(see illustration)**. Hold the cable in this position, then have a helper insert a narrow screwdriver through the lockout spring on the side of the parking brake lever and pry it down **(see illustration)**. Release the front cable slowly and verify that there is slack in the cable (if

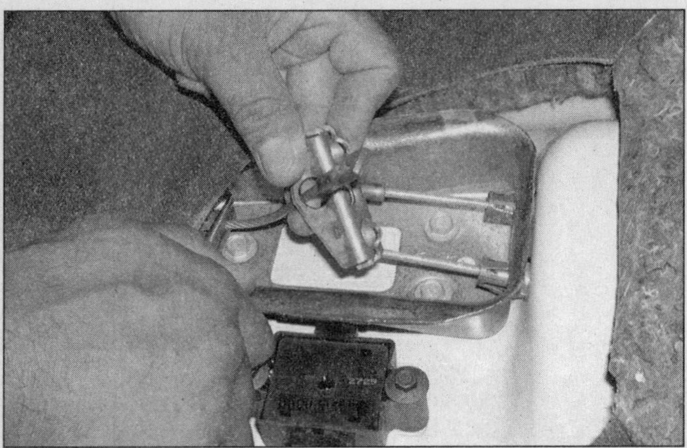

15.23 Detach the front cable from the equalizer by passing it through the slot

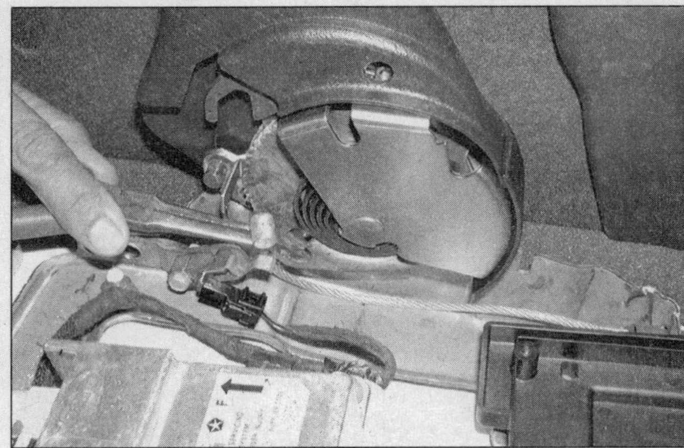

15.24 Unhook the front end of the cable from the tensioner on the parking brake lever

there isn't, the lockout spring was not properly engaged).

23 Pass the front cable through the slot in the equalizer **(see illustration).**

24 Detach the end of the cable from the adjuster on the parking brake lever **(see illustration).**

25 Unscrew the nuts securing the cable brackets to the floor pan **(see illustration).**

26 Depress the tangs on the cable retainers and detach the ends of the cable from the brackets **(see illustration).** Remove the cable.

27 Installation is the reverse of removal. Operate the parking brake lever several times to release the lockout spring and adjust the cables.

Rear cables

Refer to illustrations 15.30, 15.31, 15.34 and 15.35

28 Remove the center console and the rear seat cushion (see Chapter 11).

29 Perform Step 22.

30 Pass the cable through the slot in the equalizer **(see illustration).**

31 Depress the tangs on the cable retainer and detach the cable from the bracket **(see illustration).**

32 Loosen the rear wheel lug nuts, raise the rear of the vehicle and support it securely on jackstands. Block the front wheels to prevent the vehicle from rolling. Remove the rear wheels.

33 Remove the rear brake caliper, mounting bracket and disc (see Sections 4 and 5). Don't disconnect the brake hose from the caliper. Also, don't let the caliper hang by the brake hose - hang it with a piece of wire from the coil spring.

34 Unclip the ABS sensor wiring harness from the parking brake cable, detach the cable

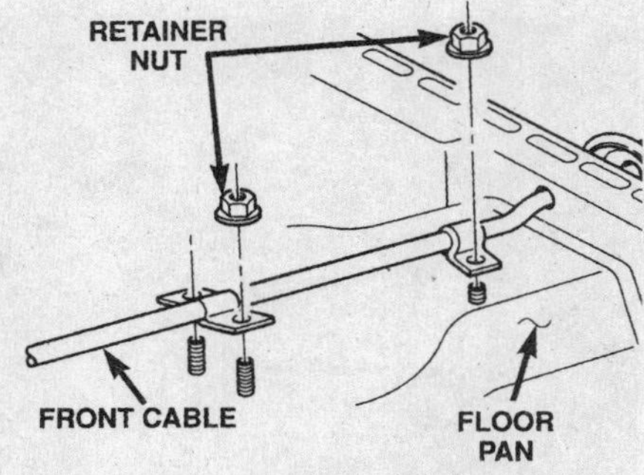

15.25 Remove the nuts securing the cable to the floorpan

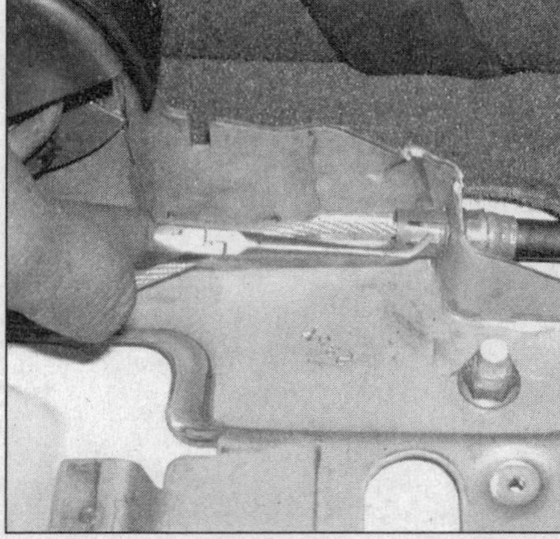

15.26 Depress the tangs on the retainer and pass the cable through the bracket

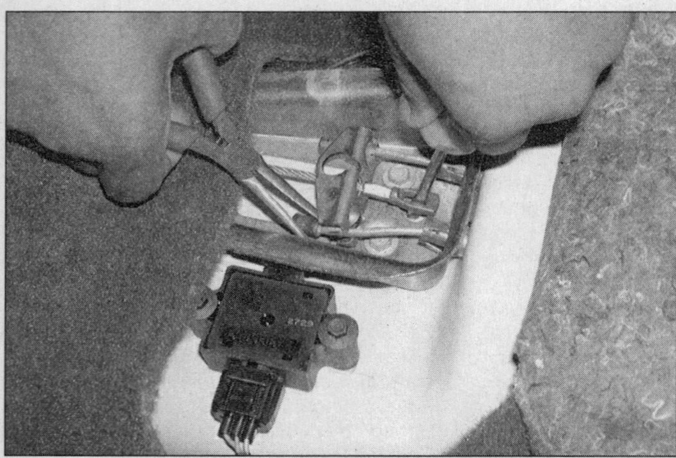

15.30 Separate the cable from the equalizer

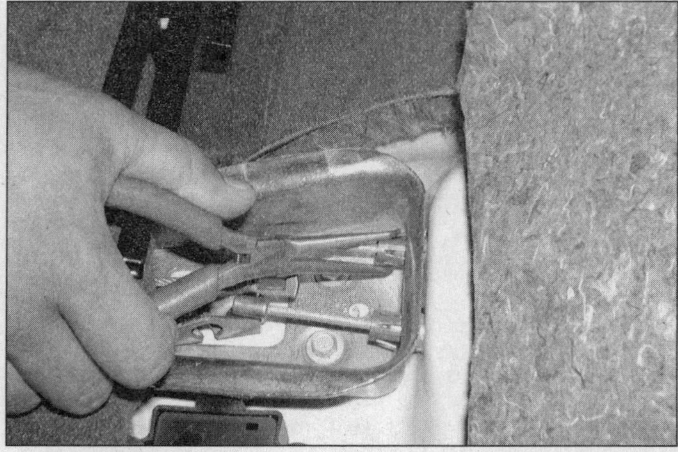

15.31 Depress the tangs on the retainer and pass the cable through the equalizer bracket

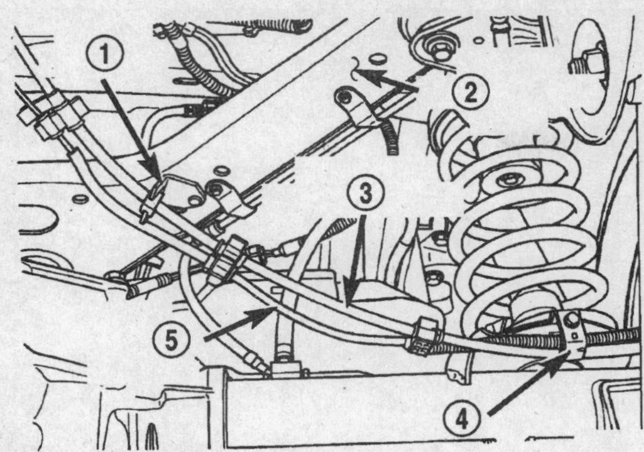

15.34 Rear parking brake cable details (left side shown)

1 Cable bracket
2 Upper control arm
3 Parking brake cable
4 Cable retainer
5 ABS sensor wiring harness

15.35 Using a screwdriver, push the cable in, lift it up and detach it from the parking brake actuator

from the bracket on the suspension upper control arm, then unbolt the cable retainer from the axle housing **(see illustration)**.

35 Detach the cable from the parking brake actuator **(see illustration)** and remove the cable.

35 Installation is the reverse of removal. Be sure to tighten the wheel lug nuts to the torque listed in the Chapter 1 Specifications. Operate the parking brake lever several times to release the lockout spring and adjust the cables.

16 Parking brake shoes (rear disc brakes) - replacement

Refer to illustrations 16.3a, 16.3b and 16.4a through 16.4o

Warning: *Dust created by the brake system is harmful to your health. Never blow it out with compressed air and don't inhale any of it. An approved filtering mask should be worn when working on the brakes. Do not, under any circumstances, use petroleum-based solvents to clean brake parts. Use brake system cleaner only!*

1 Loosen the wheel lug nuts, release the parking brake, raise the rear of the vehicle and support it securely on jackstands. Block the front wheels to keep the vehicle from rolling. Remove the rear wheels.

2 Remove the rear brake caliper (don't disconnect the hose) and hang it with a length of wire. On 1999 and later models, also remove the caliper mounting bracket (see Section 5). Also on 1999 and later models, refer to **illustrations 15.22a and 15.22b** to provide some slack in the parking brake cables.

3 Cut off the pressed-metal retainers, if equipped **(see illustration 5.6)**, and remove the brake disc. **Note:** *If the brake disc cannot be easily pulled off the axle and shoe assembly, make sure that the parking brake is completely released, then apply some penetrating oil at the hub-to-disc joint. Allow the oil to soak in and try to pull the disc off. If the disc still cannot be pulled off, the parking brake shoes will have to be retracted. This is accomplished by first removing the plug from the backing plate. With the plug removed, turn the adjusting wheel with a narrow screwdriver or brake adjusting tool, moving the shoes away from the braking surface* **(see illustrations)**. *The disc should now come off.*

4 Clean the parking brake shoe assembly with brake system cleaner, then follow the accompanying illustrations **(16.4a through 6.4o)** for the parking brake shoe replacement procedure. Be sure to stay in order and read

16.3a Remove the plug from the hole in the brake backing plate, insert a brake adjuster tool through the hole . . .

16.3b . . . position the tool at the top of the adjuster star wheel and rotate the wheel down (clockwise, as viewed from the rear of the vehicle, facing forward), moving the parking brake shoes away from the disc hub (which is already removed in this photo for the sake of clarity

16.4a Pushing on the pin from the backing plate side with your finger, pry the front hold-down clip loose, then pull the pin out

9

16.4b Remove the rear hold-down clip and pin the same way

16.4c Disengage the lower spring from the parking brake shoes

the caption under each illustration. **Note:** *All four parking brake shoes must be replaced at the same time, but to avoid mixing up parts, work on only one brake assembly at a time.*

5 Before reinstalling the disc, check the parking brake surfaces of the disc hub for cracks, score marks, deep scratches and hard spots, which will appear as small discolored areas. If hard spots or any of the other conditions listed above cannot be removed with sandpaper or emery cloth, the disc must be replaced.

6 Once all of the new parking brake shoes are in place, install the discs.

7 Remove the rubber plugs from the brake backing plates, insert a narrow screwdriver or brake adjusting tool through the adjustment hole and turn the star wheel until the shoes drag slightly as the disc is turned **(see illustrations 16.3a and 16.b)**. Turn the star wheel in the opposite direction until the disc turns freely. Install the backing plate plugs.

8 Mount the wheel, install the lug nuts and lower the vehicle. Tighten the lug nuts to the torque listed in the Chapter 1 Specifications.

16.4d Remove the adjuster

16.4e Disengage the upper spring from the rear parking brake shoe

9 If you're working on a 1999 or later model, operate the parking brake lever several times to release the lockout spring and adjust the cables. If you're working on a 1998 or earlier model, adjust the parking brake cable as described in Section 14.

10 Check brake operation before driving the vehicle in traffic.

16.4f Remove both parking brake shoes

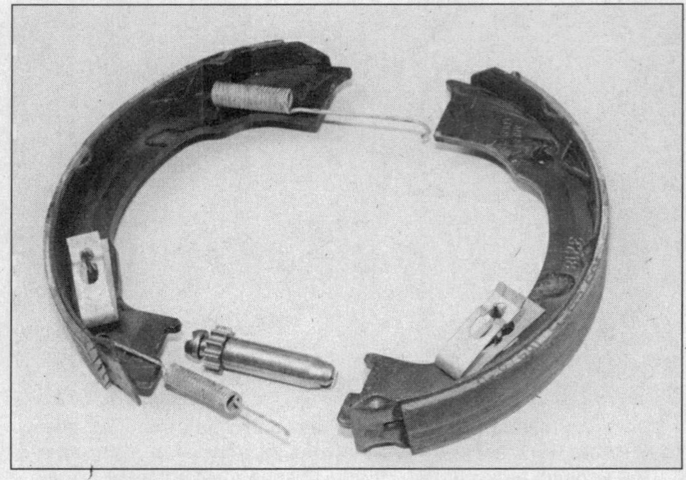

16.4g Here's how the shoes, springs and adjuster go together

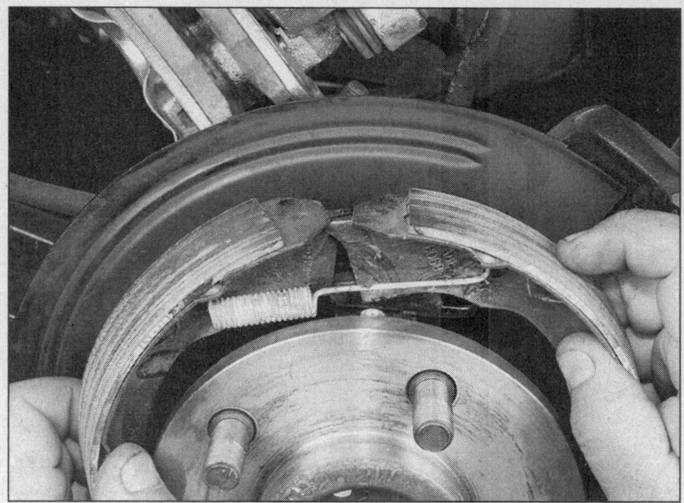

16.4h Holding them together with the upper spring, install the new shoes

16.4i Spread the upper ends of the shoes apart and engage them with the anchor as shown

16.4j Install the adjuster

16.4k Make sure the adjuster is engaged with the shoes as shown

16.4l Install the lower spring

16.4m Make sure the coils of the lower spring are facing down, away from the adjuster, as shown

16.4n Install the rear pin and hold-down clip as shown, with the head of the pin firmly seated into the lower, smaller part of the hole in the clip

16.4o Install the front pin and hold-down clip the same way

Notes

Chapter 10
Suspension and steering systems

Contents

Specifications

Torque Specifications

	Ft-lbs (unless otherwise indicated)
Front suspension	
Shock absorber	
Upper nut	
1993	168 in-lbs
1994 through 1998	17
1999 and later	26
Lower nuts	
1993 and 1994	168 in-lbs
1995 and later	20
Stabilizer bar	
Bushing clamp bolts	
1993 through 1998	40
1999 and later	68
Stabilizer bar-to-link nut	
1993 through 1995	27
1996 through 1998	45
1999 and later	78
Stabilizer link-to-axle bolt/nut	
1993 through 1998	70
1999 and later	78
Upper suspension arm-to-frame bolts/nuts	
1993 through 1999	55
2000	45
Upper suspension arm-to-axle bolts/nuts	
1993 through 1999	55
2000	45
Lower suspension arm-to-frame bolts/nuts	
1993 and 1994	130
1995 and 1996	85
1997 through 1999	130
2000	115
Lower suspension arm-to-axle bolts/nuts	
1993 and 1994	130
1995 through 1998	85
1999	130
2000	120
Coil spring retainer bolt (1998 and earlier)	16
Track bar ballstud-to-frame bracket nut (1998 and earlier)	60
Track bar-to-frame bracket bolt/nut (1999 and later)	80

Torque Specifications

	Ft-lbs (unless otherwise indicated)
Front suspension (continued)	
Track bar-to-axle housing bolt/nut	
1993	74
1994 through 1998	55
1999 and later	74
Balljoint nuts	
Upper	75
Lower	
1993 and 1994	100
1995 and later	80
Rear suspension	
Shock absorber	
Upper nut	
1993 through 1998	52
1999	68
2000	80
Lower nut	
1993	44
1994 through 1999	68
2000	85
Stabilizer bar	
Bushing clamp bolts	40
Stabilizer bar-to-link nut	
1993 through 1998	27
1999 and later	40
Stabilizer link-to-bracket nut	
1993 through 1998	27
1999 and later	68
Upper suspension arm-to-frame bolts/nuts	
1993 through 1998	55
1999 and later	74
Upper suspension arm-to-axle bolts/nuts (1998 and earlier models)	55
Upper suspension arm ballstud nut (1999 and later models)	
1999	90
2000	105
Upper suspension arm balljoint-to-axle housing bolts (1999 and later)	100
Lower suspension arm-to-frame bolt/nut	
1993 through 1999	130
2000	115
Lower suspension arm-to-axle bolt/nut	
1993 through 1999	130
2000	120
Track bar (1998 and earlier models)	
Track bar-to-frame bracket bolt/nut	74
Track bar-to-axle tube bracket bolt/nut	74
Coil spring retainer bolt (1998 and earlier models)	16
Steering	
Airbag module mounting nuts (1998 and earlier)	90 in-lbs
Airbag module mounting screws (1999 and later)	90 in-lbs
Steering wheel-to-steering shaft nut	45
Steering gear mounting bolts	
1993	100
1994 through 1998	65
1999 and later	80
Pitman arm-to-steering gear nut	185
Pitman arm-to-drag link nut	
1993 through 1995	55
1996 through 1998	60
1999 and later	65
Drag link-to-steering knuckle nut	35
Tie-rod-to-drag link nut (1998 and earlier)	
1993 and 1994	55
1995 through 1997	35
1998	55

Torque Specifications

	Ft-lbs (unless otherwise indicated)
Tie-rod-to-steering knuckle nut	35
Tie-rod adjuster clamp nuts	
1993	25
1994 and 1995	20
1996 through 1998	
Six-cylinder engine	20
V8 engine	36
1999	50
2000	30
Drag link adjuster clamp nuts	
1993 through 1995	36
1996	
Six-cylinder engine	20
V8 engine	36
1997 and 1998	36
1999	50
2000	30
Steering damper-to-linkage nut	
1993	35
1994 through 1998	55
1999	50
2000	30
Steering damper-to-axle bracket bolt	
1993 through 1998	55
1999 and 2000	65
Wheel lug nuts	See Chapter 1

1.1a Front suspension and steering components (1998 and earlier 2WD models)

1 *Stabilizer bar*
2 *Coil spring*
3 *Pitman arm*
4 *Adjuster tube*
5 *Drag link*
6 *Tie-rod*
7 *Track bar*
8 *Tube axle*
9 *Lower suspension arm (upper arms not visible in this photo)*
10 *Axle yoke*
11 *Left tie-rod end*
12 *Lower yoke-to-steering knuckle balljoint (upper balljoints not visible in this photo)*

1.1b Front suspension and steering components (1998 and earlier 4WD models)

1 *Stabilizer bar*
2 *Stabilizer bar link*
3 *Coil spring*
4 *Adjuster tube*
5 *Drag link*
6 *Steering damper*
7 *Track bar*
8 *Lower suspension arm (upper arms not visible in this photo)*
9 *Differential*
10 *Axle tube*
11 *Axle tube yoke*
12 *Lower yoke-to-steering knuckle balljoint (upper balljoints not visible in this photo)*
13 *Axleshaft CV joint*
14 *Left tie-rod end*
15 *Adjuster tube*
16 *Tie-rod*

1 General information

Front suspension

Refer to illustrations 1.1a, 1.1b and 1.1c

All of the vehicles covered by this manual utilize a solid front axle. A simple tube axle is used on 2WD models; the axle on 4WD models consists of a cast-iron or aluminum housing for the differential with a pair of steel tubes for the axleshafts welded or pressed to either side (for more information on the innards of the front axle, see Chapter 8). The front axle is suspended by a pair of shock absorbers and a two coil springs. Four control arms allow the axle to move vertically; lateral movement is prevented by a track bar. A stabilizer bar controls body roll. Each steering knuckle is positioned by a pair of balljoints pressed into the upper and lower ends of a yoke welded to the end of the axle **(see illustrations)**.

Rear suspension

Refer to illustrations 1.2a and 1.2b

The rear axle consists of a cast-iron or aluminum center housing for the differential with a pair of steel tubes for the axleshafts welded or pressed to either side (for more information on the innards of the rear axle, see Chapter 8). The rear axle is suspended by two shock absorbers and two coil springs. Four (1998 and earlier) or three (1999 on) control arms allow the axle to move vertically; lateral movement is prevented by a track bar on 1998 and earlier models; on 1999 and later models, the upper control arm performs this function. A stabilizer bar controls body roll **(see illustrations)**.

Steering

The power steering system uses a recirculating-ball type steering gearbox which transmits turning force through the steering linkage (Pitman arm, drag link and tie-rod) to the steering knuckle arms. A steering damper between the axle and the drag link (1998 and earlier models) or between the axle and tie-rod (1999 and later models) reduces unwanted “bump steer” (the slight turning or steering of a wheel away from its normal direction of travel as it moves through its suspension travel). A small U-joint connects the steering column to the steering gearbox. The steering column is designed to collapse in the event of an accident. **Warning:** *The following recall bulletin was issued by the National Highway Traffic Safety Administration (NHTSA) regarding a critical component in the steering column of 115,000 Jeep Grand*

1.1c Front suspension and steering components (1999 and later models - 2WD shown)

1. *Stabilizer bar bushing clamp*
2. *Stabilizer bar*
3. *Pitman arm*
4. *Drag link*
5. *Tie-rod*
6. *Track bar*
7. *Steering damper*
8. *Coil spring*
9. *Tie-rod end*
10. *Steering knuckle*
11. *Lower balljoint*
12. *Lower control arm*
13. *Upper control arm*
14. *Axle tube*

1.2a Rear suspension components (1998 and earlier models)

1. *Shock absorber*
2. *Coil spring*
3. *Track bar*
4. *Axle tube*
5. *Differential*
6. *Stabilizer bar*
7. *Lower suspension arm (upper arms not visible in this photo)*

1.2b Rear suspension components (1999 and later models)

1 Shock absorber
2 Stabilizer bar link
3 Coil spring
4 Stabilizer bar
5 Stabilizer bar bushing clamp
6 Lower suspension arm
7 Upper suspension arm
8 Rear axle assembly

Cherokees manufactured between December 1992 and July 1993: "The molded plastic pin which connects the upper steering column shaft and the lower column shaft may have been sheared prior to installation of the steering column assembly into the vehicle. If this occurred, the upper steering shaft may separate from the lower steering shaft." Consequently, "if the upper and lower steering shafts separate, complete loss of steering control will occur, which could result in an accident." If your vehicle is affected, take it to a Chrysler/Jeep dealer and have it repaired. A restraining sleeve will be added to the intermediate shaft to limit the amount of steering shaft movement available. If you have problems obtaining the needed repair, contact the Chrysler zone office in your area. It's listed under "Service Assistance" in your owner's manual. A zone representative will arrange for prompt attention to your vehicle. Or, you can contact the NHTSA's Auto Safety Hotline at 1-800-424-9393. Another 800 number, (800) 424-9153, is available for the hearing impaired.

Frequently, when working on the suspension or steering system components, you may come across fasteners which seem impossible to loosen. These fasteners on the underside of the vehicle are continually subjected to water, road grime, mud, etc., and can become rusted or "frozen," making them extremely difficult to remove. In order to unscrew these stubborn fasteners without damaging them (or other components), be sure to use lots of penetrating oil and allow it to soak in for a while. Using a wire brush to clean exposed threads will also ease removal of the nut or bolt and prevent damage to the threads. Sometimes a sharp blow with a hammer and punch is effective in breaking the bond between a nut and bolt threads, but care must be taken to prevent the punch from slipping off the fastener and ruining the threads. Heating the stuck fastener and surrounding area with a torch sometimes helps too, but isn't recommended because of the obvious dangers associated with fire. Long breaker bars and extension, or "cheater," pipes will increase leverage, but never use an extension pipe on a ratchet - the ratcheting mechanism could be damaged. Sometimes, turning the nut or bolt in the tightening (clockwise) direction first will help to break it loose. Fasteners that require drastic measures to unscrew should always be replaced with new ones.

Since most of the procedures that are dealt with in this chapter involve jacking up the vehicle and working underneath it, a good pair of jackstands will be needed. A hydraulic floor jack is the preferred type of jack to lift the vehicle, and it can also be used to support certain components during various operation. **Warning:** *Never, under any circumstances, rely on a jack to support the vehicle while working on it.*

Whenever any of the suspension or steering fasteners are loosened or removed they must be inspected and, if necessary, be replaced with new ones of the same part number or of original equipment quality and design. Torque specifications must be followed for proper reassembly and component retention. Never attempt to heat or straighten any suspension or steering components. Instead, replace any bent or damaged part with a new one.

2 Stabilizer bar and bushings (front) - removal and installation

Removal

Refer to illustrations 2.2a, 2.2b, 2.2c and 2.3

1 Apply the parking brake. Raise the front of the vehicle and support it securely on jackstands.

2.2a To disconnect the stabilizer bar link from the bar on 1998 and earlier models, remove the upper link nut (arrow); it isn't necessary to remove the lower link nut and bolt (arrows) unless you're planning to replace the link itself

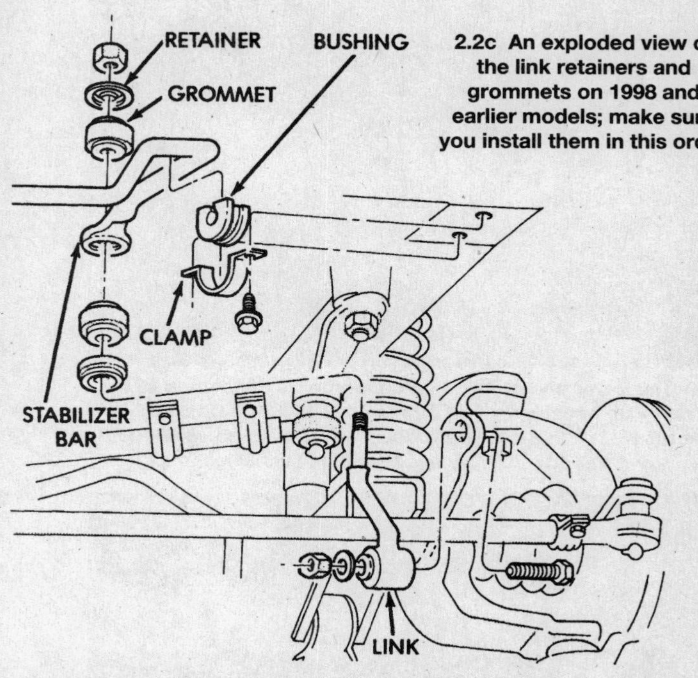

2.2c An exploded view of the link retainers and grommets on 1998 and earlier models; make sure you install them in this order

2.2b On 1999 and later models the stabilizer bar link is retained by a nut and bolt at each end

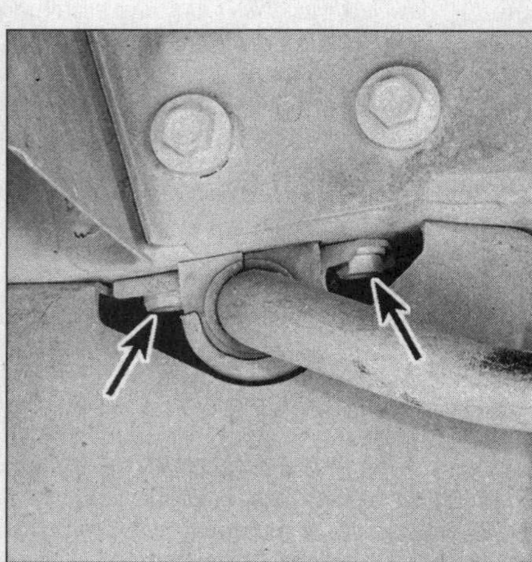

2.3 To disconnect the stabilizer bar from the frame, remove the bushing clamp bolts (arrows); inspect the rubber bushings and replace them if they're hard, cracked or otherwise deformed

2 Detach the upper ends of the links from the stabilizer bar **(see illustrations)**. On 1998 and earlier models, note the order in which the retainers and grommets are installed on the links **(see illustration)**. If it is necessary to remove the links, simply unbolt them from the axle brackets.

3 Remove the retaining bolts from the stabilizer bar bushing clamps **(see illustration)** and detach the bar from the vehicle.

4 Pull the bushings off the stabilizer bar and inspect them for cracks, hardness and other signs of deterioration. If the bushings are damaged, replace them. Inspect the bushings in the lower ends of the links, replacing them if necessary.

3.2 When removing the shock absorber stem nut (arrow), it may be necessary to hold the stem with an open end wrench or locking pliers to prevent it from turning

Installation

5 Position the stabilizer bar bushings on the bar. Don't use any grease or oil on the bushings.

6 Push the brackets over the bushings and raise the bar up to the frame. Install the bracket bolts but don't tighten them completely at this time.

7 Connect the link to the stabilizer bar and the axle. On 1998 and earlier models, be sure to assemble the grommets and retainers in the proper order **(see illustration 2.2c)**. Tighten the fasteners to the torque listed in this Chapter's Specifications.

8 Tighten the bracket bolts to the torque listed in this Chapter's Specifications.

3 Shock absorber (front) - removal and installation

Removal

Refer to illustrations 3.2 and 3.3

1 Apply the parking brake. Loosen the wheel lug nuts, raise the vehicle and support it securely on jackstands, then remove the wheel. Support the axle with a floor jack placed under the axle tube nearest the shock absorber to be changed.

2 Remove the shock absorber upper stem nut **(see illustration)**. Use an open end wrench to keep the stem from turning. If the nut won't loosen because of rust, squirt some penetrating oil on the stem threads and allow it to soak in for awhile. It may be necessary to keep the stem from turning with a pair of locking pliers, since the flats provided for a wrench are quite small.

3.3 The lower end of the shock absorber is connected to the front axle housing by two bolts and nuts (not visible in this photo). Note: *1999 and later models only have nuts - the bottom of the shock absorber has two studs connected to it*

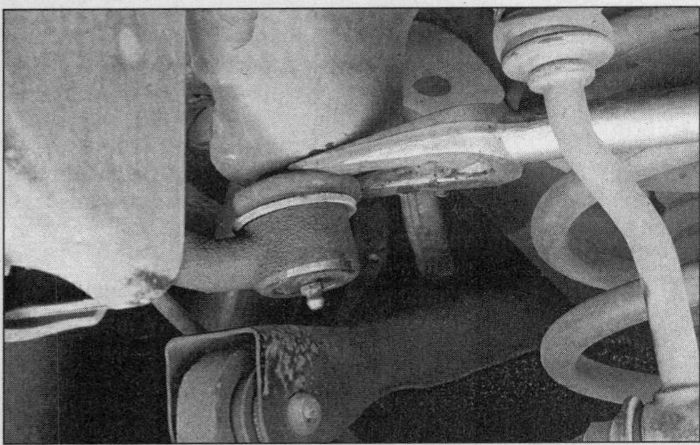

4.2b If the ballstud at the upper end of the track bar is difficult to separate from the frame bracket, use a picklefork to separate it (1998 and earlier models)

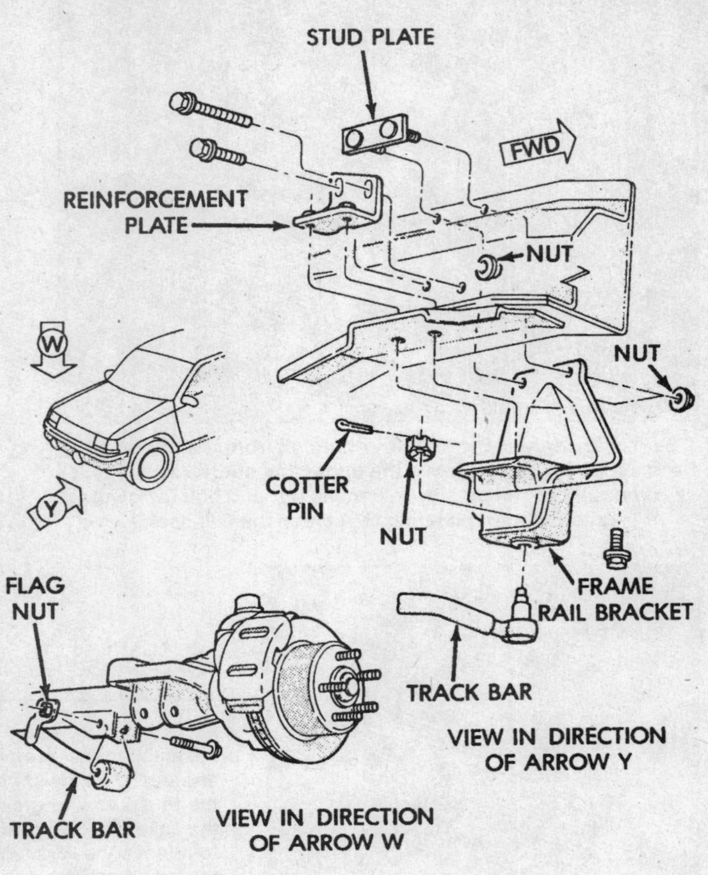

4.2a Track bar mounting details - 1998 and earlier models

3 Remove the two lower mounting nuts and, on 1998 and earlier models, bolts **(see illustration)** (the shock absorbers on 1999 and later models have two studs on the bottom). Remove the shock absorber out from the wheel well, then remove the washers and rubber grommets from the top of the shock absorber.

4.3 On 1999 and later models the upper end of the track bar is attached to the frame bracket with a bolt and nut

Installation

4 Extend the new shock absorber as far as possible. Position a new washer and rubber grommet on the stem and guide the shock up into the upper mount.

5 Install the upper rubber grommet and washer and wiggle the stem back-and-forth to ensure the grommets are centered in the mount. Tighten the stem nut to the torque listed in this Chapter's Specifications.

6 Connect the lower end of the shock to the axle, then install the lower mounting bolts (1998 and earlier models only) and nuts, tightening them to the torque listed in this Chapter's Specifications.

4 Track bar (front) - removal and installation

Refer to illustrations 4.2a, 4.2b, 4.3 and 4.4

1 Raise the front of the vehicle and support it securely on jackstands.

2 If you're working on a 1998 or earlier model, remove the cotter pin and retaining nut from the ballstud at the frame rail bracket **(see illustration)**. Discard the cotter pin. The ballstud should fall out of the frame bracket when the nut is removed. If it doesn't, use a picklefork type balljoint separator to separate the stud from the frame bracket **(see illustration)**. Grease the tool and the rubber boot to minimize damage to the boot.

3 If you're working on a 1999 or later model, remove the nut and bolt securing the upper end (driver's side) of the track bar to the frame bracket **(see illustration)**.

4 Remove the bolt and nut connecting the track bar to the bracket at the front right (passenger's side) end of the axle housing **(see illustration)** and remove the bar.

5 Installation is the reverse of removal. Before tightening the fasteners to the torque listed in this Chapter's Specifications, lower the vehicle (or support it with jackstands placed under the axle tubes); this is to prevent the bushings from becoming distorted, which can cause them to fail prematurely.

4.4 Remove the bolt (arrow) and nut (not visible in this photo) and separate the lower end of the track bar from the axle housing

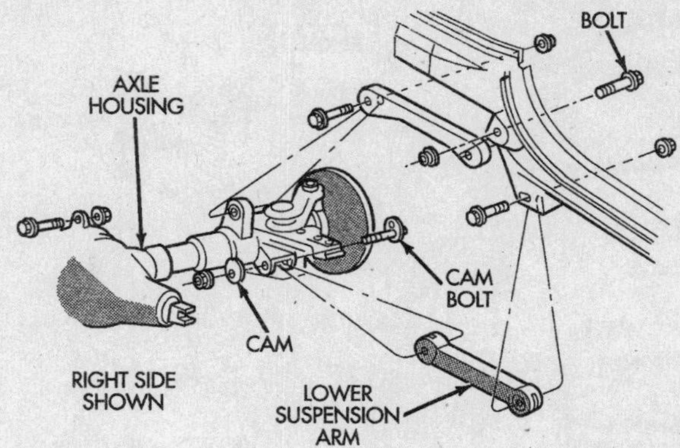

5.2a An exploded view of the right front suspension arms (4WD model shown; 2WD models similar)

5.2b To disconnect the upper suspension arm from the axle housing, remove the nut (not visible in this photo) and bolt (arrow), then slide the bolt out of the bracket

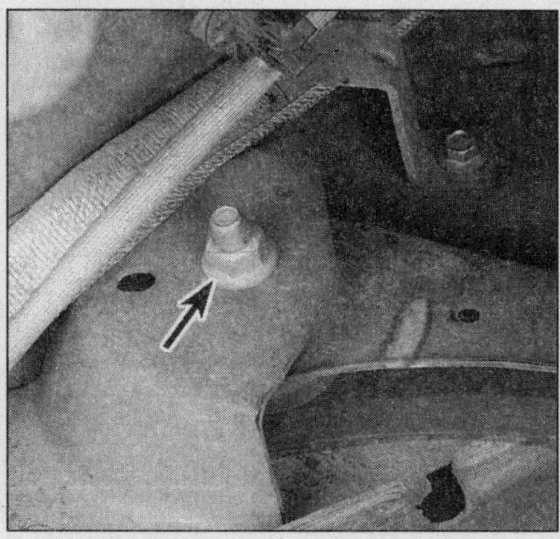

5.3 To disconnect the upper suspension arm from the frame, remove this nut (arrow) and slide out the bolt (the bolt head is accessed through a hole in the frame that faces out, into the wheel well)

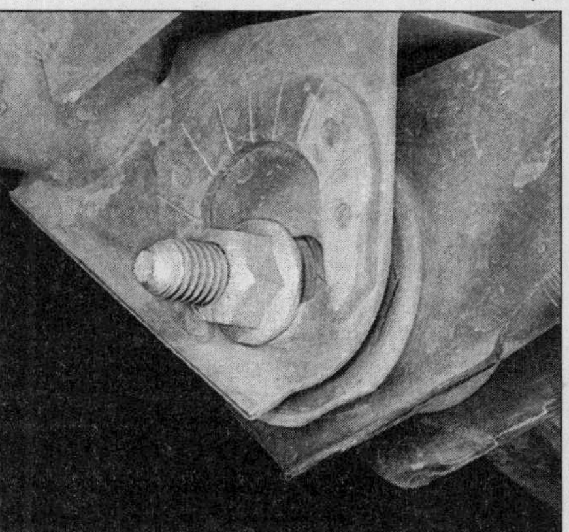

5.7 Before disconnecting the lower suspension arm from the axle housing bracket, mark the relationship of the adjuster cam (if equipped) or nut to the bracket (1998 and earlier models only)

5 Suspension arms (front) - removal and installation

Note: *Remove and install only one suspension arm at a time to avoid the possibility of the axle housing shifting out of position, which would make reassembly much more difficult. If it is absolutely necessary to remove more than one at a time, support the axle with a floor jack.*

1 Raise the front of the vehicle and support it securely on jackstands placed under the subframe. Support the axle with a floor jack placed under the axle tube on the side from which the arm is to be removed.

Upper suspension arm

Refer to illustrations 5.2a, 5.2b and 5.3

2 Remove the nut and bolt securing the upper suspension arm to the axle housing bracket **(see illustrations)**.

3 Remove the bolt and nut connecting the upper suspension arm to the frame bracket and remove the arm from the vehicle **(see illustration)**.

4 Check the arm for distortion and cracks. If the arm is damaged, replace it.

5 Inspect the bushing in the axle housing (upper arm only) for cracking, hardness and general deterioration. If it is in need of replacement, reinstall the suspension arm and take the vehicle to a dealer service department or an automotive machine shop to have it replaced (special tools are required).

6 Installation is the reverse of removal. Before tightening the fasteners to the torque listed in this Chapter's Specifications, lower the vehicle (or support it with jackstands placed under the axle tubes); this is to prevent the bushings from becoming distorted, which can cause them to fail prematurely.

Lower suspension arm

Refer to illustrations 5.7 and 5.9

7 If you're working on a 1998 or earlier model, make alignment marks on the adjuster cams (if equipped) or nuts and the axle housing brackets to insure proper reassembly **(see illustration)**. **Note:** *The caster adjuster cams are not installed at the factory but are available at your dealer parts department.*

8 Remove the lower suspension arm nut, the cam (on models so equipped) and bolt from the bracket on the axle housing.

9 Remove the nut and bolt from the frame rail bracket **(see illustration)** and remove the lower suspension arm.

10 Installation is the reverse of removal. On 1998 and earlier models, be sure to re-align the marks you made between the cam

5.9 To disconnect the lower suspension arm from the frame bracket, remove the nut (not visible in this photo) and pull out this bolt

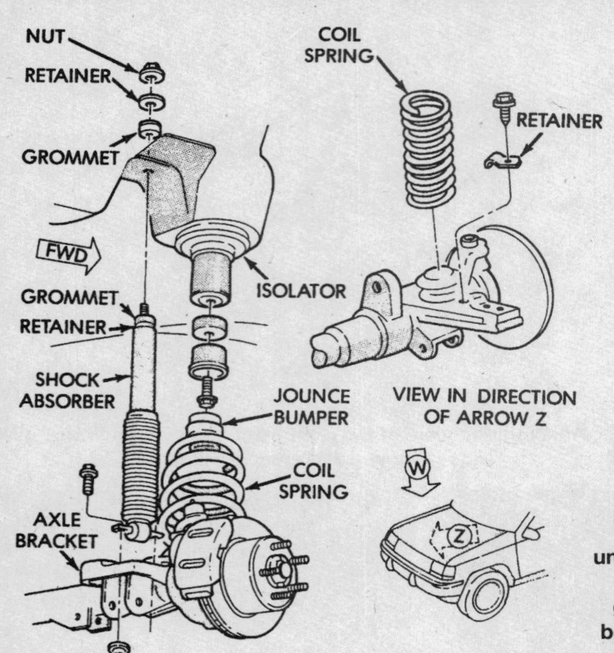

6.6 Coil spring mounting details (1998 and earlier model shown) - once the spring is fully extended, unscrew the bolt and remove the spring retainer, then guide the spring out from between the axle and frame

adjuster or nut and the bracket on the axle. Before tightening the fasteners to the torque listed in this Chapter's Specifications, lower the vehicle (or support it with jackstands placed under the axle tubes); this is to prevent the bushings from becoming distorted, which can cause them to fail prematurely.

6 Coil spring (front) - removal and installation

Removal

Refer to illustration 6.6

1 Loosen the front wheel lug nuts, raise the front of the vehicle and support it securely on jackstands placed under the subframe rails. Remove the wheels.

2 Mark and disconnect the front driveshaft from the front differential (see Chapter 8), then hang the driveshaft out of the way with a piece of wire.

3 Support the axle assembly with two floor jacks, one at each end of the axle.

4 Unbolt the stabilizer bar links and the shock absorbers at the front axle housing (see Sections 2 and 3).

5 Disconnect the track bar at the axle bracket (see Section 4). If you're working on a 1998 or earlier model, unbolt the lower control arms from the axle housing (see Section 5).

6 Slowly lower the axle assembly until the coil springs are fully extended. If you're working on a 1998 or earlier model, remove the spring retainer **(see illustration)**. Remove the spring from the vehicle.

7 Check the spring for deep nicks and corrosion, which could cause premature failure of the spring. Replace the spring if these or any other questionable conditions are evident.

8 Check the upper and lower isolators for damage, replacing them as necessary.

Installation

Refer to illustration 6.9

9 Install the spring isolators. Position the coil spring on the axle housing and, on 1998 and earlier models, place the spring retainer over the bottom coil of the spring and tighten the retainer bolt to the torque listed in this Chapter's Specifications. If you're working on a 1999 or later model, position the end of the spring against the locator on the lower isolator **(see illustration)**.

10 Raise the axle up into position and, on 1998 and earlier models, connect the lower suspension arms to the axle housing (see Section 5). Don't tighten the fasteners yet.

11 Connect the track bar to its bracket on the axle, but don't tighten the fasteners yet (see Section 4).

12 Connect the stabilizer bar links and the shock absorbers to the front axle housing (see Sections 2 and 3).

13 Connect the front driveshaft to the differential pinion shaft yoke (see Chapter 8). Tighten the fasteners to the torque listed in the Chapter 8 Specifications.

14 Install the wheels and lug nuts and lower the vehicle. Tighten the lug nuts to the torque listed in the Chapter 1 Specifications.

15 Tighten the track bar fasteners and, on 1998 and earlier models, the lower control arm fasteners, to the torque values listed in this Chapter's Specifications. If you can't reach the fasteners with the vehicle on the ground, raise it again and support it with jackstands placed under the axle tube(s); this is to prevent the bushings from becoming distorted, which can cause them to fail prematurely.

6.9 When installing the coil spring on a 1999 or later model, make sure the end of the spring contacts the locator on the lower isolator

7 Steering knuckle - removal and installation

Removal

Refer to illustrations 7.5 and 7.6

1 Loosen the wheel lug nuts, raise the vehicle and support it securely on jackstands. Remove the wheel.

2 Disconnect the tie-rod from the steering knuckle (see Section 15). If you're removing the right (passenger's side) steering knuckle, also detach the drag link from the knuckle (on 1998 and earlier models, only the drag link is connected to the steering knuckle).

3 Remove the disc brake caliper and disc (see Chapter 9).

4 If you're working on a 4WD model, remove the front axle hub and bearing assembly and remove the axleshaft (see Chapter 8).

5 Unplug the electrical connector for the ABS wheel sensor and remove the sensor **(see illustration)**.

6 Remove the cotter pins and loosen the castle nuts on the balljoint studs. Using a brass hammer, tap the steering knuckle at the top to separate it from the balljoint studs; if either end of the knuckle hangs up on a

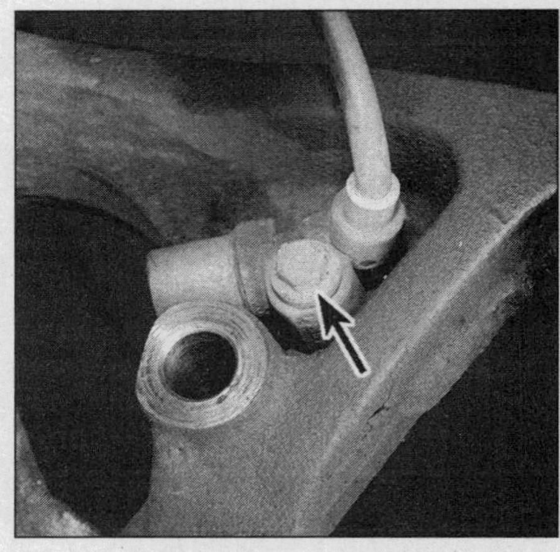

7.5 Unplug the electrical connector for the ABS wheel sensor, remove the sensor retaining bolt (arrow) and remove the sensor

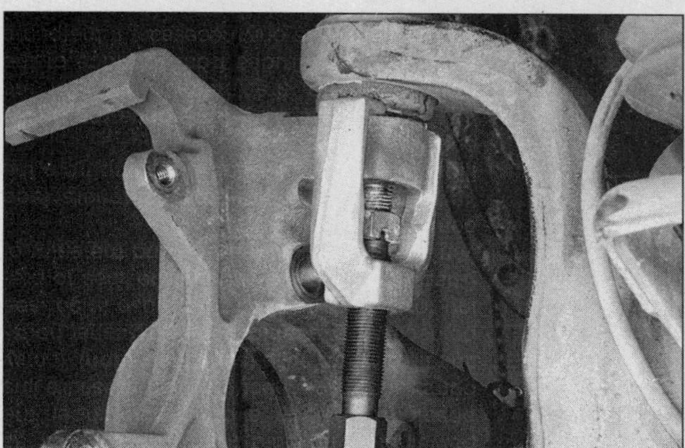

7.6 If striking the top of the steering knuckle doesn't separate it from the balljoint studs, install a small puller and press it off (puller shown in place on upper ballstud; separation procedure for lower ballstud similar)

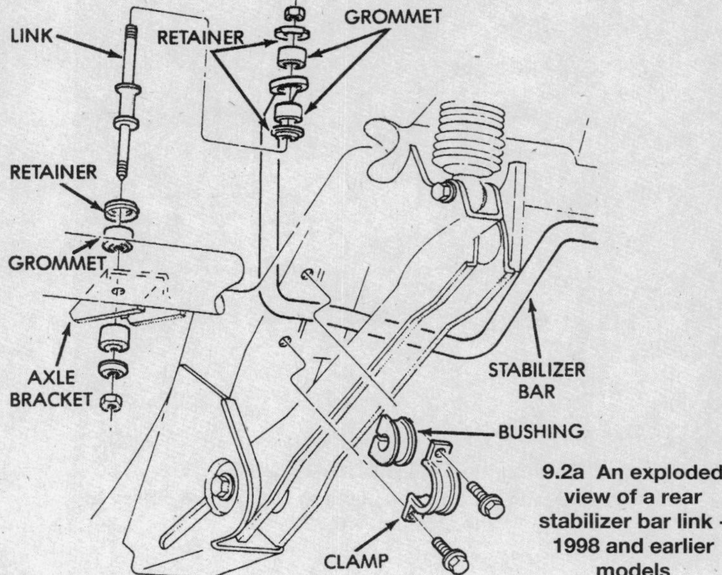

9.2a An exploded view of a rear stabilizer bar link - 1998 and earlier models

ballstud, press it off with a small puller **(see illustration)**. Remove the nuts and detach the knuckle from the axle housing.

7 Carefully check the steering knuckle for cracks, especially around the steering arm and spindle mounting area. Check for elongated balljoint stud holes. Replace the steering knuckle if any of these conditions are found.

Installation

8 Position the steering knuckle on the axle housing yoke and insert the balljoint studs into the holes in the knuckle. Install the nuts and tighten them to the torque listed in this Chapter's Specifications. Use new cotter pins to secure the nuts.

9 If you're working on a 4WD model, install the axleshaft (see Chapter 8).

10 Install the front axle hub and bearing assembly (see Chapter 8).

11 Install the brake disc and caliper (see Chapter 9).

12 Connect the tie-rod (and/or drag link, if you're working on the right side) to the steering knuckle (see Section 15). Tighten the mounting nut(s) to the torque listed in this Chapter's Specifications.

13 Install the wheel and lug nuts. Lower the vehicle and tighten the lug nuts to the torque listed in the Chapter 1 Specifications.

8 Balljoints - replacement

The balljoints are a press fit in the front axle housing yokes, which necessitates the use of a special tool to remove and install them. Since the tool is not normally available to the home mechanic, it is recommended that the vehicle be taken to a dealer service department or other qualified repair shop to have the balljoints replaced.

9 Stabilizer bar and bushings (rear) - removal and installation

Removal

Refer to illustrations 9.2a, 9.2b, 9.2c, 9.3a and 9.3b

1 Raise the rear of the vehicle and support it securely on jackstands.

2 Disconnect the stabilizer bar links from the stabilizer bar **(see illustrations)**.

3 Unbolt the stabilizer bar bushing clamps **(see illustrations)** and remove the bar from the vehicle. On 1998 and earlier models the

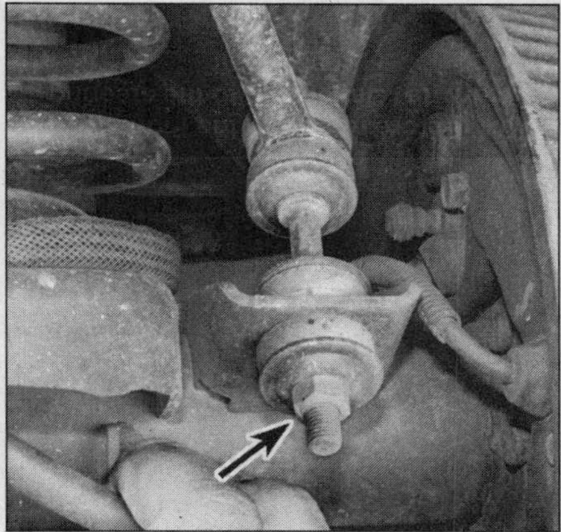

9.2b To disconnect the rear stabilizer bar link from the axle tube bracket on 1998 and earlier models, remove this nut (arrow)

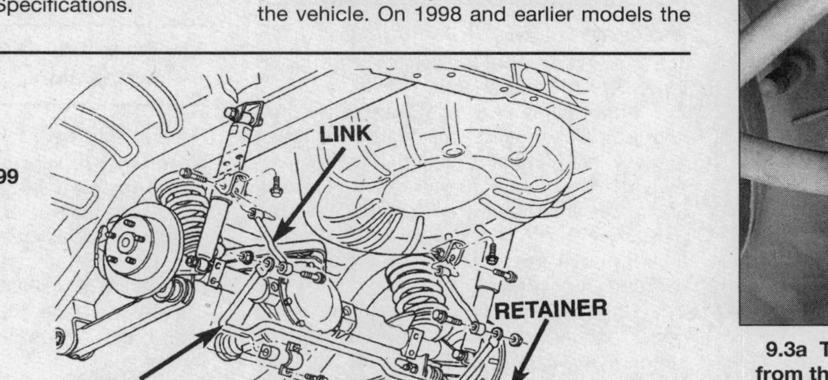

9.2c Stabilizer mounting details - 1999 and later models

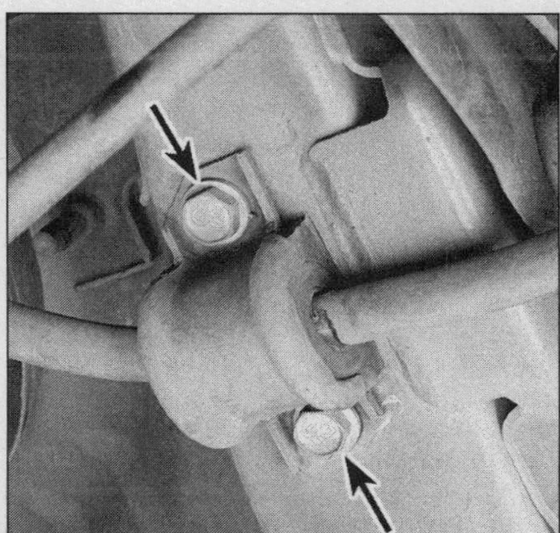

9.3a To detach the rear stabilizer bar from the frame rails on 1998 and earlier models, remove the bushing clamp bolts (arrows) from both clamps

9.3b On 1999 and later models the bushing clamps are bolted to the axle housing

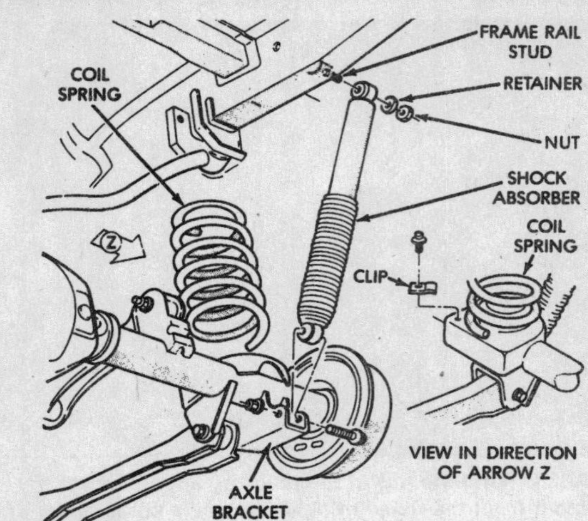

10.3a Shock absorber and coil spring installation details - 1998 and earlier models

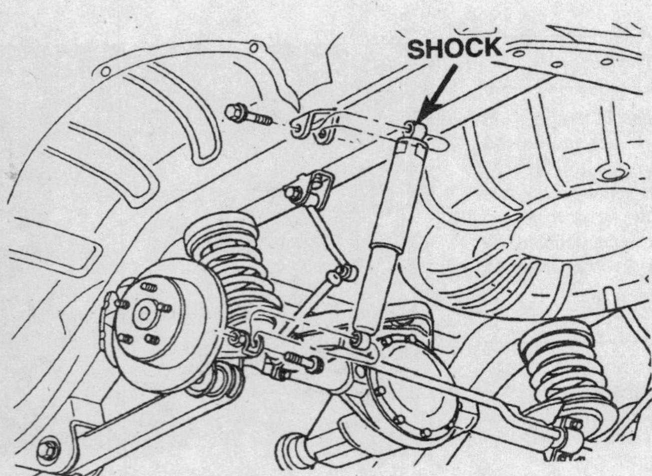

10.3b Shock absorber installation details - 1999 and later models

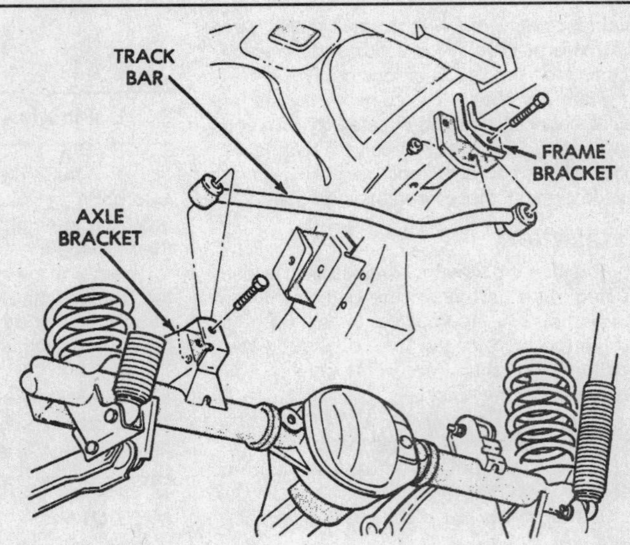

11.2a Track bar installation details (1998 and earlier models)

clamps are bolted to the underbody. On 1999 and later models the clamps are bolted to the rear axle housing.

4 Pull the clamps and bushings off the bar and inspect them for wear, cracking and general deterioration. Replace them if necessary.

Installation

5 Push the bushings and clamps onto the stabilizer bar. Position the bar against the frame (1998 and earlier models) or axle (1999 and later models), install the clamp bolts and tighten them to the torque listed in this Chapter's Specifications.

6 Connect the stabilizer links and tighten the nuts to the torque listed in this Chapter's Specifications.

7 Remove the jackstands and lower the vehicle.

10 Shock absorber (rear) - removal and installation

Refer to illustrations 10.3a and 10.3b

1 Raise the rear of the vehicle and support it securely on jackstands.

2 Support the rear axle assembly with a floor jack placed under the differential. Raise the jack just enough to take the spring pressure off the shock absorbers (the shock absorbers limit downward travel of the suspension).

3 Remove the lower mounting bolt and nut **(see illustrations)**.

4 On 1998 and earlier models, remove the upper mounting nut and washer and slide the shock eye off the mounting stud. On 1999 and later models, remove the nut and bolt from the upper mounting bracket and remove the shock absorber.

5 Installation is the reverse of the removal procedure. Tighten the fasteners to the torque listed in this Chapter's Specifications.

11 Track bar (rear) - removal and installation

Refer to illustrations 11.2a, 11.2b and 11.3

Note: *This procedure applies to 1998 and earlier models only.*

1 Raise the rear of the vehicle and support it securely on jackstands.

2 Remove the bolt and nut from the frame rail bracket **(see illustrations)**.

3 Remove the bolt and nut from the axle tube bracket **(see illustration)** and remove the bar.

4 Installation is the reverse of removal. Before tightening the fasteners to the torque listed in this Chapter's Specifications, lower

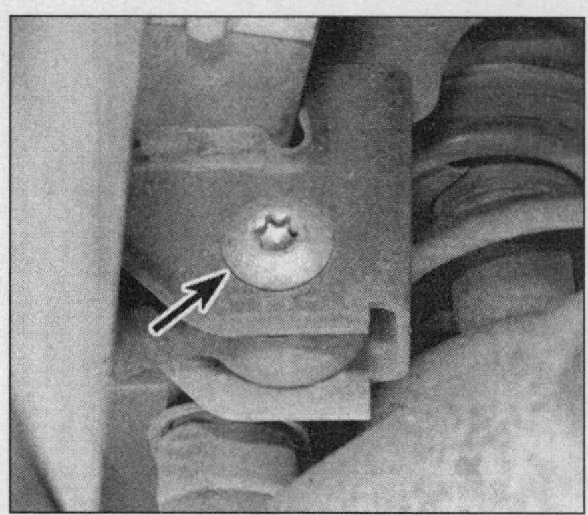

11.2b To disconnect the upper end of the track bar from the frame rail bracket, remove this bolt (arrow) and nut (not visible in this photo)

11.3 To disconnect the lower end of the track bar from the axle tube bracket, remove this bolt (arrow) and nut (not visible in this photo)

the vehicle (or support it with jackstands placed under the axle tubes); this is to prevent the bushings from becoming distorted, which can cause them to fail prematurely.

12 Suspension arms (rear) - removal and installation

Note: *Remove and install only one suspension arm at a time to avoid the possibility of the axle housing shifting out of position, which would make reassembly much more difficult. If it is absolutely necessary to remove more than one at a time, support the axle with a floor jack.*

1 Raise the rear of the vehicle and support it securely on jackstands placed under the subframe rails (not under the rear axle housing).

Upper suspension arm

1998 and earlier models

Refer to illustrations 12.2a, 12.2b, 12.3

2 Support the axle with a floor jack placed under the axle tube on the side from which the arm is to be removed. Remove the nut and bolt securing the upper suspension arm to the axle housing bracket **(see illustrations)**.

3 Remove the nut and bolt connecting the upper suspension arm to the frame rail bracket and remove the arm from the vehicle **(see illustration)**.

4 Check the arm for distortion and cracks. Inspect the bushings for tears and cracks. If the arm or either bushing is damaged, replace the arm.

5 Installation is the reverse of removal. Before tightening the fasteners to the torque listed in this Chapter's Specifications, lower the vehicle (or support it with jackstands placed under the axle tubes); this is to prevent the bushings from becoming distorted, which can cause them to fail prematurely.

1999 and later models

Refer to illustrations 12.8 and 12.9

6 Support the axle with a floor jack placed under the differential.

7 Detach the brake hose and the parking brake cables from the arm.

8 Loosen the balljoint nut a few turns, then separate the arm from the balljoint using a balljoint separator **(see illustration)**. Remove the nut.

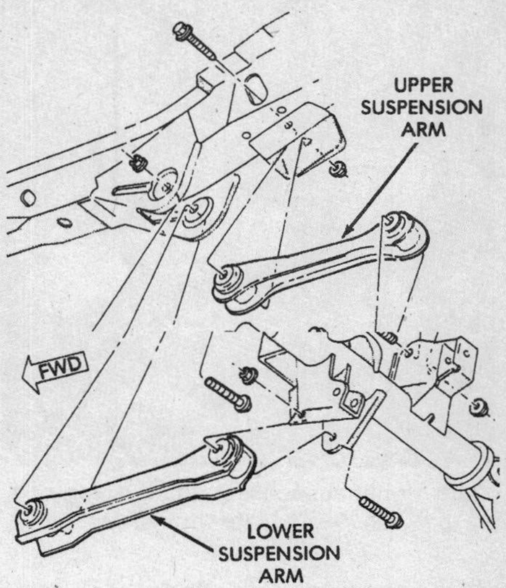

12.2a Rear suspension arm installation details (1998 and earlier models)

12.2b To disconnect the upper rear suspension arm from the axle housing bracket, remove this nut (arrow) and bolt (1998 and earlier models)

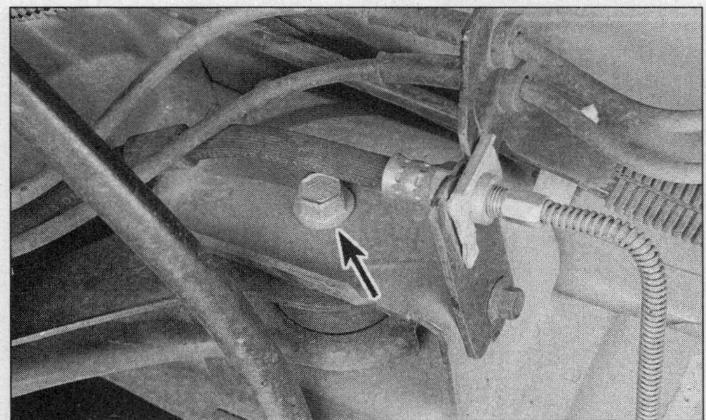

12.3 To disconnect the upper rear suspension arm from the frame rail bracket, remove the nut (accessible through a hole in the other side of the frame rail) and this bolt (arrow) (1998 and earlier models)

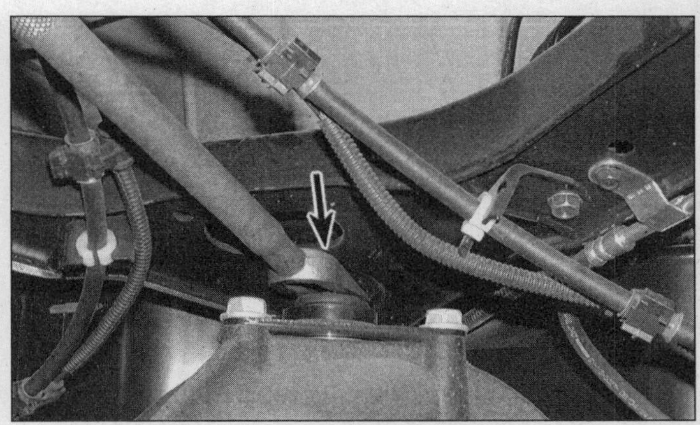

12.8 Using a "picklefork" type balljoint separator (arrow) to detach the upper suspension arm from the balljoint stud (1999 and later models)

12.9 Upper suspension arm bushing bolt (1999 and later models)

12.11 To disconnect the lower rear suspension arm from the axle housing bracket, remove this nut and bolt (arrows)

12.12 To disconnect the lower rear suspension arm from the frame rail bracket, remove the nut (not visible in this photo) and this bolt

9 Unscrew the bushing bolts and remove the arm **(see illustration)**. **Note:** *If the balljoint shows signs of wear, it can be unbolted from the axle housing and replaced. Signs of wear include a damaged boot, looseness, or rough movement.*

10 Installation is the reverse of removal. Be sure to tighten the fasteners to the torque listed in this Chapter's Specifications.

Lower suspension arm

Refer to illustrations 12.11 and 12.12

11 Support the axle with a floor jack placed under the axle tube on the side from which the arm is to be removed. Remove the nut and bolt connecting the lower suspension arm to the axle bracket **(see illustration)**.

12 Remove the nut and bolt connecting the lower suspension arm to the frame rail bracket **(see illustration)** and remove the lower suspension arm.

13 Installation is the reverse of removal. Before tightening the fasteners to the torque listed in this Chapter's Specifications, lower the vehicle (or support it with jackstands placed under the axle tubes); this is to prevent the bushings from becoming distorted, which can cause them to fail prematurely.

13 Coil spring (rear) - removal and installation

1 Loosen the wheel lug nuts, raise the rear of the vehicle and support it securely on jackstands. Remove the wheels.

2 Support the axle with a floor jack placed under the axle tube nearest the coil spring to be removed.

3 Unbolt the stabilizer bar link from the stabilizer bar (see Section 9).

4 Unbolt the shock absorber from the axle (see Section 10).

5 On 1998 and earlier models, disconnect the track bar at the axle bracket (see Section 11).

6 Slowly lower the axle until the coil spring is fully extended. Unbolt the spring retainer (1998 and earlier models) and remove the spring from the vehicle.

7 Check the spring for deep nicks and corrosion, which could cause premature failure of the spring. Replace the springs as a set if these or any other questionable conditions are evident.

8 Also check the spring isolators for damage, replacing them as necessary.

9 Installation is the reverse of removal. If you're working on a 1998 or earlier model, make sure the coil spring is seated as shown in **illustration 10.3a**. If you're working on a 1999 or later model, position the end of the spring against the locator on the lower isolator **(see illustration 6.9)**. Tighten all fasteners to the torque values listed in this Chapter's Specifications.

14 Steering wheel - removal and installation

Refer to illustrations 14.2a, 14.2b, 14.2c, 14.2d, 14.3, 14.4a, 14.4b, 14.5a and 14.5b

Warning: *These models are equipped with airbags. Always disable the airbag system before working in the vicinity of any airbag system component to avoid the possibility of accidental deployment of the airbag(s), which could cause personal injury* (see Chapter 12).

1 Park the vehicle with the wheels pointing straight ahead. Disconnect the cable from the negative terminal of the battery.

2 Remove the four airbag module retaining nuts (1998 and earlier models) or the two screws (1999 and later models), detach the module from the steering wheel and unplug the electrical connectors from the module **(see illustrations)**. **Warning:** *Whenever handling the airbag module, always keep the airbag opening (the trim side) pointed away from your body. Never set the airbag down with the airbag opening facing the surface. Always place the airbag module in a safe location with the airbag opening facing up.*

3 Remove the steering wheel retaining nut, then mark the relationship of the steering shaft to the hub (if marks don't already exist or don't line up) to simplify installation and ensure steering wheel alignment **(see illustration)**. **Caution:** *Don't allow the steering*

14.2a To remove the air bag module on a 1998 or earlier model, remove the upper nuts . . .

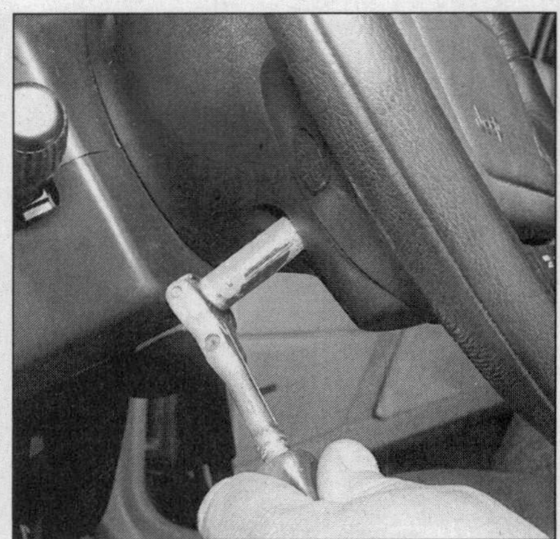

14.2b . . . and the lower nuts

14.2c To remove the airbag module on a 1999 or later model, remove the screw from each side of the steering wheel

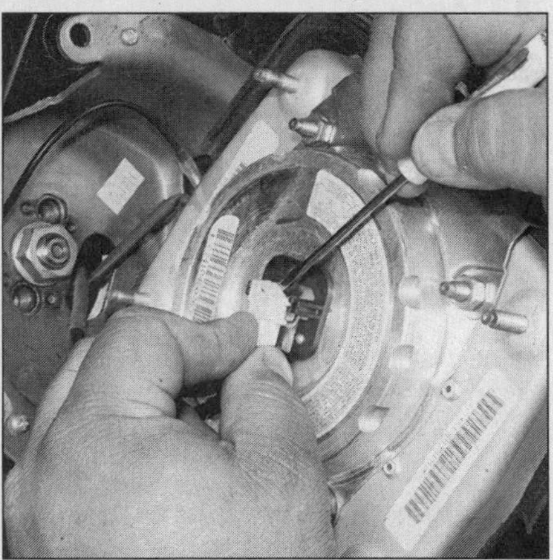

14.2d Disengage the locking tangs on the airbag module electrical connector and unplug it from the module

14.3 Before removing the steering wheel, look for alignment marks between the steering wheel and the steering shaft; if there are none, use a sharp scribe or white paint to make your own marks

14.4a On 1998 and earlier models, remove the wheel from the shaft with a puller that uses bolts threaded into the hub of the wheel - DO NOT HAMMER ON THE SHAFT!

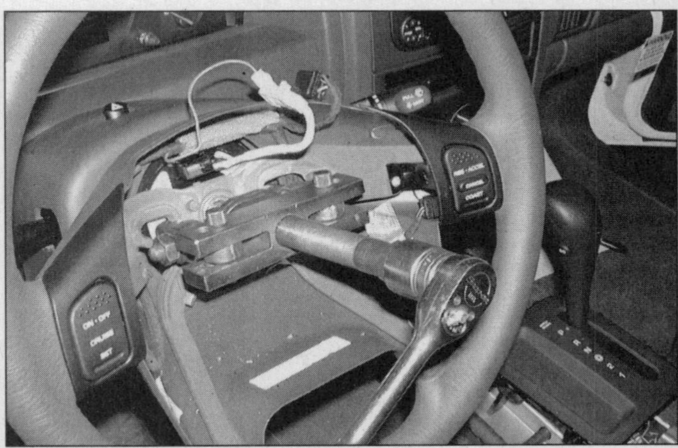

14.4b On 1999 and later models, a jaw type puller will have to be used; make sure the jaws of the puller seat completely in the pockets in the steering wheel hub

shaft to turn while the steering wheel is removed.

4 Use a puller to detach the steering wheel from the shaft **(see illustrations)**. Don't hammer on the shaft to dislodge the steering wheel.

5 If the front wheels remained in the straight-ahead position throughout the procedure, the steering wheel can be installed now. If, however, the wheels were turned, the clockspring for the airbag will have to be centered. To do this, first make sure the front wheels are in the straight-ahead position. If you're working on a 1999 or later model, remove the clockspring from the steering column. Turn the clockspring hub in the clockwise direction until it won't turn any more (don't apply excessive force). **Note:** *On 1998 and earlier models you'll have to depress the locking tabs in the center of the clockspring.* Now, turn the hub counterclockwise 2-1/2 turns **(see illustrations)**.

6 To install the wheel, align the mark on the steering wheel hub with the mark on the

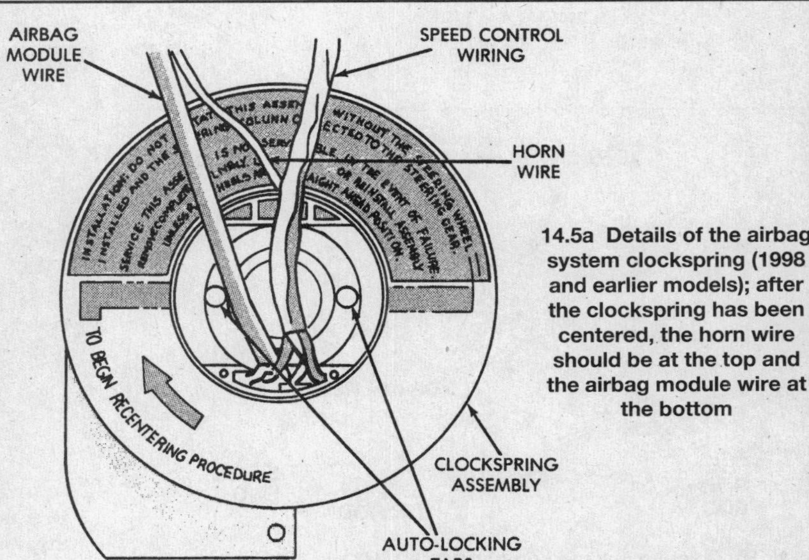

14.5a Details of the airbag system clockspring (1998 and earlier models); after the clockspring has been centered, the horn wire should be at the top and the airbag module wire at the bottom

shaft and slip the wheel onto the shaft. Install the nut and tighten it to the torque listed in this Chapter's Specifications.

7 Install the airbag module and tighten the fasteners to the torque listed in this Chapter's Specifications.

8 Connect the negative battery cable.

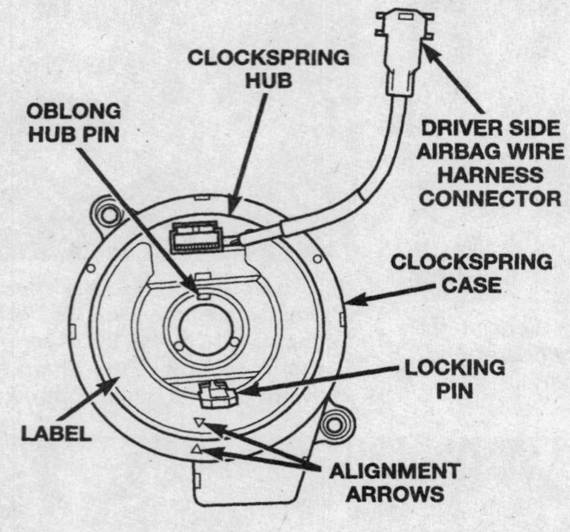

14.5b Details of the airbag system clockspring (1999 and later models); when properly centered, the alignment arrows should be pointing at each other, and the oblong hub pin should be at the top

15 Steering linkage - inspection, removal and installation

Inspection

Refer to illustrations 15.1a and 15.1b

1 The steering linkage **(see illustrations)** connects the steering gear to the front wheels and keeps the wheels in proper relation to each other. The linkage consists of the Pitman arm, fastened to the steering gear shaft, which moves the drag link back and forth. The back-and-forth motion of the drag link is transmitted to the right steering knuckle. A tie-rod transmits motion to the left steering knuckle. The tie-rod is made up of a tube, clamps and two tie-rod ends. A steering damper, connected between the drag link (1998 and earlier models) or tie-rod (1999 and later models) and the front axle housing

PITMAN ARM
WASHER
COTTER PIN
NUT
DAMPENER
TIE ROD
CENTER LINK

15.1a Steering linkage components - 1998 and earlier models

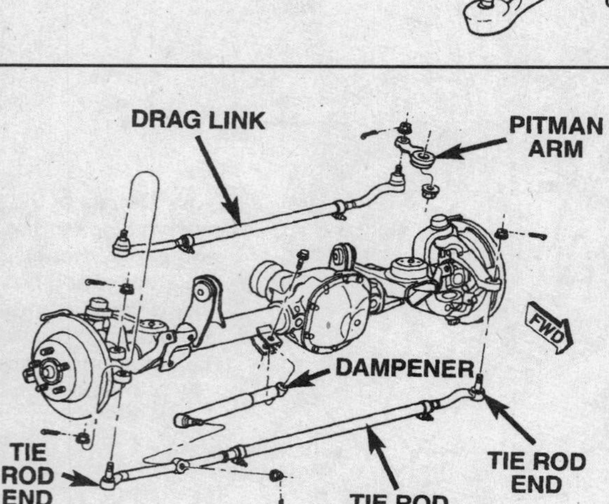

15.1b Steering linkage components - 1999 and later models

15.8 Use a two-jaw puller to detach the tie-rod end from the steering knuckle - notice the nut has been loosened but not removed; this will prevent the components from separating violently

15.9 If a tie-rod end must be replaced, count the number of threads visible at the end of the adjuster tube and install the new tie-rod end with the same number of threads showing

reduces shimmy and unwanted forces to the steering gear.

2 Set the wheels in the straight ahead position and lock the steering wheel.

3 Raise one side of the vehicle until the tire is approximately 1-inch off the ground.

4 Mount a dial indicator with the needle resting on the outside edge of the wheel. Grasp the front and rear of the tire and using light pressure, wiggle the wheel back-and-forth and note the dial indicator reading. The gauge reading should be less than 0.108-inch. If the play in the steering system is more than that, inspect each steering linkage pivot point and ballstud for looseness and replace parts if necessary.

5 Raise the vehicle and support it on jackstands. Check for torn ballstud boots, frozen joints and bent or damaged linkage components.

Removal and installation

Tie-rod

Refer to illustrations 15.8 and 15.9

6 Loosen the wheel lug nuts, raise the vehicle and support it securely on jackstands. Apply the parking brake. Remove the wheel.

7 Remove the cotter pins and loosen, but do not remove, the castle nuts from the ballstuds on each end of the tie-rod.

8 Using a small puller, separate the tie-rod end from the steering knuckle (both knuckles on 1999 and later models) and, on 1998 and earlier models, the drag link **(see illustration)**. **Note:** *On 1999 and later models the steering damper must also be detached from the tie-rod.* Remove the castle nuts and pull the tie-rod from the knuckle and drag link. **Caution:** *The use of a picklefork-type balljoint separator most likely will cause damage to the balljoint boot.*

9 If a tie-rod end must be replaced, count the number of threads showing and jot down this number to maintain correct toe-in during reassembly **(see illustration)**. Loosen the adjuster tube clamp and unscrew the tie-rod end.

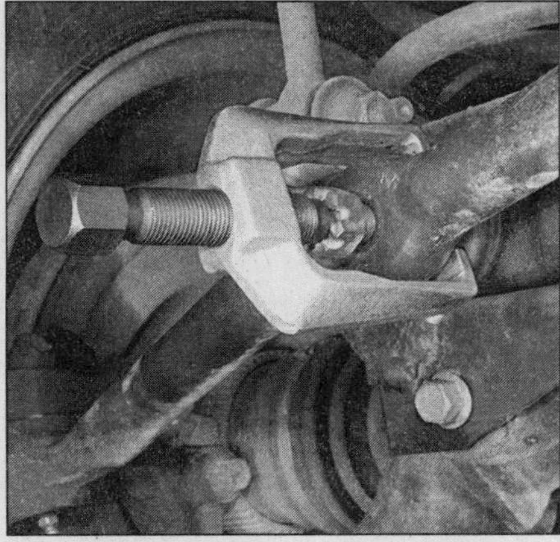

15.16 To disconnect the drag link from the tie-rod, use a small puller

10 Lubricate the threaded portion of the tie-rod end with chassis grease. Screw the new tie-rod end into the adjuster tube and adjust the distance from the tube to the ballstud by threading the tie-rod into the adjuster tube until the same number of threads are showing as before (the number of threads showing on both sides of the adjuster tube should be within three threads of each other). Don't tighten the adjuster tube clamps yet.

11 To install the tie-rod assembly, reverse the removal procedure. Install the nuts and tighten them to the torque listed in this Chapter's Specifications. If a ballstud spins when attempting to tighten the nut, force it into the tapered hole with a large pair of pliers.

12 Install new cotter pins. If necessary, tighten the nut slightly to align a slot in the nut with the hole in the ballstud.

13 Tighten the adjuster tube clamp nuts to the torque listed in this Chapter's Specifications. The adjuster tube clamp bolts should be nearly horizontal.

14 Install the wheel and lug nuts, lower the vehicle and tighten the lug nuts to the torque listed in the Chapter 1 Specifications. Drive the vehicle to an alignment shop to have the front end alignment checked and, if necessary, adjusted.

Drag link

Refer to illustration 15.16

15 If you're working on a 1999 or later model, loosen the left front wheel lug nuts. Raise the front of the vehicle and support it securely on jackstands. Apply the parking brake. Remove the wheel (1999 and later models).

16 Loosen, but do not remove, the nuts securing the drag link ballstuds to the tie-rod (1998 and earlier) or, on 1999 and later models, the steering knuckle. Separate the drag link from the steering knuckle **(see illustration 15.8)** and, on 1998 and earlier models, from the tie-rod **(see illustration)** with a two-jaw puller, then remove the nuts.

17 Separate the drag link from the Pitman arm using the same technique. If the tie-rod end on the end of the drag link is in need of replacement, follow Steps 9 and 10 of this Section.

18 Installation is the reverse of the removal procedure. If the ballstuds spin when attempting to tighten the nuts, force them into the tapered holes with a large pair of pliers. Be sure to tighten all fasteners to the torque listed in this Chapter's Specifications.

Pitman arm

19 Refer to Section 16 of this Chapter for the Pitman arm removal procedure.

Steering damper

20 Raise the front of the vehicle and support it securely on jackstands.

21 Separate the steering damper from the drag link (1998 and earlier models) or the tie-rod (1999 and later models) using the technique described in Step 16.

22 Unbolt the damper from the axle housing and remove it from the vehicle.

23 Installation is the reverse of removal. Be sure to tighten the steering damper fasteners to the torque listed in this Chapter's Specifications.

16 Steering gear - removal and installation

Warning: *Make sure the steering shaft is not turned while the steering gear is removed or you could damage the airbag system clockspring. One way to prevent the shaft from turning is to place the ignition key in the LOCK position or run the seat belt through the steering wheel and clip the seat belt into place.*

Removal

Refer to illustrations 16.2, 16.4, 16.5, 16.6a and 16.6b

1 Raise the front of the vehicle and support it securely on jackstands. Apply the parking brake.

2 Place a drain pan under the steering

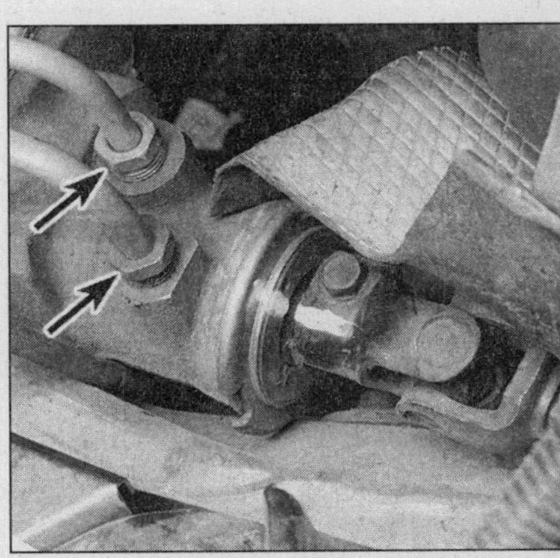

16.2 Loosen these fittings (arrows) and disconnect the power steering fluid lines from the steering gearbox, then mark the relationship of the intermediate shaft U-joint to the steering gear input shaft and loosen the pinch bolt

10

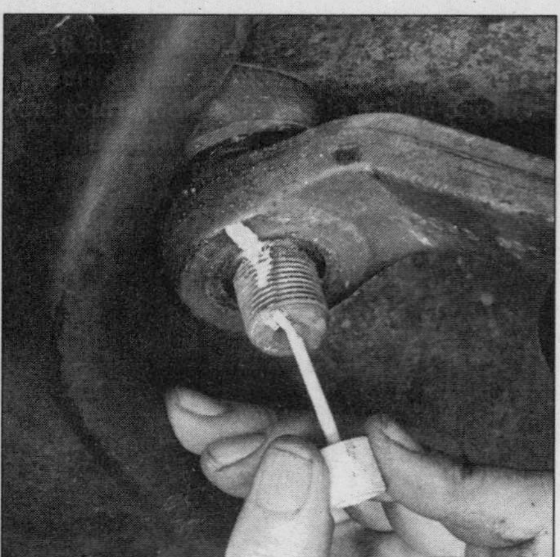

16.4 Before removing the Pitman arm, mark its relationship to the steering gear shaft

16.5 Use a puller to separate the Pitman arm from the steering gear shaft

16.6a On 1998 and earlier models, the steering gear is mounted to the frame rail with three bolts (arrows)

gear. Remove the hoses/lines and cap the ends to prevent excessive fluid loss and contamination **(see illustration)**. If available, use a flare-nut wrench to remove the hoses/lines.

3 Mark the relationship of the intermediate shaft lower universal joint to the steering gear input shaft. Remove the intermediate shaft lower pinch bolt.

4 Remove the Pitman arm nut and washer. Mark the relationship of the Pitman arm to the shaft so it can be installed in the same position **(see illustration)**.

5 Separate the Pitman arm from the shaft with a two-jaw puller **(see illustration)**.

6 Support the steering gear and remove the mounting bolts **(see illustrations)**. **Note:** *If you're working on a 1999 or later model, you'll have to unbolt and reposition the windshield washer fluid reservoir for access to the two forward bolts.* Lower the unit, separate the intermediate shaft from the steering gear input shaft and remove the steering gear from the vehicle.

16.6b On 1999 and later models, the steering gear is mounted with four bolts (the fourth bolt, not visible in this photo, is behind the windshield washer fluid reservoir)

Installation

7 Raise the steering gear into position and connect the intermediate shaft, aligning the marks.

8 Install the mounting bolts and washers and tighten them to the torque listed in this Chapter's Specifications.

9 Slide the Pitman arm onto the shaft. Make sure the marks are aligned. Install the washer and nut and tighten the nut to the torque listed in this Chapter's Specifications.

10 Install the intermediate shaft lower pinch bolt and tighten it to the torque listed in this Chapter's Specifications.

11 Connect the power steering hoses/lines to the steering gear and fill the power steering pump reservoir with the recommended fluid (see Chapter 1).

12 Lower the vehicle and bleed the steering system (see Section 18).

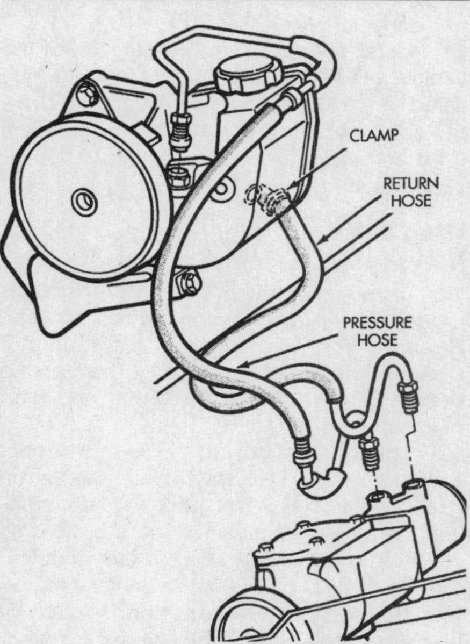

17.2 Power steering pump line and hose connections (typical)

17 Power steering pump - removal and installation

Refer to illustrations 17.2, 17.3a, 17.3b, 17.3c, 17.4, 17.5a and 17.5b

1 Loosen the pump drivebelt and slip the belt over the pulley (see Chapter 1).

2 Using a suction gun, suck out as much power steering fluid from the reservoir as possible. Position a drain pan under the pump and disconnect the high pressure line and fluid return hose **(see illustration)**. It may be necessary to remove the air cleaner housing for access to the return hose (if so, see Chapter 4). Cap the ends of the lines to prevent excessive fluid leakage and the entry of contaminants.

3 Remove the mounting fasteners **(see illustrations)** and detach the pump from the engine.

4 If it is necessary to remove the pulley from the pump, first measure how far the pump shaft protrudes from the face of the pulley hub (this is only necessary if the pump has a plastic pulley). Remove the pulley from the shaft with a special power steering pump pulley removal tool **(see illustration)**. This tool can be purchased at most auto parts stores.

5 A special pulley installation tool is also available for pressing the pulley back onto the pump shaft **(see illustration)**, but an alternate tool can be fabricated from a long bolt, nut, washer and a socket of the same diameter as the pulley hub **(see illustration)**. If the pulley is plastic, push it onto the shaft until the shaft protrudes from the hub the previously recorded amount. If the pulley is metal, press it on until it is flush with the end of the shaft.

6 Installation of the pump is the reverse of the removal procedure. Be sure to tighten the fasteners securely, and bleed the power steering system following the procedure in Section 18.

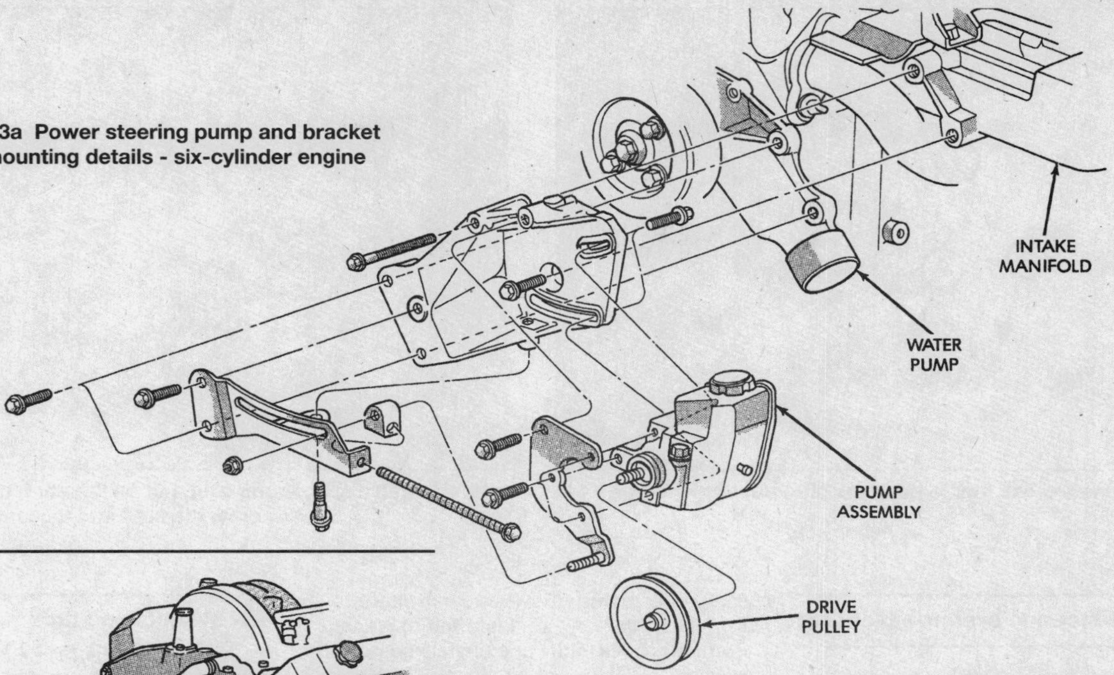

17.3a Power steering pump and bracket mounting details - six-cylinder engine

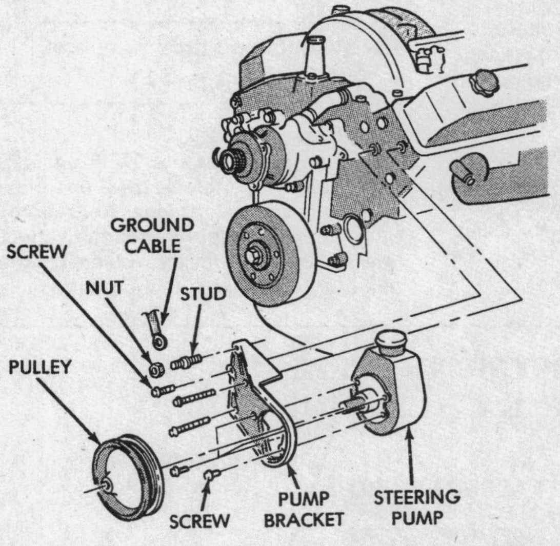

17.3b Power steering pump and bracket mounting details - 5.2L V8 engine

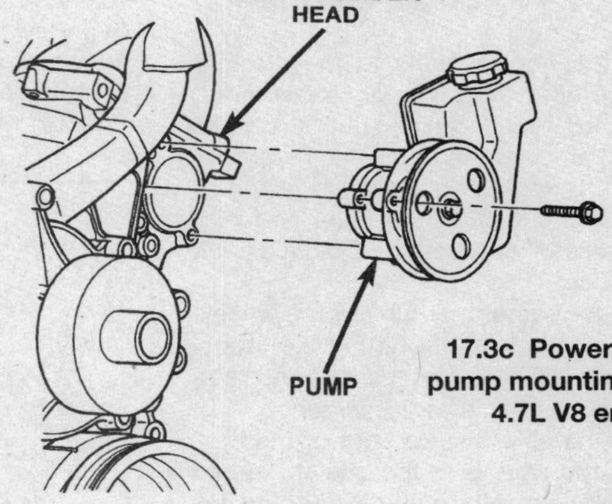

17.3c Power steering pump mounting details - 4.7L V8 engine

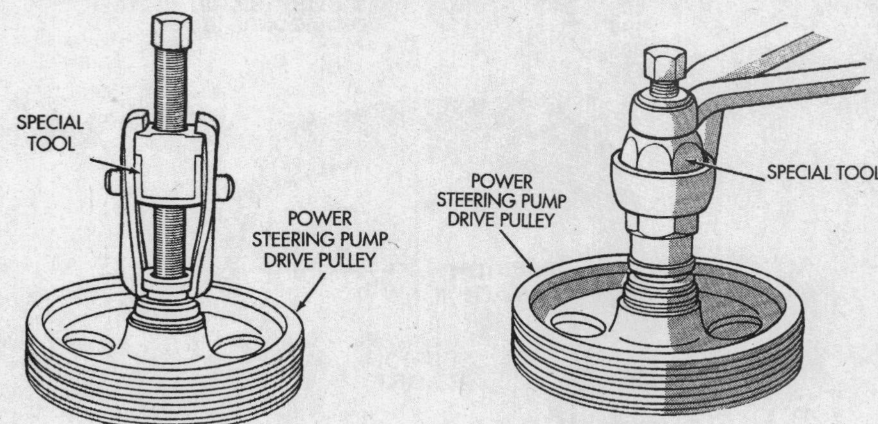

17.4 This special tool, designed for removing power steering pump pulleys, is available at most auto parts stores

17.5a There's also a special tool available for installing power steering pump pulleys, but . . .

17.5b . . . a long bolt with the same thread pitch as the internal threads of the power steering pump shaft, a nut, washer and a socket that is the same diameter as the pulley hub can also be used to install the pulley on the shaft

19.3 Use a press tool to push the stud out of the flange

1 Hub flange 2 Lug nut on stud 3 Press tool

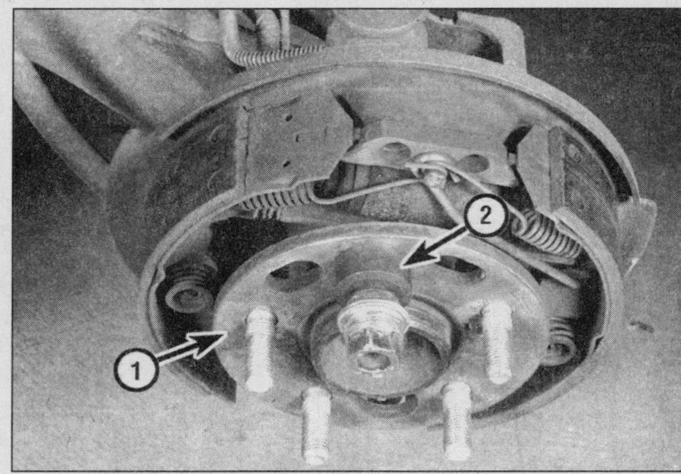

19.4 Install a spacer and a lug nut on the stud, then tighten the nut to draw the stud into place

1 Hub flange 2 Spacer

18 Power steering system - bleeding

1 Following any operation in which the power steering fluid lines have been disconnected, the power steering system must be bled to remove all air and obtain proper steering performance.

2 With the front wheels in the straight ahead position, check the power steering fluid level and, if low, add fluid until it reaches the Cold (C) mark on the dipstick.

3 Start the engine and allow it to run at fast idle. Recheck the fluid level and add more if necessary to reach the Cold (C) mark on the dipstick.

4 Bleed the system by turning the wheels from side-to-side, without hitting the stops. This will work the air out of the system. Keep the reservoir full of fluid as this is done.

5 When the air is worked out of the system, return the wheels to the straight ahead position and leave the vehicle running for several more minutes before shutting it off.

6 Road test the vehicle to be sure the steering system is functioning normally and noise-free.

7 Recheck the fluid level to be sure it is up to the Hot (H) mark on the dipstick while the engine is at normal operating temperature. Add fluid if necessary (see Chapter 1).

19 Wheel studs - replacement

Refer to illustrations 19.3 and 19.4

Note: *This procedure applies to both the front and rear wheel studs.*

1 Loosen the wheel lug nuts, raise the vehicle and support it securely on jackstands. Remove the wheel.

2 Remove the brake disc or drum (see Chapter 9).

3 Install a lug nut part way onto the stud being replaced. Push the stud out of the hub flange with a press tool **(see illustration)**.

4 Insert the new stud into the hub flange from the back side and install some flat washers and a lug nut on the stud **(see illustration)**.

5 Tighten the lug nut until the stud is seated in the flange.

6 Reinstall the brake drum or disc. Install the wheel and lug nuts. Lower the vehicle and tighten the lug nuts to the torque listed in the Chapter 1 Specifications.

20 Wheels and tires - general information

Refer to illustration 20.1

The vehicles covered by this manual are equipped with metric-sized fiberglass or steel-belted radial tires **(see illustration)**. Use of other size or type of tires may affect the ride and handling of the vehicle. Don't mix different types of tires, such as radials and

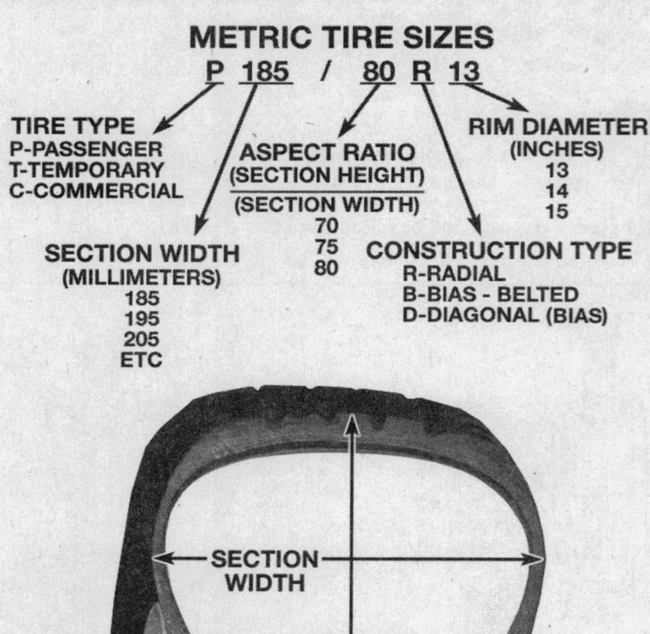

20.1 Metric tire size code

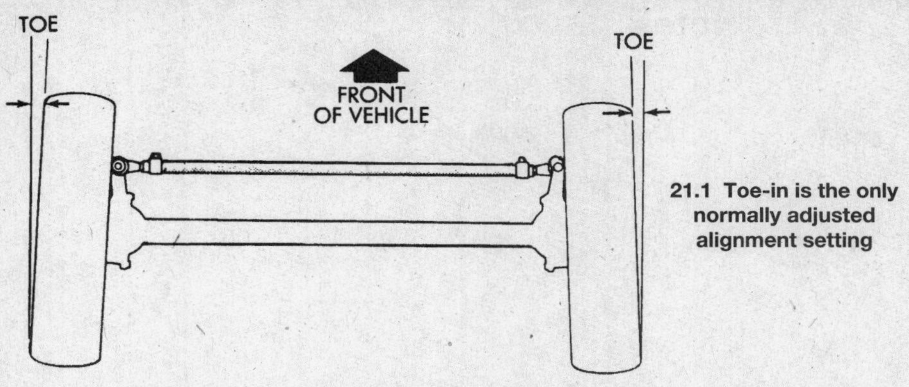

21.1 Toe-in is the only normally adjusted alignment setting

bias belted, on the same vehicle as handling may be seriously affected. It's recommended that tires be replaced in pairs on the same axle, but if only one tire is being replaced, be sure it's the same size, structure and tread design as the other.

Because tire pressure has a substantial effect on handling and wear, the pressure on all tires should be checked at least once a month or before any extended trips (see Chapter 1).

Wheels must be replaced if they are bent, dented, leak air, have elongated bolt holes, are heavily rusted, out of vertical symmetry or if the lug nuts won't stay tight. Wheel repairs that use welding or peening are not recommended.

Tire and wheel balance is important to the overall handling, braking and performance of the vehicle. Unbalanced wheels can adversely affect handling and ride characteristics as well as tire life. Whenever a tire is installed on a wheel, the tire and wheel should be balanced by a shop with the proper equipment.

21 Front end alignment - general information

Refer to illustrations 21.1 and 21.2

A front end alignment refers to the adjustments made to the front wheels so they are in proper angular relationship to the suspension and the ground. Front wheels that are out of proper alignment not only affect steering control, but also increase tire wear. The only front end adjustments possible on 1998 and earlier models are caster and toe-in; on 1999 and later models the only adjustment possible is toe-in.

Getting the proper front wheel alignment is a very exacting process, one in which complicated and expensive machines are necessary to perform the job properly. Because of this, you should have a technician with the proper equipment perform these tasks. We will, however, use this space to give you a basic idea of what is involved with front end alignment so you can better understand the process and deal intelligently with the shop that does the work.

Toe-in is the turning in of the front wheels **(see illustration)**. The purpose of a toe specification is to ensure parallel rolling of the front wheels. In a vehicle with zero toe-in, the distance between the front edges of the wheels will be the same as the distance between the rear edges of the wheels. The actual amount of toe-in is normally only a fraction of an inch. Toe-in adjustment is controlled by the tie-rod end position on the tie-rod. Incorrect toe-in will cause the tires to wear improperly by making them scrub against the road surface.

Caster is the tilting of the top of the front steering axis from vertical **(see illustration)**. A tilt toward the rear is positive caster and a tilt toward the front is negative caster. This angle is adjusted on 1998 and earlier models by adding caster adjustment cams (available at a dealer parts department) at the front of the lower control arm (this angle is not adjustable on 1999 and later models).

Camber (the tilting of the front wheels from vertical when viewed from the front of the vehicle) is factory present at 0-degrees and cannot be adjusted. If the camber angle isn't correct, the components causing the problem must be replaced. **Caution:** *Never attempt to adjust the camber angle by heating or bending the axle or any other suspension component!*

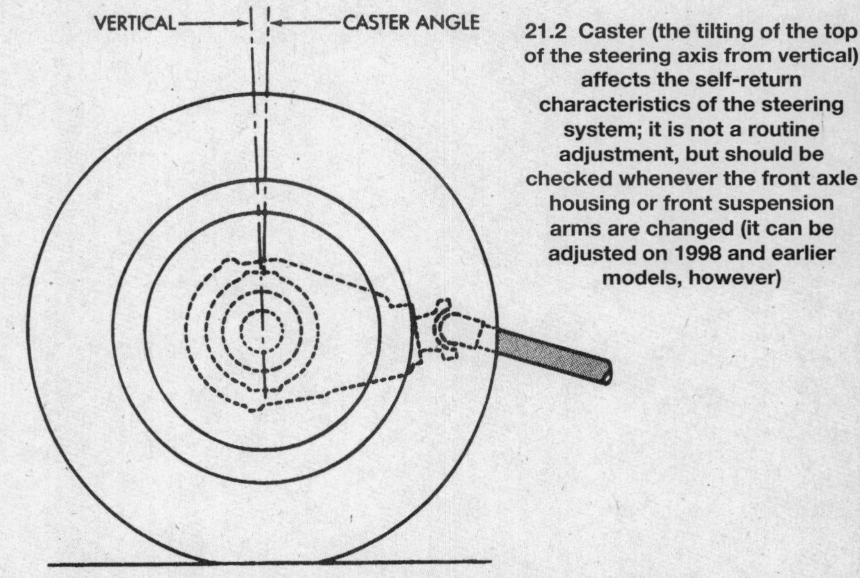

21.2 Caster (the tilting of the top of the steering axis from vertical) affects the self-return characteristics of the steering system; it is not a routine adjustment, but should be checked whenever the front axle housing or front suspension arms are changed (it can be adjusted on 1998 and earlier models, however)

Notes

Chapter 11 Body

Contents

1 General information

These models feature a "unibody" layout, using a floor pan with front and rear frame side rails which support the body components, front and rear suspension systems and other mechanical components. Certain components are particularly vulnerable to accident damage and can be unbolted and repaired or replaced. Among these parts are the body moldings, bumpers, doors, the hood and liftgate and all glass. Only general body maintenance practices and body panel repair procedures within the scope of the do-it-yourselfer are included in this chapter.

2 Body - maintenance

1 The condition of your vehicle's body is very important, because the resale value depends a great deal on it. It's much more difficult to repair a neglected or damaged body than it is to repair mechanical components. The hidden areas of the body, such as the wheel wells, the frame and the engine compartment, are equally important, although they don't require as frequent attention as the rest of the body.

2 Once a year, or every 12,000 miles, it's a good idea to have the underside of the body steam cleaned. All traces of dirt and oil will be removed and the area can then be inspected carefully for rust, damaged brake lines, frayed electrical wires, damaged cables and other problems. The front suspension components should be greased after completion of this job.

3 At the same time, clean the engine and the engine compartment with a steam cleaner or water-soluble degreaser.

4 The wheel wells should be given close attention, since undercoating can peel away and stones and dirt thrown up by the tires can cause the paint to chip and flake, allowing rust to set in. If rust is found, clean down to the bare metal and apply an anti-rust paint.

5 The body should be washed about once a week. Wet the vehicle thoroughly to soften the dirt, then wash it down with a soft sponge and plenty of clean soapy water. If the surplus dirt is not washed off very carefully, it can wear down the paint.

6 Spots of tar or asphalt thrown up from the road should be removed with a cloth soaked in solvent.

7 Once every six months, wax the body and chrome trim. If a chrome cleaner is used to remove rust from any of the vehicle's plated parts, remember that the cleaner also removes part of the chrome, so use it sparingly.

3 Vinyl trim - maintenance

Don't clean vinyl trim with detergents, caustic soap or petroleum-based cleaners. Plain soap and water works just fine, with a soft brush to clean dirt that may be ingrained. Wash the vinyl as frequently as the rest of the vehicle. After cleaning, application of a high-quality rubber and vinyl protectant will help prevent oxidation and cracks. The protectant can also be applied to weatherstripping, vacuum lines and rubber hoses, which often fail as a result of chemical degradation, and to the tires.

4 Upholstery and carpets - maintenance

1 Every three months remove the floormats and clean the interior of the vehicle (more frequently if necessary). Use a stiff whiskbroom to brush the carpeting and loosen dirt and dust, then vacuum the upholstery and carpets thoroughly, especially along seams and crevices.

2 Dirt and stains can be removed from carpeting with basic household or automotive carpet shampoos available in spray cans. Follow the directions and vacuum again, then use a stiff brush to bring back the "nap" of the carpet.

3 Most interiors have cloth or vinyl upholstery, either of which can be cleaned and maintained with a number of material-specific cleaners or shampoos available in auto supply stores. Follow the directions on the product for usage, and always spot-test any upholstery cleaner on an inconspicuous area (bottom edge of a back seat cushion) to ensure that it doesn't cause a color shift in the material.

4 After cleaning, vinyl upholstery should be treated with a protectant. **Note:** *Make sure the protectant container indicates the product can be used on seats - some products may make a seat too slippery.* **Warning:** *Do not use protectant on vinyl-covered steering wheels.*

5 Leather upholstery requires special care. It should be cleaned regularly with saddlesoap or leather cleaner. Never use alcohol, gasoline, nail polish remover or thinner to clean leather upholstery.

6 After cleaning, regularly treat leather upholstery with a leather conditioner, rubbed in with a soft cotton cloth. Never use car wax on leather upholstery.
7 In areas where the interior of the vehicle is subject to bright sunlight, cover leather seating areas of the seats with a sheet if the vehicle is to be left out for any length of time.

5 Body repair - minor damage

Repair of scratches

1 If the scratch is superficial and does not penetrate to the metal of the body, repair is very simple. Lightly rub the scratched area with a fine rubbing compound to remove loose paint and built up wax. Rinse the area with clean water.
2 Apply touch-up paint to the scratch, using a small brush. Continue to apply thin layers of paint until the surface of the paint in the scratch is level with the surrounding paint. Allow the new paint at least two weeks to harden, then blend it into the surrounding paint by rubbing with a very fine rubbing compound. Finally, apply a coat of wax to the scratch area.
3 If the scratch has penetrated the paint and exposed the metal of the body, causing the metal to rust, a different repair technique is required. Remove all loose rust from the bottom of the scratch with a pocketknife, then apply rust inhibiting paint to prevent the formation of rust in the future. Using a rubber or nylon applicator, coat the scratched area with glaze-type filler. If required, the filler can be mixed with thinner to provide a very thin paste, which is ideal for filling narrow scratches. Before the glaze filler in the scratch hardens, wrap a piece of smooth cotton cloth around the tip of a finger. Dip the cloth in thinner and then quickly wipe it along the surface of the scratch. This will ensure that the surface of the filler is slightly hollow. The scratch can now be painted over as described earlier in this section.

Repair of dents

See photo sequence

4 When repairing dents, the first job is to pull the dent out until the affected area is as close as possible to its original shape. There is no point in trying to restore the original shape completely as the metal in the damaged area will have stretched on impact and cannot be restored to its original contours. It is better to bring the level of the dent up to a point that is about 1/8-inch below the level of the surrounding metal. In cases where the dent is very shallow, it is not worth trying to pull it out at all.
5 If the backside of the dent is accessible, it can be hammered out gently from behind using a soft-face hammer. While doing this, hold a block of wood firmly against the opposite side of the metal to absorb the hammer blows and prevent the metal from being stretched.
6 If the dent is in a section of the body which has double layers, or some other factor makes it inaccessible from behind, a different technique is required. Drill several small holes through the metal inside the damaged area, particularly in the deeper sections. Screw long, self-tapping screws into the holes just enough for them to get a good grip in the metal. Now pulling on the protruding heads of the screws with locking pliers can pull out the dent.
7 The next stage of repair is the removal of paint from the damaged area and from an inch or so of the surrounding metal. This is easily done with a wire brush or sanding disk in a drill motor, although it can be done just as effectively by hand with sandpaper. To complete the preparation for filling, score the surface of the bare metal with a screwdriver or the tang of a file or drill small holes in the affected area. This will provide a good grip for the filler material. To complete the repair, see the section on *filling and painting*.

Repair of rust holes or gashes

8 Remove all paint from the affected area and from an inch or so of the surrounding metal using a sanding disk or wire brush mounted in a drill motor. If these are not available, a few sheets of sandpaper will do the job just as effectively.
9 With the paint removed, you will be able to determine the severity of the corrosion and decide whether to replace the whole panel, if possible, or repair the affected area. New body panels are not as expensive as most people think and it is often quicker to install a new panel than to repair large areas of rust.
10 Remove all trim pieces from the affected area except those which will act as a guide to the original shape of the damaged body, such as headlight shells, etc. Using metal snips or a hacksaw blade, remove all loose metal and any other metal that is badly affected by rust. Hammer the edges of the hole on the inside to create a slight depression for the filler material.
11 Wire brush the affected area to remove the powdery rust from the surface of the metal. If the back of the rusted area is accessible, treat it with rust inhibiting paint.
12 Before filling is done, block the hole in some way. This can be done with sheet metal riveted or screwed into place, or by stuffing the hole with wire mesh.
13 Once the hole is blocked off, the affected area can be filled and painted. See the following subsection on *filling and painting*.

Filling and painting

14 Many types of body fillers are available, but generally speaking, body repair kits which contain filler paste and a tube of resin hardener are best for this type of repair work. A wide, flexible plastic or nylon applicator will be necessary for imparting a smooth and contoured finish to the surface of the filler material. Mix up a small amount of filler on a clean piece of wood or cardboard (use the hardener sparingly). Follow the manufacturer's instructions on the package, otherwise the filler will set incorrectly.
15 Using the applicator, apply the filler paste to the prepared area. Draw the applicator across the surface of the filler to achieve the desired contour and to level the filler surface. As soon as a contour that approximates the original one is achieved, stop working the paste. If you continue, the paste will begin to stick to the applicator. Continue to add thin layers of paste at 20-minute intervals until the level of the filler is just above the surrounding metal.
16 Once the filler has hardened, the excess can be removed with a body file. From then on, progressively finer grades of sandpaper should be used, starting with a 180-grit paper and finishing with 600-grit wet-or-dry paper. Always wrap the sandpaper around a flat rubber or wooden block, otherwise the surface of the filler will not be completely flat. During the sanding of the filler surface, the wet-or-dry paper should be periodically rinsed in water. This will ensure that a very smooth finish is produced in the final stage.
17 At this point, the repair area should be surrounded by a ring of bare metal, which in turn should be encircled by the finely feathered edge of good paint. Rinse the repair area with clean water until all of the dust produced by the sanding operation is gone.
18 Spray the entire area with a light coat of primer. This will reveal any imperfections in the surface of the filler. Repair the imperfections with fresh filler paste or glaze filler and once more smooth the surface with sandpaper. Repeat this spray-and-repair procedure until you are satisfied that the surface of the filler and the feathered edge of the paint are perfect. Rinse the area with clean water and allow it to dry completely.
19 The repair area is now ready for painting. Spray painting must be carried out in a warm, dry, windless and dust free atmosphere. These conditions can be created if you have access to a large indoor work area, but if you are forced to work in the open, you will have to pick the day very carefully. If you are working indoors, dousing the floor in the work area with water will help settle the dust that would otherwise be in the air. If the repair area is confined to one body panel, mask off the surrounding panels. This will help minimize the effects of a slight mismatch in paint color. Trim pieces such as chrome strips, door handles, etc., will also need to be masked off or removed. Use masking tape and several thickness of newspaper for the masking operations.
20 Before spraying, shake the paint can thoroughly, then spray a test area until the spray painting technique is mastered. Cover the repair area with a thick coat of primer. The thickness should be built up using several thin layers of primer rather than one thick one. Using 600-grit wet-or-dry sandpaper, rub down the surface of the primer until it is very smooth. While doing this, the work area should be thoroughly rinsed with water and

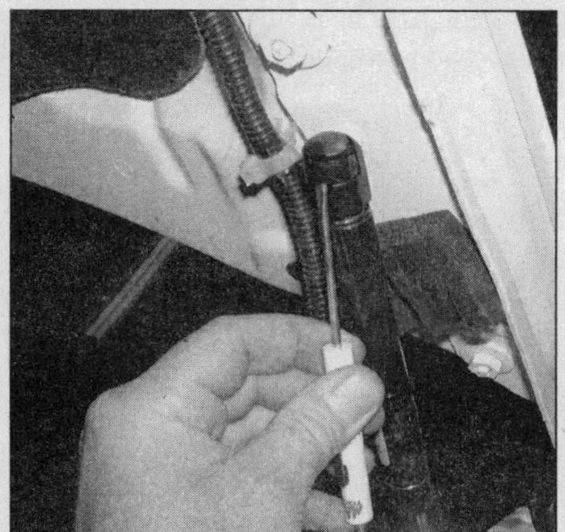

9.3 Pry the out the retaining clip and pull the strut from its ball-stud mount - bottom of strut is similarly mounted

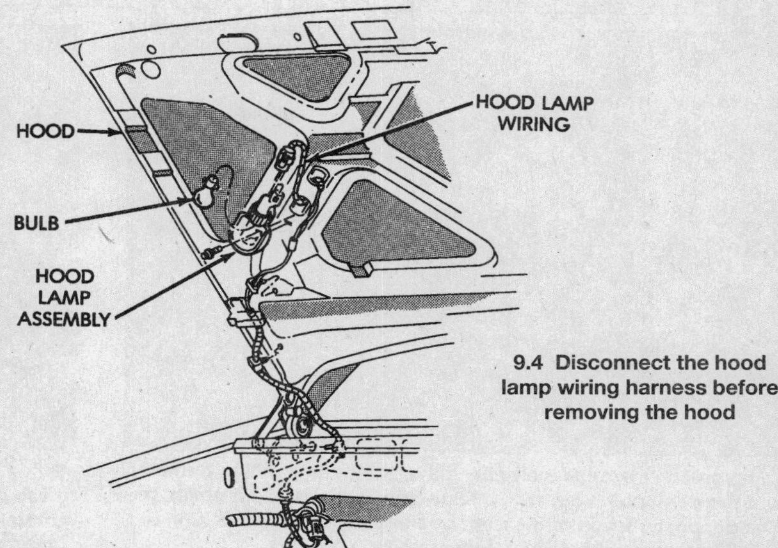

9.4 Disconnect the hood lamp wiring harness before removing the hood

the wet-or-dry sandpaper periodically rinsed as well. Allow the primer to dry before spraying additional coats.

21 Spray on the top coat, again building up the thickness by using several thin layers of paint. Begin spraying in the center of the repair area and then, using a circular motion, work out until the whole repair area and about two inches of the surrounding original paint is covered. Remove all masking material 10 to 15 minutes after spraying on the final coat of paint. Allow the new paint at least two weeks to harden, then use a very fine rubbing compound to blend the edges of the new paint into the existing paint. Finally, apply a coat of wax.

6 Body repair - major damage

1 Major damage must be repaired by an auto body shop specifically equipped to perform unibody repairs. These shops have the specialized equipment required to do the job properly.

2 If the damage is extensive, the body must be checked for proper alignment or the vehicle's handling characteristics may be adversely affected and other components may wear at an accelerated rate.

3 Due to the fact that all of the major body components (hood, fenders, etc.) are separate and replaceable units, any seriously damaged components should be replaced rather than repaired. Sometimes the components can be found in a wrecking yard that specializes in used vehicle components, often at considerable savings over the cost of new parts.

7 Hinges and locks - maintenance

Once every 3000 miles, or every three months, the hinges and latch assemblies on the doors, hood and trunk should be given a few drops of light oil or lock lubricant. The door latch strikers should also be lubricated with a thin coat of grease to reduce wear and ensure free movement. Lubricate the door and trunk locks with spray-on graphite lubricant.

8 Windshield and fixed glass - replacement

Replacement of the windshield and fixed glass requires the use of special fast-setting adhesive/caulk materials and some specialized tools and techniques. These operations should be left to a dealer service department or a shop specializing in glass work.

9 Hood - removal, installation and adjustment

Refer to illustrations 9.3, 9.4 and 9.11

Note: *The hood is heavy and somewhat awkward to remove and install - at least two people should perform this procedure.*

Removal and installation

1 Use blankets or pads to cover the cowl area of the body and the fenders. This will protect the body and paint as the hood is lifted off.

2 Scribe alignment marks around the hinge plates to insure proper alignment during installation (paint or a permanent-type felt-tip marker also will work for this).

3 On 1999 and later models, secure the hood in the open position with a prop, and disconnect the hood support struts **(see illustration)**.

4 Disconnect any cables or wire harnesses that will interfere with removal **(see illustration)**. **Note:** *On models with an underhood light, the light and harness can be left attached to the hood by disconnecting the connector located under the cowl grille* (see Section 28).

9.11 Thread the rubber bumper in-or-out to make fine adjustments to the hood closed height

5 Have an assistant support the weight of the hood. Remove the hinge-to-hood nuts.

6 Lift off the hood.

7 Installation is the reverse of removal.

Adjustment

8 Fore-and-aft and side-to-side adjustment of the hood is done by moving the hood in relation to the hinge plate after loosening the bolts.

9 Scribe or trace a line around the entire hinge plate so you can judge the amount of movement.

10 Loosen the nuts and move the hood into correct alignment. Move it only a little at a time. Tighten the hinge bolts or nuts and carefully lower the hood to check the alignment.

11 Adjust the hood bumpers on the radiator support so the hood is flush with the fenders when closed **(see illustration)**.

These photos illustrate a method of repairing simple dents. They are intended to supplement *Body repair - minor damage* in this Chapter and should not be used as the sole instructions for body repair on these vehicles.

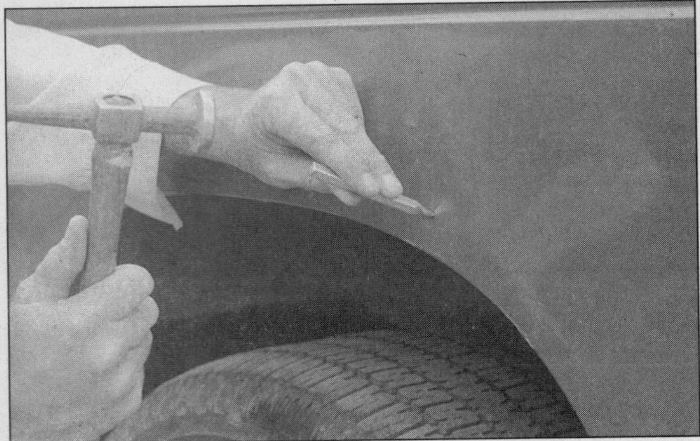

1 If you can't access the backside of the body panel to hammer out the dent, pull it out with a slide-hammer-type dent puller. In the deepest portion of the dent or along the crease line, drill or punch hole(s) at least one inch apart . . .

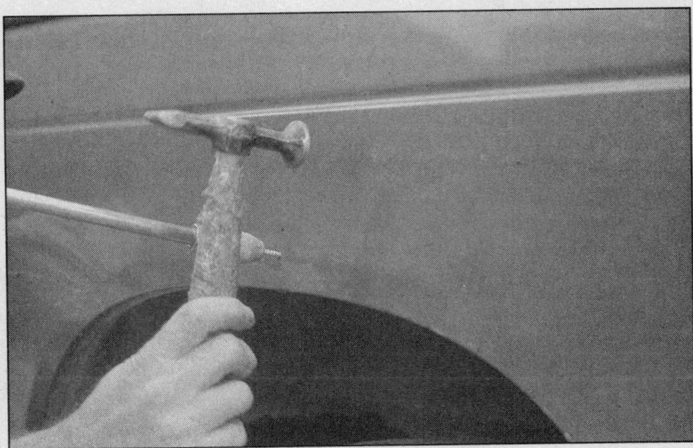

2 . . . then screw the slide-hammer into the hole and operate it. Tap with a hammer near the edge of the dent to help 'pop' the metal back to its original shape. When you're finished, the dent area should be close to its original contour and about 1/8-inch below the surface of the surrounding metal

3 Using coarse-grit sandpaper, remove the paint down to the bare metal. Hand sanding works fine, but the disc sander shown here makes the job faster. Use finer (about 320-grit) sandpaper to feather-edge the paint at least one inch around the dent area

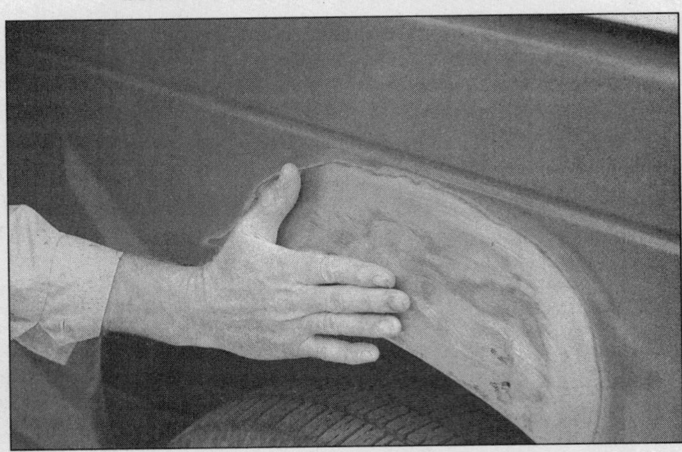

4 When the paint is removed, touch will probably be more helpful than sight for telling if the metal is straight. Hammer down the high spots or raise the low spots as necessary. Clean the repair area with wax/silicone remover

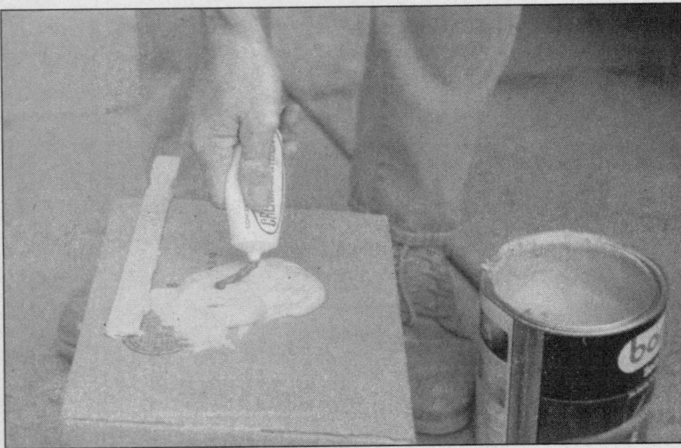

5 Following label instructions, mix up a batch of plastic filler and hardener. The ratio of filler to hardener is critical, and, if you mix it incorrectly, it will either not cure properly or cure too quickly (you won't have time to file and sand it into shape)

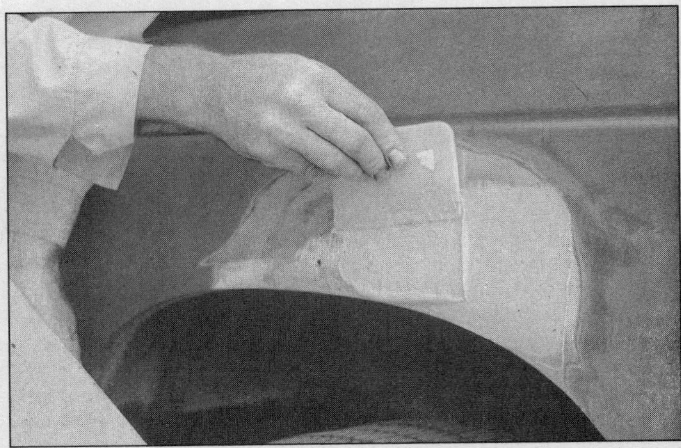

6 Working quickly so the filler doesn't harden, use a plastic applicator to press the body filler firmly into the metal, assuring it bonds completely. Work the filler until it matches the original contour and is slightly above the surrounding metal

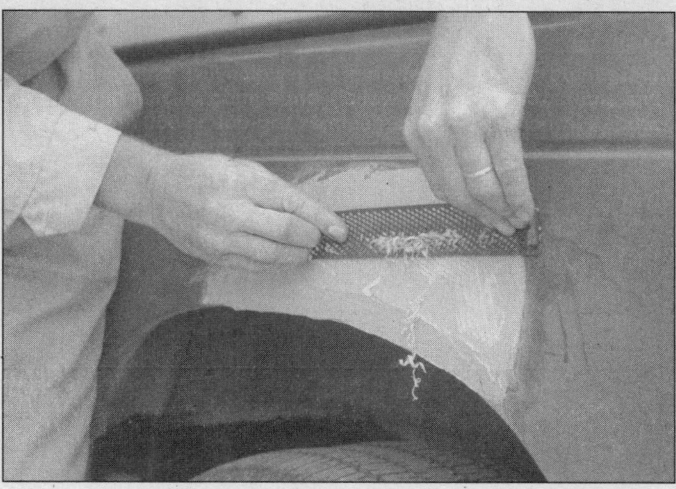

7 Let the filler harden until you can just dent it with your fingernail. Use a body file or Surform tool (shown here) to rough-shape the filler

8 Use coarse-grit sandpaper and a sanding board or block to work the filler down until it's smooth and even. Work down to finer grits of sandpaper - always using a board or block - ending up with 360 or 400 grit

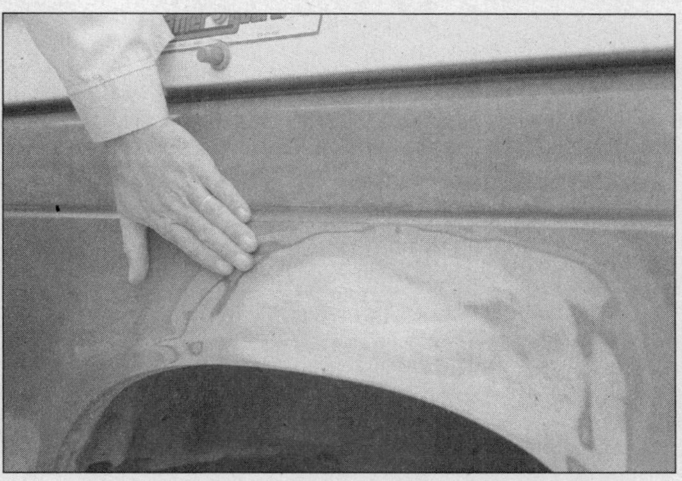

9 You shouldn't be able to feel any ridge at the transition from the filler to the bare metal or from the bare metal to the old paint. As soon as the repair is flat and uniform, remove the dust and mask off the adjacent panels or trim pieces

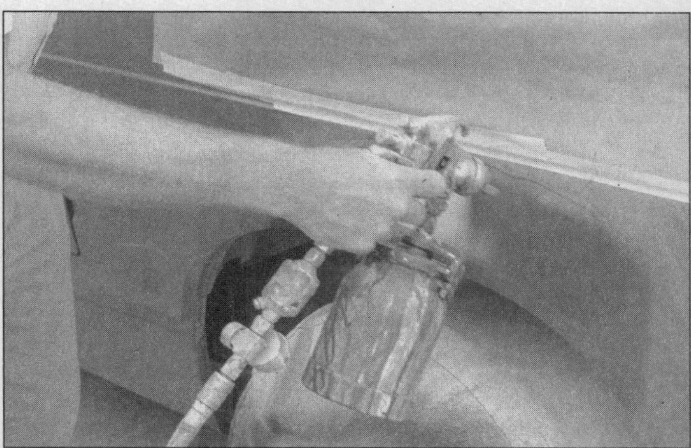

10 Apply several layers of primer to the area. Don't spray the primer on too heavy, so it sags or runs, and make sure each coat is dry before you spray on the next one. A professional-type spray gun is being used here, but aerosol spray primer is available inexpensively from auto parts stores

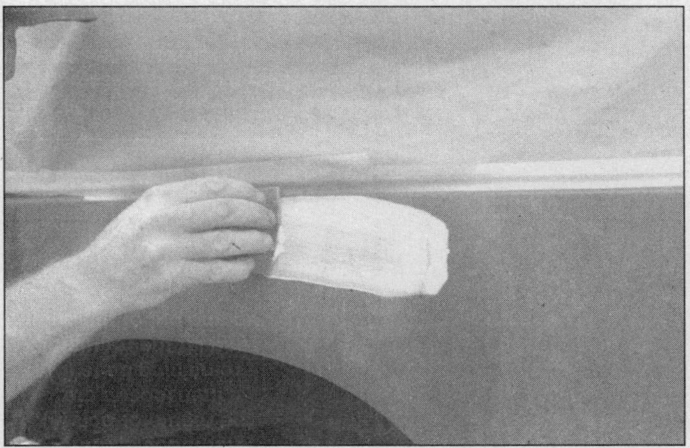

11 The primer will help reveal imperfections or scratches. Fill these with glazing compound. Follow the label instructions and sand it with 360 or 400-grit sandpaper until it's smooth. Repeat the glazing, sanding and respraying until the primer reveals a perfectly smooth surface

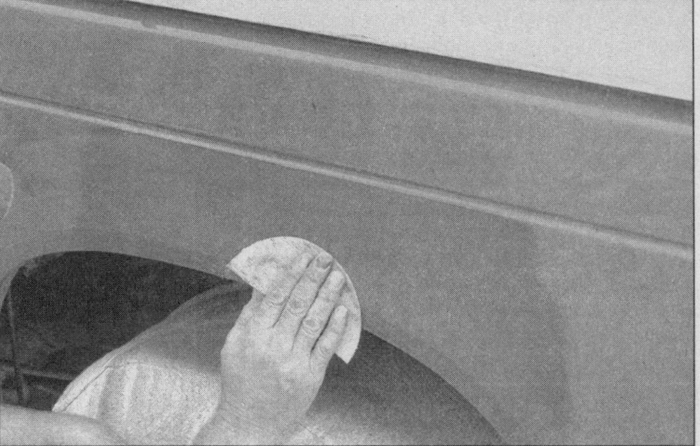

12 Finish sand the primer with very fine sandpaper (400 or 600-grit) to remove the primer overspray. Clean the area with water and allow it to dry. Use a tack rag to remove any dust, then apply the finish coat. Don't attempt to rub out or wax the repair area until the paint has dried completely (at least two weeks)

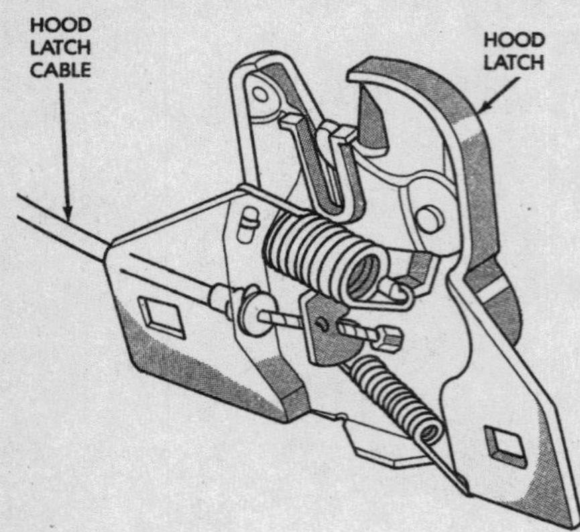

10.1 Detach the hood latch cable from the hood latch

12 The safety catch assembly on the hood itself can also be adjusted fore-and-aft and side-to-side after loosening the bolts.

13 The hood latch assembly, as well as the hinges, should be periodically lubricated with white lithium-base grease to prevent sticking and wear.

10 Hood latch and cable - removal and installation

Refer to illustrations 10.1 and 10.4

Latch

1 Remove the bolts, detach the latch assembly, then use a screwdriver to detach the cable end **(see illustration)**.

2 Installation is the reverse of removal.

Cable

3 Disconnect the cable from the latch **(see illustration 10.1)**.

4 In the passenger compartment, remove the kick panel. Remove the screws and detach the hood release cable and handle assembly **(see illustration)**.

5 Connect a string or piece of wire to the engine compartment end of the cable, then detach the cable and pull it through the firewall into the passenger compartment.

6 Connect the string or wire to the new cable and pull it through the firewall into the engine compartment.

7 The remainder of installation is the reverse of removal.

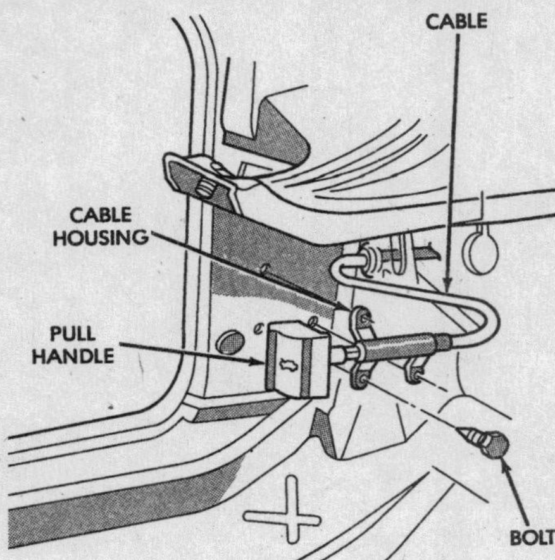

10.4 Hood release cable and pull handle installation details

11 Radiator grille - removal and installation

Refer to illustration 11.2

Warning: *These models are equipped with airbags. The airbag is armed and can deploy (inflate) anytime the battery is connected. To prevent accidental deployment (and possible injury), disconnect the negative battery cable whenever working near airbag components. After the battery is disconnected, wait at least 2 minutes before beginning work (the system has a back-up capacitor that must fully discharge). For more information see Chapter 12.*

1 Disconnect the negative cable from the battery.

2 Remove the mounting screws, detach the lower clips and detach the grille assembly **(see illustration)**. **Note:** *The radiator grille on 1999 and later models is an integral part of the front bumper fascia. See Section 13 for the radiator grille/bumper fascia replacement procedures on these models.*

3 Installation is the reverse of removal.

12 Front fender - removal and installation

Refer to illustrations 12.5, 12.7a and 12.7b

Warning: *These models are equipped with airbags. The airbag is armed and can deploy (inflate) anytime the battery is connected. To prevent accidental deployment (and possible injury), disconnect the negative battery cable whenever working near airbag components. After the battery is disconnected, wait at least 2 minutes before beginning work (the system has a back-up capacitor that must fully discharge). For more information see Chapter 12.*

1 Disconnect the negative cable from the battery.

2 Raise the vehicle, support it securely on jackstands and remove the front wheel.

3 Remove the headlight housing assembly, side marker and turn signal lamps (see Chapter 12).

4 On models through 1998, remove the front bumper fascia assembly (see Section 13).

5 Remove the inner fender liner **(see illustration)**.

6 Disconnect the antenna and all light bulb wiring harness connectors and other components that would interfere with fender removal.

7 Remove the fender mounting bolts and nuts **(see illustrations)**. **Note:** *On 1999 and later models, the lower front fender mounting fasteners can be removed by simply pulling the corner of the bumper fascia out. The fascia does not have to be removed.*

8 Detach the fender. It's a good idea to have an assistant support the fender while

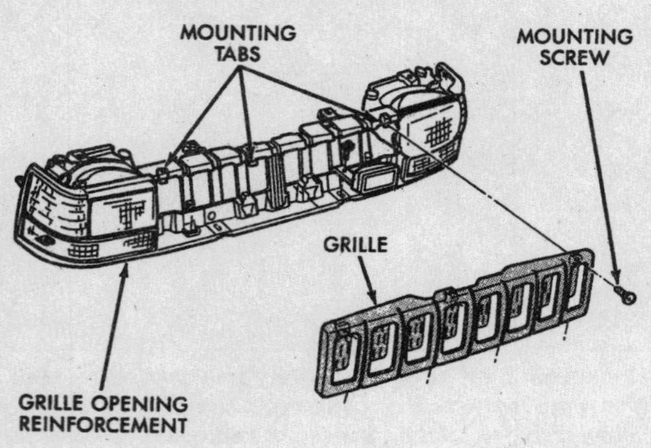

11.2 Radiator grille installation details

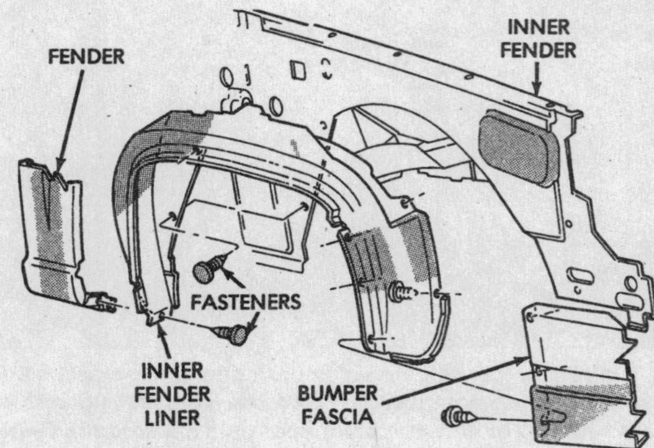

12.5 Inner fender liner installation details

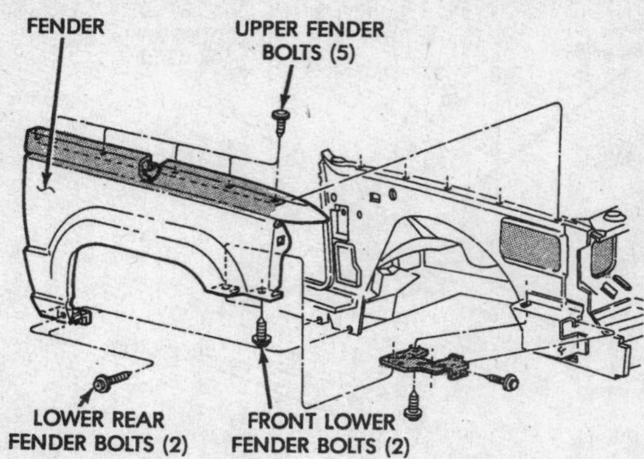

12.7a Remove the bolts and detach the fender - typical

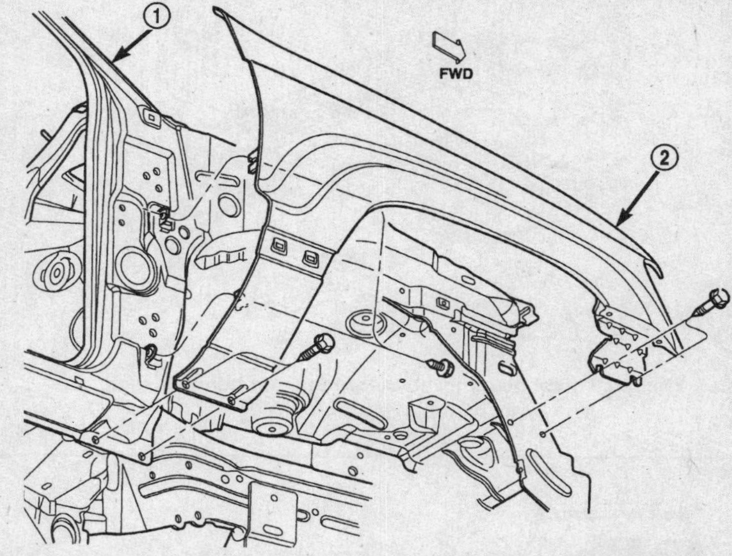

12.7b Lower front fender mounting details on 1999 and later models

1 Body *2 Fender*

it's being moved away from the vehicle to prevent damage to the surrounding body panels.

9 Installation is the reverse of removal.

10 Tighten all nuts, bolts and screws securely.

13 Bumpers - removal and installation

Warning: *These models are equipped with airbags. The airbag is armed and can deploy (inflate) anytime the battery is connected. To prevent accidental deployment (and possible injury), disconnect the negative battery cable whenever working near airbag components. After the battery is disconnected, wait at least 2 minutes before beginning work (the system has a back-up capacitor that must fully discharge). For more information see Chapter 12.*

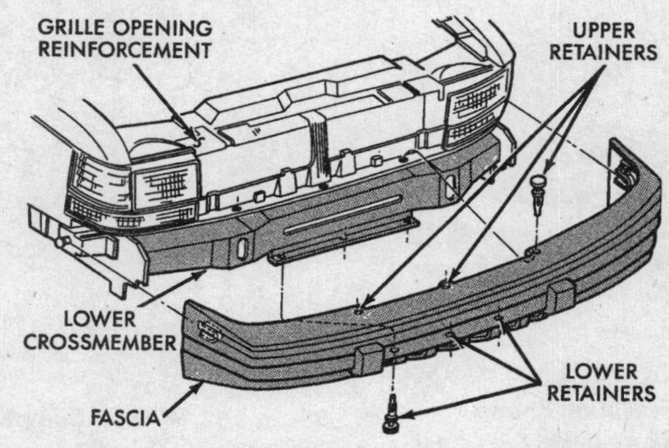

13.3a Remove the upper and lower fascia retainers (1998 and earlier shown, 1999 and later similar)

Front

Refer to illustrations 13.3a, 13.3b and 13.4

1 The front bumper on these models consists of a fascia, which covers the lower crossmember that functions as the actual bumper. Disconnect the cable from the negative battery terminal and remove the grille (see Section 11). **Note:** *On some models, the grille is part of the front bumper fascia and, once unclipped at the top from the body, will come off with the fascia.*

2 On models through 1998, remove the right and left headlight housing assemblies, side marker and turn signal lamps (see Chapter 12).

3 Remove the front fascia retainers **(see illustrations)**.

4 Remove the plastic rivets from the front of the inner fender liners **(see illustration)**.

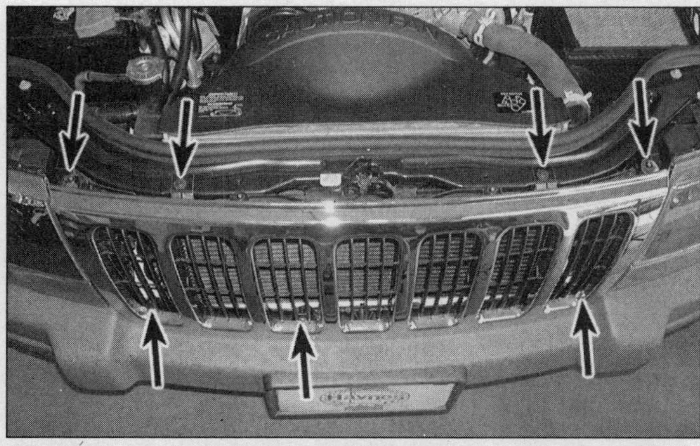

13.3b On 1999 and later models, the upper retainers (arrows) are located in the grille area

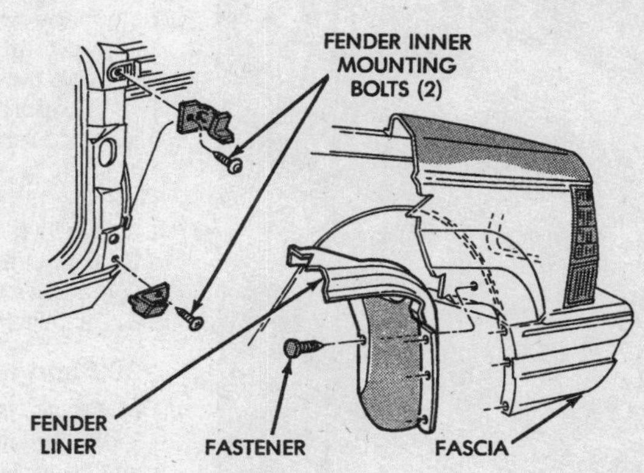

13.4 Remove the fasteners retaining the fender liner to the fascia

11

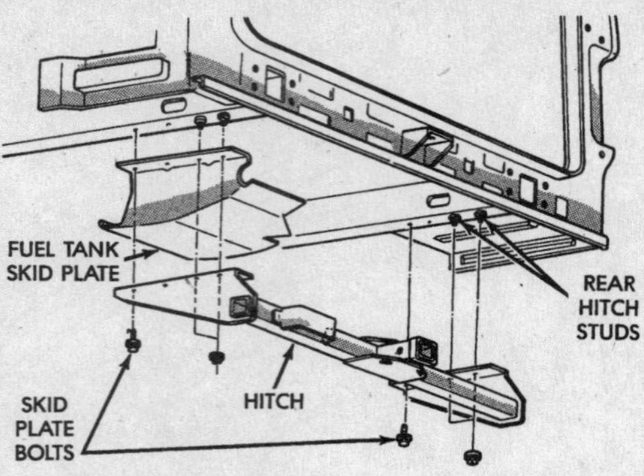

13.8 The hitch must be removed before the rear bumper can be removed

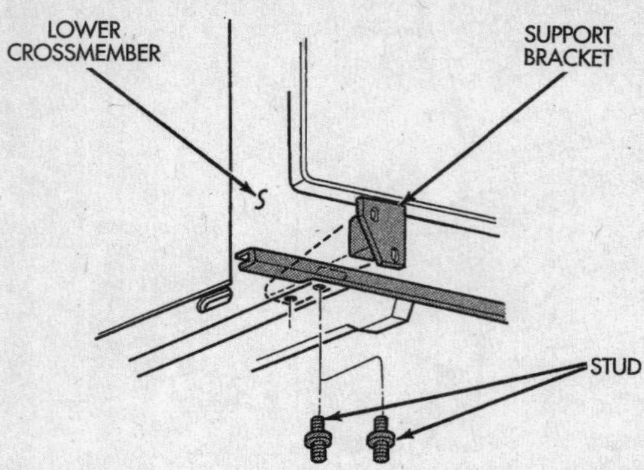

13.10 Remove the bumper support bracket bolts

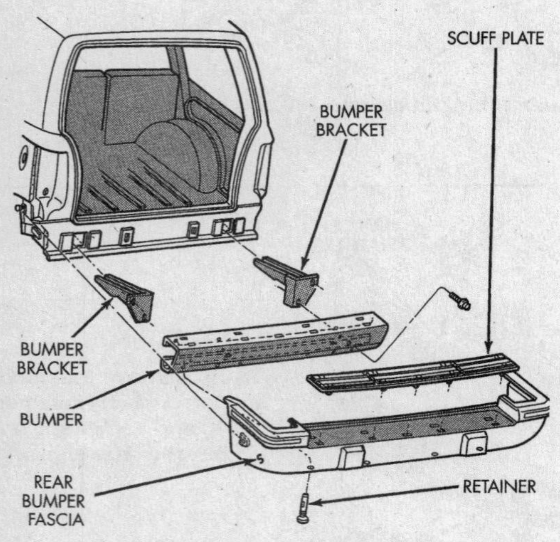

13.11 Rear bumper installation details - 1998 and earlier models

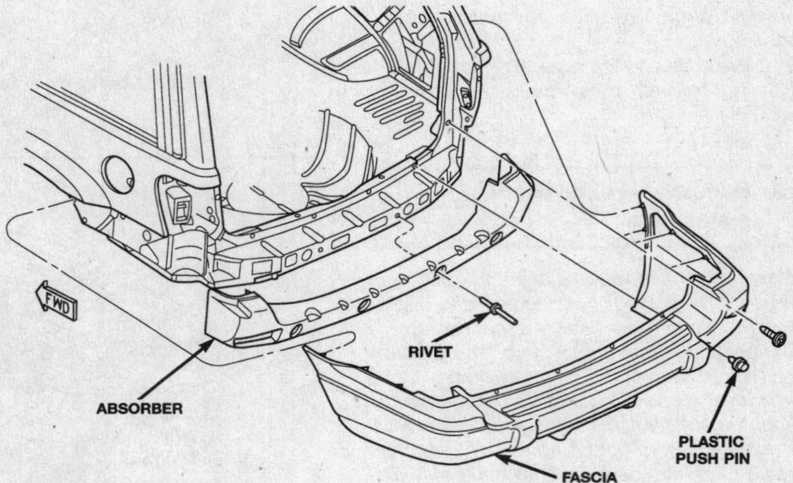

13.16 Rear bumper installation details - 1999 and later models

5 Slide the fascia off the retaining pegs, using a small screwdriver to detach the locating tangs under the turn signal mounts.
6 Installation is the reverse of removal.

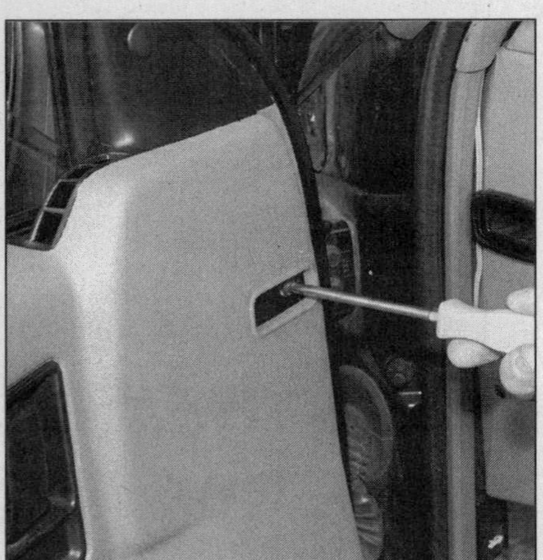

14.2a Use a Phillips screwdriver to remove the door trim panel screws

Rear

1998 and earlier

Refer to illustrations 13.8, 13.10 and 13.11

7 Raise the rear of the vehicle and support it securely on jackstands.
8 On models equipped with a trailer hitch, detach the wiring harness, support the hitch, remove retaining nuts and lower the hitch assembly **(see illustration)**.
9 Support the bumper and remove the inner fender retainers.
10 Remove the bumper support bolts and slide the bumper beam/fascia assembly off the vehicle **(see illustration)**.
11 Remove the support brackets from the bumper (**see illustration)**.
12 Installation is the reverse of removal.

1999 and later models

Refer to illustration 13.16

13 On these models, the bumper reinforcement beam is riveted to the chassis, with the plastic fascia covering it.
14 Remove the plastic push pins and rivets securing the fascia to the wheelwell liners and the rear skid plate (under the fuel tank).
15 In each fenderwell, remove the nuts securing the front edges of the fascia to the body.
16 Open the liftgate and remove the screws, pushpins and retainer clips securing the fascia to the liftgate opening (**see illustration**). **Note:** *Remove the plastic trim at the rear body pillars and pull back the rubber weatherstrip to access the fascia retaining clips.*

14 Door trim panel - removal and installation

Refer to illustrations 14.2a, 14.2b, 14.2c, 14.2d, 14.5 and 14.6

1 Disconnect the negative cable from the battery.
2 Remove the door trim panel retaining screws and the door pull/armrest assembly **(see illustrations)**.
3 On manual window models, remove the

14.2b Remove the door trim panel screw located in the armrest

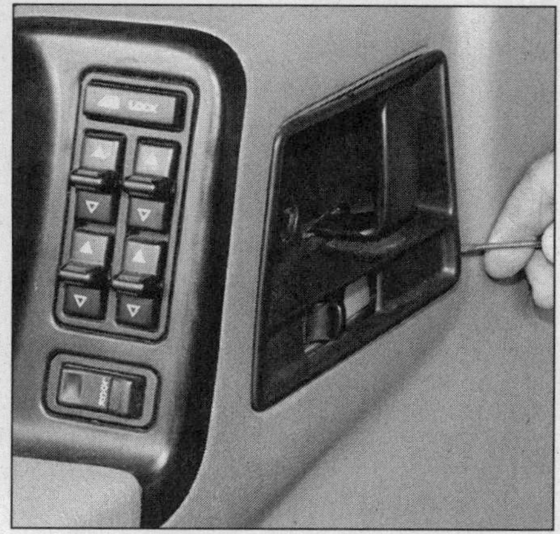
14.2c Remove the screw, then pry off the door handle cover

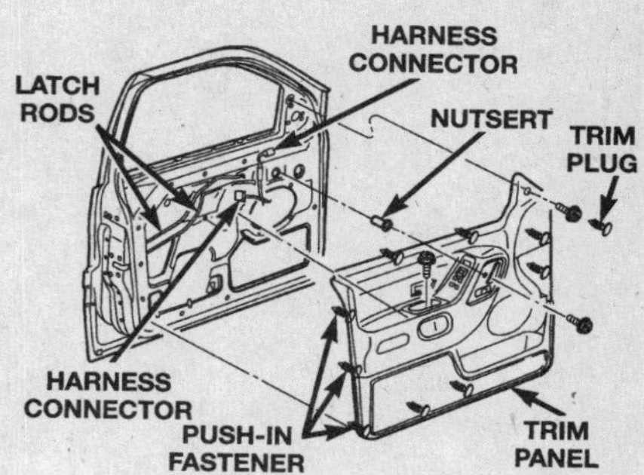

14.2d Door trim panel screw locations on 1999 and later models - a plastic plug in the outside mirror area must be removed to access the upper front screw

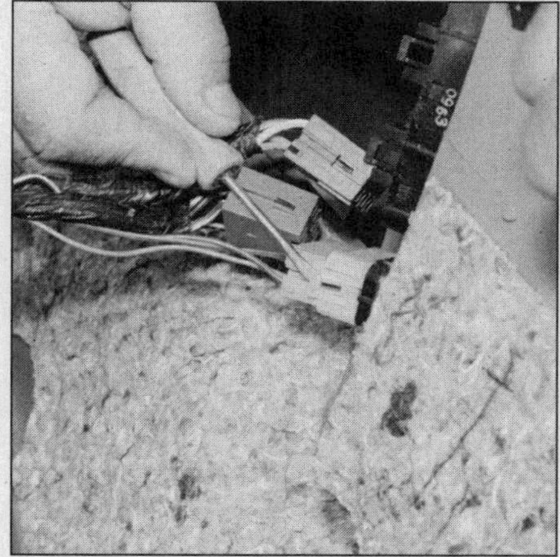
14.5 Pull the panel away from the door and disconnect the electrical connectors

14.6 Peel the water deflector carefully away from the door, taking care not to tear or distort it

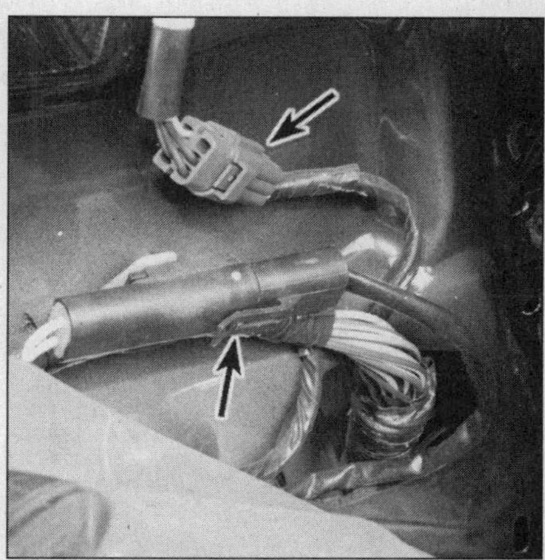
15.1 Disconnect the door electrical connectors (arrows) and remove the harness from the door

window crank.

4 Insert a special door panel clip removal tool (available at auto parts stores) or a putty knife between the trim panel and the door and disengage the retaining clips. Work around the outer edge until the panel is free.

5 Once all of the clips are disengaged, detach the trim panel, disconnect any wire harness connectors and remove the trim panel from the vehicle **(see illustration)**. **Note:** *On some models, disconnect the latch rod from the inner door handle*.

6 For access to the inner door, carefully peel back the plastic watershield **(see illustration)**.

7 Prior to installation of the door panel, be sure to reinstall any clips in the panel which may have come out during the removal procedure and remain in the door itself.

8 Connect the wire harness connectors and place the panel in position in the door. Press the trim panel into place until the clips are seated.

9 Install the armrest/door pulls and the window crank. Connect the negative battery cable.

15 Door - removal, installation and adjustment

Refer to illustrations 15.1 and 15.4

1 Remove the door trim panel. Disconnect any wire harness connectors and push them through the door opening so they won't interfere with door removal **(see illustration)**. On 1999 and later models, disconnect the single large connector at the doorjamb edge of the door.

2 Place a jack under the door or have an assistant on hand to support it when the hinge bolts are removed. **Note:** *If a jack is used, place a rag between it and the door to protect the door's painted surfaces*.

3 Remove the bolts securing the door check strap to the doorjamb on 1999 and later models.

4 Scribe around the mounting bolt heads with a marking pen, remove the bolts and carefully lift off the door **(see illustration)**.

5 Installation is the reverse of removal, making sure to align the hinge with the marks made during removal before tightening the bolts.

15.4 Use a marking pen to mark the bolt alignment locations before removing the door

6 Following installation of the door, check the alignment and adjust it if necessary by moving the door lock striker.

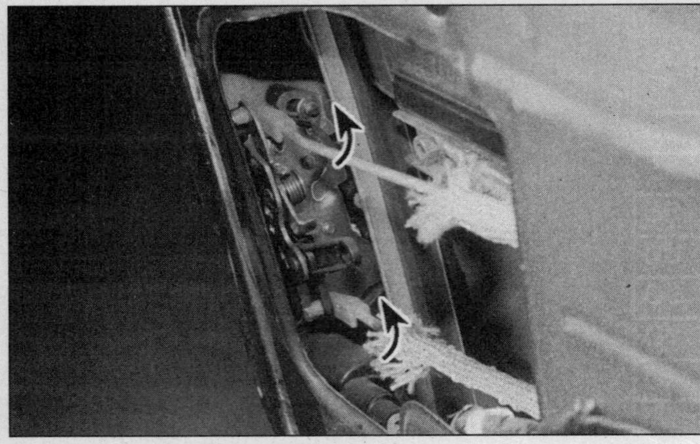

16.2 Rotate the plastic clip up and disconnect the latch rod links

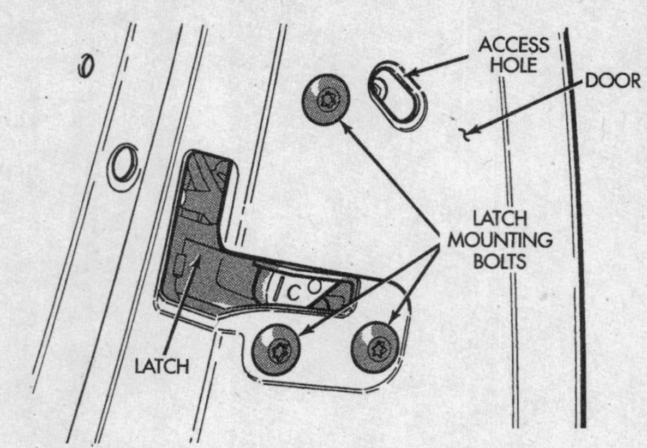

16.3 Remove the three door latch bolts

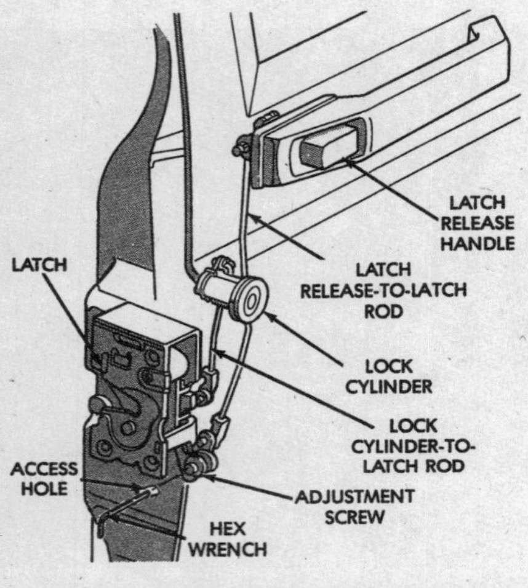

16.7 Insert a hex-head wrench into the adjustment hole, loosen the screw and move the wrench up in the slot to remove slack from the linkage

16.9 Remove the Torx screw (arrow) and the retainer plate to separate the lock cylinder from the outside door handle on 1999 and later models

16 Door latch, lock cylinder and handles - removal and installation

Refer to illustrations 16.2, 16.3, 16.7, 16.9, 16.11, 16.13a and 16.13b

Latch

1 Raise the window completely and remove the door trim panel and watershield (see Section 14).

2 Disconnect the link rods from the latch **(see illustration)**.

3 Remove the three Torx-head mounting bolts **(see illustration)**. It may be necessary to use an impact-driver to loosen them. **Note:** *On 1999 and later models, remove the screw in the rear edge of the door, just above the latch, to allow moving the bottom of the rear glass run channel out of the way for latch removal.*

4 Remove the latch from the door.

5 Place the latch in position and install the mounting bolts. Tighten the bolts securely.

6 Connect the link rods to the latch.

7 Check the door to make sure it closes properly. Readjust the latch (by loosening the bolts and moving it) as necessary until the door closes smoothly (with the door handle flush with the door). After installation, loosen the linkage adjusting screw by inserting a 5/32-inch hex-head wrench through the adjustment hole in the end of the door **(see illustration)**. Loosen the adjustment screw and push it up in the slot to remove all slack from the linkage, then tighten it.

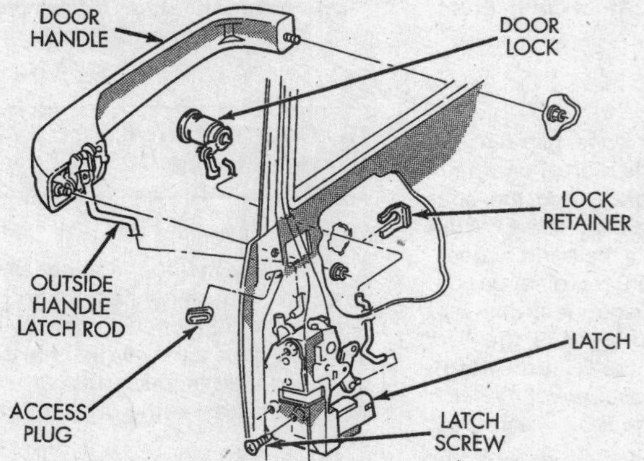

16.11 Front door outside handle and lock cylinder installation details - 1998 and earlier models

Lock cylinder

8 Remove the outside door handle (see below). On 1998 and earlier models disconnect the link, then use a screwdriver to push the key lock cylinder retainer off and withdraw the lock cylinder from the door.

9 On 1999 and later models, the lock cylinder is part of the outside handle assembly. Remove the retainer to extract the lock cylinder from the handle, then disconnect the link **(see illustration)**.

10 Installation is the reverse of removal.

Outside handle

11 Disconnect the outside handle (remote control) link from the latch, remove the mounting nuts and detach the handle from

16.13a On 1998 and earlier models, remove the bolt (arrow), then rotate the inside handle forward and disconnect the rods

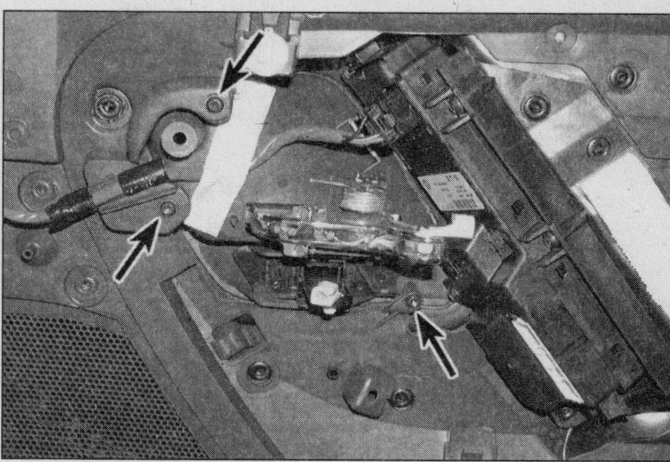
16.13b Inside door handle retaining screws (arrows) - 1999 and later models

the door **(see illustration)**. On 1999 and later models, there are two handle mounting screws in the door jamb area, and two nuts accessed from inside the door.

12 Place the handle in position, attach the link and install the nuts. Tighten the nuts securely.

Inside handle

13 Remove the retaining bolt, rotate the handle forward and disconnect the links, then lift the handle off the door **(see illustrations)**.

17 Door window glass - removal and installation

Refer to illustrations 17.4, 17.5a and 17.5b

1 Remove the door trim panel and water-shield (see Section 14).

2 Lower the glass.

3 Pry the weatherstripping and moldings (inside and outside at the beltline window opening) out of the door glass opening. On most models, there is a screw retaining the outer beltline molding to the rear edge of the door, and a screw in the latch-end of the door retaining the rear edge of the glass run channel.

4 On rear windows, remove the screws, tilt the stationary glass forward and lift it out of the door **(see illustration)**.

5 Remove the two nuts or two clips retaining the glass to the window track and slide the glass until the studs line up with the holes in the track, then detach the glass **(see illustrations)**.

6 Lift the glass up and out of the door through the glass opening.

7 To install, lower the glass into the door, slide it into position and install the nuts or clips.

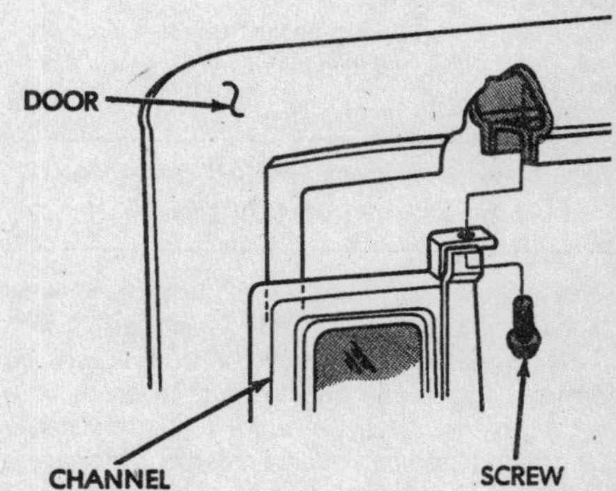

17.4 Remove the screws and detach the rear door stationary glass from the channel

18 Door window regulator - removal and installation

Refer to illustrations 18.3a, 18.3b and 18.3c

1 Remove the door trim panel and water-shield (see Section 14).

2 Remove the door window glass (see Section 17).

3 Loosen or remove the window regula-

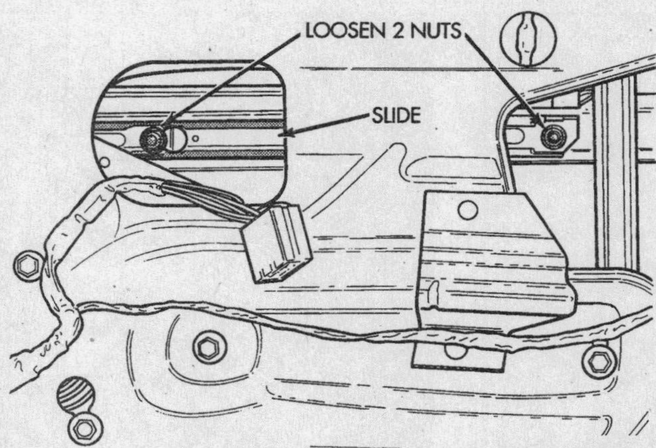

17.5a On 1998 and earlier models, loosen the nuts located at each end, slide the track off and lift the glass out of the door

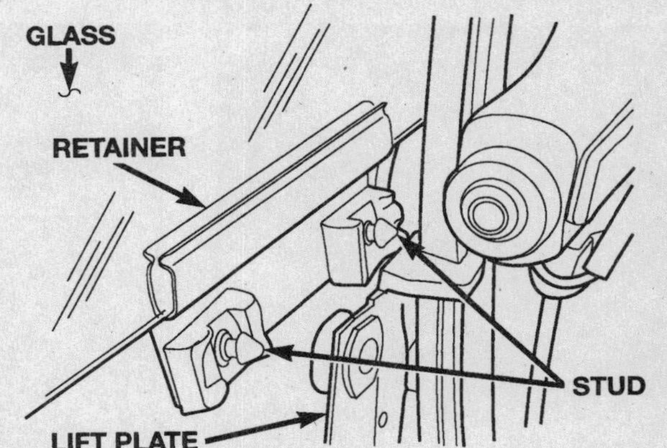

17.5b On 1999 and later models, the glass is retained to the lift plate with clips that go over studs on the glass mount - use a long hooked tool to remove the clips

11

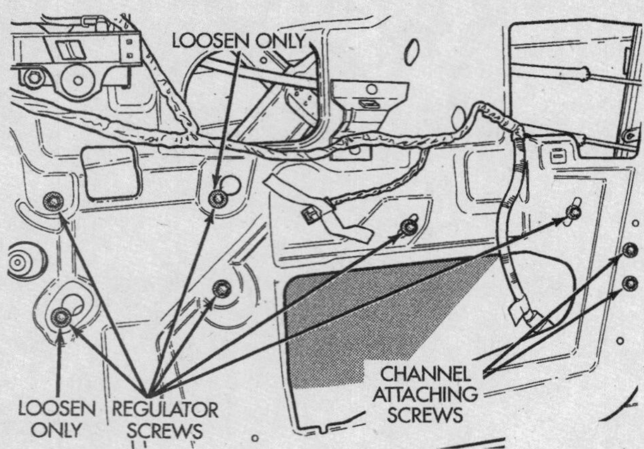

18.3a Front door window regulator details - 1998 and earlier models

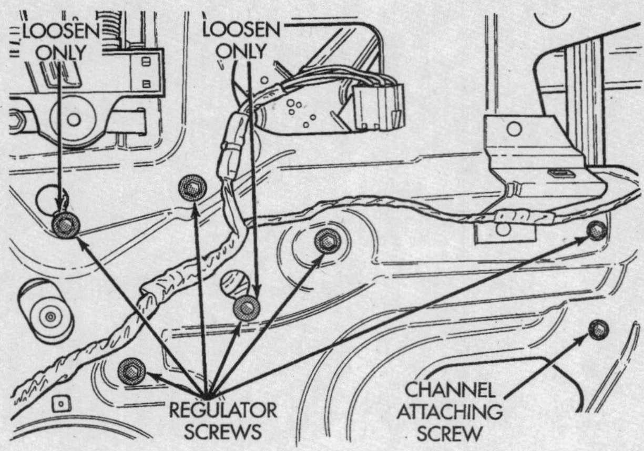

18.3b Rear door window regulator details - 1998 and earlier models

tor-to-door and channel attaching screws **(see illustrations)**.

4 Remove the regulator from the door.

5 Installation is the reverse of removal.

18.3c On 1999 and later models, remove the regulator bolt (A), then loosen the remaining bolts (B) and lift the regulator up and out of the door

19 Center and overhead consoles - removal and installation

Warning: *These models are equipped with airbags. The airbag is armed and can deploy (inflate) anytime the battery is connected. To prevent accidental deployment (and possible injury), disconnect the negative battery cable whenever working near airbag components. After the battery is disconnected, wait at least 2 minutes before beginning work (the system has a back-up capacitor that must fully discharge). For more information see Chapter 12.*

1 Disconnect the negative cable from the battery.

Center console

Refer to illustrations 19.2a, 19.2b, 19.2c, 19.4a, 19.4b and 19.4c

2 On 1998 and earlier models, detach the 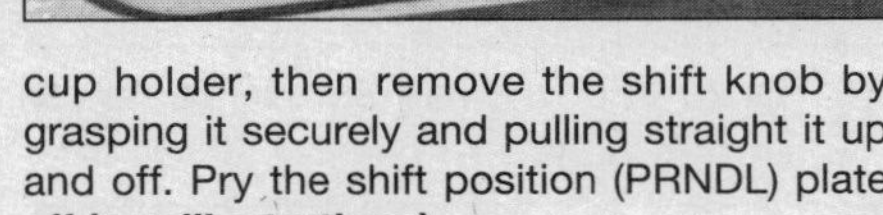 cup holder, then remove the shift knob by grasping it securely and pulling straight it up and off. Pry the shift position (PRNDL) plate off **(see illustrations)**.

3 On 1999 and later models, position the shift levers in the fully rearward position, then remove the front and rear storage bins.

4 Remove the console mounting screws/bolts **(see illustrations)**. On 1999 and later models, it will be necessary to pull the rear cupholder outward to access the rear console screws.

5 Lift the console upward slightly and disconnect any electrical connectors.

6 Remove the console from the vehicle by pulling the console up past the shift lever(s)

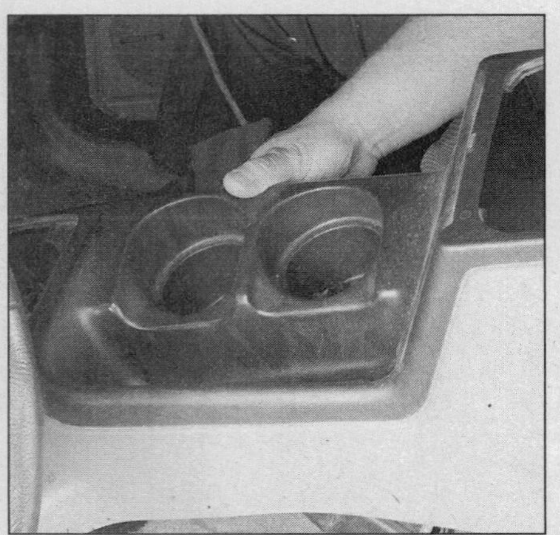

19.2a On 1998 and earlier models, detach the cup holder from the console . . .

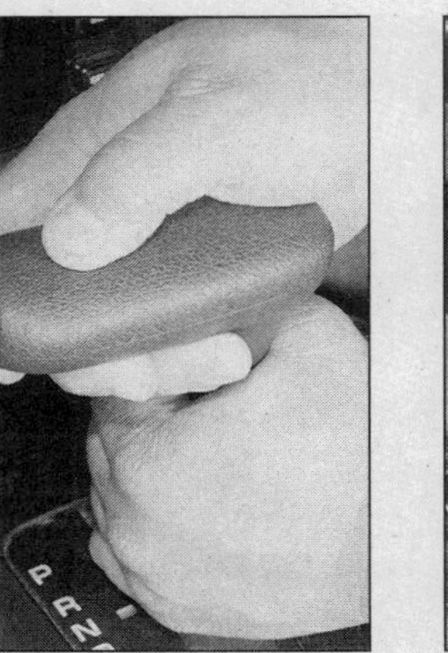

19.2b . . . grasp the shift knob securely and pull it straight up sharply to remove it . . .

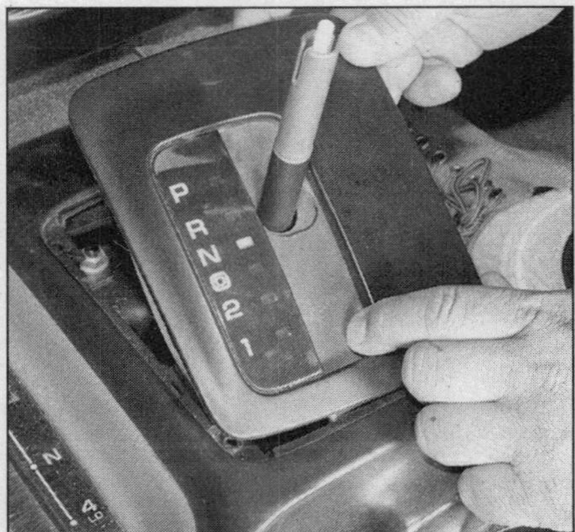

19.2c . . . then detach the shift plate

19.4a Full floor console installation details - 1998 and earlier models

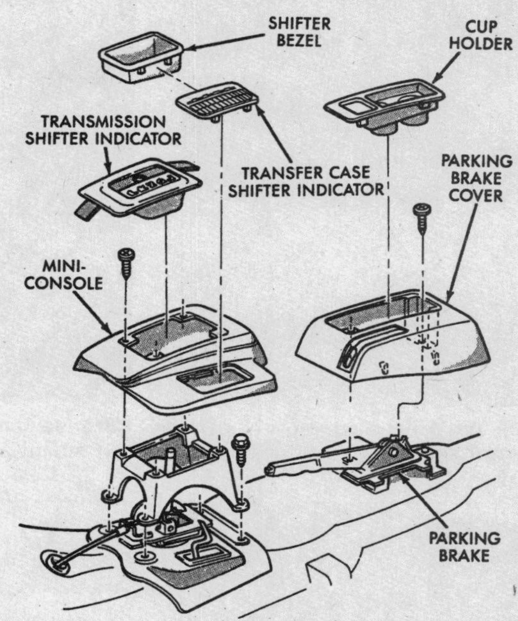

19.4b Mini console installation details - 1998 and earlier models

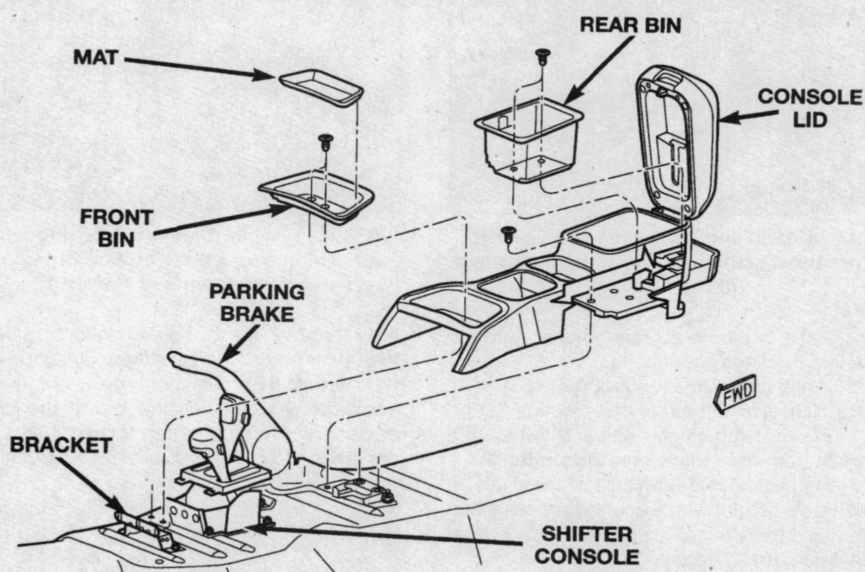

19.4c Floor console installation details - 1999 and later models

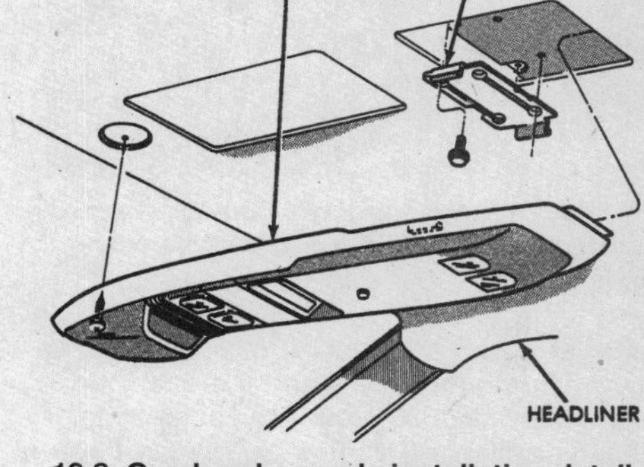

19.8 Overhead console installation details - 1998 and earlier models

and towards the rear of the vehicle.

7 Installation is the reverse of removal.

Overhead console

Refer to illustration 19.8

8 Remove the front console retaining screw and disconnect the trip computer electrical connector **(see illustration)**.

9 Slide the console forward, detach it from the bracket and lower it sufficiently to disconnect the electrical connectors, then remove it from the vehicle.

10 On 1998 and earlier models, remove the retaining screws or clips and detach the rear half of the console.

11 Installation is the reverse of removal.

20 Liftgate trim panel - removal and installation

Refer to illustrations 20.1, 20.2, 20.4 and 20.5

1 Remove the liftgate trim panel retaining screws **(see illustration)**.

2 Carefully pry loose the retaining clips with a special trim panel removal tool (avail-

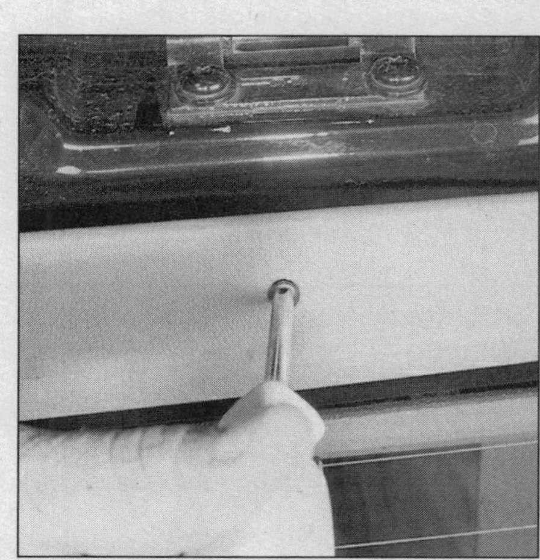

20.1 Use a Phillips screwdriver to remove the liftgate panel screws

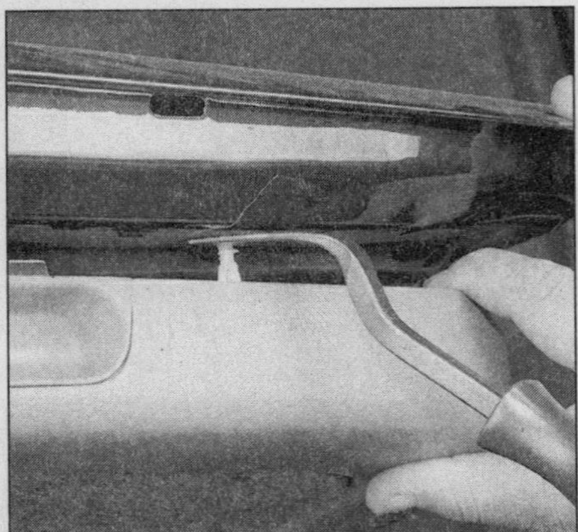

20.2 Use a trim panel clip removal tool to pry the plastic retainers out of the liftgate

20.4 Be careful not to tear or stretch the sound deadener as you peel it off

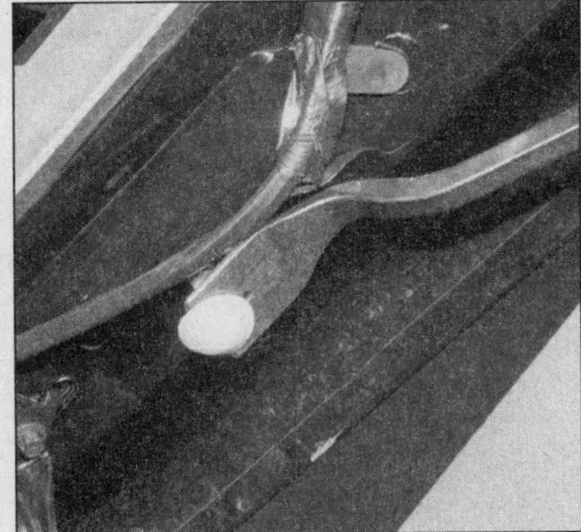

20.5 Remove any retainers that have come loose and reinstall them in the trim panel

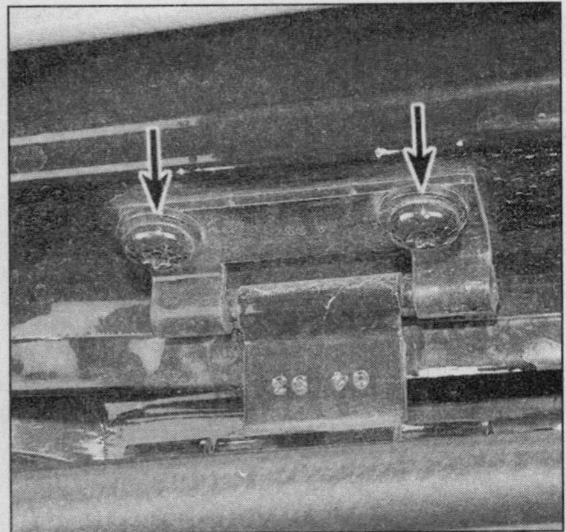

21.5 A Torx head tool will be necessary to remove the liftgate hinge bolts (arrows)

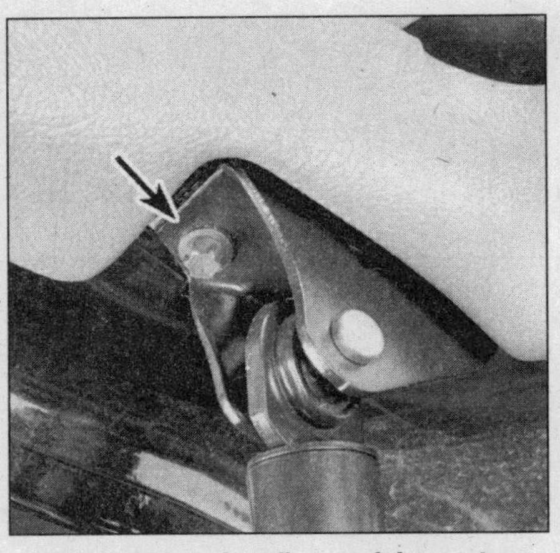

22.2a 1998 and earlier models use two Torx head screws to retain the upper end of the strut . . .

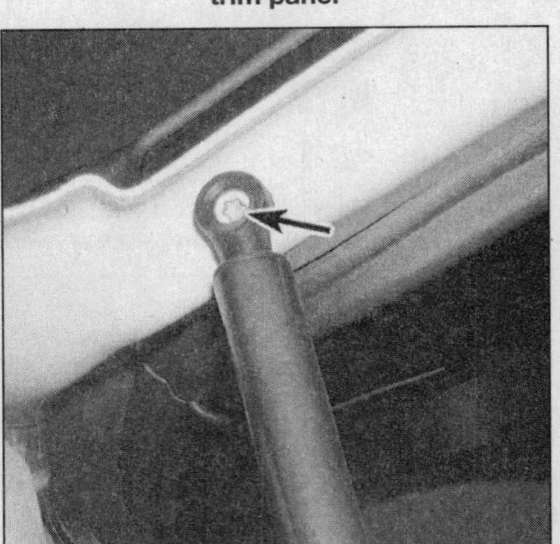

22.2b . . . while 1999 and later models, use a single torx head bolt on the upper and lower ends of the strut

able at auto parts stores) or a putty knife between the trim panel and the door. Work slowly and carefully around the outer edge of the trim panel until it's free **(see illustration)**.

3 Once all of the clips are disengaged, pull the trim panel up, disconnect any wire harness connectors and remove the panel.

4 For access to the liftgate inner panel, carefully peel back the plastic sound deadener **(see illustration)**.

5 Prior to installation of the trim panel, be sure to reinstall any clips in the panel which may have remained in the door when you removed the panel **(see illustration)**.

21 Liftgate - removal, installation and adjustment

Refer to illustration 21.5

1 Open the liftgate and cover the upper body area around the opening with pads or cloths to protect the painted surfaces when the liftgate is removed.

2 Remove the trim panel, disconnect all electrical connectors and pull the wiring harness out of the liftgate.

3 Paint or scribe alignment marks around the liftgate hinge flanges.

4 While an assistant supports the liftgate, detach the support struts (see Section 22).

5 Remove the hinge bolts and detach the liftgate from the vehicle **(see illustration)**.

6 Installation is the reverse of removal. Be extremely careful with the wire harness when threading it back into the liftgate or you could cut it on a metal edge.

7 After installation, close the liftgate and make sure it's in proper alignment with the surrounding body panels.

8 If the liftgate needs to be adjusted, loosen the hinge bolts slightly, gently close the liftgate and verify that it's centered (the striker should center it). Then carefully open the liftgate and retighten the hinge bolts.

22 Liftgate support strut - replacement

Refer to illustrations 22.2a, 22.2b and 22.3

1 Open and support the liftgate. **Note:** *If you are replacing both struts, replace one first, then the other.*

2 Remove the two Torx head screws at the upper end of the strut **(see illustrations)**.

3 On 1998 and earlier models, use a small screwdriver to detach the clip at the lower end of each strut, then grasp the strut securely and pull it out sharply to detach it

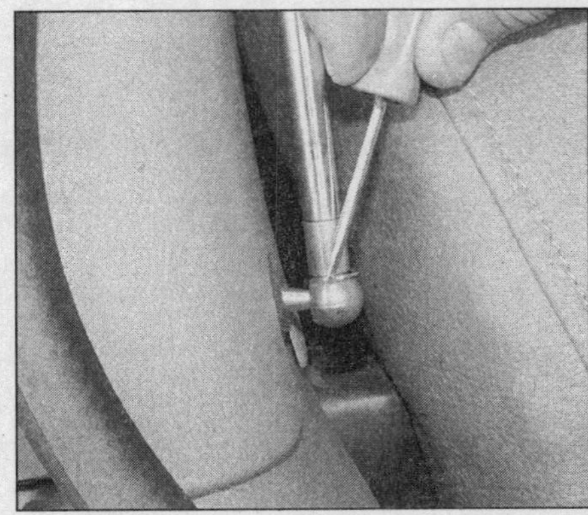

22.3 On 1998 and earlier models, use a small screwdriver to detach the clip at the lower end of the strut, then pull out sharply to detach it from the lower ballstud

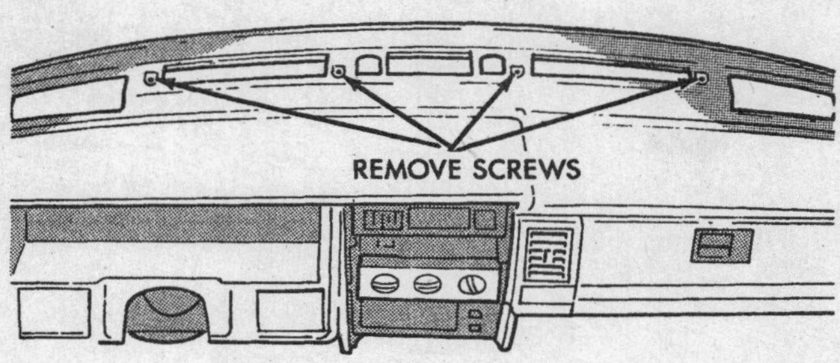

23.5 Remove the defroster duct screws (arrows) - 1995 and earlier models

23.7 Remove the cluster bezel-to-dash pad screws (arrows) - 1995 and earlier models

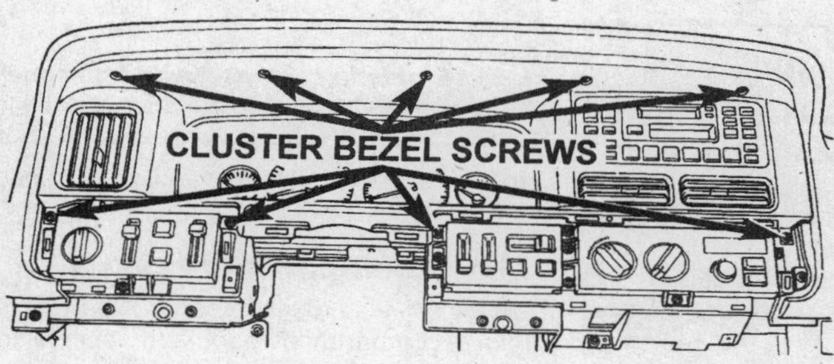

23.10 On 1996 through 1998 models, remove the screws (arrows) and detach the instrument cluster bezel

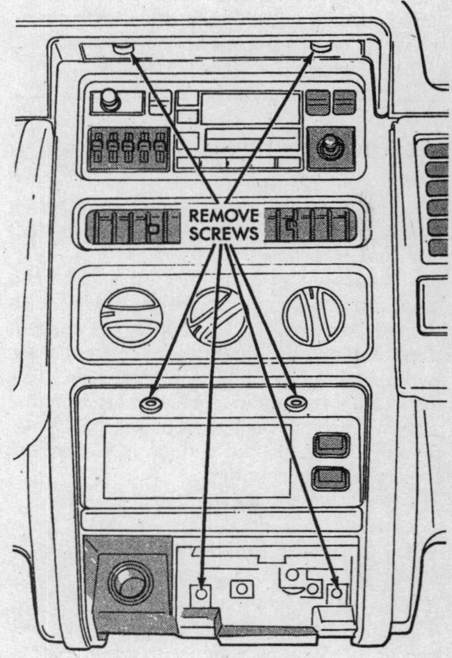

24.2 Remove the six center bezel retaining screws - 1995 and earlier models

from the ballstud **(see illustration)**. On 1999 and later models, remove the Torx bolt from the bottom of the strut.

4 Installation is the reverse of removal. Tighten all fasteners securely.

23 Instrument cluster bezel - removal and installation

Warning: *These models are equipped with airbags. The airbag is armed and can deploy (inflate) anytime the battery is connected. To prevent accidental deployment (and possible injury), disconnect the negative battery cable whenever working near airbag components. After the battery is disconnected, wait at least 2 minutes before beginning work (the system has a back-up capacitor that must fully discharge). For more information see Chapter 12.*

1 Disconnect the negative cable from the battery.

1995 and earlier models

Refer to illustrations 23.5 and 23.7

2 Remove the center instrument panel bezel (see Section 24).

3 Remove the two screws behind the top of the center bezel retaining the dash pad.

4 Pry the defroster grille out, disconnect any sensors and set the grille aside.

5 Remove the four screws in the defroster duct that retain the dash pad **(see illustration)**.

6 Open the glove box, remove the two screws retaining the dash pad and pull up on the pad to detach the end clips.

7 Remove the three cluster bezel-to-dash pad retaining screws and detach the bezel **(see illustration)**.

8 Installation is the reverse of removal.

1996 through 1998 models

Refer to illustration 23.10

9 Disconnect the negative battery cable, then use a trim tool to remove the two snap-fit bezels on the switch pods located just below the instrument cluster.

10 Remove all of the screws and remove the instrument cluster bezel **(see illustration)**.

11 Installation is the reverse of the removal procedure.

1999 and later models

12 Disconnect the negative battery cable.

13 If equipped with a tilt steering column, lower it to the lowest position.

14 Use a flat trim tool to pry in the four corners of the instrument cluster bezel to release the four clips, then rotate the bezel back and out over the steering column.

15 Installation is the reverse of the removal procedure. Make sure the clips are aligned with the receptacles on the instrument panel before pushing the bezel in.

24 Instrument panel trim bezels - removal and installation

Warning: *These models are equipped with airbags. The airbag is armed and can deploy (inflate) anytime the battery is connected. To prevent accidental deployment (and possible injury), disconnect the negative battery cable whenever working near airbag components. After the battery is disconnected, wait at least 2 minutes before beginning work (the system has a back-up capacitor that must fully discharge). For more information see Chapter 12.*

Center instrument panel bezel

1995 and earlier models

Refer to illustration 24.2

1 Remove the ashtray.

2 Remove the six screws and detach the center bezel **(see illustration)**.

3 Installation is the reverse of removal.

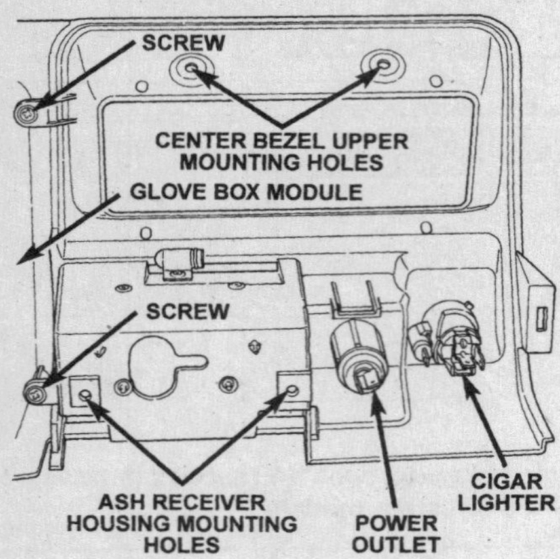

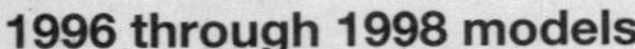

24.5 On 1996 through 1998 models, the two screws at the left in this illustration must be reached from behind, though the glove box panel opening

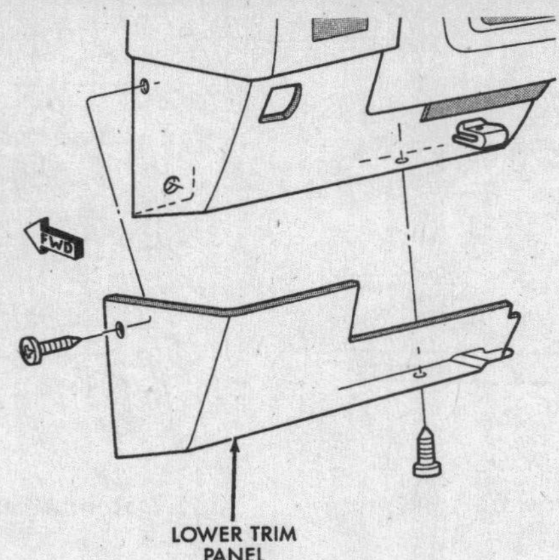

24.13 Lower trim panel installation details

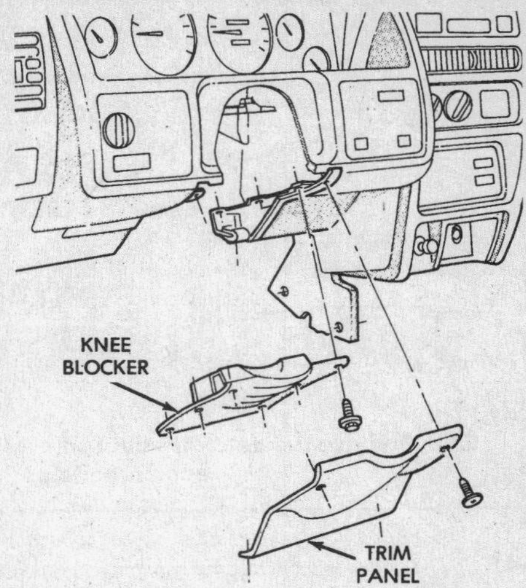

24.14 Steering column lower trim panel and knee blocker installation details

1996 through 1998 models

Refer to illustration 24.5

4 On these models, remove the glove compartment (and its surrounding panel) from the dashboard.

5 Reaching through the left side of the glove box opening, remove the two screws on the backside of the center instrument panel bezel **(see illustration)**. Remove the two upper screws from the front side, then move the bezel out to release its tab from the slot on the instrument panel.

6 Pull the bezel out enough to disconnect the electrical connectors.

7 Installation is the reverse of removal.

1999 and later models

8 There are two center bezels on these models, an upper and a lower.

9 Each bezel is attached with clips, and can be removed by prying with a trim tool.

10 To release the bottom bezel, remove the ashtray and disconnect the electrical connectors behind the bezel.

11 Installation is the reverse of removal.

Switch pod bezel(s)

1995 and earlier models

Refer to illustrations 24.13, 24.14 and 24.15

12 Remove the center bezel **(see illustration 24.2)** and the instrument cluster bezel (see Section 23).

13 Open the driver's door for access and remove the left lower trim panel **(see illustration)**.

14 Remove the lower steering column trim panel and knee blocker **(see illustration)**.

15 Remove the steering column nuts and lower the steering column, if necessary. Remove the bezel retaining screws and detach the switch pod bezel. Pull it out far enough to disconnect the electrical connectors and remove it. Turn the bezel over for access to the switch pods **(see illustration)**.

16 Installation is the reverse of removal.

1996 through 1998 models

Refer to illustration 24.17

17 The switch pod bezels on these models are secured to the instrument panel with spring clips. Removal of the bezels is easily accomplished by simply prying outward at each of the clips with a trim removal tool or a small screwdriver **(see illustration)**.

18 Installation is the reverse of removal. Be sure to align the clips with the holes in the instrument panel before snapping them back into place.

Glove compartment panel

19 On 1993 through 1995 models, open the glove compartment door and remove the screws to the relay center cover, which gives access to the screws securing the glove box hinge to the instrument panel. Remove the screws and take off the panel surrounding the glove compartment, if necessary for access in other repair work.

20 On 1996 through 1998 models, open the glove box and remove the screws that attach the glove compartment "bin" or liner, then remove the glove box hinge screws.

21 On 1999 and later models, open the glove box and push in on the sides with both hands, until the sides clear the stop bumpers and the glove box can come down fully.

22 The glove box hinges on 1999 and later models are attached to the instrument panel with pins, rather than screws. With the glove box door down, grasp both ends of the door and rock it side-to-side to loosen it from the hinge hooks. Then lift the door to a more vertical position and twist it counterclockwise.

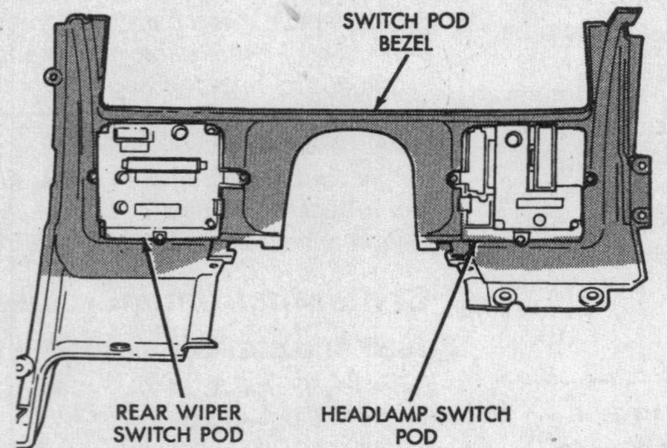

24.15 Turn the switch pod bezel over for access to the switches

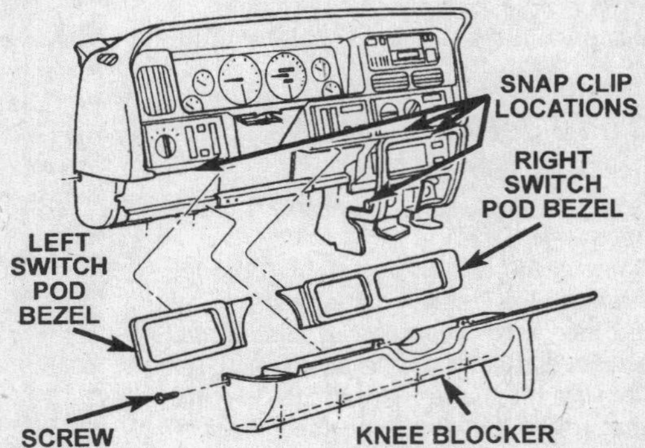

24.17 Switch pod bezel installation details - 1996 through 1998 models

With some twisting, the door will come loose from the hooks, but don't use excessive force.
23 Installation on all models is the reverse of removal.

25 Instrument panel - removal and installation

Refer to illustrations 25.11a, 25.11b, 25.11c and 25.11d

Warning: *These models are equipped with airbags. The airbag is armed and can deploy (inflate) anytime the battery is connected. To prevent accidental deployment (and possible injury), disconnect the negative battery cable whenever working near airbag components. After the battery is disconnected, wait at least 2 minutes before beginning work (the system has a back-up capacitor that must fully discharge). For more information see Chapter 12.*

Note: *This procedure is lengthy and difficult, even for an experienced mechanic. Due to the number of electrical connections, fasteners used, and the various safety systems involved, we don't recommend instrument panel removal for the home mechanic.*

1 Disconnect the negative battery cable.
2 Remove the defroster grille, instrument cluster bezel, steering column covers, knee blocker, instrument panel bezels and the switch pod bezel if equipped (see Sections 23, 24 and 26).
3 Disconnect the steering column wiring harness connector, remove the steering column nuts (three nuts at the firewall, two at the column-to-dash mount), lower the column and support it on the front seat.
4 Remove the left and right side cowl panels and kick panels. Disconnect the electrical connector and radio antenna cable at the right side kick panel.
5 Remove the hood release handle (see Section 10).
6 Remove the center floor console (see Section 19). On 1997 and later models, pull the carpeting on the transmission tunnel back until you can access and remove the bolts securing the center instrument panel brace to the floor.
7 Remove the radio (see Chapter 12) and heater/air conditioning control panel (see Chapter 3).
8 Disconnect the instrument panel bulkhead electrical connector and instrument cluster wiring harness at the lower left side of the instrument panel.
9 Disconnect the heater/air conditioning electrical and vacuum lines. Also detach the lower floor duct if equipped.
10 Remove the upper dash defroster grilles, or the upper instrument panel dash pad strip to access some of the fasteners.
11 Remove the four upper instrument panel retaining nuts. Remove the left and right hand instrument panel mounting screws **(see illus-**

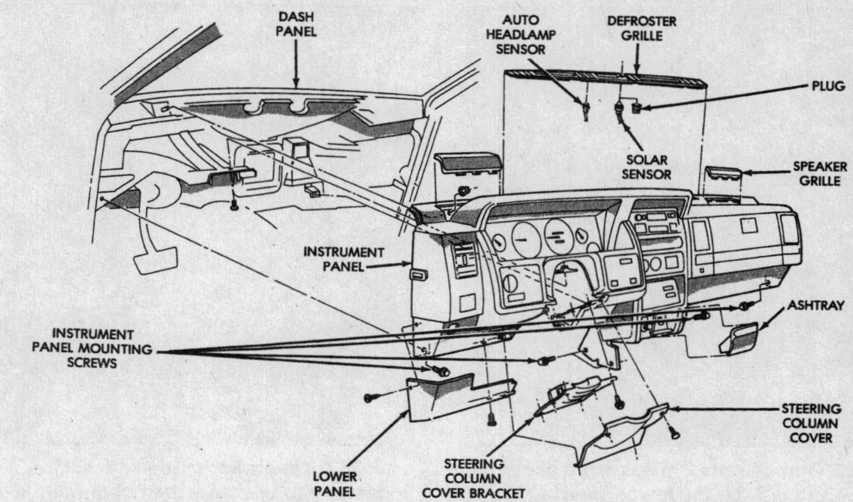

25.11a Instrument panel installation details - 1995 and earlier model shown, 1996 through 1998 models similar

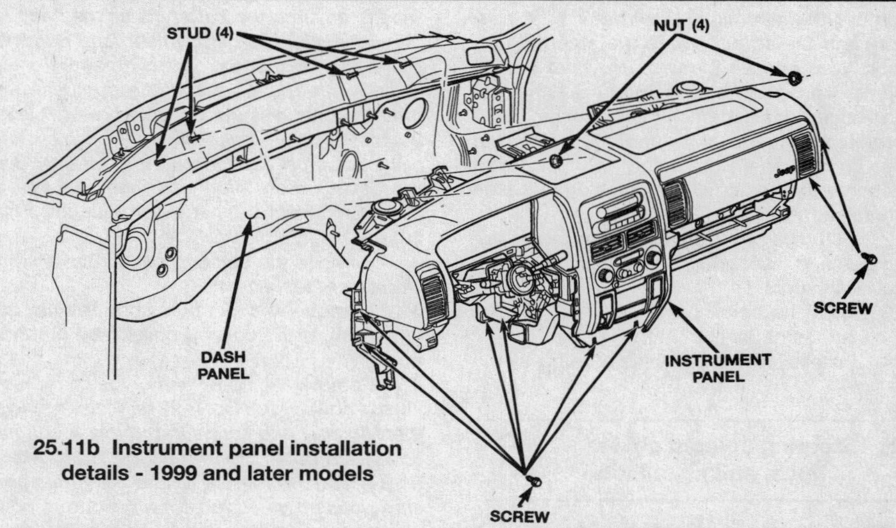

25.11b Instrument panel installation details - 1999 and later models

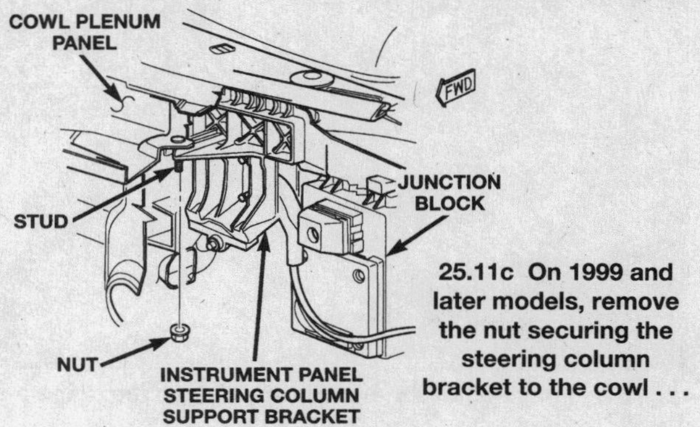

25.11c On 1999 and later models, remove the nut securing the steering column bracket to the cowl . . .

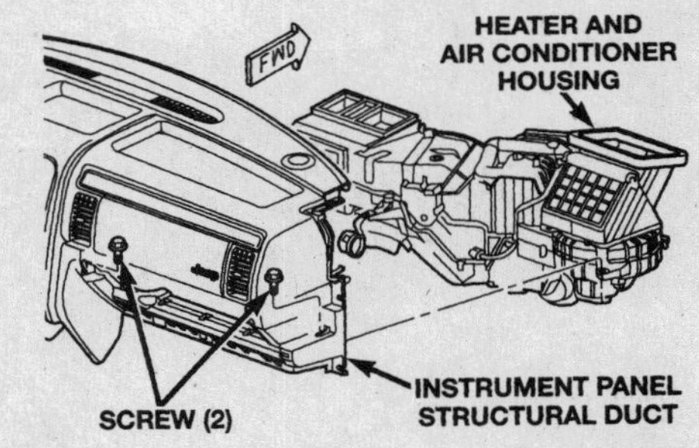

25.11d . . . and the screws securing the instrument panel to the heating/air conditioning unit

11

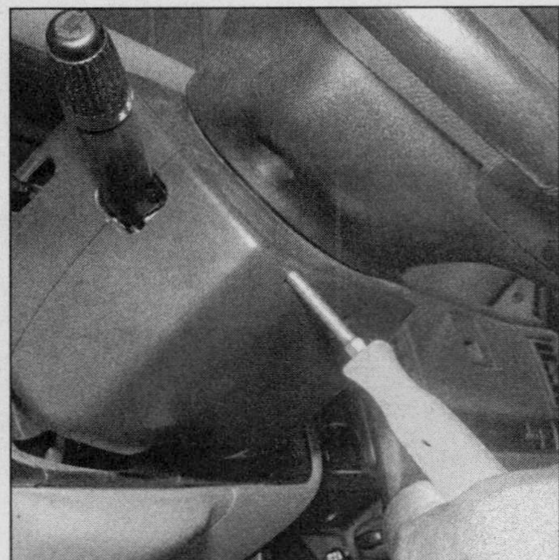

26.3 Remove the Phillips head screws and detach the lower steering column cover

26.4 On earlier models, it will be necessary to unscrew the column tilt lever to lift the upper column cover off

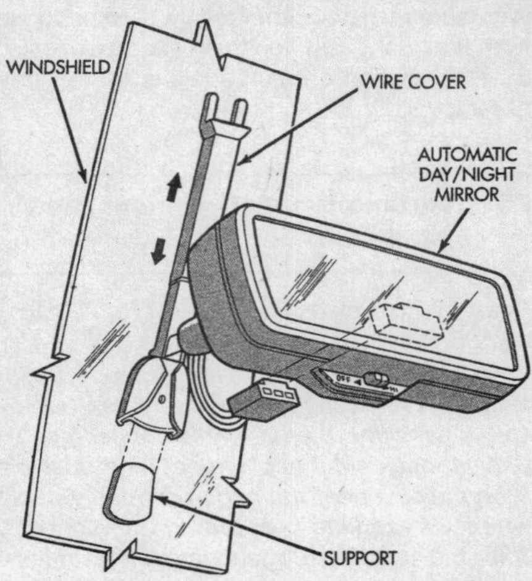

27.1 Windshield mounted rear view mirror details

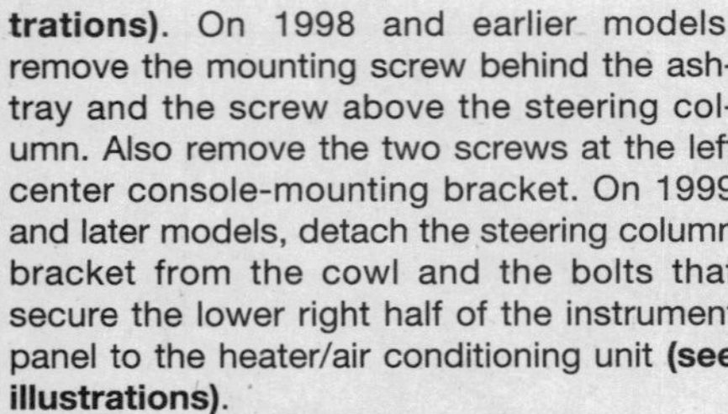

trations). On 1998 and earlier models, remove the mounting screw behind the ashtray and the screw above the steering column. Also remove the two screws at the left center console-mounting bracket. On 1999 and later models, detach the steering column bracket from the cowl and the bolts that secure the lower right half of the instrument panel to the heater/air conditioning unit **(see illustrations)**.

12 Pull the instrument panel back for access and disconnect any remaining electrical connectors.

13 With the help of an assistant, remove the instrument panel from the vehicle.

14 Installation is the reverse of removal.

26 Steering column covers - removal and installation

Refer to illustrations 26.3 and 26.4

Warning: *These models are equipped with airbags. The airbag is armed and can deploy (inflate) anytime the battery is connected. To prevent accidental deployment (and possible injury), disconnect the negative battery cable whenever working near airbag components. After the battery is disconnected, wait at least 2 minutes before beginning work (the system has a back-up capacitor that must fully discharge). For more information see Chapter 12.*

1 Disconnect the negative cable from the battery.

2 Remove the steering column lower trim panel (see Section 24).

3 Remove the screws and detach the lower half of the column cover **(see illustration)**.

4 Remove the upper half of the cover **(see illustration)**. **Note:** *On 1999 and later models, there is only one screw in the lower column cover, the upper cover must be squeezed in on the sides to release it, and the turn signal stalk does not need to be removed to remove either upper or lower column covers.*

5 Installation is the reverse of removal.

27 Rear view mirrors - removal and installation

Refer to illustrations 27.1, 27.7a, 27.7b and 27.8

Interior mirror

1 Grasp the lower section of the wire cover securely remove it by sliding it up and off **(see illustration)**. **Note:** *1999 and later models do not have a wire cover.*

2 Disconnect the electrical connector, if equipped with automatic day/night mirror.

3 Use a Phillips screwdriver to remove the setscrew, then slide the mirror up off the button on the windshield.

4 Installation is the reverse of removal.

5 If the support button for the mirror has come off the windshield, it can be reattached with a special mirror adhesive kit available at auto parts stores. Clean the glass and support base thoroughly and follow the directions on the adhesive package, allowing the base to bond overnight before attaching the mirror.

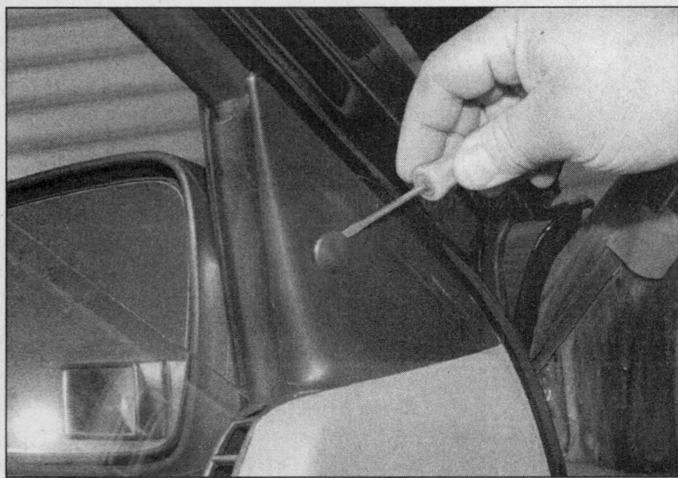

27.7a Use a small screwdriver to pry out the screw cover . . .

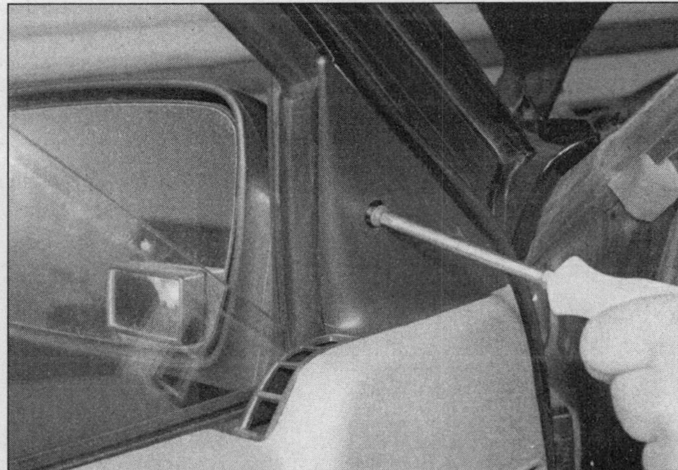

27.7b . . . then remove the screw and detach the triangle-shaped trim panel

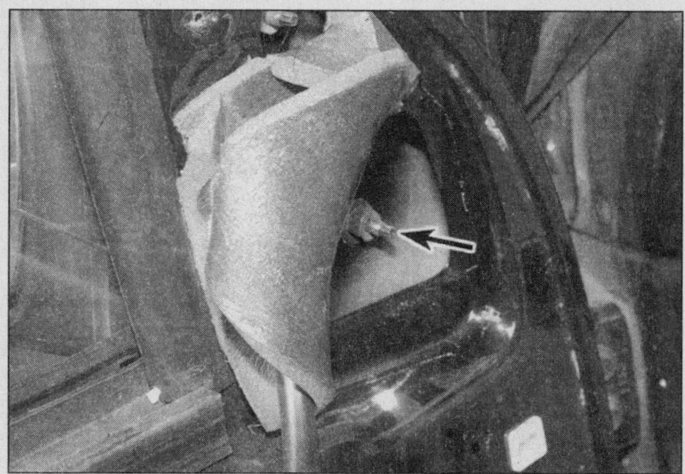

27.8 Peel back the sound deadener for access to the three mirror nuts (arrow)

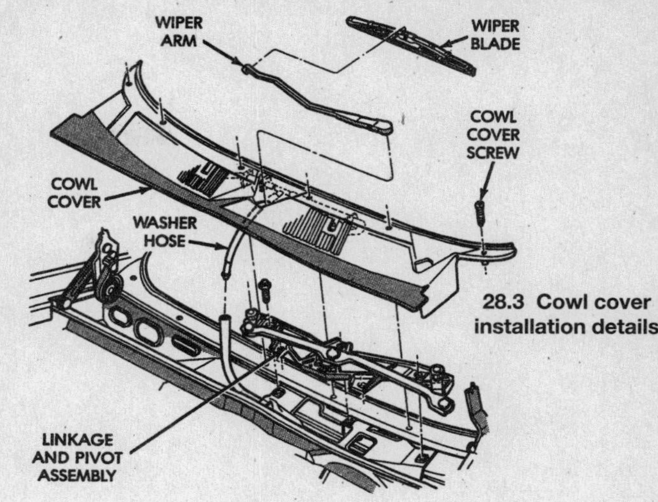

28.3 Cowl cover installation details

Exterior mirror

6 Remove the door trim panel (see Section 14).

7 Remove the screws and detach the mirror knob (if equipped) and trim panel **(see illustrations)**. On power mirrors, disconnect the electrical connector.

8 Remove the nuts and detach the mirror from the door **(see illustration)**.

9 Installation is the reverse of removal.

28 Cowl cover - removal and installation

Refer to illustration 28.3

1 Mark the position of the windshield wiper blades on the windshield with a wax marking pen.

2 Remove the wiper arms.

3 Remove the cowl cover retaining screws or plastic nuts, disconnect the windshield washer hoses and detach the cowl cover from the vehicle **(see illustration)**. **Note:** *On 1999 and later models, carefully remove the rubber cowl weatherstripping/hood seal before lifting the cowl cover off.*

4 Installation is the reverse of removal. Make sure to align the wiper blades with the marks made during removal.

29 Seats - removal and installation

Refer to illustrations 29.2 and 29.5

Front

1 Move the seat all the way forward. Remove the plastic track covers, then remove the rear seat track-to-floor bolts,

2 Move the seat rearward and remove the bolts at the front **(see illustration)**. Disconnect any electrical connectors attached to the seat.

3 Lift the seat from the vehicle.

4 Installation is the reverse of removal.

Rear

5 Remove the seat cushion-to-floor mounting bolts, then detach the cushion and lift it out **(see illustration)**.

6 Remove the bolts and lift seat back out of the vehicle.

7 Installation is the reverse of removal.

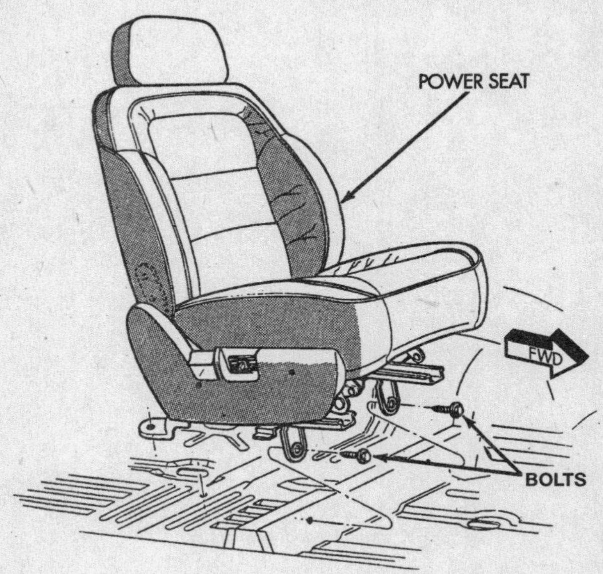

29.2 Front seat installation details

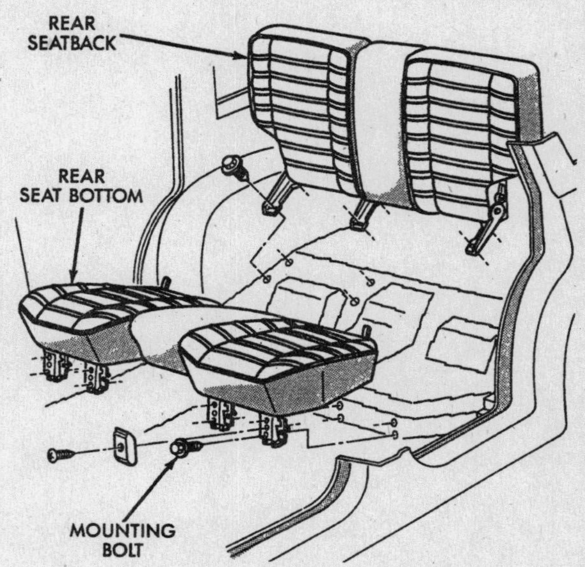

29.5 Rear seat installation details

Notes

Chapter 12
Chassis electrical system

Contents

1 General information

The electrical system is a 12-volt, negative ground type. Power for the lights and all electrical accessories is supplied by a lead/acid-type battery that is charged by the alternator.

This Chapter covers repair and service procedures for the various electrical components not associated with the engine. Information on the battery, alternator, distributor and starter motor can be found in Chapter 5. **Warning:** *When working on the electrical system, disconnect the negative battery cable from the battery to prevent electrical shorts and/or fires.*

2 Electrical troubleshooting - general information

Refer to illustrations 2.5a, 2.5b, 2.6, 2.9 and 2.15

A typical electrical circuit consists of an electrical component, any switches, relays, motors, fuses, fusible links or circuit breakers related to that component and the wiring and connectors that link the component to both the battery and the chassis. To help you pinpoint an electrical circuit problem, wiring diagrams are included at the end of this Chapter.

Before tackling any troublesome electrical circuit, first study the appropriate wiring diagrams to get a complete understanding of what makes up that individual circuit. Trouble spots, for instance, can often be narrowed down by noting if other components related to the circuit are operating properly. If several components or circuits fail at one time, chances are the problem is in a fuse or ground connection, because several circuits are often routed through the same fuse and ground connections.

Electrical problems usually stem from simple causes, such as loose or corroded connections, a blown fuse, a melted fusible link or a failed relay. Visually inspect the condition of all fuses, wires and connections in a problem circuit before troubleshooting the circuit.

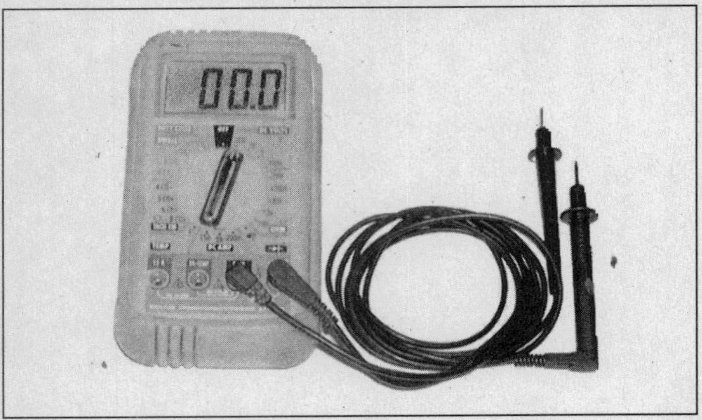

2.5a The most useful tool for electrical troubleshooting is a digital multimeter that can check volts, amps, and test continuity

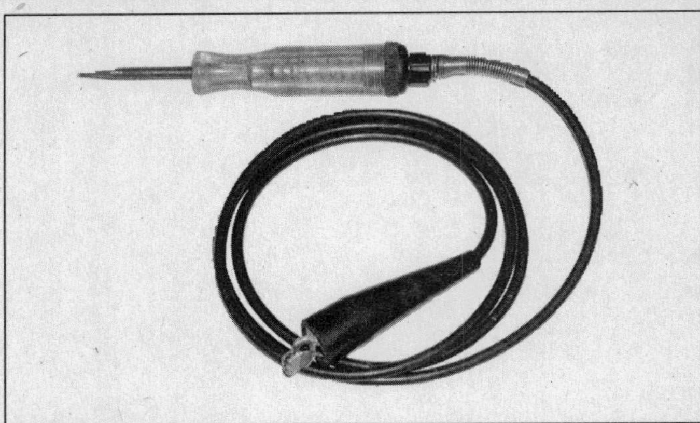

2.5b A simple test light is very handy, especially when testing for voltage

If test equipment and instruments are going to be utilized, use the diagrams to plan ahead of time where you will make the necessary connections in order to accurately pinpoint the trouble spot.

The basic tools needed for electrical troubleshooting include a circuit tester or voltmeter (a 12-volt bulb with a set of test leads can also be used), a continuity tester, which includes a bulb, battery and set of test leads, and a jumper wire, preferably with a circuit breaker incorporated, which can be used to bypass electrical components **(see illustrations)**. Before attempting to locate a problem with test instruments, use the wiring diagram(s) to decide where to make the connections.

Voltage checks

Voltage checks should be performed if a circuit is not functioning properly. Connect one lead of a circuit tester to either the negative battery terminal or a known good ground. Connect the other lead to a connector in the circuit being tested, preferably nearest to the battery or fuse **(see illustration)**. If the bulb of the tester lights, voltage is present, which means that the part of the circuit between the connector and the battery is problem free. Continue checking the rest of the circuit in the same fashion. When you reach a point at which no voltage is present, the problem lies between that point and the last test point with voltage. Most of the time the problem can be traced to a loose connection. **Note:** *Keep in mind that some circuits receive voltage only when the ignition key is in the Accessory or Run position.*

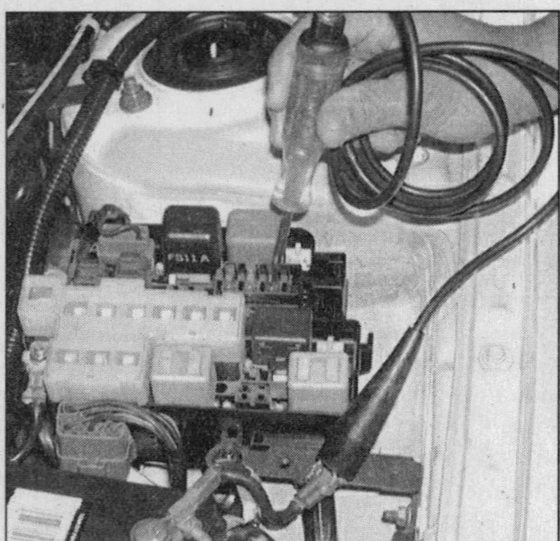

2.6 In use, a basic test light's lead is clipped to a known good ground, then the pointed probe can test connectors, wires or electrical sockets - if the bulb lights, battery voltage is present at the test point

Finding a short

One method of finding shorts in a circuit is to remove the fuse and connect a test light or voltmeter in place of the fuse terminals. There should be no voltage present in the circuit. Move the wiring harness from side-to-side while watching the test light. If the bulb goes on, there is a short to ground somewhere in that area, probably where the insulation has rubbed through. The same test can be performed on each component in the circuit, even a switch.

Ground check

Perform a ground test to check whether a component is properly grounded. Disconnect the battery and connect one lead of a continuity tester or multimeter (set to the ohms scale), to a known good ground. Connect the other lead to the wire or ground connection being tested. If the resistance is low (less than 5 ohms), the ground is good. If the bulb on a self-powered test light does not go on, the ground is not good.

Continuity check

A continuity check is done to determine if there are any breaks in a circuit - if it is passing electricity properly. With the circuit off (no power in the circuit), a self-powered continuity tester or multimeter can be used to check the circuit. Connect the test leads to both ends of the circuit (or to the "power" end and a good ground), and if the test light comes on the circuit is passing current properly **(see illustration)**. If the resistance is low (less than 5 ohms), there is continuity; if the reading is 10,000 ohms or higher, there is a break somewhere in the circuit. The same procedure can be used to test a switch, by connecting the continuity tester to the switch terminals. With the switch turned On, the test light should come on (or low resistance should be indicated on a meter).

Finding an open circuit

When diagnosing for possible open circuits, it is often difficult to locate them by sight because the connectors hide oxidation or terminal misalignment. Merely wiggling a connector on a sensor or in the wiring har-

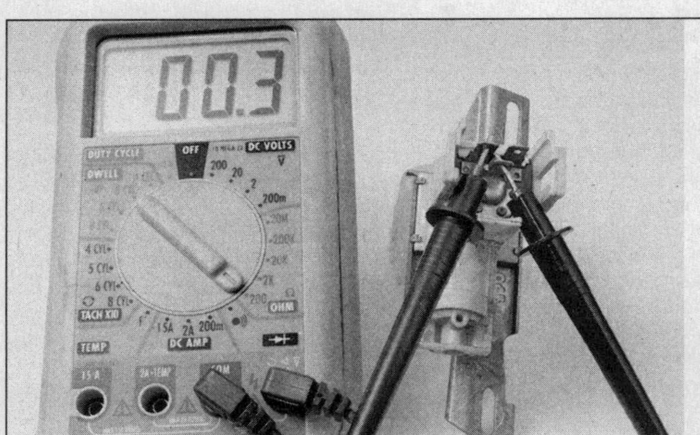

2.9 With a multimeter set to the ohms scale, resistance can be checked across two terminals - when checking for continuity, a low reading indicates continuity, a very high or infinite reading indicates lack of continuity

2.15 To backprobe a connector, insert a small, sharp probe (such as a straight-pin) into the back of the connector alongside the desired wire until it contacts the metal terminal inside; connect your meter leads to the probes - this allows you to test a functioning circuit

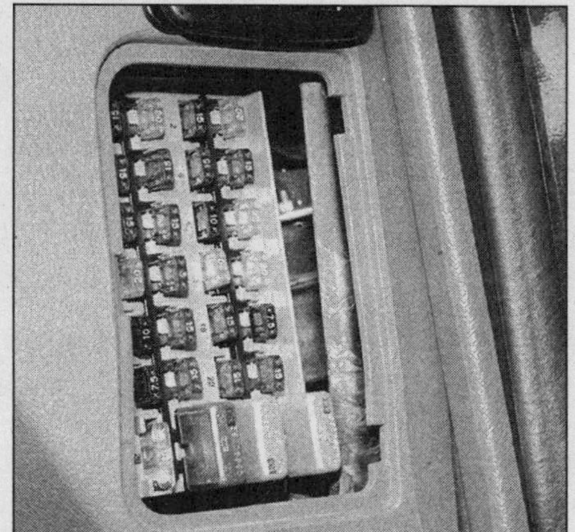

3.1a The fuse block on 1998 models is located in the right end of the instrument panel under a cover

3.1b Pull down on the fuse panel cover at the left end of the instrument panel to access the interior fuse panel on 1999 and later models

ness may correct the open circuit condition. Remember this when an open circuit is indicated when troubleshooting a circuit. Intermittent problems may also be caused by oxidized or loose connections.

Electrical troubleshooting is simple if you keep in mind that all electrical circuits are basically electricity running from the battery, through the wires, switches, relays, fuses and fusible links to each electrical component (light bulb, motor, etc.) and to ground, from which it is passed back to the battery. Any electrical problem is an interruption in the flow of electricity to and from the battery.

Connectors

Most electrical connections on these vehicles are made with multiwire plastic connectors. The mating halves of many connectors are secured with locking clips molded into the plastic connector shells. The mating halves of large connectors, such as some of those under the instrument panel, are held together by a bolt through the center of the connector.

To separate a connector with locking clips, use a small screwdriver to pry the clips apart carefully, then separate the connector halves. Pull only on the shell, never pull on the wiring harness as you may damage the individual wires and terminals inside the connectors. Look at the connector closely before trying to separate the halves. Often the locking clips are engaged in a way that is not immediately clear. Additionally, many connectors have more than one set of clips.

Each pair of connector terminals has a male half and a female half. When you look at the end view of a connector in a diagram, be sure to understand whether the view shows the harness side or the component side of the connector. Connector halves are mirror images of each other, and a terminal shown on the right side end-view of one half will be on the left side end view of the other half.

It is often necessary to take circuit voltage measurements with a connector connected. Whenever possible, carefully insert a small straight pin (not your meter probe) into the rear of the connector shell to contact the terminal inside, then clip your meter lead to the pin. This kind of connection is called "backprobing" **(see illustration)**. When inserting a test probe into a male terminal, be careful not to distort the terminal opening. Doing so can lead to a poor connection and corrosion at that terminal later. Using the small straight pin instead of a meter probe results in less chance of deforming the terminal connector.

3 Fuses - general information

Refer to illustrations 3.1a, 3.1b, 3.3a and 3.3b

The electrical circuits of the vehicle are protected by a combination of fuses, circuit breakers and cartridge fuses. The interior fuse block on 1993 through 1997 models is located in the right-side kick-panel, behind a cover. On 1998 models, it's located under a cover at the right end of the instrument panel **(see illustration)**. On 1999 and later models, the fuse/relay panel is located under the left side of the instrument panel, behind a cover **(see illustration)**. There is also a fuse and

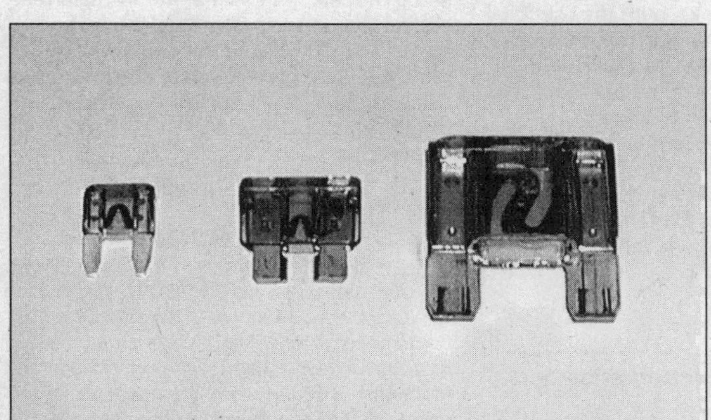

3.3a All three of these fuses are of 30-amp rating, yet are different sizes, at left is a small fuse, the center is a medium, and at right is a large - make sure you get the right amperage and size when purchasing replacement fuses

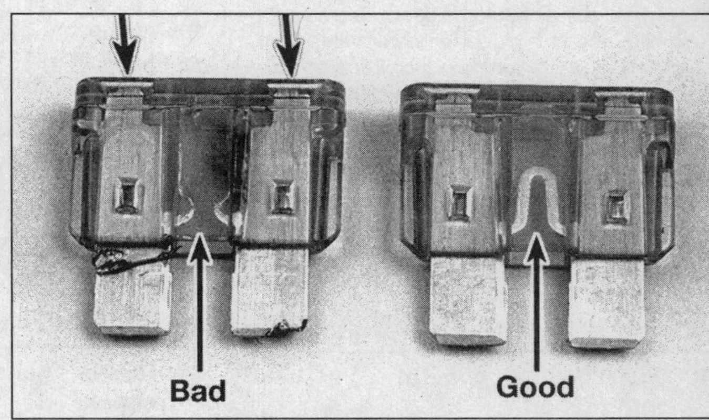

3.3b When a fuse blows, the element between the terminals melts - the fuse on the left is blown, the one on the right is good - these fuse can be tested on top (arrows) with a test light without removing the fuse

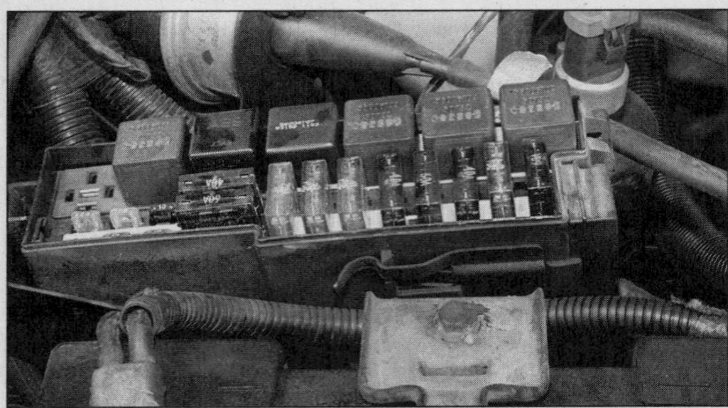

6.2 The power distribution center in the engine compartment contains fuses and relays

6.3a The relay center on some models is accessible after removing the cover in the glove compartment

relay block, called the power distribution center, located on the right side of the engine compartment (see Section 6).

Each of the fuses is designed to protect a specific circuit, and the various circuits are identified on the fuse panel itself.

Several sizes of fuses are employed in the fuse blocks. There are small, medium and large sizes of the same design, all with the same blade terminal design, as well as five "metal" fuses located in the engine compartment fuse panel. The medium and large fuses can be removed with your fingers, but the small fuses require the use of pliers or the small plastic fuse-puller tool found in most fuse boxes. The metal fuses are for heavy loads, and if the metal strip melts due to an overload, it is easily seen, although the battery should be disconnected while replacing this type fuse (see the **Cautions** in Section 1). If an electrical component fails, always check the fuse first. The best way to check the fuses is with a test light. Check for power at the exposed terminal tips of each fuse **(see illustration)**. If power is present at one side of the fuse but not the other, the fuse is blown. A blown fuse can also be identified by visually inspecting it **(see illustration)**.

Be sure to replace blown fuses with the correct type. Fuses of different ratings are physically interchangeable, but only fuses of the proper rating should be used. Replacing a fuse with one of a higher or lower value than specified is not recommended. Each electrical circuit needs a specific amount of protection. The amperage value of each fuse is molded into the fuse body.

If the replacement fuse immediately fails, don't replace it again until the cause of the problem is isolated and corrected. In most cases, the cause will be a short circuit in the wiring caused by a broken or deteriorated wire.

4 Fusible links - general information

Some circuits are protected by fusible links. The links are used in circuits that are not ordinarily fused, such as the ignition circuit.

The fusible links on these models are similar to fuses in that they can be visually checked to determine if they are melted. The fusible links are located in the power distribution block in the engine compartment (see Section 6).

To replace a fusible link, first disconnect the negative cable from the battery. Disconnect the burned-out link and replace it with a new one (available from your dealer or auto parts store). Always determine the cause for the overload that melted the fusible link before installing a new one.

5 Circuit breakers - general information

Circuit breakers protect certain circuits, such as the power windows or heated seats. Depending on the vehicle's accessories, there may be one or two 25-amp circuit breakers, located in the underhood power distribution center (see Section 6).

Because the circuit breakers reset automatically, an electrical overload in a circuit-breaker-protected system will cause the circuit to fail momentarily, then come back on. If the circuit does not come back on, check it immediately.

For a basic check, pull the circuit breaker up out of its socket on the fuse panel, but just far enough to probe with a voltmeter. The breaker should still contact the sockets.

With the voltmeter negative lead on a good chassis ground, touch each end prong of the circuit breaker with the positive meter probe. There should be battery voltage at each end. If there is battery voltage only at one end, the circuit breaker must be replaced.

6 Relays - general information and check

Refer to illustrations 6.2, 6.3a and 6.3b

1 Many electrical accessories in the vehicle utilize relays to transmit current to the component. If the relay is defective, the component won't operate properly.

6.3b On later models, relays (arrows) are combined in the interior fuse/relay box in the kick panel or instrument panel

2 Some relays are located in the power distribution center, which is a fuse and relay box located in the engine compartment **(see illustration)**.

3 Other relays are located in a relay center in the glove compartment under a cover on some models, and in the interior fuse/relay box on later models **(see illustrations)**.

4 If a faulty relay is suspected, it can be removed and tested using the procedure below or by a dealer service department or a repair shop. Defective relays must be replaced as a unit.

Testing

Refer to illustrations 6.5a, 6.5b and 6.5c

5 Most of the relays used in these vehicles are of a type often called "ISO" relays, which refers to the International Standards Organization. The terminals of ISO relays are numbered to indicate their usual circuit connections and functions. There are two basic layouts of terminals on the relays used in the vehicles covered by this manual **(see illustrations)**.

6 Refer to the wiring diagram for the circuit to determine the proper connections for the relay you're testing. If you can't determine

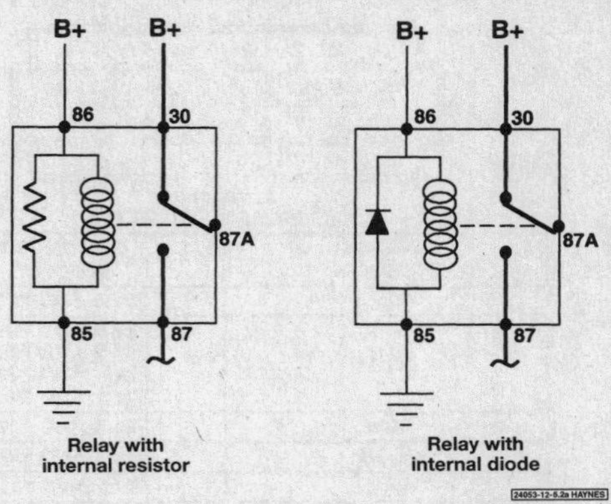

6.5a Typical ISO relay designs, terminal numbering and circuit connections

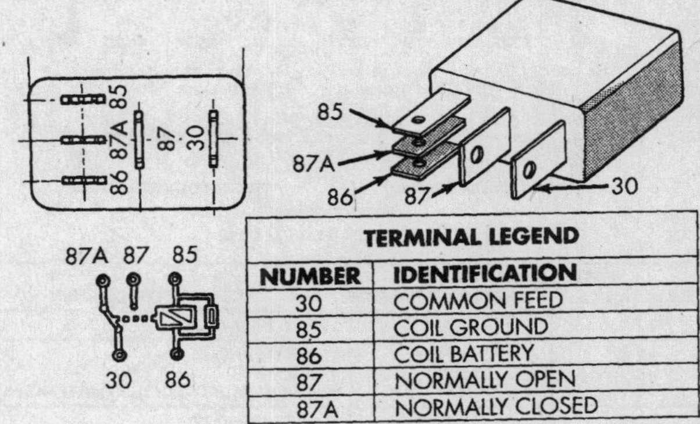

TERMINAL LEGEND	
NUMBER	IDENTIFICATION
30	COMMON FEED
85	COIL GROUND
86	COIL BATTERY
87	NORMALLY OPEN
87A	NORMALLY CLOSED

6.5b Some relays (starter relay shown) are called "micro-relays" and have a different terminal arrangement

6.5c Most relays are marked on the outside to easily identify the control circuit and power circuits - this one is of the four-terminal type

the correct connection from the wiring diagrams, however, you may be able to determine the test connections from the information that follows.

7 Two of the terminals are the relay control circuit and connect to the relay coil. The other relay terminals are the power circuit. When the relay is energized, the coil creates a magnetic field that closes the larger contacts of the power circuit to provide power to the circuit loads.

8 Terminals 85 and 86 are normally the control circuit. If the relay contains a diode, terminal 86 must be connected to battery positive (B+) voltage and terminal 85 to ground. If the relay contains a resistor, terminals 85 and 86 can be connected in either direction with respect to B+ and ground.

9 Terminal 30 is normally connected to the battery voltage (B+) source for the circuit loads. Terminal 87 is connected to the ground side of the circuit, either directly or through a load. If the relay has several alternate terminals for load or ground connections, they usually are numbered 87A, 87B, 87C, and so on.

10 Use an ohmmeter to check continuity through the relay control coil.

a) *Connect the meter according to the polarity shown in the illustration for one check; then reverse the ohmmeter leads and check continuity in the other direction.*

b) *If the relay contains a resistor, resistance will be indicated on the meter, and should be the same value with the ohmmeter in either direction.*

c) *If the relay contains a diode, resistance should be higher with the ohmmeter in the forward polarity direction than with the meter leads reversed.*

d) *If the ohmmeter shows infinite resistance in both directions, replace the relay.*

11 Remove the relay from the vehicle and use the ohmmeter to check for continuity between the relay power circuit terminals. There should be no continuity between terminal 30 and 87 with the relay de-energized.

12 Connect a fused jumper wire to terminal 86 and the positive battery terminal. Connect another jumper wire between terminal 85 and ground. When the connections are made, the relay should click.

13 With the jumper wires connected, check for continuity between the power circuit terminals. Now, there should be continuity between terminals 30 and 87.

14 If the relay fails any of the above tests, replace it.

7 Turn signal/hazard flasher - check and replacement

Refer to illustration 7.1

Warning: *These models are equipped with airbags. The airbag is armed and can deploy (inflate) anytime the battery is connected. To prevent accidental deployment (and possible injury), disconnect the negative battery cable whenever working near airbag components. After the battery is disconnected, wait at least 2 minutes before beginning work (the system has a back-up capacitor that must fully discharge). For more information see Section 27.*

1 The turn signal/hazard flasher is a small

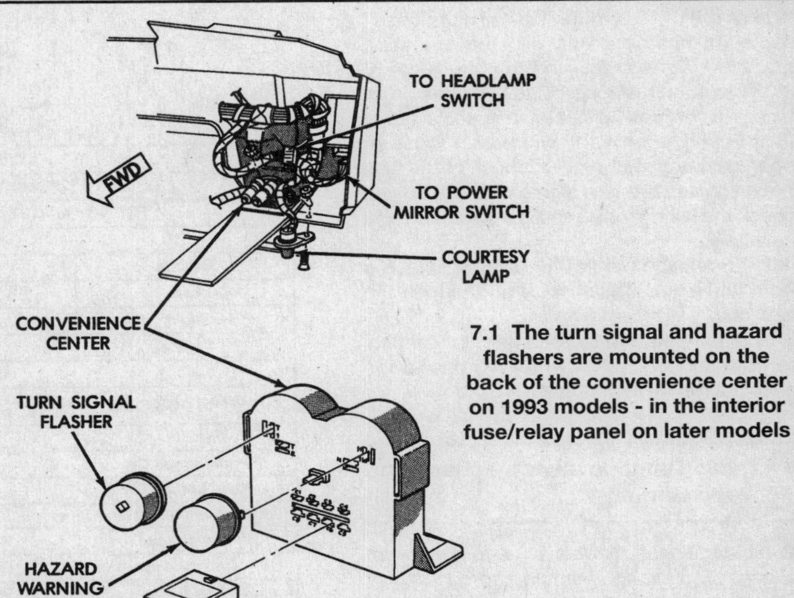

7.1 The turn signal and hazard flashers are mounted on the back of the convenience center on 1993 models - in the interior fuse/relay panel on later models

12

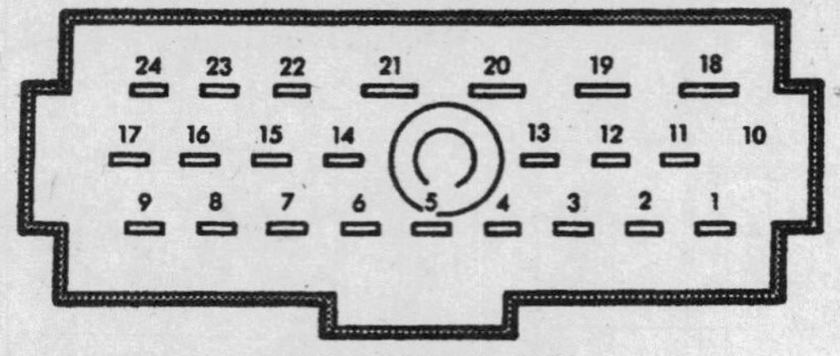

SWITCH POSITIONS		CONTINUITY BETWEEN
TURN SIGNAL	HAZARD WARNING	
NEUTRAL	OFF	12 AND 14 AND 15
LEFT LEFT LEFT	OFF OFF OFF	15 AND 16 AND 17 12 AND 14 22 AND 23 WITH OPTIONAL CORNER LAMPS
RIGHT RIGHT RIGHT	OFF OFF OFF	11 AND 12 AND 17 14 AND 15 23 AND 24 WITH OPTIONAL CORNER LAMPS
NEUTRAL	ON	11 AND 12 AND 13 AND 15 AND 16

8.3a Turn signal and hazard flasher switch terminal guide and continuity chart - 1993 through 1998 models

SWITCH POSITION	CONTINUITY BETWEEN
OFF	PIN 6 AND PIN 7
DELAY	PIN 8 AND PIN 9 PIN 2 AND PIN 4 PIN 1 AND PIN 2 PIN 1 AND PIN 4
LOW	PIN 4 AND PIN 6
HIGH	PIN 4 AND PIN 5
WASH	PIN 3 AND PIN 4
*RESISTANCE AT MAXIMUM DELAY POSITION SHOULD BE BETWEEN 270,000 OHMS AND 330,000 OHMS.	
*RESISTANCE AT MINIMUM DELAY POSITION SHOULD BE ZERO WITH OHMMETER SET ON HIGH OHM SCALE.	

8.3b Windshield wiper/washer terminal guide and continuity chart - 1993 through 1998 models

canister-shaped relay that flashes the turn signals and hazard warning lights when either switch is activated. On 1993 models, two separate units are mounted on the back of the convenience center located under the dash on the driver's side **(see illustration)**. All 1994 and later models use a combination flasher unit to perform both functions. On 1994 through 1996 models, the combination flasher is located in the glove compartment relay box. On 1997 and 1998 models, the combination flasher is located in the interior fuse/relay box in the right end of the instrument panel. On 1999 and later models, the combination flasher is located in the fuse/relay box just left of the steering column.

2 When the flasher unit is functioning properly, an audible click can be heard during its operation. If the turn signals fail on one side or the other and the flasher unit doesn't make its characteristic clicking sound, a faulty turn signal bulb is indicated.

3 If both turn signals fail to blink, the problem may be due to a blown fuse, a faulty flasher unit, a broken switch or a loose or open connection. If a quick check of the fuse box indicates the turn signal fuse has blown, check the wiring for a short before installing a new fuse.

4 To replace the flasher, simply detach it from its electrical connector and plug in the new one.

5 Make sure the replacement is identical to the original. Compare the old one to the new one before installing it.

8 Multi-function switch - check and replacement

Warning: *These models are equipped with airbags. The airbag is armed and can deploy (inflate) anytime the battery is connected. To prevent accidental deployment (and possible injury), disconnect the negative battery cable*

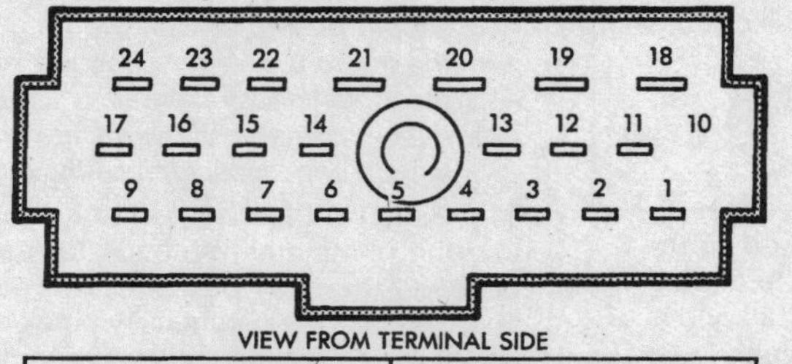

SWITCH POSITION	CONTINUITY BETWEEN
LOW BEAM	18 AND 19
HIGH BEAM	19 AND 20
OPTICAL HORN	20 AND 21

8.3c Headlight dimmer switch terminal guide and continuity chart - 1993 through 1998 models

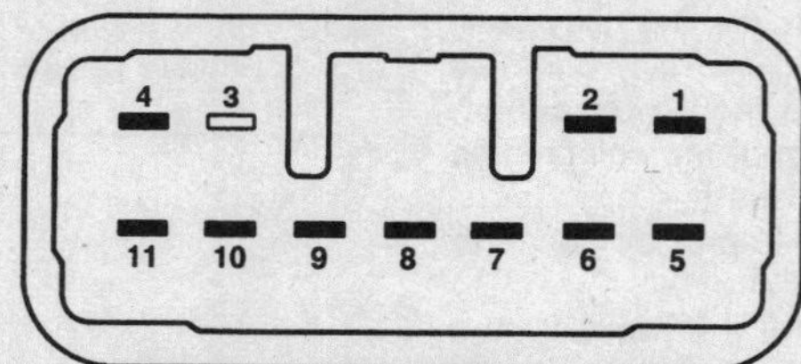

LEFT (LIGHTING) MULTI-FUNCTION SWITCH			
TURN SIGNAL AND HAZARD WARNING SWITCH TESTS			
SWITCH POSITION		CONTINUITY BETWEEN	
TURN	HAZARD		
Neutral	Off	No Related Continuity	
Left	Off	Pins 2 & 8	
Right	Off	Pins 2 & 7	
Neutral	On	Pins 2 & 9	
EXTERIOR LIGHTING SWITCH TESTS			
SWITCH POSITION	CONTINUITY BETWEEN	RESISTANCE BETWEEN	RESISTANCE RANGE (OHMS)
Off	–	Pins 4 & 11	3743 - 3824
Park Lamps On	–	Pins 4 & 11	901 - 926
Head Lamps On	–	Pins 4 & 11	345 - 358
Auto Headlamps On	–	Pins 4 & 11	74 - 81
Fog Lamps	Pins 1 & 2	–	–
Optical Horn	Pins 2 & 5	–	–
High Beam	Pins 2 & 6	–	–

8.3d Headlight switch terminal guide and continuity chart - 1999 and later models

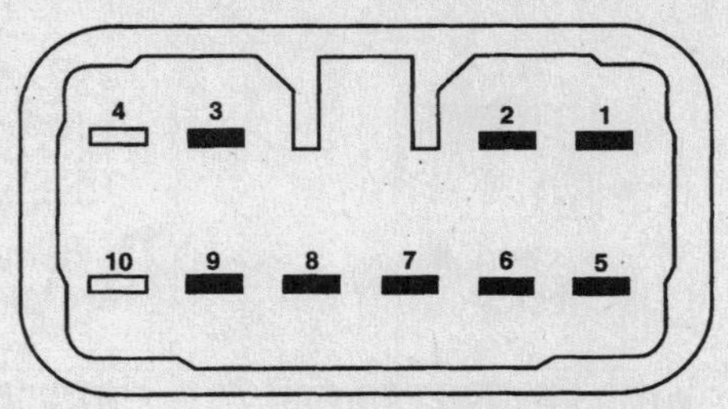

RIGHT (WIPER) MULTI-FUNCTION SWITCH			
FRONT WIPERS SWITCH TESTS			
SWITCH POSITION	CONTINUITY BETWEEN	RESISTANCE BETWEEN	RESISTANCE RANGE (OHMS)
Off	—	Pins 7 & 8	4286-4379
Intermittent Wipe Position 1	—	Pins 7 & 8	1445-1480
Intermittent Wipe Position 2	—	Pins 7 & 8	847- 870
Intermittent Wipe Position 3	—	Pins 7 & 8	556- 573
Intermittent Wipe Position 4	—	Pins 7 & 8	367- 380
Intermittent Wipe Position 5	—	Pins 7 & 8	218-229
Low Speed	—	Pins 7 & 8	99-106
High Speed	Pins 1 & 9	Pins 7 & 8	99-106
Mist	—	Pins 7 & 8	49-56
Wash	Pins 1 & 3	—	—

REAR WIPER SWITCH TESTS			
SWITCH POSITION	CONTINUITY BETWEEN	RESISTANCE BETWEEN	RESISTANCE RANGE (OHMS)
Off	—	—	—
Delay	Pins 1 & 6	—	—
On	Pins 1 & 5	—	—
Wash	Pins 1 & 5 & 6	—	—

8.3e Windshield wiper switch terminal guide and continuity chart - 1999 and later models

whenever working near airbag components. After the battery is disconnected, wait at least 2 minutes before beginning work (the system has a back-up capacitor that must fully discharge). For more information see Section 27.

1 The multi-function switch on 1993 through 1998 models is located on the left side of the steering column. It incorporates the turn signal, headlight dimmer and windshield wiper/washer functions into one switch. On 1999 and later models, there is a multi-function switch/lever on each side of the steering column. On the left is the switch for lighting operations, and the right switch is for the wiper/washer functions.

Check

Refer to illustrations 8.3a, 8.3b, 8.3c, 8.3d and 8.3e

2 Remove the multi-function switch(es) (see Step 4).

3 Using an ohmmeter or self-powered test light and the accompanying diagrams, check for continuity between the indicated switch terminals with the switch in each of the indicated positions **(see illustrations)**. If the continuity isn't as specified, replace the switch.

Replacement

Refer to illustrations 8.5a and 8.5b

4 Disconnect the negative battery cable.

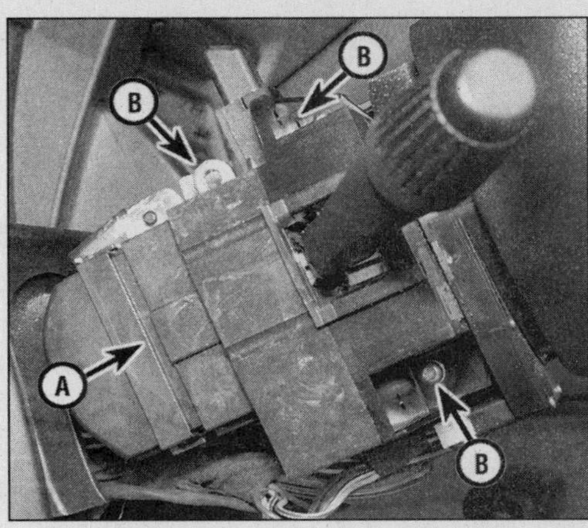

8.5a Disconnect the multi-function switch electrical connector (A), remove the three Torx-head tamper-proof screws (B) and detach the switch

8.5b On 1999 and later models, there are two multi-function switches, each mounted with two screws (arrows) (wiper switch shown)

5 Remove the steering column upper and lower covers and knee blocker (see Chapter 11). Disconnect the electrical connector, remove the bolts (this will require a special Torx-head tool), then detach the switch from the steering column **(see illustrations)**.

6 Installation is the reverse of removal.

9 Ignition switch/key lock cylinder - check and replacement

Warning: *These models are equipped with airbags. The airbag is armed and can deploy (inflate) anytime the battery is connected. To prevent accidental deployment (and possible injury), disconnect the negative battery cable whenever working near airbag components. After the battery is disconnected, wait at least 2 minutes before beginning work (the system has a back-up capacitor that must fully discharge). For more information see Section 27.*

1 The ignition switch is located on right side of the steering column and is held in place by three Torx screws (one screw on

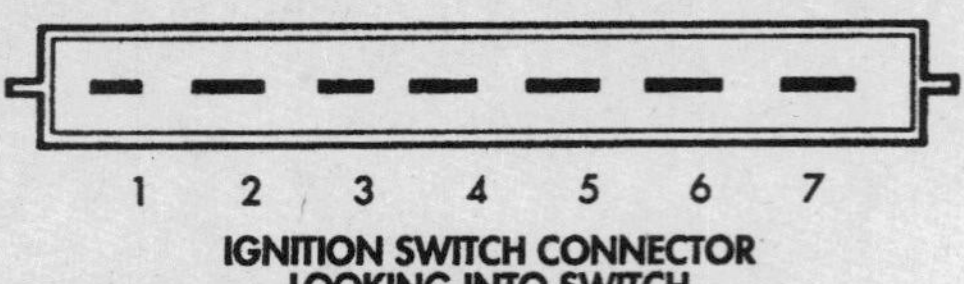

SWITCH POSITION	CONTINUITY BETWEEN
START	PIN 3 AND GROUND PIN 1 AND PIN 7 PIN 2 AND PIN 7
RUN	PIN 4 AND PIN 5 PIN 2 AND PIN 7 PIN 6 AND PIN 7
ACC	PIN 6 AND PIN 7

9.3a Ignition switch terminal guide and continuity chart - 1998 and earlier models

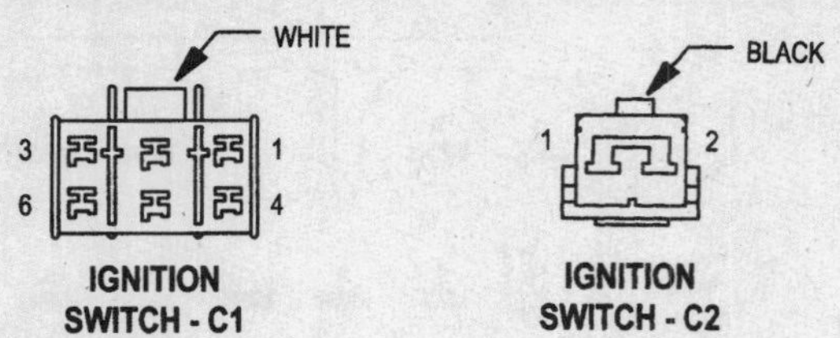

IGNITION SWITCH POSITION	CONTINUITY BETWEEN TERMINALS
Off or Lock	None
Start	C1-2 to C1-1, C1-4 to C1-6
Run	C1-2 to C1-3, C1-4 to C1-5, C1-4 to C1-6
Acc	C1-4 to C1-5

9.3b Ignition switch terminal guide and continuity chart - 1999 and later models

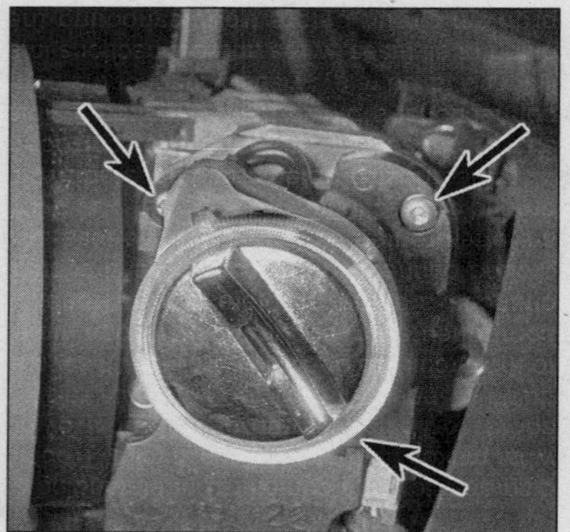

9.7a The ignition switch on most models is held in place by three Torx-head tamper-proof screws (arrows) (the lower screw is located under the cover)

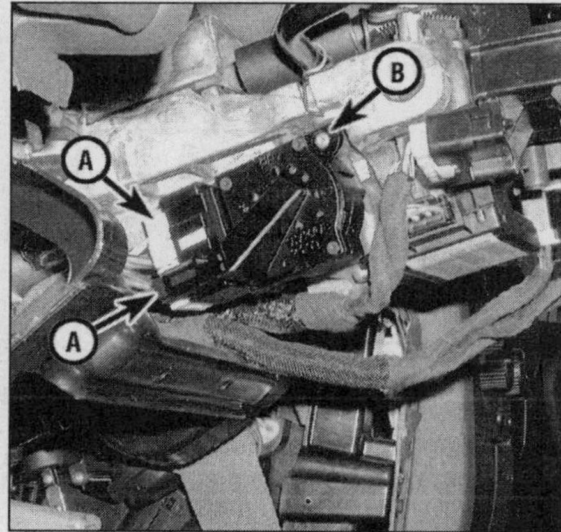

9.7b On 1999 and later models, disconnect the electrical connectors (A), remove mounting screw (B) . . .

1999 and later models) which require a special tool for removal (available at auto parts stores).

Check

Refer to illustrations 9.3a and 9.3b

2 Remove the switch (see Steps 5 through 7).

3 Use an ohmmeter or self-powered test light and check for continuity between the switch terminals **(see illustrations)**.

4 If the switch does not have correct continuity, replace it.

Replacement

Refer to illustrations 9.7a, 9.7b, 9.7c, 9.8, 9.10 and 9.11

5 Disconnect the negative battery cable.

6 Remove the steering column upper and lower covers, and on some models remove the driver's knee bolster (see Chapter 11).

7 Remove the screw(s), detach the switch from column, then disconnect the electrical connector(s) and remove the switch from the steering column **(see illustrations)**. **Note:** *On later models with Smart Key Immobilizer system, remove the one screw from below to detach the immobilizer module before removing the ignition switch.*

8 Installation is the reverse of removal. On 1998 and earlier models, make sure the column lock flag on the switch is parallel with the electrical connectors before the switch is

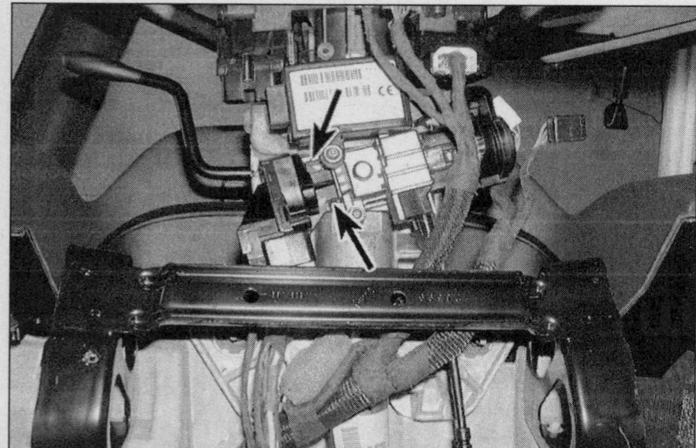

9.7c . . . and squeeze the tabs (arrows) with pliers to release the ignition switch

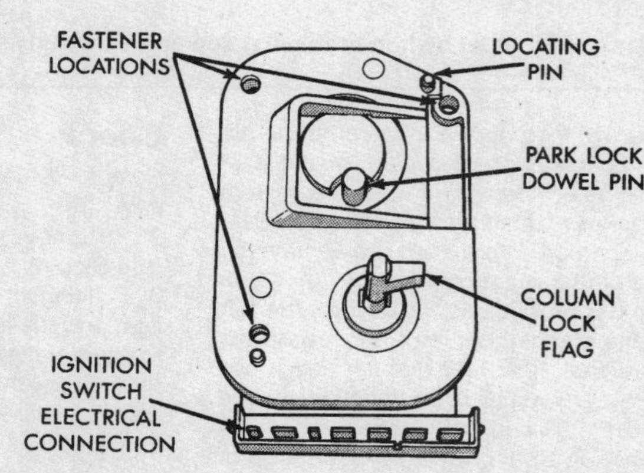

9.8 On 1998 and earlier models, make sure the switch column lock flag is parallel with the electrical connection before installation

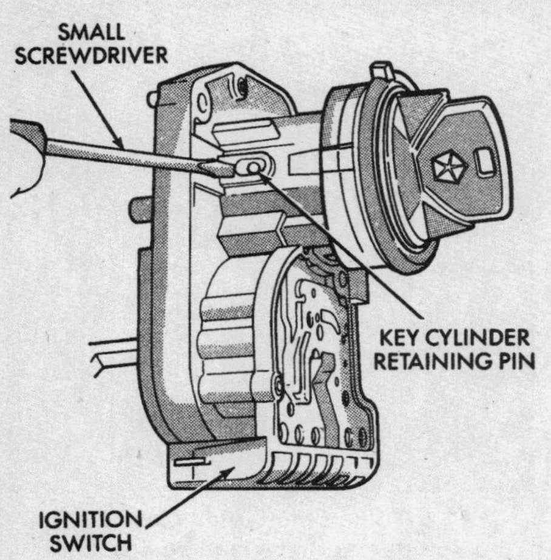

9.10 Push the retaining pin in to unseat the lock cylinder - on 1995 and 1996 models, remove the tamper-proof screw

engaged with the column **(see illustration)**. On 1999 and later models, it may be necessary to wiggle the key lock cylinder back and forth in order to engage the ignition switch with the key lock cylinder. Once the switch is engaged with the lock cylinder, push the switch into the housing until the plastic locking tabs at the bottom of the switch have engaged with the steering column housing. In either case do not force the ignition switch into the column.

9 If it's necessary to remove the key lock cylinder, first insert the ignition key into the lock cylinder.

10 On 1997 and later models, turn the ignition key to the Run position. Push inward on the lock cylinder retaining pin with a small screwdriver until the lock cylinder can be removed from the housing **(see illustration)**. When installing the lock cylinder, be sure to reposition the key in the Run position before installing it in the housing.

11 On 1996 and earlier models, turn the key to the Lock position. Press the retaining pin in with a small screwdriver until it is flush with the surface **(see illustration 9.10)**. **Note:** *On 1995 and 1996 models, the retaining pin was replaced with a theft-proof screw which must be removed to release the lock cylinder.* Turn the key clockwise to the Off position, which will unseat the lock cylinder, but don't try to remove it yet **(see illustration)**. With the cylinder unseated, rotate the key counterclockwise to the Lock position, remove the key, then remove the lock cylinder from the ignition switch.

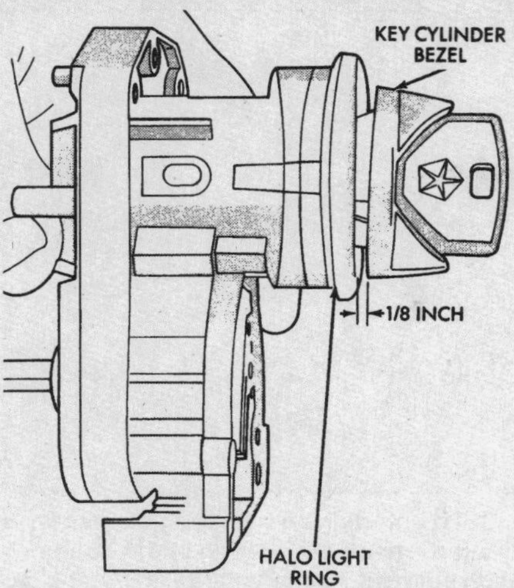

9.11 On 1996 and earlier models, the lock cylinder will protrude about 1/8-inch from the switch once it's unseated - don't try to remove it until you rotate it to the Lock position and remove the key

12 Insert the lock cylinder in the Lock position until it bottoms in the switch. While pushing the lock cylinder in, insert the key and turn it clockwise to the Run position. On 1995 and 1996 models, install the theft-proof retaining screw.

10 Rear window defogger switch - check and replacement

Refer to illustration 10.3

Warning: *These models are equipped with airbags. The airbag is armed and can deploy (inflate) anytime the battery is connected. To prevent accidental deployment (and possible injury), disconnect the negative battery cable whenever working near airbag components. After the battery is disconnected, wait at least 2 minutes before beginning work (the system has a back-up capacitor that must fully discharge). For more information see Section 27.*

1 Remove the instrument panel trim bezels and the switch pod bezel if equipped (see Chapter 11).

2 On 1998 and earlier models, remove the screws and detach the defogger switch from the bezel. On 1999 and later models, the switch is part of the air conditioning/heating controls panel. Refer to Chapter 3 for removal of the control panel.

3 On 1998 and earlier models, connect an ohmmeter or self-powered test light between terminals 2 and 3 of the switch and press the switch button **(see illustration)**. The ohmmeter should read less than 1 ohm. If it doesn't, replace the switch.

4 On 1999 and later models, the defogger circuit includes the Body Control Module (BCM), a computer module that controls a number of body system communication tasks. If the fuse, switch and relay are OK on one of these models, but the defogger still doesn't work, take the vehicle to a dealer service department where they can use a scan tool to check for a Diagnostic Trouble Code (DTC) in the BCM.

11 Rear window defogger - check and repair

1 The rear window defogger consists of a number of horizontal heating elements baked onto the inside surface of the glass. Power is supplied through a relay and fuse from the interior fuse/relay box. The heater is controlled by the instrument panel switch.

2 Small breaks in the element can be repaired without removing the rear window.

Check

Refer to illustrations 11.5, 11.6 and 11.8

3 Turn the ignition and defogger switches to the ON position.

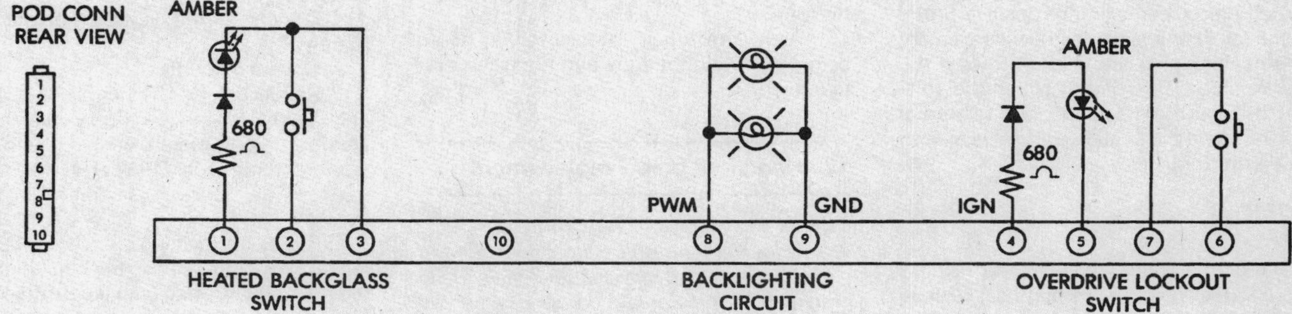

10.3 Defogger switch and circuit details - 1998 and earlier models

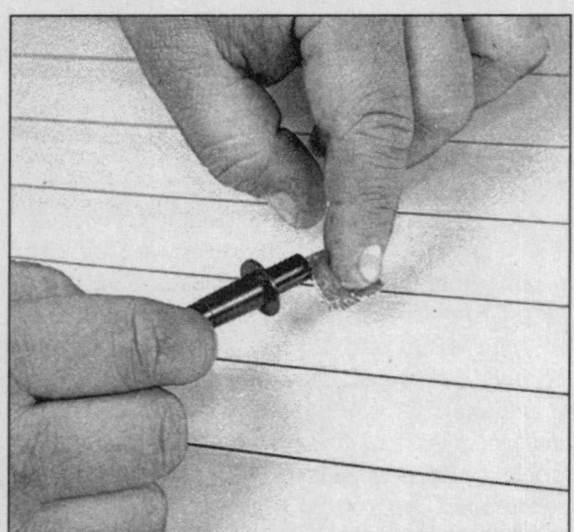

11.5 When measuring the voltage at the rear window defogger grid, wrap a piece of aluminum foil around the positive probe of the voltmeter and press the foil against the wire with your finger

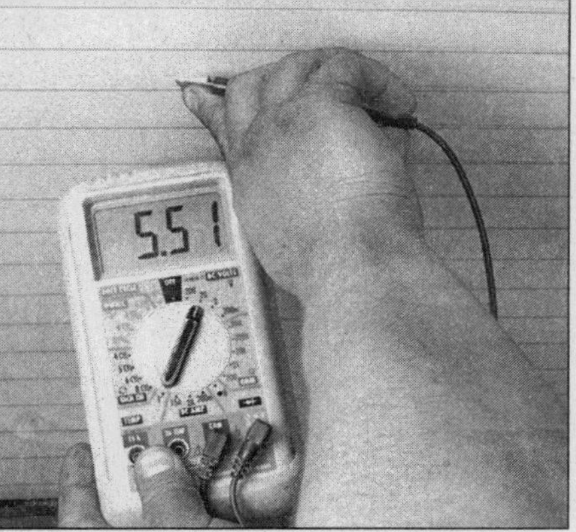

11.6 To determine if a heating element has broken, check the voltage at the center of each element - if the voltage is 6-volts, the element is unbroken

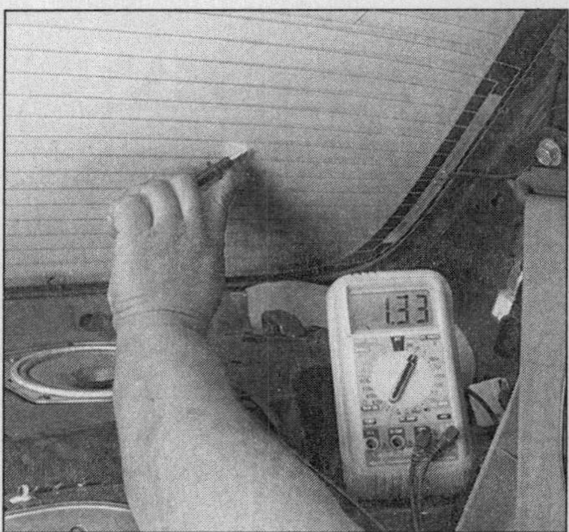

11.8 To find the break, place the voltmeter negative lead against the defogger ground terminal, place the voltmeter positive lead with the foil strip against the heat wire at the positive terminal end and slide it toward the negative terminal end - the point at which the voltmeter deflects from several volts to zero volts is the point at which the wire is broken

4 Using a voltmeter, place the positive probe against the defogger grid positive terminal and the negative probe against the ground terminal. If battery voltage is not indicated, check the fuse, defogger switch, defogger relay and related wiring. If voltage is indicated, but all or part of the defogger doesn't heat, proceed with the following tests.

5 When measuring voltage during the next two tests, wrap a piece of aluminum foil around the tip of the voltmeter positive probe and press the foil against the heating element with your finger **(see illustration)**. Place the negative probe on the defogger grid ground terminal.

6 Check the voltage at the center of each heating element **(see illustration)**. If the voltage is 5 to 6 volts, the element is okay (there is no break). If the voltage is 0 volts, the element is broken between the center of the element and the positive end. If the voltage is 10 to 12 volts, the element is broken between the center of the element and the ground side. Check each heating element.

7 If none of the elements are broken, connect the negative probe to a good chassis ground. The voltage reading should stay the same, if it doesn't the ground connection is bad.

8 To find the break, place the voltmeter negative probe against the defogger ground terminal. Place the voltmeter positive probe with the foil strip against the heating element at the positive side and slide it toward the negative side. The point at which the voltmeter deflects from several volts to zero is the point where the heating element is broken **(see illustration)**.

Repair

Refer to illustration 11.14

9 Repair the break in the element using a repair kit specifically for this purpose, such as Dupont paste No. 4817 (or equivalent). The kit includes conductive plastic epoxy.

10 Before repairing a break, turn off the system and allow it to cool for a few minutes.

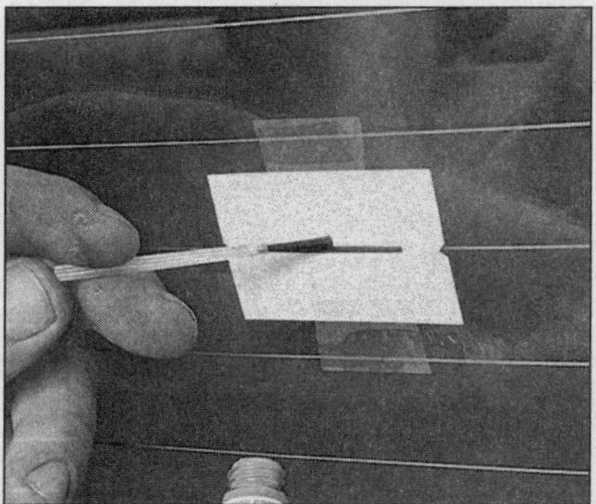

11.14 To use a defogger repair kit, apply masking to the inside of the window at the damaged area, then brush on the special conductive coating

11 Lightly buff the element area with fine steel wool; then clean it thoroughly with rubbing alcohol.

12 Use masking tape to mask off the area being repaired.

13 Thoroughly mix the epoxy, following the kit instructions.

14 Apply the epoxy material to the slit in the masking tape, overlapping the undamaged area about 3/4-inch on either end **(see illustration)**.

15 Allow the repair to cure for 24 hours before removing the tape and using the system.

12 Headlight bulb - replacement

Refer to illustrations 12.3a, 12.3b and 12.4

Warning: *Halogen bulbs are gas-filled and under pressure and may shatter if the surface is scratched or the bulb is dropped. Wear eye protection and handle the bulbs carefully, grasping only the base whenever possible. Don't touch the surface of the bulb with your fingers because the oil from your skin could cause it to overheat and fail prematurely. If you do touch the bulb surface, clean it with rubbing alcohol.*

12.3a Disconnect the headlight bulb electrical connector - 1998 and earlier models

1 Disconnect the negative cable from the battery.

2 Open the hood.

3 Reach behind the headlight assembly and disconnect the electrical connector **(see illustrations)**. If you are unable to access the rear of the headlight (1999 and later models), remove the housing assembly (see Section 14).

4 Rotate the lockring or bulb socket counterclockwise and remove the bulbholder from the headlight housing **(see illustration)**.

5 Insert the new bulb into the holder, secure it into the housing with the lockring and connect the electrical connector.

12.3b On 1999 and later models, access to the bulbholder (arrow) is with the housing pulled forward

12.4 Rotate the bulb lockring counterclockwise and remove the bulb assembly from the housing

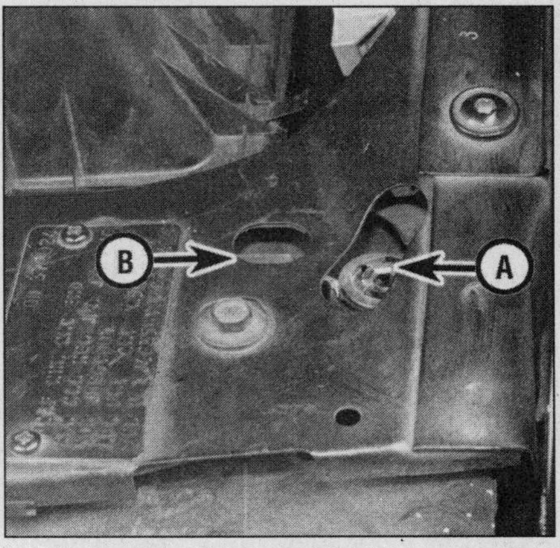

13.1a On 1998 and earlier models, use an Allen wrench to turn the adjuster screws - the outer screw (A) is used to adjust the headlights up-and-down and the inner screw (B) adjusts them left-and-right

13.1b On 1999 and later models, the headlight adjustment screw is accessed through the hole (arrow) in the radiator support panel - a long torx driver will be needed to reach the adjustment screw

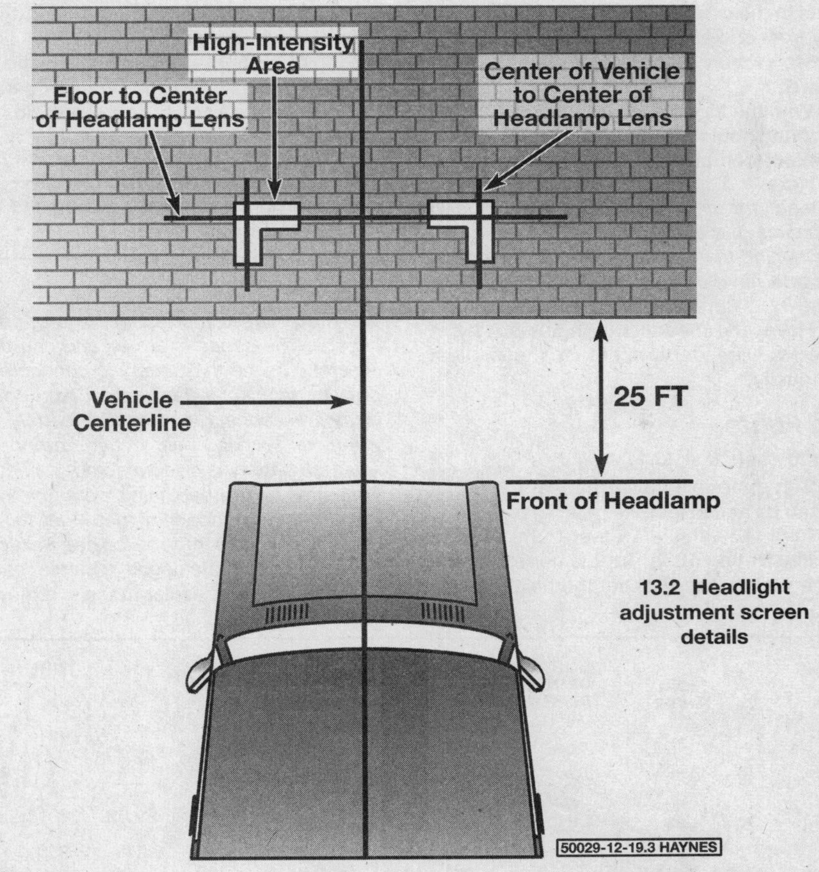

13.2 Headlight adjustment screen details

6 The remainder of installation is the reverse of the removal procedure.

13 Headlights - adjustment

Warning: *The headlights must be aimed correctly. If adjusted incorrectly, they could temporarily blind the driver of an oncoming vehicle and cause an accident or seriously reduce your ability to see the road. The headlights should be checked for proper aim every 12 months and any time a new headlight is installed or front-end bodywork is performed. The following procedure is only an interim step to provide temporary adjustment until the headlights can be adjusted by a properly equipped shop.*

Headlights

Refer to illustrations 13.1a, 13.1b and 13.2

1 Headlights have two spring loaded adjusting screws, located at the rear of the housing. The outer adjuster controls up-and-down movement and the inner adjuster controls left and-right movement **(see illustrations)**.

2 There are several methods of adjusting the headlights. The simplest method requires an open area with a blank wall and a level floor **(see illustration)**.

3 Position masking tape vertically on the wall in reference to the vehicle centerline and the centerlines of both headlights.

4 Position a horizontal tapeline in reference to the centerline of all the headlights. **Note:** *It may be easier to position the tape on the wall with the vehicle parked only a few inches away.*

5 Adjustment should be made with the vehicle parked 25 feet from the wall, sitting level, the gas tank half-full and no unusually heavy load in the vehicle.

6 Starting with the low beam adjustment,

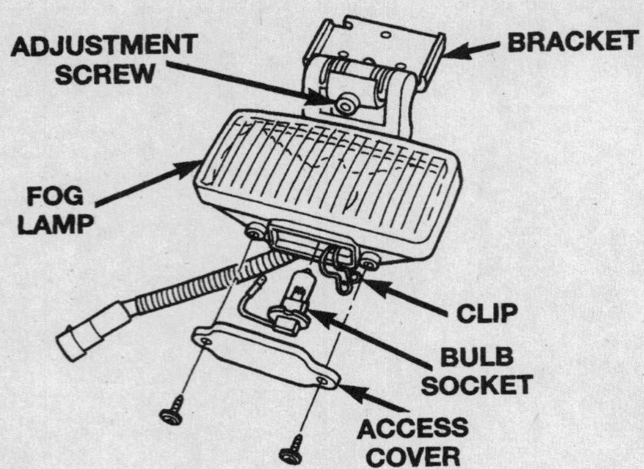

13.11 Use the adjustment screw on the foglamp bracket to adjust the foglamp beam (1998 model shown, the screw is on the exterior of the vehicle on 1999 and later models)

14.2 Remove this jackscrew (arrow) on 1999 and later headlight housings, then pull the housing straight out to disengage it from the ball-studs

position the high intensity zone so it is two inches below the horizontal line and two inches to the side of the vertical headlight line away from oncoming traffic. Twist the adjustment screws until the desired level has been achieved.

7 With the high beams on, the high intensity zone should be vertically centered with the exact center just below the horizontal line. **Note:** *It may not be possible to position the headlight aim exactly for both high and low beams. If a compromise must be made, keep in mind that the low beams are the most used and have the greatest effect on driver safety.*

8 Have the headlights adjusted by a dealer service department at the earliest opportunity.

Fog lights

Refer to illustration 13.11

9 Some models have optional fog lights that can be aimed just like headlights.

10 Park the vehicle 25 feet from the wall **(see illustration 13.2)**. Tape a horizontal line on the wall that represents the height of the fog lamps, and another tapeline four inches below that line.

11 Using the adjusting screws on the fog lamps, adjust the pattern on the wall so that the top of the fog lamp beam meets the lower line on the wall, and the beam is centered horizontally in front of the fog lamps **(see illustration)**.

14 Headlight housing - removal and installation

Refer to illustrations 14.2 and 14.3

Warning: *These models are equipped with airbags. The airbag is armed and can deploy (inflate) anytime the battery is connected. To prevent accidental deployment (and possible injury), disconnect the negative battery cable whenever working near airbag components. After the battery is disconnected, wait at least 2 minutes before beginning work (the system has a back-up capacitor that must fully discharge). For more information see Section 27.*

1 Disconnect the negative battery cable.

2 On 1998 and earlier models, remove the radiator grille (Chapter 11). On 1999 and later models, remove the jackscrew at the top of the headlight housing **(see illustration)**.

3 Grasp the lower edge of the headlight housing and pull it straight back sharply to detach it from the lower adjuster pivots **(see illustration)**. Grasp the upper edge and pull it straight back to detach the housing from the upper pivots.

4 Disconnect the electrical connectors and remove the housing assembly from the vehicle.

5 Installation is the reverse of removal.

15 Bulb replacement

Side marker lights

Refer to illustrations 15.1 and 15.2

1 On 1998 and earlier models, the side marker light housing is located next to the headlight housing. Remove the screws and pull out the housing to access the bulbholder on the back of the housing **(see illustration)**.

2 On 1999 and later models, the park and turn signal bulbs are located in the bottom of

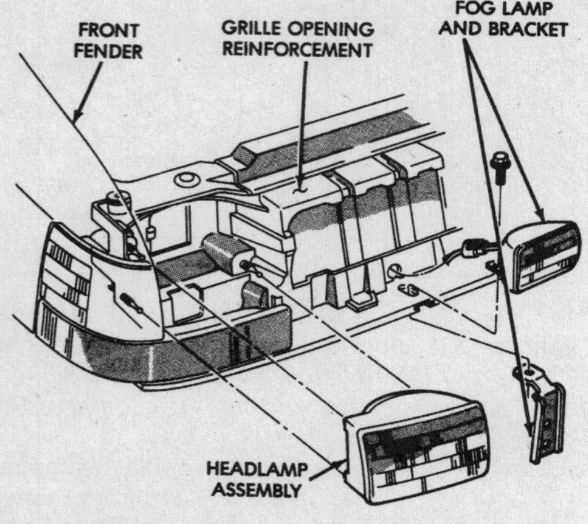

14.3 Headlight housing installation details - 1998 and earlier models

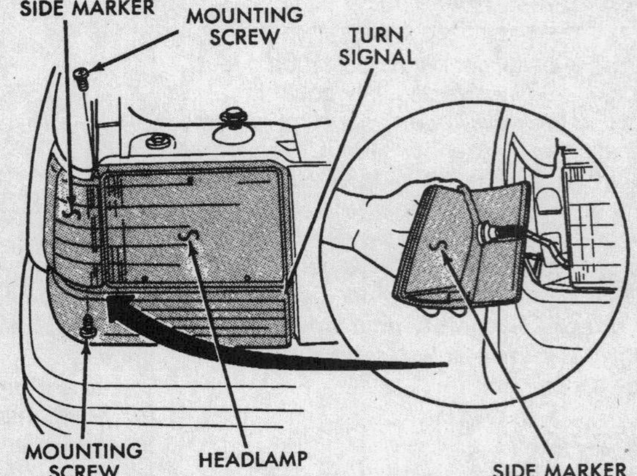

15.1 Side marker bulb installation details - 1998 and earlier models

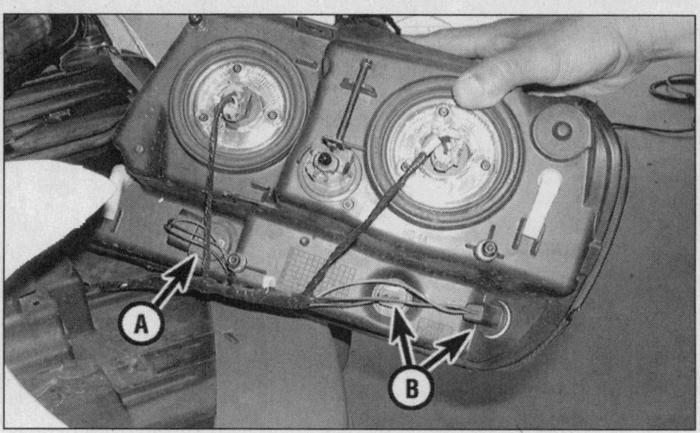

15.2 On 1999 and later models, the turn signal (A) and park light bulbs (B) are located in the headlight housing

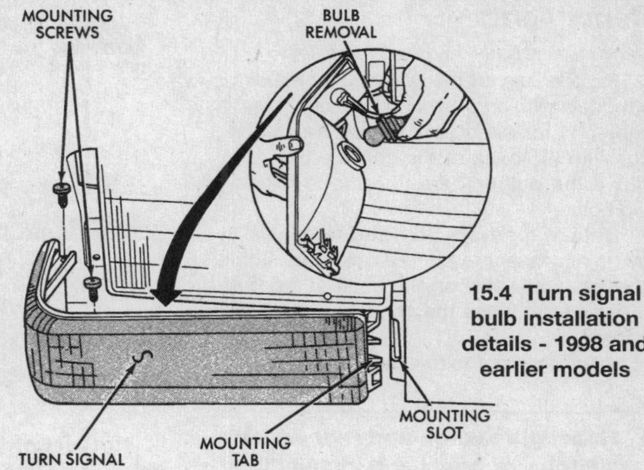

15.4 Turn signal bulb installation details - 1998 and earlier models

15.6 Open the liftgate, remove the screws and detach the tail light housing by pulling it straight back - 1998 and earlier shown, 1999 and later similar

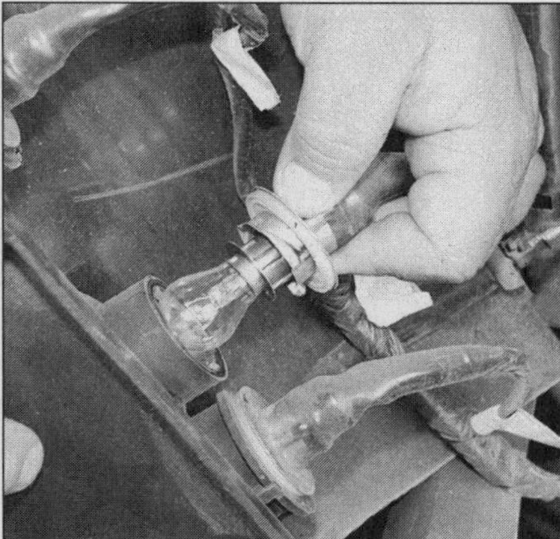
15.7 Rotate the bulb holder and pull it out - push in and rotate the bulb to remove it

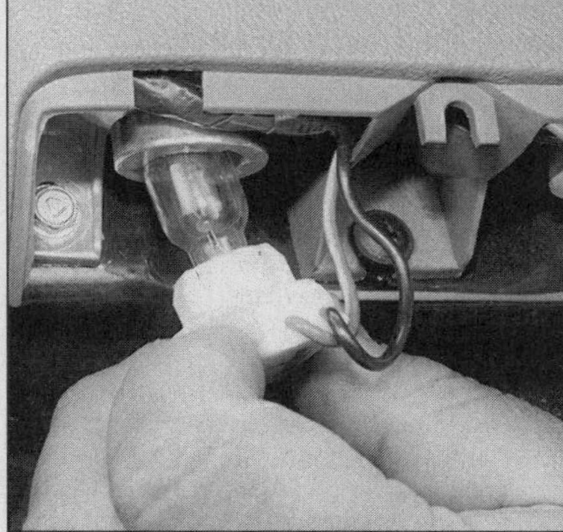
15.9 Remove bulb holder, then grasp the bulb and pull it straight out of the holder

the headlight housing. Refer to Section 14 for removal. Twist the bulbholders out to replace the park or turn signal bulbs **(see illustration)**.

3 Installation is the reverse of the removal procedure.

Front turn signal lights

Refer to illustration 15.4

4 Remove the screws that secure the turn signal housing, lift the housing out, rotate the bulb holder counterclockwise and pull the bulb out **(see illustration)**. On 1999 and later models, see Step 2.

5 Installation is the reverse of removal.

Rear tail light/brake light/turn signal

Refer to illustrations 15.6 and 15.7

6 Remove bolts and detach the taillight housing **(see illustration)**.

7 Turn the bulb holder counterclockwise and remove it from the back of the housing **(see illustration)**. Push in and rotate the turn signal, brake light and tail light bulbs counterclockwise to remove them from the holder. On most models, the backup lights are also in the taillight housing.

8 Installation is the reverse of removal.

High-mounted brake light

Refer to illustration 15.9

9 Open the tailgate. Remove the screw and pry off the housing cover. Pull the bulb holder out and replace the bulb by pulling it straight out **(see illustration)**. On later models, pry off the cover, remove the two bolts and lift the housing out and turn it over, then rotate bulb holder counterclockwise to remove it and pull the bulb straight out. **Note:** *On 1999 and later models, the light is mounted to the exterior of the liftgate. Remove the two screws to pull out the housing and replace the bulb.*

10 Installation is the reverse of removal.

Instrument cluster lights

11 To gain access to the instrument cluster illumination lights, the instrument cluster will have to be removed (see Section 20). The bulbs can then be removed and replaced from the rear of the cluster.

12 Rotate the bulb counterclockwise to remove it.

13 Installation is the reverse of removal.

License plate light

Refer to illustration 15.14

14 On 1998 and earlier models, remove the screws and detach the license plate bracket, then remove the bulbs from the back **(see illustration)**. On 1999 and later models, the light is part of a long plastic panel above the license plate. Remove the screws securing the panel to the liftgate, then pull to panel out to disengage the clips at each end. Replacement of the bulb can then be accomplished from the back of the panel.

15 Installation is the reverse of removal.

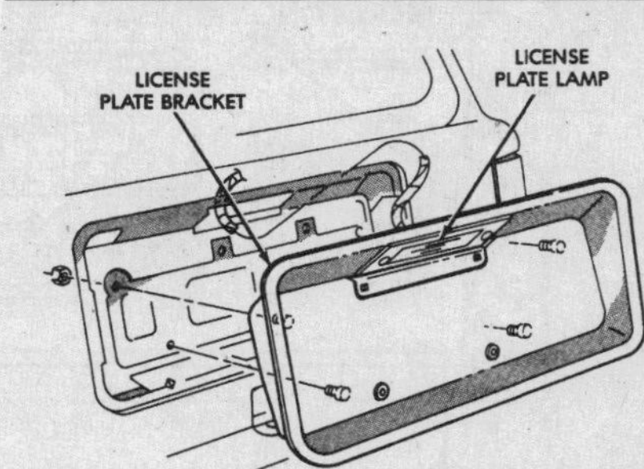

15.14 License plate lamp bulb installation details - 1998 and earlier models

Interior lights

Refer to illustrations 15.17a and 15.17b

16 Pry the lens off the interior light housing. On models with an overhead console, refer to Chapter 11 for removal of the overhead console. With the back of the console exposed, remove the bulbholders to replace reading lamp bulbs.

17 Detach the bulb from the terminals. It may be necessary to pry the bulb out - if this is the case, pry only on the ends of the bulb (otherwise the glass may shatter **(see illustrations)**.

18 Installation is the reverse of removal.

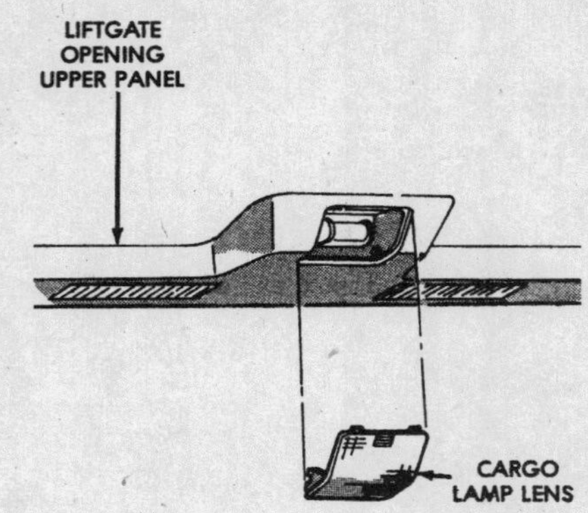

15.17a Cargo lamp bulb installation details - on later models the lens is round, but bulb replacement is the same

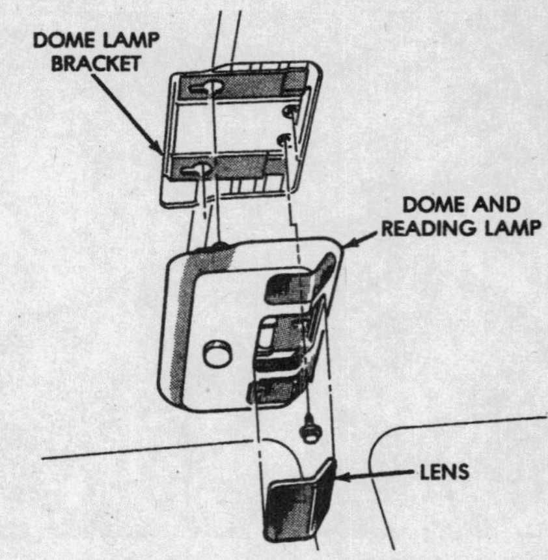

15.17b Dome lamp bulb installation details - typical

16 Headlight switch and rear wiper switch - removal and installation

Warning: *These models are equipped with airbags. The airbag is armed and can deploy (inflate) anytime the battery is connected. To prevent accidental deployment (and possible injury), disconnect the negative battery cable whenever working near airbag components. After the battery is disconnected, wait at least 2 minutes before beginning work (the system has a back-up capacitor that must fully discharge). For more information see Section 27.*

1995 and earlier models

Refer to illustration 16.4

1 Disconnect the negative battery cable.

2 Remove the center trim panel and the instrument cluster bezel (see Chapter 11).

3 Remove the switch pod bezel retaining screws and pull the bezel out to disconnect the electrical connectors from the switches, then remove the bezel and switch assembly from the instrument panel.

4 From the back of the bezel, remove the switch pod mounting screws for either the headlight or rear-wiper switches **(see illustration)**.

5 Installation is the reverse of removal.

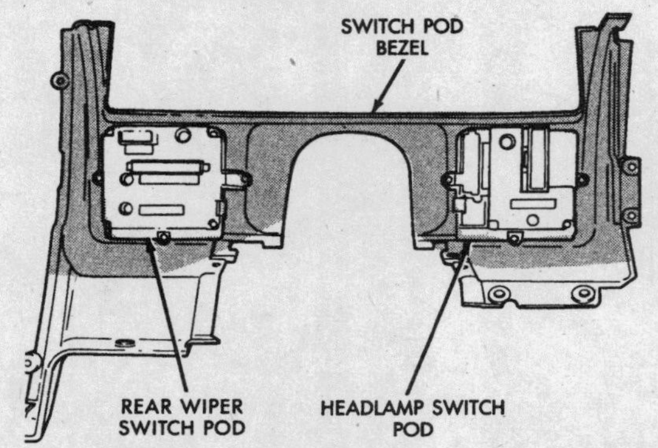

16.4 On 1995 and earlier models, remove the switches from the back of the switch pod bezel

1996 through 1998 models

Refer to illustration 16.8

6 On these models, the left (headlight switch) and right (wiper) switch pods are separate units.

7 Pry off the switch pod bezels from the instrument panel (see Chapter 11).

8 Remove the mounting screws and pull the switch pod out far enough from the instrument panel to disconnect the electrical connectors from the rear of the switch **(see illustration)**.

9 Installation is the reverse of removal.

1999 and later models

10 On these models, the rear wiper switch function is incorporated into the right-hand multi-function lever/switch, and the headlight switch function is incorporated into the left multi-function lever/switch (see Section 8).

17 Radio and speakers - removal and installation

Warning: *These models are equipped with airbags. The airbag is armed and can deploy (inflate) anytime the battery is connected. To prevent accidental deployment (and possible injury), disconnect the negative battery cable whenever working near airbag components. After the battery is disconnected, wait at least 2 minutes before beginning work (the system has a back-up capacitor that must fully discharge). For more information see Section 27.*

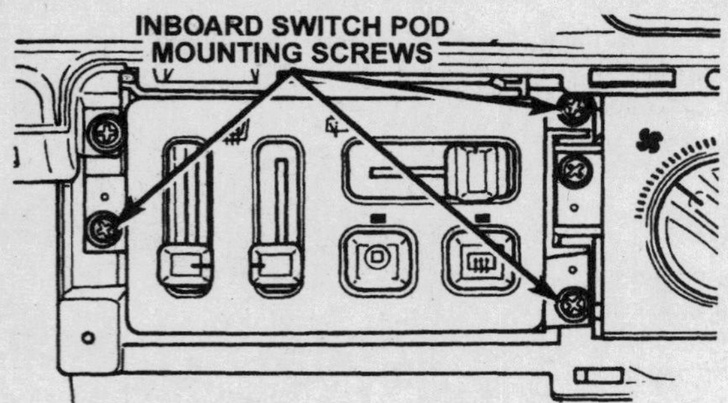

16.8 On 1996 through 1998 models, the switch pods on either side of the steering column are mounted directly to the instrument panel

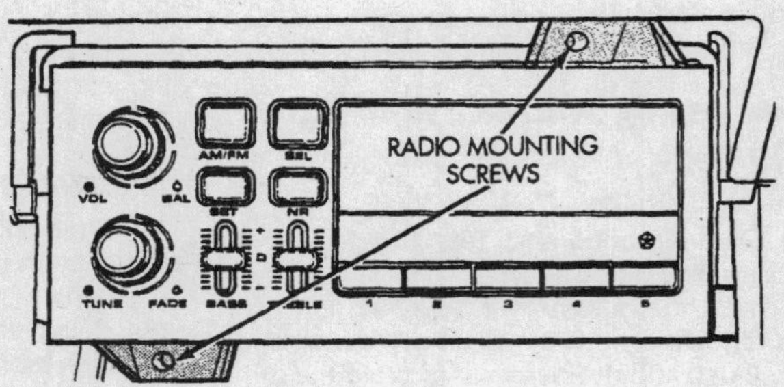

17.3 Remove the screws and pull the radio out - typical

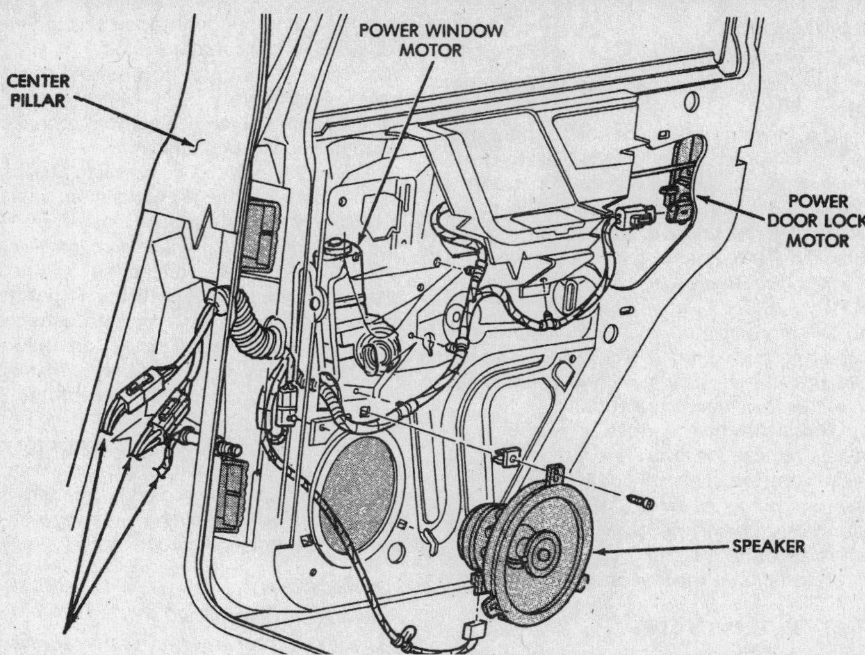

17.6 Door-mounted speaker installation details (typical rear door installation shown)

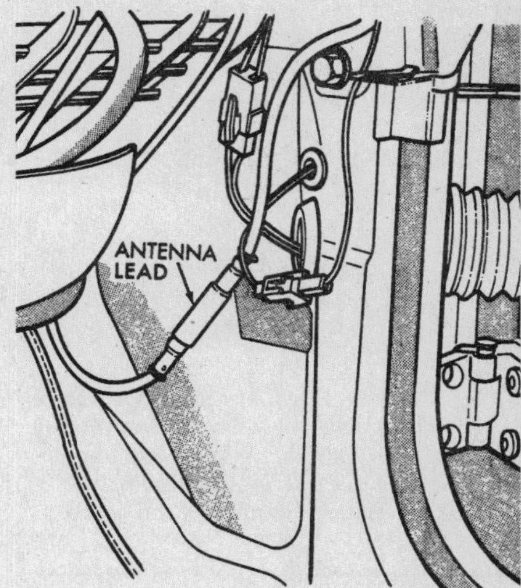

18.4 Disconnect the antenna lead by pulling on the metal connectors - DO NOT pull on the cable

Radio

Refer to illustration 17.3

1 Disconnect the negative battery cable.

2 On 1995 and earlier models, remove the instrument panel center bezel (see Chapter 11). On 1996 through 1998 models, remove the instrument cluster bezel (see Chapter 11). On 1999 and later models, remove the radio trim bezel only.

3 Remove the mounting screws, pull the radio out of the instrument panel far enough to disconnect the electrical connectors and antenna, then remove it from the vehicle **(see illustration)**.

4 Installation is the reverse of removal.

Speakers

Door

Refer to illustration 17.6

5 Remove the door trim panel (see Chapter 11).

6 Remove the screws and detach the speaker **(see illustration)**. Pull the speaker out of the door, disconnect the electrical connector and remove the speaker from the vehicle.

Instrument panel

7 Remove the dash pad on 1995 and earlier models (see Chapter 11). On 1996 and later models, remove only the upper cowl top trim panel (see Chapter 11).

8 Disconnect the connector, remove the screws and detach the speaker.

9 Installation is the reverse of removal.

18 Antenna - removal and installation

Refer to illustrations 18.4, 18.5a and 18.5b

1 Disconnect the negative battery cable.

2 If equipped with a fixed antenna mast, use a small wrench to remove the mast.

3 Remove the plastic rivets from the rear edge of the right side inner fender splash shield and pull the shield away for access to the antenna.

4 Remove the passenger side kick panel and disconnect the antenna lead **(see illustration)**. If equipped with a power antenna, disconnect the harness connector. Pull the grommet out of the kick panel and feed the antenna lead and harness through the hole.

5 Remove the capnut and escutcheon. If equipped with a power antenna, remove the mounting bolts. Remove the antenna from inside the wheel well **(see illustrations)**.

6 Installation is the reverse of removal.

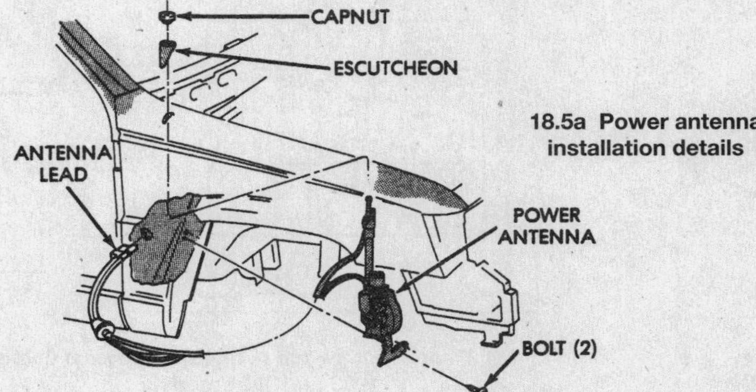

18.5a Power antenna installation details

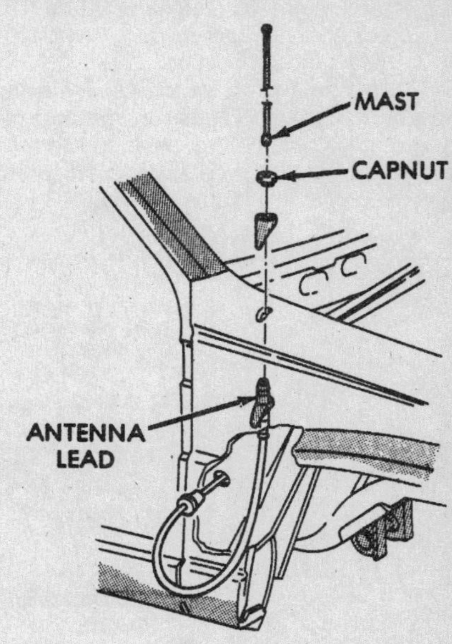

18.5b Fixed mast antenna installation details

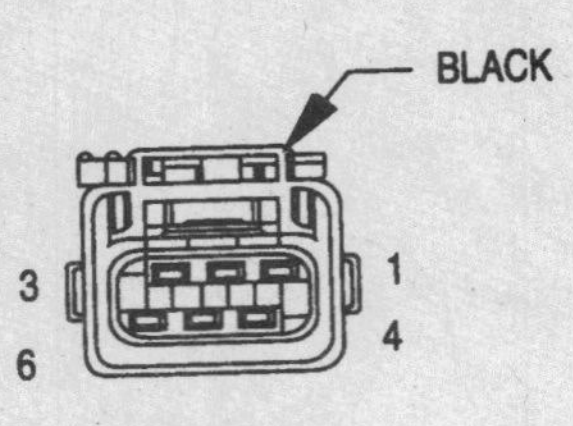

19.5 Windshield wiper motor connector terminal details - 1998 and later model connector shown, on 1997 and earlier models the connector is oval, but terminals are numbered the same

19 Wiper motor - check and replacement

Front wiper motor

Check

Refer to illustration 19.5

1 If the wiper motor does not run at all, first check the fuse block for a blown fuse (see Section 3).

2 Check the wiper switch (see Section 8).

3 Remove the circuit breaker from the fuse block (see Section 3). Use an ohmmeter to check for continuity across the terminals. If an open circuit is indicated, replace the circuit breaker.

4 Remove the electrical connector from the wiper motor (see Replacement below for access to the connector).

5 With wiper switch Off, measure the resistance between terminal 4 of the motor connector and ground **(see illustration)**. The ohmmeter should read zero. If it doesn't, repair the ground connection.

6 Turn the ignition switch On. With the wiper switch in the Low position, measure the voltage at terminal 5. The meter should read battery voltage. If it does and the motor still doesn't operate, the motor is faulty.

Replacement

Refer to illustration 19.10

7 Disconnect the negative cable from the battery.

8 Use a wax marking pencil to mark the wiper blade position on the windshield, then, remove the nuts and detach the wiper arms. **Note:** *If the wiper arm is stuck firmly, don't use a prying tool under it. Use a battery cable terminal puller to release it without damage.*

9 Remove the cowl cover (see Chapter 11).

10 Disconnect the wiper motor electrical connector and remove the five mounting bolts and lift the motor and linkage assembly out of the cowl **(see illustration)**.

11 Hold the motor drive crank with a wrench, remove the crank nut and detach the crank. Turn the assembly over, remove the three attaching screws and separate the motor from the linkage assembly **(see illustration 19.10)**.

12 Installation is the reverse of removal.

Rear wiper motor

Check

Refer to illustration 19.14

13 Check the wiper fuse in the fuse block (see Section 3).

14 Remove the instrument panel switch pod bezel and remove the rear wiper switch (see Section 16). Connect the switch and backprobe the power terminal with the ignition switch On **(see illustration)**. The voltmeter should read battery voltage. If it doesn't, there is an open circuit between the switch and the fuse block. **Note:** *Check for power in at terminal 1 on 1993 and 1994 models, at terminal 4 on 1995 and 1996 models, and at terminal 2 on 1997 and 1998 models. For testing power out (with the switch turned On), test at terminal 3 on 1993 and 1994 models, at terminal 2 on 1995 and 1996 models, and at terminal 1 on 1997 and 1998 models. For 1999 and later models, the switch function is performed by the right multi-function switch* (see Section 8).

15 If the switch fails the test, replace the switch. If there **is** battery voltage at the corresponding terminals, and nothing is wrong with the wiring or connections and the motor still won't run, the motor is faulty.

16 Open the liftgate and remove the trim panel (see Chapter 11).

17 Disconnect the electrical connector from the wiper motor assembly.

18 With the ignition Off, check to make sure there is zero resistance between terminal 3 of the wiper motor and ground, except on 1999 and later models, on which terminal 1 is the ground. If there is resistance, repair the ground connection. Close the liftgate and make sure there is zero resistance between terminal 6 and ground, except on 1999 and later models, on which you test terminal 3. If there is resistance, repair the liftgate ajar switch and wiring.

19 Turn the ignition On and check for battery voltage at the motor connector with the wiper switch in each position. If battery voltage is present and nothing is wrong with the switch or wiring, replace the motor.

Replacement

Refer to illustration 19.20

20 Detach the wiper arm and remove the nut, bezel and gasket **(see illustration). Caution:** *Place masking tape around the bezel, and use a plastic trim tool to pry the bezel up without damaging the paint.*

21 Open the liftgate, remove the liftgate trim panel (see Chapter 11) and disconnect the electrical connector.

22 Support the motor with one hand while removing the two retaining bolts, then lower the wiper motor from the vehicle.

23 Installation is the reverse of removal.

20 Instrument cluster - removal and installation

Refer to illustration 20.3

Warning: *These models are equipped with airbags. The airbag is armed and can deploy (inflate) anytime the battery is connected. To prevent accidental deployment (and possible injury), disconnect the negative battery cable whenever working near airbag components. After the battery is disconnected, wait at least*

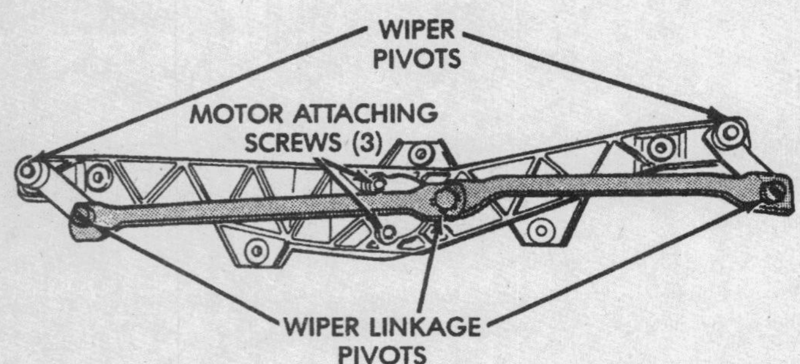

19.10 Windshield wiper motor and linkage assembly details

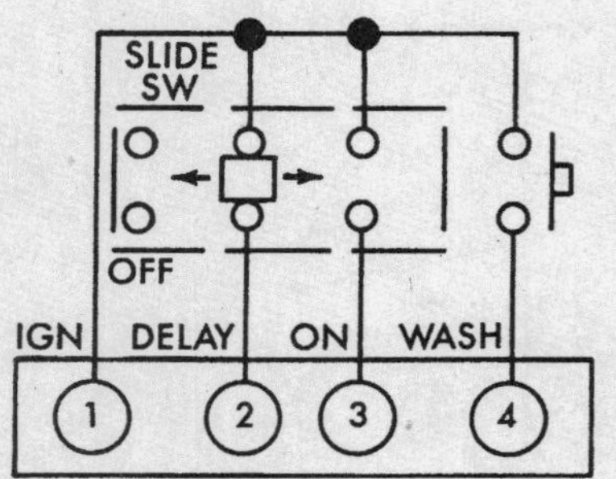

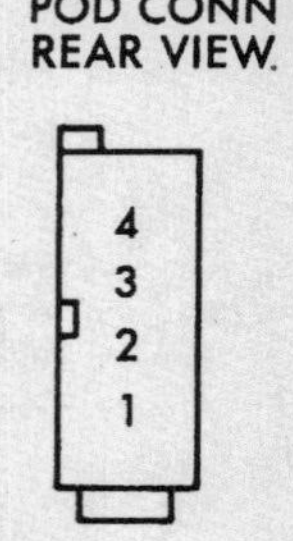

19.14 Rear wiper switch connector terminal details, 1993 and 1994 models

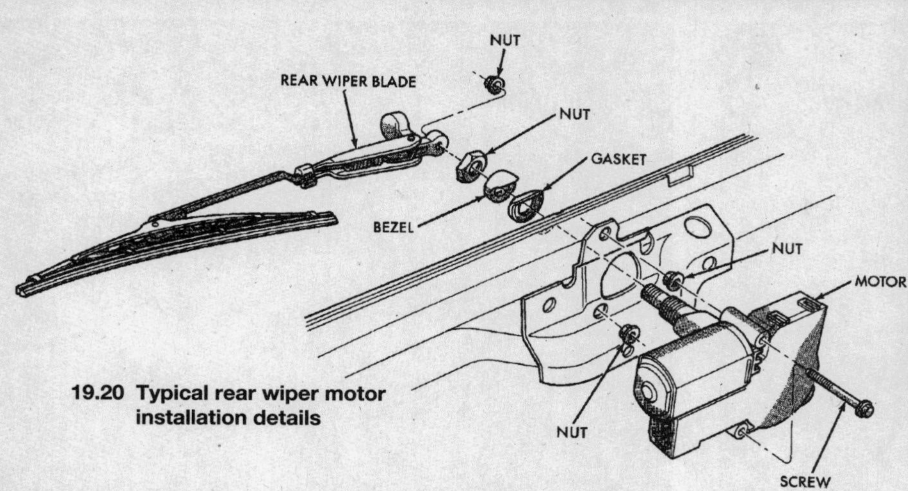

19.20 Typical rear wiper motor installation details

20.3 Typical instrument cluster retaining screw locations

2 minutes before beginning work (the system has a back-up capacitor that must fully discharge). For more information see Section 27.

1 Disconnect the negative battery cable.

2 Remove the instrument cluster bezel (see Chapter 11).

3 Remove the cluster mounting screws **(see illustration)**. Pull the assembly up and back, then disconnect the electrical connector(s).

4 Installation is the reverse of removal.

21 Horn - check and replacement

Refer to illustration 21.4

Note: *Check the fuse before beginning electrical diagnosis.*

1 On 1998 and earlier models, remove the lower air deflector from below the radiator to access the horns. On 1999 and later models, remove the lower screws from the right front wheel well housing, then pull the lower edge of the wheelwell housing back to access the horns.

2 Disconnect the electrical connector from the horn. Test the horn by connecting battery voltage to the two terminals on the horn with a pair of jumper wires.

3 If the horn doesn't sound, replace it. It does sound, the problem lies in the switch, relay or the wiring between components. See Section 6 for relay testing.

4 To replace the horn, disconnect the electrical connector(s) and remove the mounting bolt **(see illustration)**.

5 Installation is the reverse of removal.

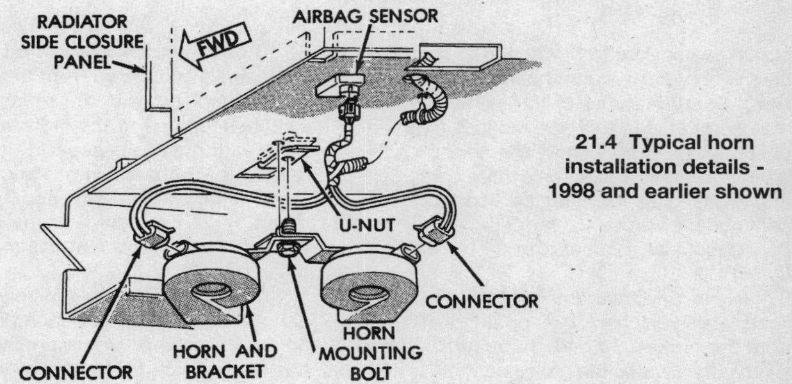

21.4 Typical horn installation details - 1998 and earlier shown

22 Cruise control system - description and check

Refer to illustrations 22.1, 22.5a and 22.5b

1 The cruise control system maintains vehicle speed with servomotor connected to the throttle linkage by a cable. The system consists of the servo motor, brake switch,

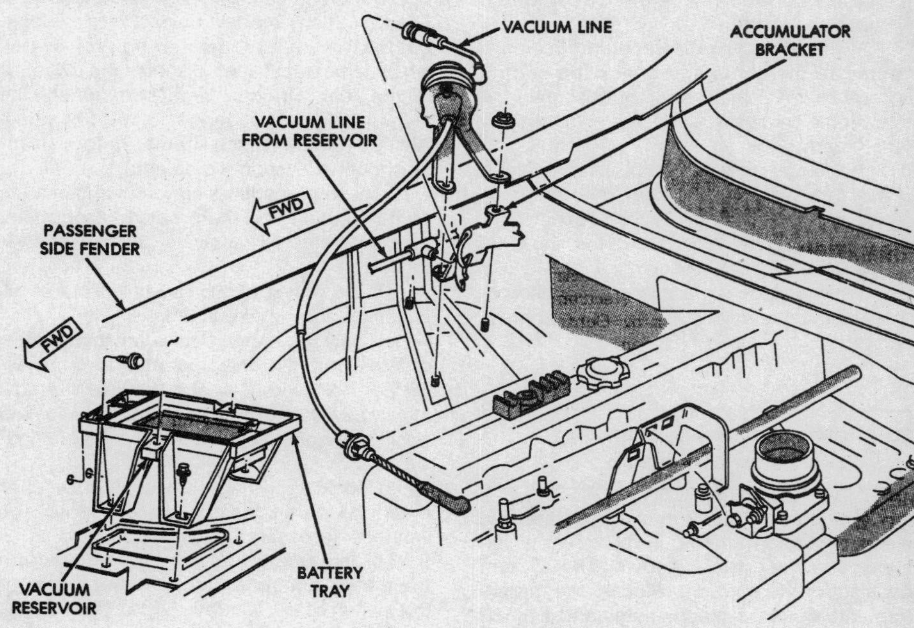

22.1 Typical cruise control system details

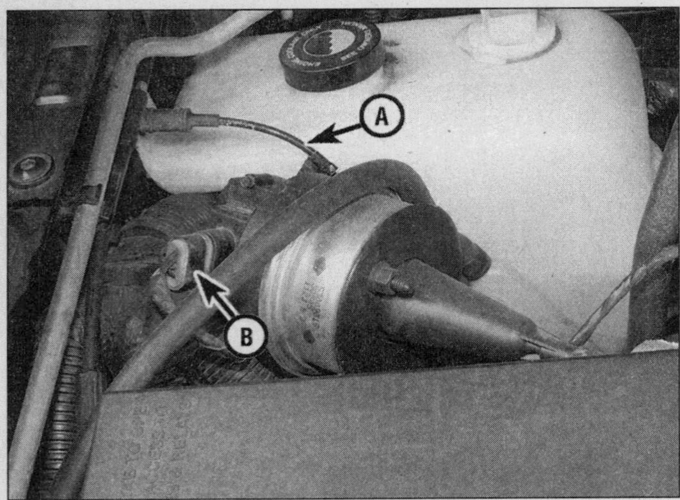

22.5a Check the cruise control vacuum (A) and electrical (B) connections

22.5b Check the cruise control connection to the throttle linkage (arrow)

clutch switch, control switches and associated wiring and vacuum hoses **(see illustration)**. Some features of the system require the use of a scan tool and diagnostic procedures which are beyond the scope of the home mechanic. Listed below are some general procedures that may be used to locate common problems.

2 Locate and check the fuse (see Section 3).

3 Have an assistant operate the brake lights while you check their operation (voltage from the brake light and, if equipped, clutch switch deactivates the cruise control).

4 If the brake lights don't come on or don't shut off, correct the problem and retest the cruise control. Check the clutch switch (Chapter 8).

5 Visually inspect the vacuum hose connected to the servo and check the control linkage between the cruise control servo and the throttle linkage and replace as necessary **(see illustrations)**.

6 If the speedometer is inoperative along with the cruise control system, refer to Chapter 6 and check the vehicle speed sensor.

7 Test-drive the vehicle to determine if the cruise control is now working. If it isn't, take it to a dealer service department or an automotive electrical specialist for further diagnosis and repair.

23 Power window system - description and check

1 The power window system operates the electric motors mounted in the doors which lower and raise the windows. The system consists of the control switches, the motors (regulators), glass mechanisms and associated wiring. When the ignition is On, power flows from the fuse in the power distribution center to the circuit breaker in the fuse block, then to the master switch.

2 Power windows are wired so they can be lowered and raised from the master control switch by the driver or by remote switches located at the individual windows. Each window has a separate motor that is reversible. The position of the control switch determines the polarity and therefore the direction of operation. Some systems are equipped with relays that control current flow to the motors.

3 Some vehicles are equipped with a separate circuit breaker for each motor in addition to the fuse or circuit breaker protecting the whole circuit. This prevents one stuck window from disabling the whole system.

4 The power window system will only operate when the ignition switch is ON. In addition, many models have a window lockout switch at the master control switch which, when activated, disables the switches at the rear windows and, sometimes, the switch at the passenger's window also. Always check these items before troubleshooting a window problem.

5 These procedures are general in nature, so if you can't find the problem using them, take the vehicle to a dealer service department.

6 If the power windows don't work at all, check the fuse or circuit breaker.

7 If only the rear windows are inoperative, or if the windows only operate from the master control switch, check the rear window lockout switch for continuity in the unlocked position. Replace it if it doesn't have continuity.

8 Check the wiring between the switches and fuse panel for continuity. Repair the wiring, if necessary.

9 If only one window is inoperative from the master control switch, try the other control switch at the window. **Note:** *This doesn't apply to the drivers door window*.

10 If the same window works from one switch, but not the other, check the switch for continuity.

11 If the switch tests OK, check for a short or open in the wiring between the affected switch and the window motor.

12 If one window is inoperative from both switches, remove the trim panel from the affected door and check for voltage at the switch and at the motor while the switch is operated. First check for voltage at the connectors. With the ignition key On and connectors in place on the switch, backprobe at the designated wire (see the wiring diagrams at the end of this Chapter) with a grounded test light. Pushing the driver's window switch Down, there should be voltage at one terminal. Pushing the same switch Up, there should be voltage at another terminal. If these voltage are OK, disconnect the electrical connector at the driver's motor, and check for voltage there when the switch is operated.

13 If voltage is reaching the motor and the switch is OK, disconnect the glass from the regulator (see Chapter 11). Move the window up and down by hand while checking for binding and damage. Also check for binding and damage to the regulator. If the regulator is not damaged and the window moves up and down smoothly, replace the motor. If there's binding or damage, lubricate, repair or replace parts, as necessary.

14 If voltage isn't reaching the motor, check the wiring in the circuit for continuity between the switches and motors. You'll need to consult the wiring diagram for the vehicle. Some power window circuits are equipped with relays. If equipped, check that the relays are grounded properly and receiving voltage from the switches. Also check that each relay sends voltage to the motor when the switch is turned on. If it doesn't, replace the relay.

15 Test the windows after you are done to confirm proper repairs. If the main power window switch is to be replaced, pry the switch unit out of the door panel, then disconnect the two connectors.

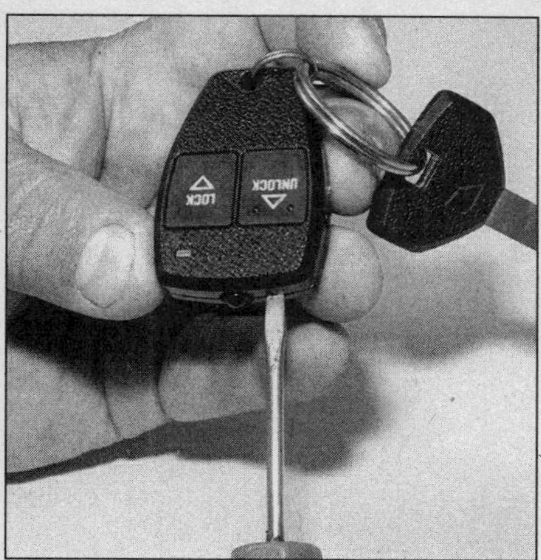

24.15 Use a small screwdriver to separate the transmitter halves

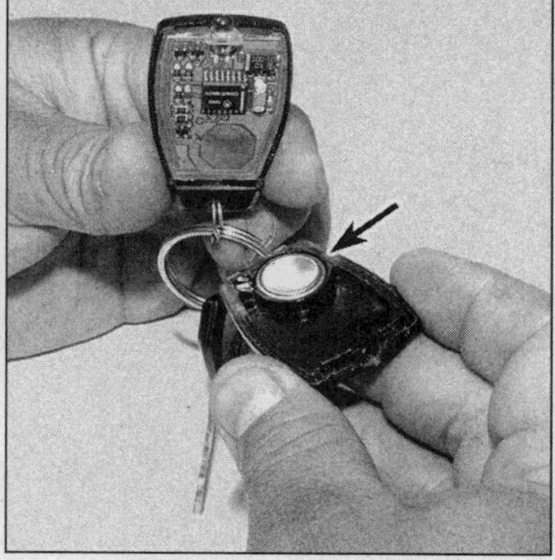

24.16 Replace the lithium batteries (arrow)

24 Power door lock and keyless entry system - description and check

1 The power door lock system operates the door lock actuators mounted in each door. The system consists of the switches, actuators, relays and associated wiring. Diagnosis can usually be limited to simple checks of the wiring connections and actuators for minor faults that can be easily repaired.

2 Power door lock systems are operated by bi-directional solenoids located in the doors. The lock switches have two operating positions: Lock and Unlock. These switches activate a relay, which in turn connects voltage to the door lock solenoids. Depending on which way the relay is activated, it reverses polarity, allowing the two sides of the circuit to be used alternately as the feed (positive) and ground side.

3 Some vehicles may have anti-theft systems incorporated into the power locks. If you are unable to locate the trouble using the following general steps, consult your dealer service department.

4 Always check the circuit protection (fuses and relays) first.

5 Operate the door lock switches in both directions (Lock and Unlock) with the engine off. Listen for the faint click of the relay operating.

6 If there's no click, check for voltage at the switches. If no voltage is present, check the wiring between the fuse block and the switches for shorts and opens.

7 If voltage is present but no click is heard, test the switch for continuity. Replace it if there's no continuity in either switch position.

8 If the switch has continuity but the relay doesn't click, check the wiring between the switch and relay for continuity. Repair the wiring if there's not continuity.

9 If the relay is receiving voltage from the switch but is not sending voltage to the solenoids, check for a bad ground at the relay case. If the relay case is grounding properly, replace the relay.

10 If all but one lock solenoids operate, remove the trim panel from the affected door (see Chapter 11) and check for voltage at the solenoid while the lock switch is operated. One of the wires should have voltage in the Lock position; the other should have voltage in the unlock position.

11 If the inoperative solenoid is receiving voltage, replace the solenoid.

12 If the inoperative solenoid isn't receiving voltage, check for an open or short in the wire between the lock solenoid and the relay. **Note:** *It's common for wires to break in the portion of the harness between the body and door (opening and closing the door fatigues and eventually breaks the wires).*

Keyless entry system

Refer to illustrations 24.15 and 24.16

13 The keyless entry system consists of a remote control transmitter that sends a coded infrared signal to a receiver located in the overhead console or dome lamp housing that operates the door lock system.

14 Replace the transmitter batteries when the red LED light on the side of the case doesn't light when the button is pushed.

15 Use a small screwdriver to carefully separate the case halves **(see illustration)**.

16 Replace the two 3-volt batteries **(see illustration)**. On 1993 to 1995 models, the batteries are CR1616, on 1996 through 1998 models they are DL2016, and 1999 and newer models use CR2016 batteries.

17 Snap the case halves together.

25 Power seats - description and check

1 Power seats allow you to adjust the position of the seat with little effort. These models feature a six-way seat that goes forward and backward, up and down and tilts forward and backward. The seats are powered by three reversible motors mounted in one housing that are controlled by switches on the side of the seat. Each switch changes the direction of seat travel by reversing polarity to the drive motor. Newer models have optional ten-way seats that have extra motors for power lumbar support and power seatback recline.

2 Diagnosis is a simple matter, using the following procedures.

3 Look under the seat for any object which may be preventing the seat from moving.

4 If the seat won't work at all, check the circuit breaker in the fuse block.

5 With the engine off to reduce the noise level, operate the seat controls in all directions and listen for sound coming from the seat motors.

6 If the motor doesn't work or make noise, check for voltage at the motor while an assistant operates the switch.

7 If the motor is getting voltage but doesn't run, test it off the vehicle with jumper wires. If it still doesn't work, replace it. On most power seat systems, the individual components are not available separately. The whole power-seat track must be purchased as an assembly.

8 If the motor isn't getting voltage, remove the seat side panel to access the switch and check for voltage. If there's no voltage at the switch, check the wiring between the fuse block and the switch. If there's battery voltage at the switch, check the circuit breaker. Check the black wire and replace the switch if there's no continuity. If the switch is OK, check for a short or open in the wiring between the switch and motor.

9 Test the completed repairs. **Note:** *On newer models with driver's seat "position memory" option, these tests may be insufficient to determine the problem. On these models, have the dealership service department test the system with a scan tool.*

26 Electric rear view mirrors - description and check

Refer to illustration 26.9

Side mirror

1 The electric rear view mirrors use two motors to move the glass; one for up and down adjustments and one for left-right adjustments. Some models also have a heating element for defogging. The newest models also have a "position memory" option that returns the mirrors to the driver's preferred positions.

2 The control switch has a selector portion which sends voltage to the left or right side mirror. With the ignition switch in the ACC position, roll down the windows and operate the mirror control switch through all functions (left-right and up-down) for both the

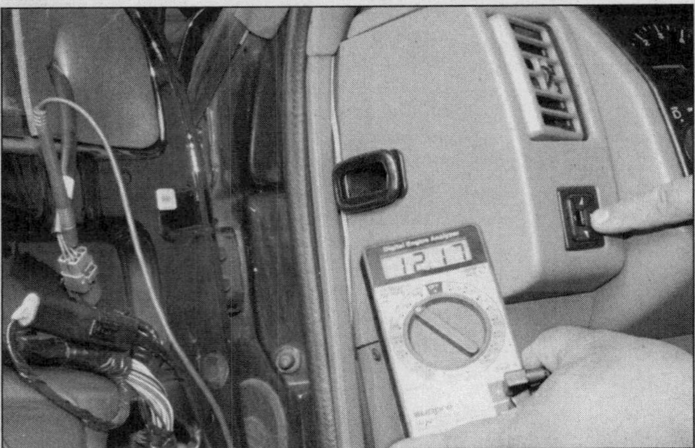

26.9 Check the mirror switch-to-motor connector voltage with the switch in each position

27.6 The airbag control module (arrow) is located under the floor console - be careful not to disconnect the yellow connectors attached to it

left and right side mirrors.

3 Listen carefully for the sound of the electric motors running in the mirrors.

4 If the motors can be heard but the mirror glass doesn't move, the problem is usually with the drive mechanism inside the mirror, which will necessitate replacement of the mirror.

5 If the mirrors don't operate and no sound comes from the mirrors, check the fuse in the fuse block located in the right side of the dash (see Section 3).

6 If the fuse is OK, remove the switch bezel (see Chapter 11) for access to the back of the mirror control switch, without disconnecting the wires attached to it. Turn the ignition ON and check for voltage at the switch. There should be voltage at one terminal. If there's no voltage at the switch, check for an open or short in the wiring between the fuse panel and the switch.

7 If there's voltage at the switch, disconnect it. Refer to the wiring diagrams at the end of this Chapter and check the switch for continuity in all its operating positions. If the switch does not have continuity, replace it.

8 Re-connect the switch. Locate the wire going from the switch to ground. Leaving the switch connected, connect a jumper wire between this wire and ground. If the mirror works normally with this wire in place, repair the faulty ground connection.

9 If the mirror still doesn't work, remove the mirror and check the wires at the mirror for voltage. Check with the ignition ON and the mirror selector switch on the appropriate side. Operate the mirror switch in all its positions **(see illustration)**. There should be voltage at one of the switch-to-mirror wires in each switch position (except the neutral "off" position).

10 If voltage is not present in each switch position, check the wiring between the mirror and control switch for opens and shorts.

11 If there's voltage, remove the mirror and test it off the vehicle with jumper wires. Replace the mirror if it fails this test.

Automatic/day night mirror

12 To reduce glare, the automatic day/night mirror adjusts the amount of light reflected according to conditions. This is achieved with two photocell sensors, one facing forward and one facing rearward, that darken or lighten the thin layer of electrochromic material incorporated into the mirror glass.

13 With the ignition switch On, place the mirror switch on non-sunroof models, in the High or Low position or on sunroof equipped models, in the On position.

14 Cover the forward facing sensor, shine a light into the rear facing sensor, make sure the glass darkens, then shift the transmission into reverse and make sure it lightens. If it doesn't, disconnect the electrical connector. With the ignition On, cavity 1 should have battery voltage, cavity 2 should have continuity to ground and cavity 3 should have continuity with the transmission in reverse.

15 If it fails any of these tests, replace the mirror (see Chapter 11).

27 Airbag - general information

Refer to illustration 27.6

These models are equipped with a Supplemental Restraint System (SRS), more commonly called an airbag system. On later models, there are two airbags, one for the driver and one for the front seat passenger. The SRS system is designed to protect the driver (and on 1996 and later models, the passenger as well) from serious injury in the event of a head-on or frontal collision. On models through 1996, there are two crash sensors mounted at the front of the vehicle and an airbag control module that also contains a crash sensor located inside the passenger compartment. On later models, there is only one crash sensor, the one located in the airbag control module.

Airbag module

The airbag module contains a housing incorporating the cushion (airbag) and inflator unit. The inflator assembly is mounted on the back of the housing over a hole through which gas is expelled, inflating the bag almost instantaneously when an electrical signal is sent from the system. On the driver's airbag, the specially wound wire that carries this signal to the module is called a clockspring. The clockspring is a flat, ribbon-like electrically conductive tape that winds and unwinds as the steering wheel is turned so it can transmit an electrical signal regardless of wheel position.

On 1996 and later models, a second air bag assembly for the front seat passenger is located in the top of the dashboard, above the glove compartment.

Sensors

Early systems had three sensors: two crash sensors at the front of the vehicle behind the bumper and a safing sensor in the airbag control module located between the front seats, in the center console. Later models have only the sensor in the control module.

The front crash sensors are basically pressure sensitive switches that complete an electrical circuit during an impact of sufficient G force. The electrical signal from the crash sensors is sent to the safing sensor in the airbag control module, which then completes the circuit and inflates the airbag(s).

Airbag control module

The ASDM contains the safing sensor and an on-board microprocessor that monitors the operation of the system **(see illustration)**. It checks this system every time the vehicle is started, causing the AIRBAG light to go on, then off, if the system is operating properly. If there is a fault in the system, the light will go on and stay on and the airbag control module will store fault codes indicating the nature of the fault. If the AIRBAG light doesn't come on, or goes on and stays on, or the light blinks at all times, the vehicle should be taken to your dealer immediately for service.

Disabling the system

Whenever working in the vicinity of the steering wheel, steering column, floor console or near other components of the airbag system, the system should be disarmed. To do this perform the following steps:

a) Turn the ignition switch to the Off position.

	LEGEND OF SYMBOLS USED ON WIRING DIAGRAMS		
+	POSITIVE	→≻—	CONNECTOR
−	NEGATIVE	→	MALE CONNECTOR
	GROUND	≻—	FEMALE CONNECTOR
	FUSE		DENOTES WIRE CONTINUES ELSEWHERE
	GANG FUSES WITH BUSS BAR		DENOTES WIRE GOES TO ONE OF TWO CIRCUITS
	CIRCUIT BREAKER		SPLICE
	CAPACITOR	J2 2	SPLICE IDENTIFICATION
Ω	OHMS		THERMAL ELEMENT
	RESISTOR	TIMER	TIMER
	VARIABLE RESISTOR		MULTIPLE CONNECTOR
	SERIES RESISTOR		OPTIONAL WIRING WITH / WIRING WITHOUT
	COIL		"Y" WINDINGS
	STEP UP COIL	88:88	DIGITAL READOUT
	OPEN CONTACT		SINGLE FILAMENT LAMP
	CLOSED CONTACT		DUAL FILAMENT LAMP
	CLOSED SWITCH		L.E.D. – LIGHT EMITTING DIODE
	OPEN SWITCH		THERMISTOR
	CLOSED GANGED SWITCH		GAUGE
	OPEN GANGED SWITCH		SENSOR
	TWO POLE SINGLE THROW SWITCH		FUEL INJECTOR
	PRESSURE SWITCH	#36	DENOTES WIRE GOES THROUGH BULKHEAD DISCONNECT
	SOLENOID SWITCH	#19 STRG COLUMN	DENOTES WIRE GOES THROUGH STEERING COLUMN CONNECTOR
	MERCURY SWITCH	INST PANEL #14	DENOTES WIRE GOES THROUGH INSTRUMENT PANEL CONNECTOR
	DIODE OR RECTIFIER	ENG #7	DENOTES WIRE GOES THROUGH GROMMET TO ENGINE COMPARTMENT
	BY-DIRECTIONAL ZENER DIODE		DENOTES WIRE GOES THROUGH GROMMET
	MOTOR		HEATED GRID ELEMENTS
	ARMATURE AND BRUSHES		

28.4a Wiring diagram symbol legend - typical

b) *Disconnect the cable from the negative battery terminal.*

c) *Wait at least two minutes for the backup power supply to be depleted before beginning work.*

Enabling the system

To enable the airbag system, perform the following steps:

a) *Turn the ignition switch to the Off position.*

b) *Connect the cable to the negative battery terminal.*

c) *Without putting your body in front of either airbag, turn the ignition switch to the On position. Observe that the airbag warning light should glow for 6 to 8 seconds, then go out, indicating the system is functioning properly.*

28 Wiring diagrams - general information

Refer to illustrations 28.4a and 28.4b

Since it isn't possible to include all wiring diagrams for every year covered by this manual, the following diagrams are those that are typical and most commonly needed.

Prior to troubleshooting any circuits, check the fuse and circuit breakers (if equipped) to make sure they are in good condition. Make sure the battery is properly charged and has clean, tight cable connections (see Chapter 1).

When checking the wiring system, make sure that all electrical connectors are clean, with no broken or loose pins. When disconnecting an electrical connector, do not pull on the wires, only on the connector housings themselves.

Refer to the **accompanying illustrations** for the legend and wire color codes applicable to your vehicle.

COLOR CODE	COLOR	STANDARD TRACER COLOR
BL	BLUE	WT
BK	BLACK	WT
BR	BROWN	WT
DB	DARK BLUE	WT
DG	DARK GREEN	WT
GY	GRAY	BK
LB	LIGHT BLUE	BK
LG	LIGHT GREEN	BK
OR	ORANGE	BK
PK	PINK	BK or WT
RD	RED	WT
TN	TAN	WT
VT	VIOLET	WT
WT	WHITE	BK
YL	YELLOW	BK
*	WITH TRACER	

28.4b Wire diagram color code chart

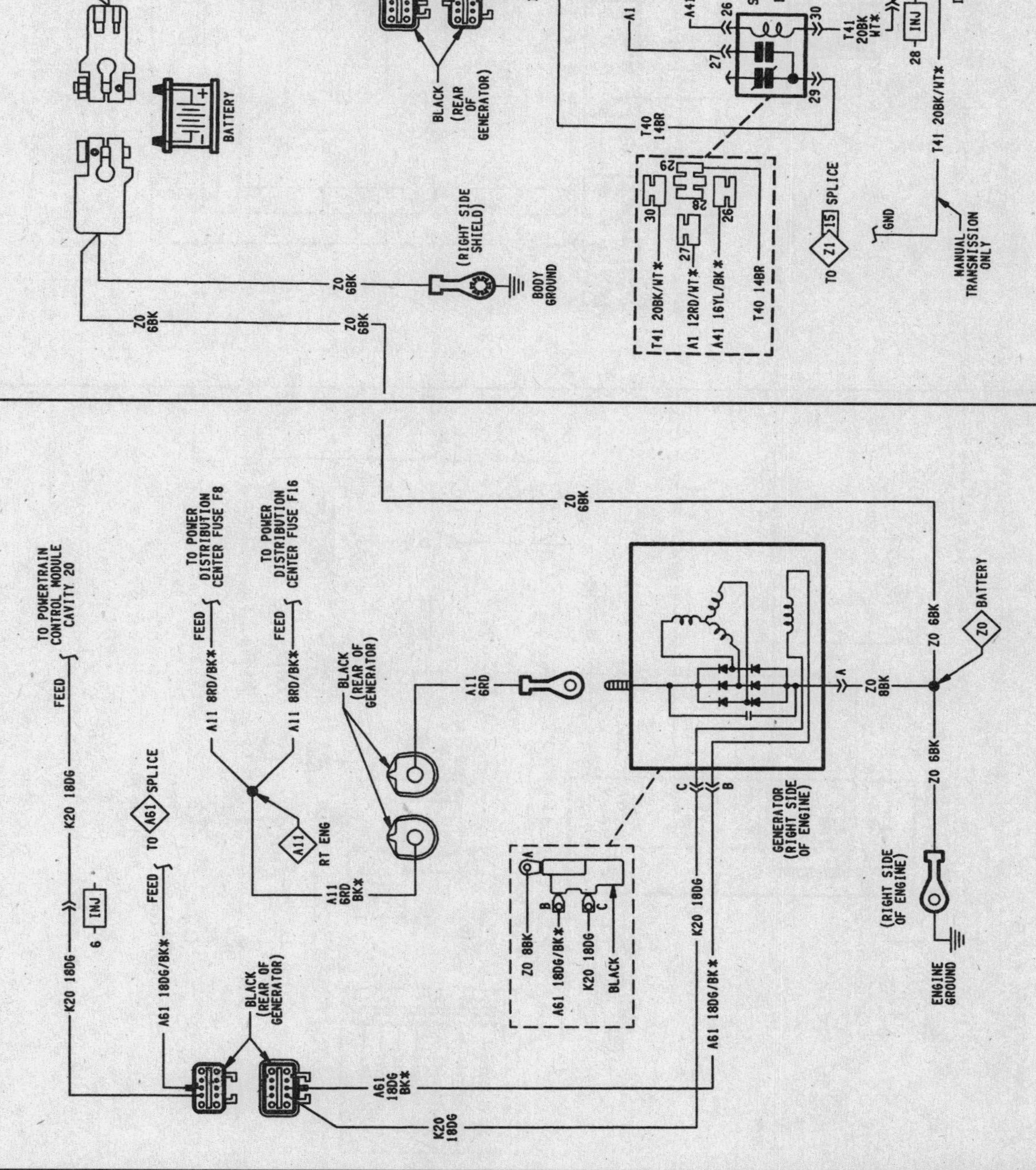

Typical starting system wiring diagram

Typical charging system wiring diagram

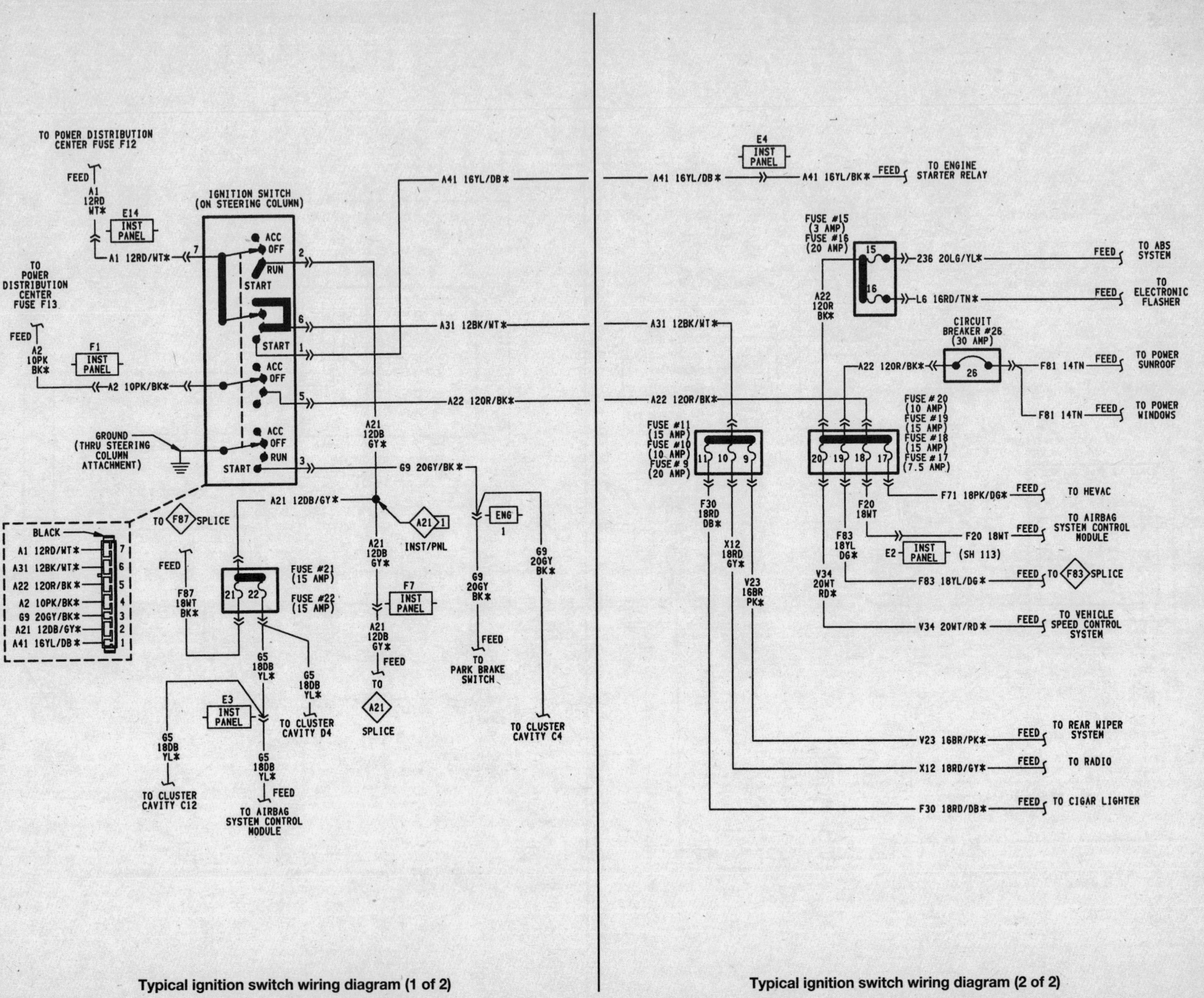

Typical ignition switch wiring diagram (1 of 2)

Typical ignition switch wiring diagram (2 of 2)

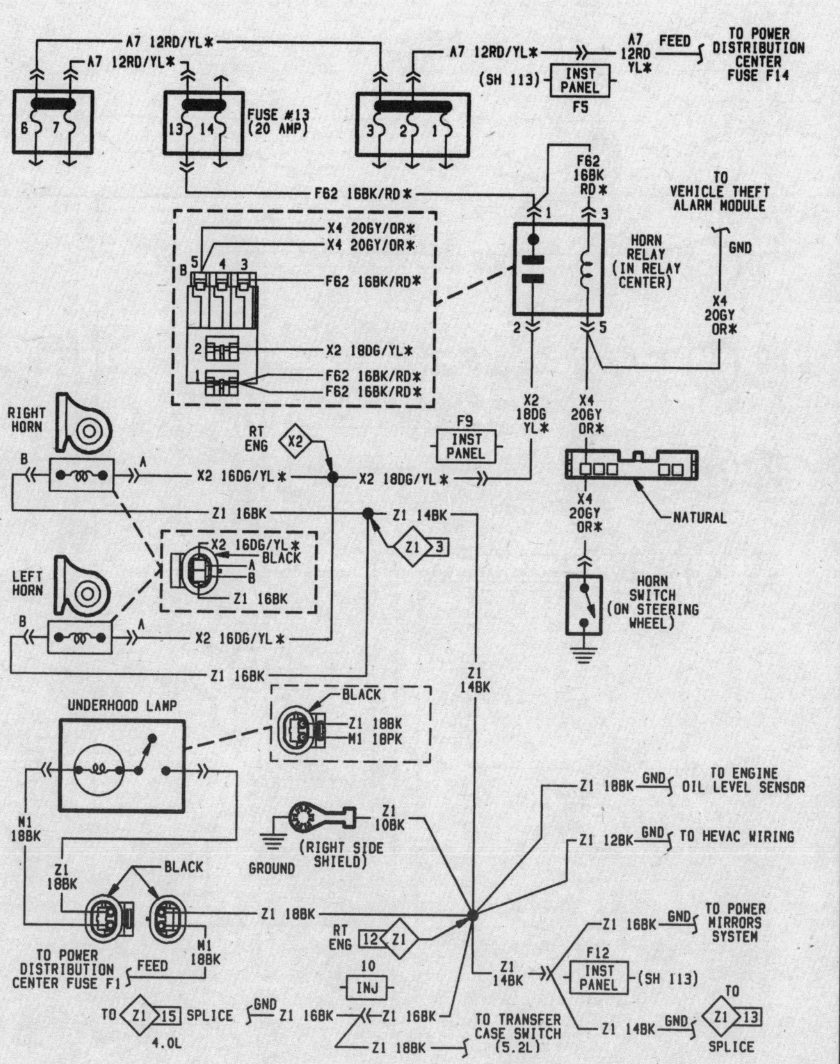

Typical underhood lamp/horn system wiring diagram

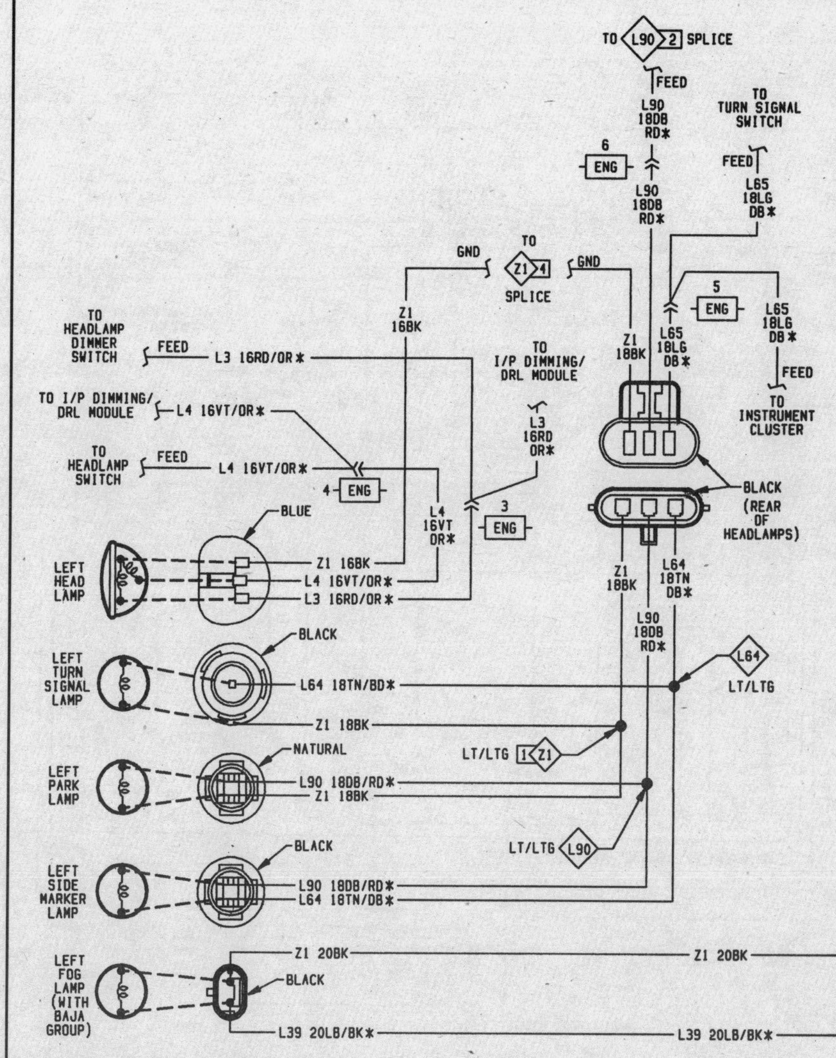

Typical front end lighting system wiring diagram (1 of 2)

Typical headlight switch wiring diagram (1 of 2)

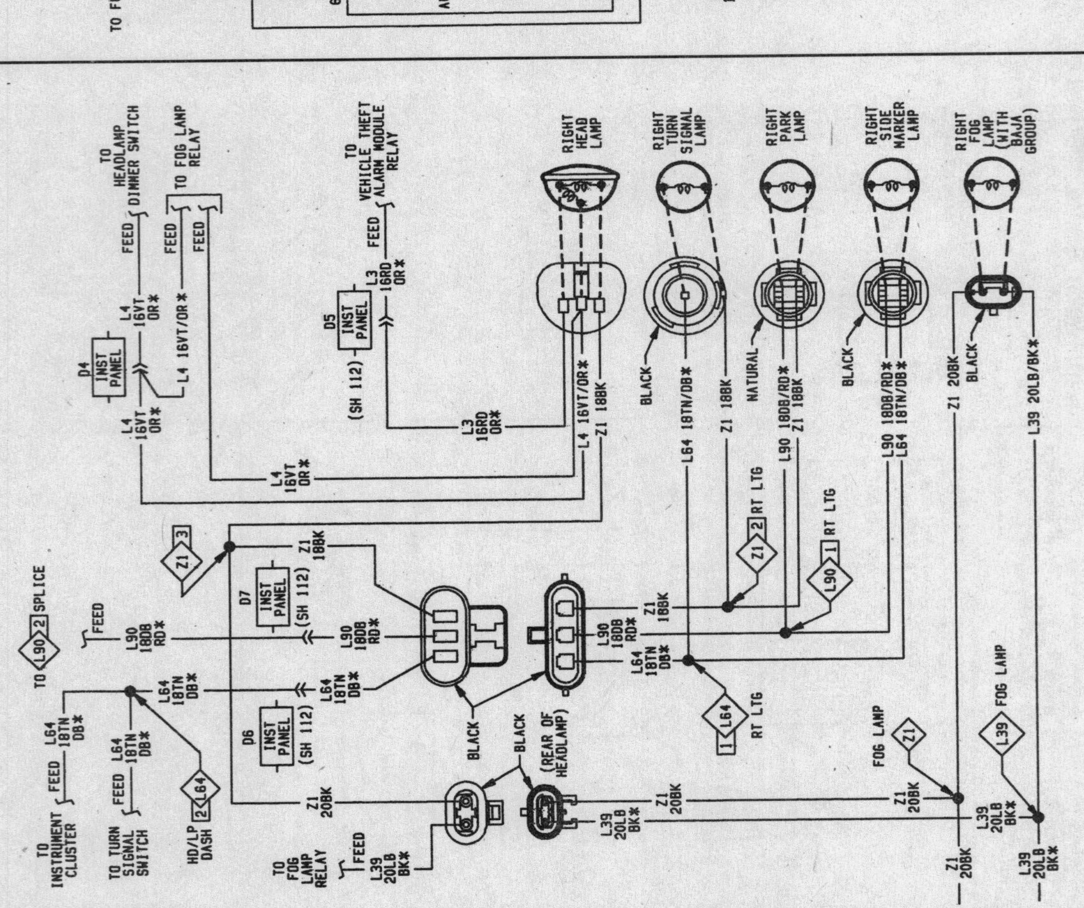

Typical front end lighting system wiring diagram (2 of 2)

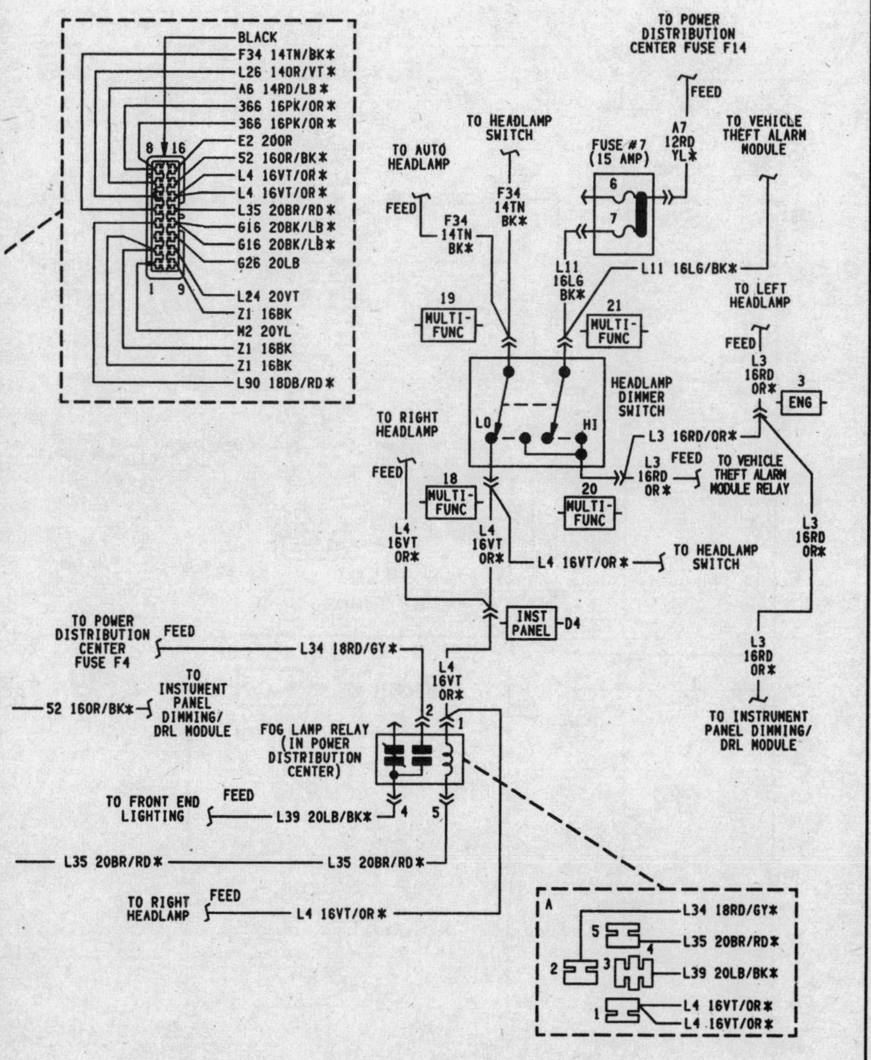

Typical headlight switch wiring diagram (2 of 2)

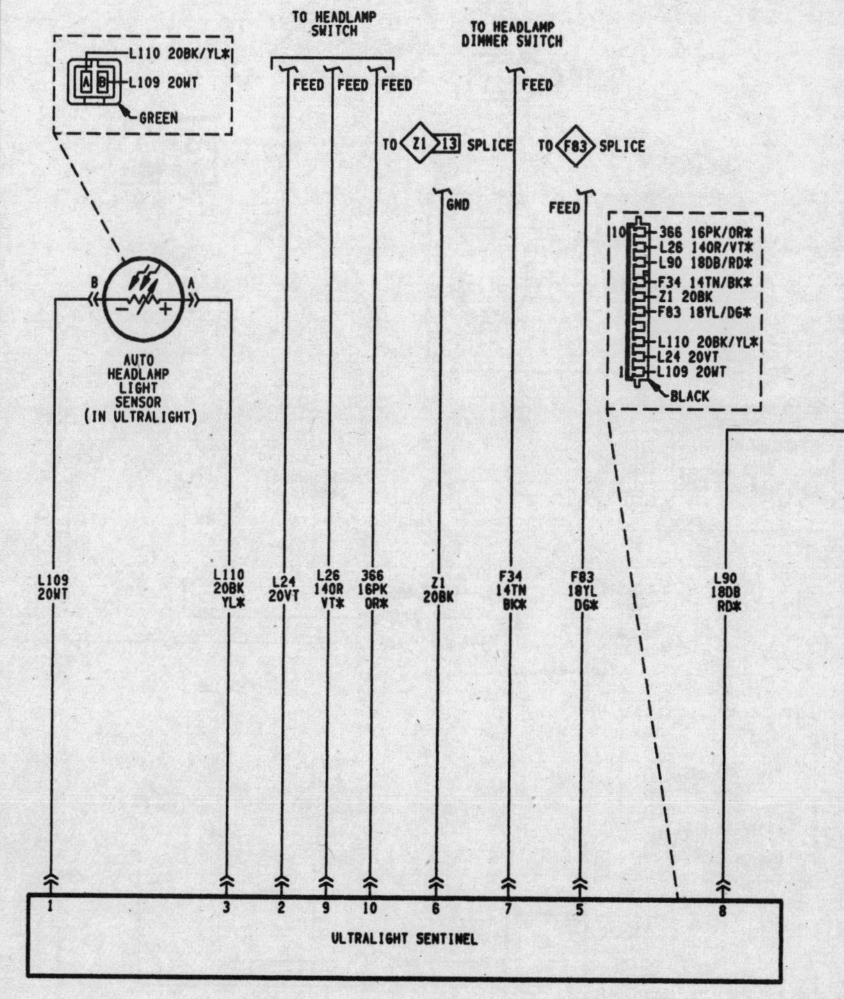

Typical automatic headlight switch system wiring diagram (1 of 2)

Typical day/night mirror and cigar lighter circuit wiring diagram

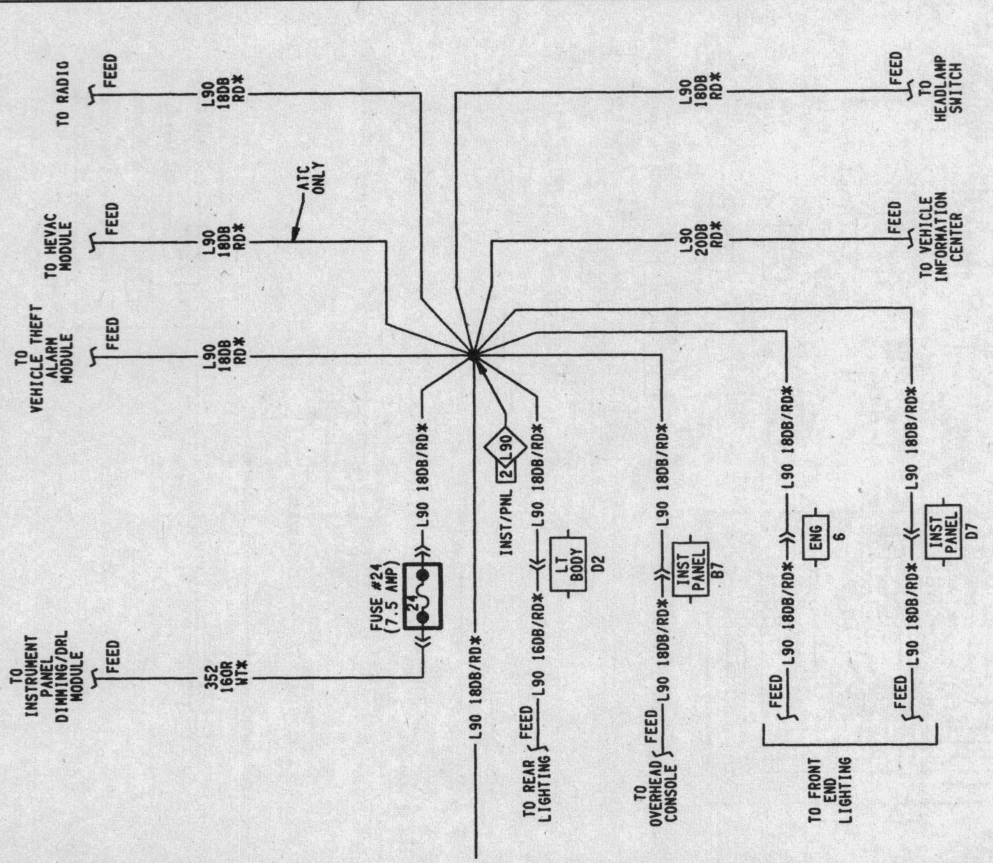

Typical automatic headlight switch system wiring diagram (2 of 2)

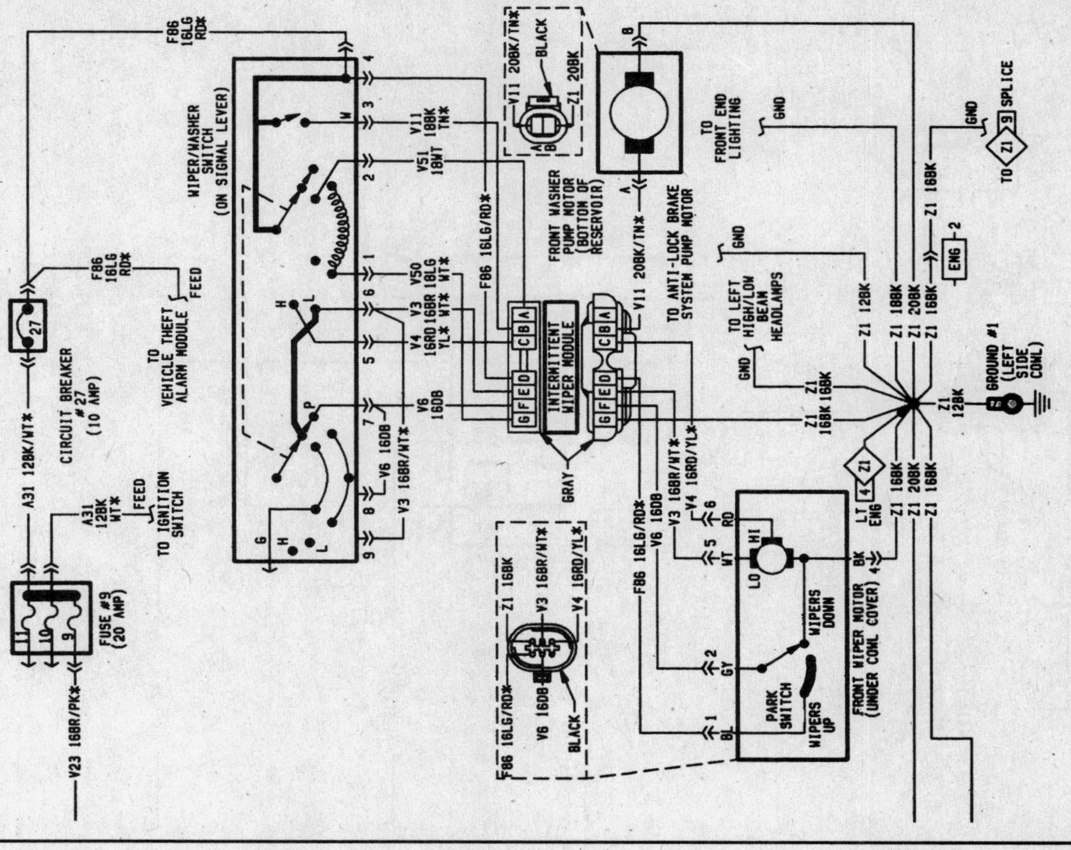

Typical wiper/washer circuit wiring diagram (2 of 2)

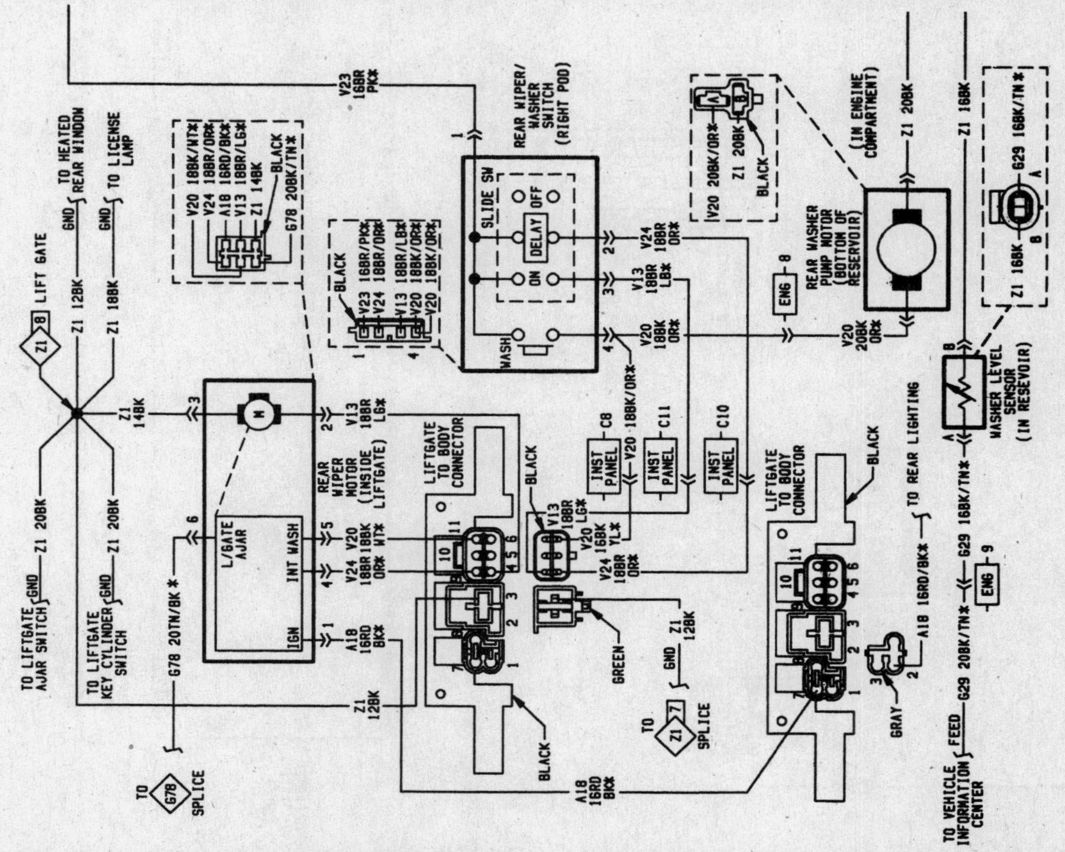

Typical wiper/washer circuit wiring diagram (1 of 2)

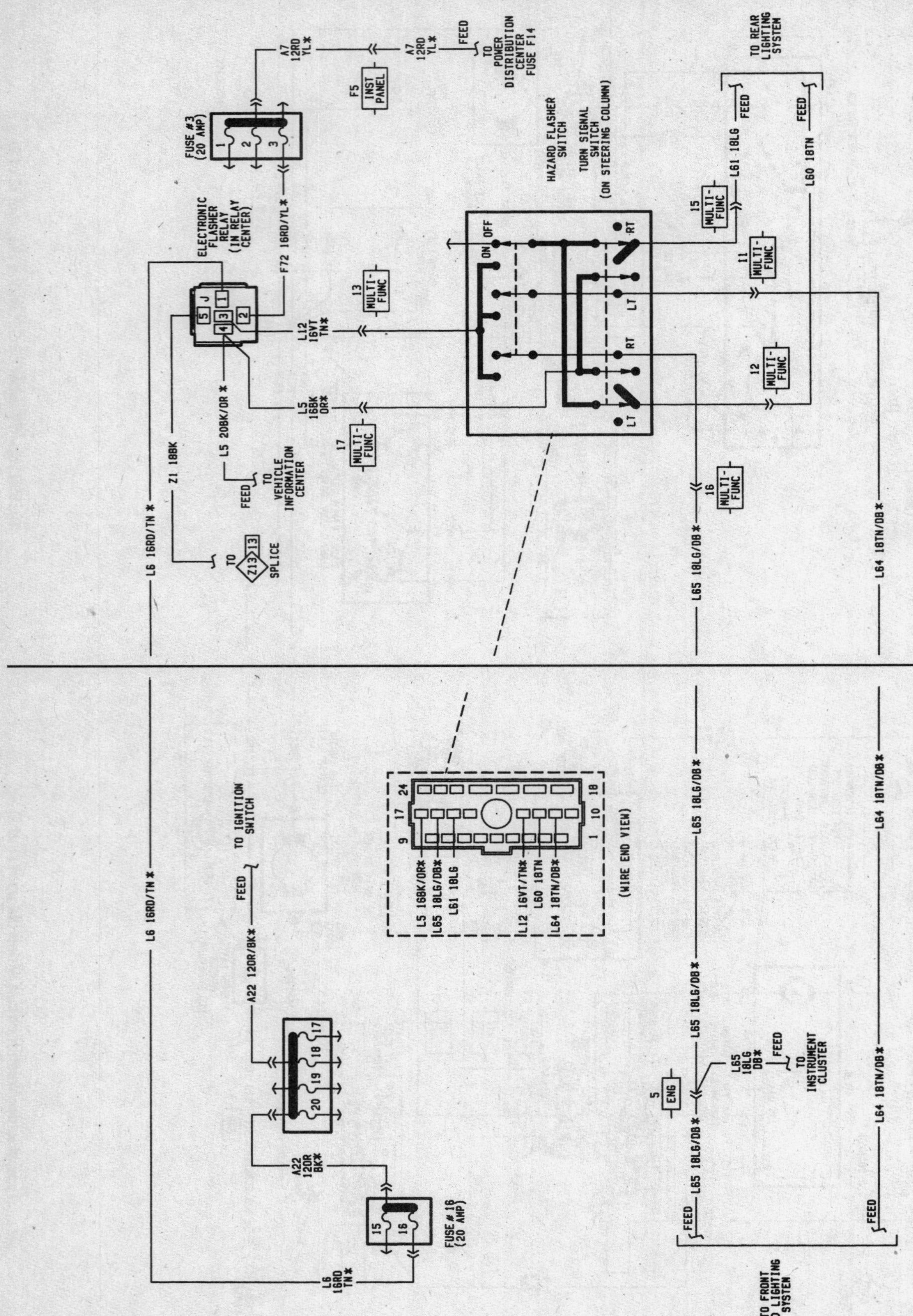

Typical turn signal and hazard flasher circuit wiring diagram (1 of 2)

Typical turn signal and hazard flasher circuit wiring diagram (2 of 2)

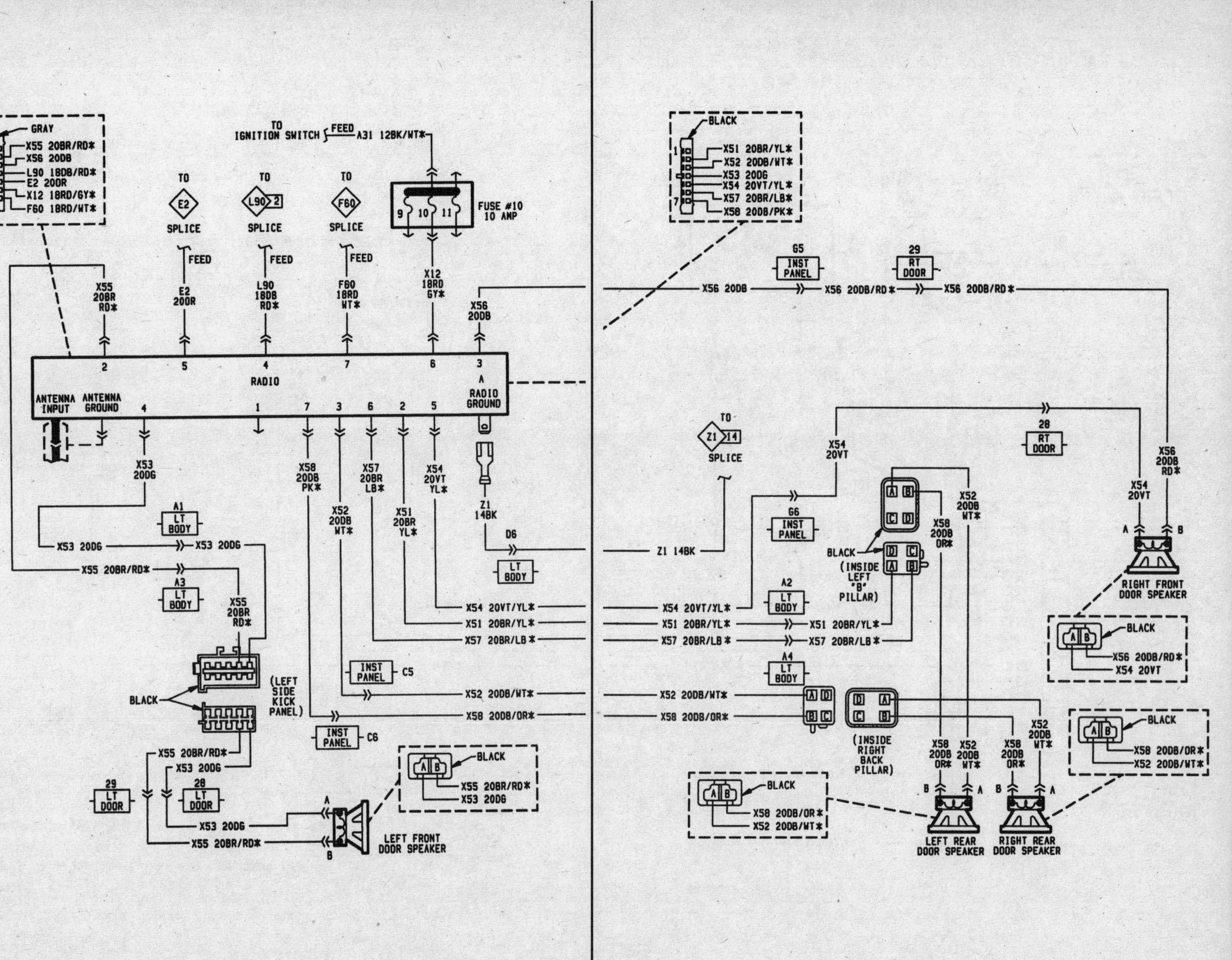

Typical radio circuit wiring diagram (1 of 2)

Typical radio circuit wiring diagram (2 of 2)

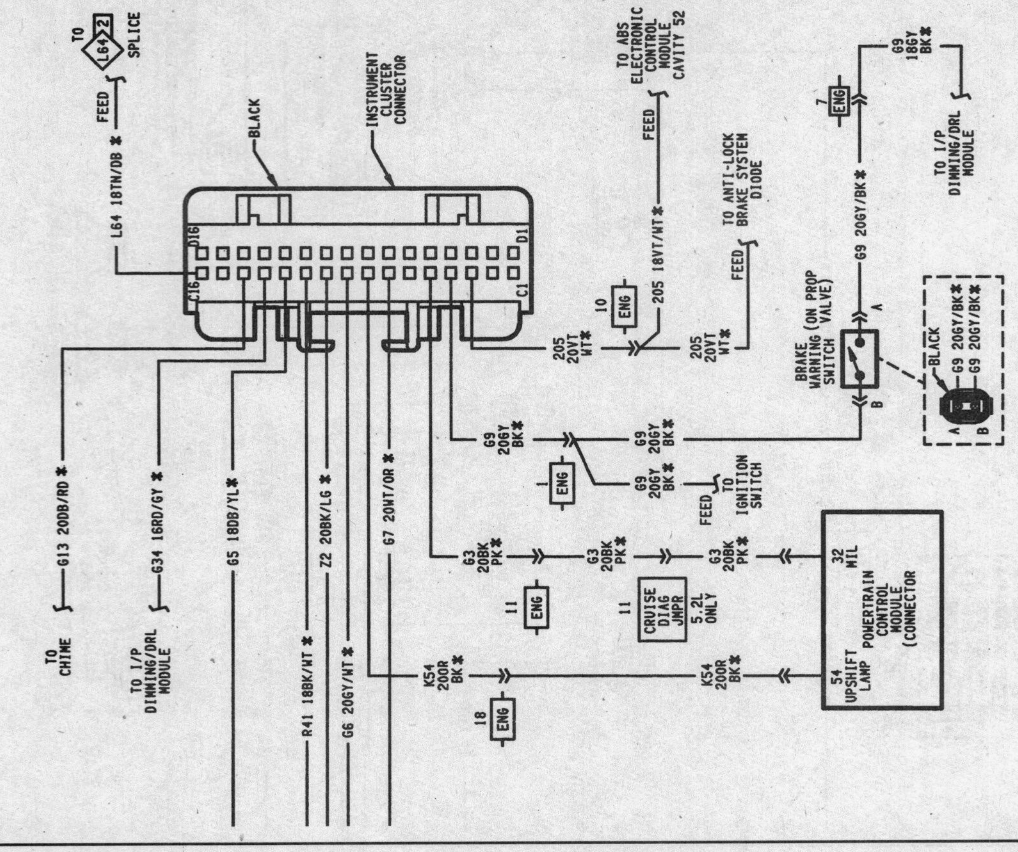

Typical instrument cluster circuit wiring diagram (2 of 4)

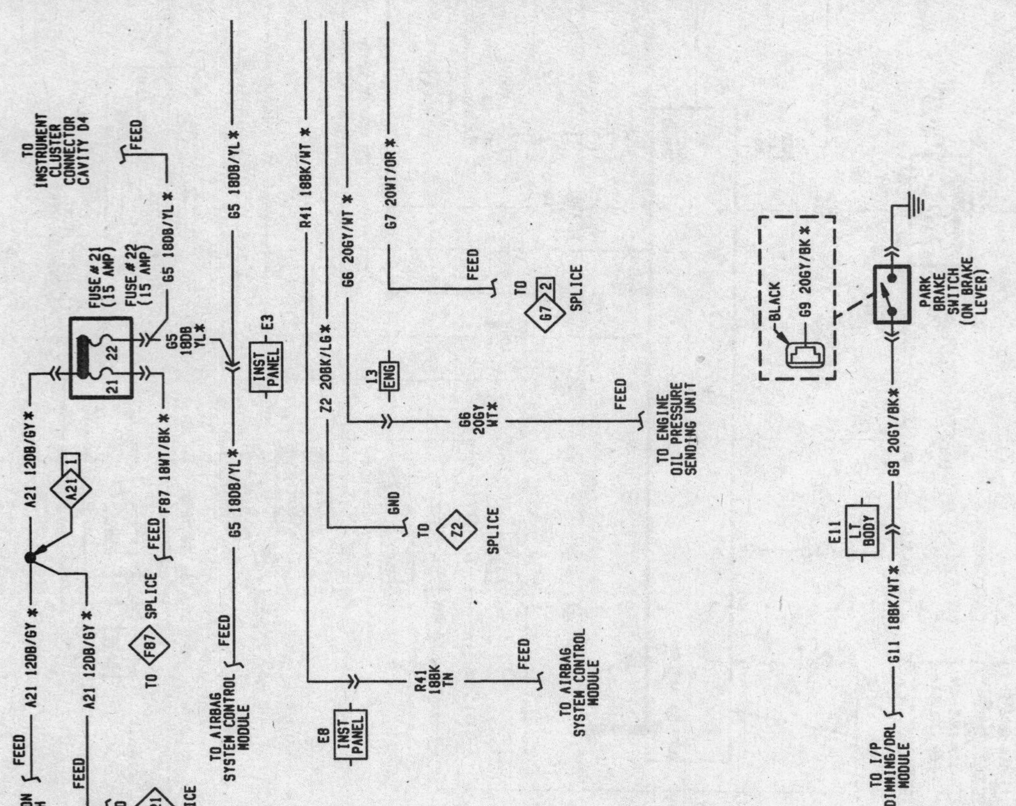

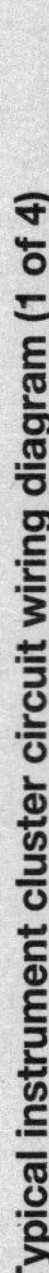

Typical instrument cluster circuit wiring diagram (1 of 4)

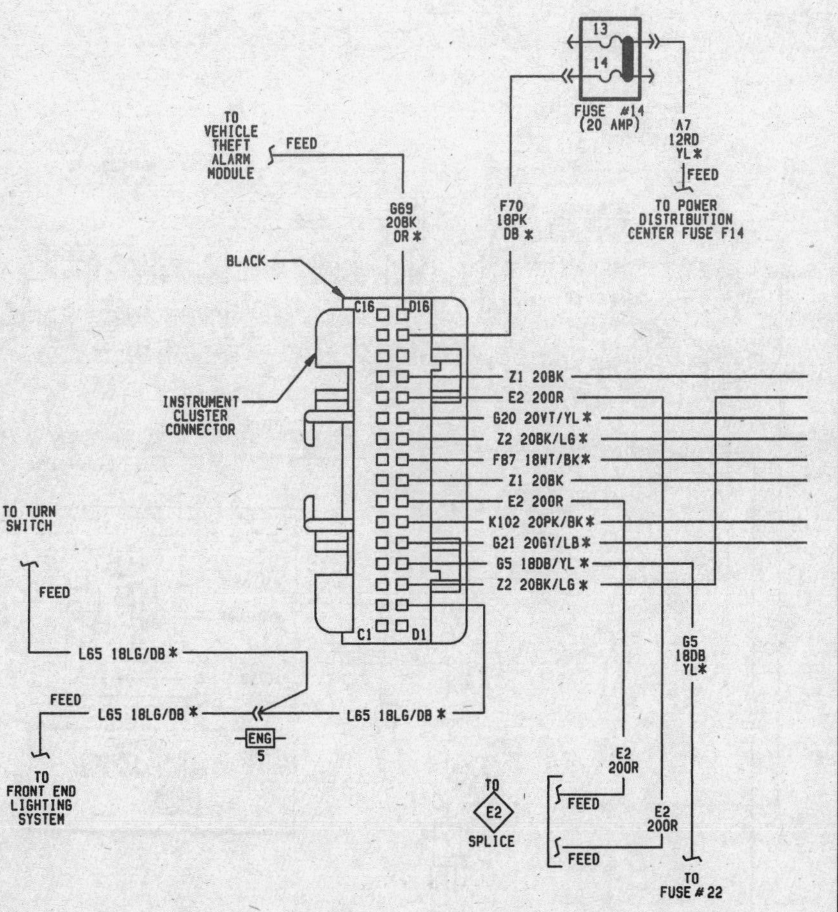

Typical instrument cluster circuit wiring diagram (3 of 4)

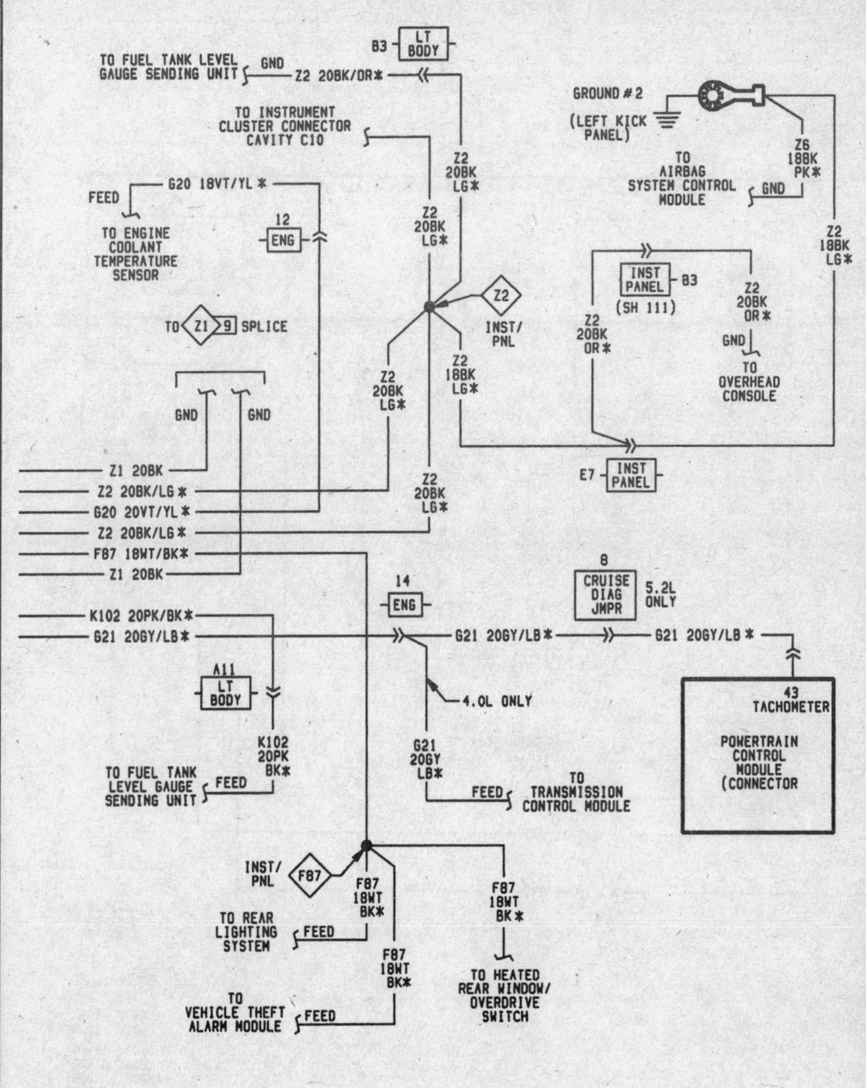

Typical instrument cluster circuit wiring diagram (4 of 4)

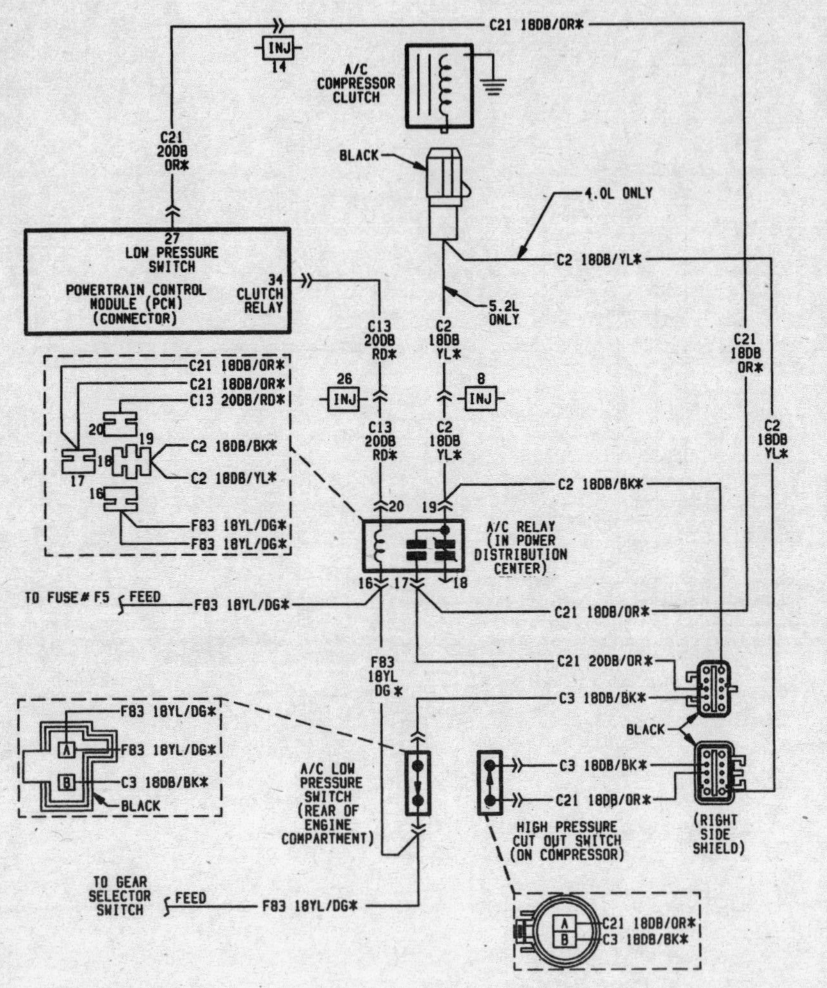

Typical air conditioning relay circuit diagram

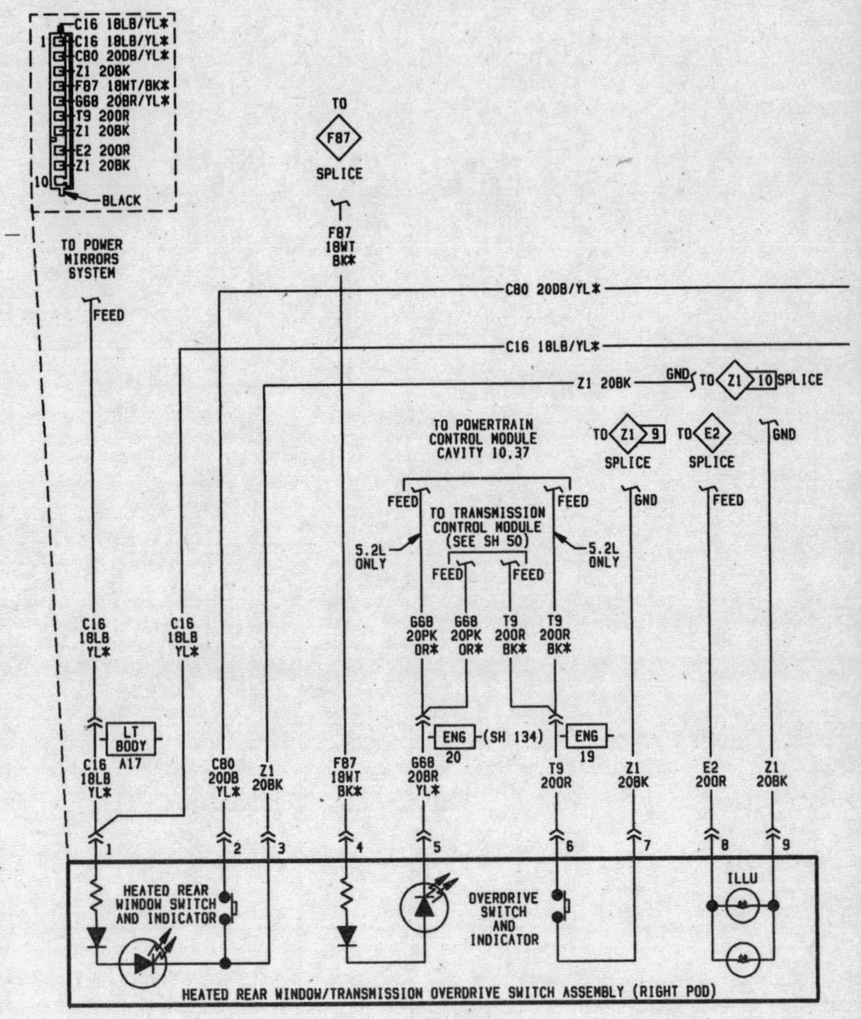

Typical rear window defogger circuit wiring diagram (1 of 2)

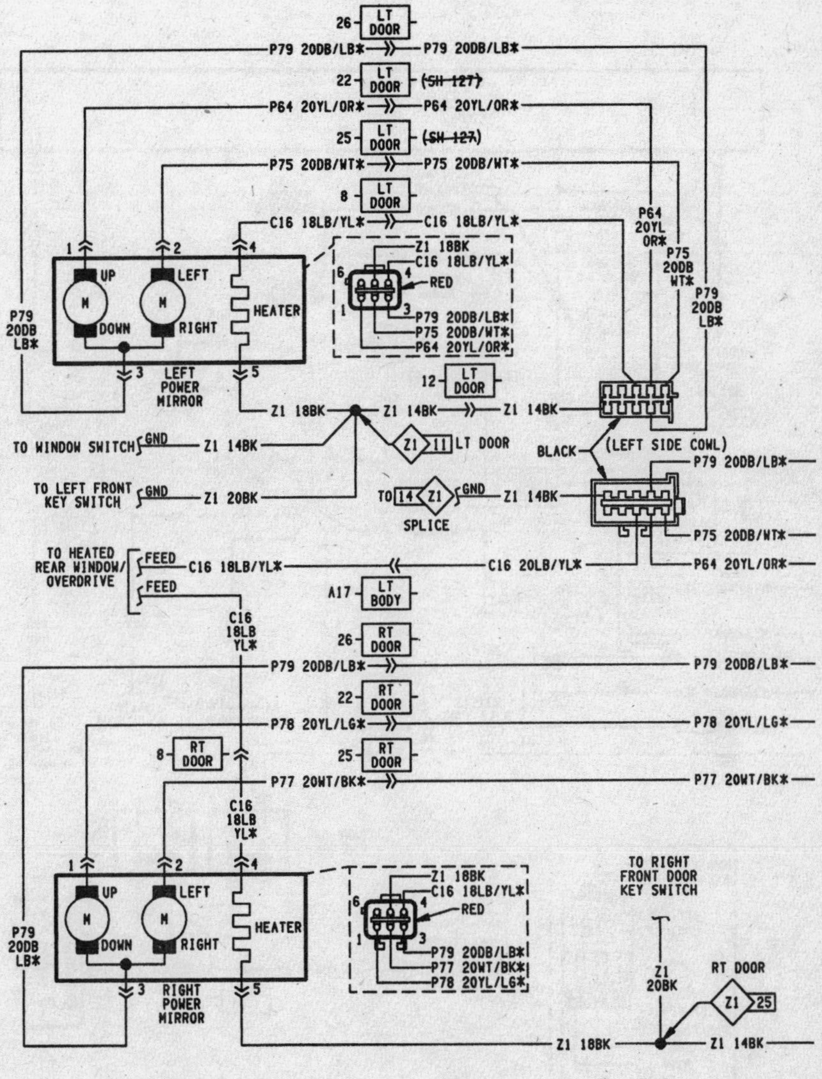

Typical power mirror circuit wiring diagram (1 of 2)

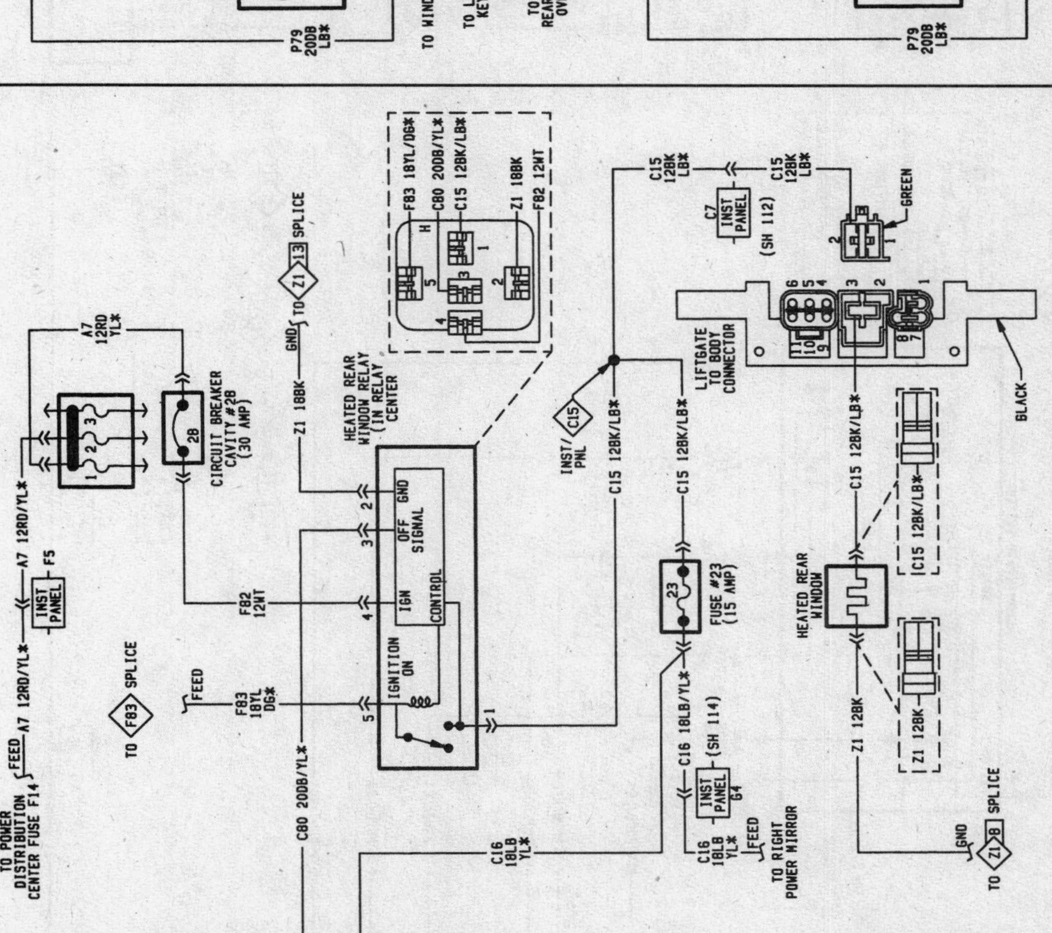

Typical rear window defogger circuit wiring diagram (2 of 2)

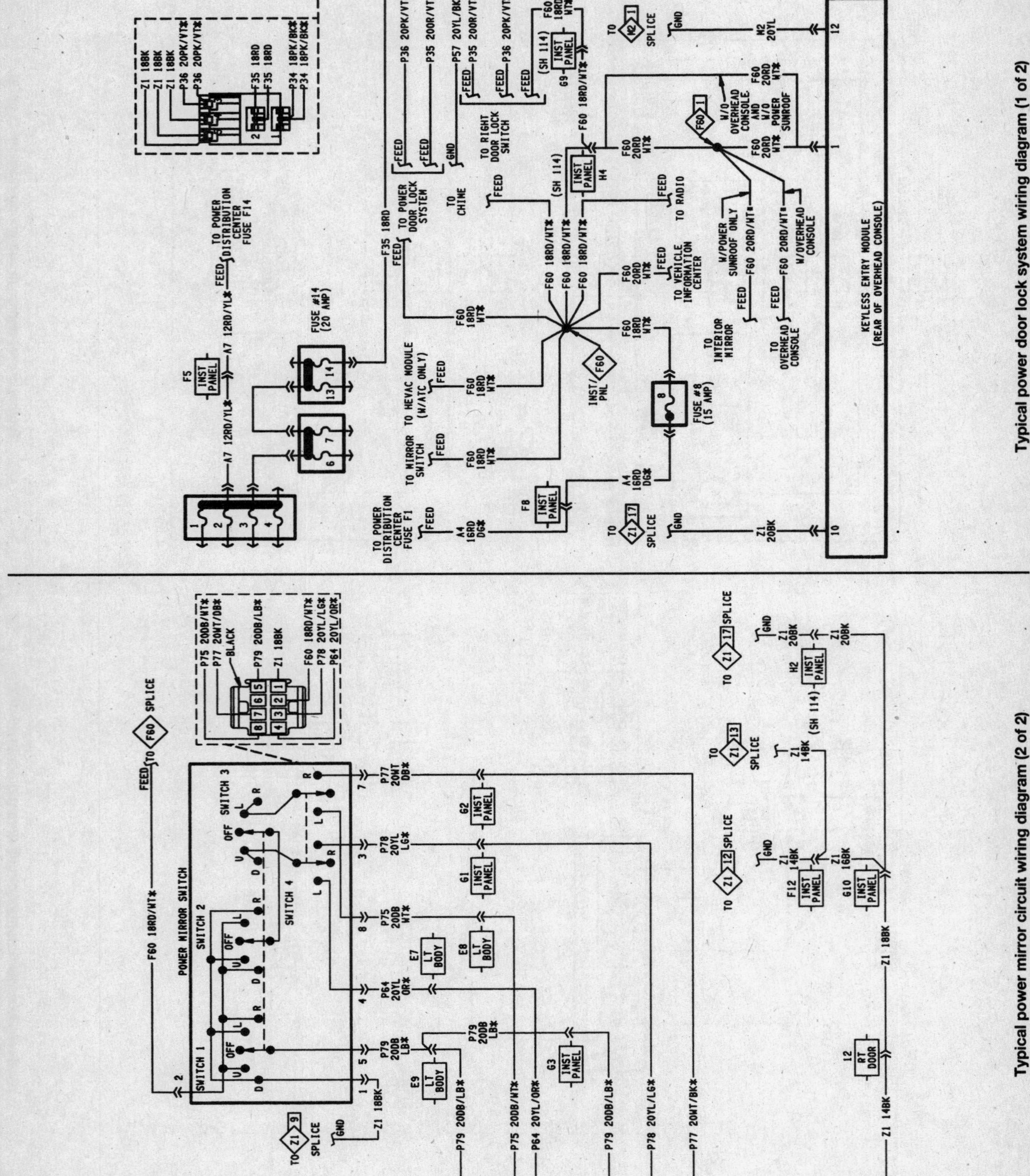

Typical power door lock system wiring diagram (1 of 2)

Typical power mirror circuit wiring diagram (2 of 2)

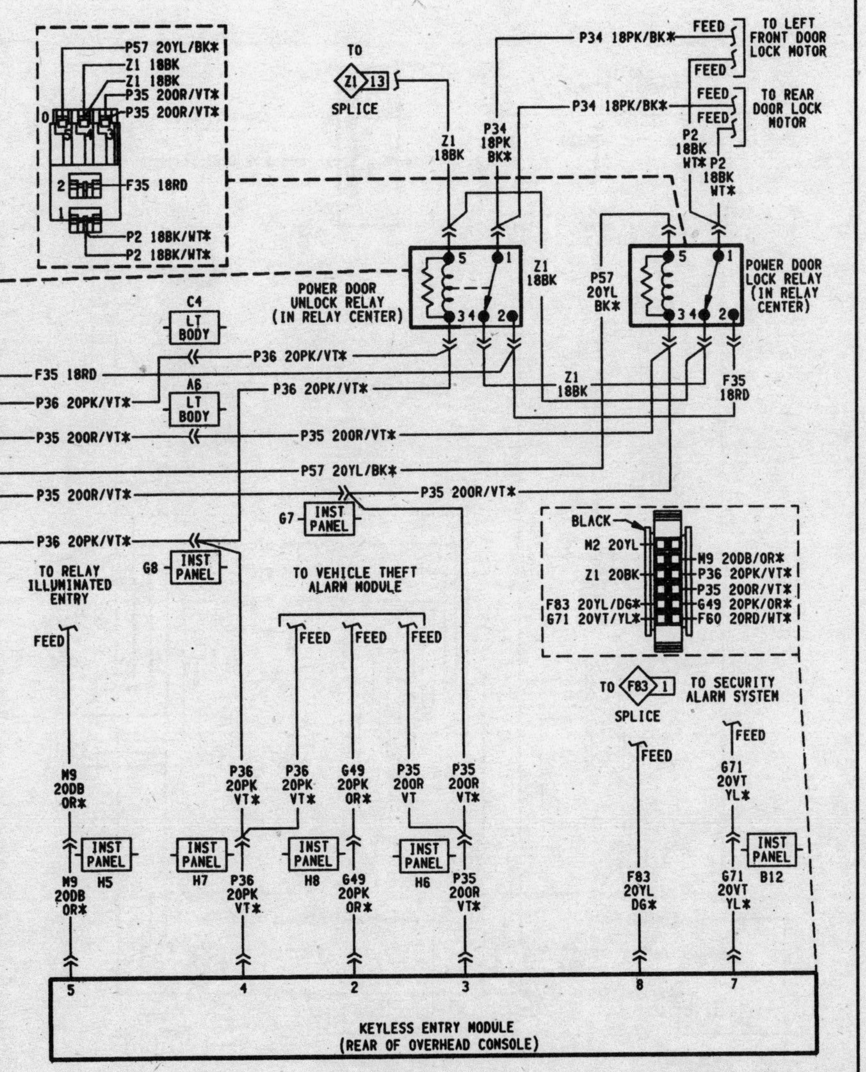

Typical power door lock system wiring diagram (2 of 2)

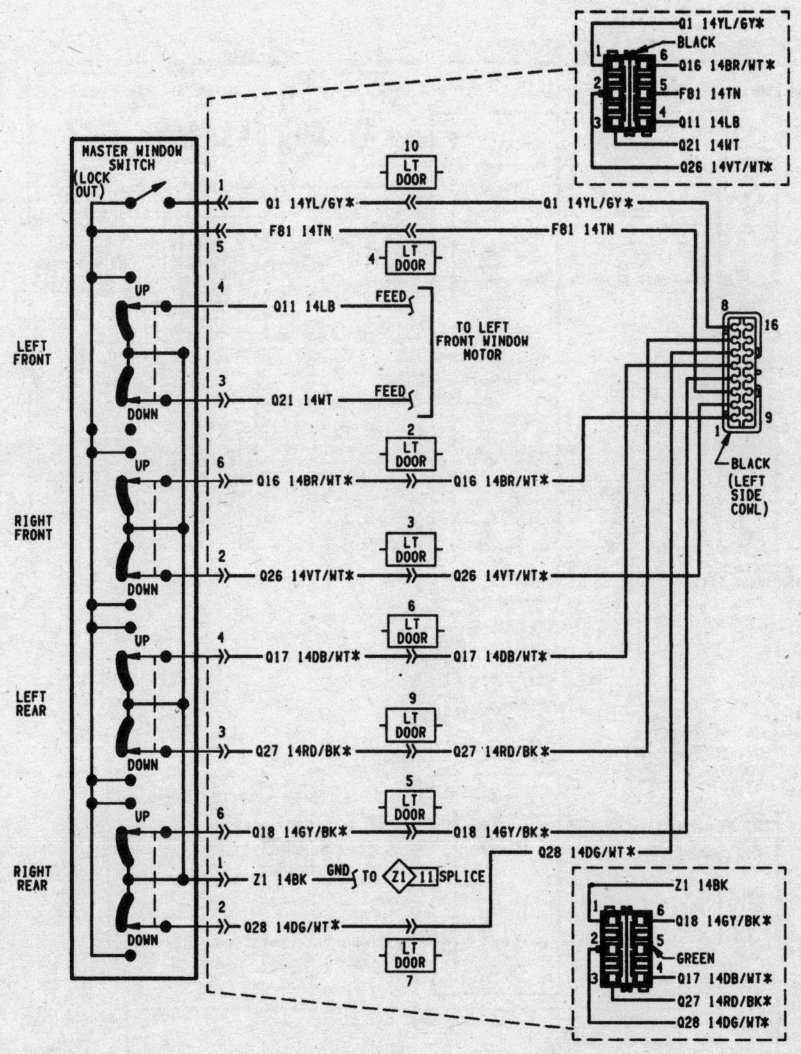

Typical power window system wiring diagram (1 of 3)

12

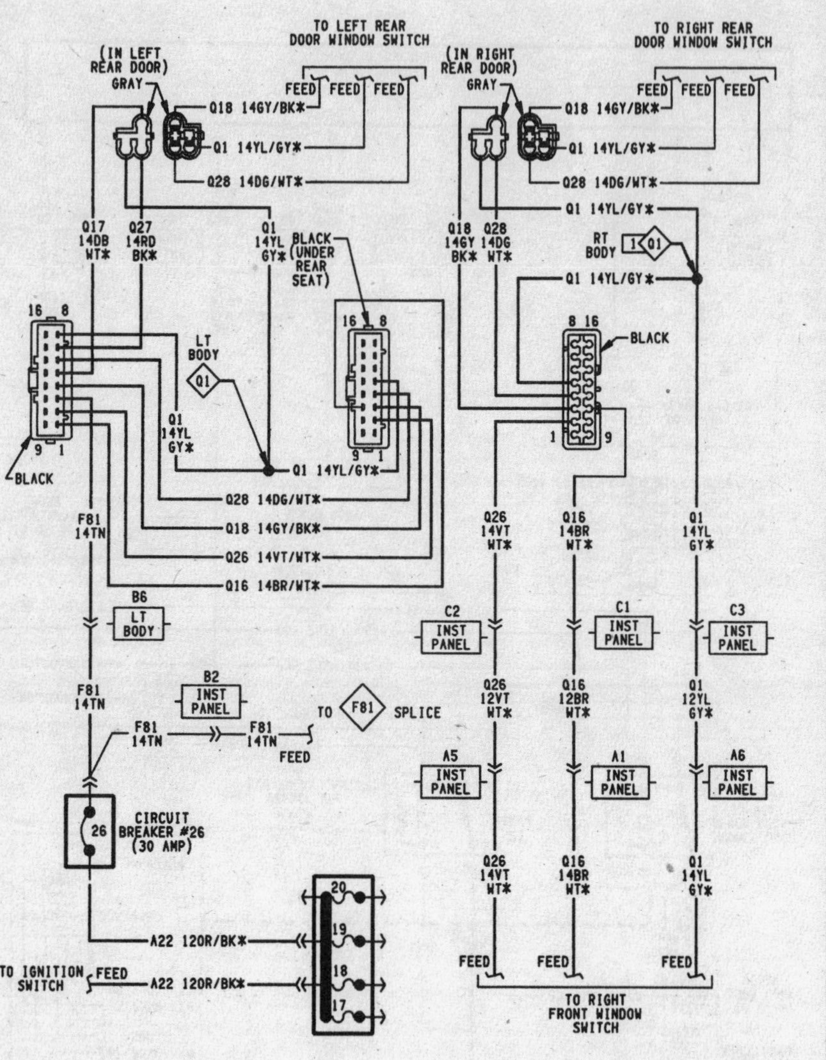

Typical power window system wiring diagram (2 of 3)

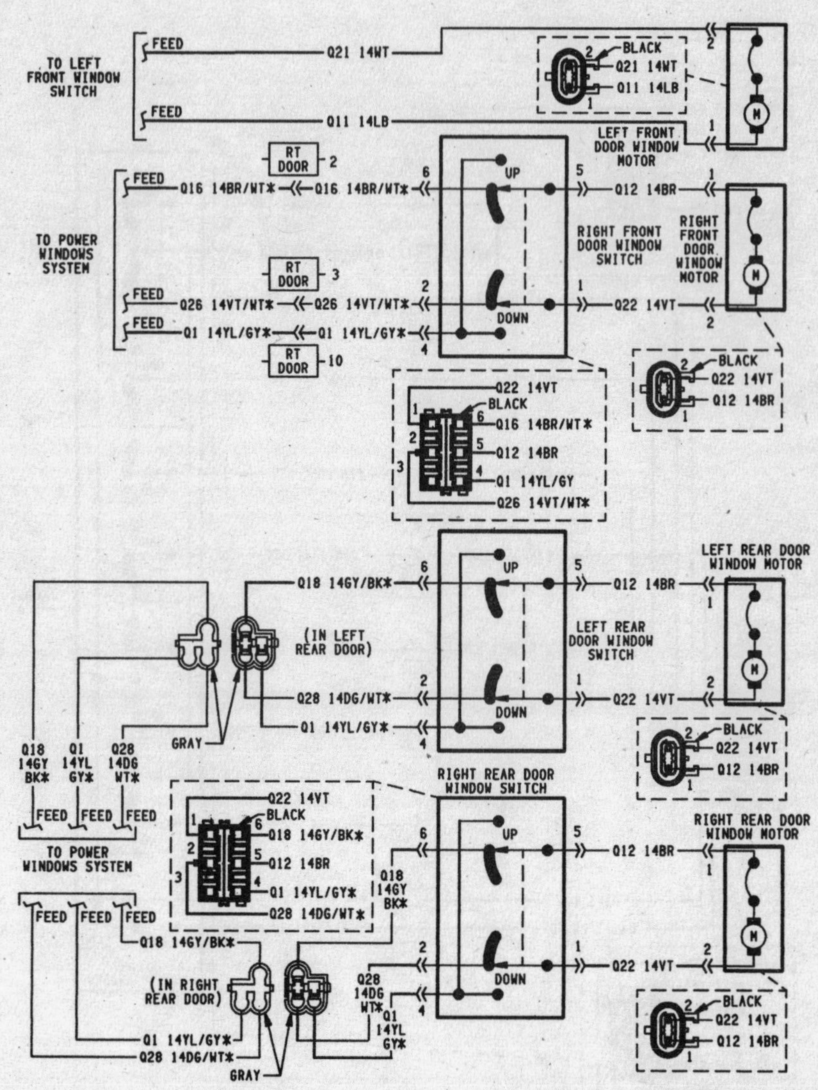

Typical power window system wiring diagram (3 of 3)

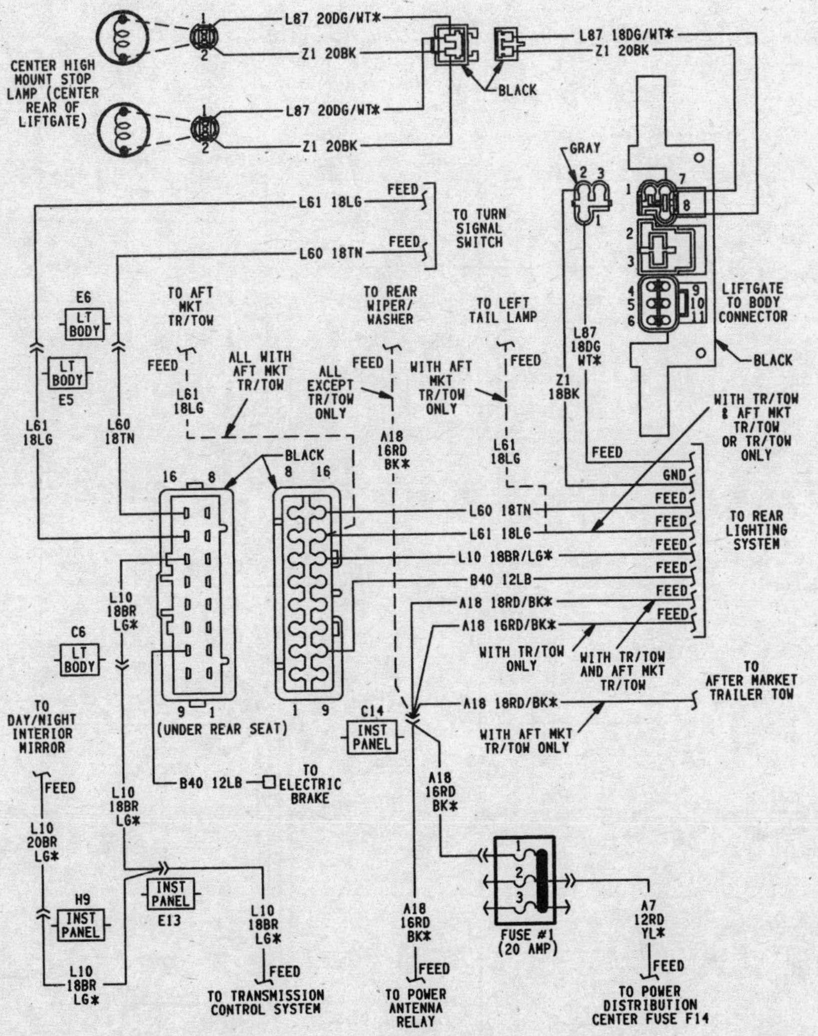

Typical rear lighting system wiring diagram (1 of 3)

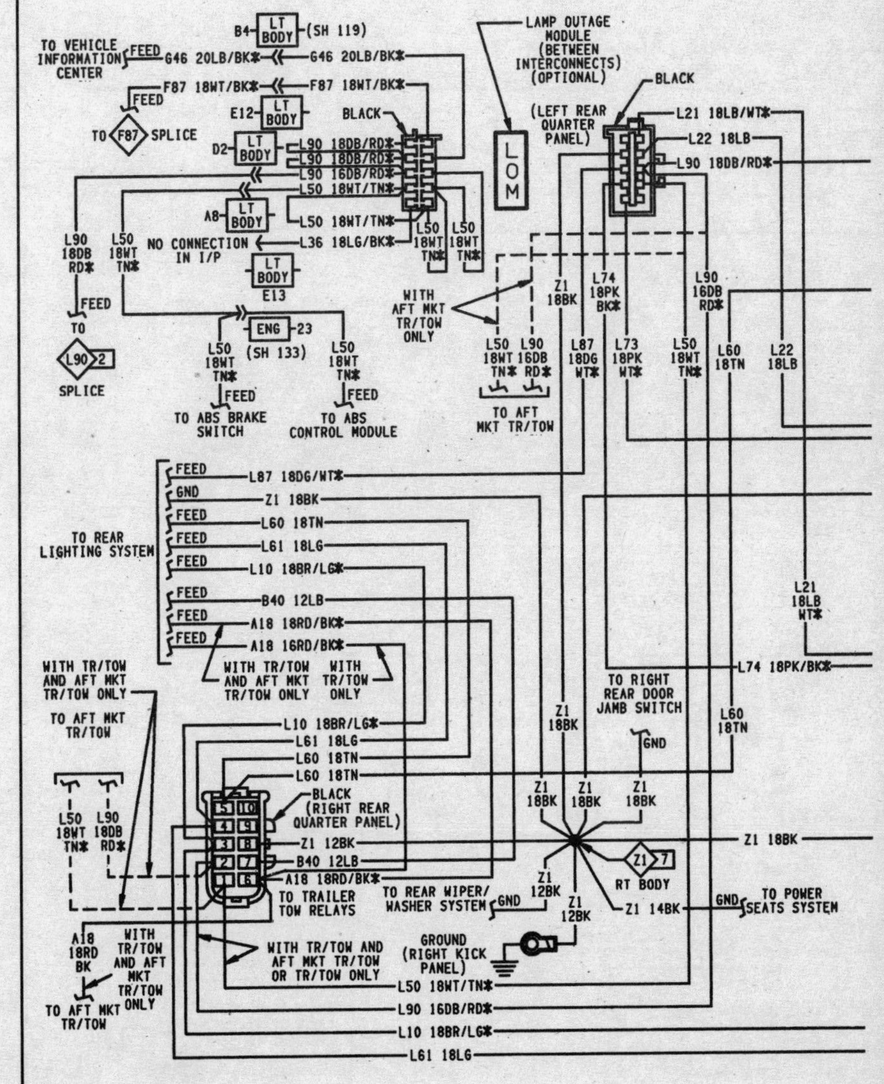

Typical rear lighting system wiring diagram (2 of 3)

Typical fuel injection system wiring diagram - 1998 and earlier models (1 of 6)

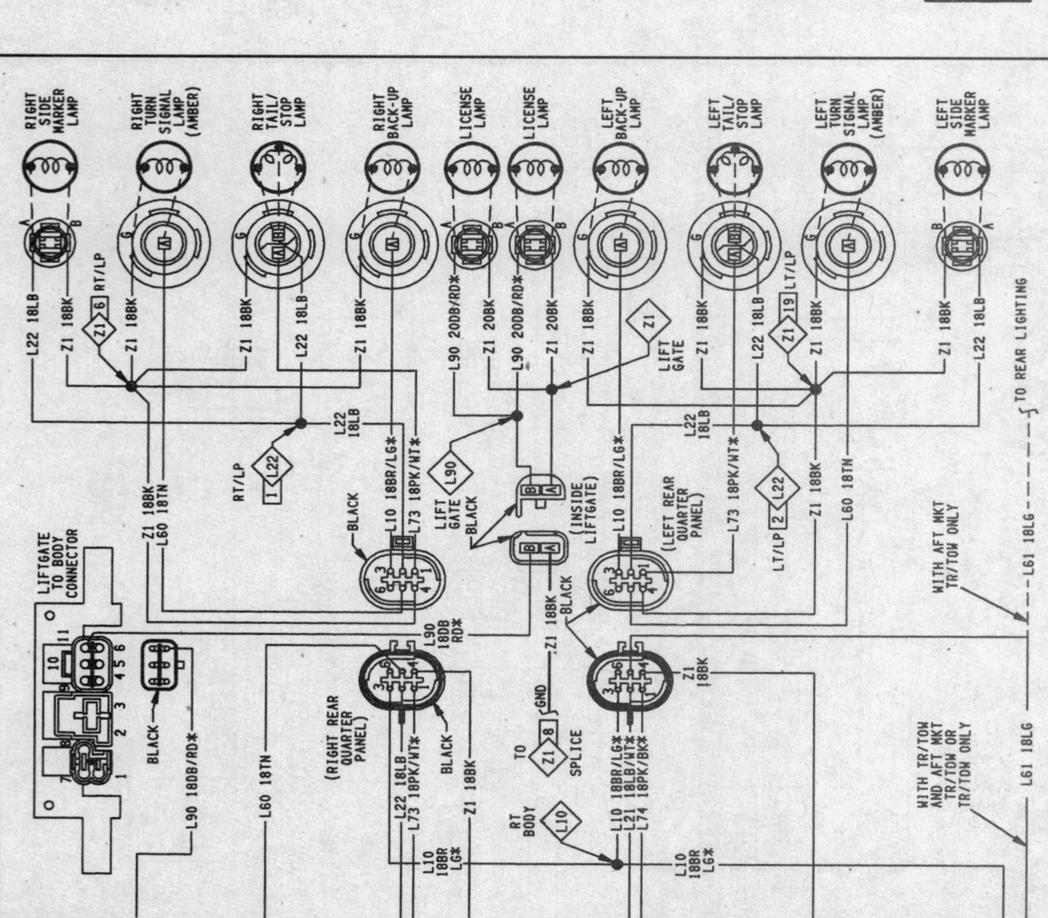

Typical rear lighting system wiring diagram (3 of 3)

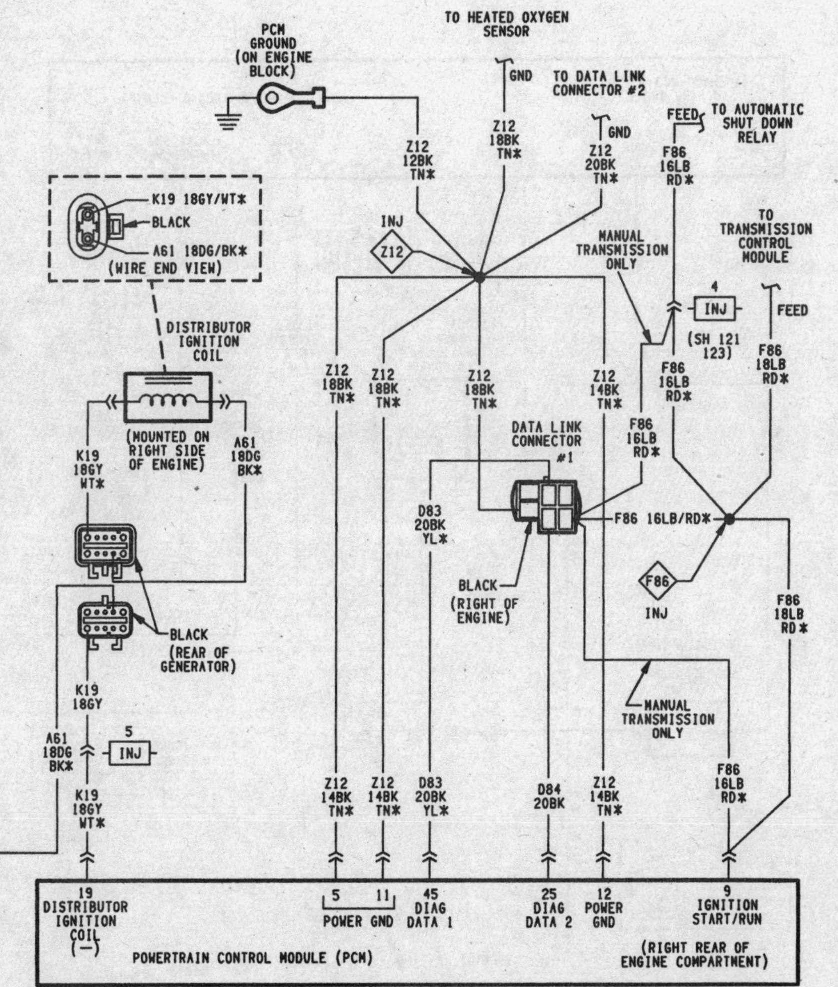

Typical fuel injection system wiring diagram - 1998 and earlier models (2 of 6)

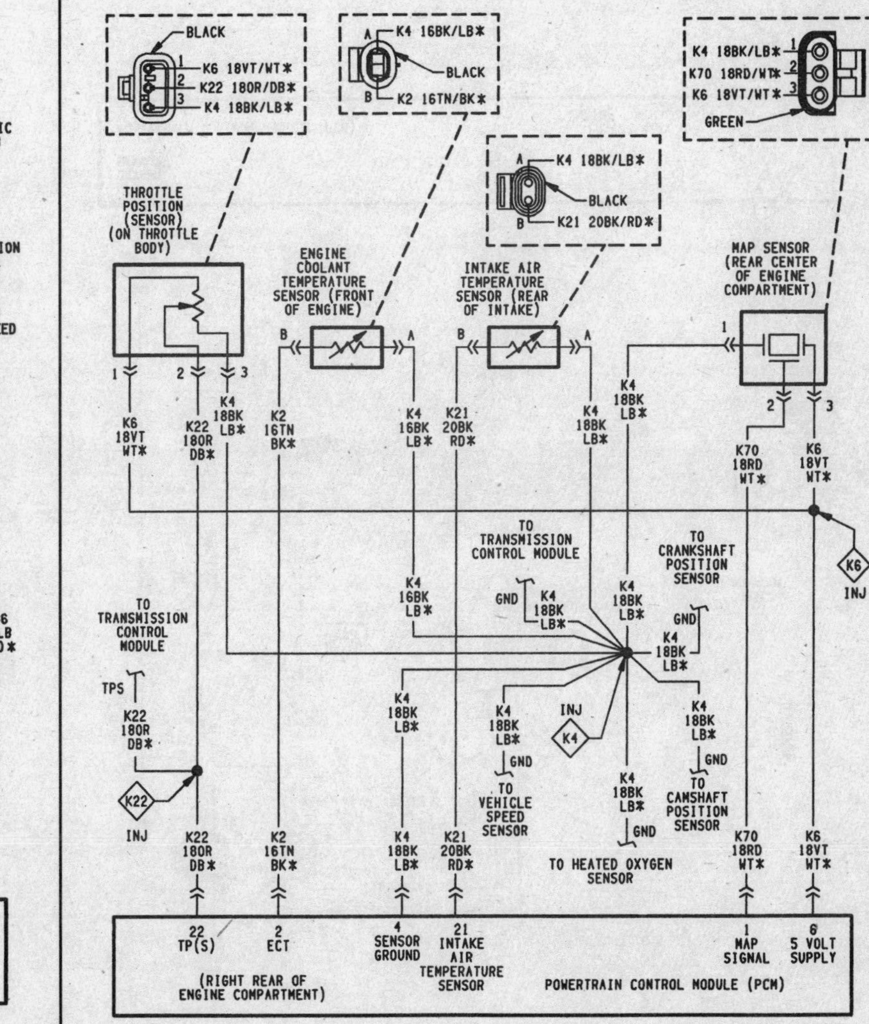

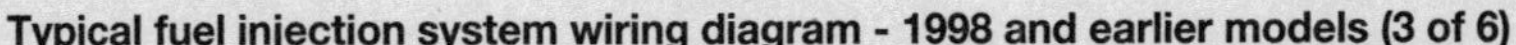
Typical fuel injection system wiring diagram - 1998 and earlier models (3 of 6)

12

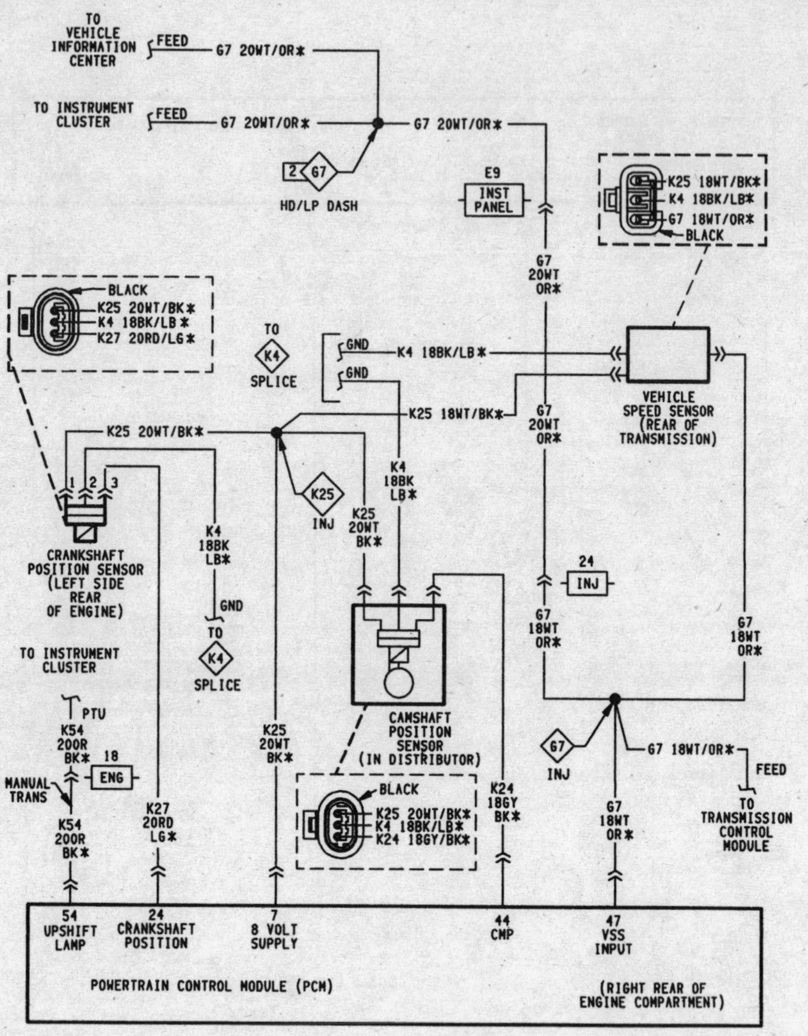

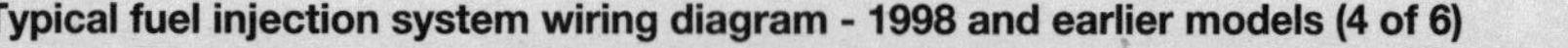
Typical fuel injection system wiring diagram - 1998 and earlier models (4 of 6)

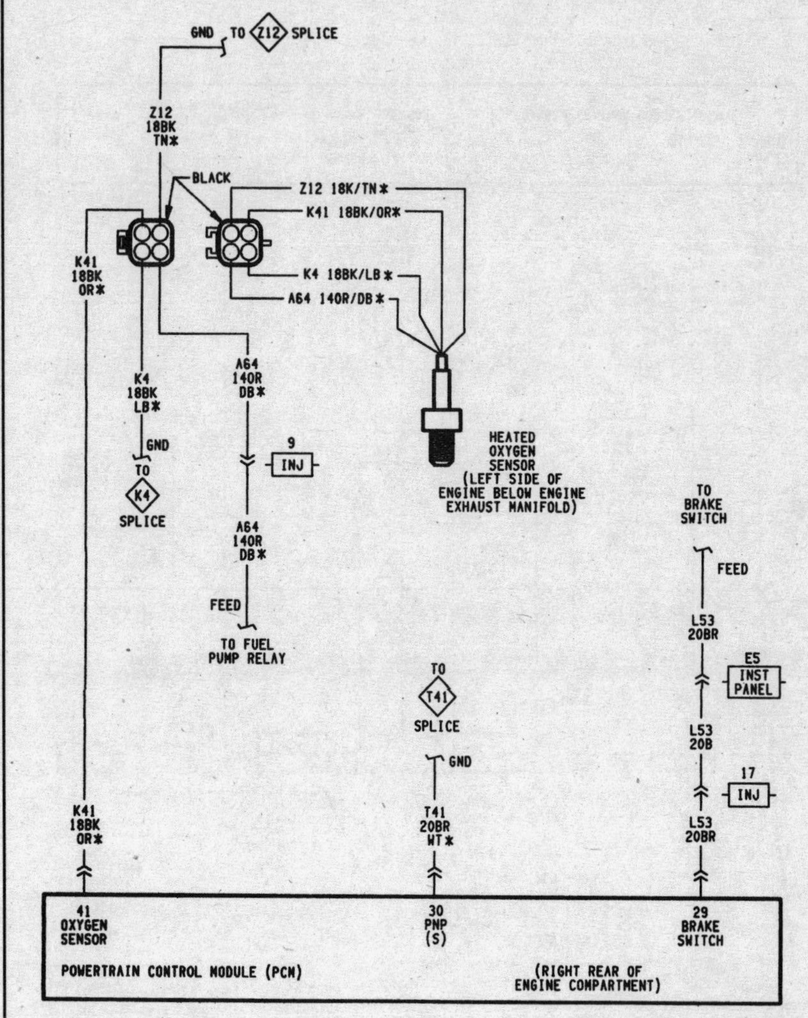

Typical fuel injection system wiring diagram - 1998 and earlier models (5 of 6)

Typical fuel injection system wiring diagram - 1999 and later models (1 of 17)

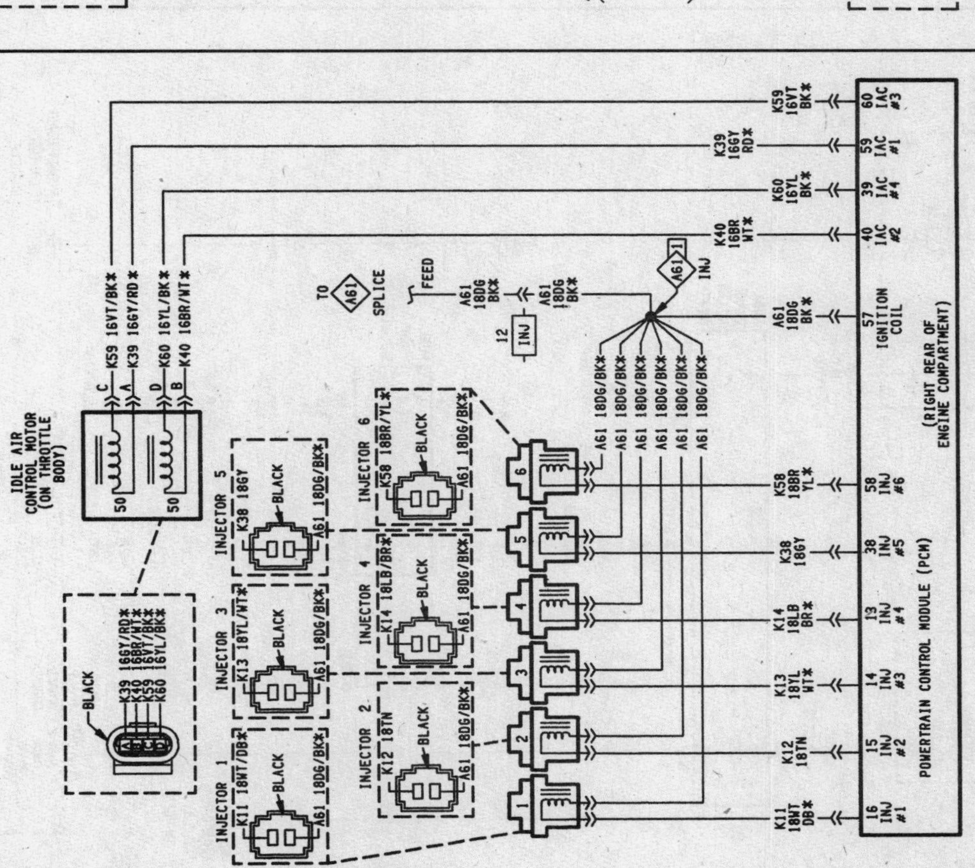

Typical fuel injection system wiring diagram - 1998 and earlier models (6 of 6)

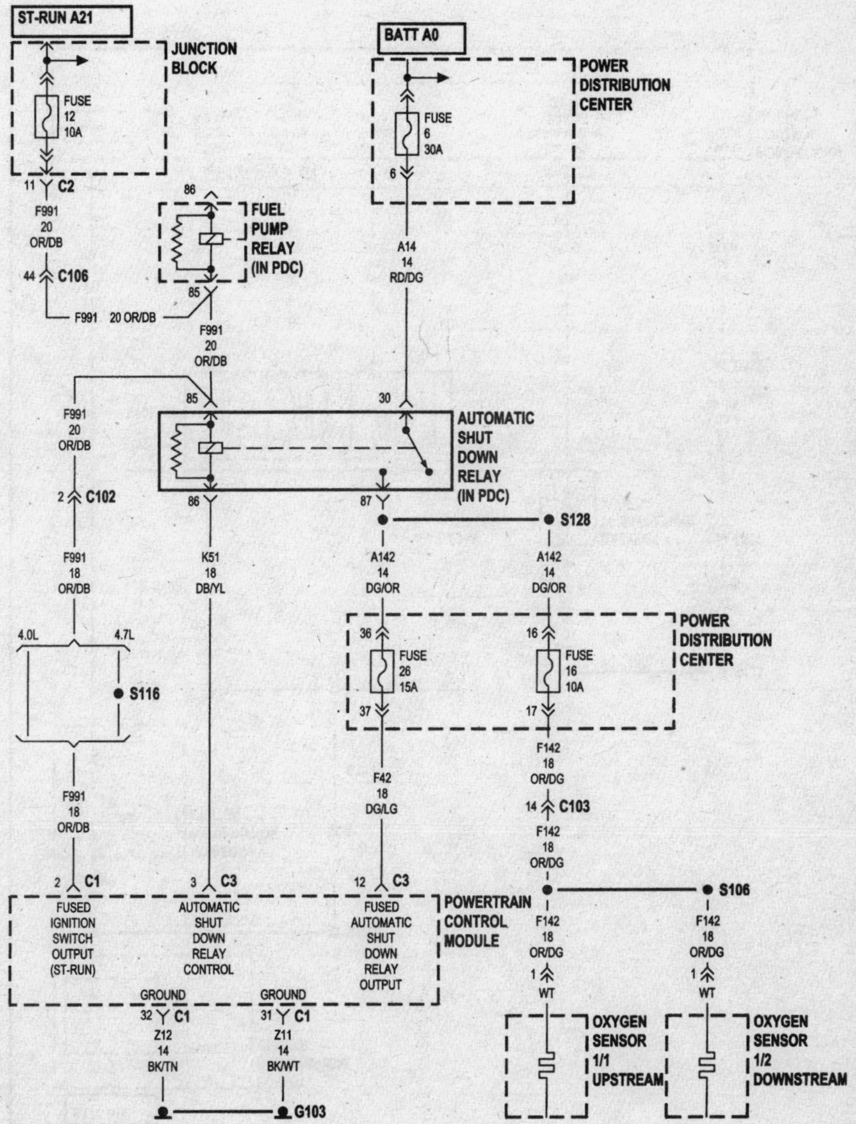

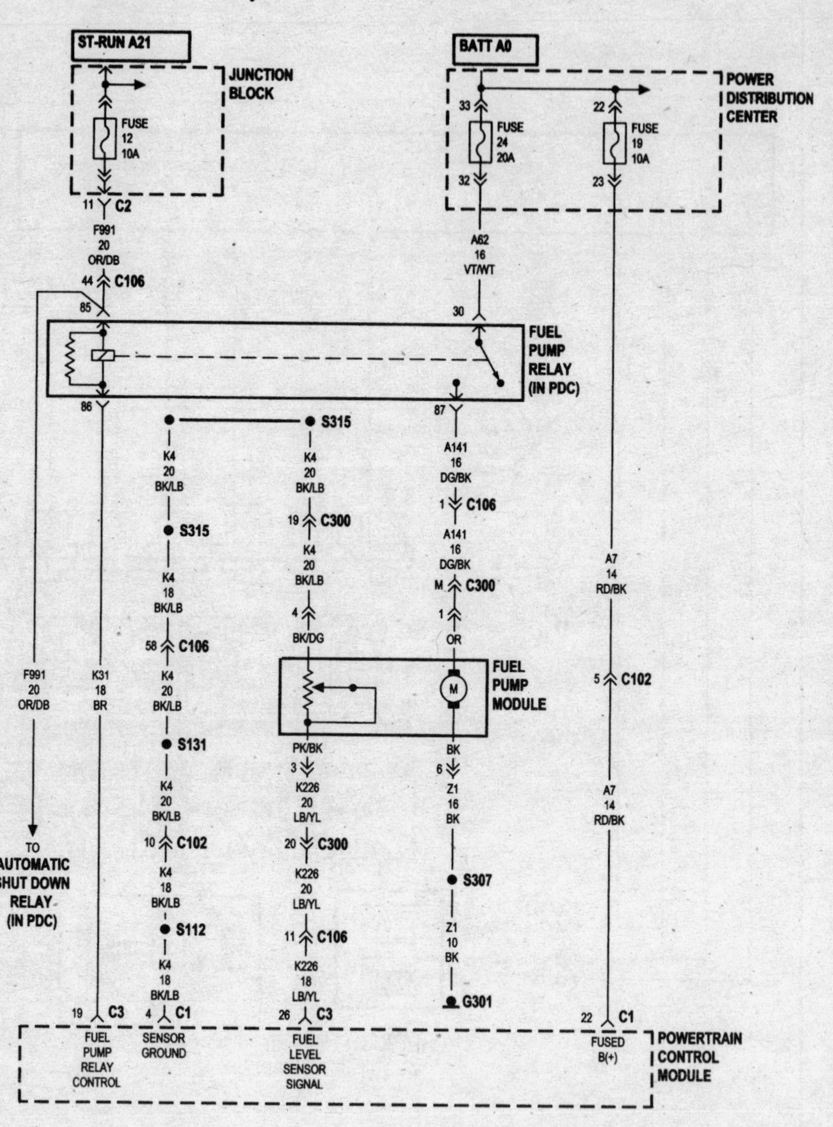

Typical fuel injection system wiring diagram - 1999 and later models (2 of 17)

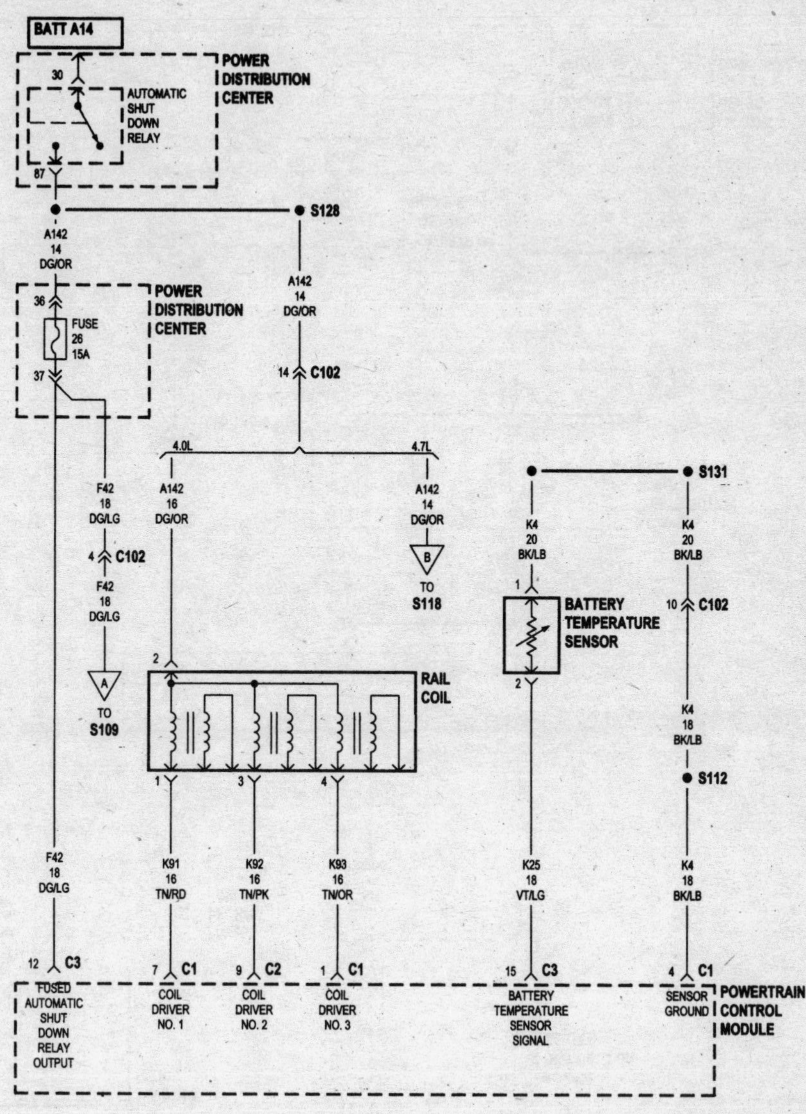

Typical fuel injection system wiring diagram - 1999 and later models (3 of 17)

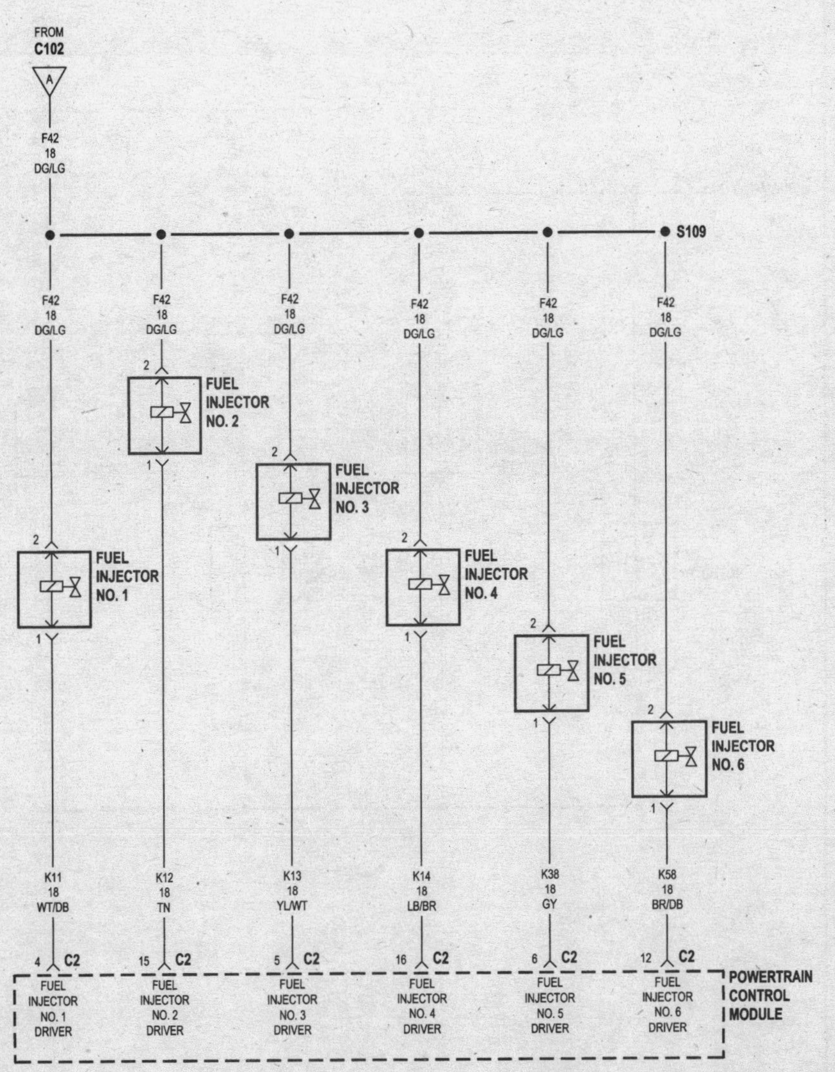

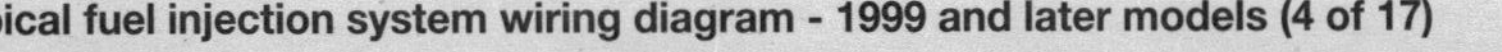
Typical fuel injection system wiring diagram - 1999 and later models (4 of 17)

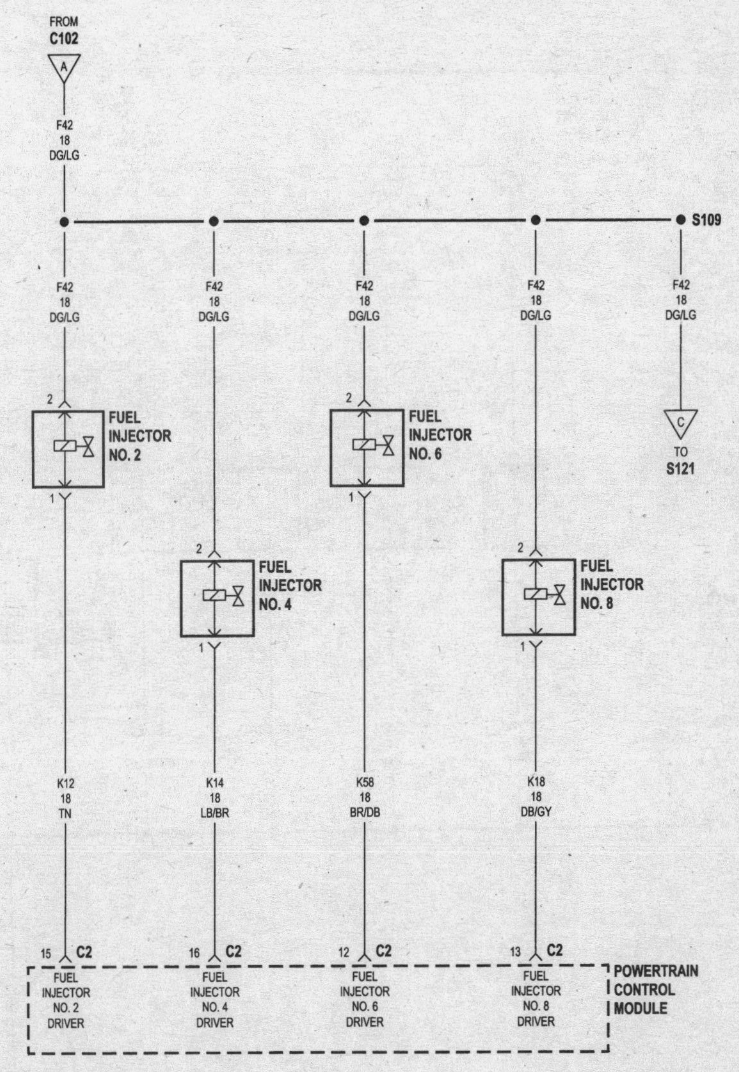

Typical fuel injection system wiring diagram - 1999 and later models (5 of 17)

Typical fuel injection system wiring diagram - 1999 and later models (7 of 17)

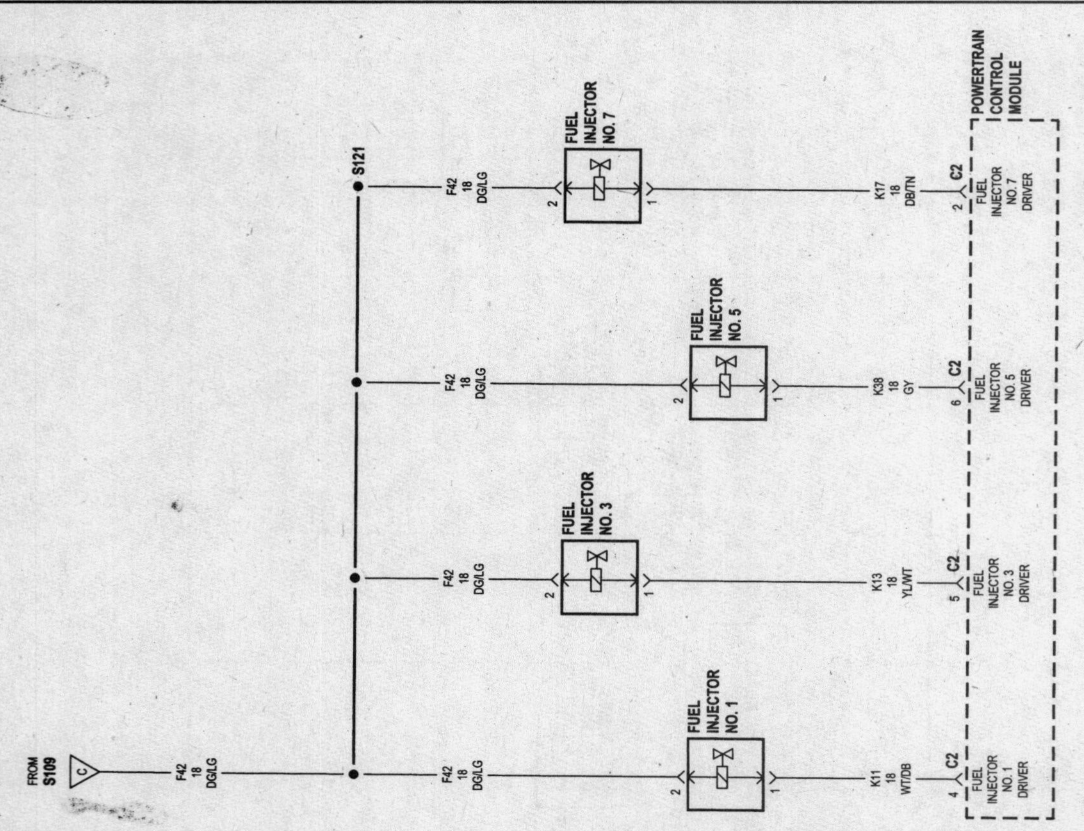

Typical fuel injection system wiring diagram - 1999 and later models (6 of 17)

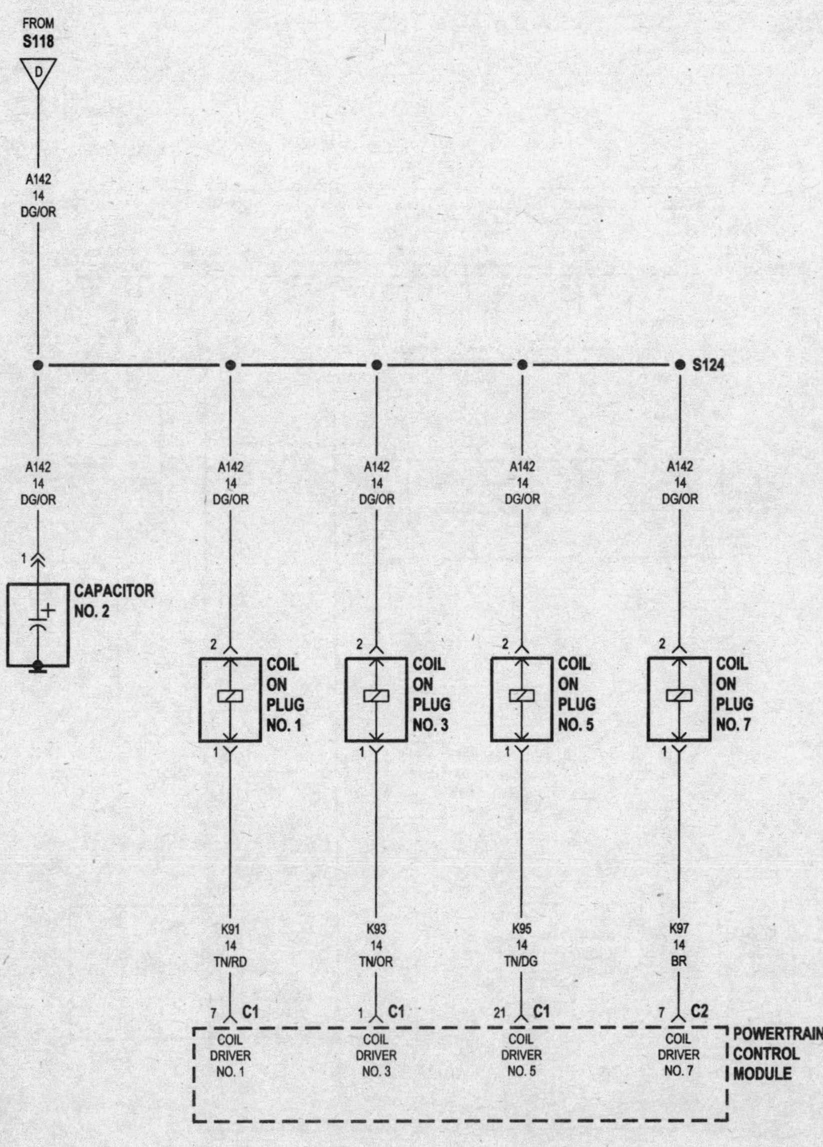

Typical fuel injection system wiring diagram - 1999 and later models (8 of 17)

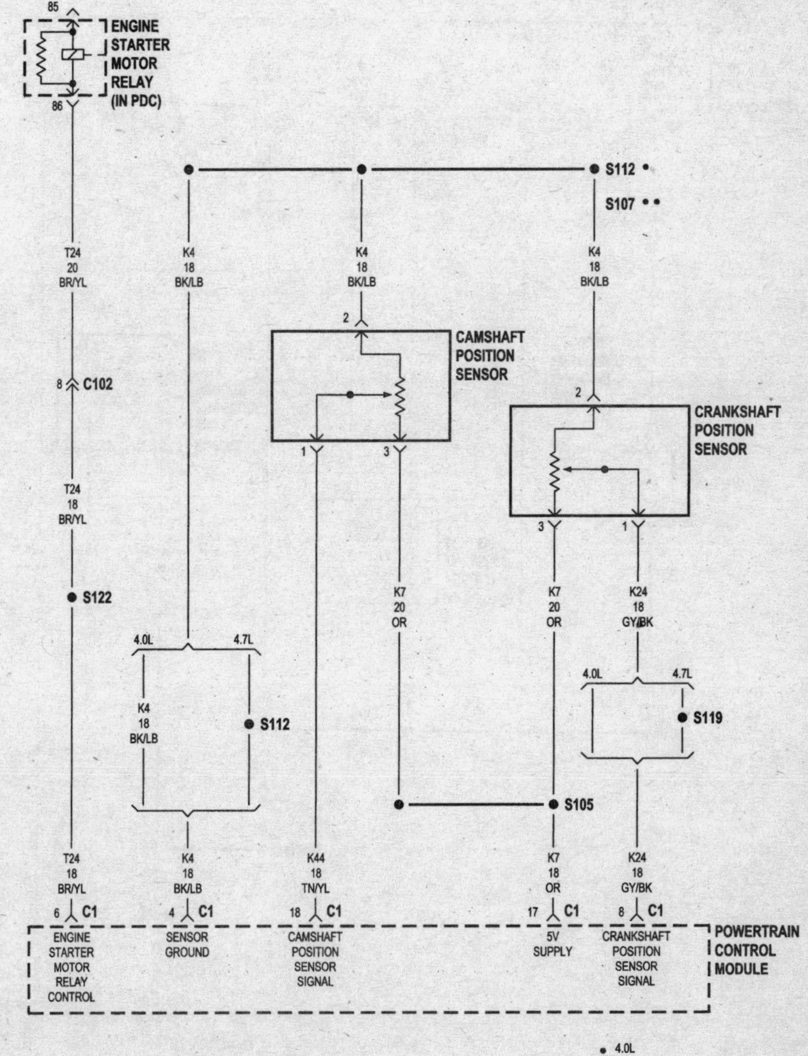

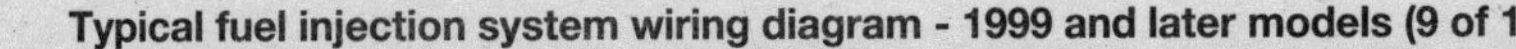
Typical fuel injection system wiring diagram - 1999 and later models (9 of 17)

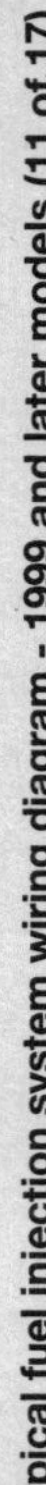

Typical fuel injection system wiring diagram - 1999 and later models (11 of 17)

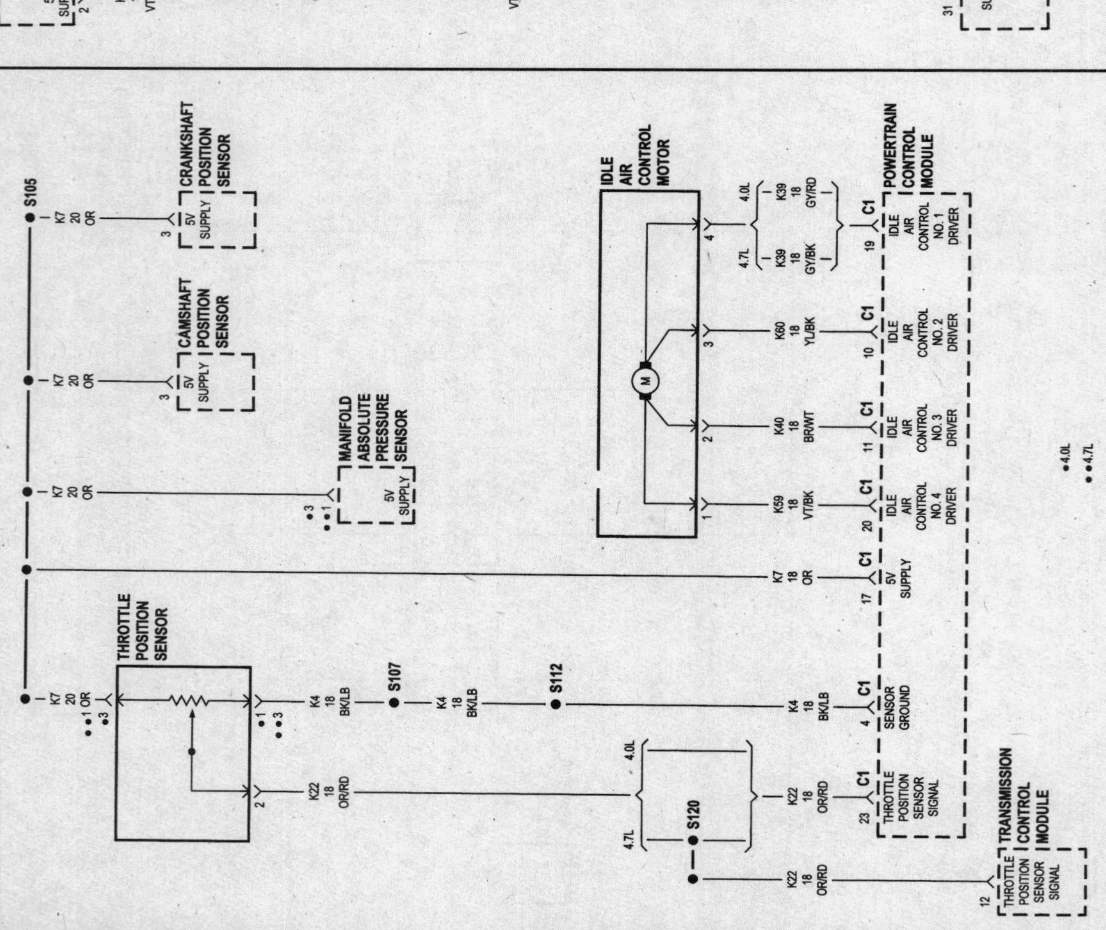

Typical fuel injection system wiring diagram - 1999 and later models (10 of 17)

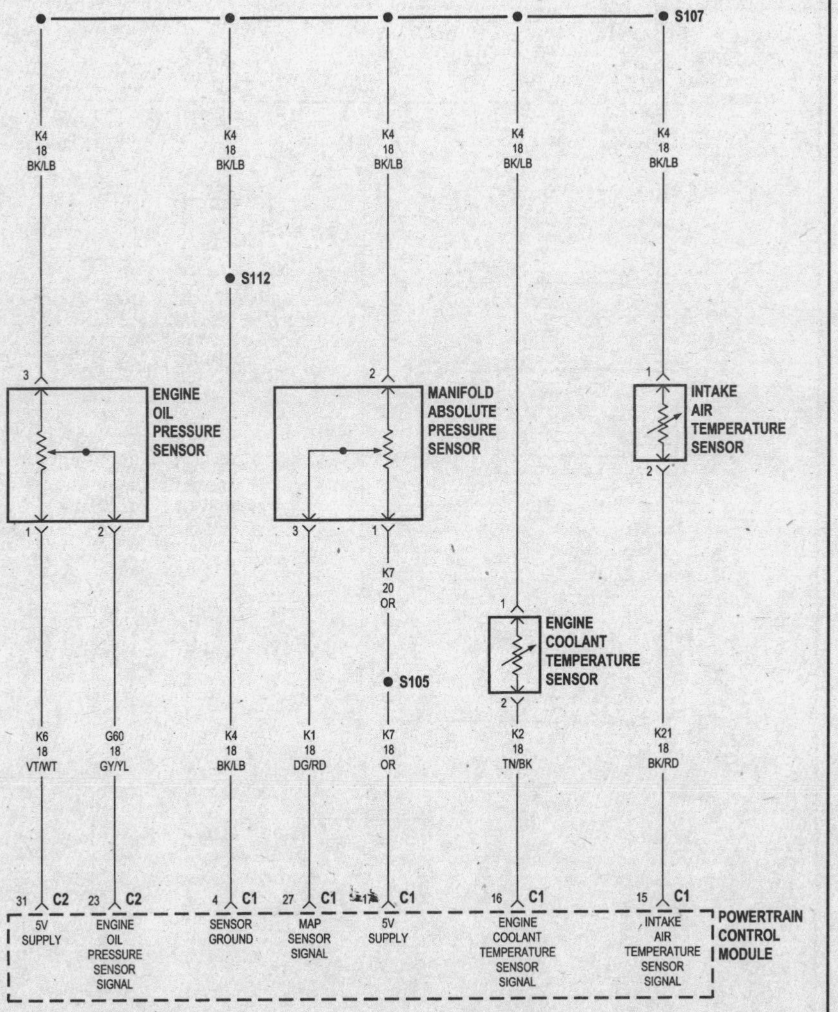

Typical fuel injection system wiring diagram - 1999 and later models (12 of 17)

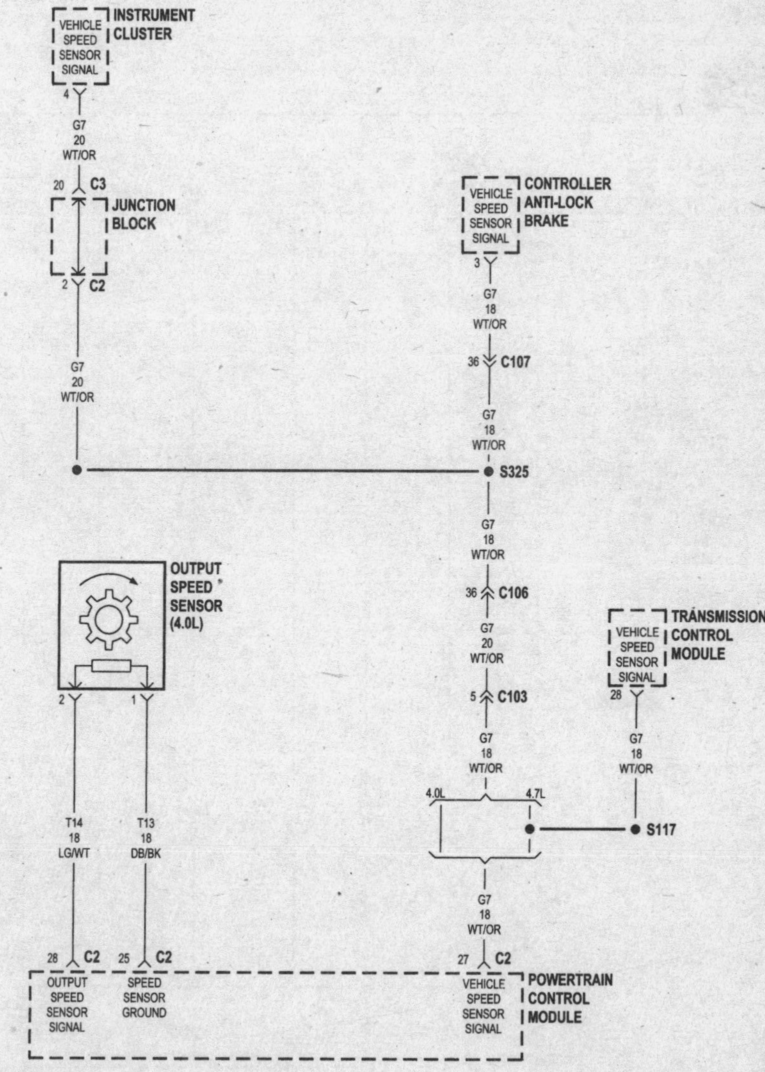

Typical fuel injection system wiring diagram - 1999 and later models (13 of 17)

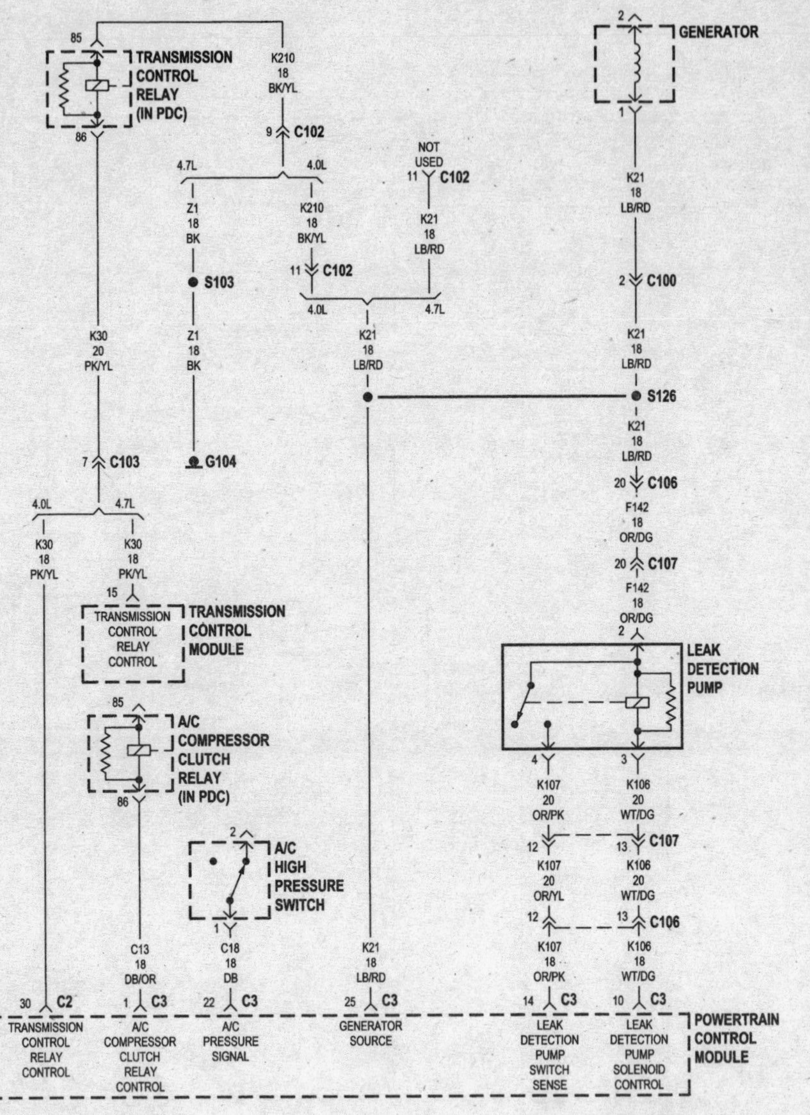

Typical fuel injection system wiring diagram - 1999 and later models (14 of 17)

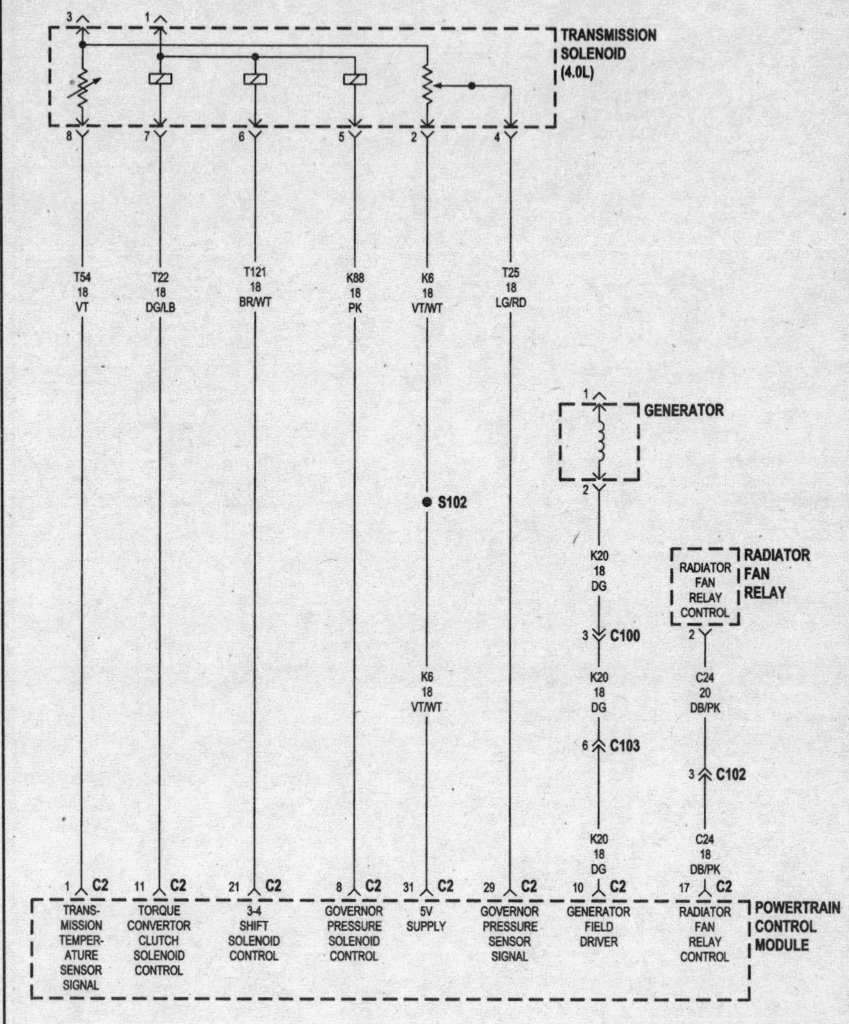

Typical fuel injection system wiring diagram - 1999 and later models (15 of 17)

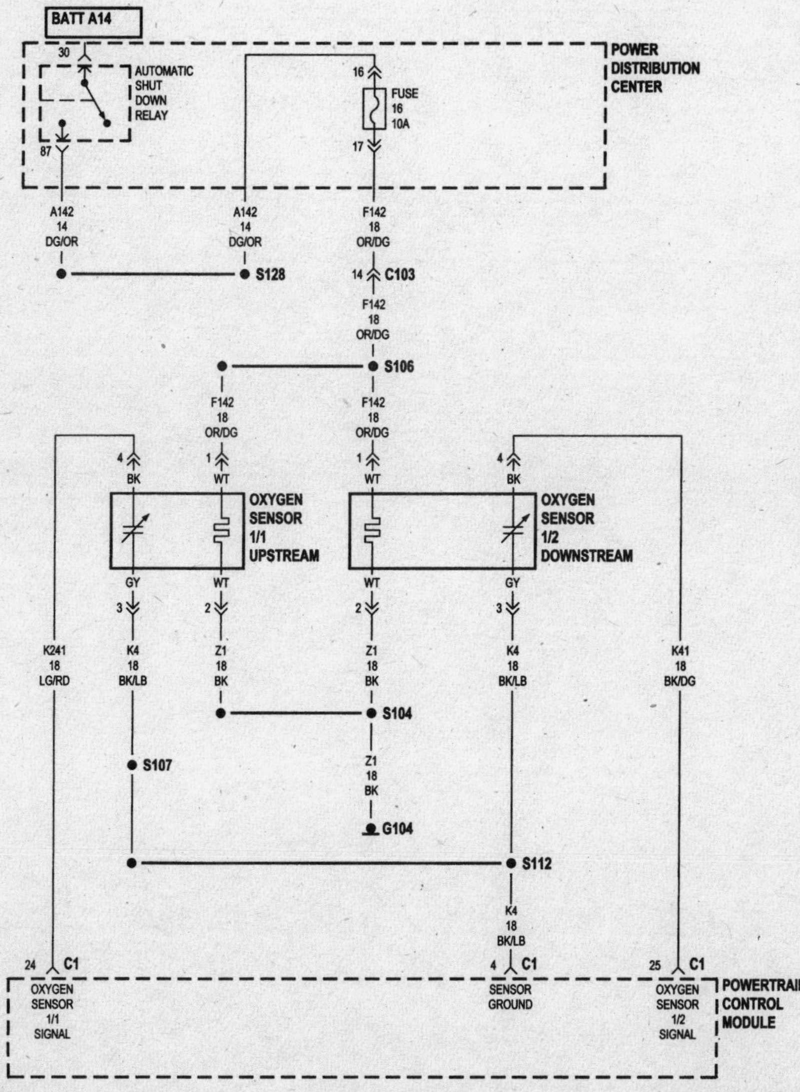

Typical fuel injection system wiring diagram - 1999 and later models (16 of 17)

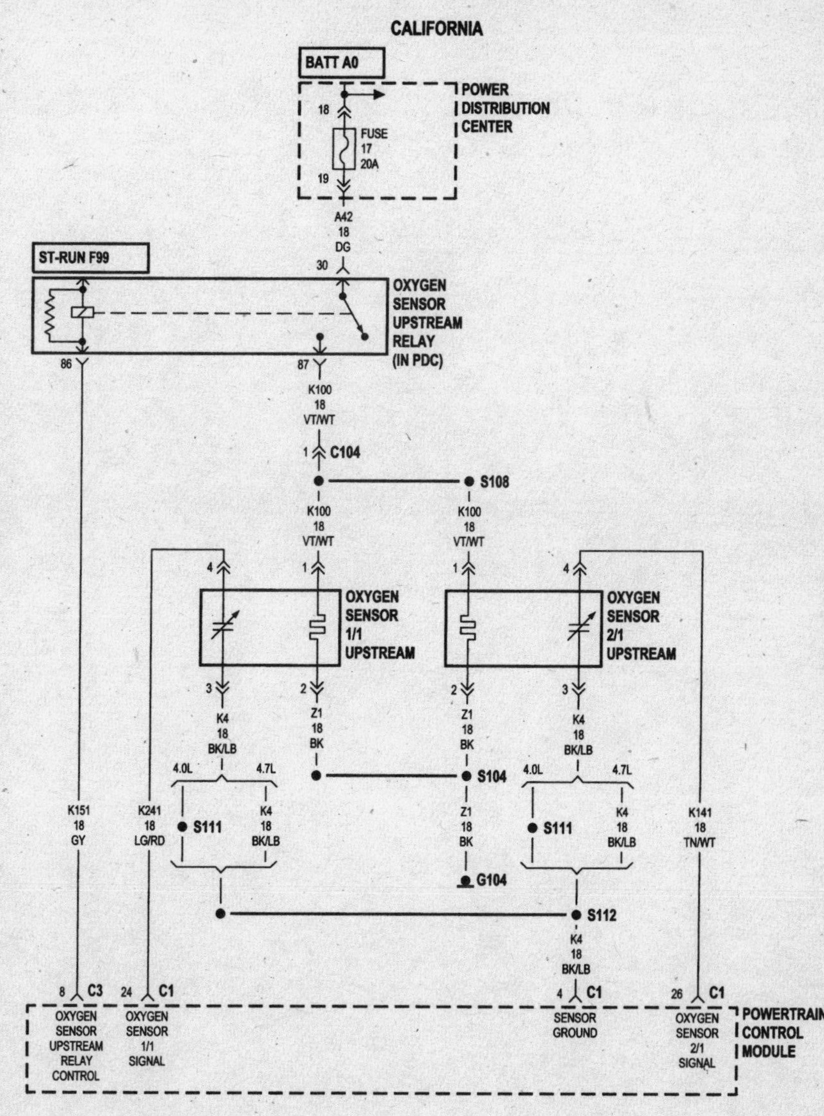

Typical fuel injection system wiring diagram - 1999 and later models (17 of 17)

12

Notes

Index

D

E

F

G

H

I

J

K

L

M

O

P

U

V

W

Haynes Automotive Manuals

NOTE: New manuals are added to this list on a periodic basis. If you do not see a listing for your vehicle, consult your local Haynes dealer for the latest product information.

ACURA
- 12020 Integra '86 thru '89 & Legend '86 thru '90
- 12021 Integra '90 thru '93 & Legend '91 thru '95

AMC
- Jeep CJ - *see JEEP (50020)*
- 14020 Concord/Hornet/Gremlin/Spirit '70 thru '83
- 14025 (Renault) Alliance & Encore '83 thru '87

AUDI
- 15020 4000 all models '80 thru '87
- 15025 5000 all models '77 thru '83
- 15026 5000 all models '84 thru '88

AUSTIN
- Healey Sprite - *see MG Midget (66015)*

BMW
- *18020 3/5 Series '82 thru '92
- *18021 3 Series including Z3 models '92 thru '98
- 18025 320i all 4 cyl models '75 thru '83
- 18050 1500 thru 2002 except Turbo '59 thru '77

BUICK
- Century (FWD) - *see GM (38005)*
- *19020 Buick, Oldsmobile & Pontiac Full-size (Front wheel drive) '85 thru '00
- 19025 Buick Oldsmobile & Pontiac Full-size (Rear wheel drive) '70 thru '90
- 19030 Mid-size Regal & Century '74 thru '87
- Regal - *see GENERAL MOTORS (38010)*
- Skyhawk - *see GM (38030)*
- Skylark - *see GM (38020, 38025)*
- Somerset - *see GENERAL MOTORS (38025)*

CADILLAC
- 21030 Cadillac Rear Wheel Drive '70 thru '93
- Cimarron, Eldorado & Seville - *see GM (38015, 38030, 38031)*

CHEVROLET
- 10305 Chevrolet Engine Overhaul Manual
- *24010 Astro & GMC Safari Mini-vans '85 thru '98
- 24015 Camaro V8 all models '70 thru '81
- 24016 Camaro all models '82 thru '92
- Cavalier - *see GM (38015)*
- Celebrity - *see GM (38005)*
- *24017 Camaro & Firebird '93 thru '00
- 24020 Chevelle, Malibu, El Camino '69 thru '87
- 24024 Chevette & Pontiac T1000 '76 thru '87
- Citation - *see GENERAL MOTORS (38020)*
- 24032 Corsica/Beretta all models '87 thru '96
- 24040 Corvette all V8 models '68 thru '82
- *24041 Corvette all models '84 thru '96
- 24045 Full-size Sedans Caprice, Impala, Biscayne, Bel Air & Wagons '69 thru '90
- 24046 Impala SS & Caprice and Buick Roadmaster '91 thru '96
- Lumina '90 thru '94 - *see GM (38010)*
- *24048 Lumina & Monte Carlo '95 thru '01
- Lumina APV - *see GM (38035)*
- 24050 Luv Pick-up all 2WD & 4WD '72 thru '82
- Malibu - *see GM (38026)*
- 24055 Monte Carlo all models '70 thru '88
- Monte Carlo '95 thru '01 - *see LUMINA*
- 24059 Nova all V8 models '69 thru '79
- 24060 Nova/Geo Prizm '85 thru '92
- 24064 Pick-ups '67 thru '87 - Chevrolet & GMC, all V8 & in-line 6 cyl, 2WD & 4WD '67 thru '87; Suburbans, Blazers & Jimmys '67 thru '91
- 24065 Pick-ups '88 thru '98 - Chevrolet & GMC, all full-size models '88 thru '98; C/K Classic '99 & '00; Blazer & Jimmy '92 thru '94; Suburban '92 thru '99; Tahoe & Yukon '95 thru '99
- *24066 Pick-ups '99 thru '01 - Chevrolet Silverado & GMC Sierra '99 thru '01; Suburban/Tahoe/Yukon/Yukon XL '00 & '01
- 24070 S-10 & GMC S-15 Pick-ups '82 thru '93
- *24071 S-10, Gmc S-15 & Jimmy '94 thru '01
- *24075 Sprint & Geo Metro '85 thru '94
- *24080 Vans - Chevrolet & GMC '68 thru '96

CHRYSLER
- 10310 Chrysler Engine Overhaul Manual
- 25015 Chrysler Cirrus, Dodge Stratus, Plymouth Breeze, '95 thru '98
- 25020 Full-size Front-Wheel Drive '88 thru '93
- K-Cars - *see DODGE Aries (30008)*
- Laser - *see DODGE Daytona (30030)*
- 25025 Chrysler LHS, Concorde & New Yorker, Dodge Intrepid, Eagle Vision, '93 thru '97
- *25026 Chrysler LHS, Concorde, 300M, Dodge Intrepid '98 thru '01
- 25030 Chrysler/Plym. Mid-size '82 thru '95
- Rear-wheel Drive - *see DODGE (30050)*
- *25040 Chrysler Sebring/Dodge Avenger '95 thru '02

DATSUN
- 28005 200SX all models '80 thru '83
- 28007 B-210 all models '73 thru '78
- 28009 210 all models '78 thru '82
- 28012 240Z, 260Z & 280Z Coupe '70 thru '78
- 28014 280ZX Coupe & 2+2 '79 thru '83
- 300ZX - *see NISSAN (72010)*
- 28016 310 all models '78 thru '82
- 28018 510 & PL521 Pick-up '68 thru '73
- 28020 510 all models '78 thru '81
- 28022 620 Series Pick-up all models '73 thru '79
- 720 Series Pick-up - *NISSAN (72030)*
- 28025 810/Maxima all gas models, '77 thru '84

DODGE
- 400 & 600 - *see CHRYSLER (25030)*
- 30008 Aries & Plymouth Reliant '81 thru '89
- 30010 Caravan & Ply. Voyager '84 thru '95
- *30011 Caravan & Ply. Voyager '96 thru '99
- 30012 Challenger/Plymouth Saporro '78 thru '83
- Challenger '67-'76 - *see DART (30025)*
- 30016 Colt/Plymouth Champ '78 thru '87
- 30020 Dakota Pick-ups all models '87 thru '96
- *30021 Durango '98 & '99, Dakota '97 thru '99
- 30025 Dart, Challenger/Plymouth Barracuda & Valiant 6 cyl models '67 thru '76
- 30030 Daytona & Chrysler Laser '84 thru '89
- Intrepid - *see Chrysler (25025, 25026)*
- *30034 Dodge & Plymouth Neon '95 thru '99
- *30035 Omni & Plymouth Horizon '78 thru '90
- 30040 Pick-ups all full-size models '74 thru '93
- *30041 Pick-ups all full-size models '94 thru '01
- *30045 Ram 50/D50 Pick-ups & Raider and Plymouth Arrow Pick-ups '79 thru '93
- 30050 Dodge/Ply./Chrysler RWD '71 thru '89
- 30055 Shadow/Plymouth Sundance '87 thru '94
- 30060 Spirit & Plymouth Acclaim '89 thru '95
- *30065 Vans - Dodge & Plymouth '71 thru '99

EAGLE
- Talon - *see MITSUBISHI (68030, 68031)*
- Vision - *see CHRYSLER (25025)*

FIAT
- 34010 124 Sport Coupe & Spider '68 thru '78
- 34025 X1/9 all models '74 thru '80

FORD
- 10355 Ford Automatic Transmission Overhaul
- 10320 Ford Engine Overhaul Manual
- 36004 Aerostar Mini-vans '86 thru '97
- Aspire - *see FORD Festiva (36030)*
- 36006 Contour/Mercury Mystique '95 thru '00
- 36008 Courier Pick-up all models '72 thru '82
- *36012 Crown Victoria & Mercury Grand Marquis '88 thru '00
- 36016 Escort/Mercury Lynx '81 thru '90
- 36020 Escort/Mercury Tracer '91 thru '00
- Expedition - *see FORD Pick-up (36059)*
- *36024 Explorer & Mazda Navajo '91 thru '01
- 36028 Fairmont & Mercury Zephyr '78 thru '83
- 36030 Festiva & Aspire '88 thru '97
- 36032 Fiesta all models '77 thru '80
- *36034 Focus all models '00 and '01
- 36036 Ford & Mercury Full-size '75 thru '87
- 36040 Granada & Mercury Monarch '75 thru '80
- 36044 Ford & Mercury Mid-size '75 thru '86
- 36048 Mustang V8 all models '64-1/2 thru '73
- 36049 Mustang II 4 cyl, V6 & V8 '74 thru '78
- 36050 Mustang & Mercury Capri '79 thru '86
- *36051 Mustang all models '94 thru '00
- 36054 Pick-ups and Bronco '73 thru '79
- 36058 Pick-ups and Bronco '80 thru '96
- *36059 Pick-ups, Expedition & Lincoln Navigator '97 thru '99
- *36060 Super Duty Pick-up, Excursion '97 thru '02
- 36062 Pinto & Mercury Bobcat '75 thru '80
- 36066 Probe all models '89 thru '92
- 36070 Ranger/Bronco II gas models '83 thru '92
- *36071 Ford Ranger '93 thru '00 & Mazda Pick-ups '94 thru '00
- 36074 Taurus & Mercury Sable '86 thru '95
- *36075 Taurus & Mercury Sable '96 thru '01
- 36078 Tempo & Mercury Topaz '84 thru '94
- 36082 Thunderbird/Mercury Cougar '83 thru '88
- 36086 Thunderbird/Mercury Cougar '89 thru '97
- 36090 Vans all V8 Econoline models '69 thru '91
- *36094 Vans full size '92 thru '01
- *36097 Windstar Mini-van '95 thru '01

GENERAL MOTORS
- 10360 GM Automatic Transmission Overhaul
- 38005 Buick Century, Chevrolet Celebrity, Olds Cutlass Ciera & Pontiac 6000 '82 thru '96
- *38010 Buick Regal, Chevrolet Lumina, Oldsmobile Cutlass Supreme & Pontiac Grand Prix front wheel drive '88 thru '99
- 38015 Buick Skyhawk, Cadillac Cimarron, Chevrolet Cavalier, Oldsmobile Firenza Pontiac J-2000 & Sunbird '82 thru '94
- *38016 Chevrolet Cavalier/Pontiac Sunfire '95 thru '00
- 38020 Buick Skylark, Chevrolet Citation, Olds Omega, Pontiac Phoenix '80 thru '85
- 38025 Buick Skylark & Somerset, Olds Achieva, Calais & Pontiac Grand Am '85 thru '98
- *38026 Chevrolet Malibu, Olds Alero & Cutlass, Pontiac Grand Am '97 thru '00
- 38030 Cadillac Eldorado & Oldsmobile Toronado '71 thru '85, Seville '80 thru '85, Buick Riviera '79 thru '85
- *38031 Cadillac Eldorado & Seville '86 thru '91, Deville & Buick Riviera '86 thru '93, Fleetwood & Olds Toronado '86 thru '92
- *38035 Chevrolet Lumina APV, Oldsmobile Silhouette & Pontiac Trans Sport '90 thru '95
- *38036 Chevrolet Venture, Olds Silhouette, Pontiac Trans Sport & Montana '97 thru '01
- General Motors Full-size Rear-wheel Drive - *see BUICK (19025)*

GEO
- Metro - *see CHEVROLET Sprint (24075)*
- Prizm - *see CHEVROLET (24060) or TOYOTA (92036)*
- 40030 Storm all models '90 thru '93
- Tracker - *see SUZUKI Samurai (90010)*

GMC
- Vans & Pick-ups - *see CHEVROLET*

HONDA
- 42010 Accord CVCC all models '76 thru '83
- 42011 Accord all models '84 thru '89
- 42012 Accord all models '90 thru '93
- 42013 Accord all models '94 thru '97
- *42014 Accord all models '98 and '99
- 42020 Civic 1200 all models '73 thru '79
- 42021 Civic 1300 & 1500 CVCC '80 thru '83
- 42022 Civic 1500 CVCC all models '75 thru '79
- 42023 Civic all models '84 thru '91
- 42024 Civic & del Sol '92 thru '95
- *42025 Civic '96 thru '00, CR-V '97 thru '00, Acura Integra '94 thru '00
- Passport - *see ISUZU Rodeo (47017)*
- *42040 Prelude CVCC all models '79 thru '89

HYUNDAI
- *43010 Elantra all models '96 thru '01
- 43015 Excel & Accent all models '86 thru '98

ISUZU
- Hombre - *see CHEVROLET S-10 (24071)*
- *47017 Rodeo '91 thru '97, Amigo '89 thru '94, Honda Passport '95 thru '97
- 47020 Trooper '84 thru '91, Pick-up '81 thru '93

JAGUAR
- 49010 XJ6 all 6 cyl models '68 thru '86
- 49011 XJ6 all models '88 thru '94
- 49015 XJ12 & XJS all 12 cyl models '72 thru '85

JEEP
- 50010 Cherokee, Comanche & Wagoneer Limited all models '84 thru '00
- 50020 CJ all models '49 thru '86
- *50025 Grand Cherokee all models '93 thru '00
- 50029 Grand Wagoneer & Pick-up '72 thru '91
- *50030 Wrangler all models '87 thru '00

LEXUS
- ES 300 - *see TOYOTA Camry (92007)*

LINCOLN
- Navigator - *see FORD Pick-up (36059)*
- *59010 Rear Wheel Drive all models '70 thru '01

MAZDA
- 61010 GLC (rear wheel drive) '77 thru '83
- 61011 GLC (front wheel drive) '81 thru '85
- *61015 323 & Protegé '90 thru '00
- *61016 MX-5 Miata '90 thru '97
- 61020 MPV all models '89 thru '94
- Navajo - *see FORD Explorer (36024)*
- 61030 Pick-ups '72 thru '93
- Pick-ups '94 on - *see Ford (36071)*
- 61035 RX-7 all models '79 thru '85
- 61036 RX-7 all models '86 thru '91
- 61040 626 (rear wheel drive) '79 thru '82
- *61041 626 & MX-6 (front wheel drive) '83 thru '91
- *61042 626 '93 thru '01, & MX-6/Ford Probe '93 thru '97

MERCEDES-BENZ
- 63012 123 Series Diesel '76 thru '85
- 63015 190 Series 4-cyl gas models, '84 thru '88
- 63020 230, 250 & 280 6 cyl sohc '68 thru '72
- 63025 280 123 Series gas models '77 thru '81
- 63030 350 & 450 all models '71 thru '80

MERCURY
- 64200 Villager & Nissan Quest '93 thru '01
- *All other titles, see FORD listing.*

MG
- 66010 MGB Roadster & GT Coupe '62 thru '80
- 66015 MG Midget & Austin Healey Sprite Roadster '58 thru '80

MITSUBISHI
- 68020 Cordia, Tredia, Galant, Precis & Mirage '83 thru '93
- 68030 Eclipse, Eagle Talon & Plymouth Laser '90 thru '94
- *68031 Eclipse '95 thru '01, Eagle Talon '95 thru '98
- 68040 Pick-up '83 thru '96, Montero '83 thru '93

NISSAN
- 72010 300ZX all models incl. Turbo '84 thru '89
- 72015 Altima all models '93 thru '01
- 72020 Maxima all models '85 thru '92
- *72021 Maxima all models '93 thru '01
- 72030 Pick-ups '80 thru '97, Pathfinder '87 thru '95
- *72031 Frontier Pick-up '98 thru '01, Xterra '00 & '01, Pathfinder '96 thru '01
- 72040 Pulsar all models '83 thru '86
- 72050 Sentra all models '82 thru '94
- 72051 Sentra & 200SX all models '95 thru '99
- 72060 Stanza all models '82 thru '90

OLDSMOBILE
- *73015 Cutlass '74 thru '88
- *For other OLDSMOBILE titles, see BUICK, CHEVROLET or GM listings.*

PLYMOUTH
- *For PLYMOUTH titles, see DODGE.*

PONTIAC
- 79008 Fiero all models '84 thru '88
- 79018 Firebird V8 models except Turbo '70 thru '81
- 79019 Firebird all models '82 thru '92
- 79040 Mid-size Rear-wheel Drive '70 thru '87
- *For other PONTIAC titles, see BUICK, CHEVROLET or GM listings.*

PORSCHE
- 80020 911 Coupe & Targa models '65 thru '89
- 80025 914 all 4 cyl models '69 thru '76
- 80030 924 all models incl. Turbo '76 thru '82
- 80035 944 all models incl. Turbo '83 thru '89

RENAULT
- Alliance, Encore - *see AMC (14020)*

SAAB
- *84010 900 including Turbo '79 thru '88

SATURN
- *87010 Saturn all models '91 thru '99

SUBARU
- 89002 1100, 1300, 1400 & 1600 '71 thru '79
- 89003 1600 & 1800 2WD & 4WD '80 thru '94

SUZUKI
- *90010 Samurai/Sidekick/Geo Tracker '86 thru '01

TOYOTA
- 92005 Camry all models '83 thru '91
- 92006 Camry all models '92 thru '96
- *92007 Camry/Avalon/Solara/Lexus ES 300 '97 thru '01
- 92015 Celica Rear Wheel Drive '71 thru '85
- 92020 Celica Front Wheel Drive '86 thru '99
- 92025 Celica Supra all models '79 thru '92
- 92030 Corolla all models '75 thru '79
- 92032 Corolla rear wheel drive models '80 thru '87
- 92035 Corolla front wheel drive models '84 thru '92
- *92036 Corolla & Geo Prizm '93 thru '01
- 92040 Corolla Tercel all models '80 thru '82
- 92045 Corona all models '74 thru '82
- 92050 Cressida all models '78 thru '82
- 92055 Land Cruiser FJ40/43/45/55 '68 thru '82
- *92056 Land Cruiser FJ60/62/80/FZJ80 '80 thru '96
- 92065 MR2 all models '85 thru '87
- 92070 Pick-up all models '69 thru '78
- 92075 Pick-up all models '79 thru '95
- *92076 Tacoma '95 thru '00, 4Runner '96 thru '00, T100 '93 thru '98
- 92080 Previa all models '91 thru '95
- *92082 RAV4 all models '96 thru '02
- 92085 Tercel all models '87 thru '94

TRIUMPH
- 94007 Spitfire all models '62 thru '81
- 94010 TR7 all models '75 thru '81

VW
- 96008 Beetle & Karmann Ghia '54 thru '79
- *96009 New Beetle '98 thru '00
- 96016 Rabbit, Jetta, Scirocco, & Pick-up gas models '74 thru '91 & Convertible '80 thru '92
- 96017 Golf & Jetta '93 thru '97
- *96018 Golf & Jetta '98 thru '01
- 96020 Rabbit, Jetta, Pick-up diesel '77 thru '84
- *96023 Passat '98 thru '01, Audi A4 '96 thru '01
- 96030 Transporter 1600 all models '68 thru '79
- 96035 Transporter 1700, 1800, 2000 '72 thru '79
- 96040 Type 3 1500 & 1600 '63 thru '73
- 96045 Vanagon air-cooled models '80 thru '83

VOLVO
- 97010 120, 130 Series & 1800 Sports '61 thru '73
- 97015 140 Series all models '66 thru '74
- 97020 240 Series all models '76 thru '93
- 97025 260 Series all models '75 thru '82
- 97040 740 & 760 Series all models '82 thru '88

TECHBOOK MANUALS
- 10205 Automotive Computer Codes
- 10210 Automotive Emissions Control Manual
- 10215 Fuel Injection Manual, 1978 thru 1985
- 10220 Fuel Injection Manual, 1986 thru 1999
- 10225 Holley Carburetor Manual
- 10230 Rochester Carburetor Manual
- 10240 Weber/Zenith/Stromberg/SU Carburetor
- 10305 Chevrolet Engine Overhaul Manual
- 10310 Chrysler Engine Overhaul Manual
- 10320 Ford Engine Overhaul Manual
- 10330 GM and Ford Diesel Engine Repair
- 10340 Small Engine Repair Manual
- 10345 Suspension, Steering & Driveline
- 10355 Ford Automatic Transmission Overhaul
- 10360 GM Automatic Transmission Overhaul
- 10405 Automotive Body Repair & Painting
- 10410 Automotive Brake Manual
- 10415 Automotive Detailing Manual
- 10420 Automotive Eelectrical Manual
- 10425 Automotive Heating & Air Conditioning
- 10430 Automotive Reference Dictionary
- 10435 Automotive Tools Manual
- 10440 Used Car Buying Guide
- 10445 Welding Manual
- 10450 ATV Basics

SPANISH MANUALS
- 98903 Reparación de Carrocería & Pintura
- 98905 Códigos Automotrices de la Computadora
- 98910 Frenos Automotriz
- 98915 Inyección de Combustible 1986 al 1999
- 99040 Chevrolet & GMC Camionetas '67 al '87
- 99041 Chevrolet & GMC Camionetas '88 al '98
- 99042 Chevrolet Camionetas Cerradas '68 al '95
- 99055 Dodge Caravan/Ply. Voyager '84 al '95
- 99075 Ford Camionetas y Bronco '80 al '94
- 99077 Ford Camionetas Cerradas '69 al '91
- 99083 Ford Modelos de Tamaño Grande '75 al '87
- 99088 Ford Modelos de Tamaño Mediano '75 al '86
- 99091 Ford Taurus & Mercury Sable '86 al '95
- 99095 GM Modelos de Tamaño Grande '70 al '90
- 99100 GM Modelos de Tamaño Mediano '70 al '88
- 99110 Nissan Camionetas '80 al '96, Pathfinder '87 al '95
- 99118 Nissan Sentra '82 al '94
- 99125 Toyota Camionetas y 4-Runner '79 al '95

** Listings shown with an asterisk (*) indicate model coverage as of this printing. These titles will be periodically updated to include later model years - consult your Haynes dealer for more information.*

Nearly 100 Haynes motorcycle manuals also available

7-02

Haynes North America, Inc., 861 Lawrence Drive, Newbury Park, CA 91320 • (805) 498-6703